Climate Change Impacts in Texas

This book focuses on climate change impacts and adaptation in Texas, a vital economic and ecologically important region in the United States. Texas spans twelve distinct ecoregions and a GDP greater than that of many countries; however, it is the only large state in the United States that does not have an active climate action or adaptation plan. This book analyzes the ways in which every aspect of life in Texas could be affected by climate change and the severe risks posed to the future well-being of Texans. It discusses the effects of climate change on the natural environment, human-environmental systems, and a diversity of approaches used for climate adaptation and mitigation, whether intentional or not. Finally, this book provides a different perspective for both the climate enthusiast and the climate agnostic.

Features

- Offers a contemporary perspective on the changing climate in Texas and why it matters—especially with respect to impacts, points of vulnerability, and adaptation.
- Includes contributions from the state climatologist and disciplinary experts who have a solid understanding of the ecological, social, economic, and policy issues for Texas.
- Provides a perspective on the political and economic motivations of climate mitigation in Texas, or lack thereof.
- Discusses the multi-faceted and interconnected nature of impacts and links projected impacts to adaptation and mitigation needs.
- Extends the discussion of climate change to relate to the broader science of planetary limitations.

This book serves as a foundational resource for graduate students, industry professionals, government officials, environmental practitioners, researchers, and academics dealing with climate change everywhere, not just in Texas, and gives readers a sense of urgency for changes that need to be made.

Environmental Assessment and Management

Series Editors

Glenn W. Suter II,

U.S. Environmental Protection Agency, Cincinnati, OH, USA

Mona Wells

Ronin Institute, USA

Published Titles

Ecological Causal Assessment

Edited by Susan B. Norton, Susan M. Cormier, Glenn W. Suter II

Multi-Criteria Decision Analysis: Case Studies in Engineering and the Environment, Second Edition

Igor Linkov, Emily Moberg, Benjamin D. Trump, Boris Yatsalo, and Jeffrey M. Keisler

Management of Water Resources Using Electromechanical Methods

Edited by Guoshuai Liu, Yong Jiang, and Changyong Zhang

Environmental Bioavailability Handbook: Methods and Assessments

Xiaokai Zhang and Mona Wells

Climate Change Impacts in Texas: Integrated Analysis and Adaptation

Edited by Mona Wells and Ryan A. McManamay

For more about this series, please visit https://www.routledge.com/Environmental-Assessment-and-Management/book-series/CRCENVASSANDMAN

Climate Change Impacts in Texas

Integrated Analysis and Adaptation

Edited by
Mona Wells and Ryan A. McManamay

CRC Press
Taylor & Francis Group
Boca Raton New York London

CRC Press is an imprint of the
Taylor & Francis Group, an **informa** business

Designed cover image: © Dylan Baddour, https://dylanbaddour.com/, the image is of the Rio Grande River at Santa Elena Canyon, Big Bend, going dry.

First edition published 2026
by CRC Press
2385 NW Executive Center Drive, Suite 320, Boca Raton FL 33431

and by CRC Press
4 Park Square, Milton Park, Abingdon, Oxon, OX14 4RN

CRC Press is an imprint of Taylor & Francis Group, LLC

ISBN: 978-1-032-52416-0 (hbk)
ISBN: 978-1-032-52420-7 (pbk)
ISBN: 978-1-003-40660-0 (ebk)

DOI: 10.1201/9781003406600

Typeset in Times
by KnowledgeWorks Global Ltd.

Contents

Preface

Some truths are too loud to hear. That's how it felt in the summer of 2023, the hottest summer on record for large cities in Texas including Austin, Dallas, El Paso, Houston, and San Antonio. Having spent most of my adult life away from Texas prior to 2023, I stepped outside to find the afternoon temperature in my front yard registering 117°F—a moment that carried me back nearly forty years to the first scholarly lecture I saw on climate change when I was a graduate student at Texas A&M University. Many important US Federal Agencies came together for a closed-door conference, attendance by invitation only, hosted at Texas A&M to discuss the ten greatest existential threats to humanity and the United States in particular. My professor was invited, and I was beside myself to wangle an invitation through him, an aspiration he bemusedly indulged. I recall clearly the session on climate change—that is, anthropogenically induced climate change. Back then, in the mid-1980s, after seeing the data presented, alongside the fundamental theory, it seemed obvious to me. It was obvious to others as well, and I do not mean my contemporaries or the decades of young scientists who have come along since. It was obvious to Svante Arrhenius, later one of the first Nobel Prize winners, who, in 1896, calculated that an eventual human-driven doubling of Earth's atmospheric CO_2 would lead to a global temperature increase of approximately 5–6°C (9–11°F)—a remarkably close estimate to modern climate models. After the conference, I asked my professor why it was closed session. He explained that the science of the topic and how scientists communicate about it differed greatly to wider communications with national and corporate interests involved. Decades later, I see: some things are too loud to hear, and these days it is difficult to find a rational, solely fact-driven conversation to be had on the topic of climate change, at any point in the political spectrum, and particularly in Texas.

While some are "deniers," others are practically acolytes. Unhappily, recycling or driving an electric car, as one energy analyst that I know says, will not "move the needle" for dealing with climate change. Energy use is inextricably tied to resource use, little to none of which is carbon neutral, despite the best attempts of greenwashing efforts to convince otherwise, and in a purely scientific framework, other impacts of resource use are currently viewed as being more quantitatively threatening to the human endeavor as we know it than climate change. But, as the Texas country music star Lyle Lovett sings "Look, I understand too little too late ... But what would you be if you didn't even try? You have to try." Which brings me, fast-forwarding to the end of the summer of 2023, to another conference, after which I sat around with a group of scientists for another sort of closed session discussion. We all agreed that we felt an extreme sense of urgency. The analogy of the modern response to climate change to the Roman Emperor Nero playing his lyre (fiddling) while Rome burned was mentioned by one, and rapidly adopted by the group as apt. It was agreed that some effort should be made by the group to raise our facts-based voices in an attempt to retain some vestige of rationality within the overwhelming miasma of non-fact-based opinion, speculation, dissimulation, and downright effulgent mendacity that has now begun to dominate the mainstream climate change narrative. While our initial plans did not work out as envisaged, this book did (Figure 0.1).

The intention of this book is to present an integrated analysis of climate change impacts in Texas, focusing on both natural ecosystems and human-environmental systems. Climate change is often discussed in isolation—rising temperatures, sea level shifts, or extreme weather events—but its effects are deeply interconnected, influencing agriculture, water management, infrastructure, and economic resilience—in fact, touching every facet of the basic planetary resources on which we depend. Although a regional study on climate change in Texas, the book is not fundamentally about Texas. By framing Texas as a case study, this book illustrates how climate change unfolds in a region with vast environmental diversity and economic dependencies. The progression of topics—from foundational climate science to specific ecological and societal consequences—is intended to

FIGURE 0.1 The Emperor Nero "fiddles" while Rome burns. (While the story is thought to be more legend than fact, there are thought to be some aspects of historical accuracy, and the analogy is apt regarding the urgency of response needed to deal with climate change versus the dilatory and lackluster actual responses occurring.)

provide a representative overview of both the science and its implications. Texas, with its history of extreme weather and central role in global energy markets, serves as a microcosm for broader climate adaptation challenges, with challenge and a call to action being key themes. The book's structure also reflects a logical progression from impact to action. The first section establishes a scientific foundation, explaining fundamental climate concepts and shifts in Texas's environment. From there, in the second section we include assessments of how climate change affects freshwater systems, coastal regions, land use, and ecosystems—essential elements in the balance of life and industry in Texas. The third section explores direct human consequences, from food production disruptions to urban heat stress and economic shifts. Finally, the book transitions into adaptation and mitigation strategies, outlining practical considerations for policymakers, environmental managers, and industries working to navigate a changing climate.

A project of this scale—connecting scientific foundations, ecological impacts, human systems, and adaptation strategies—required the effort of many individuals and organizations whose expertise and support made this book possible. First and foremost, this book would not exist without the dedication and vision of my co-editor, Ryan McManamay, whose relentless enthusiasm and leadership transformed an ambitious idea into reality. His ability to navigate challenges, inspire collaboration, and push forward when momentum stalled was invaluable. Without his energy and commitment, this project would never have come together. I am also deeply grateful to the Meadows Foundation, whose vision and commitment to advancing environmental research and policy led to the support provided for my work in Texas as Director of Climate Science for one of their research centers, a role that catalyzed the development of the proposal for this book. Finally, Ryan and I extend our sincere appreciation to the authors, publisher, and editorial team who enabled us to bring this complex and lengthy endeavor to final fruition.

This book is not just an analysis of climate change in Texas—it is a call to engage with the realities of the global environmental crisis, to look beyond fragmented and rebarbative debates, to recognize the interconnected forces shaping our future, and to take action. Ryan and I hope this book sparks meaningful discussions, deepens awareness, and empowers readers to take action—while any single individual may not be able to move the needle, large numbers of us changing our habits and overly consumptive lifestyles would. Climate change was a distant concern to Arrhenius, becoming considered an existential threat requiring closed-door discussions by the 1980s. Now the state of play is urgent; each informed voice and, most importantly, each individual agent for change matters.

Mona Wells

About the Editors

Mona Wells is an interdisciplinary environmental scientist specialized in Environmental Impact Assessment, Environmental Risk Assessment, and Environmental Management. Her roles have included posts in different locales internationally as an academic and as an environmental practitioner, on projects spanning a broad portfolio including technical project work, research, commercialization, program development, policy advisement, science communication, public consultation, pedagogical development, and continuous improvement. She recently served as Director of Climate Science for one of the Meadows Foundation Research Institutes in Texas.

Ryan A. McManamay is an Associate Professor of Environmental Science at Baylor University in Waco, Texas, where he teaches classes on energy and environment, water management, and ecology. His research explores past and future anthropogenic pressures, from land use change to infrastructure development, on ecosystems with the goal of balancing ecological and societal needs. Previously, he was a research scientist at Oak Ridge National Laboratory, serving in both administrative and technical leadership roles and working in numerous multi-institutional research projects across several U.S. Department of Energy programs. His research has influenced natural resource management along multiple dimensions, including informing environmental flow management for several state and federal agencies, decision-support and training for hydropower mitigation and basin planning, and long-term integrative planning for renewable energy and urban infrastructure. He has held several editorial roles, currently serving as Associate Editor of *Earth's Future* (American Geophysical Union).

Contributors

Srinivasulu Ale
Texas A&M AgriLife Research
Vernon, Texas

Melissa R. Allen-Dumas
Oak Ridge National Laboratory
Oak Ridge, Tennessee

Nafissatou Ba
Baylor University
Waco, Texas

Mahesh Bade
Baylor University
Waco, Texas

Dominic Boyer
Rice University
Houston, Texas

Susan Power Bratton
Department of Environmental Science
Baylor University
Waco, Texas

Tirthankar Chakraborty (TC)
Pacific Northwest National Laboratory
Richland, Washington

Forrest Cobb
Texas A&M Natural Resources Institute
Texas A&M University
San Antonio, Texas

Francisco Contreras
Department of Civil, Environmental, and Architectural Engineering
College of Engineering and Applied Science
University of Colorado Boulder
Boulder, Colorado

Aparajita Datta
University of Houston
Houston, Texas

Jiabi Du
Department of Marine and Coastal Environmental Sciences
College of Marine Sciences and Maritime Studies
Texas A&M University
Galveston, Texas

Stephen Ferencz
Energy and Environment Directorate
Pacific Northwest National Laboratory
Richland, Washington

Michelle A. Hummel
Department of Civil Engineering
University of Texas at Arlington
Arlington, Texas

Nilkamal Jaisawal
Baylor University
Waco, Texas

Jordan T. Jatko
Baylor University
Waco, Texas

Mark Keeley
City of Waco
Waco, Texas

Julie A. King
Baylor University
Waco, Texas

Kritika Kothari
Indian Institute of Technology Roorkee
Roorkee, Uttarakhand, India

Ramanan Krishnamoorti
University of Houston
Houston, Texas

Mukesh Kumar
Waco Metropolitan Planning Organization
Waco, Texas

Yang Li
Baylor University
Waco, Texas

Roel Lopez
Texas A&M Natural Resources Institute
Texas A&M University
College Station, Texas

Alison Lund
Texas A&M Natural Resources Institute
Texas A&M University
College Station, Texas

Ryan A. McManamay
Department of Environmental Science
Baylor University
Waco, Texas

John W. Nielsen-Gammon
Texas A&M University
College Station, Texas

Perpetua I. Okoye
Baylor University
Waco, Texas

Alejandra Rodriguez
Texas A&M Natural Resources Institute
Texas A&M University
College Station, Texas

Sayantan Samanta
Department of Agricultural and Biological Engineering
Mississippi State University
Starkville, Mississippi

Alla Semenova
St. Mary's College of Maryland
St. Mary's City, Maryland

Bhupinder Singh
Maricopa Agricultural Center
University of Arizona
Maricopa, Arizona

Levi T. Sweet-Breu
Baylor University
Waco, Texas

Cristina Torres-Machi
Department of Civil, Environmental, and Architectural Engineering
College of Engineering and Applied Science
University of Colorado Boulder
Boulder, Colorado

Matthew J. Troia
The University of Texas at San Antonio
San Antonio, Texas

Mona Wells
Ronin Institute
USA

Joseph D. White
Baylor University
Waco, Texas

Lili Yao
Energy and Environment Directorate
Pacific Northwest National Laboratory
Richland, Washington

Section I

Introduction

I.1 Why Texas?

Ryan A. McManamay

So, you ask "Why a book on climate change in Texas?" Perhaps this is a compulsive response that should be expected when any global topic is applied to a non-obvious or seemingly mundane place. A counter argument would, of course, simply start by pointing to the virtue of place-based analyses, which provide context to understand complex environmental issues. In other words, lessons can best be learned through case studies and real-world applications. This is true, but why not focus on fragile ecosystems on the brink of collapse, such as boreal tundra, sea-ice ecosystems, or the rainforests of the Congo? Of any place for climate change application, why Texas?

The simple and most immediate answer is that Texas is a mesocosm of the world. While it doesn't have charismatic fauna such as the mountain gorilla or the Asian elephant, it reflects the condition of our planet that most people will experience. It represents the extreme endpoints of humanity's conflict with climate change. Texas provides a scaled-down example of the global spectra of stressors that come with climate change and extremes, and the diversity of political ideologies, technological hurdles, the resilience and vulnerabilities, and socioeconomic solutions, or lack thereof, related to climate adaptation and mitigation. Due in part to its sheer size and location, Texas possesses an immense diversity of climates, ecosystems, and socioeconomic systems. It is a place of gradients, from deserts to subtropics, from high plains to oceans, urban epicenters to remote wilderness, and extremes ranging from hurricanes to drought to wildfire. Most of the environmental and socioeconomic challenges identified in various parts of the world are very relevant and very real in Texas. Of course, being focused on Texas there are issues that this book will not cover, such as ice cap melt, glacial recession, and permafrost thawing, to name a few. Nonetheless, you will find that Texas is also an ideal case because it demands a practical outlook on the changing climate and why it matters. Appropriate perspectives are critical to solving problems. Texas has plenty of environmental and social challenges in the absence of climate change: water overuse, electricity outages, health concerns from elevated airborne particulate matter, excessive pollution and deoxygenation in rivers and estuaries, rapid urban expansion, and rural land fragmentation are all critical issues. We cannot blame these issues, at least directly, on global warming or extremely high atmospheric CO_2 concentrations. Our society's attempts to fix some environmental problems cannot ignore other interrelated ones, or else, we face a grim situation with no solutions, or even worse, a future of harmful unanticipated consequences. Hopefully, the reasoning behind the geographical focus of this book will become evidently clear as you explore the sections of this book – the natural ecosystems, the social fabric, the economy, the energy landscape, and yes, the climate context, of the Lone Star state.

Texas is the world's eighth largest economy, grossing $2.7 trillion in 2023 (BEA, 2025) and is home to 31 million people, roughly 10% of the total U.S. population (U.S. Census Bureau, 2024). It is the largest state in terms of area for the conterminous U.S. (Figure I.1.1), second only to Alaska for the entire nation, and second to California in total population. Five of the U.S.'s largest metropolitan areas are found in Texas. Houston, Dallas, Fort Worth, Austin, and San Antonio collectively comprise over 7 million residents. Most conservative projections suggest that population in Texas could increase by 40% by 2050 and increase by 70% by 2080 (US Census Bureau, 2023). However, some future population estimates exceed 60 million Texans by 2080, a 100% increase (Figure I.1.2). Part of the attraction and influx of increasing population is the affordability of Texas, but also the business friendly, ultra-pro-capitalism culture that fosters economic growth. Ultimately, this has led to significant land and water resource demands. The American Farmland Trust reported that Texas farmlands are the most threatened in the U.S. from urbanization, fragmentation, and loss

DOI: 10.1201/9781003406600-2

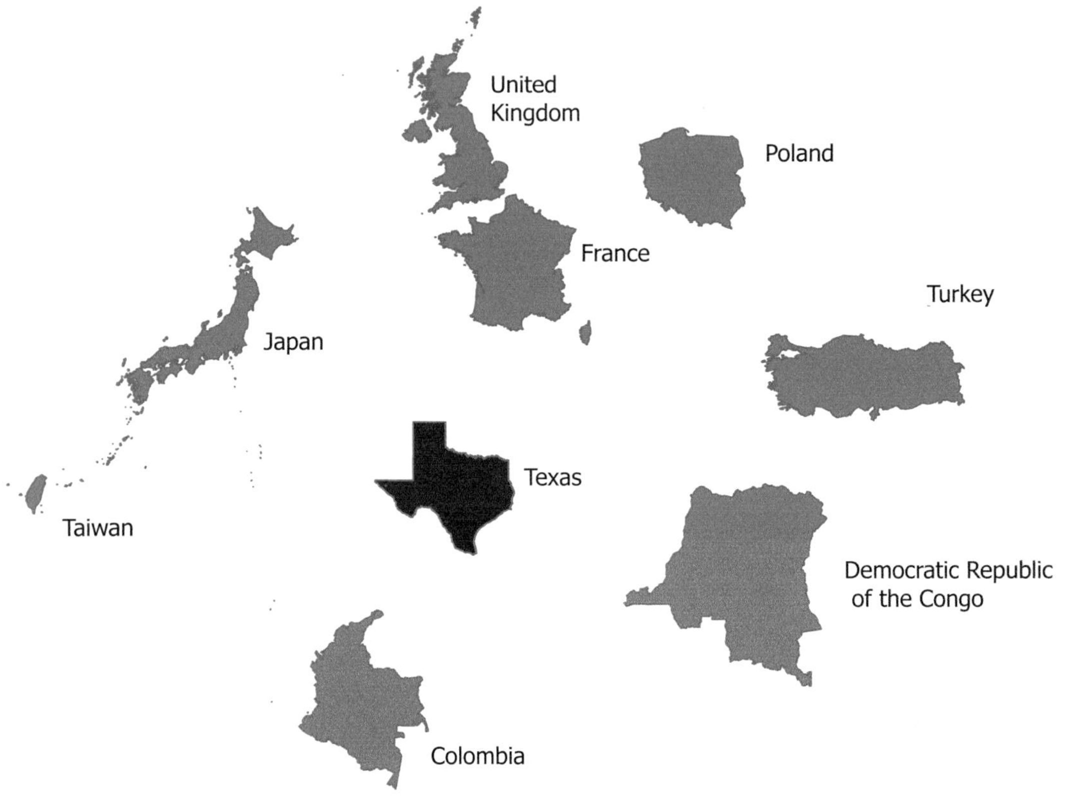

FIGURE I.1.1 Comparison of the size of Texas relative to other countries.

(Freedgood et al., 2020). From 1997 to 2002, 3.7 million acres of working lands were converted to non-agricultural uses with 1.8 million of those acres being lost within the last five years (TAMNRI, 2025). The impetus for this change has primarily been economical. Land market values have mirrored losses in farms with average appraisal values of rural land increasing over 500% from 1997 to 2023 (TAMNRI, 2025).

Texas landscapes cover almost 700,000 km^2, supporting a diversity of habitats and extraordinary biodiversity. Ten ecoregions support at least 800 habitat types (TPWD, 2025). Topographic diversity is significant, with western mountain ranges reaching over 2,700 m. At the southeastern portion of the state, there is 590 km of coastline where 21 river basins terminate. Despite the immense diversity of natural landscapes and resources, Texas lands are predominantly privately owned and managed – only 4% of land is designated as protected through federal, state, or local means (USGS, 2024). Hence, increasing the resilience and mitigation capacity of natural ecosystems across Texas has and will require careful coordination with private landowners. According to Texas Parks and Wildlife, Texas hosts tens of thousands of native animal and plant species (TXNDD, 2025), including 145 mammals (Schmidly and Bradley, 2016), 672 birds (TBRC, 2025), 191 freshwater fish (Birdsong et al., 2021), 5,000 plant species (NPST, 2025), and 29,000 insects (Abbott and Abbott, 2020). However, 1,100 of these species are listed as being of "greatest conservation need" due to having some risk of major loss or extinction in the state (TPWD, 2025).

Freshwater in Texas ranges from being an abundant to a scarce resource, both in surface waters and deep aquifers. The 100th meridian, running through the heart of Texas, is near to the exact dividing line for the U.S., splitting major continental climate regimes into the humid eastern and arid western portions (Figure I.1.3). Within Texas, annual precipitation ranges from 59 inches

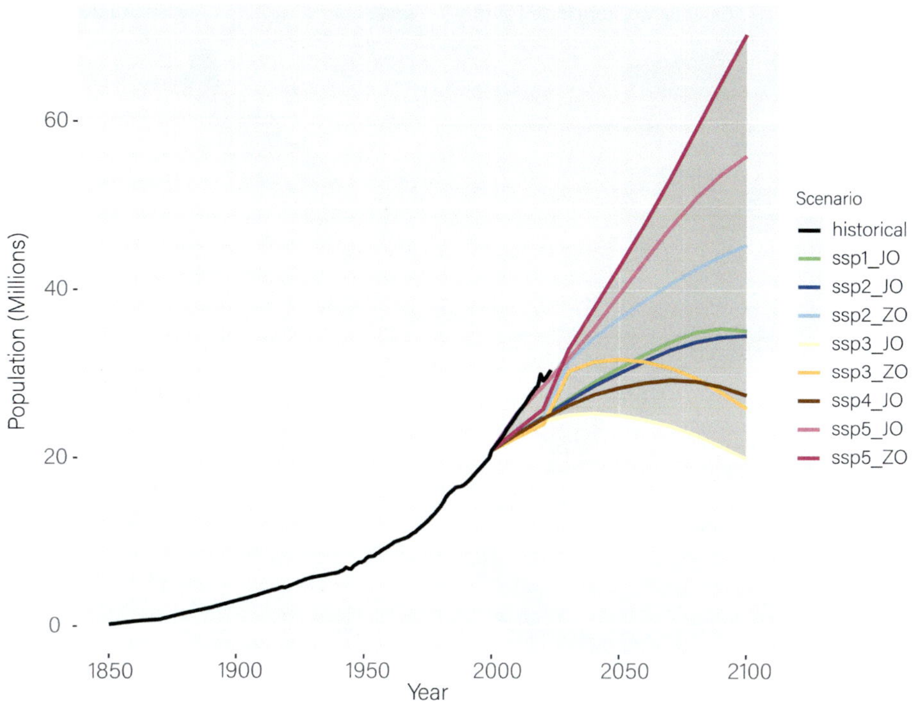

FIGURE I.1.2 Population growth and projections in Texas. Census population records (1850–2017) obtained from Texas State Library and Archives Commission (2017), whereas recent population (2018–2023) obtained from the U.S. Census Bureau (2025). Future projections for different Shared Socioeconomic Pathways (SSPs) provided by JO (Jones and O'Neill, 2013) and ZO (Zoraghein and O'Neill, 2020).

in Orange in the eastern portion to less than 8 inches in El Paso (see Chapter I.2 by Nielsen-Gammon). This climate gradient presents a range of challenges that will require unique adaptation strategies as climate continues to evolve. Water availability also follows seasonal extremes, with most precipitation in the late spring and early fall, generally followed by dry or extremely dry conditions at other times of the year. Not surprisingly, adaptation to this boom or bust synchrony has been to store surface water in impoundments, which currently exceed 43,000 in number, more than any other state, including Minnesota (Cheruvelil et al., 2021). This is an astounding number considering the vast majority of "lakes" in Texas are man-made. Virtually all major rivers originate in the central or western arid portions of the state and flow eastward, eventually joining other river systems or flowing directly to the gulf. This orientation is not necessarily exceptional – many rivers in the western U.S. originate in mountainous arid regions; however, the juxtaposition of drainage pattern relative to the most densely populated megaregions, particularly the high rates of surface water use in the east-central half of the state, create vulnerable situations of over-use and groundwater depletion (Figure I.1.3). Indeed, during extreme droughts, several of the major perennial rivers have dried up in recent years, including the Rio Grande in 2022, the Frio River in 2024, the Guadalupe in 2023, the Medina River in 2021, and the Brazos in 2011 (Figure I.1.4). A series of droughts has placed water reserves at critically low levels, which may result in river managers cutting flows to the estuaries for some systems. Unfortunately, projections suggest combinations of increasing drought severities and growing demands could create critical situations. The Texas Water Development Board projects 16.5% increases in total water demand by 2050 and over 20% increases by 2080 (TWDB, 2022), although this can be locally challenging based on intense use for some regions. For instance, municipal water demand projections suggest over 70% increases in demand by 2080, generally tracking expected population growth. Twenty-three major reservoirs

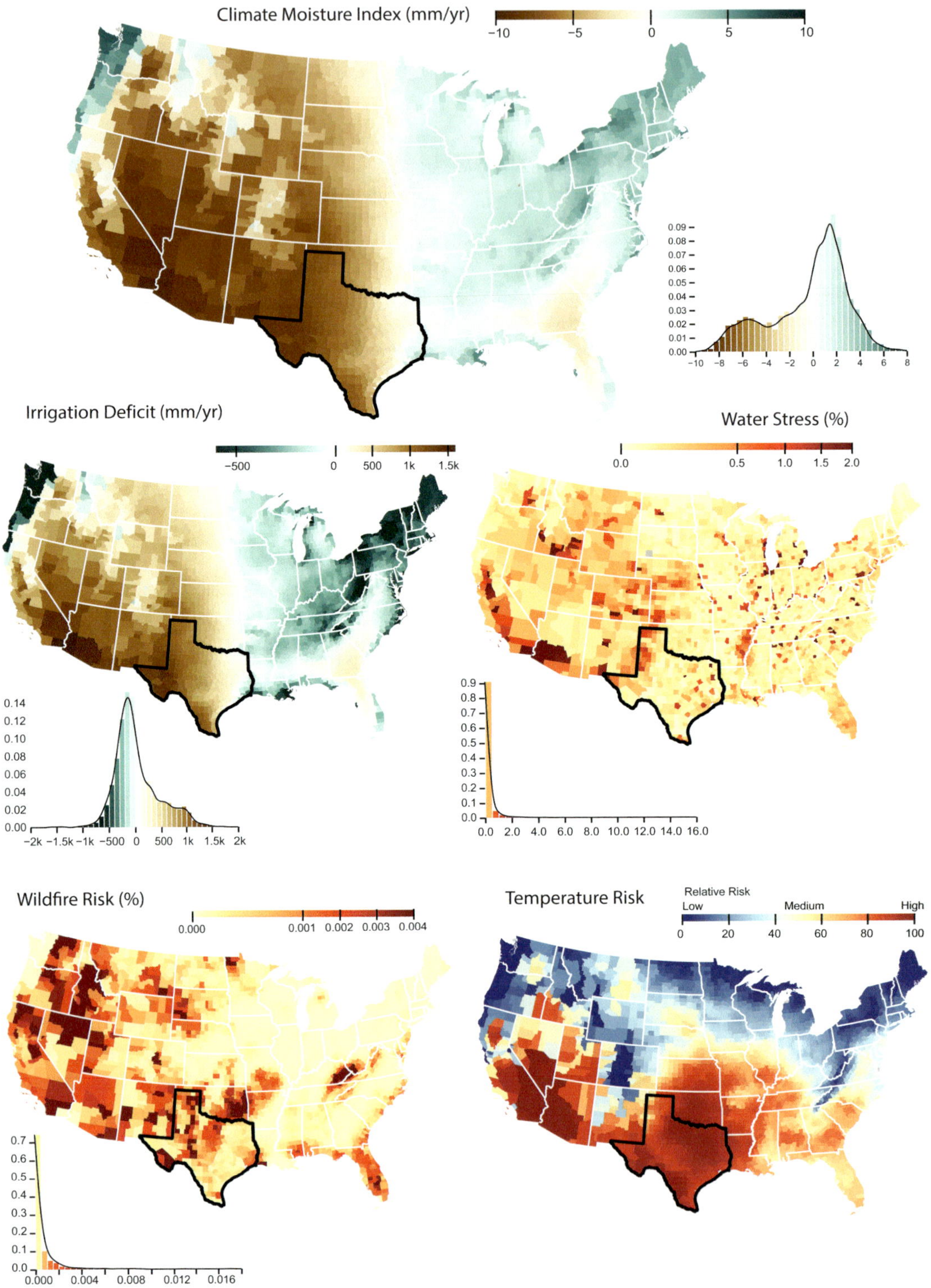

FIGURE I.1.3 Indicators of climate-related stressors in Texas relative to the remainder of the United States. Metrics taken from Massachusetts Institute of Technology, Joint Program on the Science and Policy of Global Change STRESS (System for the Triage of Risks from Environmental and Socio-economic Stressors) Platform (MIT, 2025).

FIGURE I.1.4 The Brazos River ran dry in Knox County, Texas during the drought of 2011. Provided by Kennedy (2011), National Oceanic and Atmospheric Administration, Climate.gov.

over 5,000 acre-ft in size are currently being planned for development in Texas to expand water storage (TWDB, 2022).

The energy context of Texas is simultaneously a marvel and a paradox. Compared to the remainder of the U.S. interconnected electricity grid system, Texas's electric grid, the Texas Interconnection, is isolated and managed separately by the Electric Reliability Council of Texas (ERCOT). As we will see later in the book, this can create its own criticalities. Texas has the highest U.S. energy consumption, the highest energy production, and the highest CO_2 emissions, more than double that of California (EIA, 2025). By far, Texas leads the nation in natural gas production and crude oil production (5.86 million barrels per day (mbd)), roughly equivalent to production for the remainder of the U.S. and exceeding total oil production rates for Canada (4.80 mbd) (EIA, 2025). Yet, simultaneously, Texas leads the nation in wind energy generation and is only second to California in utility and residential solar energy generation. Even so, generation from renewable sources comprises only 28% of total net generation, indicating significant continued reliance on fossil fuel sources for electricity beyond primary heat and transportation uses (EIA, 2025). On one hand, growth in renewables has been primarily a result of aggressive legislation and both state- and federal-production tax credits. However, this appears to be purely driven from an economic, rather than environmental, standpoint. As Dr. Dominic Boyer elegantly states in his chapter on Texan Energopolitics (Chapter IV.4), "in this logic [that of John Locke and the American manifestation in Texas], the whole world is nothing more than a bundle of resources to be developed and transformed by humans as a matter of divine right and dispensation." In other words, renewables in Texas are not really a climate mitigation strategy, but a means to profit for the electricity sector. In part, this strategy is seemingly nurtured by the extreme deregulation of the electricity system in Texas, which is among the most aggressive deregulated energy markets globally. On the other hand, the deregulation of wholesale and retail electricity markets was supposedly established to prevent monopolies in power infrastructure and generation and foster competition, allowing penetration by renewables. Indeed, in the current system, that has worked and supported an explosion of independent power producers. The only unforeseen outcome was that many of these producers are fueled by natural gas, providing over 50% of Texas's electricity supply. As long as the price of fossil fuels remains low and storage technologies prevent wind and solar from competing in the peak market, natural gas will continue to outcompete the renewable strategies in Texas.

Of course, in addition to all the above issues, the climate is changing. Weather extremes – heat waves, drought, winter storms, and tropical cyclones – are not new to Texas (TAM OTSC, 2025). For instance, the Great Galveston Storm, a hurricane in September 1900, remains one of the U.S.'s deadliest weather events, completely inundating the island and killing 8,000 people. A seven-year extreme drought in the 1950s led to $36 billion in losses to the state's agricultural industry alone. Extreme weather and climate events in Texas are estimated to have cost over $400 billion USD in infrastructure, property, and asset losses since 1980 – this estimate only includes the major events exceeding $1 billion each in losses. Despite the history of extremes, the public has increasingly become more aware, perhaps more sensitized, to the noticeable and recent shifts in climate – longer and hotter heat waves, more frequent droughts, intense precipitation events (e.g., Hurricane Harvey in 2017), and complete abnormalities (e.g., Winter Storm Uri in 2021). The impacts of climate-related stressors are multifaceted and dependent on the vulnerability and exposure of humans and infrastructures. Texas sits at a critical geographical and socioeconomical juxtaposition, as it experiences many of these multifaceted stressors represented across the remainder of the U.S., including water stress, wildfire risk, and temperature-related health risks (Figure I.1.3). An important consideration is that, although many modern climate extremes are not necessarily unprecedented events, the current socioeconomic context is new and continues to evolve. Population, resource demands, and urban development have increased, simultaneously increasing exposure to extremes. Sophisticated infrastructures to support human livelihoods have dramatically expanded, placing additional stress on natural systems. In theory, this means that climate events of similar frequency, duration, and magnitude to historical conditions could have more severe health and infrastructural impacts to society simply due to increased exposure and the nature of our intense use and dependence on resources. For instance, the 1950s drought is considered the 'drought-of-record' for the state due to the length of time in drought severity; however, the more recent 2011 and 2022 droughts could be considered more severe based on the total areas impacted and the complete loss of streamflow in multiple rivers. In their chapter on climate extremes (Chapter I.4), Yao and others describe that when assessing the severity and impact of drought, multiple interacting factors, not just the duration of zero precipitation, are important considerations – these include the compounding effects of low precipitation, high temperatures, and high evapotranspiration, the hydrologic conditions of the soil, and the current degree of human use of water. Unfortunately, the climate issues that have plagued Texas will not only continue but will likely worsen. According to the National Climate Assessment (NCA, 2025), under a 2°C global warming scenario, the drying trends in Texas will persist and intensify, and remaining precipitation events, when they occur, will be more extreme bouts, expected to increase in intensity from 10% to 42%. Heat waves have increased over time and that trend is expected to continue. Since 1900, the number of days exceeding 100°F have increased, on average, from 10 to 25 (a 150% increase). The number of days exceeding 95°F could increase by one to two months, whereas days exceeding 105°F could conservatively increase by over 20 days.

Equally important to the above factors is the sociopolitical, cultural, and ideological context of Texas. In many ways, Texan sociopolitics is not necessarily an anomaly, but rather singularly represents the diversity of ideologies represented across the remainder of the U.S. Interestingly, 70% of Texans agree that global warming is real and happening, very close to the U.S. average, 72% (Figure I.1.5a); however, fewer than 58% of Texans believe climate change is caused mostly by human activities, again approximating the U.S. national average, 59% (Figure I.1.5b). Of course, one difference is that a significant portion of Texas's economy and workforce is related to the fossil fuel

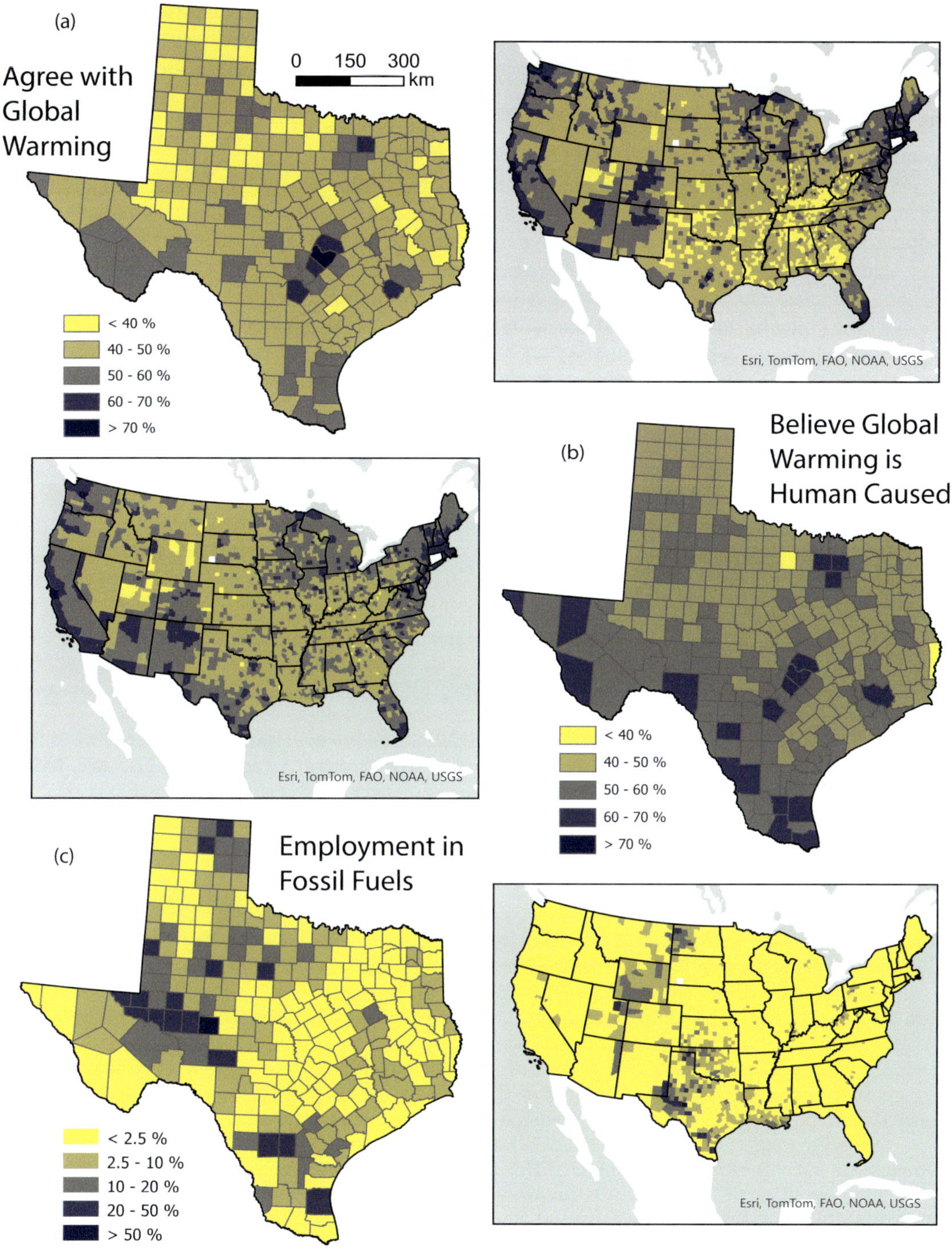

FIGURE I.1.5 Climate opinions and demographic trends summarized for respondents for United States counties. Outcomes of the Yale Climate Opinion survey include (a) % of respondents who agree that global warming is real, (b) % of respondents that believe climate change is caused by human activities. Additionally, for reference, (c) the % of people employed in fossil fuels are mapped. Climate opinion data obtained from Marlon et al., 2025, which was derived from Howe et al. (2015) and Marlon et al. (2022). Employment in fossil fuel data provided by MIT (2025).

industry (Figure I.1.5c). There are, however, complex ideologies not easily captured in these results. Although Texans are supportive of diverse incentives to stimulate economic development including carbon neutral strategies, they also have a strong sense of independence and land conservation ethic, which can, at times, align with but also conflict with federal climate policies. Understanding the ideologies of "green laws in red states" (Hess et al., 2016) can provide important lessons on social dynamics of climate mitigation and policy adoption in the future.

Perhaps the saying, "As California Goes, So Goes the Nation" is true, at least in shaping future sociopolitical philosophies and social movements associated with technological innovation (e.g. Apple). So why not a book on climate change in California? In response, consider the fact that Texas now leads the nation in most Fortune 500 companies headquartered within the state (e.g., Hewlett Packard, Oracle); many of these companies moved to Texas for practical economic reasons. Also, consider the following companies: Exxon Mobil, Dr. Pepper, JC Penney, HEB, Whole Foods Market, Southwest Airlines, Dell Technologies, AT&T, GameStop, USAA, Whataburger, Taco Cabana, Blue Bell ice cream, Scholtzky's, Jason's Deli, Fuddruckers, Six Flags, and yes, Buc-ee's. All these are major franchises started in Texas, and while none are necessarily credited with inducing social transformation or global innovation, these companies all reflect a simple mission: meeting the practical needs of society.

So, why Texas? In short, Texas is practical. It is our collective reality. Everyone can identify with climate change through the lens of Texas. And besides, who doesn't like Buc-ee's?

REFERENCES

Abbott JC, Abbott K, 2020. Common Insects of Texas and Surrounding States: A Field Guide. University of Texas Press, Austin, TX. 448 pg.

BEA (Bureau of Economic Analysis), 2025. GDP by State. https://www.bea.gov/data/gdp/gdp-state (accessed May 8, 2025).

Birdsong T, Bean M, Mayes K, Robertson S, Curtis S, 2021. Conservation status of Texas freshwater fishes and protected species recommendations. Journal of Southeastern Association of Fish and Wildlife Agencies. 8:40–52.

Cheruvelil KS, Soranno PA, McCullough IM, Webster KE, Rodriguez LK, Smith NJ, 2021. LAGOS-US LOCUS v1.0: Data module of location, identifiers, and physical characteristics of lakes and their watersheds in the conterminous U.S.. Limnology & Oceanography Letters 6: 270–292.

EIA (Energy Information Administration), 2025. U.S. States. https://www.eia.gov/state/ (accessed May 8, 2025).

Freedgood J, Hunter M, Dempsey J, Sorensen A, 2020. Farms Under Threat: The State of the States. Washington, DC., American Farmland Trust. https://farmlandinfo.org/publications/farms-under-threat-the-state-of-the-states/ (accessed May 8, 2025).

Hess DJ, Mai QD, Brown KP, 2016. Red states, green laws: Ideology and renewable energy legislation in the United States. Energy Research & Social Science. 11:19–28.

Howe P, Mildenberger M, Marlon J, Leiserowitz A, 2015. Geographic variation in opinions on climate change at state and local scales in the USA. Nature Climate Change. 5. doi:10.1038/nclimate2583.

Jones B, O'Neill BC, 2013. Historically grounded spatial population projections for the continental United States. Environmental Research Letters. 8(4):044021.

Kennedy C, 2011. Brazos River Runs Dry During Texas Drought. National Oceanic and Atmospheric Administration. Climate.gov. https://www.climate.gov/news-features/featured-images/brazos-river-runs-dry-during-texas-drought (accessed June 5, 2025).

Marlon J, Goddard E, Howe P, Mildenberger M, Jefferson M, Carman J, Rosenthal S, Fine E, Gillreath-Brown A, Leiserowtiz A, 2025. Yale climate opinion maps 2024. Yale Program on Climate Change Communication. https://climatecommunication.yale.edu/visualizations-data/ycom-us/ (accessed May 8, 2025).

Marlon JR, Wang X, Bergquist P, Howe P, Leiserowitz A, Maibach E, Mildenberger M, Rosenthal S, 2022. Change in US state-level public opinion about climate change: 2008–2020. Environmental Research Letters. 17(12):124046.

MIT (Massachusetts Institute of Technology), 2025. Joint Program on the Science and Policy of Global Change, 2025. STRESS Platform: System for the Triage of Risks from Environmental and Socio-Economic Stressors. https://est.mit.edu/ (accessed May 8, 2025).

National Climate Assessment (NCA), 2025. The Fifth National Climate Assessment. https://nca2023.globalchange.gov/ (accessed May 8, 2025).

NPST (Native Plant Society of Texas), 2025. Native Plant Database. https://www.npsot.org/resources/native-plants/native-plants-database/ (accessed May 8, 2025).

Schmidly DJ, Bradley RD, 2016. The Mammals of Texas, 7th ed. University of Texas Press, Austin, TX.

TAMNRI (Texas A&M Natural Resources Institute), 2025. Texas Land Trends: A Database of Compiled and Analyzed Values for Working Lands in Texas. College Station, TX. USA. http://txlandtrends.org

TAM OTSC (Texas A&M, Office of the Texas State Climatologist), 2025. Texas Severe Weather. https://climatexas.tamu.edu/products/severe-weather-summaries/index.html (accessed May 8, 2025).

TBRC (Texas Bird Records Committee), 2025. https://www.texasbirdrecordscommittee.org/home (accessed May 8, 2025).

Texas State Library and Archives Commission, 2017. United States and Texas Populations 1850–2017. https://www.tsl.texas.gov/ref/abouttx/census.html (accessed May 8, 2025).

TPWD (Texas Parks and Wildlife Department), 2025. Wildlife Conservation Program: What We Do. https://tpwd.texas.gov/wildlife/wildlife-diversity/about/ (accessed May 8, 2025).

TWDB (Texas Water Development Board), 2022. 2022 State Water Plan. https://www.twdb.texas.gov/waterplanning/swp/ (accessed May 8, 2025).

TXNDD (Texas Natural Diversity Database), 2025. Element Occurrence Data Export. Wildlife Diversity Program of Texas Parks & Wildlife Department. (accessed May 8, 2025).

US Census Bureau, 2023. https://www.census.gov/programs-surveys/popproj.html (accessed May 8, 2025).

US Census Bureau, 2024. https://www.census.gov/data/tables/time-series/demo/popest/2020s-state-total.html (accessed May 8, 2025).

US Census Bureau, 2025. Population and Housing Unit Estimates Datasets. https://www.census.gov/programs-surveys/popest/data/data-sets.html (accessed May 8, 2025).

USGS (U.S. Geological Survey Gap Analysis Project), 2024. Protected Areas Database of the United States (PAD-US) 4.1: U.S. Geological Survey Data Release. https://doi.org/10.5066/P96WBCHS (accessed May 8, 2025).

Zoraghein H, O'Neill BC, 2020. US state-level projections of the spatial distribution of population consistent with shared socioeconomic pathways. Sustainability. 12(8):3374.

1.2 The Changing Climate of Texas

John W. Nielsen-Gammon

1.2.1 CLIMATE ACROSS TEXAS

Texas is a land of climate variations both subtle and stark. According to the conventional Köppen climate classification system, Texas includes parts of five climate zones. Most of the state, east of Childress, Abilene, and Uvalde, is a humid subtropical climate, a classification shared by most of the south-central and southeast United States. Farther west, as the climate gets progressively drier, semi-arid and desert climates are found, subdivided into hot and cold semi-arid and desert. Hot semi-arid is the second most common Texas climate, found in much of west-central Texas as well as southwestern Texas from Del Rio to Brownsville. Farther upstream, the Rio Grande and part of the Pecos River flow (when they flow) through desert climates, with the mountains between the two rivers able to catch enough moisture to rise above the desert, both figuratively and literally.

Texas's climate is determined by its distance from the equator and by the surrounding land and water. Texas lies between 25.8°N and 36.5°N. This puts it in the transition zone between the tropics and the mid-latitudes. During the cooler times of the year, the weather patterns in Texas move from west to east, as they do in the mid-latitudes, with storm systems carried and steered by the jet stream. During the warmest times of the year, the weather can be stagnant, almost unchanging from day to day, with an occasional tropical weather system moving inland from the coast.

The Gulf of Mexico (also known as the Gulf of America) is one of the three key geographical features affecting Texas's climate. Winds from the Gulf dominate Texas weather during the summertime, bringing warm and humid air inland. The Gulf serves as a moderating influence, with its relatively steady sea surface temperatures keeping coastal temperatures within reasonable bounds. Moisture from the Gulf (and farther upwind from the Caribbean and the open Atlantic) is the source of most of the precipitation that falls in Texas.

The Western North American Cordillera (including the Rocky Mountains and the Sierra Madre) is the second key geographic feature. These mountains block moisture from the Pacific Ocean, meaning that western Texas must rely on moisture carried long distances from the Gulf. Especially in wintertime, the normal midlatitude wind direction is west to east, so copious rainfall is rare in western Texas. When strong west winds do develop, they descend the eastern slope of the mountains, causing the air to become warm and dry. Intense versions of this weather pattern can lead to dust storms and wildfires.

The final key geographic feature is the Great Plains, stretching from Texas north into Canada. Unlike the Gulf, which imposes its temperature on the air above it, the Great Plains serve as a vast north-south expressway, allowing Arctic air to penetrate southward into Texas and sometimes beyond to Mexico and Central America. The western mountains serve as one bank of a channel, forcing air southward that might otherwise have spread westward and causing the cold fronts (the leading edges of the Arctic air masses) to become unusually sharp. Texans call these cold air outbreaks "blue northers."

People moving to Texas from the northeastern United States often complain about the lack of autumn. Technically, Texas has three months of fall just like anywhere else, but it tends to lack the colorful leaves and crisp clear air of northeastern autumn. Texas makes up for this by having a memorable spring season, when colorful wildflowers combine with dramatic thunderstorms, while the northeastern United States is tolerating mud season.

The temperature patterns in Texas are well summarized by the normal daily maximum and minimum temperatures in January and July (Figure I.2.1). The January temperature patterns are

DOI: 10.1201/9781003406600-3

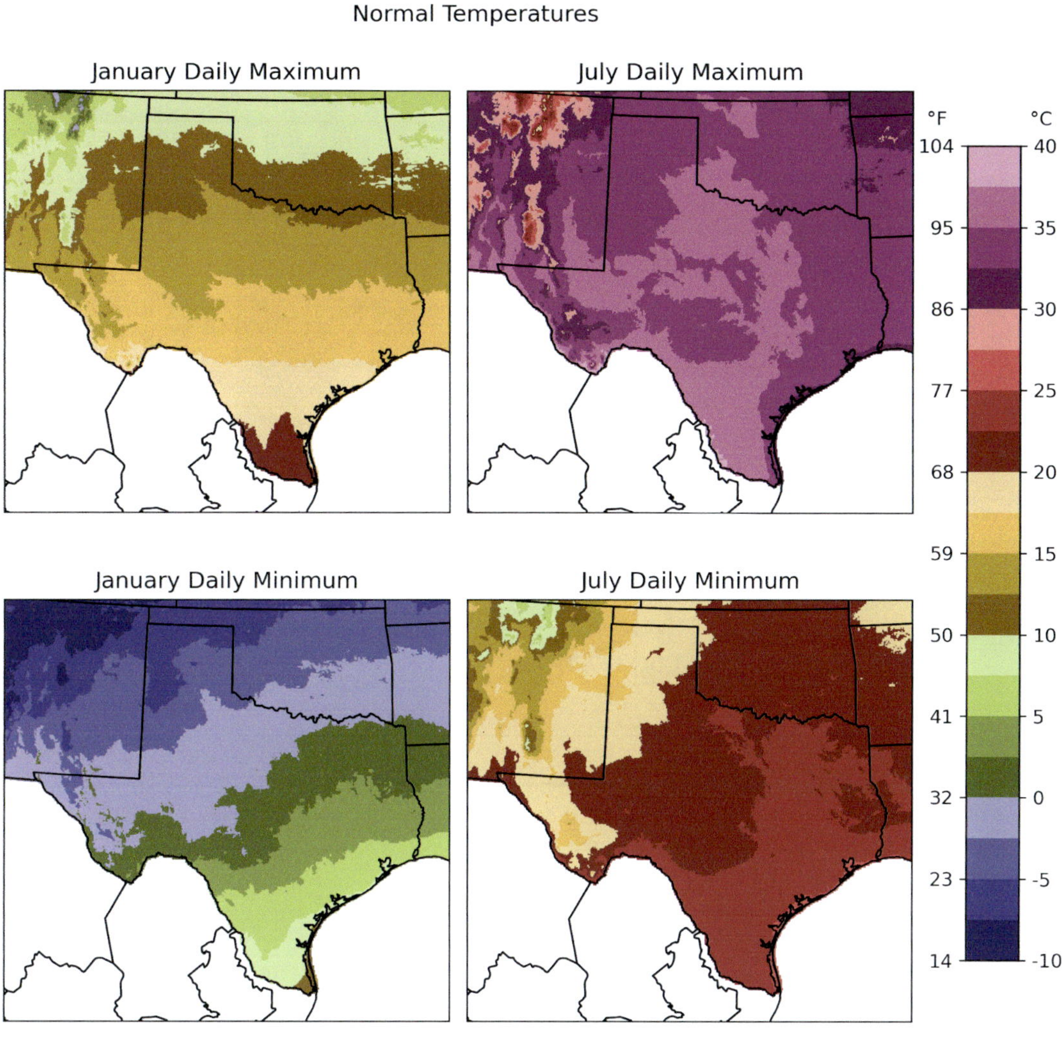

FIGURE I.2.1 Maps of the normal January and July daily maximum and minimum temperatures in Texas and surrounding areas. (Data from https://prism.oregonstate.edu/.)

simple to remember: colder to the north, warmer to the south. Normal daily maximum temperatures in January vary by 14°C (25°F) from the northwestern Panhandle to the Lower Valley, and normal daily minimum temperatures in January vary by more than 17°C (30°F). This temperature difference is caused in part by the normal variation of temperature with latitude: southern latitudes receive much more warming sunlight in winter than northern latitudes. The Gulf also serves as a heat source in winter, and the temperature gradient reflects that northern Texas gets more winter air from the Great Plains and southeastern Texas gets more winter air from the Gulf, simply due to proximity. Departures from this pattern are mainly in far west Texas, where large changes of elevation have a substantial influence on local temperatures.

Summertime maximum temperatures are much more uniform across the state. In part this is because the summer sun is fairly intense at all Texas latitudes; in part the role of the Gulf is reversed from wintertime, now serving as a cooling influence. The highest maximum temperatures are typically found along the Rio Grande in the Big Bend region, with Laredo, Midland, and Wichita Falls near the center of other patches of relatively high temperatures. Nighttime temperatures in

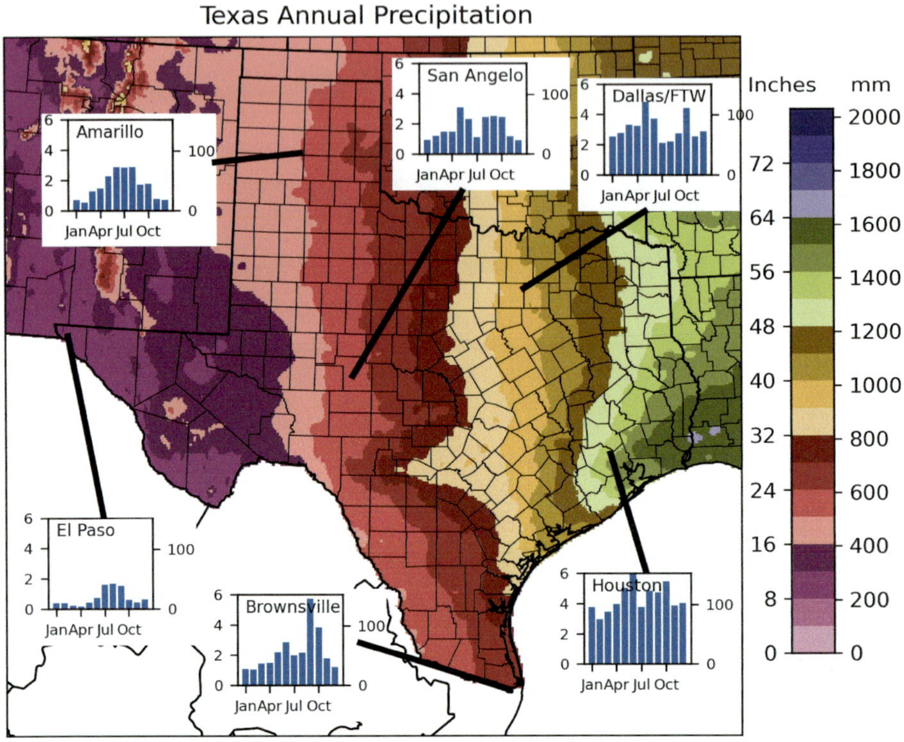

FIGURE I.2.2 Map of normal annual precipitation for the state of Texas. Also shown are the monthly precipitation amounts for six cities around the state. Precipitation amounts are given in inches (left) and mm (right). (Data for map from https://prism.oregonstate.edu/; data for bar charts from the Applied Climate Information System.)

summer have more geographical variation, which is attributable to differences in humidity. In eastern Texas, and especially along the coast, dew point temperatures are typically in the low to mid 20s °C (70s °F) in July, while in western Texas they are more commonly in the upper single digits °C (40s °F). This allows the ground and air to cool more at night in western Texas. Also, due to its relatively high altitude compared to its surroundings, the Hill Country (Edwards Plateau) provides some refuge from high daytime and nighttime temperatures.

While temperature varies mainly from north to south across Texas, precipitation varies mainly from west to east (Figure I.2.2). The desert areas of the state average as little as eight inches of precipitation a year, while the southeast corner of the state typically receives over sixty inches. Variations in that pattern are due to topography. As mentioned earlier, the mountains of west Texas capture more rainfall than the lowlands. A tongue of relatively high average precipitation extends from east to west along the Balcones Escarpment just north of San Antonio. The escarpment is the first sizable obstacle humid Gulf air encounters, and the resulting upward motion can trigger thunderstorms or enhance existing precipitating systems. There is also a tendency for enhanced precipitation just inland from the coast, partly due to the sea breeze.

I.2.2 TEXAS CLIMATE THROUGH THE YEAR

The sequence of weather and climate patterns throughout the year is hinted at in the annual precipitation figure. In stations in western Texas (El Paso and Amarillo), precipitation is strongly seasonal, with most of the precipitation occurring during the warm half of the year and little from November through February. This is in keeping with the seasonal wind patterns. While winds from the west keep moisture away from western Texas during the wintertime, low pressure develops over the desert southwest during summer, drawing moist air northwestward from the Gulf. This change in wind patterns fits the technical definition of a monsoon and the resulting rainfall in July and August fits the popular definition of a monsoon as well, so this weather pattern is known as the Southwest Monsoon. Most of the summertime rainfall comes from thunderstorms, while the wintertime precipitation is a mix of rain and snow.

In most of the rest of Texas, there are two seasonal peaks of rainfall. The first comes in May and June, with May the preferred month for squall lines and other mesoscale convective systems that form near the dry line in western Texas and march eastward and southward across the state. In June, such organized convection can be supplemented by disturbances that form in the Gulf and move inland. Sometimes these are the first named storms of hurricane season; storms that form in the Gulf are more common during June than at any other time during hurricane season. These storms can be triggered by weak upper-level disturbances breaking off from the southern edge of the jet stream.

By July the jet stream is far enough north that disturbances become rare. July and, to a lesser extent, August, produce less rain than May and June due to the lack of disturbances and the reliance on ordinary thunderstorms. Rainfall in eastern Texas in July is barely larger than rainfall in western Texas. As hurricane season ramps up, average precipitation increases across most of the state in September. In eastern Texas the rain is fed by storms from the Atlantic, while in western Texas the rain is fed by storms from the Pacific that cross northern Mexico. Also, in October the jet stream returns south and plays a more active role in the weather, and since sea surface temperatures have reached their peak, the moisture available to weather systems can be plentiful.

Tornadoes are most common in April and May. Hail and strong thunderstorm winds are also favored during the springtime, but often occur during the summer too. Especially in eastern Texas, a secondary severe weather season develops in the fall, peaking in November. Severe weather in general is most likely in northern Texas.

With temperatures varying from north to south, the agricultural season varies too. Crops are grown throughout the year in the Lower Rio Grande Valley, where freezing temperatures are relatively rare. The beginning of the growing season progresses northward from late February to early April for most crops, with corn planted earliest and cotton planted latest. Winter wheat, grown mostly in the Panhandle and west-central Texas, is off-cycle, being planted in the fall and coming to maturity in late spring.

1.2.3 HOW CLIMATE CHANGE BROADLY AFFECTS TEXAS

Global climate changes, by whatever cause, affect the weather and climate of Texas. In the past 20,000 years, western Texas has seen massive sand dune fields (Holliday, 2001), and the coastline of Texas was at least 50 miles farther southeast than its present position (Simms et al., 2007). At the same time, not all historical changes are entirely caused by global climate change. Local changes in land use and the random variations of the weather both contribute to what Texas has experienced over the past century or more.

It can be useful to separate climate change impacts into direct impacts caused specifically by rising temperatures and indirect impacts caused by changes in weather patterns and storm structure. The latter are also caused ultimately by rising temperatures, but through a longer and more complex chain of cause and effect. In general, the direct effects are much better understood and easier to quantify.

The magnitude of the effects of climate change can be assessed through a combination of three different approaches. The first approach relies on information regarding historical trends. This approach is reliable when the historical data is extensive and of high quality and when the year-to-year variability is small compared to the long-term trend. Examples of high-quality, high signal-to-noise trend data are temperature records and sea level records. An example of low-quality historical data is the time series of tornado frequency and intensity: in the not-too-distant past, people tended to flee tornadoes rather than racing toward them with video cameras, so we don't have good information on changes in the frequency of tornadoes over time.

The second approach relies on computer model simulations. This approach is reliable if the phenomenon is easily and accurately simulated by global climate models. An example of this is, again, temperature changes, which tend to be large-scale and are simulated directly by models. An example of a phenomenon for which models don't help much is, again, tornadoes, which are much too small to be simulated by a global model. About the best we can do is try to identify the key atmospheric features that lead to tornado formation and examine how those features are simulated to change.

The third approach is good old basic physical understanding. If rising temperatures obviously ought to cause a certain effect, we can have confidence that the trends we see in observations or model output are real. Once more, temperatures lead the way: if temperatures globally increase, we would expect temperatures locally to increase, too. Tornadoes again provide a good counterexample: the process by which tornadoes form is very complex and is not directly related to temperature. So scientists have an excellent basis for inferring trends in temperatures and very little basis for inferring trends in tornadoes. Other types of extreme weather lie somewhere between these two extremes.

Figure I.2.3 shows four temperature records: the average annual temperature for Texas and the globe, the average of the hottest days of June, July, August, and September, and the average of the coldest days of December, January, and February. The Texas values are averages from a representative station in each county, while the global values assume a historic global mean temperature of about 14.0°C.

The historical trend of temperatures in Texas is clear for hot extremes, cold extremes, and everything in between: a gradual warming trend during the first half of the 20th century, then cooling into the 1970s, and warming since then. Based on this historical data as well as climate model simulations and basic understanding, temperatures are expected to continue rising in Texas at a similar pace to the past half-century for another several decades at least. According to climate models, the rate of change of temperature in Texas ought to be about 20%–30% larger than the rate of change of the global mean surface temperature. This is similar to what has been experienced in recent decades and is also similar to other land areas. It is due to the more gradual response of oceans to the Earth's energy imbalance, as well as carbon dioxide reducing the amount of transpiration by plants.

Warmer temperatures would lead to more heat extremes, and indeed the annual number of triple digit days (in Fahrenheit, fortunately) has more than doubled over the past 50 years (Nielsen-Gammon et al., 2024). Conversely, warmer temperatures would lead to fewer cold extremes, and indeed cold extremes have become less common historically and are projected to continue to become less common. However, as the chart illustrates, extremes are more erratic than average temperatures. The average annual temperature in Texas may not in the foreseeable future be below its long-term average anymore, while both hot and cold extremes are still frequently below their respective averages. In particular, the warming trend of extreme cold temperatures is much smaller than the typical difference in extreme cold from one year to the next.

The recent downturn in the graph of extreme cold temperatures is due to the period 1990–2020 having remarkably few extreme cold events, followed by a more historically representative period recently. At least one study claimed to show that the weather patterns associated with extreme cold in Texas were becoming more frequent because of global warming and resulting Arctic ice loss, but that does not mean extreme cold itself becomes less frequent. The overall trend is due to a combination of the weather pattern trend (indirect) and the temperature trend in the Arctic (direct), and all evidence indicates that the Arctic warming trend is easily strong enough to overwhelm any weather pattern changes and make the Arctic air that reaches Texas milder.

Note that here we have an example of a type of extreme weather that is expected to become less common in Texas. Other related examples are snowstorms and ice storms. Because the factor that

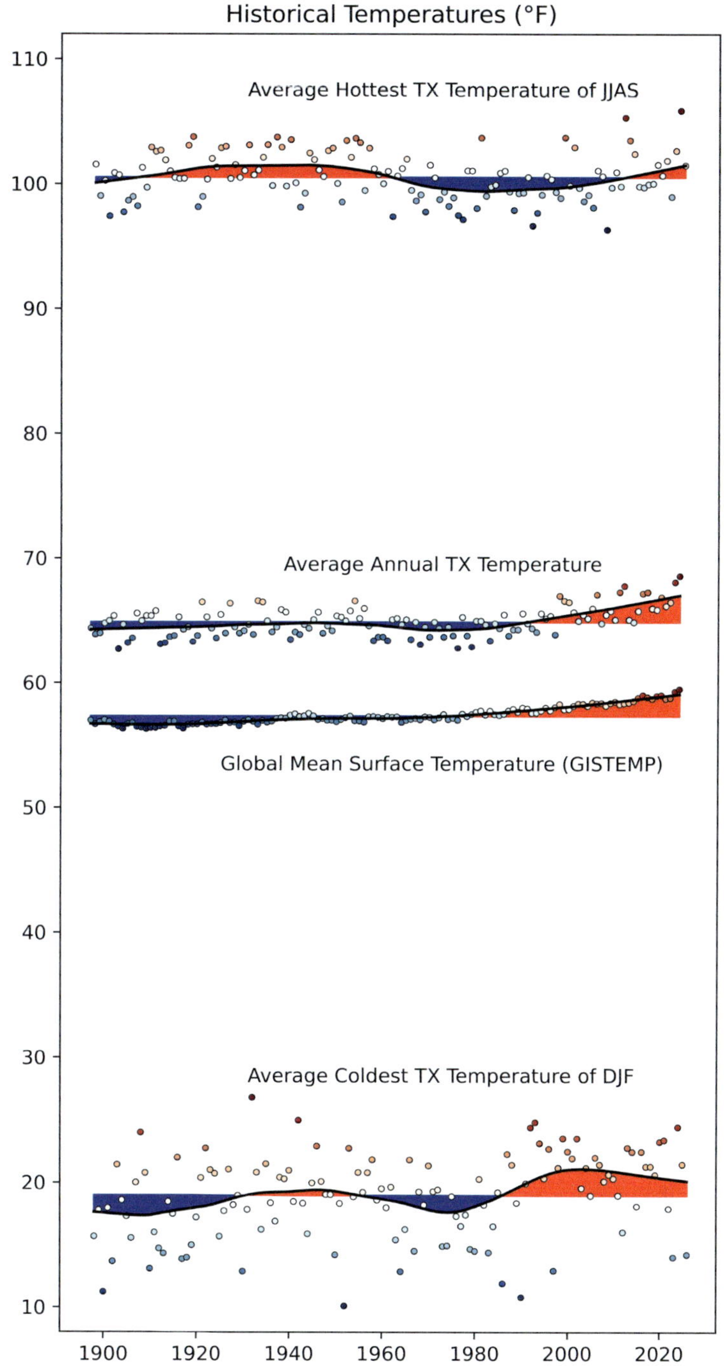

FIGURE I.2.3 Annual values and smoothed graphs of surface temperatures. From top to bottom: the annual average hottest monthly temperature in Texas for the months of June through September, the annual average temperature in Texas, the global mean annual average temperature, and the annual average coldest monthly temperature in Texas for the months of December through February. Shading indicates departures above (red) and below (blue) the overall average, while dot color intensity indicates the magnitude of the individual departure from the overall average. Texas values based on averages individual composite stations (one per county) with station data from the Applied Climate Information System, while global values are from the GISTEMP data set hosted by the National Aeronautics and Space Administration's Goddard Institute for Space Studies.

limits the amount of snow and ice in Texas is the winter temperature, warmer winter temperatures should directly lead to less winter weather.

The Texas temperature trend, while similar to what was expected in recent decades, was unusual during the period 1950–1980. Global temperatures during that period stayed relatively stable, while in Texas temperatures plummeted to their lowest historical value. Scientists do not know why; it is possible that changing land use patterns, such as surface reservoirs and increasing use of irrigation, may have reduced temperatures locally, especially in the summertime. Extreme heat is only now on a par with typical values from the early 20th century.

One way of thinking about the temperature trends is to translate them into geographical patterns. For example, Texas temperatures are about 1.4°C (2.5°F) above their 20th century average. So the climate at various locations in Texas is becoming more like the climate farther south. How much farther south? In wintertime, the total difference between temperatures at the northern Texas border and the southern Texas border is close to 17°C (30°F). A warming of 1.4°C (2.5°F) is about one-twelfth of that, so it corresponds to moving south one-twelfth the height of Texas, or about 105 km (65 miles). For comparison, a typical Texas county is about 50 km (30 miles) across, so the climate has gotten two counties warmer in wintertime. In summer, the climate at any given location in Texas is becoming unlike anything previously found within the state of Texas.

Another type of weather extreme that is directly affected by temperatures is rainfall intensity. The intensity of extreme rainfall has increased historically in Texas (Jorgensen and Nielsen-Gammon 2024), and the increase is expected to continue simply because warmer air can hold more water vapor, producing greater amounts of condensation if the air rises at the same rate as before. While indirect effects on weather patterns may change the intensity and frequency of upward motion, it seems that the direct effect of higher temperatures is the dominant factor over most mid-latitude land areas, including Texas.

So if rainfall intensity is expected to increase, does that mean that southeast Texas will see more rain events like Hurricane Harvey? Probably not. The intense multi-day rainfall associated with Hurricane Harvey has been estimated to have an annual likelihood of about 1 in 10,000 of occurring at any given location. While the risk of a Harvey is increasing, it is not increasing fast enough to make another Harvey likely.

Also, just because Harvey hit southeast Texas in 2017 does not mean that a storm like Harvey cannot hit, say, Brownsville or Victoria. There has historically been a large increase in extreme rainfall in the Houston area because they have been nailed by storms like Allison, Harvey, and the Tax Day Flood. That trend is larger than what would be expected from climate change, so the trend is unlikely to continue at the same pace. Conversely, the places like Victoria that have been spared extreme rainfall recently have seen a long-term trend that is smaller than what would be expected from climate change, so the trend there is likely to increase substantially. Except for things like temperature which are increasing everywhere, local trends in weather extremes are generally not a reliable measure of the magnitude of the effect of climate change.

Overall precipitation amounts in Texas do not necessarily track changes in extreme precipitation. On a global scale, we know they are different: extreme rainfall trends are typically much larger than average rainfall trends, and the chain of cause and effect is very different for those two aspects of precipitation. Also, the weather in Texas, especially during the period October through May, is strongly affected by the presence of El Niño (typically cool and wet in Texas) and La Niña (typically warm and dry in Texas). This remote influence makes rainfall from year to year much more variable than at places in the northern United States with similar annual precipitation. Hence it is necessary to examine trends over a longer period of time if possible.

The long-term trend of precipitation in Texas (Figure I.2.4) presents a mixed picture. In central and eastern Texas, precipitation generally increased by 1%–2% per decade, which over a century amounts to around a 20% increase. Meanwhile, in western Texas precipitation trends were flat or even declined by as much as 1% per decade.

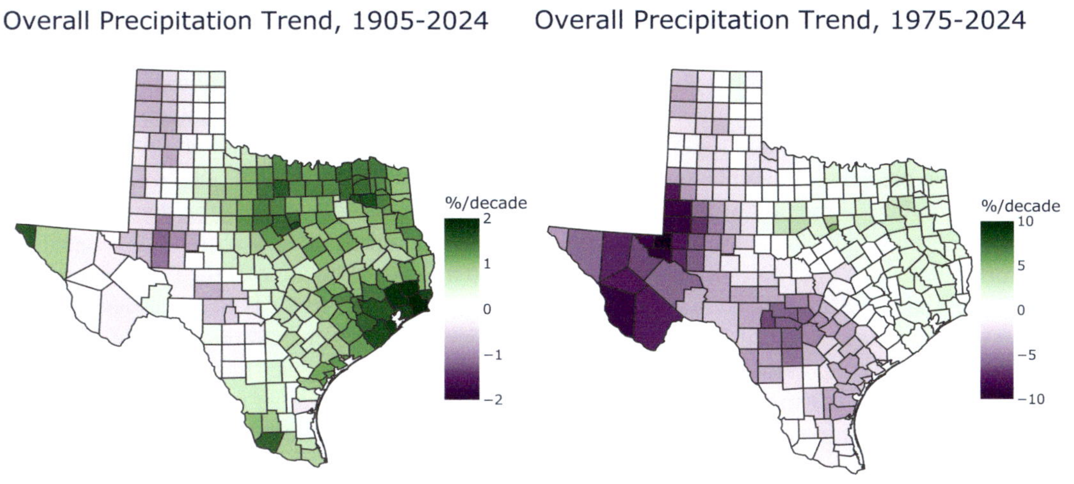

FIGURE I.2.4 Precipitation trends (percent per decade) from linear regression over the period 1905–2024 (left) and 1975–2024 (right). Values based the nClimDiv county-scale data set provided by the National Centers for Environmental Information.

More recent trends are larger. Over the past half-century, precipitation in far west Texas has declined by as much as 10% per decade. The longer-term rise in central Texas and the Coastal Bend region hides a shorter-term decline of a few percent per decade. In northern and eastern Texas, the positive trend persists and has even amplified in places (note the different color scale for the longer-term and shorter-term trends).

As mentioned earlier, natural variability can play a large role in these swings. Computer model projections, on average, project steady rainfall or a slight decline (1% per decade), with Texas occupying the middle ground between systematic declines in Mexico and systematic increases in the northern United States and Canada. Individual climate models present a diverse set of possibilities for precipitation, from large declines to moderate increases. However, no climate model projects anything near the large spatial variation in trends seen in Texas over the past half-century, suggesting that those trends are dominated by natural variability rather than the changing large-scale climate.

Knowing, or at least suspecting, that these trends are largely temporary does not make them any less real. Recent years have seen devastating droughts in western and southern Texas; the historic trends demonstrate that precipitation in those areas has indeed become less reliable, at least temporarily. Conversely, if the substantial increases in northern and eastern Texas are also temporary, it may be wise to prepare for a drying trend there.

I.2.4 THE FUTURE OF CLIMATE CHANGE IN TEXAS

The question of what the future holds for climate in Texas is a tough one. To figure out what will happen, and what could happen, we actually need to answer three separate questions.

I.2.4.1 Question #1: How Much Will the Global Climate Change?

The global-scale rise in temperatures, seen in surface air temperatures, sea surface temperatures, atmospheric temperatures aloft, and deep ocean temperatures, is unequivocally positive. The magnitude of temperature change in these various parts of the climate system is also well known to within relatively small uncertainty ranges. But that is the past. Predictions are hard, especially about the future.[1]

While forecasting the weather is distinctly the province of meteorologists, forecasting the climate is well beyond the skill set of climatologists. The future climate depends on technological advances in clean energy, on global and regional climate change policies, and on economic prosperity. Such things are so patently impossible to predict that climate scientists use several potential scenarios to simulate possible future climates. The scenarios, ranging from high carbon emissions to negative carbon emissions, get plugged into climate models to simulate the potential climate consequences.

You actually don't need a fancy climate model to get a basic expectation of the future magnitude of climate change. All you need is a decent estimate of *climate sensitivity*, typically specified as the amount of warming expected from a doubling of carbon dioxide concentrations. There are plenty of other things driving climate change, such as different greenhouse gases (more warming) and air pollution in the form of tiny airborne particles (less warming). Those other two sets of drivers rank second and third behind carbon dioxide over the past several decades, and they are nearly equal and opposite, so the climate response to carbon dioxide alone is not far off from the overall change in global temperature caused by humans.

As recently as a couple of decades ago, the main tools for estimating the magnitude of climate sensitivity were the global climate models themselves: run them and see how hot the simulations get. But there are so many details that matter, such as the amount of low-level cloud cover in the southeastern Pacific Ocean, that the climate models only get you in the right ballpark. Nowadays, there are three other mostly independent ways of estimating climate sensitivity that seem to be more accurate collectively: looking at the amount of change so far, looking at the size and drivers of shorter-term natural climate fluctuations, and looking at estimates of global temperatures in the distant past, such as during the last big Northern Hemisphere glaciation about 20,000 years ago. The resulting estimate is about 2.5°C to 4.0°C of eventual warming if carbon dioxide concentrations are doubled and stay doubled.

Without knowing the ultimate carbon dioxide concentration, we cannot say how much the climate will warm. The various scenarios give temperatures in 2100 ranging from less than 1.5°C of warming (the massive reductions in greenhouse gas emissions required probably won't happen) to more than 5.0°C of warming (the massive increases in greenhouse gas emissions required probably won't happen either). A middle-of-the-road projection might be something like 2.5°C (optimistic) to 4.0°C (pessimistic).

I.2.4.2 Question #2: How Much Will the Local Climate Change in Response to a Change in the Global Climate?

Based on past climate changes as well as climate model simulations of the future, the temperature increases over land are larger than the temperature increases over the ocean. As mentioned above, Texas temperatures are expected to rise about 20% more than global temperatures. That means that a middle-of-the-road global projection such as 3.0°C corresponds to a middle-of-the-road local projection of 3.6°C. But Texas uses Fahrenheit, and 3.6°C translates to 6.5°F, which sounds like it ought to be a high-end projection rather than a middle-of-the-road projection.

Just as Texas temperature changes are proportional to global temperature changes, so too are many other aspects of climate change. That means that, instead of estimating future climate based on a scenario that probably won't happen, we can skip that part and just estimate the local climate change that corresponds to a given amount of global change. So, for example, a global warming level of 2.0°C will correspond to Texas warming of about 2.5°C (4.5°F) (Figure I.2.5). Even if a climate model gets its overall sensitivity wrong, we can still use its output to get the relative amount of climate change at different global warming levels. Depending on the scenario, the globe will surpass 2.0°C warmer than preindustrial (and Texas about 2.5°C (4.5°F) warmer) sometime around 2040–2050, unless there are massive and lasting reductions in emissions.

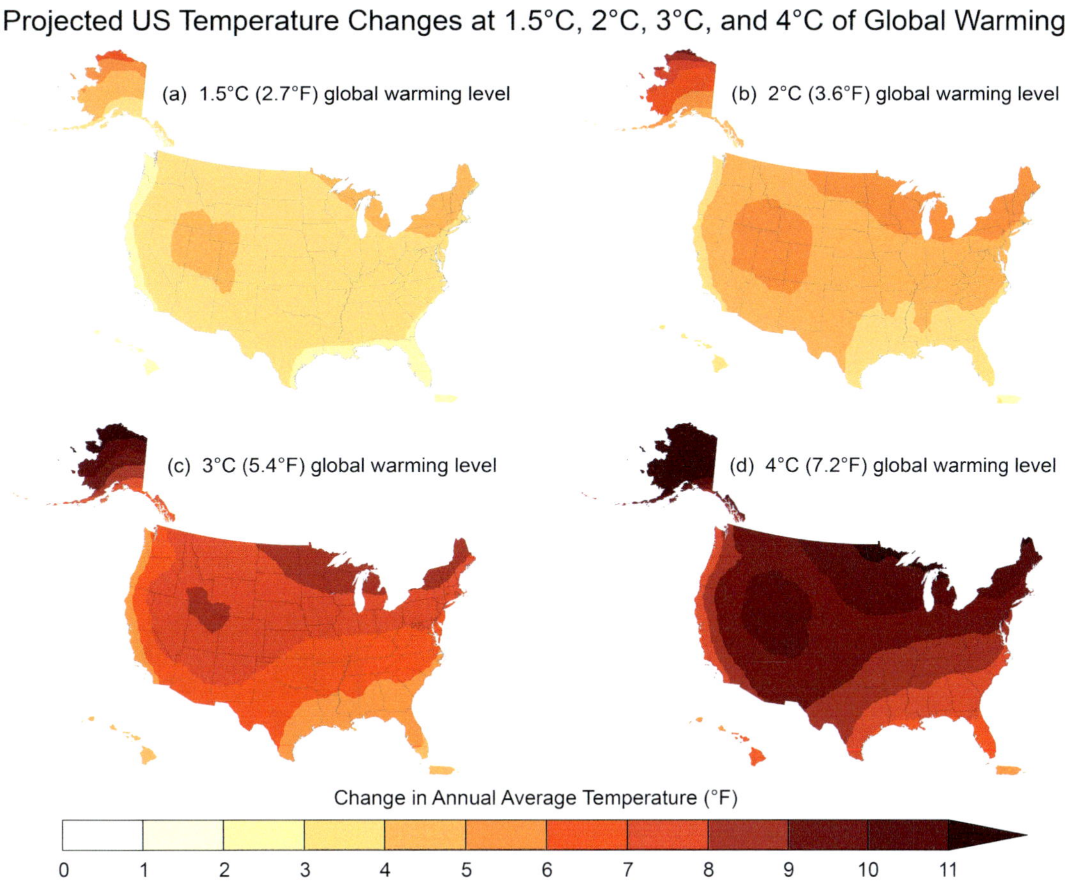

FIGURE I.2.5 (a)–(d) Projected changes in average temperatures for various levels of global warming. (Figure 2.9 in the Fifth National Climate Assessment (Marvel et al., 2023).)

Under the higher emissions scenarios, global temperatures would pass the 4.0°C mark (and Texas temperatures would approach 5.5°C (10°F) warmer than preindustrial) between about 2080 and 2120.

The projected temperature changes in Texas are smaller than those in most of the northern and western United States. This is because of Texas's proximity to water, which unfortunately also means that humidity is expected to increase too.

So how much will rainfall in Texas change if the globe is on average 2.0°C warmer? We don't know, as noted above. Generally speaking, if climate models agree on the local change of some particular climate parameter, we suspect that they are probably right. If they don't agree, we will have to rely more on historical trends or other lines of evidence. With precipitation, they don't generally agree on whether precipitation will increase, decrease, or stay the same. There is a general tendency (but not a consensus) for models to project relatively wetter trends toward Northeast Texas and relatively drier trends toward South and Southwest Texas (Figure I.2.6).

As with temperatures, the projected changes in precipitation are somewhat smaller in Texas than in most of the rest of the United States. However, that is only the average of the projections. In models that project a drying trend for most of the United States, that drying trend is largest in Texas and the rest of the Southwest. The lack of hatching over Texas in Figure I.2.6 indicates that at least a fifth of the models disagree on whether precipitation will go up or go down. That is not surprising;

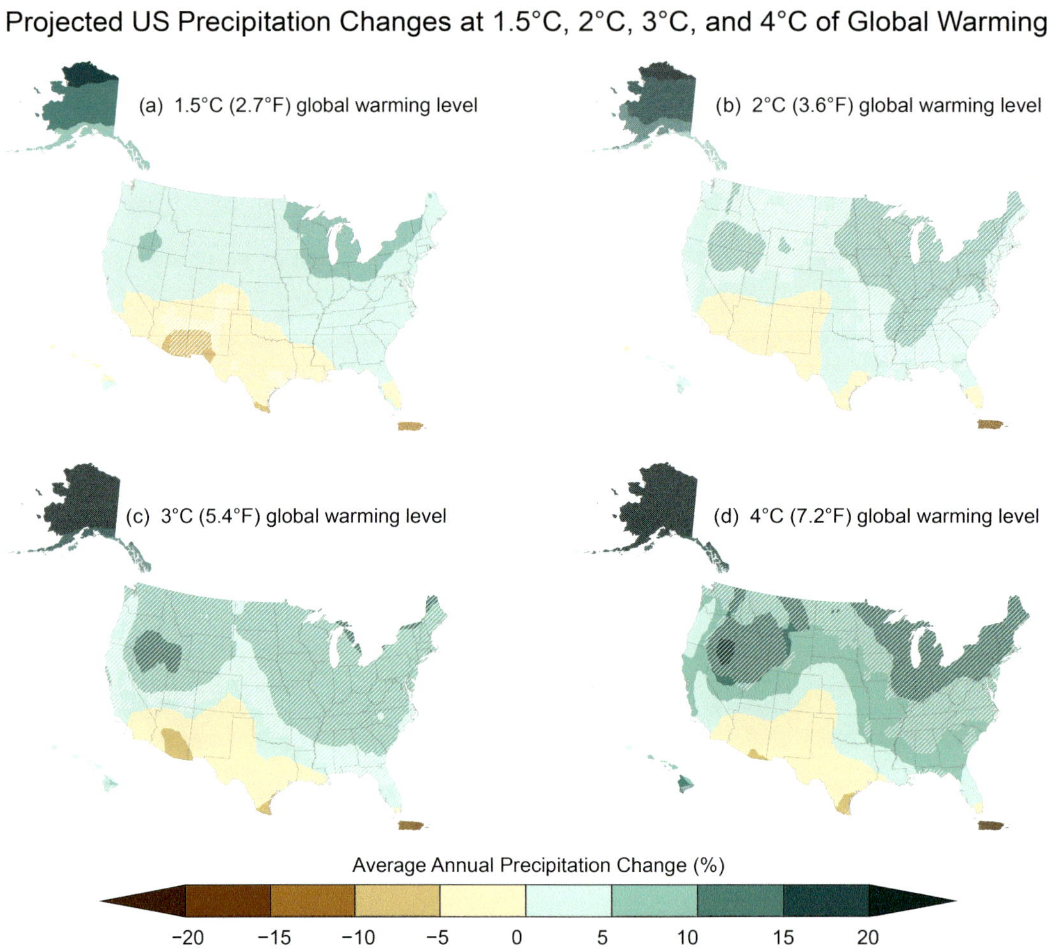

FIGURE I.2.6 (a)–(d) Projected changes in annual precipitation for various levels of global warming. White hatching indicates where at least 80% of model projections agree on the sign of the change. (Figure 2.10 in the Fifth National Climate Assessment (Marvel et al., 2023).)

with the average projected trend being zero across parts of northern Texas, there must be about half of the models projecting an increase and half projecting a decrease.

I.2.4.3 Question #3: What Else Will Cause Texas Climate to Change?

Texas climate has always been changing. We live in an era in which that change is remarkably rapid and for which we have sufficient evidence to understand the cause. But we know that other things besides greenhouse gases have changed the climate in the past. Some of those are global in their impacts, while some are more local.

On the global side, catastrophic volcanic eruptions are possible. Prehistoric evidence shows that really intense volcanic eruptions can lower global temperatures by a degree or more for a few years. But we cannot predict them, and climate change probably has not affected the risk of a super-eruption. So we tend to ignore that risk, just as we always have.

On the local side, changes in the Texas landscape affect the climate. The urban heat island, for example, elevates urban temperatures, and within major urban areas in the state the urban heat island temperature increase can be as large or larger than the climate change-driven increase in overall Texas temperatures. As shown above, though, on a county scale it is not really possible to pick out the big Texas metropolitan areas on the basis of recent temperature trends. In rural areas, irrigation of crops raises humidity and lowers daytime temperatures, much like a gigantic swamp cooler. There may also be urban and rural influences on rainfall, through their effects on temperature and moisture and even (in the case of urban areas) winds. All of these will help determine future Texas climate.

Also mattering is good old-fashioned natural variability. Besides El Niño and La Niña, other variations in the Pacific, Atlantic, and even the Indian Ocean affect Texas weather. That is partly why there is so much difference in precipitation from year to year and even decade to decade. Because these are all short-term natural cycles, they do not fall into the category of climate change. But with projections of changes in Texas precipitation being relatively small, natural variability will hide changes in long-term precipitation trends for a few decades.

The bottom line is that human-driven global climate change is likely to have a large effect on Texas climate. It will not be the only thing having an effect, but it will be a dominant driver of local climate change and we have good ways of estimating its future impact on many aspects of Texas climate. Temperature, for example, is fairly easy. Most other things, as discussed in section I.2.5, are harder.

I.2.5 THE DETAILS OF CLIMATE CHANGE IN TEXAS: IT'S COMPLICATED

Texas experiences a laundry list of extreme weather: droughts, floods, severe thunderstorms, hurricanes, and wildfires, in addition to the extreme heat, extreme cold, and extreme rainfall already discussed. For each of these phenomena, projections of future impacts in Texas are complicated by the complexity of the weather involved, the challenge of making projections, and other local factors that will also have large effects on their human, environmental, and economic impacts.

Droughts are so complicated that there is no generally accepted definition! Generally speaking, it is a meteorological condition that leads to an abnormal and impactful lack of water. Different types of droughts are distinguished by their impacts: agricultural drought, hydrologic drought, socioeconomic drought. Because of this, it is hard to make a blanket statement about drought trends. But some aspects of drought trends are clear. First, large changes in total rainfall can overwhelm just about anything else affecting droughts. If rainfall does not change much, the higher temperatures will lead to increased evaporation rates, meaning water does not last as long. A robust projection of climate models is that soil moisture will decrease, even in places where rainfall increases slightly. This could have adverse effects on agriculture and would also lead to reduced replenishment of aquifers. Higher evaporation is also not good for reservoir water supplies.

In eastern Texas, normal rainfall is adequate to refill reservoirs, but in western Texas it takes a flood. The competing effects of drier soils and more intense rainfall means that normal stream flows are likely to decrease, but the most intense floods might become worse, making surface water supplies more erratic. In urban areas, soil moisture has little effect on flash floods because much of the ground surface is pavement or is otherwise impermeable to moisture. Thus, more intense rainfall is likely to drive more frequent and intense urban flooding. The actual risk of flooding will depend also on what sorts of countermeasures are taken to avoid or prevent floods. Flood control measures might keep up with, or even outpace, the risk of flooding.

Trends in severe thunderstorms are particularly unclear, due to a lack of long-term data, the challenge in simulating them, and the myriad of factors that can affect them. Perhaps in five or ten years we will have enough climate-scale simulations at high enough resolution to simulate severe thunderstorms directly, but until then it is more a matter of waiting and seeing. For example, hail might become more frequent as instability in the atmosphere increases, but hail might become less frequent because it is more prone to melting in a warmer climate.

Hurricanes involve many different trends. Intense rainfall from hurricanes is likely to increase, partly for the same reason that intense rainfall in general is increasing. The intensity of hurricanes themselves is expected to change, with weaker hurricanes becoming weaker and stronger hurricanes becoming stronger. The crossover point, above which hurricanes become more frequent, is thought to be somewhere around Category 3 or Category 4 on the Saffir-Simpson scale. So, like floods, hurricanes are likely to become less frequent, except for the really devastating ones. Hurricane storm surge depends on the strength of hurricanes but also depends upon sea level rise. Along the Texas Gulf coast, sea level rise is partly due to coastal subsidence and partly due to rising global sea levels. By the middle of the 21st century, the overall risk of storm surge may have doubled from sea level rise alone.

Wildfires involve fuel availability, fuel dryness, and suitable weather conditions. Fuel availability is expected to stay the same or increase, partly due to increased carbon dioxide levels and partly due to increases in brush cover on managed lands. Fuel dryness is expected to become more common due to the higher temperatures allowing dead branches and leaves to dry out faster. It is not known whether or to what extent the number of hot, dry, and windy days will change, but the other two factors point to a general increase in the extent of wildfires due to an increase in the number of days with fuel ready to burn. Overall trends in wildfires will also depend on how well human-caused ignitions are avoided.

1.2.6 SUMMARY

The Texas climate is diverse and is changing in many different ways, along with climates elsewhere. The impacts of these changes also involve local human-driven factors, such as agricultural practices and flood protection. The eventual magnitude of these changes will depend on both the extent of global climate change and how the local climate responds to those global changes. Some changes are clear and are obviously already happening, while other changes are still mostly a matter of speculation. Where changes are clear, plans can be made; where changes are unclear, contingency plans are called for.

NOTE

1. First overheard in Danish Parliament, 1937–1938, according to https://quoteinvestigator.com/2013/10/20/no-predict/

REFERENCES

Holliday VT, 2001. Stratigraphy and geochronology of upper quaternary eolian sand on the Southern high plains of Texas and New Mexico, United States. GSA Bulletin. 13(1):88–108. https://doi.org/10.1130/0016-7606(2001)113<0088:SAGOUQ>2.0.CO;2

Jorgensen S, Nielsen-Gammon JW, 2024. Nonstationarity in extreme precipitation return values along the United States Gulf and southeastern coasts. Journal of Hydrometeorology. 25:771–788. https://doi.org/10.1175/JHM-D-22-0157.1

Marvel K, Su W, Delgado R, Aarons S, Chatterjee A, Garcia ME, Hausfather Z, Hayhoe K, Hence DA, Jewett EB, Robel A, Singh D, Tripati A, Vose RS, 2023. Ch. 2. Climate trends. In Fifth National Climate Assessment, Crimmins AR, Avery CW, Easterling DR, Kunkel KE, Stewart BC, and Maycock TK, eds. U.S. Global Change Research Program, Washington, DC. https://doi.org/10.7930/NCA5.2023.CH2

Nielsen-Gammon JW, Holman S, Buley A, Jorgensen S, Escobedo J, Ott C, Dedrick D, Van Fleet A, 2024. Assessment of historic and future trends in extreme weather in Texas, 1900–2036: 2024 update. Document OSC-202101, Office of the State Climatologist, Texas A&M University, College Station. https://texas2036.org/weather

Simms AR, Lambeck K, Purcell A, Anderson JB, Rodriguez AB, 2007. Sea-level history of the Gulf of Mexico since the last glacial maximum with implications for the melting history of the Laurentide ice sheet. Quaternary Science Reviews. 26(7–8):920–940. https://doi.org/10.1016/j.quascirev.2007.01.001

1.3 Alarming Sea Level Rise Along Texas Coast

Causes, Impacts, and Mitigation Strategies

Jiabi Du

1.3.1 INTRODUCTION

1.3.1.1 Why Do We Care About Sea Level Rise?

Sea level rise (SLR) is a significant concern for coastal communities worldwide. As of 2018, approximately 38% of the global population lived within 100 km of the shoreline, and this percentage is expected to increase in future as more people settle along coastlines (Cosby et al., 2024). The population residing in low-elevation coastal zones (elevation <10 m) is projected to grow from 50% in 2000 to 71% by 2050 (Merkens et al., 2016). This growing coastal population amplifies the socioeconomic consequences of increased flooding risks associated with SLR. Texas exemplifies this challenge, with alarmingly high SLR rates recorded at tidal gauging stations. At Pier 21 in Galveston, for instance, long-term monitoring (1904 to present) reveals an SLR rate exceeding 6 $mm \cdot yr^{-1}$ (Liu et al., 2020), one of the highest rates observed along the U.S. coast. For comparison, the average SLR rate along the U.S. coastline is approximately 3 $mm \cdot yr^{-1}$.

Flooding risk is one of the most pressing consequences of SLR, with wide-ranging socioeconomic implications. However, the impact of SLR extends far beyond flooding. It accelerates shoreline erosion (Anderson et al., 2014), contributes to the loss of salt marshes (Ravens et al., 2009), and slows the renewal and weakens the flushing of coastal waters (Hong and Shen, 2012). SLR also exacerbates saltwater intrusion, alters sedimentation patterns, affects geological evolution (Anderson et al., 2014), and degrades water quality (Hong and Shen, 2012).

SLR is expected to amplify the negative effects of a warming climate, which is itself a key driver of SLR. Significant research has been conducted on the impacts and mitigation strategies for SLR along the Texas coast. Now is an opportune time to synthesize this knowledge and present a comprehensive overview of SLR's challenges and implications for Texas.

1.3.1.2 What Is Relative Sea Level Rise?

To better understand sea level rise (SLR), it is essential to differentiate between key concepts associated with it.

1.3.1.2.1 Eustatic Sea-Level Rise (ESLR)

Eustatic SLR (ESLR) refers to global sea-level changes caused by variations in ocean volume. These changes are driven by two factors: mass and density. Water density decreases with increasing temperature, a process known as thermal expansion. Consequently, ESLR results from two primary processes: thermal expansion due to rising water temperatures and the addition of water from sources such as melting glaciers, ice sheets, and land-based water storage.

DOI: 10.1201/9781003406600-4

1.3.1.2.2 Relative Sea-Level Rise (RSLR)

Relative SLR (RSLR), observed at specific locations, combines ESLR with local vertical land motion (uplift or subsidence). Vertical land motion occurs due to natural processes, such as tectonic plate movement, or human activities, like groundwater extraction. The following are key contributors to vertical land motion, leading to either uplift or subsidence:

1. Groundwater/Oil/Gas Withdrawal. Excessive withdrawal causes land subsidence as aquifer materials compact when water is removed.
2. Land Reclamation. Subsidence can occur on reclaimed land due to soil compaction or when wetlands are enclosed by levees and drained to create polders.
3. Natural Soil Compaction. This is especially common in salt marshes, where organic material, such as decaying plant roots, compacts and causes the land to sink.
4. Glacial Isostatic Adjustment (GIA). The Earth's crust adjusts slowly to changes in glacier mass. When glaciers melt, previously glaciated areas experience land uplift, while surrounding areas may subside. GIA also affects ocean levels through gravitational and deformational forces. Large glaciers, such as those in Alaska, Greenland, and Antarctica, exert a gravitational pull that attracts ocean water, raising local sea levels. Melting ice reduces this pull, redistributing water and lowering RSLR in nearby regions. For example, GIA reduces the rate of RSLR in northern Washington (NAS, 2012). Conversely, in previously glaciated regions, such as Canada and Northwestern Europe, relative sea levels continue to fall due to postglacial rebound, with rates exceeding 10 mm·yr^{-1} (Peltier, 1999).
5. Plate Tectonic Movements. Plate interactions cause land to rise or subside, affecting local sea levels. For example, subduction along the Cascadia Subduction Zone causes uplift in Washington, Oregon, and northernmost California at rates of 1.5–3 mm·yr^{-1} (NAS, 2012). Combined with strong GIA effects, the U.S. West Coast has experienced relatively low rates of RSLR.

In regions with significant subsidence due to groundwater or oil/gas extraction and isostatic adjustment, RSLR often far exceeds ESLR. While the global rate of ESLR has been approximately 2 mm·yr‾1 over the past century (Church et al., 2013), RSLR in some coastal and riverine systems has been reported at rates an order of magnitude higher (Al Mukaimi et al., 2018). For instance, the RSLR in the Mississippi Delta is approximately 10 mm·yr^{-1} (Penland and Ramsey, 1990).

1.3.1.3 How Do We Measure Sea-Level Rise?

Before satellite altimetry became available in 1993, sea-level measurements were collected primarily through tide gauge stations. In the United States, these stations are mainly maintained by National Oceanic and Atmospheric Administration (NOAA), with additional support from state agencies. Tide gauge stations initially used mechanical pen-and-ink recorders and later advanced to acoustic sensors. These instruments measure the distance from a fixed point on land to the water surface, capturing relative SLR. By analyzing long-term water level records, RSLR rates can be estimated. However, special efforts are required to convert RSLR rates to ESLR rates (Church and White, 2006).

In 1993, the launch of two satellites, TOPEX/Poseidon, followed by the Jason-1, Jason-2, and Jason-3 missions, revolutionized SLR monitoring. These satellites use altimetry techniques to measure global sea level from space with high precision. Satellite data provides a comprehensive global perspective and has been instrumental in estimating SLR rates (e.g., Church et al., 2013; Kuo, 2006).

1.3.1.4 What Is Past and Current Global Sea Level Rise Rate?

Globally, ESLR was estimated at 1.7 $\pm$ 0.2 mm·yr^{-1} over the last century based on tide gauge data (Church and White, 2006, 2011). In contrast, satellite altimetry data indicate a faster rate of

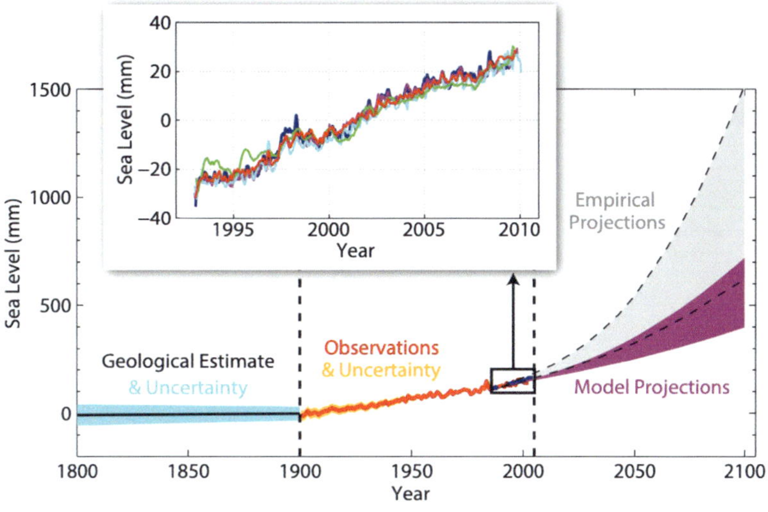

FIGURE I.3.1 Current knowledge of estimated, observed, and projected global sea level rise from 1800 to 2100. The tide gauge record is shown from 1900 to 2005 (red, uncertainty bounds shaded yellow, from Church and White, 2006), and the satellite altimetry record is shown from 1985 to 2005 (blue, from Kuo, 2006). The projected 21st century SLR of 26–59 cm (pink envelop, from IPCC, 2007) is based on coupled climate models using the A1F1 scenario assuming heavy reliance on fossil fuels. The projected of 21st century SLR of 40–150 cm is from Rahmstorf (2007) using an empirical method. The inset shows estimates of globally averaged sea level rise from different selected studies (Kuo, 2006; Leuliette and Scharroo, 2010; Michaël et al., 2009; Nerem et al., 2010), with the averaged trend being 3.2 ± 0.5 mm·yr^{-1}. Here, the seasonal signals have been removed and all the corrections are applied, including the small impact of net sinking across the ocean floors (also known as glacial isostatic adjustment, or GIA). Figure from Willis et al. (2010).

3.3 ± 0.4 mm·yr^{-1} between 1993 and 2010 (Church et al., 2013; Nicholls and Cazenave, 2010). This significant increase reflects a clear acceleration in SLR, which began as early as the 1960s (Figure I.3.1; Dangendorf et al., 2019; Frederikse et al., 2020). Recent studies have investigated the drivers of SLR acceleration. Frederikse et al. (2020) attributed the increase since the 1970s to a combination of ocean thermal expansion and accelerated ice mass loss, particularly from Greenland. Dangendorf et al. (2019) suggested an earlier onset of acceleration in the 1960s, linked to intensified Southern Hemispheric westerlies and increased ocean heat uptake. More recently, Yin (2023) identified a slowdown in the Atlantic Meridional Overturning Circulation during 2009–2010 as a contributing factor to the acceleration observed from 2010–2022. Additionally, Rahmstorf (2007) found a positive empirical relationship between SLR rates and global ocean temperatures, highlighting the role of warming oceans in driving faster SLR. Despite differences in these explanations, there is a broad consensus that SLR is accelerating and will continue to do so throughout the 21st century.

ESLR is driven by two primary processes: thermal expansion and the addition of water, primarily from melting glaciers. Research shows that added water has contributed approximately twice as much to SLR since 1900 compared to thermal expansion (Figure I.3.2; Frederikse et al., 2020).

It is important to note that ESLR is not uniform across the globe. Regional variations in SLR are influenced by large-scale ocean-atmosphere dynamics, ice melt, and freshwater redistribution. For example, trend analyses using global altimetry data (1993–2023) show an SLR rate of ~8 mm·yr^{-1} in the subtropical Pacific Ocean but only 3–4 mm·yr^{-1} in the North Atlantic at the same latitude (Figure I.3.2). Along the U.S. West Coast, the rate decreases to nearly 0 mm·yr^{-1} and, in subpolar regions near Antarctica, some areas even experience negative rates. These spatial differences arise from sterodynamic sea-level change, which reflects combined effects of local circulation, heat redistribution, and salinity changes.

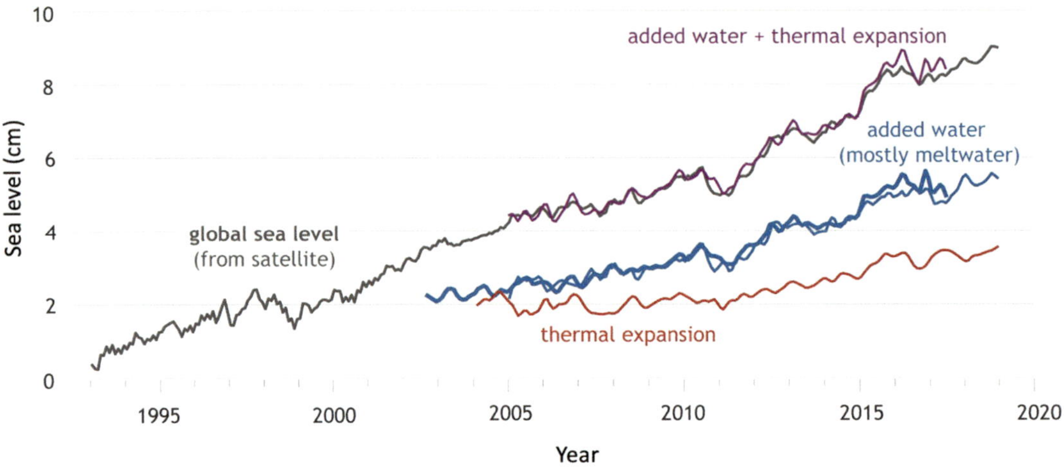

FIGURE I.3.2 Contribution from thermal expansion and added ocean water mass to the global SLR. Adapted from Lindsey (2023) and Figure 3.15a in State of the Climate in 2018 (https://www.ametsoc.org/index.cfm/ams/publications/bulletin-of-the-american-meteorological-society-bams/state-of-the-climate/). Combination of thermal expansion (red line) and added ocean water mass (thick blue line) approximate the global SLR with some minor difference. The thin blue line is the inferred global ocean water mass, calculated by subtracting global mean steric sea level (i.e., due to thermal expansion) from global mean sea level.

I.3.2 SEA LEVEL RISE ALONG TEXAS COAST

I.3.2.1 What Is the Texas SLR Rate?

Tide gauge data reveal that RSLR rates also vary significantly across regions (Figure I.3.3). On the U.S. East Coast, all stations report positive RSLR rates, while most stations on the West Coast report negative rates, largely due to glacial isostatic adjustment (GIA) and tectonic activity (NAS, 2012). In the Gulf of Mexico, RSLR rates are notably high. Two stations, Galveston, Texas, and the Mississippi Delta, show the highest rates, exceeding 6 mm·yr^{-1}. Long-term records along the Texas coast indicate RSLR rates between 3–7 mm·yr^{-1} (Figure I.3.4). Near Galveston Bay, the rate is the second highest in the Gulf of Mexico, following the Mississippi Delta.

Texas has four NOAA-maintained tide gauge stations with over 30 years of records. These data reveal a decrease in RSLR rates from 6.6 mm·yr^{-1} at Galveston Bay to 4.3 mm·yr^{-1} at Port Isabel (Figure I.3.5). However, these rates vary depending on the temporal span of the data. For instance, Penland and Ramsey (1990) reported a rate of 6.3 mm·yr^{-1} at Galveston (1908–1988) and 3.1 mm·yr^{-1} at Port Isabel (1944–1979). The increasing rates observed in longer records suggest an acceleration of SLR in recent decades.

I.3.2.2 Why Is the SLR Rate Different along the Texas Coast?

The significant spatial variability in sea level rise along the Texas coast is strongly linked to differences in local vertical land motion. Over the last century, extensive groundwater extraction to support petrochemical industries and a growing population has caused significant land subsidence (Al Mukaimi et al., 2018).

On average, the Texas coastline experienced a natural land subsidence rate of 1.4 mm·yr^{-1} between 1970 and 2010 (Zhou et al., 2021). However, this rate varies considerably across regions, with particularly high subsidence near the Houston area. For example, at Galveston Pier 21, subsidence rates were estimated at 3.5 mm·yr^{-1} from 1909 to 1937, increased to 6.1 mm·yr^{-1} from 1937 to 1983, and

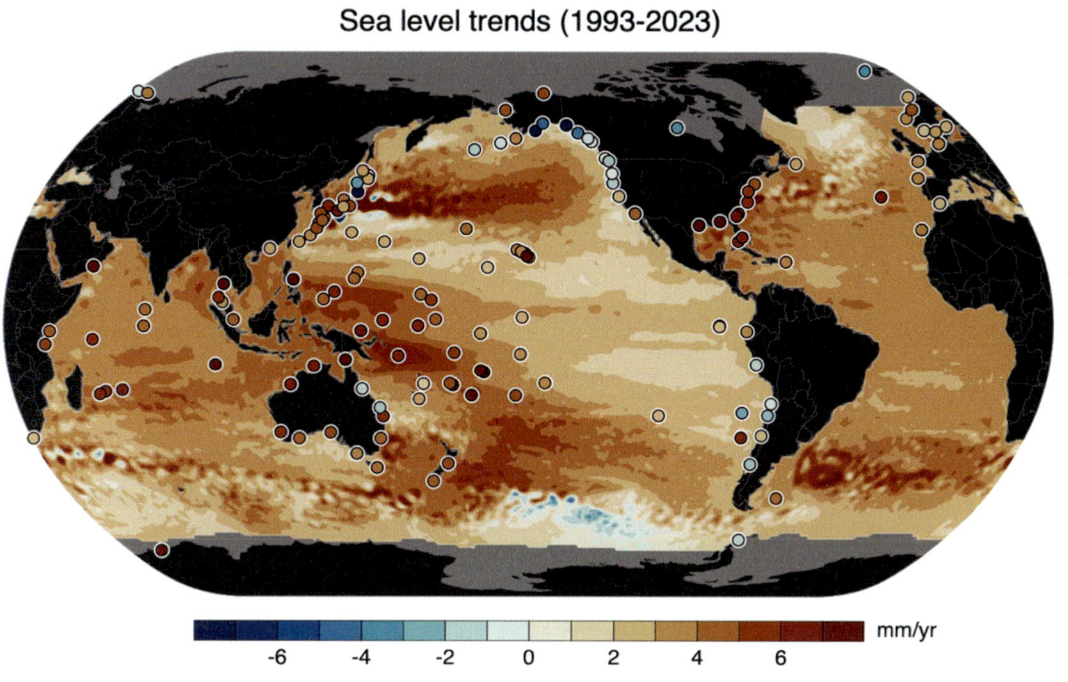

FIGURE I.3.3 Global sea level rise trend from satellite data (color shades) and tide gauge data (colored circles). Tide gauge data are from the University of Hawai'i Sea Level Center (UHSLC) Fast-Delivery database (https://uhslc.soest.hawaii.edu), while satellite altimeter data were produced and distributed by the E.U. Copernicus Marine and Environment Monitoring Service (CMEMS). (Figure source: https://uhslc.soest.hawaii.edu; accessed on 12/11/2024.)

returned to 3.5 mm·yr^{-1} after 1983 (Liu et al., 2020). Historically, Houston has experienced extreme subsidence rates, reaching as high as 60 mm·yr^{-1} in some areas (Gabrysch and Bonnet, 1975). This significant subsidence, even without considering ESLR, is projected to inundate 76 square kilometers in the Galveston Bay area by 2100 (Miller and Shirzaei, 2021).

Subsidence rates along the Texas coast generally decrease from north to south (Younas et al., 2023). Based on data from global navigation satellite systems, rates range from 2 to 26 mm·yr^{-1},

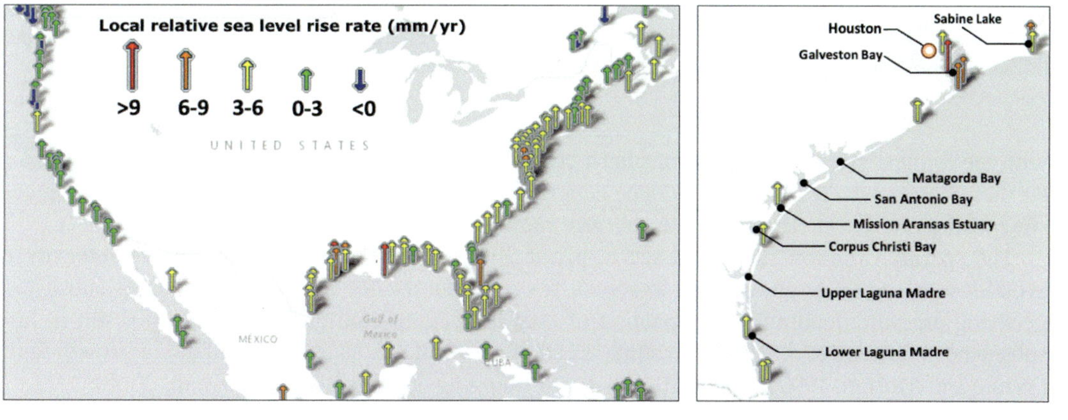

FIGURE I.3.4 Local relative sea level rise rate. Figure adapted from NOAA's sea level trend product available at https://tidesandcurrents.noaa.gov/sltrends/sltrends.html.

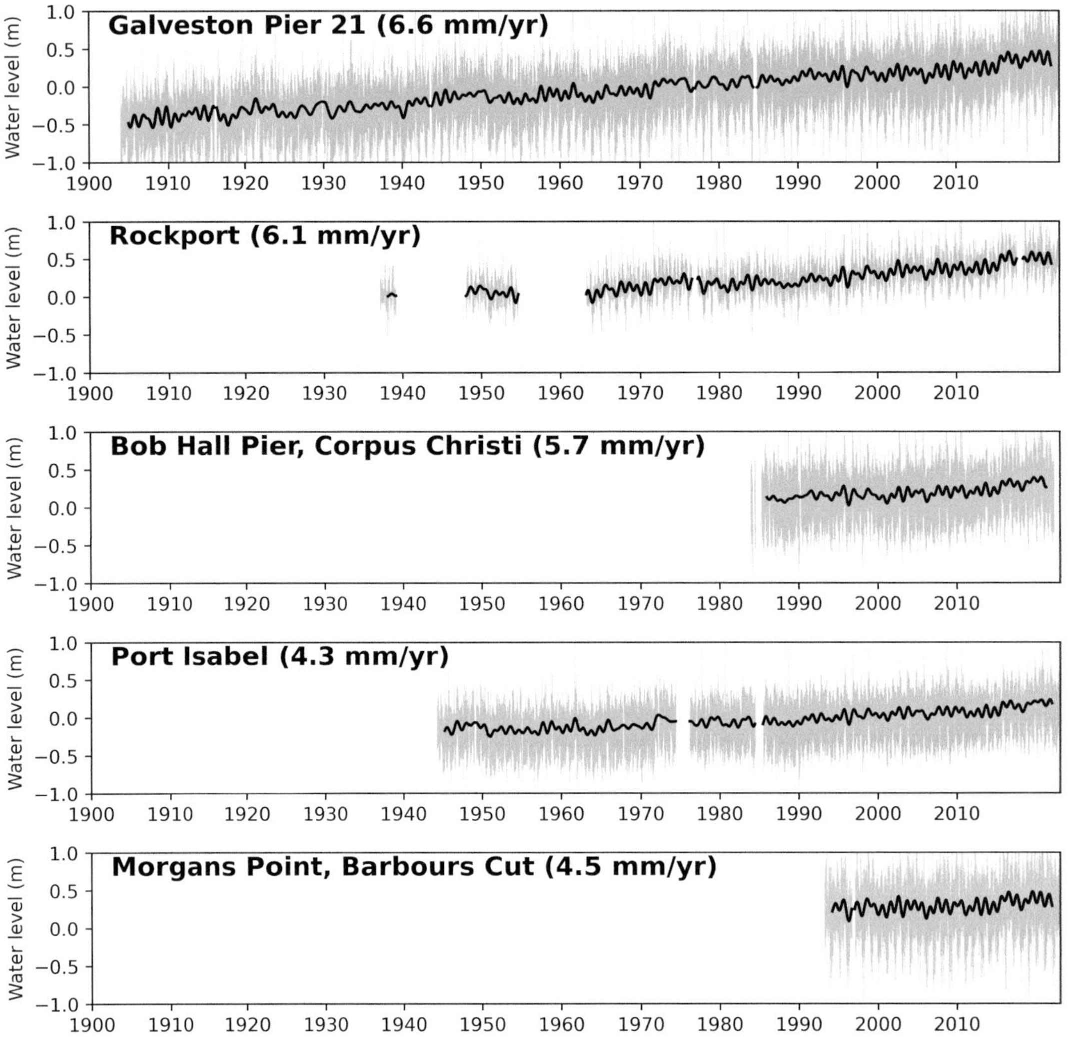

FIGURE I.3.5 Hourly water level (grey lines) and low pass filtered signal (cut-off period of 365 days) at NOAA tide gauge stations in Texas with at least 30-year records. Relative sea level trend based on linear regression of the low-pass filtered signal (with a cut-off period of 30 days) is shown in the text following the station name. Hourly water level data in the North American Vertical Datum of 1988 is available at https://tidesandcurrents.noaa.gov.

with the highest rates observed in northern counties such as Harris, Montgomery, Fort Bend, and Karnes (Younas et al., 2023). Subsidence generally decreases toward the southwest based on recent study using interferometric synthetic aperture radar data (Figure I.3.6; Haley et al., 2022).

The spatial variation in subsidence is primarily attributed to differences in the intensity of groundwater and oil or gas extraction across Texas. Groundwater withdrawal increases subsurface geostatic pressure, leading to compaction of clay lattices and fault slipping, which result in land subsidence (Bawden et al., 2012). In some areas, this subsidence is exacerbated along growth faults (White and Morton, 1997), as detected by Lidar mapping (Engelkemeir and Khan, 2008).

Natural processes such as tectonic activity, glacial isostatic adjustments, and local sediment compaction also contribute to subsidence. However, these natural rates are much smaller than the rates observed in recent decades. For example, based on the differential elevations of modern shorelines

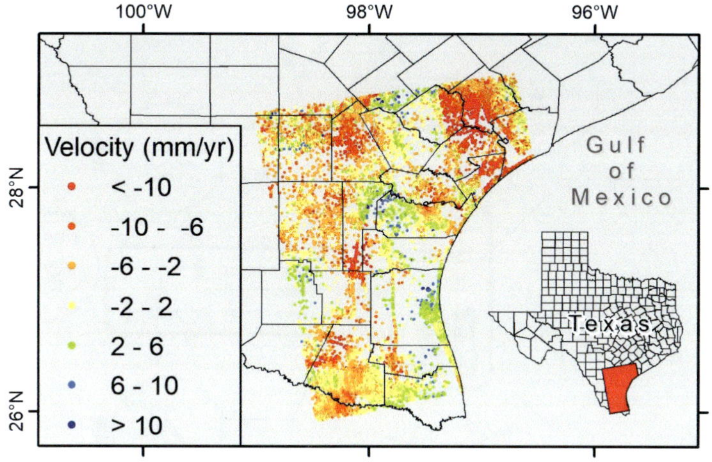

FIGURE I.3.6 InSAR-derived land subsidence rates (vertical component; in mm/yr) generated over the Texas Coastal Bend during the period from October 2016 to July 2019. Figure from Haley et al. (2022).

and last interglacial geomorphic features, Simms et al. (2013) estimated subsidence rates over the past 120k years ranging from 0.03 mm·yr^{-1} inland to 0.09 mm·yr^{-1} near the Brazos and Colorado River Delta. Paine (1993) similarly found that long-term (~10^5 years) historical subsidence rates were 0.05 mm·yr^{-1} or less. These rates are two orders of magnitude lower than modern rates of SLR.

Other coastal processes also influence local sea levels, including alongshore wind, river discharge, and dynamics in the deeper Gulf of Mexico. Strong downcoast winds (blowing from Louisiana to Texas) can push coastal waters against the shoreline due to Ekman transport, a phenomenon in which water moves at a right angle to wind direction in the northern hemisphere. While this process contributes to short-term changes like storm surge, its long-term impact on SLR is minimal. River discharge can also temporarily raise local water levels, but there is no persistent trend in river flow rates along the Texas coast to suggest a significant contribution to SLR.

In contrast, deep ocean dynamics play a more significant role in long-term coastal sea level changes, especially near strong boundary currents like the Gulf Stream. An increase in water mass in the Gulf of Mexico due to SLR can affect shallow coastal areas, depending on how the added water is redistributed (Steinberg et al., 2024). Long-term variations in coastal sea levels in the Gulf of Mexico are linked to the presence of loop current eddies and the strength of water flow through the Yucatan Channel and Florida Strait (Ezer, 2022).

I.3.3 IMPACTS OF SEA LEVEL RISE

SLR will impact the coastal water in terms of hydrodynamic, wave, water quality, and sediment dynamics, as well as ecosystem. A quick summary of these impacts is summarized in Figure I.3.7.

I.3.3.1 Flooding

The Texas coast is highly susceptible to storm-induced flooding due to its extensive low-lying areas. Flooding from major hurricanes, such as Hurricane Ike in 2008 and Hurricane Harvey in 2017, has caused billions of dollars in economic losses. Hurricane Ike is notable for its extremely high storm surge, reaching up to 3 meters (Rego and Li, 2010), while Hurricane Harvey is remembered for unprecedented rainfall of up to 1,200 mm over a five-day period (Du et al., 2019). Hurricane Harvey exemplifies compound flooding, which results from multiple factors including tide, wave action, storm surge, precipitation, and river flow (Huang et al., 2021).

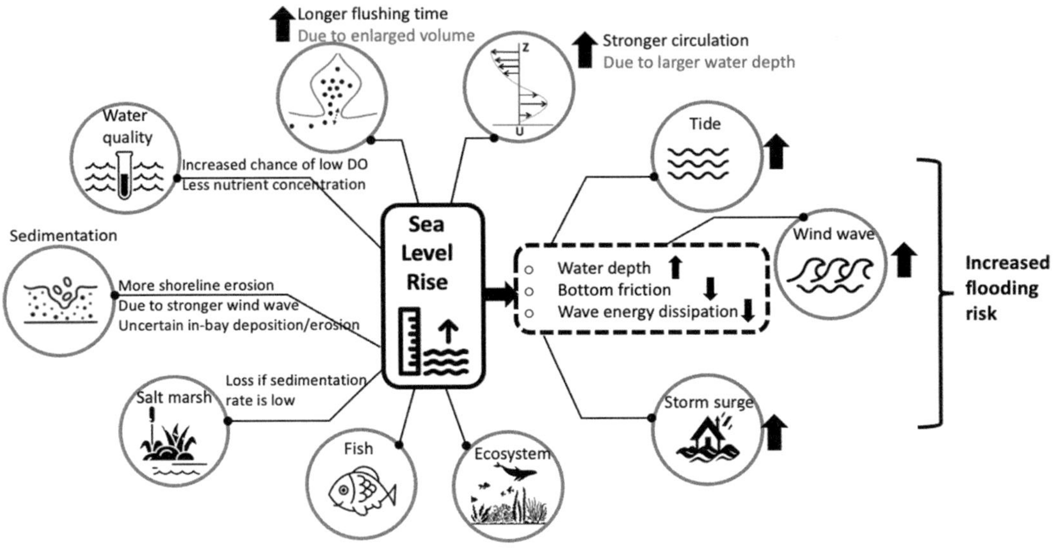

FIGURE I.3.7 Sketch diagram showing impacts from SLR. SLR will have similar impacts on wave energy for tidal wave, wind wave, and storm surge, due to a larger water depth, smaller bottom friction, and thus less energy dissipation.

With SLR, the risk of flooding along the Texas coast is expected to increase significantly. However, this increase is not solely due to the static "bathtub" effect, which considers only the rise in mean sea level (Figure I.3.8). SLR will disproportionally amplify flooding risks through dynamic changes in waves, tides, and storm surges (Yang et al., 2014), as well as alterations to landforms such as salt marshes, mangroves, sand dunes, and barrier islands. Greater water depths result in reduced bottom friction, allowing wave energy to dissipate less and leading to increased wave heights (Figure I.3.7). Storm surges, which can be considered long-period waves, will also grow in magnitude under SLR. Flooding driven by interactions among water level, wave action, and tides is referred to as "dynamic flooding." In contrast, static or "bathtub" models often fail to account for critical hydrodynamic effects such as tidal wave amplification, damping, and reflection (Khojasteh et al., 2021). Studies have shown that SLR can lead to earlier peak surge arrival and prolonged inundation (Bilskie et al., 2016). Furthermore, dynamic modeling reveals a threefold increase in exposed coastal populations compared to static models that consider only SLR (Barnard et al., 2019).

SLR will also impact tidal amplitude and phase, key factors in coastal flooding. While tidal amplitude often increases with SLR (Passeri et al., 2016), it may decrease in short estuaries due to wave reflection or the inundation of low-lying lands (Du et al., 2018). Since tidal amplitudes are relatively small (around 0.1 m for major tidal constituents like M2, S2, O1, and Q1), even a 10% change in amplitude contributes minimally (~1 mm) to overall flooding risk. Nevertheless, tides remain a critical component of flooding dynamics.

Waves, however, have a much greater influence on flooding risk. Across the Texas bays, the current average water depth is about 2.5 meters. With a 1-meter sea level rise, the depth would increase by 40%, substantially reducing bottom friction and increasing wave heights. SLR will alter mean wave heights and, more critically, extreme wave heights (Chini et al., 2010). Wave energy has already increased globally in association with climate change (Reguero et al., 2019), and wave contributions to flooding are widely recognized as significant (Vousdoukas et al., 2018).

Storm surge impacts will also grow with SLR, particularly for smaller storms. While major hurricanes like Ike (2008) produced maximum surges of around 3 meters, smaller storms such as Cindy (1963), Alicia (1983), and Rita (2005), with surges of approximately 1 meter, are expected to experience relatively larger impacts from SLR (Warner and Tissot, 2012).

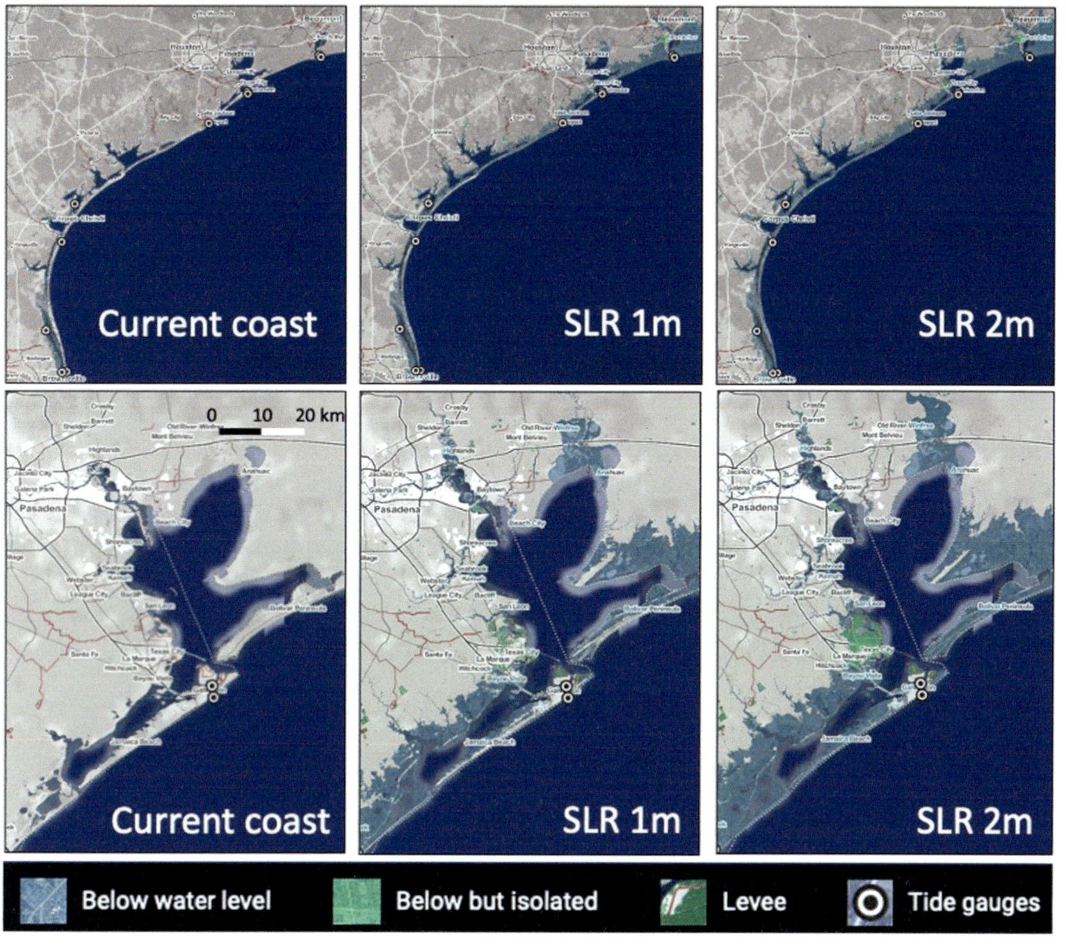

FIGURE I.3.8 Flooded area under current water level, SLR of 1 m and SLR of 2 m along Texas coast (upper panels) and in Galveston Bay (lower panels). Original maps are from ss2.climatecentral.org (last accessed August 27, 2025).

SLR will further influence shorelines by affecting the extent and migration of salt marshes and mangroves, which serve as critical buffers against wave energy. These ecosystems effectively attenuate wave energy, trap sediments, and protect shorelines. While salt marshes may adapt to SLR if sediment supply is sufficient, many have been lost globally due to insufficient sediment, coastal eutrophication, and SLR (Deegan et al., 2012). Along the Texas coast, salt marsh losses have been documented (Ravens et al., 2009; White et al., 2002) and are expected to continue with accelerated SLR, leading to stronger wind wave and storm surge impacts in the future.

SLR not only affects ocean water levels but also influences flooding dynamics on land. Urban drainage systems, for instance, may become less efficient as sewer pipes are inundated and groundwater levels rise, exacerbating urban flooding during heavy precipitation events.

In the northern Gulf of Mexico, Texas is not the most vulnerable state to flooding under SLR. Louisiana's coast, particularly near the Mississippi Delta, is at greater risk due to higher land subsidence rates of ~10 $mm{\cdot}yr^{-1}$. According to the Coastal Vulnerability Assessment by the U.S. Geological Survey (Pendleton et al., 2010), the region between Sabine and Galveston Bay is among the most vulnerable to flooding, with vulnerability decreasing to the west.

I.3.3.2 Salt Intrusion

SLR is commonly projected to enhance salt intrusion into coastal bays and aquifers. Salt intrusion results from tidal and subtidal exchanges between fresher bay waters and saltier coastal ocean water. Both tidal and subtidal exchanges are expected to intensify under SLR. Larger water depths associated with rising sea levels tend to amplify tidal amplitude and currents. Subtidal exchange, primarily driven by estuarine circulation, is also expected to strengthen. Estuarine circulation typically features a two-layer flow, with seaward flow at the surface and landward flow at the bottom. Its strength is heavily influenced by water depth and is proportional to the cube of the depth (H^3, where H represents water depth) (MacCready, 2004). For Texas bays, which have a mean water depth of 2.5 meters, a 1-meter SLR would increase estuarine circulation strength to 274% of its current value.

The phenomenon of enhanced salt intrusion under SLR has been documented in numerous coastal bays, such as Chesapeake Bay (Hong and Shen, 2012). Similar effects are anticipated in Texas bays. Numerical experiments using a well-calibrated ocean model (Du et al., 2019) indicate that mean salinity, averaged over a year, is expected to rise by 1–4 in most Texas bays. However, exceptions are observed in Upper and Lower Laguna Madre, where mean salinity is projected to decrease (Figure I.3.9). These exceptions are attributed to the hypersaline conditions in these bays,

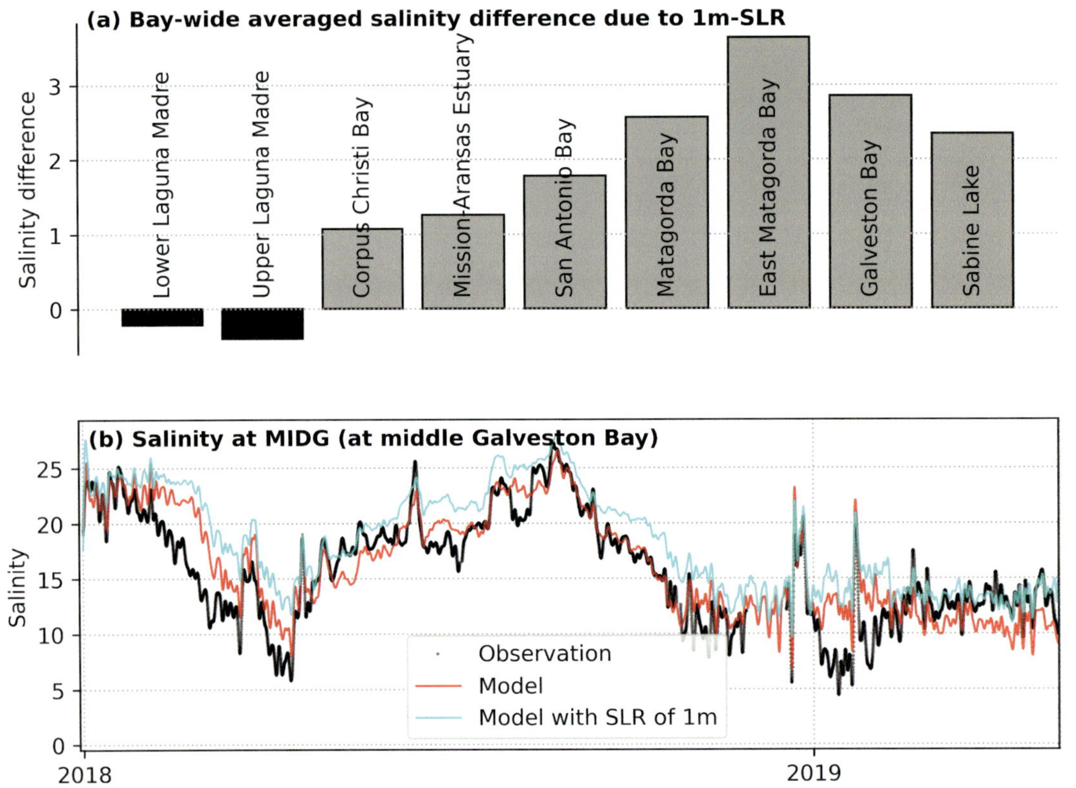

FIGURE I.3.9 Influence of SLR on salinity in major Texas bays. (a) Mean of bay-wide averaged salinity difference due to 1 m-SLR, with positive values indicating larger salinities under SLR. The bay-wide averaged salinity difference differs from day to day and the mean over a one-year period (from day 30 to day 395 since January 1, 2018) is plotted. (b) Observed and modeled salinity at station MIDG (a long-term monitoring station maintained by Texas Water Development Board). Observation data is available at https://waterdatafortexas.org/coastal.

where salinity levels are higher than those of adjacent coastal ocean waters. Under SLR, strengthened estuarine circulation transports more coastal ocean water (with lower salinity) into the bays, resulting in reduced salinity levels in these regions.

Beyond open water salinity, SLR can also affect salt intrusion into coastal aquifers, altering the freshwater-saltwater interface and threatening groundwater resources. However, the extent of this impact is limited, as salt intrusion into aquifers is primarily governed by freshwater influx from inland sources toward the coast (Uddameri et al., 2014).

1.3.3.2.1 Water Quality and Water Exchange

SLR tends to decrease dissolved oxygen levels in bottom waters due to reduced replenishment caused by a thicker water column. Hypoxia, defined as dissolved oxygen levels below 2 mg/L, can severely impact benthic organisms, shellfish, and fish species, particularly benthic ones that are immobile or slow-moving. Hypoxia often occurs in stratified systems like Chesapeake Bay and Mobile Bay. Mobile Bay, which has a similar depth to Texas bays, provides a relevant comparison. Hypoxia in Mobile Bay is typically triggered during calm weather after substantial freshwater influx, which rapidly stratifies the water column and creates salinity differences of 10–20 between surface and bottom layers. The stratification inhibits vertical mixing and replenishment of dissolved oxygen in bottom waters (Park et al., 2007). Although Texas bays generally receive less freshwater input than Mobile Bay (~1,700 m^3/s mean flow), episodic heavy precipitation events, such as those associated with hurricanes, could cause temporary stratification and hypoxic conditions. SLR and climate warming may exacerbate this issue. Rising sea levels will further hinder vertical oxygen replenishment, while higher water temperatures will reduce oxygen solubility at the surface and increase oxygen demand in biogeochemical processes.

SLR will not only impact the vertical mixing and bottom water replenishment, but also the overall horizontal exchange with the adjacent coastal ocean. While tidal and estuarine circulation may enhance water exchange with adjacent coastal waters, the enlarged water volume due to SLR increases flushing time (or residence time), meaning nutrients and pollutants discharged into bays will linger longer before being exported (Hong and Shen, 2012). Longer retention times exacerbate the effects of eutrophication in bays already burdened with excessive nutrients. However, increased water volume under SLR (projected to rise by 40% for Texas bays with a 1-meter SLR) could dilute concentrations of terrestrial materials.

1.3.3.3 Sediment Erosion

SLR, coupled with stronger wind waves, significantly enhances sediment erosion along shorelines and plays a central role in global shoreline retreat (Leatherman, 1984). While the effects of SLR on sediment erosion along coastal margins are evident, its impact on sedimentation in open waters, such as the middle of coastal bays, remains less understood. SLR increases wind wave energy, which amplifies bottom shear stress, yet the simultaneous deepening of the water column reduces this stress. The overall effect on sediment dynamics depends on the interplay between the enhanced wind wave energy and the mitigative influence of greater water depths. This dual effect introduces complexities in understanding and predicting sediment transport patterns in response to rising sea levels.

The impacts of SLR on sediment erosion are especially pronounced during storm events, particularly in low-energy coastal regions like the northern Gulf of Mexico, where weak tidal forces dominate. Storm events such as Hurricane Rita (2005) have demonstrated the vulnerability of shorelines to erosion, with sediment often being transported offshore or upland to maintain the equilibrium bottom profile (Leatherman, 1984; Park and Edge, 2011). SLR amplifies storm surge and wind wave energy, exacerbating sediment redistribution during such events. Waves overtopping dunes can result in extensive sand transport and reshaping of coastal landscapes (Stockdon et al., 2012). Furthermore, SLR influences the bottom profile, which determines how coastal storms interact with the sediment, highlighting the multifaceted ways in which rising sea levels exacerbate erosion and reshape coastal systems.

1.3.3.4 Salt Marsh

Salt marsh loss due to sea level rise (SLR) and insufficient sediment supply has been a persistent challenge for the Texas coast over the past century (Osland et al., 2017). For instance, the Galveston Island State Park marsh experienced a dramatic reduction, losing 405 of its original 445-hectare area between 1930 and 1994 (Glass and Hollingsworth, 1999). Similarly, about 12% of the salt marshes in Galveston Bay were lost between 1950 and 1989, largely attributed to RSLR and wave-induced erosion (White et al., 1993). From the 1950s to 2002, Galveston Island alone saw a 32% loss of its estuarine marsh (White et al., 2004). While moderate SLR paired with adequate sediment supply might allow salt marshes to migrate and maintain their spatial extent, the sedimentation rates along most of the Texas coast fall short of matching the regional RSLR. For example, sedimentation rates in Trinity Bay range between 1.6 $mm{\cdot}yr^{-1}$ and 5 $mm{\cdot}yr^{-1}$, while West Galveston Bay records a rate of approximately 2 $mm{\cdot}yr^{-1}$, both insufficient to counterbalance SLR (Ravens et al., 2009; White et al., 2002; Yeager et al., 2007). Moreover, larger wave heights exacerbated by SLR further destabilize salt marshes, with stability often compromised when 20% exceedance wave height surpasses 0.14 m (Shafer et al., 2003). Projections indicate that from 1990 to 2080, SLR could cause the loss of up to 22% of global coastal wetlands. When combined with anthropogenic factors, this figure could rise to 70%, although uncertainties remain (Nicholls et al., 1999).

While SLR poses severe threats to salt marshes, it may simultaneously support the expansion of mangrove habitats due to a projected decrease in the frequency and intensity of coastal freezes under a warming climate (Comeaux et al., 2012). Along the Texas coast, where mangroves and marshes coexist, periodic freeze events have historically limited mangrove expansion. However, milder winters in recent decades have enabled significant mangrove proliferation. Between 1990 and 2010, mangrove area expanded by 16.1 km^2, a 74% increase, while salt marsh area decreased by 77.8 km^2, resulting in a 24% net loss (Armitage et al., 2015). Notably, only 6% of salt marsh loss during this period was directly linked to mangrove expansion, with the majority attributed to conversion to tidal flats or open water (Armitage et al., 2015). These contrasting dynamics underscore the complex interplay of SLR, sediment supply, and climate in shaping coastal wetland ecosystems.

1.3.3.5 FISH AND ECOSYSTEM

The Gulf of Mexico's Texas coastline is a vital hub for marine fisheries, contributing approximately 29% of the total fisheries value in the Gulf (NOAA Fisheries). SLR indirectly impacts fisheries and ecosystems through changes in salinity, sediment erosion, habitat loss (e.g., marshes and mangroves), and prolonged retention of terrestrial materials and pollutants in coastal waters. A 33-year gillnet survey of eight major Texas bays revealed increasing fish diversity over time, likely driven by the expansion of tropical species into these bays as the climate warms (Fujiwara et al., 2019; Pawluk et al., 2021). However, rising salinity and temperature force fish species adapted to moderate salinity to migrate to upper estuaries where these conditions are more favorable. Unlike fish populations on the U.S. East Coast, which can migrate to higher latitudes, those in the northern Gulf of Mexico face constraints due to land boundaries. While higher biodiversity can benefit ecosystems, it may also bring unforeseen functional consequences.

Texas coastal bays provide critical nursery habitats for juvenile fish and invertebrate species, such as brown shrimp (*Farfantepenaeus aztecus*), white shrimp (*Litopenaeus setiferus*), and blue crab (*Callinectes sapidus*) (Rozas et al., 2007). The loss of wetlands associated with SLR poses a significant threat to the abundance of these species, which depend on such habitats during their early life stages. Wetland degradation and habitat reduction could undermine the sustainability of these economically and ecologically important fisheries (Yoskowitz et al., 2017).

I.3.4 FUTURE SEA LEVEL RISE AND MITIGATION STRATEGIES

Estimates of the magnitude and rates of future SLR are based on potential carbon emission scenarios and are constructed from various assessments of ocean warming and ice sheet loss. Regardless of the methods or emission scenarios used to estimate future sea levels, it is generally accepted that sea levels are rising, and that the rate of increase is expected to accelerate during the 21st century (Church et al., 2013). Projections suggest that global sea levels will rise by 0.4 to 1.5 meters above 1990 levels by 2100 (Figure I.3.1), based on a semi-empirical relationship between global sea level and global mean surface temperature (Rahmstorf, 2007). An even higher rise is also possible (DeConto and Pollard, 2016).

The actual magnitude of future SLR remains uncertain, primarily due to uncertainties surrounding the contributions of water from the Greenland and Antarctic ice sheets, which are major contributors to global SLR. Paleoclimate studies (e.g., Dutton et al., 2015) underscore the significance of the Antarctic ice sheets. Over the past several million years, global sea levels have, at times, been 6–9 meters higher than present, during periods when polar temperatures were slightly warmer than today (by 0–2°C). Three million years ago, during the Pliocene, the average climate was about 2° to 3°C warmer, and sea levels were 25 to 35 meters higher than current levels (Dowsett et al., 1994). A recent modeling study (DeConto and Pollard, 2016) suggests that Antarctica could contribute more than 1 meter of SLR by 2100 and over 15 meters by 2500 if emissions continue unabated. These paleoclimate and modeling studies indicate that future SLR could be even higher, given the current rate of warming.

Regionally, RSLR will also depend on local vertical land motion. While it is uncertain whether land subsidence will continue at its current rate, there is little doubt that ongoing groundwater, oil, and gas withdrawals in Texas will exacerbate RSLR, making it larger than the global mean SLR.

I.3.5 SLR MITIGATION OPTIONS

There are five major adaptation strategies (Griggs and Reguero, 2021) in response to SLR-induced shoreline erosion and coastal flooding including:

1. *Do nothing (or wait and see).* It has the lowest cost upfront but the greatest risk for potential consequences. With accelerated SLR, such strategies only apply to low-populated rural areas.
2. *Managed retreat and relocation.* It has been increasingly considered in many coastal areas. Managed retreat is very complex and can be controversial as it will unevenly affect coastal communities and receive resistance from coastal residents with high stakes.
3. *Beach nourishment.* This method helps preserve the aesthetic and recreational value of protected beaches. It has been implemented in over 200 areas along the U.S. coast since the 1920s (Campbell and Benedet, 2006). The process involves adding additional sand to the sand dunes, subaerial beach, and nearshore submerged sandbars, typically sourced from the offshore continental shelf or dredged from nearby ship channels. In Texas, beach nourishment has been applied in various locations, including Galveston Island, Bryan Beach (near Freeport), Mustang Island (near Port Aransas), and South Padre Island. However,

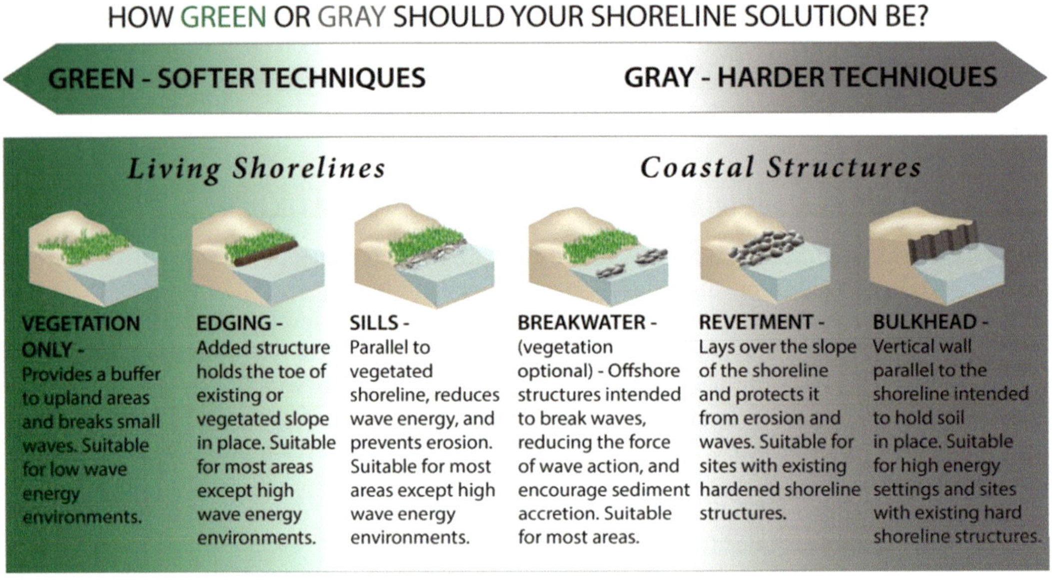

FIGURE I.3.10 Sketch diagram showing different strategies in defending coast from SLR and flooding (from NOAA, 2015; modified from SAGE, 2015).

beach nourishment efforts are only carried out on the more populated sections of the Texas coast, while the majority of Texas beaches remain untouched and subject to natural erosion. Over the period from 1930 to 2007, the average shoreline retreat along Texas coast was 1.26 meters per year (Paine et al., 2014).

4. *Soft protection*. This approach involves sediment relocation, sand dune and beach nourishment, and the restoration and installation of natural infrastructures such as salt marshes, mangroves, oyster reefs, and vegetation (both emerged and submerged) (Figure I.3.10). Several salt marsh restoration projects near Galveston Island have been completed historically, supported by the Texas General Land Office, Texas Parks & Wildlife Department, and the Galveston Bay Foundation. These restored salt marshes not only provide essential natural habitats for numerous fish and wildlife species, but also act as natural buffer zones that reduce wave impact and slow shoreline erosion.
5. *Armoring or hardening the shoreline*. This approach involves installing infrastructure, such as dikes, levees, and storm surge gates, to defend against elevated water levels and waves. An ongoing project, approved by the U.S. Congress and currently in the design phase, is the Galveston Bay Storm Surge Barrier System. The flood mitigation system includes multiple lines of defense to reduce storm-induced flooding damage to critical petrochemical and refinery complexes in and around Galveston Bay (Figure I.3.11). Storm surge barrier systems have been widely used in Europe, and one system has already been constructed in New Orleans, Louisiana, by the U.S. Army Corps of Engineers. The impacts of storm surge barrier systems on hydrodynamics, salt intrusion, and wave dynamics are currently unknown. However, it is believed that the impact will likely be notable only locally (e.g., near the gates) and is unlikely to affect the open water of the bay significantly, as the gates are not permanently closed. It is important to note that the storm surge barrier system does not address precipitation-induced flooding, which has become more frequent in recent years. Therefore, concurrent improvements to urban sewage systems should be considered a key component of flood mitigation strategies.

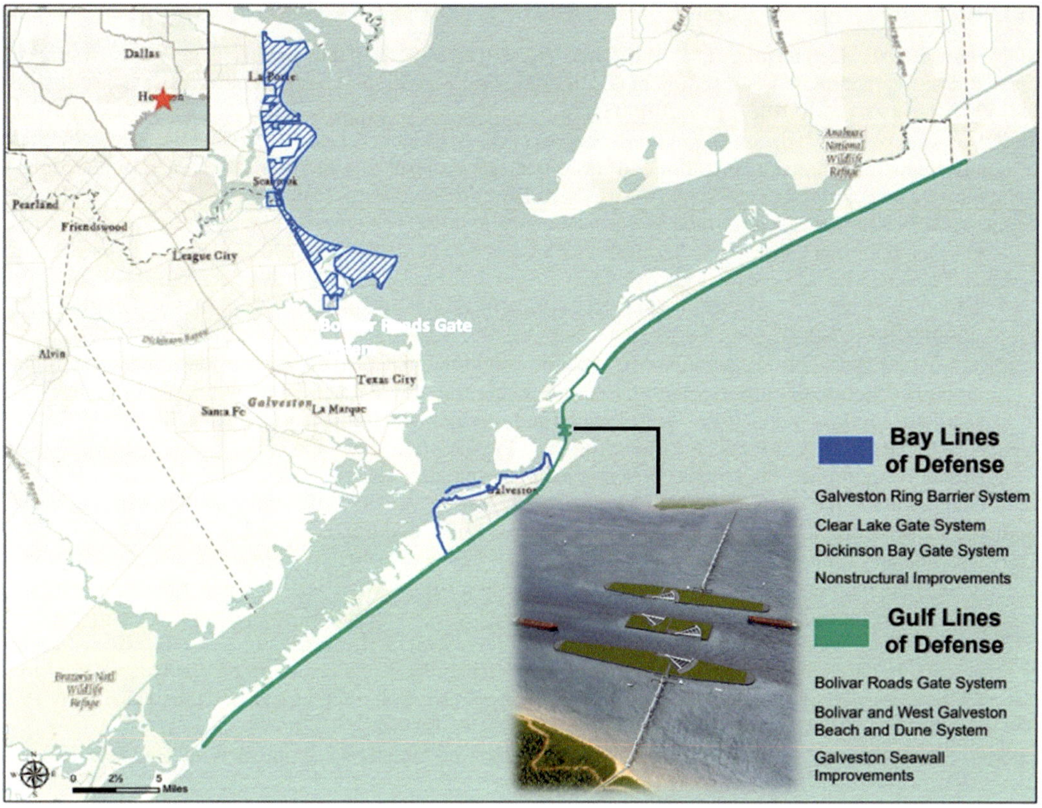

FIGURE I.3.11 Galveston Bay Storm Surge Barrier System. The System includes the Bolivar Road Gate System across the Bay entrance, 43 miles of beach and dune segments on Bolivar Peninsular and West Galveston Island, improvements to the existing 10-mile seawall on Galveston Island, an 18-mile Galveston Ring Barrier System, and Gate System near on the perimeter of Clear Lake and Dickinson Bay. (Figure adapted from USACE & GLO, 2021.)

I.3.6 SUMMARY

In this chapter, we discuss the causes, impacts, and mitigation strategies for the Texas coast by synthesizing existing studies focused on the Texas coastline, as well as other global coastal regions. Findings from these studies are often transferable (e.g., on waves, shoreline erosion, salt intrusion, and salt marshes), but it is important to consider the shallow nature of Texas bays, which currently have a mean depth of 2.5 meters. This shallow characteristic may make the impacts of SLR much more profound than those observed in relatively deeper coastal estuaries, such as Chesapeake Bay.

The Texas coast has experienced a concerning rise in sea level over the past century, largely due to regional vertical land motion resulting from extensive extraction of groundwater and hydrocarbons (oil and gas). The resulting SLR and increased flooding risk have already prompted various management efforts, including both shoreline softening and hardening strategies.

As sea levels rise, ecosystems in Texas bays will be exposed to greater depths, stronger tides and waves, larger storm surges, continuous loss of salt marshes, and increased coastal ocean water intrusion (leading to higher salinity in most bays except for Upper and Lower Laguna Madre). Combined with ongoing ocean warming, SLR could exacerbate water quality issues, such as hypoxia. While many mitigation efforts currently focus on shoreline protection and flooding prevention, it is crucial to plan ahead for the potential water quality and ecosystem challenges.

REFERENCES

Al Mukaimi, ME, Dellapenna, TM, Williams, JR, 2018. Enhanced land subsidence in Galveston Bay, Texas: Interaction between sediment accumulation rates and relative sea level rise. Est Coast Shelf Sci, 207:183–193, doi:10.1016/j.ecss.2018.03.023

Anderson, JB, Wallace, DJ, Simms, AR, Rodriguez, AB, Milliken, KT, 2014. Variable response of coastal environments of the northwestern Gulf of Mexico to sea-level rise and climate change: Implications for future change. Mar Geol, 352:348–366, doi:10.1016/j.margeo.2013.12.008

Armitage, AR, Highfield, WE, Brody, SD, Louchouarn, P, 2015. The contribution of mangrove expansion to salt marsh loss on the Texas Gulf coast. PLoS One, 10(5):e0125404, doi:10.1371/journal.pone.0125404

Barnard, PL, Erikson, LH, Foxgrover, AC, Hart, JAF, Limber, P, O'Neill, AC, et al., 2019. Dynamic flood modeling essential to assess the coastal impacts of climate change. Sci Rep, 9(1), 4309, doi:10.1038/s41598-019-40742-z

Bawden, GW, Johnson, MR., Kasmarek, MC, Brandt, J, Middleton, CS, 2012. Investigation of land subsidence in the Houston-Galveston Region of Texas by using the global positioning system and interferometric synthetic aperture radar, 1993–2000 (Scientific Investigations Report No. 2012–5211).

Bilskie, MV, Hagen, SC, Alizad, K, Medeiros, SC, Passeri, DL, Needham, HF, Cox, A, 2016. Dynamic simulation and numerical analysis of hurricane storm surge under sea level rise with geomorphologic changes along the northern Gulf of Mexico. Earth Future, 4(5):177–193, doi:10.1002/2015EF000347

Campbell, T, Benedet, L, 2006. Beach nourishment magnitudes and trends in the U.S. J Coast Res, 39:57–64.

Chini, N, Stansby, P, Leake, J, Wolf, J, Roberts-Jones, J, Lowe, J, 2010. The impact of sea level rise and climate change on inshore wave climate: A case study for East Anglia (UK). Coast Eng, 57(11):973–984, doi:10.1016/j.coastaleng.2010.05.009

Church, JA, Clark, PU, Cazenave, A, Gregory, JM, Jevrejeva, S, Levermann, A, et al., 2013. Sea level change. In Climate Change 2013: The Physical Science Basis. Contribution of Working Group I to the Fifth Assessment Report of the Intergovernmental Panel on Climate Change. Cambridge Univ Press, Cambridge.

Church, JA, White, NJ, 2006. A 20th century acceleration in global sea-level rise. Geophys Res Lett, 33(1):L01602, doi:10.1029/2005GL024826

Church, JA, White, NJ, 2011. Sea-level rise from the late 19th to the early 21st century. Surv Geophys, 32 (4–5):585–602, doi:10.1007/s10712-011-9119-1

Comeaux, RS, Allison, MA, & Bianchi, TS, 2012. Mangrove expansion in the Gulf of Mexico with climate change: Implications for wetland health and resistance to rising sea levels. Est Coast Shelf Sci, 96:81–95, doi:10.1016/j.ecss.2011.10.003

Cosby, AG, Lebakula, V, Smith, CN, Wanik, DW, Bergene, K, Rose, AN, et al., 2024. Accelerating growth of human coastal populations at the global and continent levels: 2000–2018. Sci Rep, 14(1):22489, doi:10.1038/s41598-024-73287-x

Dangendorf, S, Hay, C, Calafat, FM, Marcos, M, Piecuch, CG, Berk, K, Jensen, J, 2019. Persistent acceleration in global sea-level rise since the 1960s. Nat Clim Chang, 9(9):705–710, doi:10.1038/s41558-019-0531-8

DeConto, RM, Pollard, D, 2016. Contribution of Antarctica to past and future sea-level rise. Nature, 531(7596):591–597, doi:10.1038/nature17145

Deegan, LA, Johnson, DS, Warren, RS, Peterson, BJ, Fleeger, JW, Fagherazzi, S, Wollheim, WM, 2012. Coastal eutrophication as a driver of salt marsh loss. Nature, 490(7420):388–392, doi:10.1038/nature11533

Dowsett, H, Thompson, R, Barron, J, Cronin, T, Fleming, F, Ishman, S, et al., 1994. Joint investigations of the Middle Pliocene climate I: PRISM paleoenvironmental reconstructions. Global Planet Chang, 9(3):169–195, doi:10.1016/0921-8181(94)90015-9

Du, J, Park, K, Shen, J, Zhang, YJ, Yu, X, Ye, F, et al., 2019. A hydrodynamic model for Galveston Bay and the shelf in the Northern Gulf of Mexico. Ocean Sci, 15(4):951–966, doi:10.5194/os-15-951-2019

Du, J, Shen, J, Zhang, YJ., Ye, F, Liu, Z, Wang, Z, et al., 2018. Tidal response to sea-level rise in different types of estuaries: The importance of length, bathymetry, and geometry. Geophys Res Lett, 45(1):227–235, doi:10.1002/2017GL075963

Dutton, A, Carlson, AE, Long, AJ, Milne, GA, Clark, PU, DeConto, R, et al., 2015. Sea-level rise due to polar ice-sheet mass loss during past warm periods. Science, 349(6244):aaa4019, doi:10.1126/science.aaa4019

Engelkemeir, RM, Khan, SD, 2008. Lidar mapping of faults in Houston, Texas, USA. Geosphere, 4(1), 170–182, doi:10.1130/GES00096.1

Ezer, T, 2022. Sea level variability in the Gulf of Mexico since 1900 and its link to the Yucatan Channel and the Florida Strait flows. Ocean Dyn, 72(11):741–759, doi:10.1007/s10236-022-01530-y

Frederikse, T, Landerer, F, Caron, L, Adhikari, S, Parkes, D, Humphrey, VW, et al., 2020. The causes of sea-level rise since 1900. Nature, 584(7821):393–397, doi:10.1038/s41586-020-2591-3

Fujiwara, M, Martinez-Andrade, F, Wells, RJD, Fisher, M, Pawluk, M, Livernois, MC, 2019. Climate-related factors cause changes in the diversity of fish and invertebrates in subtropical coast of the Gulf of Mexico. Commun Biol, 2(1):1–9, doi:10.1038/s42003-019-0650-9

Gabrysch, RK, & Bonnet, CW, 1975. Land-surface subsidence in the Houston-Galveston region, Texas (Open-File Report). U.S. Geological Survey.

Glass, P, Hollingsworth, T, 1999. Wetlands restoration at Galveston Island State Park a multi-agency project. In Galveston Bay Estuary Program State of the Bay Symposium IV (pp. 201–204). Galveston, Texas.

Griggs, G, Reguero, BG, 2021. Coastal adaptation to climate change and sea-level rise. Water, 13(16):2151, doi:10.3390/w13162151

Haley, M, Ahmed, M, Gebremichael, E, Murgulet, D, Starek, M, 2022. Land subsidence in the Texas coastal bend: Locations, rates, triggers, and consequences. Remote Sens, 14(1):192, doi:10.3390/rs14010192

Hong, B, Shen, J, 2012. Responses of estuarine salinity and transport processes to potential future sea-level rise in the Chesapeake Bay. Est Coast Shelf Sci, 104–105:33–45, doi:10.1016/j.ecss.2012.03.014

Huang, W, Ye, F, Zhang, YJ, Park, K, Du, J, Moghimi, S, et al., 2021. Compounding factors for extreme flooding around Galveston Bay during Hurricane Harvey. Ocean Model, 158:101735, doi:10.1016/j.ocemod.2020.101735

IPCC, 2007. Summary for policymakers. In Climate Change 2007: The Physical Science Basis. Contribution of Working Group I to the Fourth Assessment Report of the Intergovernmental Panel on Climate Change, S. Solomon, D. Qin, M. Manning, Z. Chen, M. Marquis, K.B. Averyt, M. Tignor, and H.L. Miller, eds, (pp. 1–18). Cambridge University Press, Cambridge, UK, and New York, NY. https://www.ipcc.ch/site/assets/uploads/2018/02/ar4-wg1-spm-1.pdf

Khojasteh, D, Glamore, W, Heimhuber, V, & Felder, S, 2021. Sea level rise impacts on estuarine dynamics: A review. Sci Total Environ, 780:146470, doi:10.1016/j.scitotenv.2021.146470

Kuo, CY, 2006. Determination and characterization of 20th century global sea level rise (No. 478). The Ohio State University. https://earthsciences.osu.edu/sites/earthsciences.osu.edu/files/report-478.pdf

Leatherman, SP, 1984. Coastal geomorphic responses to sea level rise: Galveston Bay, Texas. In Greenhouse Effect and Sea Level Rise, Barth MC, Titus JG eds. Springer US, Boston, doi:10.1007/978-1-4684-6569-3_5

Leuliette, EW, Scharroo, R, 2010. Integrating jason-2 into a multiple-altimeter climate data. Mar Geod, 33:504–517, doi:10.1080/01490419.2010.487795

Lindsey, R, 2023. Climate change: Global sea level. Last accessed on 03/13/2025 at http://www.climate.gov/news-features/understanding-climate/climate-change-global-sea-level

Liu, Y, Li, J, Fasullo, J, Galloway, DL, 2020. Land subsidence contributions to relative sea level rise at tide gauge Galveston Pier 21, Texas. Sci Rep, 10(1):17905, doi:10.1038/s41598-020-74696-4

MacCready, P, 2004. Toward a unified theory of tidally-averaged estuarine salinity structure. Estuaries, 27(4):561–570. doi:10.1007/BF02907644

Merkens, JL, Reimann, L, Hinkel, J, Vafeidis, AT, 2016. Gridded population projections for the coastal zone under the shared socioeconomic pathways. Global and Planetary Change, 145, 57–66, doi:10.1016/j.gloplacha.2016.08.009

Michaël, A, Cazenave, A, Valladeau, G, Guinehut, S, 2009. A new assessment of global mean sea level from altimeters highlights a reduction of global trend from 2005 to 2008. Ocean Sci, 5:193–201, doi:10.5194/osd-6-31-2009

Miller, MM, Shirzaei, M, 2021. Assessment of future flood hazards for southeastern Texas: Synthesizing subsidence, sea-level rise, and storm surge scenarios. Geophy Res Lett, 48(8):e2021GL092544, doi:10.1029/2021GL092544

NAS, 2012. Sea-Level Rise for the Coasts of California, Oregon, and Washington: Past, Present, and Future. National Academy of Sciences.

Nerem, RS, Chambers, DP, Choe, C, Mitchum, GT, 2010. Estimating mean sea level change from the TOPEX and Jason altimeter missions. Mar Geod, 33(supl):435–446, doi:10.1080/01490419.2010.491031

Nicholls, RJ, Cazenave, A, 2010. Sea-level rise and its impact on coastal zones. Science, 328(5985):1517–1520, doi:10.1126/science.1185782

Nicholls, RJ, Hoozemans, FMJ, Marchand, M, 1999. Increasing flood risk and wetland losses due to global sea-level rise: Regional and global analyses. Global Environ Chang, 9:S69–S87, doi:10.1016/S0959-3780(99)00019-9

NOAA, 2015. Guidance for considering the use of living shorelines. Last accessed on 05/13/2025 at https://www.habitatblueprint.noaa.gov/wp-content/uploads/2018/01/NOAA-Guidance-for-Considering-the-Use-of-Living-Shorelines_2015.pdf

Osland, MJ, Griffith, KT, Larriviere, JC, Feher, LC, Cahoon, DR, Enwright, NM, et al., 2017. Assessing coastal wetland vulnerability to sea-level rise along the northern Gulf of Mexico coast: Gaps and opportunities for developing a coordinated regional sampling network. PLoS One, 12(9):e0183431, doi:10.1371/journal.pone.0183431

Paine, JG, 1993. Subsidence of the Texas coast: Inferences from historical and late Pleistocene Sea levels. Tectonophysics, 222(3), 445–458, doi:10.1016/0040-1951(93)90363-O

Paine, JG, Caudle, T, Andrews, J, 2014. Shoreline Movement along the Texas Gulf Coast, 1930's to 2012. Final Report Prepared for the General Land Office.

Park, K, Kim, CK., Schroeder, WW, 2007. Temporal variability in summertime bottom hypoxia in shallow areas of Mobile Bay, Alabama. Estuar Coast, 30(1):54–65, doi:10.1007/BF02782967

Park, YH, Edge, BL, 2011. Beach erosion along the Northeast Texas Coast. J Coast Res, 27(3):502–514, doi:10.2112/JCOASTRES-D-09-00042.1

Passeri, DL, Hagen, SC, Plant, NG, Bilskie, MV, Medeiros, SC, Alizad, K, 2016. Tidal hydrodynamics under future sea level rise and coastal morphology in the Northern Gulf of Mexico. Earth Future, 4(5): 159–176, doi:10.1002/2015EF000332

Pawluk, M, Fujiwara, M, Martinez-Andrade, F, 2021. Climate effects on fish diversity in the subtropical bays of Texas. Est Coast Shelf Sci, 249:107121, doi:10.1016/j.ecss.2020.107121

Peltier, WR, 1999. Global sea level rise and glacial isostatic adjustment. Global Planet Chang, 20(2):93–123, doi:10.1016/S0921-8181(98)00066-6

Pendleton, EA, Barras, JA, Williams, SJ, Twichell, DC, 2010. Coastal vulnerability assessment of the Northern Gulf of Mexico to sea-level rise and coastal change. U.S. Geological Survey Open-File Report No. 2010–1146, doi:10.3133/ofr20101146

Penland, S, Ramsey, KE, 1990. Relative sea-level rise in Louisiana and the Gulf of Mexico: 1908–1988. J Coast Res. 6(2):323–342.

Rahmstorf, S, 2007. A semi-empirical approach to projecting future sea-level rise. Science, 315(5810), 368–370, doi:10.1126/science.1135456

Ravens, TM, Thomas, RC, Roberts, KA, Santschi, PH, 2009. Causes of salt marsh erosion in Galveston Bay, Texas. J Coast Res, 2009(252):265–272, doi:10.2112/07-0942.1

Rego, JL, Li, C, 2010. Storm surge propagation in Galveston Bay during Hurricane Ike. J Mar Sys, 82(4): 265–279, doi:10.1016/j.jmarsys.2010.06.001

Reguero, BG, Losada, IJ, Méndez, FJ, 2019. A recent increase in global wave power as a consequence of oceanic warming. Nature Commun, 10(1):205, doi:10.1038/s41467-018-08066-0

Rozas, LP, Minello, TJ, Zimmerman, RJ, Caldwell, P, 2007. Nekton populations, long-term wetland loss, and the effect of recent habitat restoration in Galveston Bay, Texas, USA. Mar Ecol Prog Ser, 344:119–130, doi:10.3354/meps06945

SAGE, 2015. Natural and structural measures for shoreline stabilization brochure. Last accessed on 05/13/2025 at https://coast.noaa.gov/data/digitalcoast/pdf/living-shoreline.pdf.

Shafer, DJ, Roland, R, Douglass, S, 2003. Preliminary Evaluation of Critical Wave Energy Thresholds at Natural and Created Coastal Wetlands. The U.S. Army Engineer Research and Development Center. Last accessed on 05/13/2025 at https://apps.dtic.mil/sti/citations/ADA413157

Simms, AR, Anderson, JB, DeWitt, R, Lambeck, K, Purcell, A, 2013. Quantifying rates of coastal subsidence since the last interglacial and the role of sediment loading. Global Planet Chang, 111:296–308, doi:10.1016/j.gloplacha.2013.10.002

Steinberg, JM, Piecuch, CG, Hamlington, BD, Thompson, PR, Coats, S, 2024. Influence of deep-ocean warming on coastal sea-level decadal trends in the Gulf of Mexico. J Geophys Res Oceans, 129(1), e2023JC019681, doi:10.1029/2023JC019681

Stockdon, HF, Doran, KJ, Thompson, DM, Sopkin, CL, Plant, NG, Sallenger, AH, 2012. National assessment of hurricane-induced coastal erosion hazards: Gulf of Mexico. U.S. Geological Survey Open-File Report No. 2012–1084.

Uddameri, V, Singaraju, S, & Hernandez, EA, 2014. Impacts of sea-level rise and urbanization on groundwater availability and sustainability of coastal communities in semi-arid South Texas. Environl Earth Sci, 71(6): 2503–2515, doi:10.1007/s12665-013-2904-z

USACE, GLO, 2021. Coastal Texas protection and restoration feasibility study. Last accessed on 05/13/2025 at https://www.swg.usace.army.mil/Portals/26/Coastal%20TX%20Executive%20Summary_FINAL_20210827.pdf

Vousdoukas, MI, Mentaschi, L, Voukouvalas, E, Verlaan, M, Jevrejeva, S, Jackson, LP, Feyen, L, 2018. Global probabilistic projections of extreme sea levels show intensification of coastal flood hazard. Nature Commun, 9(1):2360, doi:10.1038/s41467-018-04692-w

Warner, NN, Tissot, PE, 2012. Storm flooding sensitivity to sea level rise for Galveston Bay, Texas. Ocean Eng, 44:23–32, doi:10.1016/j.oceaneng.2012.01.011

White, WA, Morton, RA, 1997. Wetland losses related to fault movement and hydrocarbon production, southeastern Texas Coast. J Coast Res, 13(4):1305–1320.

White, WA, Morton, RA, Holmes, CW, 2002. A comparison of factors controlling sedimentation rates and wetland loss in fluvial–deltaic systems, Texas Gulf coast. Geomorphology, 44(1):47–66, doi:10.1016/S0169-555X(01)00140-4

White, WA, Tremblay, TA, Waldinger, RL, Calnan, TR, 2004. Status and trends of wetland and aquatic habitats on barrier islands, upper Texas coast, Galveston and Christmas Bays. Final Report Prepared for the Texas General Land Office and National Oceanic and Atmospheric Administration. https://hdl.handle.net/2152/125595

White, WA, Tremblay, TA, Wermund, EG, Jr, Handley, LR, 1993. Trends and Status of Wetlands and Aquatic Habitats in the Galveston Bay System, Texas. Galveston Bay National Estuary Program Publication No. 31.

Willis, JK, Chambers, DP, Kuo, CY, & Shum, CK, 2010. Global sea level rise: recent progress and challenges for the decade to come. Oceanogr, 23(4):26–35, doi:10.5670/oceanog.2010.03

Yang, Z, Wang, T, Leung, R, Hibbard, K, Janetos, T, Kraucunas, I, et al., 2014. A modeling study of coastal inundation induced by storm surge, sea-level rise, and subsidence in the Gulf of Mexico. Nat Hazards, 71(3):1771–1794, doi:10.1007/s11069-013-0974-6

Yeager, KM, Santschi, PH, Rifai, HS, Suarez, MP, Brinkmeyer, R, Hung, CC, et al., 2007. Dioxin chronology and fluxes in sediments of the Houston Ship Channel, Texas: Influences of non-steady-state sediment transport and total organic carbon. Environ Sci Technol, 41(15):5291–5298, doi:10.1021/es062917p

Yin, J, 2023. Rapid decadal acceleration of sea level rise along the U.S. East and Gulf Coasts during 2010–22 and its impact on hurricane-induced storm surge. J Climat, 36(13):4511–4529, doi:10.1175/JCLI-D-22-0670.1

Yoskowitz, D, Carollo, C, Pollack, JB, Santos, C, Welder, K, 2017. Integrated ecosystem services assessment: Valuation of changes due to sea level rise in Galveston Bay, Texas, USA. Integr Environ Assess Manag, 13(2):431–443, doi:10.1002/ieam.1798

Younas, M, Khan, SD, Tirmizi, O, Hamed, Y, 2023. Geospatial analytics of driving mechanism of land subsidence in Gulf Coast of Texas, United States. Sci Total Environ, 902:166102, doi:10.1016/j.scitotenv.2023.166102

Zhou, X, Wang, G, Wang, K, Liu, H, Lyu, H, Turco, MJ, 2021. Rates of natural subsidence along the Texas Coast derived from GPS and tide gauge measurements (1904–2020). J Surv Eng, 147(4), 04021020, doi:10.1061/(ASCE)SU.1943-5428.0000371

I.4 Scorched Earth and Shifting Sands

Drought, Dust, and Wildfire in a Changing Landscape

Lili Yao, Mahesh Bade, Nilkamal Jaisawal, Nafissatou Ba, and Yang Li

I.4.0 OVERVIEW

Climate change is reshaping meteorological patterns, intensifying extreme environmental conditions such as drought, dust, and wildfire across large regions of Texas. These events are interconnected through complex feedback mechanisms involving atmospheric processes, vegetation dynamics, and land use changes (Figure I.4.1). Dry soils and stressed vegetation increase susceptibility to both wind-driven dust emissions and wildfire outbreaks, while human activities and land management add further complexity. Research has explored changes in these extreme events, with studies available across different databases (Table I.4.1). Drought and dust events in Texas have been widely studied, whereas fire events have received less attention with fewer publications. This chapter reviews the changing characteristics of drought, dust, and wildfire, highlighting their key drivers and potential consequences for Texas.

I.4.1 DROUGHT

I.4.1.1 Introduction

Drought is a complex and widespread natural hazard that poses significant challenges to both the environment and human societies. It is marked by a prolonged period of water shortage, involving deficits in precipitation (meteorological drought), soil moisture (agricultural drought), and surface or groundwater resources (hydrological drought) (AghaKouchak et al., 2021; Hao and Singh, 2015; Mishra and Singh, 2011, 2010; Van Loon, 2015). These conditions can ultimately lead to insufficient water supply to meet human demands, adversely affecting socioeconomic activities (socioeconomic drought). Unlike other natural disasters, droughts typically develop gradually and can persist for months or even years. The impacts of droughts can be extensive and severe, affecting ecosystems, agriculture, water resources, and economies, making them among the costliest natural disasters in the United States, with damages totaling $367.5 billion from 1980 to 2024 (National Centers for Environmental Information, 2025). Furthermore, droughts can intensify risk for other environmental hazards, such as wildfires, flash flooding, landslides, and desertification, creating a cascade of negative effects on both natural and human systems (AghaKouchak et al., 2018; Deb et al., 2020). Recent research suggests that climate change is reshaping drought dynamics, potentially leading to more frequent and severe events, while amplifying the difficulties of mitigation and adaptation (Gu et al., 2023; Rastogi et al., 2024; Ullah et al., 2023; Zhou et al., 2023).

 DOI: 10.1201/9781003406600-5

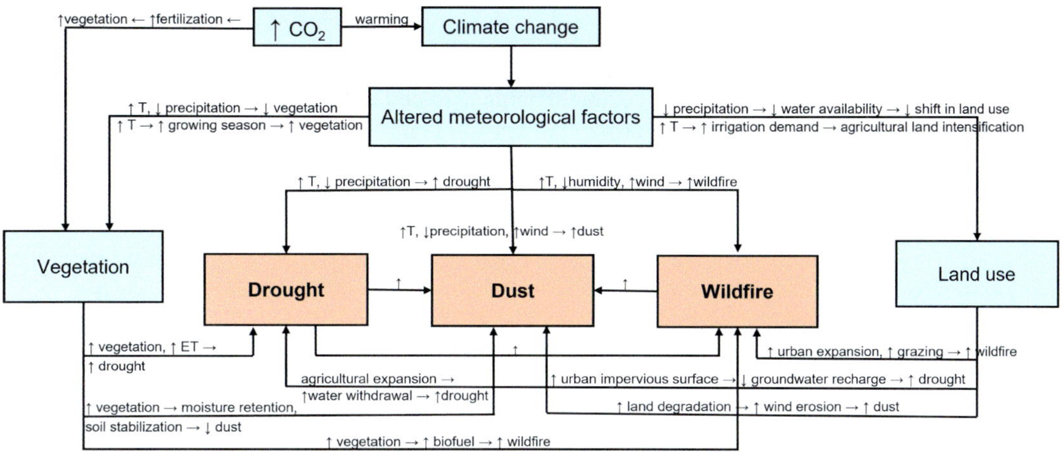

FIGURE I.4.1 Drivers and interconnections of drought, dust, and wildfires under a changing climate.

TABLE I.4.1
The Number of Scholarly Sources Available for Texas Based on Keyword Search

Topic	Keywords	Academic Search Complete	GeoRef	Web of Science	PubMed	Scopus
Drought	Texas AND drought	1068	540	1204	107	1286
	Texas AND drought AND (change OR trend)	209	213	505	25	439
	Texas AND drought AND (change OR trend) AND (historical OR past)	36	54	96	4	90
	Texas AND drought AND (change OR trend) AND (future OR projection OR prediction)	50	61	139	10	123
Dust	Texas AND dust	236	188	394	85	452
	Texas AND dust AND (change OR trend)	25	31	74	5	78
	Texas AND dust AND (change OR trend) AND (historical OR past)	4	3	11	2	10
	Texas AND dust AND (change OR trend) AND (future OR projection OR prediction)	7	6	12	1	12
Wildfire	Texas AND (wildfire OR forest fire)	273	33	229	9	211
	Texas AND (wildfire OR forest fire) AND (change OR trend)	37	10	120	1	67
	Texas AND (wildfire OR forest fire) AND (change OR trend) AND (historical OR past)	5	1	34	0	21
	Texas AND (wildfire OR forest fire) AND (change OR trend) AND (future OR projection OR prediction)	7	4	24	0	19

Source: Academic Search Complete (https://about.ebsco.com/products/research-databases/academic-search-complete), GeoRef (https://about.ebsco.com/products/research-databases/georef), Web of Science (https://www.webofscience.com/wos/woscc/basic-search), PubMed (https://pubmed.ncbi.nlm.nih.gov/advanced/), and Scopus (https://www.scopus.com/pages/home?display=basic&zone=header&origin=sbrowse#basic).

1.4.1.2 Drought Identification

Drought can be quantified using a variety of indices and metrics that capture different aspects of water deficits, depending on the type of drought being assessed (Bachmair et al., 2016; Heim, 2002; Van Loon, 2015). Among these, standardized indices are currently the most used drought indicators because they transform raw hydrometeorological variables (e.g., precipitation, streamflow) into standardized values, enabling direct comparisons across different regions and timescales (Zhang et al., 2022). These indices typically involve fitting a probability distribution to the data and then converting the values into a standardized normal distribution with a mean of zero and standard deviation of one, where negative values indicated drier-than-normal conditions and positive values indicate wetter-than-normal conditions (Tijdeman et al., 2020).

A commonly used standardized index is the Standardized Precipitation Index (SPI), which quantifies meteorological drought by measuring precipitation deficits over a given period and is often calculated by fitting a gamma distribution to precipitation data (McKee et al., 1993). Another widely used index is the Standardized Precipitation Evapotranspiration Index (SPEI), derived by fitting a log-logistic distribution to the difference between precipitation and potential evapotranspiration (PET) (Vicente-Serrano et al., 2010). By incorporating temperature-driven PET, SPEI is more sensitive to warming-induced droughts, a key advantage over the SPI. Both SPI and SPEI can be computed at various temporal scales, such as 1-month, 6-month, or 12-month periods, allowing for drought condition assessments over different durations. For example, the 6-month SPI is derived by summing precipitation over six consecutive months and comparing it to historical 6-month precipitation totals. Shorter timescales, like the 1-month SPI, capture short-term meteorological fluctuations, while longer timescales, such as the 12-month SPI, are better suited for assessing prolonged droughts and their hydrological impacts.

Beyond SPI and SPEI, the standardized framework has been applied to other hydrological variables to assess different aspects of drought. For example, the Standardized Runoff Index (SRI) and Standardized Streamflow Index (SSFI) are used to evaluate hydrological drought (Rajsekhar et al., 2015b; Shukla and Wood, 2008; Vicente-Serrano et al., 2012), and the Standardized Soil Moisture Index (SSMI) is applied to agricultural drought (Samaniego et al., 2013). Alongside standardized drought indices, the Palmer Drought Severity Index (PDSI) (Palmer, 1965) is another widely used drought metric. Unlike SPI and SPEI, which are purely statistical, PDSI incorporates temperature, precipitation, and modeled soil moisture to represent multifaceted water deficits within a water balance framework. While PDSI has been valuable for assessing long-term drought characteristics, it has limitations, including complex calculation procedures, lack of transparency, lack of flexibility with time scales, and challenges in comparing values across different regions (Vicente-Serrano et al., 2010). Along with the standardized indices, the theory of runs (Yevjevich, 1967) has been commonly used to analyze drought indices by applying a threshold value to identify drought events and assess drought characteristics such as duration and severity.

In addition to standardized indices, alternative methods exist for drought detection. For instance, percentile-based methods rank hydrometeorological values (e.g., soil moisture, streamflow) and assign a percentile score, with drought conditions typically determined when values fall below a predefined threshold, such as the 20th percentile, representing the driest 20% of historical records (Hammond et al., 2022; Tanguy et al., 2023; Wuebbles et al., 2014). However, no single index can fully capture the interconnected nature of drought impacts, underscoring the need for continued innovation in drought assessment methods. Recognizing these shortcomings, researchers have developed multivariate standardized drought indices that integrate multiple variables into a single framework to provide more comprehensive drought assessment indices (Hao and AghaKouchak, 2013; Hao and Singh, 2015; Ma et al., 2015; Rajsekhar et al., 2015a; Tadesse et al., 2015).

1.4.1.3 Historical Droughts in Texas

Drought has long been a defining characteristic of Texas's climate and socio-economic landscape. Drought in Texas has been assessed using various data sources and measurement techniques that capture different aspects of drought conditions. Ground-based climate and hydrological records from weather stations, river gauges, and other in-situ networks provide direct and precise measurements of key drought variables such as precipitation, temperature, streamflow, groundwater levels, and soil moisture. However, instrumental records are often limited by the density and distribution of monitoring stations, as well as the temporal coverage and consistency of the data, which constrain the ability to assess drought over large areas and extended periods (Yao et al., 2024). To address these limitations, numerical simulation models such as the Variable Infiltration Capacity (VIC) model (Liang et al., 1994), the Soil and Water Assessment Tool (SWAT) (Arnold and Fohrer, 2005), the Distributed Hydrology Soil and Vegetation Model (DHSVM) (Wigmosta et al., 1994), and the Noah-Multiparameterization (Noah-MP) (Niu et al., 2011; Yang et al., 2011) have been used to provide more spatially and temporally continuous variables for drought analysis in Texas (Chen et al., 2021, 2019; Rajsekhar et al., 2015b; Zhao et al., 2016). Each model offers distinct strengths depending on the research objectives, available data, and study region. For example, on the one hand VIC and DHSVM focus on detailed land surface and hydrological processes (e.g., 30 m to 90 m spatial resolution), while SWAT is particularly useful for assessing the impact of land use and agricultural practices on water availability. Noah-MP, on the other hand, integrates multiple parameterization options, allowing for flexibility in representing land-atmosphere interactions.

In addition to modeling approaches, advancements in remote sensing have led to the widespread adoption of satellite-derived datasets for drought analysis in Texas, such as Vegetation Condition Indicators (VCI) from the Moderate Resolution Imaging Spectroradiometer (MODIS), total water storage estimates from the Gravity Recovery and Climate Experiment (GRACE) satellite, soil moisture availability from the Advanced Microwave Scanning Radiometer (AMSR) and Soil Moisture Active Passive (SMAP) (Abbaszadeh et al., 2021; AghaKouchak et al., 2015; Geruo et al., 2017; Gyawali et al., 2022; Long et al., 2013; Quiring and Ganesh, 2010; Yuan, 2020). These satellite products have become valuable for monitoring drought conditions in Texas, complementing traditional ground-based observations and model simulations. A plausible approach to seamlessly combining in-situ measurements, satellite data, and model simulations to generate spatiotemporal land surface variables is to leverage observations to update model states and parameters. This method, known as data assimilation, integrates model simulations with observational data and is widely used in drought studies (Ahmadalipour et al., 2017; Hassani et al., 2024; Yan et al., 2015). Paleoclimate reconstructions, which aim to gain information about climate prior to the widespread availability of instrumental records, offer a broader context for understanding the range and severity of potential drought conditions (Banner et al., 2024, 2010; Cleaveland et al., 2011; Cook et al., 2015). This information is crucial for setting realistic expectations regarding future drought scenarios, particularly as climate change may exacerbate drought conditions in Texas.

Texas has experienced numerous significant droughts throughout its history. The extended drought that occurred in the 1950s, lasting from 1950 to 1957, is considered the state's drought of record. It ranks as the most severe drought statewide based on the record of precipitation measurements (Texas Water Development Board, 2022a; Vaughan et al., 2012). The 1950s drought has been widely used as a worst-case scenario for benchmarking the severity of future drought risks in Texas water management plans. Its impact was devastating, causing an estimated $18 to $28 billion agricultural loss when adjusted to 2017 dollars based on the report from the Texas Water Development Board (2022a). Streamflow was consistently below median across most parts of the state throughout the duration of the drought, indicated by the United States Geological Survey (USGS) streamflow records. However, tree ring studies have revealed evidence of even longer and more severe droughts in the past (Banner et al., 2010; Cleaveland et al., 2011; Cook et al., 2015). More recently, the

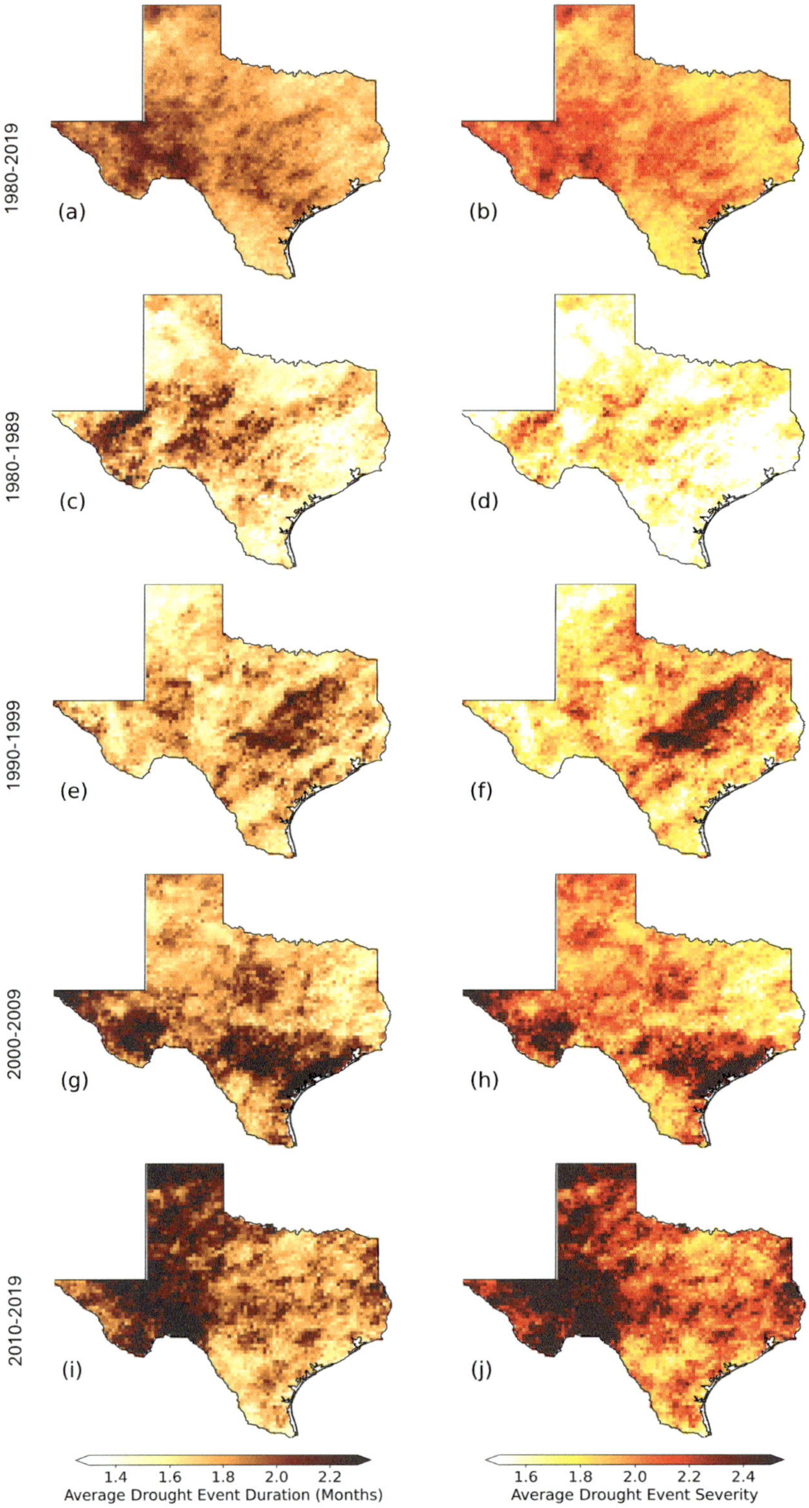

FIGURE I.4.2 Spatial distribution of average duration (left) and severity (right) of meteorological drought across Texas, derived from the 1-month SPEI for the period 1980 to 2019 (a–b) and for each decade from 1980 to 2019 (c–j).

2010–2015 drought had extensive impacts on both the natural environment and economic sectors (Ferencz et al., 2024; Hoerling et al., 2013; Long et al., 2013; Scanlon et al., 2013; Tadesse et al., 2015). This drought surpassed the 1950s drought in several key measures. For instance, the peak intensity of the 2010–2015 drought, as indicated by the PDSI, was more severe than that of the 1950s drought. Additionally, the cumulative streamflow deficit during the 2010–2015 period exceeded that of the 1950s drought, highlighting its hydrological severity. The year of 2011, in particular, stands out as the worst single-year drought on record in Texas, resulting in direct agricultural loss estimated at over $7.6 billion (Texas Water Development Board, 2022a).

Texas exhibits a generally decreasing pattern of historical drought severity from west to east, reflecting the state's climatic gradient, which transitions from arid and semi-arid conditions in western Texas to more humid and precipitation-rich environments in the east (Chen et al., 2021; Rajsekhar et al., 2013). This spatial variation in drought severity is primarily driven by differences in precipitation, evapotranspiration rates, and prevailing atmospheric circulation patterns (Nielsen-Gammon et al., 2020; Rajsekhar et al., 2015b). As a result, regions such as the Panhandle and western Texas experience longer and more severe droughts than other parts of the state. Figures I.4.2a and I.4.2b illustrate the average duration and magnitude of meteorological drought events based on 1-month SPEI from 1980 to 2019, derived from precipitation and Penman-Monteith PET using 1/8-degree Weather Research & Forecasting (WRF) climate data (Jones et al., 2023). A threshold of –0.5 was used to identify a drought event. Drought duration is measured in months, while severity is quantified as the absolute value of the accumulated SPEI (unitless) over an event. West Texas and the most northwestern Panhandle experienced the most severe and prolonged droughts, with average duration up to 2.4 months and severity up to 2.3. In contrast, the eastern and the southernmost parts of the state experienced shorter and less severe droughts, with durations around 1.7 months and severity near 1.8.

The effect of climate change on drought has already been observed in historical records. The statewide 1-month SPEI from 1980 to 2019 shows a statistically significant decreasing trend of –0.09 per decade, which reflects a long-term drying pattern in Texas (Figure I.4.3). The observed change in drought properties has not been homogeneous across the state. Figure I.4.2 (c–j) presents the average drought duration and severity for each decade from 1980 to 2019, showing that meteorological drought has generally become longer and more severe across Texas. This trend is particularly pronounced in northern and western Texas, echoing the intensifying hydrological drought in these regions detected by Rajsekhar et al. (2015b). Worsening trends in meteorological drought could result in increasing challenges for drought impacts to hydrological and agricultural systems (Gyawali et al., 2022; Lee et al., 2017; Yuan et al., 2021). Table I.4.2 lists the studies referenced in this section that directly focus on historical droughts in Texas.

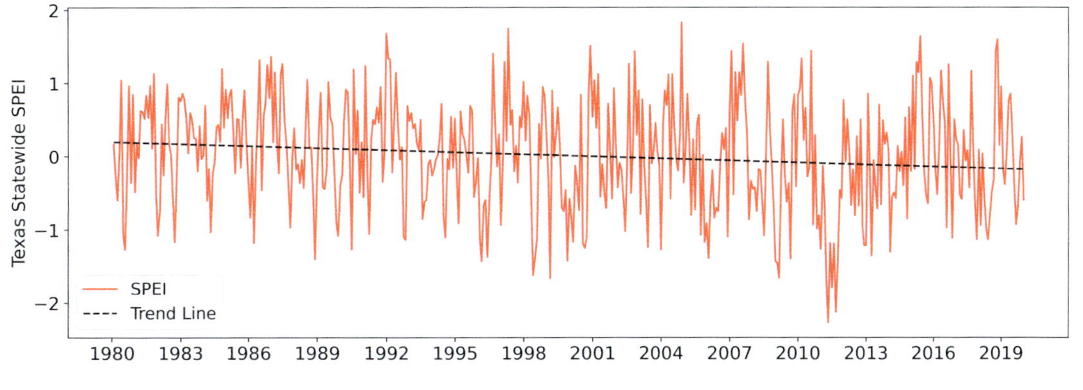

FIGURE I.4.3 Texas statewide 1-month SPEI from 1980 to 2019. The Mann-Kendall test (Mann 1945; Kendall, 1975) reveals a decreasing trend ($\tau = -0.10$, $p < 0.001$), indicating drying conditions over time.

TABLE I.4.2
Studies on Historical Droughts with Direct Texas Focus Referenced in This Section

Study	Location	Study period	Data & Method	Key findings
Cleaveland et al. (2011)	South central and western Texas	1500–2008	PDSI based on tree ring chronologies	Decadal-scale droughts have occurred in Texas at least once a century since the 1500s.
Stahle and Cleaveland (1988)	Northern and Southern Texas	1698–1980	PDSI based on tree ring chronologies	The multiyear drought of the 1950s was the most severe continuous drought since 1698, and meteorological observations from 1931 to 1980 appear to be representative of growing season climate conditions from 1968 to 1980.
Banner et al. (2024)	Austin, Texas	1844–2018	Ring width index based on tree ring chronologies	Trees in urbanized watersheds are decoupled from drought due to the sustained baseflow from municipal water networks.
Lee et al. (2017)	Texas Gulf Region (10 major river basins	1918–2011	10th percentile soil moisture based on VIC simulation	Most basins show decreasing precipitation and rising temperatures, with the Colorado River Basin and Gulf Coast regions being particularly vulnerable to drought.
Rajsekhar et al. (2015b)	Texas	1950–2000	SSFI based on VIC simulation	Hydrological severity decreases from west to east, with worsening drought trends in arid and continental regions and alleviating trends along the Gulf Coast and in humid and semi-humid regions.
Rajsekhar et al. (2013)	Texas	1950–2000	SSFI based on VIC simulation	West Texas had the most severe hydrological drought, South Texas the longest, and East Texas the least severe and shortest.
Quiring and Ganesh (2010)	Texas	1982–1999	VCI based on the National Oceanic and Atmospheric Administration (NOAA) Advanced Very High Resolution Radiometer (AVHRR) satellite data	VCI is most strongly correlated with 6-month SPI, 9-month SPI and PDSI, with a stronger correlation in western ($R^2 > 0.6$) than eastern ($R^2 < 0.1$) Texas.
Yao et al. (2024)	Texas Gulf Region (7 major river basins	The study period varies among gages, with start years ranging from 1940 to 1985 and end years ranging from 1987 to 2018.	SPI and SSI based on USGS gaged flow and naturalized flow from Water Rights Analysis Package	Water management reduced hydrological drought frequency at 81% of the gages analyzed and increased the response timescale of hydrological drought by a median of 2.5 months.
Moorhead et al. (2015)	Texas High Plains	1997–2010	SPEI based on Texas High Plains Evapotranspiration network	SPEI with potential crop evapotranspiration (ET) rather than reference ET is a better index to monitor agricultural drought for summer crops.

TABLE I.4.2 (*Continued*)
Studies on Historical Droughts with Direct Texas Focus Referenced in This Section

Study	Location	Study period	Data & Method	Key findings
Gyawali et al. (2022)	Texas Gulf Coast Aquifer	2002–2019	Groundwater storage derived from multiple data sources: GRACE-derived terrestrial water storage, Global Land Data Assimilation System and North American Land Data Assimilation System derived soil moisture storage, TWDB surface water storage	A significant decline in the coastal groundwater storage, at a rate of 0.35 ± 0.078 $km^3 \cdot yr^{-1}$.
Geruo et al. (2017)	Texas and surrounding semiarid regions	2003–2013	Total water storage, soil moisture availability, enhanced vegetation index, solar-induced fluorescence, and gross primary production derived from multiple data sources: AMSR-E & AMSR-2 satellite microwave sensors, GRACE, MODIS	Vegetation dependency on TWS weakened in shrub-dominated western Texas but strengthened in central and eastern Texas. TWS deficits lasted longer than SM deficits during droughts (7–9 months in shrubland, 4–5 months in grassland, vs. 2.7 months in shrubland and 3 months in grassland).
Ferencz et al. (2024)	Colorado Basin, Texas	2008–2015	Multisectoral analysis of drought impacts (e.g., environmental impacts) and management responses (e.g., water supply planning) based on data from local, state, and federal sources.	The region's robust economy and developed water infrastructure effectively mitigated the economic and societal impacts of the record drought.
Yuan (2020)	Coke County, Texas	2009 and 2011	Landsat Normalized Difference Vegetation Index (NDVI)	NDVI dropped significantly in 2011 compared to 2009 due to drought impact, with shrubs most affected and crops least affected.
Chen et al. (2021)	Texas	2010–2013	Soil moisture percentile based on Noah-MP simulation	The assimilation experiment predicts more severe and prolonged droughts in Texas, highlighting not just 2011 but also 2012–2013, with droughts most frequent in the north, west, and southwest.
Long et al. (2013)	Texas	2011	Total water storage derived from GRACE	Total water storage depletion was 62.3 ± 17.7 km^3 in response to the 2011 drought in Texas.
Hoerling et al. (2013)	Texas	2011	Precipitation and surface temperature from observation (NOAA's monthly Climate Division Data) and simulation (NOAA's Climate Forecast System and phase 5 of the Coupled Model Intercomparison Project (CMIP5))	The main cause of the 2011 Texas drought/heat wave was a severe rainfall deficit linked to anomalous sea surface temperatures, which increased the frequency of severe rainfall deficit from 9% to 34% compared to 1981–2010.

(*Continued*)

TABLE I.4.2 (*Continued*)
Studies on Historical Droughts with Direct Texas Focus Referenced in This Section

Study	Location	Study period	Data & Method	Key findings
Scanlon et al. (2013)	Texas	2011	Electricity and water demand and supply from state and federal agencies	During the 2011 drought, electricity generation increased by 6%, water consumption rose by 9%, but reservoir storage for cooling dropped by 30% compared to 2010.
Tadesse et al. (2015)	Southern Great Plains including Texas	2011	Vegetation drought response index (VegDRI) based on remote sensing data (e.g., AVHRR NDVI), biophysical data (e.g., USGS Land cover), and climate data (SPI and PDSI based on weather station records)	During the 2011 growing season (from April to October), the total percent area under drought in Texas ranged from 15.5% to 78.4%.

The studies are ordered by the first year of their study period.

I.4.1.4 Future Droughts in Texas

Texas is projected to undergo significant climatic changes in the 21st century, driven by rising temperatures, shifting precipitation patterns, and a likely reduction in precipitation amount (Chiang et al., 2018; Jiang and Yang, 2012; Rajsekhar et al., 2015c; Seager et al., 2007; Yoon et al., 2018). These changes are expected to have profound implications for drought conditions across the state. An overall drying trend, indicated by the soil moisture indices, is shown in Figure I.4.4, derived from multiple CMIP5 global climate model (GCM) simulations by Nielsen-Gammon et al. (2020). Understanding evolving drought dynamics is critical for managing the growing risks posed by drought. This section discusses how climate change could impact drought in Texas from various perspectives. Table I.4.3 summarizes relevant studies referenced in this section that specifically focus on future drought in Texas.

Meteorological drought can cause other types of droughts, such as agricultural drought and hydrological drought that harm crop production and reduce water availability. While meteorological drought is traditionally defined by precipitation deficits, rising temperatures under climate change are expected to exacerbate its severity by increasing atmospheric demand for moisture through

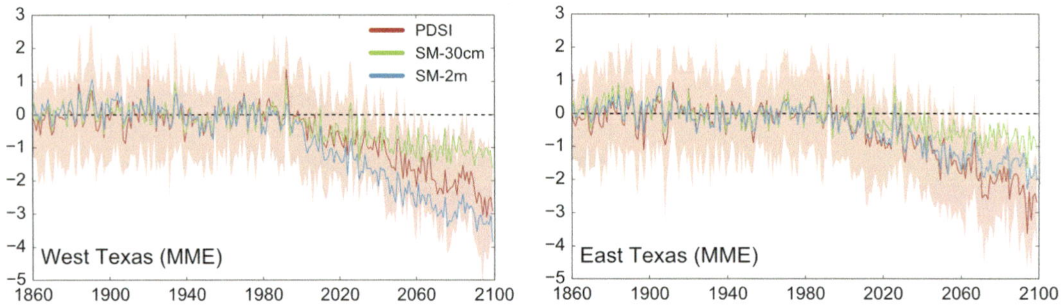

FIGURE I.4.4 Multimodel ensemble (MME) simulations of historical and projected PDSI and standardized soil moisture (SM) indices at 30 cm and 2 m depths for West Texas (left) and East Texas (right) under the Representative Concentration Pathways (RCP) 8.5 scenario. Adapted from Nielsen-Gammon et al. (2020).

TABLE I.4.3
Studies on Future Droughts with Direct Texas Focus Referenced in This Section

Study	Location	Future scenario & Study Period	Data & Method	Key Findings
Nielsen-Gammon et al. (2020)	Texas	1860–2100 under RCP 8.5 (historical + future)	PDSI and standardized soil moisture based on CMIP5 GCM	Climate models consistently project summer soil moisture decline across eastern and western Texas in the 21st century.
Loáiciga et al. (2000)	Edwards Balcones Fault Zone aquifer in Texas	Historical climatic time series (1949–1959, 1978–1989, 1975–1990) scaled to 2 × CO_2 conditions	Spring flow from TWDB's finite-difference groundwater simulation model (GWSIM IV)	In a 2 × CO_2 climate, pumping up to 140,000 acre-feet per year would minimize aquifer impacts while maintaining a base water supply.
Hernandez and Uddameri (2014)	South Texas	2000–2099 under emission scenarios A1B (higher fossil fuel use, 720 ppm CO_2) and B1 (lower fossil fuel use, 550 ppm CO_2)	SPI and SPEI based on CMIP3 GCM	Early-century (2000–2049) droughts are PET controlled, while late-century (2050–2099) droughts become more severe controlled by both precipitation and PET.
Zhao et al. (2018)	Dallas water source watersheds in Texas	2000–2099 under RCP4.5 and RCP8.5	Reservoir storage and water supply reliability based on DHSVM	Reservoir storage and water supply reliability reduced by 16.1% and 14.2% in the late century (2050–2099) compared to the near future (2000–2049) under RCP8.5.
Yoon et al. (2018)	Texas	2006–2080 under RCP8.5	Groundwater storage based on Community Earth System Model	Groundwater storage is projected to decline at a rate of approximately 1.5 cm per decade.
Wuebbles et al. (2014)	South-central United States including Texas	2006–2100 under RCP4.5 and RCP 8.5	Precipitation minus ET (net water availability) based on CMIP5 GCM	Under RCP8.5, the net water balance is projected to decline by 20% in winter, spring, and fall by the late 21st century.
Li et al. (2024)	Texas	2015–2074 under SSP245 and SSP585	Net water availability from Budyko model and Hargreaves-Samani model	Texas's water availability is expected to decline as demand rises, leading to shortages in most regions, especially under SSP585.
Rajsekhar et al. (2015c)	Texas	2015–2099 under A1B, A2, and B1.	Multivariate Drought Index based on VIC simulation	"Very High" drought risk areas are concentrated in northern, western, and southern Texas.
Shao et al. (2023)	Upper Trinity River Basin in Texas	2019–2099 under SSP245 and SSP585	Ratio of Firm Yield based on Water Availability Modeling & DHSVM	The median total firm yield in the far future (2060–2099) is 13.6% and 16.8% smaller than the baseline (1980–2018) under SSP245 and SSP585, respectively.
Valiya Veettil et al. (2024)	Texas	2020–2070 under RCP4.5 and RCP8.5	Blue water scarcity index based on USGS monthly water balance model	Both scenarios show a significant increase in drought severity. Under RCP8.5, surface blue water availability may decline by up to 20% in Texas's High Plains, East Texas, and Upper Coast, with the most severe groundwater declines projected in northern counties.

(*Continued*)

TABLE I.4.3 (*Continued*)
Studies on Future Droughts with Direct Texas Focus Referenced in This Section

Study	Location	Future scenario & Study Period	Data & Method	Key Findings
Chen et al. (2019)	Southern High Plains of Texas	2020–2099 under RCP4.5 and RCP 8.5	ET, irrigation, and crop yield from SWAT model	Irrigated corn yield decreased by 3%–22%, with ET and irrigation dropping by 8%–25% and 15%–42%, respectively. Sorghum's ET decreased by 6%–27%, irrigation by 15%, and yield by 6%–42%. Dryland sorghum saw a 10%–16% reduction in ET, with yield dropping by less than 10% and evaporation falling 15%–23%.
Chakraborty et al. (2021)	Edwards Aquifer Region in Texas	2021–2100 under RCP4.5 and RCP8.5	Groundwater level from Artificial Intelligence Model	Groundwater withdrawals would need to be cut by at least 40% under the worst-case scenario (RCP8.5) for over half of the time in three of the next eight decades under the current tiered critical period management strategy.
Chen et al. (2001)	Edwards Aquifer in Texas	2030–2090 under IPCC1992a emission scenario	Water use, springflow, and economic impacts estimated by the Edward Aquifer economic and hydrological systems model (EDSIM)	By 2030, springflows at Comal will decrease by 10–16%, and by 2090, by 20–24%. Agricultural income will drop by 16–30% by 2030 and 30–45% by 2090.
Zhang et al. (2019)	South Central Texas	2071–2100 under RCP8.5	Probabilistic multivariate drought index based on WRF and Hymod simulations	Short-term drought duration in the Guadalupe, Blanco, Mission, and Frio River Basin decreased from 29, 25, 31, and 30 months to 16, 22, 10, and 10 months, while intensities increased from 0.58, 0.65, 0.62, and 0.70 to 0.74, 0.76, 0.70, and 0.75.

The studies are ordered by the first year of their study period.

enhanced evapotranspiration in Texas. Under warming conditions, particularly along high greenhouse gas (GHG) emission scenarios, models project an increased risk of meteorological drought, especially based on drought indicators involving temperature components, such as the SPEI and PDSI (Gautam et al., 2023; Venkataraman et al., 2016). This increased risk is most pronounced in arid and semi-arid regions like El Paso and Laredo, with naturally low baseline precipitation and strong sensitivity to rising temperatures intensifying drought conditions in these regions (Nielsen-Gammon et al., 2020; Venkataraman et al., 2016). In contrast, sub-humid regions like Houston may experience milder change, as the gulf's influence—through moisture supply and thermal buffering—helps temper both temperature increase and precipitation declines (Venkataraman et al., 2016). However, projections of future meteorological drought could vary depending on indicators used. The SPI, based solely on precipitation, may show different patterns compared to the SPEI, which also accounts for PET (Hernandez and Uddameri, 2014). Using multiple indicators, such as applying both SPI and SPEI, offers a more comprehensive view of future drought risks in Texas by capturing different dimensions of drought hazards, thus providing a more robust assessment for monitoring and planning.

Agricultural drought, characterized by soil moisture deficit that reduces crop growth and productivity, is closely linked to the increased evapotranspiration and reduced soil moisture recharge. Studies using CMIP5 and CMIP6 models project significant reduction in soil moisture across Texas, particularly under high GHG emission scenarios, with more severe impacts expected in western Texas (Cook et al., 2015; Wuebbles et al., 2014; Yuan et al., 2021). In contrast, eastern Texas may experience less severe agricultural droughts due to relatively stable precipitation patterns, although rising temperatures and increasing precipitation variability can still exacerbate drought conditions. For example, Li et al. (2019) found that while annual precipitation in the Clear Creek watershed in Houston may remain constant under moderate emission scenarios, reduced soil moisture is still likely due to increased evaporation driven by higher temperatures. Additionally, climate models project that precipitation in Texas will become more variable, with longer dry periods interspersed with intense rainfall events (Li et al., 2019; Tabassum, 2023). During prolonged dry spells, soil moisture can drop rapidly, and extreme rainfall events often result in more surface runoff rather than precipitation seeping into the ground, limiting the effective replenishment of soil moisture. In addition to changes in precipitation and temperature, elevated CO_2 concentrations can also influence soil moisture by altering vegetation responses. Higher CO_2 concentrations can reduce stomatal conductance, decreasing transpiration and conserving soil moisture. This effect has been projected in Texas, where Chen et al. (2019) found that elevated CO_2 is expected to be the primary factor contributing to reduced ET and lower irrigation demand for future crops in the Southern High Plains. However, the CO_2 fertilization effect can enhance plant growth, leading to higher water consumption and greater soil moisture loss through transpiration. The net impact of elevated CO_2 on soil moisture depends on the balance between these opposing effects (Mankin et al., 2018; Zhao et al., 2022).

Hydrological drought, which affects surface and groundwater resources, is projected to intensify across Texas due to reduced precipitation, increased evaporation, and changes in streamflow patterns. Nielsen-Gammon et al. (2020) found that CMIP5 models project significant declines in reservoir supplies for both the eastern and western areas of Texas during the 21st century, particularly under high GHG emission scenarios. This finding is further supported by Valiya Veettil et al. (2024), whose projections indicate a significant decreasing trend in the statewide SRI toward 2100 under both median and high-emission scenarios. Hydrological drought could be exacerbated not only by reductions in water availability but also by increases in evaporation losses. Zhao et al. (2023) found that rising air temperatures, combined with more frequent extreme precipitation events that expand reservoir surface areas, will lead to increased open water evaporation losses, with this effect being particularly pronounced in western Texas. Additionally, changes in the seasonal distribution of streamflow driven by shifts in precipitation patterns can further intensify water stress during critical periods. For example, Shao et al. (2023) projected a future decline in surface water availability in the upper Trinity River Basin due to increased reservoir evaporation and altered streamflow regimes, with the greatest reductions in autumn. This decline coincides with peak water demand, potentially intensifying water shortages.

Groundwater resources are also at risk, including areas reliant on the Edwards Aquifer in central Texas. Studies by Chakraborty et al. (2021), Chen et al. (2001), Loáiciga et al. (2000), and Wootten et al. (2024) highlight the sensitivity of the Edwards Aquifer to changes in precipitation and temperature, with projected declines in groundwater recharge under future climate scenarios. Chakraborty et al. (2021) found that while precipitation is projected to increase in the Edwards Aquifer region under CMIP5 climate scenarios, the benefits of this increase are largely offset by higher evapotranspiration driven by rising temperatures. The study noted that although increased precipitation during extreme events could enhance focused recharge along streams, higher temperatures would reduce diffuse recharge over larger areas, leading to an overall decline in groundwater levels. Similarly, Wootten et al. (2024) found that while CMIP6 projects increases in precipitation in the Edwards Aquifer region under intermediate emissions scenarios, the impact of warming temperatures offset the benefit of increased precipitation. Additionally, more intense precipitation events in the future could lead to higher proportions of surface runoff and reduced groundwater recharge, further contributing to groundwater declines.

Socioeconomic drought, driven by the imbalance between water availability and demand for human activities, is expected to become an increasing concern in Texas (Zhao et al., 2018). Li et al. (2024) projected that while some regions in northern and eastern Texas may see slight increases in net water availability under moderate emission scenarios, most of the state will face significant challenges due to rising water demand and declining supplies. Urban centers such as Dallas, Fort Worth, Houston, Austin, and San Antonio are particularly vulnerable, as rapid population growth and urbanization drive escalating water needs. Texas's population is projected to grow by more than 70% by 2070, with much of this growth concentrated in urban areas (Texas Water Development Board, 2022b). This surge in demand, coupled with diminishing water resources, is likely to intensify water stress in cities, especially during drought periods. Addressing these challenges will require a multifaceted approach that includes improved water management practices, increased water use efficiency, and policies that promote sustainable water use (Texas Water Development Board, 2022b).

1.4.1.5 Uncertainties in Future Drought Projections

Despite advancements in climate modeling and drought projection techniques, significant uncertainties remain in simulating future dynamics of droughts in Texas. One key source of uncertainty arises from GHG emission scenarios. Future GHG emissions depend on global socio-economic trends, policy decisions, and technological advancements, which are inherently unpredictable. Different scenarios result in varying degrees of warming and changes in precipitation in Texas, directly affecting drought risk projections (Rajsekhar et al., 2015c; Wen et al., 2024). Another major source of uncertainty is related to the variability among climate models. While GCMs consistently project that Texas will become warmer, there are large discrepancies in the projected changes in precipitation, especially at regional scales, where complex factors such as climate oscillations influence regional climate patterns (Banner et al., 2010; Chen et al., 2020; Livsey et al., 2016; Mishra et al., 2011; Seager et al., 2014). Additionally, the coarse spatial resolution of GCMs (≥100 km, Wootten et al., 2024) limits their ability to capture localized climate dynamics in Texas's diverse landscape, although this limitation has been improved in CMIP6 GCMs compared to CMIP5 (Wen et al., 2024). To address the limitations of GCMs, downscaling techniques are used to improve the spatial resolution of climate projections. However, downscaling introduces its own uncertainties due to differences in the methodologies employed, such as statistical downscaling and dynamic downscaling (Wood et al., 2004). Furthermore, hydrological models, land surface models, and other impact assessment models, which rely on meteorological outputs from GCMs to simulate the effects of changing climate on hydrological processes, such as streamflow, groundwater, and soil moistures, represent another significant source of uncertainty in future drought assessment. This is particularly relevant in arid regions of Texas, where hydrological models and land surface models often struggle to capture hydrological processes such as runoff and streamflow accurately (Abbas et al., 2025; Yan et al., 2025, 2023).

A critical source of uncertainty also comes from the application of drought indicators. Different indicators capture distinct aspects of drought, leading to variations in drought characterization. Even when the same indicator is used, differences in drought definitions can introduce substantial uncertainty. For example, the threshold used to define drought conditions (e.g., SPI < –0.5 vs SPI < –1) can alter the number, severity, and duration of identified drought events. Similarly, the temporal scale of index (e.g., 3-month, 6-month, or 12-month SPI/SPEI) affects drought interpretation as well, with shorter timescales capturing seasonal droughts and longer timescales highlighting prolonged droughts. Additionally, the reference period used as the baseline for assessing future projections plays a pivotal role. Different studies often use varying baseline historical periods, which can alter the reference for measuring future anomalies, thereby affecting the estimated severity of future droughts.

The calculation of PET, a key component in indices like the SPEI and the PDSI, introduces additional layers of uncertainty (Hoerling et al., 2012; Yuan and Quiring, 2014). Various PET estimation methods, such as the Penman-Monteith, Thornthwaite, and Hargreaves equations, produce different results depending on how they treat temperature, humidity, wind speed, and solar radiation (Venkataraman et al., 2016). Notably, some PET models include the impact of elevated atmospheric CO_2 concentrations on plant physiology by reducing stomatal conductance, thereby lowering transpiration rates. However, traditional PET calculations do not account for this CO_2 effect, potentially leading to overestimations of future evapotranspiration and drought severity, especially in regions like Texas where significant warming is projected (Islam et al., 2012).

In summary, uncertainties in future drought projections arise not only from GHG emissions scenarios, climate models, and downstream impact assessment models, but also from the methodologies used to define and measure droughts. These uncertainties can lead to varying conclusions about future drought trends, making it inherently challenging to quantify and compare the magnitude of climate change impacts on different aspects of drought. This, in turn, complicates decision-making for policymakers and water resource managers, as conflicting projections can create challenges in planning for water allocation, infrastructure development, and drought mitigation strategies. Addressing these uncertainties requires a multifaceted approach that incorporates diverse models, indices, and methodological frameworks to provide a more robust and comprehensive understanding of Texas's future drought landscape. Additionally, strengthening monitoring networks will be crucial not only for tracking actual changes but also for improving drought forecasts by providing high-resolution, real-time data that improves model performance and prediction accuracy.

I.4.1.6 Conclusions

Drought is a complex and multifaceted natural hazard with far-reaching impacts on Texas's environment, economy, and society. It has been identified and evaluated from various perspectives based on in-situ measurements, satellite data, model simulations, and paleoclimate reconstructions. Historically, Texas has experienced several severe droughts, including the 1950s drought of record and the more recent 2010–2015 drought, both demonstrating the state's vulnerability to prolonged periods of water scarcity. These historical events provide critical context for understanding the potential severity of future droughts. In recent decades, increasing drought risk has already been observed. Looking to the future, model projections indicate that Texas will potentially undergo significant climatic changes in the 21st century, and drought conditions are expected to be intensified especially in arid and semi-arid regions. This will pose even greater challenges for water availability and resource management. However, projecting future drought conditions in Texas is subject to significant uncertainties due to unpredictable GHG emissions, model variability, and differences in drought assessment methods. A comprehensive understanding of future drought risk in Texas will depend on continued strengthening of monitoring networks, advancements in climate models, improved impact assessment simulations, and the development of robust, adaptive drought indicators.

I.4.2 DUST

I.4.2.1 Introduction

Airborne soil particles, commonly known as dust, are solid particles suspended in the air due to wind or other disturbances. Dust is often characterized based on particle size, with particulate matter measuring 10 μm or less in diameter (PM_{10}) and 2.5 μm or less ($PM_{2.5}$) commonly used to distinguish coarse and fine fractions, respectively. These size classifications are crucial for understanding dust transport and health impacts (Kuo and Shen, 2010; Lai and Cheng, 2008). Additionally, studies identify dust using chemical composition markers, such as high concentrations of aluminum (Al), silicon (Si), calcium (Ca), and iron (Fe), which differentiate it from other aerosols like urban pollution or biomass burning emissions (Ho et al., 2003; Offor et al., 2016; Sharma et al., 2017). Tracers, particularly iron (Fe) and calcium (Ca), have been widely used in aerosol studies to distinguish dust sources from industrial or volcanic emissions, providing insights into dust origins and transport pathways (Scheuvens and Kandler, 2014; Sharma et al., 2017). When strong, turbulent winds mobilize large quantities of these particles, they form dust storms. These storms lift fine particles from the land surface into the atmosphere (Goudie, 2014; Middleton, 2017). Dust storms are particularly common in arid and semi-arid regions, where dry, unconsolidated soils, sparse vegetation, and extreme winds make landscapes prone to dust mobilization, such as in Texas, United States (U.S.) (Goudie, 2014; Goudie and Middleton, 2006). The World Meteorological Organization (WMO) defines a dust storm as an event in which strong surface winds lift large amounts of dust, reducing visibility at eye level (1.8 meters) to less than 1,000 meters, with severe storms leading to near-zero visibility (McTainsh and Pitblado, 1987).

Dust originates from multiple natural and anthropogenic sources. Deserts, such as the Sahara Desert, are major contributors, with dust particles traveling vast distances across continents, affecting air quality and precipitation patterns (Goudie and Middleton, 2006; Middleton and Goudie, 2001). Agricultural activities, including tilling and plowing, disturb the soil, generating significant dust emissions in farming regions (Nordstrom and Hotta, 2004). Urban dust, resulting from construction, traffic, and industrial activities, further contributes to elevated dust concentrations, especially in densely populated areas (Sager, 2020; Vlasov et al., 2022). The interaction of natural and human-induced sources makes dust a pervasive and complex environmental issue (Berhane et al., 2024).

Dust storms, while naturally occurring, have increased in frequency and intensity, particularly since the mid-20th century, due to human activities such as deforestation, overgrazing, unsustainable agricultural practices, and urban expansion. These activities, especially in arid and semi-arid regions globally, disturb soil surfaces and reduce vegetation cover, making land more susceptible to wind erosion (Ghosh and Pal, 2014). Fine dust particles can rise into the troposphere, where prevailing winds transport them over long distances, affecting regions far from their sources (Pye, 2015). Dust particles can erode plant surfaces, reducing crop yields (Middleton et al., 2019), and pose respiratory health risks to humans and livestock, leading to potential complications (Howard et al., 2024; Mu et al., 2013).

Dust events are a significant component of the atmospheric system, influencing the Earth's climate, environment, and human well-being in multiple ways. Dust particles directly impact climate by scattering and absorbing solar radiation (Lau et al., 2020; Wang et al., 2009) and indirectly by acting as cloud condensation and ice nuclei (Ardon-Dryer and Levin, 2014; Chen, 2019). Additionally, they can alter the atmospheric vertical electric field (Ardon-Dryer et al., 2022), disrupt economic activities (Abdullaev and Sokolik, 2020; Al-Hemoud et al., 2019), and pose serious health risks (Ardon-Dryer et al., 2020; Bhattachan et al., 2019; Goudie, 2014). Increased dust concentrations degrade air quality, leading to respiratory issues, reduced visibility, and higher accident rates on roads and in aviation (Li et al., 2021; Tong et al., 2023).

1.4.2.2 Dust in Texas

Dust events in the U.S. contribute to about 5% of global dust emissions, with most of these events occurring in the western U.S. (Goudie, 2014; Miller et al., 2013; Rublee et al., 2020). Recent studies have noted an increasing frequency of dust events in the central and southeastern U.S. over the past decade (Hand et al., 2016; Kelley and Ardon-Dryer, 2021; Tong et al., 2017). These events are influenced by factors such as climate conditions and land cultivation (Cook et al., 2015; Prein et al., 2016; Tong et al., 2017). Texas, in particular, is vulnerable to dust events, especially in its semi-arid to arid region, such as the Panhandle and West Texas, where dust storms are frequent (Kelley and Ardon-Dryer, 2021; Novlan, 2007). The combination of low annual rainfall, seasonal fluctuations, and dry soil, especially in the summer, exacerbates this issue. Strong winds from the southwest during dry periods, particularly after harvest or tilling, easily lift soil into the air (Babcock, 2024). Using data from Modern-Era Retrospective analysis for Research and Applications, Version 2 (MERRA-2) (https://disc.gsfc.nasa.gov/), the spatial distribution of dust mass concentration across Texas from 1985 to 2024 highlights these affected regions (Figure I.4.5). This pattern is consistent with previous studies that have reported increasing dust emissions in the southwestern U.S., including Texas, New Mexico, and northern Mexico, driven by rising temperatures, prolonged droughts, and land-use changes (Kandakji et al., 2020; Pu and Ginoux, 2017). Kandakji et al. (2020) found that dust concentrations in the Great Plains increased from 2001 to 2016 due to declining precipitation and vegetation cover. Pu and Ginoux (2017) reported that variations in fine dust concentration in the southwestern U.S. from 1990 to 2015 were largely controlled by precipitation, surface bareness, and surface wind speed, with enhanced dust activity linked to climate variability. Figure I.4.6 illustrates the annual changes in dust events and the number of affected counties in Texas from 2001 to 2024, using National Centers for Environmental Information (NCEI) (https://www.ncdc.noaa.gov/stormevents/choosedates.jsp?/ statefips=48%2CTEXAS) datasets. The data show variations in both the frequency and spatial extent of these dust events across Texas.

In Texas, dust storms often involve a mix of mineral and agricultural dust particles, and the region's vulnerability is heightened by human activities. Agricultural practices, such as tilling, harvesting, and overgrazing, disturb the soil, making it more prone to wind erosion. Additionally, land disturbances from oil and gas extraction sites, particularly in the Permian Basin, contribute to dust mobilization, further intensifying dust events (Kelley and Ardon-Dryer, 2021). Historically,

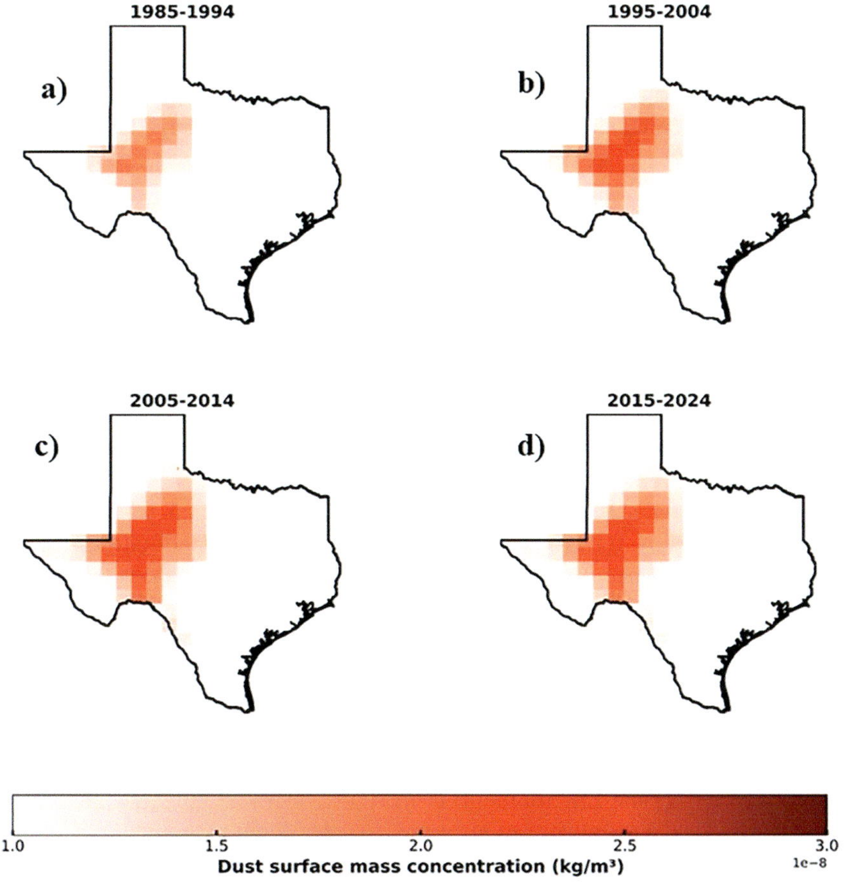

FIGURE I.4.5 Spatial distribution of dust mass concentration in Texas based on MERRA-2 reanalysis data for the periods: (a) 1985–1994, (b) 1995–2004, (c) 2005–2014, and (d) 2015–2024.

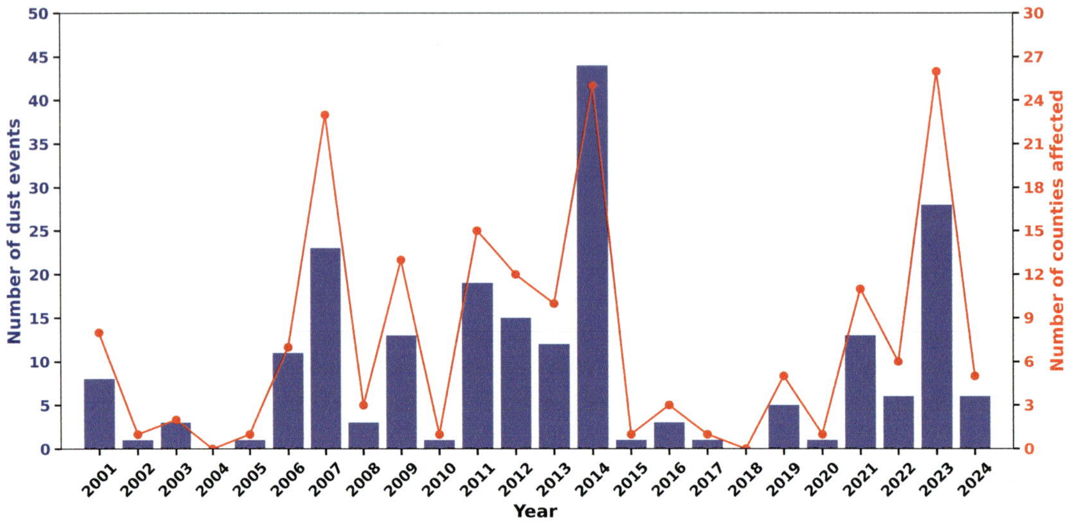

FIGURE I.4.6 The number of dust events (blue bars) and the number of counties affected (red line) from 2001 to 2024 using NCEI datasets.

the Dust Bowl of the 1930s is a stark example of how droughts combined with poor agricultural practices, like deep plowing and monoculture cropping, can lead to widespread soil erosion and catastrophic dust storms (Alexander et al., 2018; Lee and Gill, 2015; Sarafoglou et al., 2016).

The occurrence of dust events in Texas is also influenced by long-range transport of dust, especially from the Sahara Desert in North Africa (Gill et al., 2024). Saharan dust events, often driven by the trade winds, can travel thousands of miles across the Atlantic Ocean. These dust episodes, reaching the southern U.S., including Texas, typically occur in the spring and summer when the strong easterly winds from the Atlantic carry dust particles to the Gulf of America and beyond (Bozlaker et al., 2013). These Saharan dust storms can significantly affect air quality in Texas, particularly in the southern regions including populous megacities such as Houston and Dallas, by raising particulate matter concentrations in the atmosphere (Karnae and John, 2011; Prospero et al., 2005).

I.4.2.3 Historical Dust in Texas

Historical dust events have significantly driven Texas's environmental policies and agricultural practices, as well as shaping scientific research on air quality. With climate change intensifying droughts and altering wind patterns thus affecting dust occurrence, understanding dust events is crucial for mitigating future impacts, improving dust storm forecasting, and developing adaptive strategies to enhance resilience against worsening conditions. Various technologies have been employed to study dust in Texas, including ground-based monitoring networks, satellite remote sensing, and numerical modelling. Ground-based stations in Texas, such as those operated by the Texas Commission on Environmental Quality (TCEQ), provide long-term datasets on PM, which are crucial for air quality assessments (Herrera-Molina et al., 2021; Rivera, 2012). The Interagency Monitoring of Protected Visual Environments (IMPROVE) network further contributes by providing detailed composition measurements of airborne particles, helping to distinguish dust from other aerosol types (Bozlaker et al., 2019; Tong et al., 2017). Satellite observations such as from MODIS and Cloud-Aerosol Lidar and Infrared Pathfinder Satellite Observations (CALIPSO) enhance the ability to track dust transport using parameters such as Aerosol Optical Depth (AOD) and investigate vertical dust distributions (Bozlaker et al., 2019; Hennen et al., 2023; Kandakji et al., 2020;

Miller et al., 2017; Robinson et al., 2024). Additionally, atmospheric models such as Weather Research and Forecasting Model with Chemistry (WRF-Chem), and Dust Regional Atmospheric Model (DREAM) have been widely used in simulating dust distributions, transport pathways, and dispersion patterns (Rivera, 2012; Sarafoglou et al., 2016; Yin et al., 2005). WRF-Chem represents dust particles using multiple size bins, typically ranging from fine to coarse fractions (0.1–1.0 μm, 1.0–2.5 μm, 2.5–5.0 μm, and 5.0–10.0 μm) to account for their different transport and deposition behaviors. Similarly, DREAM classifies dust particles into distinct size intervals, often including bins such as 0.1–1.0 μm, 1.0–2.5 μm, and 2.5–10.0 μm, to better represent their atmospheric lifetime and removal processes. Most models categorize dust particles into different size bins (PM_{10}, $PM_{2.5}$) to better represent their transport and deposition processes, as larger particles tend to settle more quickly, while smaller particles remain suspended for longer periods and travel greater distances (Sarafoglou et al., 2016; Yin et al., 2005). The newly established Aerosol, Cloud, and Ecosystem Network (ASCENT) has expanded dust monitoring capabilities by providing more comprehensive measurements of particle properties, including size distribution, optical characteristics, and chemical composition. This network enhances the ability to study dust-related climate and air quality impacts more effectively.

Findings from these studies highlight that Texas experienced frequent dust storms, particularly during the prolonged drought periods and in regions with loose, dry soils, notably during severe droughts in the 1930s, 1990s, early 2000s, and 2011 (Heim, 1988; Opie, 1992; Sweeney et al., 2016). Analysis of ground-based and satellite data has shown that dust storms can significantly reduce visibility, degrade air quality, and exacerbate respiratory illnesses in Texas (Stout and Lee, 2003). Numerical models have provided insights into how changes in meteorological factors have influenced dust activity in Texas over past decades. For example, the frequency of windblown dust storms has surged by 240% from the 1990s to the 2000s (Tong et al., 2017). Similarly, prolonged drought conditions in the Texas Panhandle, coupled with reduced vegetation cover, have led to more intense dust outbreaks, particularly during the spring and summer months (Pu and Ginoux, 2017).

One of the most significant historical dust events was the Dust Bowl (1930s), a severe period of drought and dust storms that impacted the central U.S., particularly the Great Plains. This event highlighted the devastating consequences of prolonged arid conditions and soil erosion, ultimately shaping soil conservation policies and land management strategies (Bronnimann et al., 2009; Cook et al., 2014). Decades later, the 1998 west Texas dust storm, triggered by ongoing drought conditions and high winds, significantly reduced visibility and deteriorated air quality in areas like Lubbock and Midland (Gill et al., 2000). Another notable dust event occurred on December 15, 2003, with 146 dust sources identified, primarily from croplands and rangelands, highlighting the role of land use and soil characteristics in determining emission points (Lee et al., 2009). More recently, the 2011 dust storms in Texas, driven by a major drought, resulted in frequent and intense dust storms across Lubbock and El Paso, underscoring the persistent and severe threat of dust-related hazards in West Texas regions (Kelley and Ardon-Dryer, 2021; Robinson and Ardon-Dryer, 2024). In response to these historical events, numerous studies have examined dust storms in Texas using various observational and analytical techniques and are listed in Table I.4.4.

Studying past dust storms in Texas is challenging due to inconsistent measurement techniques and data collection methods, as well as incomplete long-term observational data records, particularly before the establishment of standardized monitoring networks such as those maintained by the Environmental Protection Agency (EPA) and NOAA. This data gap impacts trend analyses and complicates efforts to link dust events to specific climatic or human factors. Many studies rely on remote sensing, which, despite broad coverage, has resolution limits that may miss localized dust events. Satellite retrievals are also affected by cloud cover, surface reflectance variations, and algorithms, introducing uncertainties. Models like WRF-Chem and DREAM face issues with coarse resolution and imprecise parameterizations of key factors, such as soil erodibility and wind thresholds. Additionally, emission inventories may not fully capture historical variations in dust sources, adding further uncertainty to past dust assessments.

TABLE I.4.4
Historical Changes in Dust in Texas

Study	Location	Study Period	Data & Method	Key Findings
Gill et al. (2000)	Panhandle, Texas	1906–1999	United States Department of Agriculture – Agricultural Research Service (USDA-ARS) and ground-based monitoring	Dust fall magnitude analysis provided source profiles for West Texas dust, aiding receptor studies.
Stout and Lee (2003)	Southern High Plains, northwestern Texas and eastern New Mexico	1945–1995	Texas Natural Resource Conservation Commission (TNRCC) and long-term monitoring	Blowing dust declined by ~30% over 40 years; improved agricultural practices contributed to reduced wind erosion.
Stout (2001)	North Lubbock, Texas	1961–1982	USDA-ARS and long-term monitoring	TSP (Total Suspended Particles) concentrations declined by ~40% over four decades, indicating reduced wind erosion.
Eibedingil (2021)	Western US	1984-2019	Aerosol Robotic Network (AERONET), MODIS and satellite-based tracking	Satellite observations effectively tracked real-time dust events with high accuracy.
Stout (2001)	Southern High Plains, western Texas and Lubbock, Texas	1996 to 1997	Dust sampling system and field measurement	Dust events peaked in spring; wind speeds above 4 m s^{-1} under dry, bare conditions significantly increased dust concentrations.
Rivera (2012)	El Paso, Texas	2001-2005 and 2008	TCEQ CAMS, WRF-Chem modelling	Dust storm dynamics were highly sensitive to wind speed and direction; emissions peaked when wind speeds exceeded 6 m s^{-1}.
Lee et al. (2012)	West Texas and eastern New Mexico	2001–2009	MODIS and satellite-based tracking	Cultivated cropland contributed to >50% of dust emissions.
Kandakji et al. (2020)	Southern Great Plains and Chihuahuan Desert regions	2001–2016	MODIS and satellite-based tracking	Low-pressure systems drove dust transport, worsening air quality in neighboring regions.
Yin et al. (2005)	Southwestern US	December 15–17, 2003	DREAM modelling	Captured meteorological patterns and dust emissions; emphasized need for improved land surface data.
Bozlaker et al. (2013)	Houston, Texas	Spring and summer (2008)	TCEQ CAMS, CALIPSO, satellite and ground-based monitoring	Saharan dust contributed up to 85% of PM_{10} mass on peak days in spring and summer.
Herrera-Molina et al. (2021)	El Paso, Texas	2010–2014	Continuous Air Monitoring Station number-12 (CAMS-12)	PM_{10} levels ≥ 100 μg m^{-3} and wind speeds ≥ 4.5 m s^{-1} were linked to a 12% increase in hospital admissions for respiratory diseases.
Ardon-Dryer et al. (2022)	Lubbock, West Texas	Multiple events in 2019	Aerosol Research Observation Station (AEROS) and ground-based monitoring	Convective dust events resulted in PM concentrations exceeding 250 μg m^{-3}, significantly higher than synoptic dust events.

TABLE I.4.4 (*Continued*)
Historical Changes in Dust in Texas

Study	Location	Study Period	Data & Method	Key Findings
Robinson et al. (2024)	West Texas	February 26–27, 2023	TCEQ and ground-based monitoring	PM concentrations surpassed EPA's (Environmental Protection Agency) 24-hour threshold of 150 μg m^{-3}, highlighting the severity of the storm.
Ma et al. (2011)	Panhandle, Texas	Not specified	ThermoAndersonMIEDataRAM (DATARAM) AerosolMonitorDR-2000), GRIMM and optical measurements	Mass-extinction coefficient (1.74–2.97 m^2 g^{-1} at 522 nm), SSA (0.947–0.980), and Angstrom exponents for absorption (1.73–2.17) and extinction (0.110–0.168) improved dust aerosol characterization.

I.4.2.4 Future Dust in Texas

Numerous climate models and projections suggest that human-induced climate change will likely make the arid and semi-arid regions of the western U.S. hotter and drier in the coming decades. Both average and extreme drought severity are expected to intensify, with some droughts potentially surpassing those of the 1930s Dust Bowl, potentially doubling dust event frequency by the end of the century (Cayan et al., 2010). Loisel et al. (2017) projected that drought risk in the western U.S. could increase dust storm frequency by 50% by the late 21st century. Under RCP 8.5 scenario, Li et al. (2021) estimated that, by 2100, the southwestern U.S. could warm by 3.3°C, with spring precipitation decreasing by nearly 40%, leading to a ~30% increase in dust activity. Their study explores the impacts of climate change, CO_2 fertilization, and land use on vegetation dynamics and dust mobilization, finding that while CO_2 fertilization enhances plant cover in some areas, land use intensification disrupts soil stability, leading to increased dust emissions. Increased aridity and reduced precipitation could be particularly impactful in the Southern High Plains region of Texas. Studies suggest that drier soils in this semi-arid area could lead to a 25% rise in dust emissions (Salley et al., 2016). Reduced precipitation, stronger winds, and greater land surface bareness could amplify dust intensity by up to 40% during extreme wind events in the Southern Great Plains (Pu and Ginoux, 2017).

Land-use changes such as urbanization and agricultural expansion have significantly reduced native vegetation, contributing to an increase in dust storm frequency, particularly during drought periods (Villanueva, 2023). Nordstrom and Hotta (2004) found that agricultural practices such as tilling and vegetation removal for urban development could increase dust storm occurrences by up to 15% in urbanizing areas, while conservation techniques such as no-till farming and cover cropping could reduce dust emissions by 10–20% (Zobeck, 2010). Singh (2023) identified deforestation and urbanization as key contributors to rising dust storm frequency, while Mirzabaev et al. (2022) noted that climate-induced shifts in vegetation could either suppress or amplify dust emissions depending on the replacement species. Bolles et al. (2017) demonstrated that reductions in native vegetation, particularly in desert regions of the southwestern U.S., lead to increased soil erosion and dust emissions. These findings emphasize the importance of considering the combined effects of climate change, CO_2 fertilization, and land use when assessing the risks of future dust storms. Collectively, these studies highlight an alarming rise in the intensity and frequency of dust storms, as summarized in Table I.4.5.

TABLE I.4.5
Projected Changes in Dust and Drivers under Future Climate Scenarios

Study	Location	Future scenario & Study period	Data & Method	Key findings
Pu and Ginoux (2017)	Southern Great Plains, U.S.	RCP8.5 and by 2050	MODIS and Coupled Model Intercomparison Project Phase 5 (CMIP5)	Dust intensity increases by 40% during extreme wind events
Cayan et al. (2010)	U.S.	RCP8.5 and by 2100	Variable Infiltration Capacity (VIC) and global climate models (GCMs)	Dust event frequency could double
Li et al. (2021)	Southwestern U.S.	RCP8.5 and 21st century	Representative Concentration Pathways (RCPs) and GISS ModelE climate model	Dust activity projected to increase by ~30%
Salley et al. (2016)	Great Plains, U.S.	Future climate scenarios and 21st century	National Soil Information System (NASIS) and Newhall simulation model (NSM)	Dust emissions expected to rise by 25%
Nordstrom and Hotta (2004)	Texas, U.S.	Urbanization and agriculture and 21st century	Review	Dust frequency could rise by 15% in urbanizing areas
Loisel et al. (2017)	Southwestern U.S.	LIA (Little Ice Age) and 21st century	North American Drought Atlas (NADA, version 2a) and	Increased dust storm frequency by 50%
Zobeck (2010)	U.S. Midwest	No-till, cover crops and ongoing	Agricultural practices	Dust emission reductions of 10–20% in areas using conservation tillage

Uncertainties in predicting dust storms under a changing climate arise from complex interplay of atmospheric, terrestrial, and anthropogenic factors. Climate variables such as temperature, precipitation, and wind dynamics influence soil moisture and vegetation cover, which are key determinants of dust emissions. However, large variability exists across climate models in simulating these meteorological factors, as climate models differ in spatial resolution, land-atmosphere coupling, and aerosol process representations, introducing considerable uncertainty into future dust storm projections. Land-use changes driven by urbanization, deforestation, and agricultural expansion further modify dust sources and sinks, either enhancing or suppressing emissions. These alterations interact with shifting climate conditions, complicating efforts to predict future dust activity. Variability in dust emission, transport, and deposition remains challenging to model accurately due to uncertainties in source characterization, wind threshold dynamics, and meteorological influences such as convective activity and large-scale atmospheric patterns known as synoptic circulations. Climate change alters regional weather patterns, affecting the frequency, intensity, and distribution of dust storms in non-linear ways. For instance, increasing aridity may enhance dust production in some regions, while shifts in precipitation and vegetation growth could suppress emissions elsewhere. These competing influences make robust predictions difficult. Reducing uncertainties requires improved observational datasets, refined model parameterizations, and interdisciplinary research that bridge climate science, land use analysis, and atmospheric chemistry. Advancements in satellite remote sensing, high-resolution climate modeling, and process-based studies are crucial for improving dust storm projections under future climate scenarios.

1.4.2.5 Conclusion

This section focuses on the growing concerns of dust storms in Texas. Historically, dust storms were primarily associated with arid and semi-arid landscapes, and the frequency of dust events in these regions is projected to rise due to factors such as increased drought severity and higher wind speeds under climate change scenarios. Modern factors such as land-use changes, agricultural expansion, and prolonged droughts have intensified their frequency and intensity. Studies across various regions of Texas highlight that dust events, particularly during periods of high wind speeds and dry conditions, can result in PM concentrations exceeding safe levels, leading to significant health impacts, including increased hospital admissions for respiratory diseases. For example, in West Texas, convective dust events have led to PM concentrations surpassing 250 $\mu g\ m^{-3}$, which is notably higher than those from synoptic dust events. Future projections suggest that dust storm frequency and severity could substantially increase, particularly in the Southern Great Plains, due to intensified droughts, soil degradation, and vegetation loss. Climate models indicate that dust intensity might increase by up to 40% under extreme wind events by 2050, and dust storm occurrences are expected to rise by 50% due to increased drought severity. These findings underscore the need for enhanced forecasting and monitoring using satellite-based tracking, ground-based instruments, and regional climate models to inform effective policy measures. Uncertainty remains in long-term dust studies, primarily due to the complex interactions between climate, land use, and human activities. The integration of diverse data sources will be crucial for refining these predictions. Improving long-term predictive models is also essential in exploring the potential impacts of agricultural practices, such as no-till farming and cover crops, and identifying effective mitigation and adaptation strategies to reduce dust-related risks.

1.4.3 WILDFIRES

1.4.3.1 Introduction

Wildfires are unplanned and uncontrolled fires that occur due to natural causes, such as lightning, as well as human activities, including unauthorized ignitions and escaped fire management efforts (Fitzgerald-McGowan, 2024; Gould et al., 2024; Hoover and Hanson, 2009). Climate change has intensified wildfire activity by driving temperature increases, altering precipitation patterns, and exacerbating drought conditions. These climatic factors play a critical role in influencing wildfire frequency, intensity, and duration (USGCRP, 2017). Texas is particularly vulnerable to wildfires due to its extreme heat events, prolonged droughts, and diverse landscape consisting of grasslands, forests, and shrubland (Nielsen-Gammon et al., 2024). Lightning induced fires occur annually in Texas, intensifying during droughts and affecting vast areas of East Texas (Jurney et al., 2000). Texas experiences two primary wildfire seasons, with the highest risk occurring in late winter and early spring due to stronger winds, while summer wildfires are generally smaller in scale (Nielsen-Gammon et al., 2024). Wildfires have played a fundamental role in shaping Texas ecosystems, including grasslands, savannas, and forests (Stambaugh et al., 2018). Naturally occurring low-intensity fires, ignited by lightning and indigenous land management practices, historically played a crucial role in maintaining ecological balance by reducing excess vegetation and preventing severe wildfires (Kimmerer and Lake, 2001). Indigenous people in Texas actively utilized fire as a landscape management tool, influencing vegetation patterns and ecosystem dynamics. Evidence suggests that human fire use significantly contributed to maintaining and expanding the Blackland Prairies, shaping the region's fire-adapted ecosystems (Jurney et al., 2000). However, 20th century fire suppression policies, coupled with urbanization and land-use changes, have led to significant vegetation accumulation, increasing wildfire severity (Hann et al., 2008). Presently, much of Texas's natural vegetation excluding agricultural, urban, and barren lands has been altered, with many regions classified as disturbed ecosystems, thereby heightening wildfire risk (Hann et al., 2008). Exposure to wildfire smoke has been linked to various health conditions, affecting the respiratory, cardiovascular, neurological, dermatological, and gastrointestinal systems (Bruce et al., 2025; Liu et al., 2015; Manisalidis et al., 2020; Reid et al., 2016; Reisen et al., 2015).

I.4.3.2 Historical Wildfires in Texas

Texas has a long history of significant wildfires, driven by rises in temperature, prolonged drought conditions, strong winds, and abundant dry vegetation (Nielsen-Gammon et al., 2024). Several studies have examined historical wildfires in Texas using various datasets, models, and analytical methods. National Interagency Fire Center (NIFC) (https://www.nifc.gov/), Monitoring Trends in Burn Severity (MTBS) (https://www.mtbs.gov/), and Texas A&M Forest Service (https://fire-information-tfsgis.hub.arcgis.com/pages/historical-fire-statistics) maintain the historical wildfires record providing the information on fire perimeters, dates, severity, causes, casualties, suppression efforts, smoke, behavior, fuel moisture, fire weather, and post-fire recovery. Ground-based air quality monitoring networks and satellite measurements also track fire-induced changes in air quality (Giglio et al., 2013; Katuwal and Kafley, 2024; Mulder et al., 2013). Several studies have utilized models such as GCMs, the WRF model, Generalized Linear Models (GLMs), and the Energy Release Component (ERC fuel model) to analyze various wildfire drivers (Johnston et al., 2012; Junquera et al., 2005; Katuwal and Kafley, 2024; Lindley et al., 2021; Nielsen-Gammon et al., 2024).

Historical wildfire data for Texas reveal varying trends depending on the data source. Since 2014, there has been a decline in the total number of fires (Figure I.4.7), a pattern also observed

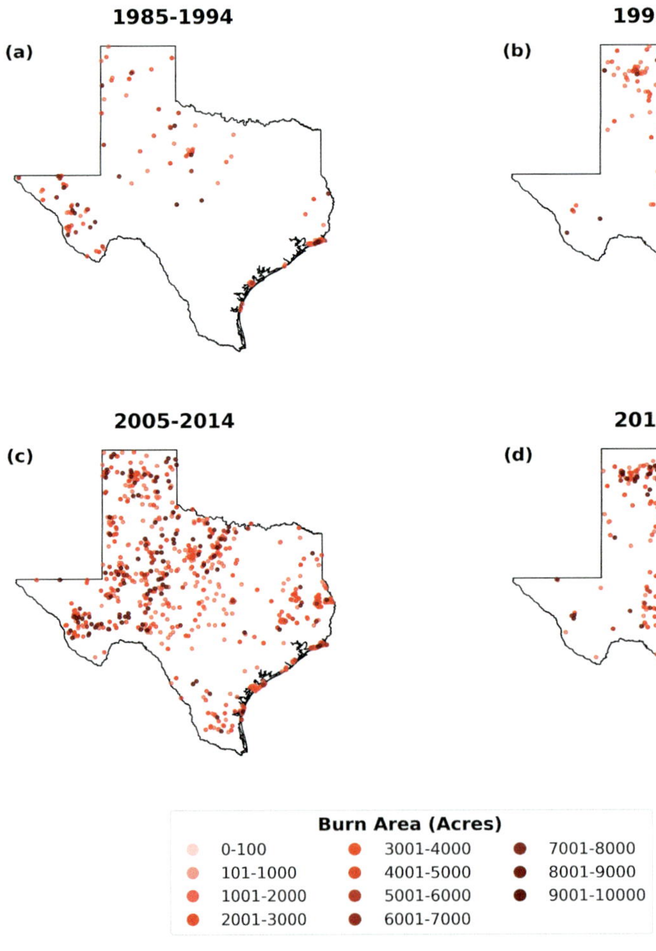

FIGURE I.4.7 Average burned areas during (a) 1985–1994, (b) 1995–2004, (c) 2005–2014, and (d) 2015–2024 due to wildfires in Texas using Monitoring Trends in Burn Severity (MTBS) datasets.

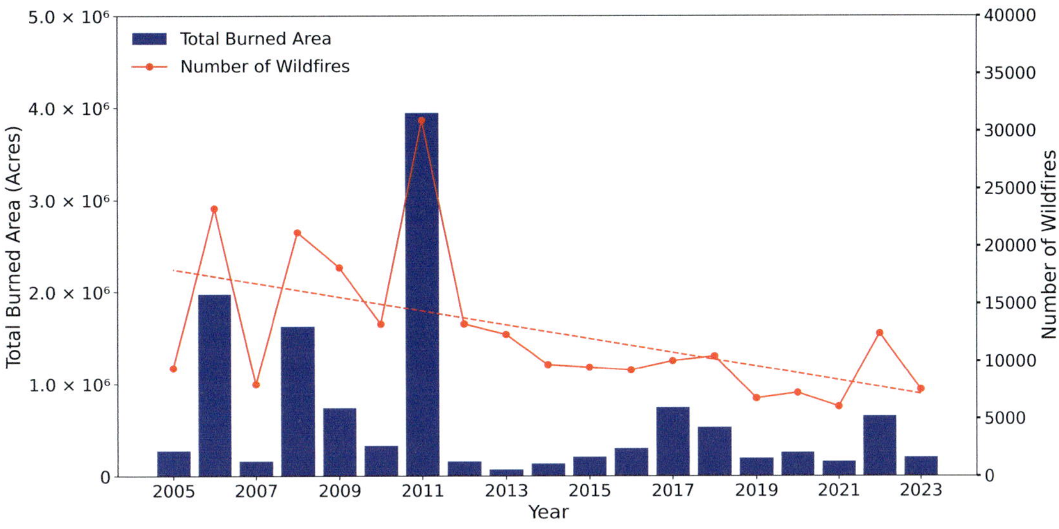

FIGURE I.4.8 Annual trends in total burned area (blue bars) and total number of wildfires incidents (red line) from 2005 to 2023 using Texas A&M Forest Service datasets.

in the trendline for wildfire occurrences (Figure I.4.8). However, the frequency of large wildfires, defined as those exceeding 405 hectares in burned area (Nagy et al., 2018), has been increasing (Figure I.4.9) (Salguero et al., 2020). This indicates a shift in wildfire dynamics, where the overall number of fires is decreasing, yet the prevalence of large fires is rising over time. Changing environmental conditions, including rising temperatures, prolonged droughts, low humidity, and strong winds, have likely contributed to this trend by altering vegetation patterns, reducing moisture availability, and increasing fuel loads (Nielsen-Gammon, 2012).

Changes in temperature and precipitation became more pronounced over the past two decades (Nielsen-Gammon et al., 2024). Wildfire risk increases with higher wind speeds and temperatures,

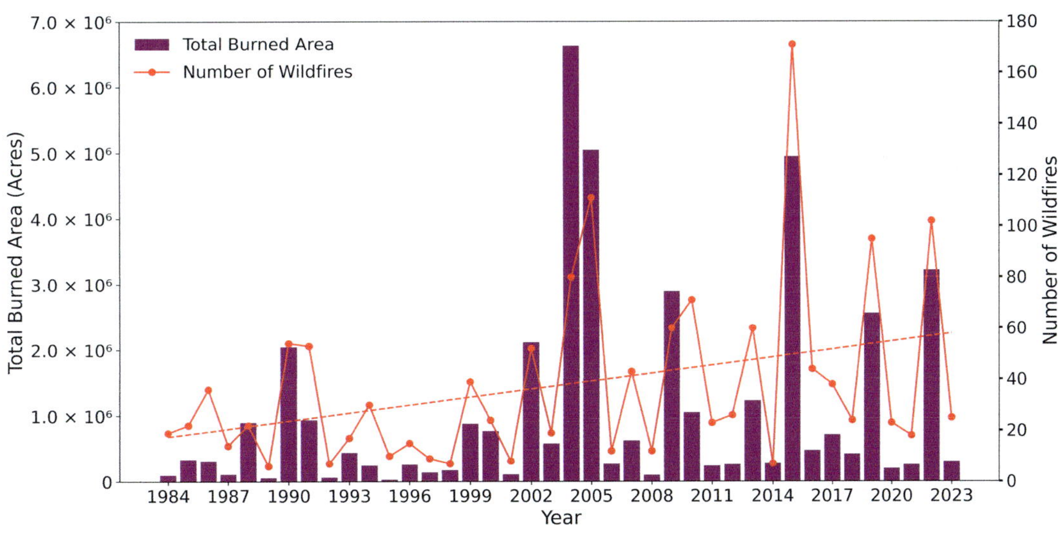

FIGURE I.4.9 Annual Trends in total burned area (purple bars) and the total number of wildfires incidents (red line) from 1984 to 2024 using Monitoring Trends in Burn Severity (MTBS) datasets.

while elevated humidity and precipitation reduce the risk of wildfires (Baijnath-Rodino et al., 2021). In western Texas, wildfire risk has risen due to increasing temperatures, whereas in eastern Texas, higher rainfall has mitigated wildfire susceptibility (Nielsen-Gammon et al., 2024). The 2011 wildfire season in Texas was among the most catastrophic in history, both in terms of fire count and total burned area, with 30,896 wildfires consuming approximately 3,943,770 acres (Texas A&M Forest Service, 2023). Early that year, 43% of Texas experienced extreme drought and record-breaking heat waves (Nielsen-Gammon, 2012). Also, the combination of a wet summer and dry fall in preceding years contributed to increased fuel loads, leading to widespread fires by April 2011, particularly in western and west-central Texas. More recently, the Smokehouse Creek Fire in the Texas Panhandles, one of the largest wildfires in state history, burned over 1.1 million acres before being contained in February 2024 (Manns, 2024). Fire exclusion has led to increased forest density in Guadalupe Mountains National Park (GMNP) and across southwestern Texas, contributing to a shift toward high-severity fire regimes. This shift was evident in the 1990 and 1994 wildfires, which burned large areas of GMNP's mixed conifer forests at high severity due to extreme weather conditions and heavy fuel accumulation (Sakulich and Taylor, 2007). Wildfires are driven by both meteorological and socioeconomic factors. Katuwal and Kafley (2024) highlighted that demographic and socioeconomic factors are key to the occurrences of arson and incendiary wildfires. Historical wildfire studies in Texas are important for assessing fire trends, understanding climate impacts, guiding land management practices, and improving risk mitigation strategies (Liu et al., 2014; Katuwal and Kafley, 2024). Historical wildfire in Texas has been investigated using a range of observational and analytical approaches, with key studies summarized in Table I.4.6.

A major challenge in historical wildfire studies is the inconsistency and incompleteness of wildfire data, particularly before standardized reporting practices were established. Spatial gaps, underreporting of smaller fires, and variability in data collection methods contribute to uncertainties in long-term trend analysis. Additionally, climate and environmental datasets exhibit inconsistency in station coverage and measurement errors, further influencing historical wildfire trend assessments. While remote sensing has been widely used to track wildfire activity, it has limitations. Sensor constraints, coarse spatial resolution, and the inability to capture smaller or less intense fire events, often due to satellite orbital period, can result in underestimations of total fire occurrences. Consequently, long-term fire histories may not fully represent actual fire frequency and intensity. Furthermore, models used for historical wildfire studies often operate at coarse spatial resolutions and rely on assumptions that may not accurately capture localized wildfire dynamics. These limitations underscore the need for improved data integration, higher-resolution modeling approaches, and advancements in remote sensing techniques to enhance the accurate and comprehensiveness of wildfire trend analyses.

I.4.3.3 Wildfires Future Trends

There are a limited number of studies available for future wildfire risk in Texas. While various tools exist for short-term fire weather forecasting, long-term predictive models emphasize the urgent need for mitigation and adaptation strategies to manage future wildfire threats effectively. Available literature supports the rise of wildfire risk in the coming decades due to climate change (Congressional Budget Office, 2024; Nielsen-Gammon et al., 2024). In Texas, wildfire risk is projected to increase due to higher temperatures, lower humidity, and lower fuel moisture (Liu et al., 2010). Rising carbon dioxide levels are anticipated to enhance plant growth, leading to a higher fuel load. Additionally, increased warming will contribute to vegetation dryness and drought conditions, elevating the risk of wildfires (Nielsen-Gammon et al., 2024; National Integrated Drought Information System, n.d.; Prestemon et al., 2016). The Intergovernmental Panel on Climate Change (IPCC) developed future climate scenarios, such as representative concentration pathways (RCPs) under the Coupled Model Intercomparison Project 5 (CMIP5) and Shared socio-economic pathways (SSPs) under the Coupled Model Intercomparison Project 6 (CMIP6), supporting the projection of wildfires under

TABLE I.4.6
Historical Wildfires Over Texas

Study	Location	Study Period	Data & Method	Key Findings
Stambaugh et al. (2018)	Southcentral Texas	1653–2011	National Climatic Data Center (NCDC)	The 2011 Bastrop County Complex Fire (BCCF) occurred during the fourth driest summer and drought is the major driver for it.
Nielsen-Gammon et al. (2024)	Texas	1900–2036	The National Centers for Environmental Information (NCEI) Ground based monitoring Modeling – GCMs	Western Texas: Increased wildfire risk due to rising temperature by 0.6 °F per decade. Eastern Texas: Reduced wildfire risk due to 15% increase in precipitation.
Jurney et al. (2000)	East Texas	1916–1997	US Forest Service, Fire and Aviation Management	The lightning fire season peaks from April to September.
Baijnath-Rodino et al. (2021)	14 western and central states	1950–2019	Texas A&M Forest Service, NIFC 2019 Wildfire Risk Report, the National Center for Environmental Information (NCEI)	90% of wildfires in Texas are caused by humans. Trends in lightning-caused fires were primarily driven by increasing numbers of large summer fires, whereas overall trends in human-caused fires were primarily driven by increasing numbers of large spring fires.
Junquera et al. (2005)	Eastern Texas	15 August 15–September 15, 2000	National Interagency Fire Management Integrated Database (NIFMID), the Texas Interagency Coordination Center, the U.S. Department of Agriculture Forest Service (USFS) Modeling – CAMx	Ozone concentrations were enhanced by more than 30 ppb at some locations due to large fires (>10,000 acres).
Poulos et al. (2020)	West Texas	2004–2018	Monitoring Trends in Burn Severity database (MTBS), Raster relative delta normalized burn ratio (rdNBR) data	Pre-fire surface fuel loading as the key driver of wildfire severity.
Lindley et al. (2021)	Eastern New Mexico, west Texas, and Oklahoma	2005–2011	TA&MFS data, Energy Release Component (ERC) data	Tendency for wildfires to occur near 2-m and 850-hPa thermal ridges when overspread by 500-hPa wind maxima.
Katuwal and Kafley (2024)	Texas	2005–2021	Fire Program Analysis fire-occurrence database (FPA FOD), the Texas A & M Forest Service (TFS) wildfire database, Bureau of Labor Statistics (BLS) data, Texas Department of Transportation (TxDOT) Modeling – GLM	1% increase in higher temperature increases arson fire by 7%. Human-caused fires tend to occur in areas with more moisture.
Nielsen-Gammon (2012)	Texas	2007–2011	U.S. Drought Monitor, Office of the State Climatologist, Texas National Climatic Data Center Historical Climatology Network, Version 2 (USHCNv2) Ground based monitoring	43 % of Texas faced extreme drought, rising wildfire risk.

future atmospheric conditions based on different anthropogenic radiative forcing and mitigation strategies (Calvin et al., 2023; IPCC, 2014).

Global climate models (GCMs) simulate atmospheric processes, land-ocean exchange, advection, and evaporation, providing a comprehensive framework for understanding climate dynamics. When combined with fire behavior and ecosystem models, GCMs have been widely used to predict future fire variability and consequences under different climate scenarios (An et al., 2015; Liu et al., 2010; McGinnis et al., 2023). Over the Southern Great Plains (SGP), a significant rise in future burning index (BI), which quantifies fire intensity by integrating both the fire spread component and energy release dynamics, was calculated based on six GCMs under both the RCP4.5 and RCP8.5 scenarios, with pronounced BI increases occurring during summer months (Fang et al., 2025). The southwestern region of the SGP, primarily in Texas, is projected to experience the greatest escalation in wildfire risk, establishing itself as a critical hotspot for future fire danger, mainly driven by declining relative humidity. The frequency of high fire danger days (HFDDs) can be determined by BI thresholds, and HFDDs with BI values of 40 or higher serve as a key metric for delineating the fire season in the SGP (Carlson et al., 2002; Fang et al., 2025). By the end of the 21st century, HFDD frequency is expected to rise substantially, exceeding a 50% increase under RCP8.5 and a 30% increase under RCP4.5 compared to historical levels (Fang et al., 2025). However, fire probability projections for Texas, calculated based on an ensemble of high-resolution downscaled GCMs in conjunction with the physical chemistry fire frequency model (PC2FM), reveal contrasting regional trends (Gao et al., 2021). In central Texas, under both the RCP4.5 and RCP8.5 scenarios, the likelihood of wildfires is expected to rise with fire activity projected to intensify throughout the 21st century. Conversely, western Texas exhibits relatively slight changes in fire probabilities, with some areas even experiencing slight declines, particularly under RCP8.5 toward the end of the 21st century. An exception to this general trend is the South-Central New Mexico Mountains, which extend into Texas and are projected to experience significant increases in fire probability under both RCP4.5 and RCP8.5, driven by increased temperatures and variations in precipitation (Gao et al., 2021).

PC2FM, along with three atmosphere-ocean general circulation models (AOGCM), was also used to project fire probabilities across the South-Central U.S., including Texas, Oklahoma, and New Mexico, under the A1B scenario with rapid economic growth, the development of efficient technology, and a population increase by mid-century followed by a decline in population growth toward the end of the century (Stambaugh et al., 2018). In the Texas High Plains, fire probabilities are expected to increase, while some decreases are observed in the Edward Plateau; however, these variations vary from mid-century to the end of century depending on the AOGCM model used.

The Ketch-Byram drought index (KBDI) and fire danger indices (FDIs) can be used to predict future wildfire risk. The KBDI is an indicator of available fuel, estimated using climate conditions including daily precipitation, daily maximum temperature, and accumulated precipitation. It ranges from 0 to 800, with 0 indicating water-saturated soil (low wildfire potential) and 800 indicating drought conditions (high wildfire risk) (Brown et al., 2021). The FDIs combine four indices: the fire potential index (FPI), which represents the combustibility of vegetation; the energy release component (ERC), which represents fuel availability; the BI, driven by the ERC, which reflects the spatial extent of wildfires; and the fire weather index (FWI), which indicates drought conditions, including fuel dryness (Yu et al., 2023). Predicted FDIs and KBDI both indicate a higher wildfire risk by the end of the 21st century, with an average increase of 10 days in the frequency of FDIs exceeding the 95th percentile in the SGP and a rise in the frequency of KBDI events exceeding 600 across all seasons in the SFP (Brown et al., 2021; Yu et al., 2023).

Smoke from future wildfires could exacerbate health risks. In the U.S., it is estimated that by mid-century, deaths resulting from exposure to wildfire $PM_{2.5}$ will increase by 51–76% by 2050, and wildfires in the Western U.S., including Northwest Texas, will lead to a 50% to 400% or higher increase in burned area by the end of the 21st century, enhancing wildfire-related illnesses (Anderegg et al., 2023; Congressional Budget Office, 2024; Spracklen et al., 2009). Table I.4.7. provides a summary of studies that predicted changes in Texas wildfire activity.

TABLE I.4.7
Comparison of Future Fire Projections for Texas, U.S.

Study	Location	Future scenario & Study Period	Data & Method	Key Findings
Stambaugh et al. (2018)	South-Central US (Texas, New Mexico, and Oklahoma)	A1B & 1900–1929 CE (baseline), 2040–2069 CE (mid-century), 2070–2099 (end of century)	PC2FM, 3 AOGCM (GFDL-CM2.1; CGCM 3.1 T47; HadCM3) driven by CMIP3 multi-model dataset, PRISM	Changes in fire probabilities will range from –47% to 165% by the mid-century and –51% to 240% by the end of the century.
Gao et al. (2021)	128 pyromes groups in CONUS	RCP4.5, RCP8.5 & 1971–2000(baseline), 2040–2069 (mid-century), 2070–2099 (late century)	Ensemble of high-resolution downscaled GCMs and PC2FM driven by MACAv2, LANDFIRE, gridMET	Central Texas: Annual fire probability will surpass 0.35–0.43 by the end of the century. Western Texas: Fire probability will increase by 1–3% by the mid-century and 3–5% by the end of the century under RCP4.5; and will span from a decrease of 0.4% to an increase of 3.2% by mid-century and decrease of 1.9–7.8% by the end of the century under RCP8.5.
Yu et al. (2023)	CONUS	RCP8.5 & 1984–2019 (historical), 2085–2094 (end of the century)	WRF derived FDIs, GCMs, NFDRS fuel model driven by MTBS, National Center for Atmospheric Research's Fire Indices, gridMET	Increase of 10 days of the averaged FDIs frequency
Fang et al. (2025)	SGP-Texas, Kansas, Oklahoma	RCP4.5, RCP8.5 & 1986–2005 (historical), 2040–2059 (mid-century), 2080–2099 (end of 21st century)	6 GCMs driven by MACAV2-METDATA, LANDFIRE, gridMET, HYDRO1K DEM	RCP4.5: 0.1 BI increase per decade, and 30% increase of HFDD in comparison to historical data; RCP8.5: 0.4 BI increase per decade, and 50% increase in HFDD in comparison to historical data.
Brown et al. (2021)	CONUS and Alaska	RCP8.5 & 1995–2004 (historical), 2045–2054 (mid-century), 2086–2094 (late century)	WRF derived KBDI driven by MTBS, MODIS, U.S. Geological Survey land use data, PRISM	Increase in KBDI magnitude in the SGP: growth of 145 in fall, 149 in summer, 134 in spring and 91 in winter from the time of 1996–2004 to 2086–2094. Increase of KBDI events frequency exceeding 600 by the late 21st century.

Predicting future wildfires involves significant uncertainties stemming from variations in modeling approaches and the datasets used to drive these models. While most studies indicate an overall increase in future fire activity, regional differences exist. For example, Gao et al. (2021) reported declining fire trends in certain regions of Texas under the RCP8.5 scenario. The reliability of these models is further influenced by the choice of input datasets, such as the MTBS and PRISM, both of which carry inherent uncertainties (Brown et al., 2021). Furthermore, discrepancies arise from differences in the time periods used for future projections, as well as variations in the historical baseline periods chosen for comparison. These factors collectively introduce complexity into wildfire trend predictions, highlighting the need for careful interpretation and cross-validation across multiple models and datasets.

1.4.3.4 Conclusion

Wildfires have long been a significant natural force in shaping Texas's ecosystems, with both natural and human-induced factors driving their occurrence. While historical fire regimes play a crucial role in maintaining ecological balance, modern influences such as urban expansion, accidental ignitions, and shifting burning seasons have led to a marked increase in the frequency and severity of wildfires. Historical wildfire trends indicate an increase in large wildfires (exceeding 405 hectares in burned area), driven by changing environmental conditions such as rising temperatures, prolonged droughts, and shifting vegetation patterns. These factors have intensified fire severity, particularly as Texas becomes increasingly vulnerable to extreme heat events. While higher rainfall in eastern Texas has mitigated wildfire risk, increasing temperatures and persistent dry conditions in the western regions have heightened fire susceptibility. Additionally, socio-economic factors, including population density, unemployment rates, wage levels, and human activities play a crucial role in influencing wildfire occurrence, highlighting the need for a more integrated understanding of fire risk in Texas that incorporates both environmental and societal considerations. Historical wildfire studies provide valuable insights into climate impacts, helping to develop mitigation measures, improve land management policies, and enhance firefighting preparedness.

Future wildfire trends indicate a significant increase in fire risk due to climate change, with rising temperatures, prolonged drought conditions, and shifts in vegetation dynamics contributing to heightened fire frequency and intensity. Projections under RCP4.5, RCP8.5, and A1B scenarios suggest a substantial rise in fire danger days, particularly in the Southern Great Plains, with the most pronounced impacts expected in Texas, Kansas, and Oklahoma. Climate and fire probability models indicate regional variations in fire risk, with central Texas facing increasing wildfire likelihood, while western Texas exhibits less variations in fire risk. Fire danger indices further support predictions of extended fire seasons and higher wildfire potential across all seasons.

REFERENCES

Abbas, S.A., Bailey, R.T., Arnold, J.G., White, M.J., Mirchi, A., 2025. Modeling agro-hydrological surface-subsurface processes in a semi-arid, intensively irrigated river basin. Journal of Hydrology: Regional Studies 57, 102188. https://doi.org/10.1016/j.ejrh.2025.102188

Abbaszadeh, P., Moradkhani, H., Gavahi, K., Kumar, S., Hain, C., Zhan, X., Duan, Q., Peters-Lidard, C., Karimiziarani, S., 2021. High-resolution SMAP satellite soil moisture product: Exploring the opportunities. Bulletin of the American Meteorological Society 102, 309–315. https://doi.org/10.1175/BAMS-D-21-0016.1

Abdullaev, S.F., Sokolik, I.N., 2020. Assessment of the Influences of Dust Storms on Cotton Production in Tajikistan. In *Landscape Dynamics of Drylands Across Greater Central Asia: People, Societies and Ecosystems*, edited by Garik Gutman, Jiquan Chen, Geoffrey M. Henebry, and Martin Kappas, 87–105. Cham: Springer International Publishing. https://doi.org/10.1007/978-3-030-30742-4_6.

AghaKouchak, A., Farahmand, A., Melton, F.S., Teixeira, J., Anderson, M.C., Wardlow, B.D., Hain, C.R., 2015. Remote sensing of drought: Progress, challenges and opportunities. Reviews of Geophysics 53, 452–480. https://doi.org/10.1002/2014RG000456

AghaKouchak, A., Huning, L.S., Chiang, F., Sadegh, M., Vahedifard, F., Mazdiyasni, O., Moftakhari, H., Mallakpour, I., 2018. How do natural hazards cascade to cause disasters? Nature 561, 458–460. https://doi.org/10.1038/d41586-018-06783-6

AghaKouchak, A., Mirchi, A., Madani, K., Di Baldassarre, G., Nazemi, A., Alborzi, A., Anjileli, H., Azarderakhsh, M., Chiang, F., Hassanzadeh, E., Huning, L.S., Mallakpour, I., Martinez, A., Mazdiyasni, O., Moftakhari, H., Norouzi, H., Sadegh, M., Sadeqi, D., Van Loon, A.F., Wanders, N., 2021. Anthropogenic drought: Definition, challenges, and opportunities. Reviews of Geophysics 59, e2019RG000683. https://doi.org/10.1029/2019RG000683

Ahmadalipour, A., Moradkhani, H., Yan, H., Zarekarizi, M., 2017. Remote Sensing of Hydrological Extremes: Vegetation, Soil Moisture, and Data Assimilation. Springer Remote Sensing/Photogrammetry. https://doi.org/10.1007/978-3-319-43744-6

Alexander, R., Nugent, C., Nugent, K., 2018. The dust bowl in the US: An analysis based on current environmental and clinical studies. The American Journal of the Medical Sciences 356(2), 90–96. https://doi.org/10.1016/j.amjms.2018.03.015

Al-Hemoud, A., Al-Dousari, A., Misak, R., Al-Sudairawi, M., Naseeb, A., Al-Dashti, H. and Al-Dousari, N., 2019. Economic impact and risk assessment of sand and dust storms (SDS) on the oil and gas industry in Kuwait. Sustainability 11(1), 200. https://doi.org/10.3390/su11010200

An, H., Gan, J., Cho, S.J., 2015. Assessing climate change impacts on wildfire risk in the United States. Forests 6, 3197–3211. https://doi.org/10.3390/f6093197

Anderegg, W.R.L., Collins, T., Grineski, S., Nicholls, S., Nolte, C., 2023. Climate change greatly escalates forest disturbance risks to US property values. Environmental Research Letters 18, 094011. https://doi.org/10.1088/1748-9326/ace639

Ardon-Dryer, K., Chmielewski, V., Bruning, E.C., Xueting, X., 2022. Changes of electric field, aerosol, and wind covariance in different blowing dust days in West Texas. Aeolian Research 54, 100762. https://doi.org/10.1016/j.aeolia.2021.100762

Ardon-Dryer, K., Levin, Z., 2014. Ground-based measurements of immersion freezing in the Eastern mediterranean. Atmospheric Chemistry and Physics 14(10), 5217–5231. https://doi.org/10.5194/acp-14-5217-2014

Ardon-Dryer, K., Mock, C., Reyes, J., Lahav, G., 2020. The effect of dust storm particles on single human lung cancer cells. Environmental Research 181, 108891. https://doi.org/10.1016/j.envres.2019.108891

Arnold, J.G., Fohrer, N., 2005. SWAT2000: Current capabilities and research opportunities in applied watershed modelling. Hydrological Processes 19, 563–572. https://doi.org/10.1002/hyp.5611

Babcock, R., 2024. Sustainable and resilient cropping practices to mitigate wind erosion and enhance soil health.

Bachmair, S., Stahl, K., Collins, K., Hannaford, J., Acreman, M., Svoboda, M., Knutson, C., Smith, K.H., Wall, N., Fuchs, B., Crossman, N.D., Overton, I.C., 2016. Drought indicators revisited: The need for a wider consideration of environment and society. WIREs Water 3, 516–536. https://doi.org/10.1002/wat2.1154

Baijnath-Rodino, J.A., Kumar, M., Rivera, M., Tran, K.D., Banerjee, T., 2021. How vulnerable are American States to wildfires? A livelihood vulnerability assessment. Fire 4, 54. https://doi.org/10.3390/fire4030054

Banner, J.L., Black, B.A., Tremaine, D.M., 2024. Positive unintended consequences of urbanization for climate-resilience of stream ecosystems. Npj Urban Sustain 4, 16. https://doi.org/10.1038/s42949-024-00144-1

Banner, J.L., Jackson, C.S., Yang, Z.-L., Hayhoe, K., Woodhouse, C., Gulden, L., Jacobs, K., North, G., Leung, R., Washington, W., Jiang, X., Castell, R., 2010. Climate change impacts on Texas water: A white paper assessment of the past, present and future and recommendations for action. Texas Water Journal 1, 1–19. https://doi.org/10.21423/twj.v1i1.1043

Berhane, S.A., Kumar, K.R., Bu, L., 2024. A 13-year space-based climatological evolution of desert dust aerosols in the Middle East and North Africa regions observed by the CALIPSO. Science of the Total Environment 948, 174793. https://doi.org/10.1016/j.scitotenv.2024.174793

Bhattachan, A., Okin, G. S., Zhang, J., Vimal, S., Lettenmaier, D. P., 2019. Characterizing the role of wind and dust in traffic accidents in California. GeoHealth 3, 328–336. https://doi.org/10.1029/2019GH000212

Bolles, K., Forman, S.L., Sweeney, M., 2017. Eolian processes and heterogeneous dust emissivity during the 1930s Dust Bowl Drought and implications for projected 21st-century megadroughts. The Holocene 27(10), 1578–1588. https://doi.org/10.1177/0959683617702235

Bozlaker, A., Prospero, J.M., Fraser, M.P., Chellam, S., 2013. Quantifying the contribution of long-range Saharan dust transport on particulate matter concentrations in Houston, Texas, using detailed elemental analysis. Environmental Science & Technology 47(18), 10179–10187. https://doi.org/10.1021/es4015663

Bozlaker, A., Prospero, J. M., Price, J., & Chellam, S., 2019. Identifying and quantifying the impacts of advected North African dust on the concentration and composition of airborne fine particulate matter in Houston and Galveston, Texas. Journal of Geophysical Research: Atmospheres 124, 12282–12300. https://doi.org/10.1029/2019JD030792

Brönnimann, S., Stickler, A., Griesser, T., Ewen, T., Grant, A.N., Fischer, A.M., Schraner, M., Peter, T., Rozanov, E., Ross, T., 2009. Exceptional atmospheric circulation during the "Dust Bowl". Geophysical Research Letters 36(8). https://doi.org/10.1029/2009GL037612

Brown, E.K., Wang, J., Feng, Y., 2021. US wildfire potential: A historical view and future projection using high-Resolution climate data. Environmental Research Letters 16, 034060. https://doi.org/10.1088/1748-9326/aba868

Bruce, E.D., Folorunsho, A., Jaisawal, N., Gaw, E., Li, Y., 2025. Intra-continental transport of Western wildfire smoke heightens health risks across North America. International Journal of Environmental Research and Public Health 22, 226. https://doi.org/10.3390/ijerph22020226

Calvin, K., Dasgupta, D., Krinner, G., et al., 2023. IPCC, 2023: Climate Change 2023: Synthesis Report. In Contribution of Working Groups I, II and III to the Sixth Assessment Report of the Intergovernmental Panel on Climate Change [Core Writing Team, H. Lee and J. Romero (eds.)]. IPCC, Geneva, Switzerland. Intergovernmental Panel on Climate Change (IPCC). https://doi.org/10.59327/IPCC/AR6-9789291691647

Carlson, J., Burgan, R., Engle, D., Greenfield, J., 2002. The Oklahoma fire danger model: An operational tool for mesoscale fire danger rating in Oklahoma. International Journal of Wildland Fire 11, 183–191. https://doi.org/10.1071/WF02003

Cayan, D.R., Das, T., Pierce, D.W., Barnett, T.P., Tyree, M. and Gershunov, A., 2010. Future dryness in the southwest US and the hydrology of the early 21st century drought. Proceedings of the National Academy of Sciences 107(50), 21271–21276. https://doi/abs/10.1073/pnas.0912391107

Chakraborty, D., Başağaoğlu, H., Gutierrez, L., Mirchi, A., 2021. Explainable AI reveals new hydroclimatic insights for ecosystem-centric groundwater management. Environmental Research Letters 16, 114024. https://doi.org/10.1088/1748-9326/ac2fde

Chen, C.-C., Gillig, D., McCARL, B.A., 2001. Effects of climatic change on a water dependent regional economy: A study of the Texas Edwards Aquifer. Climatic Change 49, 397–409.

Chen, Q., Yin, Y., Jiang, H., Chu, Z., Xue, L., Shi, R., Zhang, X., Chen, J., 2019. The roles of mineral dust as cloud condensation nuclei and ice nuclei during the evolution of a Hail Storm. Journal of Geophysical Research: Atmospheres 36(8). https://agupubs.onlinelibrary.wiley.com/doi/full/10.1029/2019JD031403

Chen, W., Huang, C., Yang, Z.-L., 2021. More severe drought detected by the assimilation of brightness temperature and terrestrial water storage anomalies in Texas during 2010–2013. Journal of Hydrology 603, 126802. https://doi.org/10.1016/j.jhydrol.2021.126802

Chen, Y., Marek, G.W., Marek, T.H., Moorhead, J.E., Heflin, K.R., Brauer, D.K., Gowda, P.H., Srinivasan, R., 2019. Simulating the impacts of climate change on hydrology and crop production in the Northern high plains of Texas using an improved SWAT model. Agricultural Water Management 221, 13–24. https://doi.org/10.1016/j.agwat.2019.04.021

Chen, Y., Marek, G.W., Marek, T.H., Porter, D.O., Moorhead, J.E., Wang, Q., Heflin, K.R., Brauer, D.K., 2020. Spatio-temporal analysis of historical and future climate data in the Texas high plains. Sustainability 12, 6036. https://doi.org/10.3390/su12156036

Chiang, F., Mazdiyasni, O., AghaKouchak, A., 2018. Amplified warming of droughts in southern United States in observations and model simulations. Science Advances 4, eaat2380. https://doi.org/10.1126/sciadv.aat2380

Cleaveland, M.K., Votteler, T.H., Stahle, D.K., Casteel, R.C., Banner, J.L., 2011. Extended chronology of drought in South Central, Southeastern, and West Texas. Texas Water Journal 2, 54–96. https://doi.org/10.21423/twj.v2i1.2049

Congressional Budget Office, 2024. The risks of climate change to the United States in the 21st Century | Congressional Budget Office [WWW Document]. https://www.cbo.gov/publication/61146 (accessed 2.4.25).

Cook, B.I., Ault, T.R., Smerdon, J.E., 2015. Unprecedented 21st century drought risk in the American Southwest and Central Plains. Science Advances 1(1), e1400082. https://doi.org/10.1126/sciadv.1400082

Cook, B.I., Seager, R., Smerdon, J.E., 2014. The worst North American drought year of the last millennium: 1934. Geophysical Research Letters 41(20), 7298–7305. https://doi.org/10.1002/2014GL061661

Deb, P., Moradkhani, H., Abbaszadeh, P., Kiem, A.S., Engström, J., Keellings, D., Sharma, A., 2020. Causes of the widespread 2019–2020 Australian Bushfire Season. Earth's Future 8, e2020EF001671. https://doi.org/10.1029/2020EF001671

Eibedingil, I.G., 2021. Drought, dust storm, and particulate matter pollution and their interaction at the cascade of spatial scales across the Western United States. Ph.D., United States – Texas: The University of Texas at El Paso.

Fang, S., Yang, J., Zou, C.B., Krueger, E.S., Ochsner, T.E., Zhang, Q., 2025. Wildfire danger under changing climates in the southern great plains throughout the 21st century. Ecological Indicators 170, 112994. https://doi.org/10.1016/j.ecolind.2024.112994

Ferencz, S.B., Sun, N., Turner, S.W.D., Smith, B.A., Rice, J.S., 2024. Multisectoral analysis of drought impacts and management responses to the 2008–2015 record drought in the Colorado Basin, Texas. Natural Hazards and Earth System Sciences 24, 1871–1896. https://doi.org/10.5194/nhess-24-1871-2024

Fitzgerald-McGowan, M., 2024. Forest fires vs wildfires [WWW Document]. https://www.nfpa.org/news-blogs-and-articles/blogs/2024/01/31/what-is-a-wildfire (accessed 2.26.25).

Gao, P., Terando, A.J., Kupfer, J.A., Morgan Varner, J., Stambaugh, M.C., Lei, T.L., Kevin Hiers, J., 2021. Robust projections of future fire probability for the conterminous United States. Science of The Total Environment 789, 147872. https://doi.org/10.1016/j.scitotenv.2021.147872

Gautam, S., Mishra, U., Scown, C.D., Ghimire, R., 2023. Increased drought and extreme events over continental United States under high emissions scenario. Scientific Reports 13, 21503. https://doi.org/10.1038/s41598-023-48650-z

Geruo, A., Velicogna, I., Kimball, J.S., Du, J., Kim, Y., Colliander, A., Njoku, E., 2017. Satellite-observed changes in vegetation sensitivities to surface soil moisture and total water storage variations since the 2011 Texas drought. Environmental Research Letters 12, 054006. https://doi.org/10.1088/1748-9326/aa6965

Ghosh, T., Pal, I., 2014. Dust storm and its environmental implications. Journal of Engineering, Computers and Applied Sciences 3, 30–37.

Giglio, L., Randerson, J.T., van der Werf, G.R., 2013. Analysis of daily, monthly, and annual burned area using the fourth-generation global fire emissions database (GFED4). Journal of Geophysical Research: Biogeosciences 118, 317–328. https://doi.org/10.1002/jgrg.20042

Gill, T.E., Mims, F.M., Chellam, S., 2024. African Dust: Occurrence, health consequences, and impacts on Texas. The Southwest Respiratory and Critical Care Chronicles 12(53), 39–47. https://doi.org/10.12746/swrccc.v12i53.1357

Gill, T.E., Reynolds, R.L. and Zobeck, T.M., 2000, June. Measurements of current and historic settled dusts in west Texas. In Proceedings of the 93rd Air and Waste Management Association Annual Conference and Exhibition, Salt Lake City, Utah. paper (No. 0175, p. 15).

Goudie, A.S., 2014. Desert dust and human health disorders. Environment International 63, 101–113. https://doi.org/10.1016/j.envint.2013.10.011

Goudie, A.S. and Middleton, N.J., 2006. Desert dust in the global system. Berlin, Heidelberg: Springer Berlin Heidelberg.

Gould, C.F., Heft-Neal, S., Johnson, M., Aguilera, J., Burke, M., Nadeau, K., 2024. Health effects of wildfire smoke exposure. Annual Review of Medicine 75, 277–292. https://doi.org/10.1146/annurev-med-052422-020909

Gu, L., Yin, J., Slater, L.J., Chen, J., Do, H.X., Wang, H., Chen, L., Jiang, Z., Zhao, T., 2023. Intensification of global hydrological droughts under anthropogenic climate warming. Water Resources Research 59, e2022WR032997. https://doi.org/10.1029/2022WR032997

Gyawali, B., Murgulet, D., Ahmed, M., 2022. Quantifying changes in groundwater storage and response to hydroclimatic extremes in a coastal aquifer using remote sensing and ground-based measurements: The Texas gulf coast aquifer. Remote Sensing 14, 612. https://doi.org/10.3390/rs14030612

Hammond, J.C., Simeone, C., Hecht, J.S., Hodgkins, G.A., Lombard, M., McCabe, G., Wolock, D., Wieczorek, M., Olson, C., Caldwell, T., Dudley, R., Price, A.N., 2022. Going beyond low flows: Streamflow drought deficit and duration illuminate distinct spatiotemporal drought patterns and trends in the U.S. during the last century. Water Resources Research 58, e2022WR031930. https://doi.org/10.1029/2022WR031930

Hand, J.L., White, W.H., Gebhart, K.A., Hyslop, N.P., Gill, T.E., Schichtel, B.A., 2016. Earlier onset of the spring fine dust season in the southwestern United States. Geophysical Research Letters 43(8), 4001–4009. https://doi.org/10.1002/2016GL068519

Hann, W.J., Shlisky, A., Havlina, D., Schon, K., Barrett, S.W., DeMeo, T.E., Pohl, K., Menakis, J.P., Hamilton, D., Jones, J., Levesque, M. 2008. Interagency Fire Regime Condition Class (FRCC) Guidebook, *Version 1.3. 0* [online], 119 p.

Hao, Z., AghaKouchak, A., 2013. Multivariate standardized drought index: A parametric multi-index model. Advances in Water Resources 57, 12–18. https://doi.org/10.1016/j.advwatres.2013.03.009

Hao, Z., Singh, V.P., 2015. Drought characterization from a multivariate perspective: A review. Journal of Hydrology 527, 668–678. https://doi.org/10.1016/j.jhydrol.2015.05.031

Hassani, F., Zhang, Y., Kumar, S.V., 2024. Improved representation of vegetation soil moisture coupling enhances soil moisture data assimilation in water-limited regimes: A case study over Texas. Water Resources Research 60, e2023WR035558. https://doi.org/10.1029/2023WR035558

Heim Jr, R.R., 1988. About that drought…. Weatherwise, *41*(5), pp. 266–271. https://doi/abs/10.1080/00431672.1988.9925276

Heim, R.R. Jr, 2002. A review of twentieth- century drought indices used in the United States. AMS Bulletin of the American Meteorological Society 83, 1149–1166. https://doi.org/10.1175/1520-0477-83.8.1149

Hennen, M., Chappell, A., Webb, N.P., 2023. Modelled direct causes of dust emission change (2001–2020) in southwestern USA and implications for management. Aeolian Research 60, 100852. https://doi.org/10.1016/j.aeolia.2022.100852

Hernandez, E.A., Uddameri, V., 2014. Standardized precipitation evaporation index (SPEI)-based drought assessment in semi-arid south Texas. Environmental Earth Sciences 71, 2491–2501. https://doi.org/10.1007/s12665-013-2897-7

Herrera-Molina, E., Gill, T.E., Ibarra-Mejia, G., Jeon, S., 2021. Associations between dust exposure and hospitalizations in El paso, Texas, USA. Atmosphere 12(11), 1413. https://doi.org/10.3390/atmos12111413

Ho, K.F., Lee, S.C., Chow, J.C., Watson, J.G., 2003. Characterization of 10PM and 2.5PM source profiles for fugitive dust in Hong Kong. Atmospheric Environment 37(8), 1023–1032. https://doi.org/10.1016/S1352-2310(02)01028-2

Hoerling, M., Kumar, A., Dole, R., Nielsen-Gammon, J.W., Eischeid, J., Perlwitz, J., Quan, X.-W., Zhang, T., Pegion, P., Chen, M., 2013. Anatomy of an extreme event. Journal of Climate 26, 2811–2832. https://doi.org/10.1175/JCLI-D-12-00270.1

Hoerling, M.P., Eischeid, J.K., Quan, X.-W., Diaz, H.F., Webb, R.S., Dole, R.M., Easterling, D.R., 2012. Is a transition to semipermanent drought conditions imminent in the U.S. Great plains? Journal of Climate 25, 8380–8386. https://doi.org/10.1175/JCLI-D-12-00449.1

Hoover, K., Hanson, L., 2009. Wildfire statistics [WWW Document]. https://hero.epa.gov/hero/index.cfm/reference/details/reference_id/7292011 (accessed 2.28.25).

Howard, M.H., Sayes, C.M., Giesy, J.P., Li, Y., 2024. Valley fever under a changing climate in the United States. Environment International 193, 109066. https://doi.org/10.1016/j.envint.2024.109066

IPCC, 2014. Technical support unit for the synthesis report.

Islam, A., Ahuja, L.R., Garcia, L.A., Ma, L., Saseendran, A.S., 2012. Modeling the effect of elevated CO_2 and climate change on reference evapotranspiration in the semi-arid central great plains. Transactions of the ASABE 55, 2135–2146. https://doi.org/10.13031/2013.42505

Jiang, X., Yang, Z., 2012. Projected changes of temperature and precipitation in Texas from downscaled global climate models. Climate Research 53, 229–244. https://doi.org/10.3354/cr01093

Johnston, F.H., Henderson, S.B., Chen, Y., Randerson, J.T., Marlier, M., DeFries, R.S., Kinney, P., Bowman, D.M.J.S., Brauer, M., 2012. Estimated global mortality attributable to smoke from landscape fires. Environmental Health Perspectives 120, 695–701. https://doi.org/10.1289/ehp.1104422

Jones, A.D., Rastogi, D., Vahmani, P., Stansfield, A.M., Reed, K.A., Thurber, T., Ullrich, P.A., Rice, J.S., 2023. Continental United States climate projections based on thermodynamic modification of historical weather. Scientific Data 10, 664. https://doi.org/10.1038/s41597-023-02485-5

Junquera, V., Russell, M.M., Vizuete, W., Kimura, Y., Allen, D., 2005. Wildfires in eastern Texas in August and September 2000: Emissions, aircraft measurements, and impact on photochemistry. Atmospheric Environment 39, 4983–4996. https://doi.org/10.1016/j.atmosenv.2005.05.004

Jurney, D.H., Ippolito, J., Bergstrom, V., 2000. The frequency of fire in East Texas forests. ITA. https://doi.org/10.21112/.ita.2000.1.21

Kandakji, T., Gill, T.E., Lee, J.A., 2020. Identifying and characterizing dust point sources in the southwestern United States using remote sensing and GIS. Geomorphology 353, 107019. https://doi.org/10.1016/j.geomorph.2019.107019

Karnae, S., John, K., 2011. Source apportionment of fine particulate matter measured in an industrialized coastal urban area of South Texas. Atmospheric Environment 45(23), 3769–3776. https://doi.org/10.1016/j.atmosenv.2011.04.040

Katuwal, H., Kafley, H., 2024. Playing with fire: Socioeconomic factors and deliberate wildfires in Texas. Journal of Environmental Management 370, 122980. https://doi.org/10.1016/j.jenvman.2024.122980

Kelley, M.C., Ardon-Dryer, K., 2021. Analyzing two decades of dust events on the Southern great plains region of West Texas. Atmospheric Pollution Research 12(6), 101091. https://doi.org/10.1016/j.apr.2021.101091

Kimmerer, R.W., Lake, F.K., 2001. The role of indigenous burning in land management. Journal of Forestry 99, 36–41. https://doi.org/10.1093/jof/99.11.36

Kuo, H.-W., Shen, H.-Y., 2010. Indoor and outdoor 2.5PM and 10PM concentrations in the air during a dust storm. Building and Environment 45(3), 610–614. https://doi.org/10.1016/j.buildenv.2009.07.017

Lai, L.-W., Cheng, W.-L., 2008. The impact of air quality on respiratory admissions during Asian dust storm periods. International Journal of Environmental Health Research 18(6), 429–450. https://doi.org/10.1080/09603120802272227

Lau, W.K.M., Kim, K.-M., Zhao, C., Leung, L.R., Park, S.-H., 2020. Impact of dust-cloud-radiation-precipitation dynamical feedback on subseasonal-to-seasonal variability of the Asian summer monsoon in global variable-Resolution simulations with MPAS-CAM5. Frontiers in Earth Science 8. https://doi.org/10.3389/feart.2020.00226

Lee, J.A., Baddock, M.C., Mbuh, M.J., Gill, T.E. 2012. Geomorphic and land cover characteristics of aeolian dust sources in West Texas and Eastern New Mexico, USA. *Aeolian Research*, 3(4), 459–466. https://doi.org/10.1016/j.aeolia.2011.08.001

Lee, J.A., Gill, T.E., 2015. Multiple causes of wind erosion in the dust bowl. Aeolian Research 19, 15–36. https://doi.org/10.1016/j.aeolia.2015.09.002

Lee, J.A., Gill, T.E., Mulligan, K.R., Acosta, M.D., Perez, A.E., 2009. Land use/land cover and point sources of the 15 December 2003 dust storm in southwestern North America. Geomorphology 105(1–2), 18–27. https://doi.org/10.1016/j.geomorph.2007.12.016

Lee, K., Gao, H., Huang, M., Sheffield, J., Shi, X., 2017. Development and application of improved long-term datasets of surface hydrology for Texas. Advances in Meteorology 2017, 1–13. https://doi.org/10.1155/2017/8485130

Li, W., Li, D., El-Askary, H., Fisher, J.B., Fang, Z.N., 2024. Evaluating future water availability in Texas through the lens of a data-driven approach leveraged with CMIP6 general circulation models. Science of the Total Environment 921, 171136. https://doi.org/10.1016/j.scitotenv.2024.171136

Li, Y., Mickley, L.J., Kaplan, J.O., 2021. Response of dust emissions in southwestern North America to 21st century trends in climate, CO_2 fertilization, and land use: Implications for air quality. Atmospheric Chemistry and Physics 21(1), 57–68. https://doi.org/10.5194/acp-21-57-2021

Li, Z., Li, X., Wang, Y., Quiring, S.M., 2019. Impact of climate change on precipitation patterns in Houston, Texas, USA. Anthropocene 25, 100193. https://doi.org/10.1016/j.ancene.2019.100193

Liang, X., Lettenmaier, D.P., Wood, E.F., Burges, S.J., 1994. A simple hydrologically based model of land surface water and energy fluxes for general circulation models. Journal of Geophysical Research 99, 14415–14428. https://doi.org/10.1029/94JD00483

Lindley, T.T., Murdoch, G.P., Guyer, J.L., Skwira, G.D., Schneider, K.J.S., Nagle, S.R., Van Speybroeck, K.M., Smith, B.R., Beierle, M.-J., 2021. Southern Great plains wildfire outbreaks. EJSSM 9, 1–43. https://doi.org/10.55599/ejssm.v9i2.54

Liu, J.C., Pereira, G., Uhl, S.A., Bravo, M.A., Bell, M.L., 2015. A systematic review of the physical health impacts from non-occupational exposure to wildfire smoke. Environmental Research 136, 120–132. https://doi.org/10.1016/j.envres.2014.10.015

Liu, Y., Goodrick, S., Heilman, W., 2014a. Wildland fire emissions, carbon, and climate: Wildfire–climate interactions. Forest Ecology and Management 317, 80–96. https://doi.org/10.1016/j.foreco.2013.02.020

Liu, Y., Prestemon, J.P., Goodrick, S.L., Holmes, T.P., Stanturf, J.A., Vose, J.M., Sun, G., 2014b. Future Wildfire Trends, Impacts, and Mitigation Options in the Southern United States. In *Climate Change Adaption and Mitigation Management Options A Guide for Natural Resource Managers in Southern Forest Ecosystems*, 85–126. CRC Press - Taylor and Francis.

Liu, Y., Stanturf, J., Goodrick, S., 2010. Trends in global wildfire potential in a changing climate. Forest Ecology and Management, Adaptation of Forests and Forest Management to Changing Climate 259, 685–697. https://doi.org/10.1016/j.foreco.2009.09.002

Livsey, D., Simms, A.R., Hangsterfer, A., Nisbet, R.A., DeWitt, R., 2016. Drought modulated by North Atlantic Sea surface temperatures for the last 3,000 years along the northwestern gulf of Mexico. Quaternary Science Reviews 135, 54–64. https://doi.org/10.1016/j.quascirev.2016.01.010

Loáiciga, H.A., Maidment, D.R., Valdes, J.B., 2000. Climate-change impacts in a regional karst aquifer, Texas, USA. Journal of Hydrology 227, 173–194. https://doi.org/10.1016/S0022-1694(99)00179-1

Loisel, J., MacDonald, G.M., Thomson, M.J., 2017. Little Ice Age climatic erraticism as an analogue for future enhanced hydroclimatic variability across the American Southwest. PLoS ONE 12(10), e0186282. https://doi.org/10.1371/journal.pone.0186282

Long, D., Scanlon, B.R., Longuevergne, L., Sun, A.Y., Fernando, D.N., Save, H., 2013. GRACE satellite monitoring of large depletion in water storage in response to the 2011 drought in Texas. Geophysical Research Letters 40, 3395–3401. https://doi.org/10.1002/grl.50655

Ma, L., Zobeck, T.M., Hsieh, D.H., Holder, D., Morgan, C.L.S., Thompson, J.E., 2011. Optical properties of aeolian dusts common to West Texas. Aeolian Research 3(2), 235–242. https://doi.org/10.1016/j.aeolia.2011.05.002

Ma, M., Ren, L., Singh, V.P., Tu, X., Jiang, S., Liu, Y., 2015. Evaluation and application of the SPDI-JDI for droughts in Texas, USA. Journal of Hydrology 521, 34–45. https://doi.org/10.1016/j.jhydrol.2014.11.074

Mankin, J.S., Seager, R., Smerdon, J.E., Cook, B.I., Williams, A.P., Horton, R.M., 2018. Blue water trade-offs with vegetation in a CO_2-Enriched climate. Geophysical Research Letters 45, 3115–3125. https://doi.org/10.1002/2018gl077051

Mann, H.B., 1945. Nonparametric tests against trend. The Economic Journal and Economics and Sociology 13 (3), 245–259. https://doi.org/10.2307/1907187

Manns, F.T., 2024. Wild Fire Hypothesis: Fanned by the Wandering Wind - Summers of Hell in Paradise, Lytton and Smokehouse Creek.

Manisalidis, I., Stavropoulou, E., Stavropoulos, A., Bezirtzoglou, E., 2020. Environmental and health impacts of air pollution: A review. Front. Public Health 8. https://doi.org/10.3389/fpubh.2020.00014

McGinnis, S., Kessenich, L., Mearns, L., Cullen, A., Podschwit, H., Bukovsky, M., 2023. Future regional increases in simultaneous large Western USA wildfires. International Journal of Wildland Fire 32, 1304–1314. https://doi.org/10.1071/WF22107

McKee, T.B., Doesken, N.J., Kleist, J., 1993. The relationship of drought frequency and duration to time scales. In: Presented at the Proceedings of the 8th Conference on Applied Climatology, pp. 179–183.

McTainsh, G.H., Pitblado, J.R., 1987. Dust storms and related phenomena measured from meteorological records in Australia. Earth Surface Processes and Landforms 12(4), 415–424. https://doi.org/10.1002/esp.3290120407.

Middleton, N., Tozer, P., Tozer, B., 2019. Sand and dust storms: Underrated natural hazards. Disasters 43(2), 390–409. https://onlinelibrary.wiley.com/doi/full/10.1111/disa.12320

Middleton, N.J., 2017. Desert dust hazards: A global review. Aeolian Research 24, 53–63. https://doi.org/10.1016/j.aeolia.2016.12.001.

Middleton, N.J. and Goudie, A.S., 2001. Saharan dust: Sources and trajectories. Transactions of the Institute of British Geographers 26(2), 165–181.

Mishra, A.K., Singh, V.P., 2010. A review of drought concepts. Journal of Hydrology 391, 202–216. https://doi.org/10.1016/j.jhydrol.2010.07.012

Mishra, A.K., Singh, V.P., 2011. Drought modeling – A review. Journal of Hydrology 403, 157–175. https://doi.org/10.1016/j.jhydrol.2011.03.049

Mishra, A.K., Singh, V.P., Özger, M., 2011. Seasonal streamflow extremes in Texas river basins: Uncertainty, trends, and teleconnections. Journal of Geophysical Research 116, D08108. https://doi.org/10.1029/2010JD014597

Miller, S.D., Bankert, R.L., Solbrig, J.E., Forsythe, J.M., Noh, Y.J., Grasso, L.W., 2017. A dynamic enhancement with background reduction algorithm: Overview and application to satellite-based dust storm detection. Journal of Geophysical Research: Atmospheres 122(23), 12938–12959. https://agupubs.onlinelibrary.wiley.com/doi/full/10.1002/2017JD027365

Miller, S.M., Wofsy, S.C., Michalak, A.M., Kort, E.A., Andrews, A.E., Biraud, S.C., Dlugokencky, E.J., et al, 2013. Anthropogenic emissions of methane in the United States. Proceedings of the National Academy of Sciences 11(50), 20018–20022. https://doi.org/10.1073/pnas.1314392110

Mirzabaev, A., L.C. Stringer, T.A. Benjaminsen, P. Gonzalez, R. Harris, M. Jafari, N. Stevens, C.M. Tirado, and S. Zakieldeen, 2022: Cross-Chapter Paper 3: Deserts, Semiarid Areas and Desertification. In: Climate Change 2022: Impacts, Adaptation and Vulnerability. Contribution of Working Group II to the Sixth Assessment Report of the Intergovernmental Panel on Climate Change [H.-O. Pörtner, D.C. Roberts, M. Tignor, E.S. Poloczanska, K. Mintenbeck, A. Alegría, M. Craig, S. Langsdorf, S. Löschke, V. Möller, A. Okem, B. Rama (eds.)]. Cambridge University Press, Cambridge, UK and New York, NY, USA, pp. 2195–2231, doi:10.1017/9781009325844.020.

Moorhead, J.E., Gowda, P.H., Singh, V.P., Porter, D.O., Marek, T.H., Howell, T.A., Stewart, B.A., 2015. Identifying and evaluating a suitable index for agricultural drought monitoring in the Texas high plains. Journal of the American Water Resources Association 51, 807–820. https://doi.org/10.1111/jawr.12275

Mu, H., Otani, S., Shinoda, M., Yokoyama, Y., Onishi, K., Hosoda, T., Okamoto, M., Kurozawa, Y., 2013. Long-term effects of livestock loss caused by dust storm on Mongolian inhabitants: A survey 1 year after the dust storm. Yonago Acta Medica 56(1), 39–42.

Mulder, V.L., de Bruin, S., Weyermann, J., Kokaly, R.F., Schaepman, M.E., 2013. Characterizing regional soil mineral composition using spectroscopy and geostatistics. Remote Sensing of Environment 139, 415–429. https://doi.org/10.1016/j.rse.2013.08.018

Nagy, R.C., Fusco, E., Bradley, B., Abatzoglou, J.T., Balch, J., 2018. Human-related ignitions increase the number of large wildfires across U.S. Ecoregions. Fire 1, 4. https://doi.org/10.3390/fire1010004

National Centers for Environmental Information, 2025. U.S. Billion-Dollar Weather and Climate Disasters. https://www.ncei.noaa.gov/access/billions/summary-stats

National Integrated Drought Information System, n.d. Drought and wildfire management impacts | Drought.gov [WWW Document]. https://www.drought.gov/sectors/wildfire-management (accessed 3.11.25).

Nielsen-Gammon, J.W., 2012. The 2011 Texas drought. Texas Water Journal 3, 59–95. https://doi.org/10.21423/twj.v3i1.6463

Nielsen-Gammon, J.W., Banner, J.L., Cook, B.I., Tremaine, D.M., Wong, C.I., Mace, R.E., Gao, H., Yang, Z., Gonzalez, M.F., Hoffpauir, R., Gooch, T., Kloesel, K., 2020. Unprecedented drought challenges for Texas water resources in a changing climate: What do researchers and stakeholders need to know? Earth's Future 8, e2020EF001552. https://doi.org/10.1029/2020EF001552

Nielsen-Gammon, J.W., Holman, S., Buley, A., Jorgensen, S., 2024. Assessment of historic and future trends of extreme weather in Texas, 1900–2036.

Niu, G.-Y., Yang, Z.-L., Mitchell, K.E., Chen, F., Ek, M.B., Barlage, M., Kumar, A., Manning, K., Niyogi, D., Rosero, E., Tewari, M., Xia, Y., 2011. The community Noah land surface model with multiparameterization options (Noah-MP): 1. Model description and evaluation with local-scale measurements. Journal of Geophysical Research 116, D12109. https://doi.org/10.1029/2010JD015139

Nordstrom, K.F., Hotta, S., 2004. Wind erosion from cropland in the USA: A review of problems, solutions and prospects. Geoderma 121(3), 157–167. https://doi.org/10.1016/j.geoderma.2003.11.012

Novlan, 2007. J3.12 a synoptic climatology of blowing dust events in El Paso, Texas from 1932–2005 (2007 - 87ANNUAL_16applied). *American Meteorological Society J* (Vol. 3). https://ams.confex.com/ams/87ANNUAL/techprogram/paper_115842.htm

Palmer, W.C., 1965. Meteorological Drought. US Department of Commerce, Weather Bureau.

Offor, I.F., Adie, G.U., Ana, G.R.E.E., 2016. Review of particulate matter and elemental composition of aerosols at selected locations in Nigeria from 1985–2015. Journal of Health and Pollution 6 (10), 1–18. https://doi.org/10.5696/2156-9614-6-10.1

Opie, J., 1992. The drought of 1988, the global warming experiment, and its challenge to irrigation in the old dust bowl region. Agricultural History 66(2), 279–306.

Poulos, H.M., Reemts, C.M., Wogan, K.A., Karges, J.P., Gatewood, R.G., 2020. Multiple wildfires with minimal consequences: Low-severity wildfire effects on West Texas piñon-juniper woodlands. Forest Ecology and Management 473, 118293. https://doi.org/10.1016/j.foreco.2020.118293

Prein, A.F., Holland, G.J., Rasmussen, R.M., Clark, M.P., Tye, M.R., 2016. Running dry: The U.S. Southwest's drift into a drier climate state. Geophysical Research Letters 43(3), 1272–1279. https://doi.org/10.1002/2015GL066727.

Prestemon, J.P., Shankar, U., Xiu, A., Talgo, K., Yang, D., Dixon, E., McKenzie, D., Abt, K.L., 2016. Projecting wildfire area burned in the south-eastern United States, 2011–60. International Journal of Wildland Fire 25, 715–729. https://doi.org/10.1071/WF15124

Prospero, J.M., Blades, E., Mathison, G., Naidu, R., 2005. Interhemispheric transport of viable fungi and bacteria from Africa to the Caribbean with Soil Dust. Aerobiologia 21(1), 1–19. https://doi.org/10.1007/s10453-004-5872-7

Pu, B., Ginoux, P., 2017. Projection of American dustiness in the late 21st century due to climate change. Scientific Reports 7(1), 5553. https://doi.org/10.1038/s41598-017-05431-9

Pye, K., 2015. *Aeolian Dust and Dust Deposits*. Elsevier.

Quiring, S.M., Ganesh, S., 2010. Evaluating the utility of the vegetation condition index (VCI) for monitoring meteorological drought in Texas. Agricultural and Forest Meteorology 150, 330–339. https://doi.org/10.1016/j.agrformet.2009.11.015

Rajsekhar, D., Mishra, A.K., Singh, V.P., 2013. Regionalization of drought characteristics using an entropy approach. Journal of Hydrologic Engineering 18, 870–887. https://doi.org/10.1061/(ASCE)HE.1943-5584.0000683

Rajsekhar, D., Singh, V.P., Mishra, A.K., 2015a. Multivariate drought index: An information theory based approach for integrated drought assessment. Journal of Hydrology 526, 164–182. https://doi.org/10.1016/j.jhydrol.2014.11.031

Rajsekhar, D., Singh, V.P., Mishra, A.K., 2015b. Hydrologic drought atlas for Texas. Journal of Hydrologic Engineering 20, 05014023. https://doi.org/10.1061/(ASCE)HE.1943-5584.0001074

Rajsekhar, D., Singh, V.P., Mishra, A.K., 2015c. Integrated drought causality, hazard, and vulnerability assessment for future socioeconomic scenarios: An information theory perspective. JGR Atmospheres 120, 6346–6378. https://doi.org/10.1002/2014JD022670

Rastogi, D., Trok, J., Depsky, N., Monier, E., Jones, A., 2024. Historical evaluation and future projections of compound heatwave and drought extremes over the conterminous United States in CMIP6. Environmental Research Letters 19, 014039. https://doi.org/10.1088/1748-9326/ad0efe

Reid, C.E., Brauer, M., Johnston, F.H., Jerrett, M., Balmes, J.R., Elliott, C.T., 2016. Critical review of health impacts of wildfire smoke exposure. Environmental Health Perspectives 124, 1334–1343. https://doi.org/10.1289/ehp.1409277

Reisen, F., Duran, S.M., Flannigan, M., Elliott, C., Rideout, K., 2015. Wildfire smoke and public health risk. International Journal of Wildland Fire 24, 1029. https://doi.org/10.1071/WF15034

Rivera, N., 2012. Analysis and modeling of dust event data from El Paso, Texas. Open Access Theses & Dissertations. https://scholarworks.utep.edu/open_etd/2173

Robinson, M.C., Ardon-Dryer, K., 2024. Characterization of 21 years of dust events across four West Texas regions. Aeolian Research 67–69, 100930. https://doi.org/10.1016/j.aeolia.2024.100930

Robinson MC, Schueth K, Ardon-Dryer K. The Spatial and Temporal Impact of the February 26, 2023, Dust Storm on the Meteorological Conditions and Particulate Matter Concentrations Across New Mexico and West Texas. EGUsphere 2024, 1–21. https://doi.org/10.5194/egusphere-2024-113

Rublee, C.S., Sorensen, C.J., Lemery, J., Wade, T.J., Sams, E.A., Hilborn, E.D., Crooks, J.L., 2020. Associations between dust storms and intensive care unit admissions in the United States, 2000–2015. GeoHealth 4(8), e2020GH000260. https://doi.org/10.1029/2020GH000260

Sager, M., 2020. Urban soils and road dust—Civilization effects and metal pollution—A review. Environments 7(11), 98. https://doi.org/10.3390/environments7110098

Sakulich, J., Taylor, A.H., 2007. Fire regimes and forest structure in a sky island mixed conifer forest, Guadalupe mountains national park, Texas, USA. Forest Ecology and Management 241, 62–73. https://doi.org/10.1016/j.foreco.2006.12.029

Salguero, J., Li, J., Farahmand, A., Reager, J.T., 2020. Wildfire trend analysis over the contiguous United States using remote sensing observations. Remote Sensing 12, 2565. https://doi.org/10.3390/rs12162565

Salley, S.W., Sleezer, R.O., Bergstrom, R.M., Martin, P.H., Kelly, E.F., 2016. A long-term analysis of the historical dry boundary for the great plains of North America: Implications of climatic variability and climatic change on temporal and spatial patterns in soil moisture. Geoderma 274, 104–113. https://doi.org/10.1016/j.geoderma.2016.03.020

Samaniego, L., Kumar, R., Zink, M., 2013. Implications of parameter uncertainty on soil moisture drought analysis in Germany. Journal of Hydrometeorology 14, 47–68. https://doi.org/10.1175/JHM-D-12-075.1

Sarafoglou, N., Kafatos, M., Sprigg, W.A., 2016. Migration, environment and public health: Theory and interdisciplinary research from a regional science perspective. International Journal of Social Science Studies 4(4), 122–135. https://doi.org/10.11114/ijsss.v4i4.1473

Scanlon, B.R., Duncan, I., Reedy, R.C., 2013. Drought and the water–energy nexus in Texas. Environmental Research Letters 8, 045033. https://doi.org/10.1088/1748-9326/8/4/045033

Scheuvens, D., Kandler, K., 2014. On Composition, Morphology, and Size Distribution of Airborne Mineral Dust. In *Mineral Dust: A Key Player in the Earth System*, edited by Peter Knippertz and Jan-Berend W. Stuut, 15–49. Dordrecht: Springer Netherlands. https://doi.org/10.1007/978-94-017-8978-3_2.

Seager, R., Goddard, L., Nakamura, J., Henderson, N., Lee, D.E., 2014. Dynamical causes of the 2010/11 Texas–Northern Mexico Drought. Journal of Hydrometeorology 15, 39–68. https://doi.org/10.1175/JHM-D-13-024.1

Seager, R., Ting, M., Held, I., Kushnir, Y., Lu, J., Vecchi, G., Huang, H.-P., Harnik, N., Leetmaa, A., Lau, N.-C., Li, C., Velez, J., Naik, N., 2007. Model projections of an imminent transition to a more arid climate in Southwestern North America. Science 316, 1181–1184. https://doi.org/10.1126/science.1139601

Shao, M., Fernando, N., Zhu, J., Zhao, G., Kao, S., Zhao, B., Roberts, E., Gao, H., 2023. Estimating future surface water availability through an integrated climate-hydrology-management modeling framework at a basin scale under CMIP6 scenarios. Water Resources Research 59, e2022WR034099. https://doi.org/10.1029/2022WR034099

Sharma, A., Singh, S., Kulshrestha, U.C., 2017. Determination of urban dust signatures through chemical and mineralogical characterization of atmospheric dust fall in East Delhi (India). The Journal of Indian Geophysical Union 21(2), 140–147.

Shukla, S., Wood, A.W., 2008. Use of a standardized runoff index for characterizing hydrologic drought. Geophysical Research Letters 35, 2007GL032487. https://doi.org/10.1029/2007GL032487

Singh, R., 2023. Impact of Climate and Land Use Land Cover Changes on Soil Erosion. In *Soil and Water Conservation Structures Design*, edited by Rajendra Singh, 415–441. Singapore: Springer Nature. https://doi.org/10.1007/978-981-19-8665-9_14.

Spracklen, D.V., Mickley, L.J., Logan, J.A., Hudman, R.C., Yevich, R., Flannigan, M.D., Westerling, A.L., 2009. Impacts of climate change from 2000 to 2050 on wildfire activity and carbonaceous aerosol concentrations in the western United States. Journal of Geophysical Research: Atmospheres 114. https://doi.org/10.1029/2008JD010966

Stahle, D.W., Cleaveland, M.K., 1988. Texas Drought history reconstructed and analyzed from 1698 to 1980. Journal of Climate 1, 59–74. https://doi.org/10.1175/1520-0442(1988)001<0059:TDHRAA>2.0.CO;2

Stambaugh, M.C., Guyette, R.P., Stroh, E.D., Struckhoff, M.A., Whittier, J.B., 2018. Future southcentral US wildfire probability due to climate change. Climatic Change 147, 617–631. https://doi.org/10.1007/s10584-018-2156-8

Stout, J.E., 2001. Dust Trends in the Southern High Plains. In *Soil Erosion* (p. 475). American Society of Agricultural and Biological Engineers. https://doi.org/10.13031/2013.4580
Stout, J.E., Lee, J.A., 2003. Indirect evidence of wind erosion trends on the Southern high plains of North America. Journal of Arid Environments 55 (1), 43–61. https://doi.org/10.1016/S0140-1963(02)00266-5
Sweeney, M.R., Zlotnik, V.A., Joeckel, R.M., Stout, J.E., 2016. Geomorphic and hydrologic controls of dust emissions during drought from yellow Lake playa, West Texas, USA. Journal of Arid Environments 133, 37–46. https://doi.org/10.1016/j.jaridenv.2016.05.007
Tabassum, S., 2023. Deciphering changes in rainfall events and the consequences over Texas under global warming. https://doi.org/10.26153/TSW/48563
Tadesse, T., Wardlow, B.D., Brown, J.F., Svoboda, M.D., Hayes, M.J., Fuchs, B., Gutzmer, D., 2015. Assessing the vegetation condition impacts of the 2011 drought across the U.S. Southern Great Plains using the Vegetation Drought Response Index (VegDRI). Journal of Applied Meteorology and Climatology 54, 153–169. https://doi.org/10.1175/JAMC-D-14-0048.1
Tanguy, M., Chevuturi, A., Marchant, B.P., Mackay, J.D., Parry, S., Hannaford, J., 2023. How will climate change affect the spatial coherence of streamflow and groundwater droughts in Great Britain? Environmental Research Letters 18, 064048. https://doi.org/10.1088/1748-9326/acd655
Texas A&M Forest Service, 2023. Historical fire statistics [WWW Document]. https://fire-information-tfsgis.hub.arcgis.com/pages/historical-fire-statistics (accessed 2.5.25).
Texas Water Development Board, 2022a. Drought in Texas: A Comparison of the 1950–1957 and 2010–2015 Droughts.
Texas Water Development Board, 2022b. Water for Texas 2022 State Water Plan.
Tijdeman, E., Stahl, K., Tallaksen, L.M., 2020. Drought characteristics derived based on the standardized streamflow index: A large sample comparison for parametric and nonparametric methods. Water Resources Research 56, e2019WR026315. https://doi.org/10.1029/2019WR026315
Tong, D.Q., Gill, T.E., Sprigg, W.A., Van Pelt, R.S., Baklanov, A.A., Barker, B.M., et al., 2023. Health and safety effects of airborne soil dust in the Americas and beyond. Reviews of Geophysics 61, e2021RG000763. https:// doi.org/10.1029/2021RG000763.
Tong, D.Q., Wang, J.X.L., Gill, T.E., Lei, H., Wang, B., 2017. Intensified dust storm activity and valley fever infection in the Southwestern United States. Geophysical Research Letters 44(9), 4304–4312. https://doi.org/10.1002/2017GL073524
Ullah, I., Zeng, X., Mukherjee, S., Aadhar, S., Mishra, A.K., Syed, S., Ayugi, B.O., Iyakaremye, V., Lv, H., 2023. Future amplification of multivariate risk of compound drought and heatwave events on South Asian Population. Earth's Future 11, e2023EF003688. https://doi.org/10.1029/2023EF003688
USGCRP, 2017. Climate Science Special Report: Fourth National Climate Assessment, Volume I [Wuebbles, D.J., D.W. Fahey, K.A. Hibbard, D.J. Dokken, B.C. Stewart, and T.K. Maycock (eds.)]. U.S. Global Change Research Program, Washington, DC, USA, 470 pp., https://doi.org/10.7930/J0J964J6
Valiya Veettil, A., Fares, A., Awal, R., Mohtar, R., 2024. Assessing Future Water Scarcity in Texas under Climate Change and Growing Water Demands. Journal of Hydrologic Engineering 29, 05024022. https://doi.org/10.1061/JHYEFF.HEENG-6253
Van Loon, A.F., 2015. Hydrological drought explained. WIREs Water 2, 359–392. https://doi.org/10.1002/wat2.1085
Vaughan, E., Crutcher, J., Weir Labatt, T. III, McMahan, L., Bradford, B., Cluck, M., 2012. Water for Texas 2012 State Water Plan. Texas Water Development Board.
Venkataraman, K., Tummuri, S., Medina, A., Perry, J., 2016. 21st century drought outlook for major climate divisions of Texas based on CMIP5 multimodel ensemble: Implications for water resource management. Journal of Hydrology 534, 300–316. https://doi.org/10.1016/j.jhydrol.2016.01.001
Vicente-Serrano, S.M., Beguería, S., López-Moreno, J.I., 2010. A multiscalar drought index sensitive to global warming: The standardized precipitation evapotranspiration index. Journal of Climate 23, 1696–1718. https://doi.org/10.1175/2009JCLI2909.1
Vicente-Serrano, S.M., López-Moreno, J.I., Beguería, S., Lorenzo-Lacruz, J., Azorin-Molina, C., Morán-Tejeda, E., 2012. Accurate computation of a streamflow drought index. Journal of Hydrologic Engineering 17, 318–332. https://doi.org/10.1061/(ASCE)HE.1943-5584.0000433
Villanueva, F., 2023. Comparison of woody riparian vegetation communities along an urban-to- rural gradient in San Antonio, Texas. M.En., United States – Texas, The University of Texas at San Antonio
Vlasov, D., Ramírez, O., Luhar, A., 2022. Road dust in urban and industrial environments: Sources, pollutants, impacts, and management. Atmosphere 13(4), 607. https://doi.org/10.3390/atmos13040607
Wang, C., Jeong, G.R., Mahowald, N., 2009. Particulate absorption of solar radiation: anthropogenic aerosols vs. dust. Atmospheric Chemistry and Physics 9(12), 3935–3945. https://doi.org/10.5194/acp-9-3935-2009

Wen, N., Marek, G.W., Srinivasan, R., Brauer, D.K., Qi, J., Wang, N., Han, Y., Zhang, X., Feng, P., Liu, D.L., Chen, Y., 2024. Assessing the impacts of long-term climate change on hydrology and yields of diversified crops in the Texas High Plains. Agricultural Water Management 302, 108985. https://doi.org/10.1016/j.agwat.2024.108985

Wigmosta, M.S., Vail, L.W., Lettenmaier, D.P., 1994. A distributed hydrology-vegetation model for complex terrain. Water Resources Research 30, 1665–1679. https://doi.org/10.1029/94WR00436

Wood, A.W., Leung, L.R., Sridhar, V., Lettenmaier, D.P., 2004. Hydrologic implications of dynamical and statistical approaches to downscaling climate model outputs. Climatic Change 62, 189–216. https://doi.org/10.1023/B:CLIM.0000013685.99609.9e

Wootten, A.M., Başağaoğlu, H., Bertetti, F.P., Chakraborty, D., Sharma, C., Samimi, M., Mirchi, A., 2024. Customized statistically downscaled CMIP5 and CMIP6 projections: Application in the Edwards Aquifer region in South-Central Texas. Earth's Future 12, e2024EF004716. https://doi.org/10.1029/2024EF004716

Wuebbles, D., Meehl, G., Hayhoe, K., Karl, T.R., Kunkel, K., Santer, B., Wehner, M., Colle, B., Fischer, E.M., Fu, R., Goodman, A., Janssen, E., Kharin, V., Lee, H., Li, W., Long, L.N., Olsen, S.C., Pan, Z., Seth, A., Sheffield, J., Sun, L., 2014. CMIP5 climate model analyses: Climate extremes in the United States. Bulletin of the American Meteorological Society 95, 571–583. https://doi.org/10.1175/BAMS-D-12-00172.1

Yan, H., DeChant, C.M., Moradkhani, H., 2015. Improving soil moisture profile prediction with the particle filter-Markov Chain Monte Carlo method. IEEE Transactions on Geoscience and Remote Sensing 53, 6134–6147. https://doi.org/10.1109/TGRS.2015.2432067

Yan, H., Sun, N., Eldardiry, H., Thurber, T.B., Reed, P.M., Malek, K., Gupta, R., Kennedy, D., Swenson, S.C., Hou, Z., Cheng, Y., Rice, J.S., 2023. Large ensemble diagnostic evaluation of hydrologic parameter uncertainty in the Community Land Model Version 5 (CLM5). Journal of Advances in Modeling Earth Systems 15, e2022MS003312. https://doi.org/10.1029/2022MS003312

Yan, H., Sun, N., Eldardiry, H., Thurber, T., Reed, P., Kennedy, D., Swenson, S., Rice, J., 2025. Ensemble-based spatially distributed CLM5 hydrological parameter estimation for the continental United States. Journal of Advances in Modeling Earth Systems 17, e2024MS004227. https://doi.org/10.1029/2024MS004227

Yang, Z.-L., Niu, G.-Y., Mitchell, K.E., Chen, F., Ek, M.B., Barlage, M., Longuevergne, L., Manning, K., Niyogi, D., Tewari, M., Xia, Y., 2011. The community Noah land surface model with multiparameterization options (Noah-MP): 2. Evaluation over global river basins. Journal of Geophysical Research 116, D12110. https://doi.org/10.1029/2010JD015140

Yao, L., Ferencz, S.B., Sun, N., Yan, H., 2024. Assessing the complex influences of water management on hydrological drought characteristics in Texas. Environmental Research Letters 19, 114034. https://doi.org/10.1088/1748-9326/ad7d23

Yevjevich, V., 1967. An Objective Approach to Definitions and Investigations of Continental Hydrologic Droughts. In: *Hydrology Paper*, 353. Fort Collins, CO: Colorado State University.

Yin, D., Nickovic, S., Barbaris, B., Chandy, B., Sprigg, W.A., 2005. Modeling wind-blown desert dust in the southwestern United States for public health warning: A case study. Atmospheric Environment 39(33), 6243–6254. https://doi.org/10.1016/j.atmosenv.2005.07.009

Yoon, J.-H., Wang, S.-Y.S., Lo, M.-H., Wu, W.-Y., 2018. Concurrent increases in wet and dry extremes projected in Texas and combined effects on groundwater. Environmental Research Letters 13, 054002. https://doi.org/10.1088/1748-9326/aab96b

Yu, G., Feng, Y., Wang, J., Wright, D.B., 2023. Performance of fire danger indices and their utility in predicting future wildfire danger over the conterminous United States. Earth's Future 11, e2023EF003823. https://doi.org/10.1029/2023EF003823

Yuan, S., Quiring, S.M., 2014. Drought in the U.S. Great Plains (1980–2012): A sensitivity study using different methods for estimating potential evapotranspiration in the Palmer Drought Severity Index. JGR Atmospheres 119. https://doi.org/10.1002/2014JD021970

Yuan, S., Quiring, S.M., Leasor, Z.T., 2021. Historical changes in surface soil moisture over the contiguous United States: An assessment of CMIP6. Geophysical Research Letters 48, e2020GL089991. https://doi.org/10.1029/2020GL089991

Yuan, X., 2020. Analyzing drought conditions using NDVI in Coke, Texas. IOP Conference Series: Earth and Environmental Science 565, 012104. https://doi.org/10.1088/1755-1315/565/1/012104

Zhang, B., Wang, S., Wang, Y., 2019. Copula-based convection-permitting projections of future changes in multivariate drought characteristics. JGR Atmospheres 124, 7460–7483. https://doi.org/10.1029/2019JD030686

Zhang, X., Hao, Z., Singh, V.P., Zhang, Y., Feng, S., Xu, Y., Hao, F., 2022. Drought propagation under global warming: Characteristics, approaches, processes, and controlling factors. Science of the Total Environment 838, 156021. https://doi.org/10.1016/j.scitotenv.2022.156021

Zhao, B., Kao, S., Zhao, G., Gangrade, S., Rastogi, D., Ashfaq, M., Gao, H., 2023. Evaluating enhanced reservoir evaporation losses from CMIP6-based future projections in the contiguous United States. Earth's Future 11, e2022EF002961. https://doi.org/10.1029/2022EF002961

Zhao, F., Wu, Y., Yin, X., Sun, K., Ma, S., Zhang, S., Liu, S., Wang, W., Chen, J., 2022. Projected changes in population exposure to drought in China under CMIP6 forcing scenarios. Atmospheric Environment 282, 119162. https://doi.org/10.1016/j.atmosenv.2022.119162

Zhao, G., Gao, H., Kao, S.-C., Voisin, N., Naz, B.S., 2018. A modeling framework for evaluating the drought resilience of a surface water supply system under non-stationarity. Journal of Hydrology 563, 22–32. https://doi.org/10.1016/j.jhydrol.2018.05.037

Zhao, G., Gao, H., Naz, B.S., Kao, S.-C., Voisin, N., 2016. Integrating a reservoir regulation scheme into a spatially distributed hydrological model. Advances in Water Resources 98, 16–31. https://doi.org/10.1016/j.advwatres.2016.10.014

Zhou, Z., Zhang, L., Chen, J., She, D., Wang, G., Zhang, Q., Xia, J., Zhang, Y., 2023. Projecting global drought risk under various SSP-RCP scenarios. Earth's Future 11, e2022EF003420. https://doi.org/10.1029/2022EF003420

Zobeck, T.M., 2010. Soil and Water Conservation Advances in the United States—Review and Assessment. Soil and Water Conservation Advances in the United States, 60, 263–292. ASA-CSSA-SSSA.

1.5 Beyond the Banks

Urbanization, Infrastructure, and the Fight Against Flooding

Michelle A. Hummel

1.5.1 INTRODUCTION

1.5.1.1 Overview of Texas's Diverse Flood Hazards

Texas is characterized by diverse hydrologic regions, spanning a distinct wet-to-dry gradient from east to west and encompassing both inland and coastal areas and rural and urban communities. As a result, watersheds across the state face unique flooding hazards from fluvial, pluvial, and coastal sources (Texas Water Development Board, 2024). Fluvial or riverine flooding occurs when runoff during rainfall events cannot be contained within a river channel, causing overflows into the adjacent floodplain. Pluvial or urban flooding results when heavy rain falling over impervious surfaces exceeds the capacity of stormwater drainage systems. Coastal flooding is generated from tides, waves, and surge that overtop natural or engineered coastal defenses and is exacerbated by sea level rise (Zscheischler et al., 2020).

Fluvial and pluvial flood hazards, the focus of this chapter, vary significantly in both frequency and magnitude due to the diverse climate and geography across Texas. Average annual rainfall across the state varies from 140 cm in East Texas to 25 cm near El Paso in West Texas (Texas Water Development Board, 2012). The Gulf of America is an important contributor to the state's climate, providing a vast source of warm, moist air that interacts with cooler air masses from the north to produce rainfall events (Texas Water Development Board, 2012). Soil types and topography also contribute to unique flood challenges across the state. In coastal watersheds, clay soils limit the infiltration capacity, meaning that a higher percentage of rainfall remains as runoff on the land surface. In addition, the flat topography limits efficient drainage, creating longer-duration flood hazards (Federal Emergency Management Agency, 2019). In the central part of the state, known as Flash Flood Alley, shallow soils limit infiltration. The region's hilly terrain creates orographic rainfall effects when changes in topography force moist air to rise and cool, enhancing rainfall formation (Baker, 1975). In drier West Texas, flash flood events interspersed with extended periods of drought can rapidly fill dry creekbeds and alluvial fans with high-velocity flows, creating unexpected flood hazards (National Research Council, 1996).

1.5.1.2 Recent Flood Events and Impacts

Texas is one of the hardest-hit states in terms of flood damages. The National Flood Insurance Program (NFIP), which provides publicly backed insurance policies for homeowners, distributed over $22 billion in inflation-adjusted payouts between 1978 and 2021 (Federal Emergency Management Agency, 2024b). This is second only to Louisiana, which received nearly $29 billion. The county with the most claims filed during this same period was Harris County, Texas, with over 171,000 claims. Harris County and other parts of Southeast Texas were impacted by Hurricane Harvey, which stalled over the Texas coast in August 2017 and dropped extreme amounts of

DOI: 10.1201/9781003406600-6

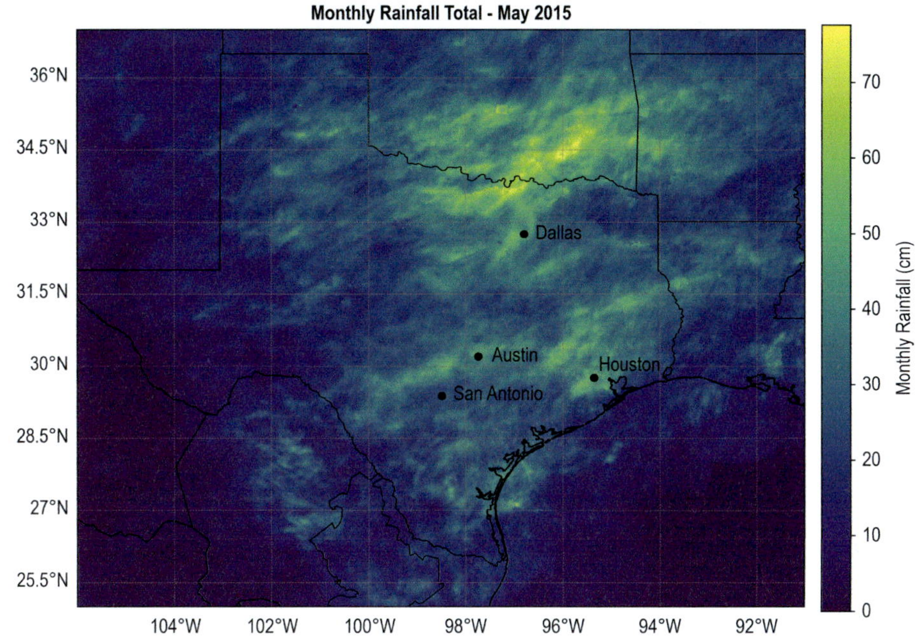

FIGURE I.5.1 May 2015 rainfall totals (in cm) for Texas. (Data from Du, 2025).

rainfall (more than 150 cm in some locations). Harvey was the second costliest hurricane in U.S. history, with over $125 billion in damages, and caused 68 deaths in Texas alone (Blake and Zelinsky, 2018).

Several years prior, in 2015, Central Texas was impacted by severe flooding over Memorial Day weekend. The flooding resulted from persistent wet conditions throughout much of May preceding a series of heavy rainfall periods on May 23–24 that produced 15–20 cm of rainfall over much of the region and upwards of 30 cm in some locations (National Weather Service, 2015). Rainfall totals over Central Texas for the month of May 2015 are shown in Figure I.5.1. Given the already saturated conditions, the rainfall during this two-day period was converted directly into runoff that caused rivers, especially the San Marcos and Blanco, to swell. Notably, water levels in the Blanco River at Wimberley, Texas, rose from 1.5 m to over 12 m in a four-hour period (National Weather Service, 2015). Damages from this storm exceeded $1.4 billion in Texas alone, and 14 deaths were reported in Central Texas (National Oceanic and Atmospheric Administration, 2025; Texas Water Development Board, 2024).

I.5.1.3 Texas State Flood Plan

In 2019, the Texas Legislature passed Senate Bill 8, which tasked the Texas Water Development Board (TWDB) with overseeing a statewide flood planning program. Given the diverse flood hazards faced across the state, a regional, watershed-centered approach was adopted. Planning groups were convened for each river basin in the state to develop region-specific flood characterization plans and to identify flood risk reduction needs. The regional plans were collated to produce the first State Flood Plan, released in 2024. The plan aims to provide a comprehensive picture of flood hazards, exposure of people and property, and vulnerability of communities, critical facilities, and infrastructure across the state, as well as to recommend strategies to enhance future resilience to flooding (Texas Water Development Board, 2024).

I.5.2 FLOOD DRIVERS: HISTORICAL PERSPECTIVES AND FUTURE TRENDS

I.5.2.1 Tropical Cyclones

Tropical cyclones that make landfall along the Texas coast bring heavy rainfall to coastal and inland watersheds. According to an analysis of gauge-based precipitation data in Texas from 1950 to 2009 (Zhu and Quiring, 2013), approximately 13% of the mean annual precipitation in Texas was due to rainfall associated with tropical cyclones. During the same historical period, extreme tropical cyclone rainfall greater than 100 mm per day occurred with nearly a one-in-three-year frequency in some parts of Southeast Texas. Although these extreme rainfall rates occurred most frequently in areas within 400 km of the coast, some inland areas were also impacted by extreme daily rainfall associated with tropical cyclones.

Of particular concern for riverine and fluvial flooding are slow-moving storms such as Tropical Storm Allison in 2001 and Hurricane Harvey in 2017, which both stalled over Southeast Texas for multiple days, dropping more than 90 cm and 150 cm of rainfall, respectively, in some locations (Blake and Zelinsky, 2018; Stewart, 2001). An analysis of storm track data over the contiguous U.S. from 1990 to 2017 found that the translation speed of tropical cyclones has decreased by 17% (Kossin, 2019), contributing to higher rainfall totals and more severe flooding in impacted areas. However, several modeling studies suggest that changes in large-scale atmospheric circulation patterns could result in an increase in tropical cyclone translation speed in the North Atlantic basin (Yamaguchi et al., 2020) and a lower probability of slow-moving tropical cyclones making landfall in Texas by the end of the 21st century (Hassanzadeh et al., 2020), although additional work is needed to further evaluate this finding.

Simulations by tropical cyclone-permitting models are inconclusive with regards to changes in the frequency of tropical cyclones but project an increase in the proportion of intense tropical cyclones of Category 4 or 5 (Knutson et al., 2020; Murakami et al., 2012; Wehner et al., 2018). Models also predict an increase in tropical cyclone rainfall intensity, as shown in Figure I.5.2. With 2°C of warming, rainfall rates are projected to increase by 12% on average globally (Knutson et al., 2020),

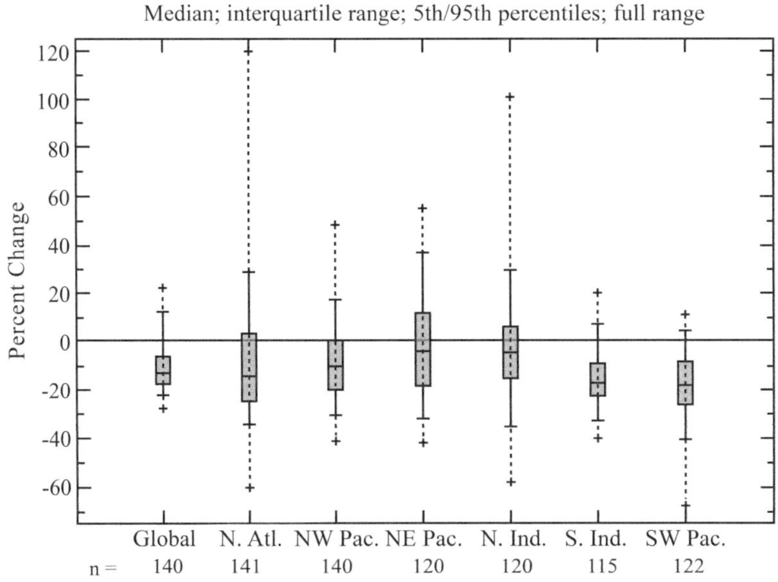

FIGURE I.5.2 Projected percentage change in tropical cyclone rainfall rates by basin (Knutson et al., 2020). © American Meteorological Society. Used with permission.

consistent with the Clausius-Clapeyron scaling that projects an approximate 7% increase in low-level water vapor per 1°C of warming (O'Gorman and Muller, 2010). More intense storms, both in terms of wind and rainfall, will result in higher damages associated with future events unless mitigation measures are implemented to reduce vulnerability.

1.5.2.2 Extratropical Cyclones

Like tropical cyclones, extratropical cyclones are rotating weather systems with low surface pressure. These systems occur in the midlatitudes, regions more than 30° latitude from the equator, and typically form due to strong horizontal temperature gradients that occur when cold air masses collide with warm air masses, forcing warm air to rise (Schultz et al., 2019). The presence of a jet stream at high altitudes enhances the strength of extratropical events (Frame et al., 2017). In Texas, extratropical storms typically bring high winds, rainfall, and thunderstorms along frontal boundaries. Rainfall is usually not as intense as that produced by tropical cyclones or convective storms but can be heavy at times. If fronts associated with extratropical cyclones stall or slow, they could drop substantial rainfall over multiple days, leading to flooding (Frame et al., 2017). With future warming, the rainfall associated with extratropical cyclones is expected to increase consistent with the Clausius-Clapeyron relationship (Bengtsson et al., 2009; Hawcroft et al., 2018; Zhang and Colle, 2017).

1.5.2.3 Convective Storms

Convective storms form when an air mass rises rapidly due to vertical stratification caused by heating of the Earth's surface or cooling at altitude (Frame et al., 2017). They may be associated with tropical or extratropical cyclones or may form individually as mesoscale convective systems (Intergovernmental Panel on Climate Change, 2021). These storms are often accompanied by heavy rainfall, thunderstorms, strong wind gusts, and tornadoes (Schumacher and Rasmussen, 2020). Convective systems comprised of multiple convective cells can take rotating or linear forms and may persist for hours as new cells are successively produced, leading to extended periods of heavy rainfall (Frame et al., 2017).

In the U.S., climate change is expected to produce higher average and maximum rainfall rates associated with convective storms (Prein et al., 2017; Schumacher and Rasmussen, 2020). In addition, models suggest that convective storms will occur more frequently in the spring months due to an increase in vertical wind shear and convective available potential energy (Diffenbaugh et al., 2013). This could result in a longer convective storm season, increasing the prevalence of heavy rainfall and severe storms associated with convective events.

1.5.2.4 Monsoons

In far West Texas, the North American monsoon is the primary driver of rainfall in the summer months (Texas Water Development Board, 2012). As air masses over the Southwestern U.S. warm in the summer, winds from the south bring moisture from the Gulf of California and, to a lesser extent, the Gulf of America (Adams and Comrie, 1997). The resulting circulation produces afternoon thunderstorms and rainfall, generally from July through September. Monsoons produce upwards of 40% of the total annual rainfall in the region (Becker, 2021).

There is some evidence that higher concentrations of atmospheric carbon have led to an increase in the intensity of monsoon rainfall since the 1970s (Intergovernmental Panel on Climate Change, 2021). However, it is not clear how climate change will impact the North American monsoon in the future. Some studies suggest that the timing of the monsoon could shift later in the year, although changes in rainfall intensity are inconclusive (Cook and Seager, 2013; Intergovernmental Panel on Climate Change, 2021).

I.5.2.5 Compound Flooding

Compound flooding occurs when two or more flood drivers or events interact, often leading to greater flood hazards and more severe impacts (Green et al., 2025). In inland watersheds, rainfall drives both fluvial and pluvial flooding. Watershed size and configuration are important determinants of subsequent flood hazards. In small watersheds and headwater systems, storms that produce localized rainfall typically result in more immediate responses in streamflow, as generated runoff is funneled into channel networks (Yang et al., 2016). In the downstream sections of larger watersheds that cover thousands of square kilometers, peak streamflow may occur in response to localized precipitation or as the flood wave resulting from rainfall in upstream portions of the watershed makes its way downstream toward the river mouth (Maymandi et al., 2022). If elevated streamflow occurs simultaneously with heavy rainfall in adjacent urban areas, backups in urban drainage systems could occur due to high river flows that inundate stormwater outlets, producing compounding effects that exacerbate both the magnitude and duration of urban flooding.

The occurrence of successive rainfall events can also contribute to preconditioned compounding of flood hazards (Zscheischler et al., 2020). As seen during the 2015 Memorial Day floods that impacted Central Texas, wet antecedent conditions that created saturated soils throughout the early part of May resulted in limited infiltration capacity during heavy rainfall that occurred at the end of the month (National Weather Service, 2015). This resulted in more runoff flowing into local river channels in a shorter period of time, creating a flashy response that produced significant increases in water levels and devastating flooding (Meitzen et al., 2018). Similar antecedent conditions were noted prior to Hurricane Harvey due to a storm that brought heavy rainfall and saturated soils across much of Houston, increasing the region's susceptibility to subsequent flooding (Valle-Levinson et al., 2020).

Fluvial and pluvial flood hazards may also interact with coastal flood drivers, most commonly during tropical cyclone events, to create compound flooding (Wahl et al., 2015). Elevated coastal water levels driven by storm surge can limit drainage of streamflow and stormwater, creating backwater effects that increase the depth and extent of flooding (Zhang et al., 2020). In addition, as sea levels rise, storm surge hazards will extend further inland (Smith et al., 2010), exposing new areas to compound flood hazards and increasing the magnitude of impacts.

I.5.2.6 Land Use Changes

In every decade since the start of the 20th century, Texas has experienced double-digit percentage increases in its population, resulting in substantial growth. Most recently, from 2023 to 2024, Texas was the fastest-growing state in the U.S., with a population increase of 563,000, for a total of 31.3 million people (U.S. Census Bureau, 2024). The state is now home to three of the ten most populous U.S. cities (Houston, San Antonio, and Dallas). As shown in Figure I.5.3, projections of future growth in the state suggest that the population could reach between 40.3 and 46.2 million people by 2060 (Texas Demographic Center, 2024).

Historical and ongoing increases in the state's population have led to significant expansion of residential, commercial, and industrial development. As a result, large swaths of land that previously consisted of agricultural, forested, or wetland coverage have been or will be converted to impervious cover such as concrete and asphalt. An analysis of satellite-derived land cover data found that Texas had the largest increase in impervious surface area between 2001 and 2011, with much of the increase centered in major cities (Bounoua et al., 2018).

Land use changes due to urbanization and expanding development can have significant impacts on fluvial flood hazards (Zhang et al., 2018; Zhao et al., 2016). Impervious surfaces have little to no infiltration capacity, meaning that any rain that falls on these surfaces remains as runoff rather than infiltrating into the subsurface. The resulting runoff is transported via engineered or natural drainage infrastructure to nearby waterways. In urbanized watersheds, more runoff enters waterways

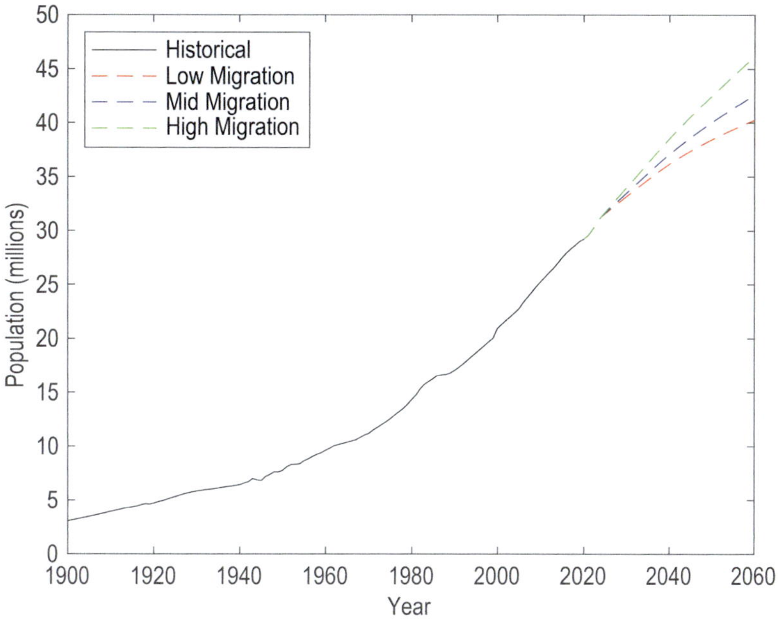

FIGURE I.5.3 Historical and projected population growth in Texas based on various migration scenarios. (Data from Texas Demographic Center, 2024.)

over a shorter time period, leading to a faster rise in river water levels and higher peak flows. For example, a modeling study of urbanization and its effects on streamflow in the San Antonio River found that the median of the annual maximum flow series for a 30-year simulation period increased by 73% when accounting for the urbanization that occurred in the basin between 1929 and 2011 (Zhao et al., 2016).

Urbanization also poses a challenge to managing pluvial flood hazards. Many stormwater drainage systems were built decades ago and have not been updated to account for increasing imperviousness and associated runoff due to continued development. As a result, these drainage systems may be unable to handle inflows during periods of heavy rain (Rosenzweig et al., 2018), leading to backups that can flood roadways and buildings and pose a threat to urban residents.

I.5.3 IMPACTS OF WATERSHED FLOODING

I.5.3.1 People and Property

Based on current flood hazard zones, an estimated 2.4 million people live or work in the 100-year floodplain in Texas, and another 2.8 million people live or work in the 500-year floodplain. This corresponds with approximately 878,000 and 786,000 buildings located within the 100-year and 500-year floodplains, respectively (Texas Water Development Board, 2024). These numbers are expected to increase as the state's population continues to grow and as climate change brings more intense rainfall events. Floodwaters can enter buildings through unsealed openings, damaging contents and posing a safety hazard for anyone remaining inside. High-velocity flows can also wash away cars or push buildings off of their foundations, mobilizing debris that produces downstream hazards (Nadal et al., 2010). Of particular concern are flash flood events, during which water levels rise rapidly with little time for residents to take protective actions or evacuate (Alipour et al., 2020), as was observed during the 2015 Memorial Day floods in Central Texas.

Along with the direct threats to life due to drowning, flooding can mobilize pollutants from nearby wastewater facilities, contaminated sites, or industrial operations, posing a health hazard to residents who come in contact with floodwaters (Quist et al., 2022). Even after floodwaters recede, residents who return to their homes without proper cleanup efforts may be exposed to indoor mold growth, which can exacerbate allergic and respiratory symptoms, particularly for those with pre-existing conditions (Institute of Medicine, 2004). A study of Houston residents conducted approximately one month after Hurricane Harvey found that nearly 80% of participants experienced at least one allergic symptom, such as sinus, eye, or throat irritation, following the storm (Oluyomi et al., 2021).

I.5.3.2 Critical Infrastructure and Facilities

Critical infrastructure and facilities represent the lifelines that support communities with essential services such as water, electricity, and mobility as well as healthcare, emergency services, and education. Flood-related disruptions to these critical assets can cascade through physical and social networks, threatening the well-being of residents, causing local and regional economic impacts, and affecting both the short- and long-term functioning of communities (Chatterjee and Mozumder, 2015). A survey of 500 Houston residents conducted in the aftermath of Hurricane Harvey (Rafi et al., 2024) found that the majority of respondents experienced loss of one or more infrastructure services, including electricity (70% of respondents), water (51%), internet (67%), phone (50%), and transportation access (>70%). On average, disruptions lasted between five and eight days, although some respondents reported service losses that lasted for multiple months. Information from the survey was used to model the macroeconomic impacts of infrastructure disruptions on business interruption losses in Houston, resulting in an estimated $6.7–9.7 billion in losses in the region (Rafi et al., 2024). Figure I.5.4 shows extensive flooding of highway infrastructure in Houston following Hurricane Harvey.

FIGURE I.5.4 Flooding of highway in Houston, Texas, after Hurricane Harvey. (Image by Revolution Messaging.)

The majority of Texas's infrastructure systems were built decades ago and were designed based on historical rainfall and streamflow data. This leaves aging infrastructure vulnerable to more extreme precipitation and flooding, increasing the likelihood of disruptions. A study of future hydraulic conditions at bridge locations in Harris and Travis Counties suggests that urbanization and increases in rainfall intensity will result in higher peak flows, reduced freeboard, and higher flow velocities at stream-crossing bridges (Pervaiz and Hummel, 2023). Such conditions could cause overtopping of bridge decks or exacerbate scour around bridge piers, leading to roadway closures. Increases in high-intensity rainfall also pose a risk to the state's wastewater treatment systems. An analysis of the performance of a wastewater treatment plant in Houston under various synthetic storm scenarios demonstrated a decrease in treatment efficiency and an increase in sewer overflows resulting from events with higher intensity, creating potential threats to human and ecosystem health (Liu et al., 2025).

1.5.3.3 Ecosystems

Flood events influence the functioning of aquatic and riparian ecosystems and the services they provide to society. Smaller floods that occur seasonally in many river systems have been shown to have positive effects on ecosystems functions and services, including increasing primary production, water storage, and recreational opportunities (Junk et al., 1989; Peters et al., 2016). Negative effects that do occur, including soil erosion and water quality impairments, are typically short-lived. In addition, because river systems are accustomed to seasonal flooding, they can absorb and recover from impacts more rapidly (Talbot et al., 2018).

In contrast, extreme floods made worse by urbanization and climate change are argued to have more negative effects across a range of ecosystems services (Talbot et al., 2018), although there is substantial variability in system responses based on watershed size, extent of urbanization, and timing of the event, among other factors. Extreme flooding that inundates large areas of developed land can mobilize a wide range of contaminants, such as fecal bacteria (LaMontagne et al., 2022), metals (Kiaghadi and Rifai, 2019), nutrients (Bende-Michl et al., 2013; Buda and DeWalle, 2009), and total suspended solids (Pizarro et al., 2014). However, the impacts to aquatic species and the services provided through drinking water and recreation depend on the mixing and dilution processes within the watershed, making generalized comparisons difficult.

High water levels and velocities associated with extreme flood events can also influence erosion and sedimentation in riverine systems and affect riparian habitat. Peak flows mobilize significant amounts of sediment, which may be re-deposited elsewhere in the watershed, providing beneficial soil formation on the floodplain or in wetlands, or washed out to receiving waters, resulting in net erosion in the watershed (Talbot et al., 2018). High flows can also cause floodplain stripping, which results when streambed material and riparian vegetation is completely removed. During the 2015 Memorial Day floods, sections of the Blanco River in Central Texas experienced substantial riparian and geomorphic disturbance due to floodplain stripping (Meitzen et al., 2018). Such extensive disturbances create immediate shifts in the habitat along river corridors and initiate the successional processes that determine the composition of the riparian habitat, a process that can take years or decades depending on the habitat that was lost. This impacts not only the types of species that are present in the area but also the aesthetic and recreational values associated with the landscape.

1.5.4 STRATEGIES FOR FLOOD RISK REDUCTION

1.5.4.1 Flood Barriers

Flood barriers include levees and floodwalls and are used to redirect floodwaters away from developed areas to prevent damages (Nunnally et al., 1987). Levees are man-made structures made of earthen materials and are meant to provide temporary protection during periods of high water.

Built along river channels, levees contain high streamflow to the channel by increasing the channel capacity, thus preventing water from flowing into the floodplain. In Texas, there are over 1,300 miles of levees that provide protection for an estimated 1.5 million people (American Society of Civil Engineers, 2025). Floodwalls serve a similar purpose but require less space to implement and are often used in more densely urbanized areas. These structures are made of manufactured materials (e.g., concrete or steel) and typically require deep foundations to provide sufficient stability against elevated water levels (U.S. Army Corps of Engineers, 2022).

In general, levees and floodwalls are an attractive solution in areas with a high concentration of residents and damageable property, where a high degree of protection is desired to prevent losses. These structures are typically designed to prevent flooding from the 100-year flood event, and they have provided substantial reduction in flooding and losses in the past (U.S. Army Corps of Engineers, 2021). However, failure of flood barriers has occurred, notably during Hurricane Harvey in 2017, when multiple levee failures occurred in the Port Arthur levee system in Jefferson County and the Columbia Lakes levee system in Brazoria County (American Society of Civil Engineers, 2025). In the 2025 Infrastructure Report Card issued by the American Society of Civil Engineers, Texas's levees received a D- grade, citing the need for improved levee safety programs that include funding to address breaches and instabilities (American Society of Civil Engineers, 2025). Continued maintenance and upgrades will be needed to Texas's system of flood barriers as infrastructure ages and rainfall intensity increases with climate change.

1.5.4.2 Stormwater Management

Stormwater management approaches, including storm sewers, channel modifications, and flood control reservoirs, are constructed in urban watersheds to help mitigate flood hazards by storing and conveying runoff (Jefferson et al., 2017). Storm sewer systems consist of inlets and pipe networks designed to collect runoff from impervious areas, such as roadways, sidewalks, and rooftops, and discharge it to nearby waterways. Given the costs associated with designing and constructing these systems, their capacities are often based on higher-frequency, lower-magnitude events such as the 25-year event (City of Arlington, 2020). Rainfall magnitudes that exceed the design event can cause backups and overflows, leading to localized pluvial flooding.

Channel modifications include changes to the geometry (width/depth), slope, roughness, and alignment of a natural channel. Increasing the width and/or depth of the channel increases the carrying capacity, while increasing the slope and/or decreasing the roughness (e.g., through vegetation removal or lining with concrete) allow for higher flow velocities (Nunnally, 1978). Taken together, these actions allow more water to be transported in a shorter period of time. Reducing the meanders of a natural channel via straightening similarly increases drainage efficiency while also reducing the spatial footprint of the channel, thus providing more space for other uses.

Despite the flood-reduction benefits of channel modifications, there is wide recognition of the negative environmental impacts associated with such approaches (Keller, 1976; Nunnally, 1978). Removing natural channel vegetation and meanders reduces the residence time of water in the channel, resulting in less water quality treatment via settling and degradation. This is a particular concern for urban runoff, which may contain high suspended sediment concentrations and chemical and biological contaminants from roadways, construction sites, and residential, commercial, and industrial properties. These contaminants are conveyed directly to receiving waterbodies with little or no treatment. In addition, modifications often result in a loss of habitat and natural hydrologic regimes, which can influence the suitability of waterways for aquatic and riparian species (Keller, 1976).

Flood control reservoirs may include small stormwater ponds used to control runoff for a single development or larger dammed reservoirs operated by flood control agencies to manage runoff at the watershed scale. Stormwater ponds may be built as detention ponds, which store water during and immediately after a flood event and then release it to the channel network once peak flows have receded, or retention ponds, which retain water for a longer period of time for aesthetic purposes or other uses (e.g., irrigation) (Jefferson et al., 2017). A combination of small ponds and larger reservoirs can be effective at managing both localized and watershed-scale flood risk.

The 2025 Infrastructure Report Card graded Texas's stormwater infrastructure as a C-, citing several ongoing challenges (American Society of Civil Engineers, 2025). First, many systems were built in the 1970s or earlier and are approaching the end of their design life (typically 50 years). Second, rapid growth and urbanization have led to higher runoff volumes that often exceed the design capacity of existing conveyance systems. Third, updates to the rainfall frequency data used by engineers to design stormwater systems have led to higher rainfall values throughout much of Texas, meaning that systems designed using the previous standards are undersized. These updated rainfall frequency values were published by the National Oceanic and Atmospheric Administration via Atlas 14 (National Oceanic and Atmospheric Administration, 2018), which updated previous estimates from the 1960–70s to include historical data observed in recent decades. The National Oceanic and Atmospheric Administration is expected to publish the next set of updates in 2027 via Atlas 15, which will include projections of rainfall changes due to climate change and will likely lead to even higher rainfall values in many areas. The Texas Flood Infrastructure Fund, established by the 86th Texas Legislature in 2019, provides financial support for communities to implement flood risk reduction strategies, but innovative approaches such as stormwater utility fees may be required to support necessary upgrades and to address funding shortfalls (American Society of Civil Engineers, 2025).

1.5.4.3 Nature-based Solutions

Nature-based solutions present an alternative to engineered stormwater management and flood control systems. These approaches leverage natural or nature-based features to reduce flood risk while simultaneously providing environmental, social, and/or economic benefits (Li et al., 2019; Prudencio and Null, 2018). Nature-based solutions may include site-scale green stormwater infrastructure (GSI), such as permeable pavements, green roofs, rain gardens, and vegetated swales implemented on a single property, or larger-scale approaches such as restoring natural river channels, floodplains, and wetlands to enhance floodwater storage across a watershed (Prudencio and Null, 2018).

Site-scale GSI solutions aim to reduce the amount of rainfall that is converted to runoff on a property. This is accomplished by replacing traditionally impervious surfaces, such as rooftops, driveways, and parking lots, with pervious materials and vegetation that allow for infiltration and evapotranspiration. In some cases, the natural soil may be amended with layers of gravel and sand to increase the infiltration capacity (Li et al., 2019). Collected water enters the groundwater system or is stored in rain barrels or buried tanks for gradual release following rainfall events. In addition to reducing the amount of runoff, these solutions also provide water quality treatment, as vegetation and soil layers remove suspended sediment and chemical and biological contaminants that would otherwise flow offsite. When implemented at high density within a city or county, site-scale GSI can significantly reduce the amount of runoff entering the conventional stormwater conveyance system, helping to alleviate concerns with undersized inlets and pipes while also providing potential cost savings compared to expensive drainage system upgrades (U.S. Environmental Protection Agency, 2015). However, although GSI systems are effective for higher-frequency, moderate-intensity rainfall events, they may not perform as well during extreme precipitation due to size and capacity limitations. In addition, ongoing maintenance is required to ensure that the systems do not become clogged and that they can continue to provide the desired infiltration rates (William et al., 2019).

At the watershed scale, restoration of natural hydrologic features often means rethinking engineered approaches that have been used for decades. For example, unlike traditional channel modification approaches, natural channel design methods focus on achieving stable hydraulic, geomorphic, and ecological functions along a river reach to reduce excessive erosion, restore habitat, and improve water quality (San Antonio River Authority, 2025). Floodplain restoration efforts such as the Dutch "room for the river" concept advocate for the removal of channel-adjacent levees to restore connectivity between the main river channel and the adjacent floodplain (Rijke et al., 2012). Levees can be relocated to the exterior edge of the floodplain, and the floodplain can be converted to natural habitat or open space that can be used for recreational purposes during the majority of the year when the floodplain is not inundated. Restoration of freshwater wetlands lost to development or hydrologic modifications can also enhance the floodwater storage capacity of a watershed, increase hydraulic residence times, and improve water quality (Kiedrzyńska et al., 2015). A major challenge to the restoration of natural hydrologic features is a lack of space due to continued development and conflicting land uses. In many Texas cities, historical decisions to implement flood barriers and channel modifications have led to encroachment of residential, commercial, and industrial land uses in the floodplain, leaving little space for restoration efforts to be implemented.

I.5.4.4 Policy-based Approaches

The above-mentioned approaches to flood risk management all focus on reducing flood hazards by modifying how water is stored and conveyed during a flood event. However, policy-based approaches also play an important role in managing flood risk by reducing exposure and vulnerability through the implementation of laws, ordinances, and codes. Some examples of policy-based approaches include land-use management, building codes, relocation, and insurance (Sebastian and Gori, 2019).

Land-use management approaches focus on controlling development in a watershed to support responsible growth while also protecting greenspace and other associated land uses that provide flood reduction benefits. Such approaches provide the greatest benefits for areas that are currently undergoing urbanization by enabling a more intentional approach to development. In communities where development has already occurred in flood-prone areas, implementation of stronger building codes with floodproofing requirements can reduce exposure. These could include retrofitting requirements for elevating homes above flood levels or using construction materials that are resistant to floodwaters. Relocating individual homes or entire neighborhoods to areas outside of the floodplain also provides significant benefits, especially in areas that are prone to repetitive flooding (Boudreault et al., 2023).

When considering these approaches to reducing flood exposure, it is important to recognize the economic and social considerations that can influence the efficacy and equity of their flood risk reduction outcomes. For example, while the benefits of stronger building codes are well-documented (Jones et al., 2006), the costs associated with bringing existing structures into compliance can be high, posing a significant financial burden for low-income homeowners (Osberghaus, 2021). Grant programs at the federal, state, and local level provide resources for homeowners who wish to implement household mitigation actions, but these funding sources will likely need to be expanded in the future if retrofits are to be implemented at scale. Relocation of homes is also highly controversial, as communities may lose their sense of place (Seebauer and Winkler, 2020). In addition, piecemeal relocation presents a challenge to municipalities as they attempt to address changing land uses while also continuing to provide critical infrastructure services, such as water deliveries and road maintenance, in shrinking neighborhoods (Faust et al., 2016).

When flooding does occur in a community, flood insurance can reduce vulnerability by alleviating the financial burden on homeowners and business owners. Federally backed flood insurance policies are provided through the NFIP coordinated by the Federal Emergency Management Agency (FEMA). Communities must adopt and enforce minimum building standards for flood hazard zones in order for residents to be eligible to purchase insurance through the NFIP (Federal Emergency Management Agency, 2025). Since it was established in 1968, the NFIP has provided billions of dollars in insurance policy payouts to Texans following flood events (Federal Emergency Management Agency, 2024b). However, critics of the program argue that it does not properly account for flood hazards in its pricing, leading to subsidized insurance rates that are not reflective of actual risks

(King, 2011). Recent adjustments to the program, which were fully implemented in April 2023, attempt to better represent actuarial risk in policy pricing. The new pricing approach, known as Risk Rating 2.0, incorporates additional flood risk variables characterizing multiple flood hazards, flood frequency, and property attributes when determining insurance rates (Federal Emergency Management Agency, 2024a).

1.5.4.5 Coordinated Planning Efforts

Although flood risk management is often undertaken at the local level, with specific projects and policies varying from one city or county to the next, the physical processes governing flood hazards occur at the watershed scale and are not restricted to jurisdictional boundaries. Flood management decisions made by upstream communities can influence flood hazards and risk for communities downstream, and vice versa (Ansari et al., 2024). Recognizing the interdependence of communities with respect to flood risk, numerous collaborative efforts have been established across Texas to manage flood risk at the regional or watershed scale. For example, the Trinity River Common Vision initiative, convened by the North Central Texas Council of Governments, was established in the 1980s to facilitate cooperative management and planning along the Trinity River corridor (North Central Texas Council of Governments, 2025). The Common Vision initiative now includes participation from nine cities, four counties, and two special districts in North Texas, as well as TWDB, FEMA, and the U.S. Army Corps of Engineers. More recently, the regional, watershed-based planning approach adopted through the state flood planning effort aims to establish similar partnerships for collaborative planning within each of the 15 major watersheds in Texas. This effort, scheduled for revision every five years, will enable ongoing coordination to understand emerging flood challenges and prioritize impactful flood risk reduction projects at the regional scale (Texas Water Development Board, 2024).

1.5.5 SUMMARY

This chapter summarizes the drivers, impacts, and risk reduction strategies associated with fluvial and pluvial flood hazards in Texas watersheds. Due to its geographic setting near the Gulf of America and unique topographic and geological characteristics, Texas has frequently experienced extreme rainfall and associated flooding, made worse by the expansion of impervious cover due to urbanization. Atmospheric warming will increase the frequency and intensity of extreme rainfall events, further stressing existing flood control infrastructure. Given the severe impacts that flooding can have on residents, property, critical infrastructure, and ecosystems, it is critical that the state develop robust flood mitigation and risk reduction strategies to enhance resilience to future hazards. The development of the first State Flood Plan in 2024 represents a positive step toward reducing the damaging effects of flooding. Continued coordination and planning at the watershed scale will ensure that risk reduction strategies are locally relevant and regionally consistent.

REFERENCES

Adams, D. K., & Comrie, A. C. (1997). The North American monsoon. *Bulletin of the American Meteorological Society*, *78*(10), 2197–2214, https://doi.org/10.1175/1520-0477(1997)078%3C2197:TNAM%3E2.0.CO;2

Alipour, A., Ahmadalipour, A., & Moradkhani, H. (2020). Assessing flash flood hazard and damages in the southeast United States. *Journal of Flood Risk Management*, *13*(2). https://doi.org/10.1111/jfr3.12605

American Society of Civil Engineers. (2025). 2025 Texas Infrastructure Report Card.

Ansari, A. H., Mejia, A., & Cibin, R. (2024). Flood teleconnections from levees undermine disaster resilience. *Npj Natural Hazards*, *1*(1). https://doi.org/10.1038/s44304-024-00002-1

Baker, V. R. (1975). Flood hazards along the Balcones Escarpment in Central Texas: Alternative approaches to their recognition, mapping, and management. Geological Circular 75-5, http://dx.doi.org/10.23867/gc7505D

Becker, E. (2021). The North American Monsoon. https://www.climate.gov/news-features/blogs/enso/north-american-monsoon

Bende-Michl, U., Verburg, K., & Cresswell, H. P. (2013). High-frequency nutrient monitoring to infer seasonal patterns in catchment source availability, mobilisation and delivery. *Environmental Monitoring and Assessment*, *185*(11), 9191–9219. https://doi.org/10.1007/s10661-013-3246-8

Bengtsson, L., Hodges, K. I., & Keenlyside, N. (2009). Will extratropical storms intensify in a warmer climate? *Journal of Climate*, *22*(9), 2276–2301. https://doi.org/10.1175/2008JCLI2678.1

Blake, E. S., & Zelinsky, D. A. (2018). Tropical cyclone report: Hurricane Harvey.

Boudreault, M., Bourdeau-Brien, M., & Milot, N. (2023). Comparison of three flood-related relocation programs with probabilistic cost-benefit analyses. *International Journal of Disaster Risk Reduction*, *96*. https://doi.org/10.1016/j.ijdrr.2023.103950

Bounoua, L., Nigro, J., Zhang, P., Thome, K., & Lachir, A. (2018). Mapping urbanization in the United States from 2001 to 2011. *Applied Geography*, *90*, 123–133. https://doi.org/10.1016/j.apgeog.2017.12.002

Buda, A. R., & DeWalle, D. R. (2009). Dynamics of stream nitrate sources and flow pathways during stormflows on urban, forest and agricultural watersheds in central Pennsylvania, USA. *Hydrological Processes*, *23*(23), 3292–3305. https://doi.org/10.1002/hyp.7423

Chatterjee, C., & Mozumder, P. (2015). Hurricane Wilma, utility disruption, and household wellbeing. *International Journal of Disaster Risk Reduction*, *14*, 395–402. https://doi.org/10.1016/j.ijdrr.2015.09.005

City of Arlington. (2020). Design criteria manual.

Cook, B. I., & Seager, R. (2013). The response of the North American monsoon to increased greenhouse gas forcing. *Journal of Geophysical Research Atmospheres*, *118*(4), 1690–1699. https://doi.org/10.1002/jgrd.50111

Diffenbaugh, N. S., Scherer, M., & Trapp, R. J. (2013). Robust increases in severe thunderstorm environments in response to greenhouse forcing. *Proceedings of the National Academy of Sciences of the United States of America*, *110*(41), 16361–16366. https://doi.org/10.1073/pnas.1307758110

Du, J. (2025). NCEP/EMC 4KM Gridded Data (GRIB) Stage IV Data. https://data.eol.ucar.edu/dataset/21.093

Faust, K. M., Abraham, D. M., & McElmurry, S. P. (2016). Water and wastewater infrastructure management in shrinking cities. *Public Works Management and Policy*, *21*(2), 128–156. https://doi.org/10.1177/1087724X15606737

Federal Emergency Management Agency. (2019). Flood insurance study: Harris County, Texas and incorporated areas.

Federal Emergency Management Agency. (2024a). NFIP's pricing approach. https://www.fema.gov/flood-insurance/risk-rating

Federal Emergency Management Agency. (2024b, September). OpenFEMA dataset: FIMA NFIP redacted claims - v2. https://www.fema.gov/openfema-data-page/fima-nfip-redacted-claims-v2

Federal Emergency Management Agency. (2025). Flood insurance. https://www.fema.gov/flood-insurance

Frame, T., Harrison, G., Hewson, T., & Roberts, N. (2017). Meteorological risk: Extra-tropical cyclones, tropical cyclones and convective storms. In K. Poljansek, M. M. Ferrer, T. De Groeve, & I. Clark (Eds.), *Science for Disaster Risk Management 2017: Knowing Better and Losing Less*. European Commission.

Green, J., Haigh, I. D., Quinn, N., Neal, J., Wahl, T., Wood, M., Eilander, D., de Ruiter, M., Ward, P., & Camus, P. (2025). Review article: A comprehensive review of compound flooding literature with a focus on coastal and estuarine regions. *Natural Hazards and Earth System Sciences*, *25*(2), 747–816. https://doi.org/10.5194/nhess-25-747-2025

Hassanzadeh, P., Lee, C. Y., Nabizadeh, E., Camargo, S. J., Ma, D., & Yeung, L. Y. (2020). Effects of climate change on the movement of future landfalling Texas tropical cyclones. *Nature Communications*, *11*(1). https://doi.org/10.1038/s41467-020-17130-7

Hawcroft, M., Walsh, E., Hodges, K., & Zappa, G. (2018). Significantly increased extreme precipitation expected in Europe and North America from extratropical cyclones. *Environmental Research Letters*, *13*(12). https://doi.org/10.1088/1748-9326/aaed59

Institute of Medicine. (2004). *Damp Indoor Spaces and Health*. National Academies Press.

Intergovernmental Panel on Climate Change. (2021). Climate Change 2021: The Physical Science Basis.

Jefferson, A. J., Bhaskar, A. S., Hopkins, K. G., Fanelli, R., Avellaneda, P. M., & McMillan, S. K. (2017). Stormwater management network effectiveness and implications for urban watershed function: A critical review. *Hydrological Processes*, *31*(23), 4056–4080. https://doi.org/10.1002/hyp.11347

Jones, C. P., Coulbourne, W. L., Marshall, J., Rogers, S. M. C., & Jones, A. (2006). Evaluation of the National Flood Insurance Program's Building Standards. http://www.fema.gov/business/nfip/nfipeval.shtm.

Junk, W. J., Bayley, P. B., & Sparks, R. E. (1989). The flood pulse concept in river-floodplain systems. *Canadian Special Publication of Fisheries and Aquatic Sciences*, *106*(1), 110–127.

Keller, E. A. (1976). Channelization: Environmental, geomorphic, and engineering aspects. In *Geomorphology and Engineering.*, 115–140, https://doi.org/10.4324/9781003028826-7
Kiaghadi, A., & Rifai, H. S. (2019). Physical, chemical, and microbial quality of floodwaters in Houston following Hurricane Harvey. *Environmental Science and Technology*, *53*(9), 4832–4840. https://doi.org/10.1021/acs.est.9b00792
Kiedrzyńska, E., Kiedrzyński, M., & Zalewski, M. (2015). Sustainable floodplain management for flood prevention and water quality improvement. *Natural Hazards*, *76*(2), 955–977. https://doi.org/10.1007/s11069-014-1529-1
King, R. O. (2011). National Flood Insurance Program: Background, Challenges, and Financial Status. Congressional Research Service, https://sgp.fas.org/crs/misc/R40650.pdf
Knutson, T., Camargo, S. J., Chan, J. C. L., Emanuel, K., Ho, C. H., Kossin, J., Mohapatra, M., Satoh, M., Sugi, M., Walsh, K., & Wu, L. (2020). Tropical cyclones and climate change assessment part II: Projected response to anthropogenic warming. *Bulletin of the American Meteorological Society*, *101*(3), E303–E322. https://doi.org/10.1175/BAMS-D-18-0194.1
Kossin, J. P. (2019). Reply to: Moon, I.-J. et al.; Lanzante, J.R. In *Nature* (Vol. 570, Issue 7759, pp. E3–E5). Nature Publishing Group. https://doi.org/10.1038/s41586-019-1222-3
LaMontagne, M. G., Zhang, Y., Guillen, G. J., Gentry, T. J., & Allen, M. S. (2022). Hurricane Harvey impacts on water quality and microbial communities in Houston, TX waterbodies. *Frontiers in Microbiology*, *13*. https://doi.org/10.3389/fmicb.2022.875234
Li, C., Peng, C., Chiang, P. C., Cai, Y., Wang, X., & Yang, Z. (2019). Mechanisms and applications of green infrastructure practices for stormwater control: A review. *Journal of Hydrology*, *568*, 626–637. Elsevier B.V. https://doi.org/10.1016/j.jhydrol.2018.10.074
Liu, L., Morrison, J., Stadler, L., Shaw, A., Vela, J. D., & Christenson, D. (2025). Modeling the resilience performance of Houston's wastewater treatment plant under wet weather conditions. *Journal of Sustainable Water in the Built Environment*, *11*(2). https://doi.org/10.1061/JSWBAY.SWENG-609
Maymandi, N., Hummel, M. A., & Zhang, Y. (2022). Compound coastal, fluvial, and pluvial flooding during historical hurricane events in the Sabine – Neches estuary, Texas. *Water Resources Research*, *58*(12). https://doi.org/10.1029/2022WR033144
Meitzen, K. M., Phillips, J. N., Perkins, T., Manning, A., & Julian, J. P. (2018). Catastrophic flood disturbance and a community's response to plant resilience in the heart of the Texas Hill Country. *Geomorphology*, *305*, 20–32. https://doi.org/10.1016/j.geomorph.2017.09.009
Murakami, H., Wang, Y., Yoshimura, H., Mizuta, R., Sugi, M., Shindo, E., Adachi, Y., Yukimoto, S., Hosaka, M., Kusunoki, S., Ose, T., & Kitoh, A. (2012). Future changes in tropical cyclone activity projected by the new high-resolution MRI-AGCM. *Journal of Climate*, *25*(9), 3237–3260. https://doi.org/10.1175/JCLI-D-11-00415.1
Nadal, N. C., Raúl, Zapata, E., Pagán, I., López, R., & Agudelo, J. (2010). Building damage due to riverine and coastal floods. https://doi.org/10.1061/ASCEWR.1943-5452.0000036
National Oceanic and Atmospheric Administration. (2018). Precipitation-frequency atlas of the United States Volume 11 Version 2.0: Texas.
National Oceanic and Atmospheric Administration. (2025). US billion-dollar weather and climate disasters. https://www.ncei.noaa.gov/access/billions/
National Research Council. (1996). Alluvial fan flooding. In *Alluvial Fan Flooding*. National Academies Press. https://doi.org/10.17226/5364
National Weather Service. (2015). Weather even summary: 2015 Memorial day weekend flooding.
North Central Texas Council of Governments. (2025). Trinity river common vision. https://www.nctcog.org/envir/watershed-management/trinity-river-common-vision
Nunnally, N. R. (1978). Stream renovation: An alternative to channelization. *Environmental Management*, *2*(5), 403–411.
Nunnally, N. R., Shields, F. D., & Hynson, J. (1987). Environmental considerations for levees and floodwalls. *Environmental Management*, *11*, 183–191.
O'Gorman, P. A., & Muller, C. J. (2010). How closely do changes in surface and column water vapor follow Clausius-Clapeyron scaling in climate change simulations? *Environmental Research Letters*, *5*(2). https://doi.org/10.1088/1748-9326/5/2/025207
Oluyomi, A. O., Panthagani, K., Sotelo, J., Gu, X., Armstrong, G., Luo, D. N., Hoffman, K. L., Rohlman, D., Tidwell, L., Hamilton, W. J., Symanski, E., Anderson, K., Petrosino, J. F., Walker, C. L., & Bondy, M. (2021). Houston Hurricane Harvey health (Houston-3H) study: Assessment of allergic symptoms and stress after Hurricane Harvey flooding. *Environmental Health: A Global Access Science Source*, *20*(1). https://doi.org/10.1186/s12940-021-00694-2

Osberghaus, D. (2021). Poorly adapted but nothing to lose? A study on the flood risk – Income relationship with a focus on low-income households. *Climate Risk Management*, *31*. https://doi.org/10.1016/j.crm.2020.100268

Pervaiz, F., & Hummel, M. A. (2023). Effects of climate change and urbanization on bridge flood vulnerability: A regional assessment for Harris County, Texas. *Natural Hazards Review*, *24*(3), 04023025-1 to 04023025-10.

Peters, D. L., Caissie, D., Monk, W. A., Rood, S. B., & St-Hilaire, A. (2016). An ecological perspective on floods in Canada. *Canadian Water Resources Journal*, *41*(1–2), 288–306). Taylor and Francis Ltd. https://doi.org/10.1080/07011784.2015.1070694

Pizarro, J., Vergara, P. M., Morales, J. L., Rodríguez, J. A., & Vila, I. (2014). Influence of land use and climate on the load of suspended solids in catchments of Andean rivers. *Environmental Monitoring and Assessment*, *186*(2), 835–843. https://doi.org/10.1007/s10661-013-3420-z

Prein, A. F., Liu, C., Ikeda, K., Trier, S. B., Rasmussen, R. M., Holland, G. J., & Clark, M. P. (2017). Increased rainfall volume from future convective storms in the US. *Nature Climate Change*, *7*(12), 880–884. https://doi.org/10.1038/s41558-017-0007-7

Prudencio, L., & Null, S. E. (2018). Stormwater management and ecosystem services: A review. In *Environmental Research Letters* (Vol. 13, Issue 3). Institute of Physics Publishing. https://doi.org/10.1088/1748-9326/aaa81a

Quist, A. J. L., Fliss, M. D., Wade, T. J., Delamater, P. L., Richardson, D. B., & Engel, L. S. (2022). Hurricane flooding and acute gastrointestinal illness in North Carolina. *Science of the Total Environment*, *809*, 151108. https://doi.org/10.1016/j.scitotenv.2021.151108

Rafi, S., Meng, S., Santos, J., & Mozumder, P. (2024). Extreme weather events and the performance of critical utility infrastructures: A case study of Hurricane Harvey. *Economics of Disasters and Climate Change*, *8*(1), 33–60. https://doi.org/10.1007/s41885-023-00139-9

Rijke, J., van Herk, S., Zevenbergen, C., & Ashley, R. (2012). Room for the river: Delivering integrated river basin management in the Netherlands. *International Journal of River Basin Management*, *10*(4), 369–382. https://doi.org/10.1080/15715124.2012.739173

Rosenzweig, B. R., McPhillips, L., Chang, H., Cheng, C., Welty, C., Matsler, M., Iwaniec, D., & Davidson, C. I. (2018). Pluvial flood risk and opportunities for resilience. *Wiley Interdisciplinary Reviews: Water*, *5*(6). John Wiley and Sons Inc. https://doi.org/10.1002/wat2.1302

San Antonio River Authority. (2025). Stream restoration. https://www.sariverauthority.org/services/sustainability/stream-restoration/

Schultz, D. M., Bosart, L. F., Colle, B. A., Davies, H. C., Dearden, C., Keyser, D., Martius, O., Roebber, P. J., Steenburgh, W. J., Volkert, H., & Winters, A. C. (2019). Extratropical cyclones: A century of research on Meteorology's centerpiece. *Meteorological Monographs*, *59*, 16.1–16.56. https://doi.org/10.1175/amsmonographs-d-18-0015.1

Schumacher, R. S., & Rasmussen, K. L. (2020). The formation, character and changing nature of mesoscale convective systems. *Nature Reviews Earth and Environment*, *1*(6), 300–314. Springer Nature. https://doi.org/10.1038/s43017-020-0057-7

Sebastian, A., & Gori, A. (2019). Flood policy and risk management in the United States. In P. B. Bedient, W. C. Huber, & B. E. Vieux (Eds.), *Hydrology and Floodplain Analysis* (6th ed.). Pearson.

Seebauer, S., & Winkler, C. (2020). Should I stay or should I go? Factors in household decisions for or against relocation from a flood risk area. *Global Environmental Change*, *60*. https://doi.org/10.1016/j.gloenvcha.2019.102018

Smith, J. M., Cialone, M. A., Wamsley, T. V., & McAlpin, T. O. (2010). Potential impacts of sea level rise on coastal surges in Louisiana. *Ocean Engineering*, *37* (1), 37–47. https://doi.org/10.1016/j.oceaneng.2009.07.008

Stewart, S. (2001). *Tropical Cyclone Report: Tropical Storm Allison.*

Talbot, C. J., Bennett, E. M., Cassell, K., Hanes, D. M., Minor, E. C., Paerl, H., Raymond, P. A., Vargas, R., Vidon, P. G., Wollheim, W., & Xenopoulos, M. A. (2018). The impact of flooding on aquatic ecosystem services. *Biogeochemistry*, *141*(3), 439–461. https://doi.org/10.1007/s10533-018-0449-7

Texas Demographic Center. (2024). Texas population projection program. https://demographics.texas.gov/Projections/2024/

Texas Water Development Board. (2012). Water for Texas.

Texas Water Development Board. (2024). 2024 State Flood Plan. www.twdb.texas.gov

US Army Corps of Engineers. (2021). Value to the Nation: Flood risk management fast facts.

US Army Corps of Engineers. (2022). Engineering and design: Floodwalls and other hydraulic retaining walls. https://www.iwr.usace.army.mil/Missions/Value-to-the-Nation/Fast-Facts/Flood-Risk-Management-Fast-Facts/

US Census Bureau. (2024). Net International Migration Drives Highest US Population Growth in Decades. https://www.census.gov/newsroom/press-releases/2024/population-estimates-international-migration.html

US Environmental Protection Agency. (2015). Flood loss avoidance benefits of green infrastructure for stormwater management. https://www.epa.gov/sites/default/files/2016-05/documents/flood-avoidance-green-infrastructure-12-14-2015.pdf

Valle-Levinson, A., Olabarrieta, M., & Heilman, L. (2020). Compound flooding in Houston-Galveston Bay during Hurricane Harvey. *Science of the Total Environment, 747.* https://doi.org/10.1016/j.scitotenv.2020.141272

Wahl, T., Jain, S., Bender, J. *et al.* (2015). Increasing risk of compound flooding from storm surge and rainfall for major US cities. *Nature Climate Change*, *5*, 1093–1097. https://doi.org/10.1038/nclimate2736

Wehner, M. F., Reed, K. A., Loring, B., Stone, D., & Krishnan, H. (2018). Changes in tropical cyclones under stabilized 1.5 and 2.0°C global warming scenarios as simulated by the community atmospheric model under the HAPPI protocols. *Earth System Dynamics*, *9*(1), 187–195. https://doi.org/10.5194/esd-9-187-2018

William, R., Gardoni, P., & Stillwell, A. S. (2019). Reliability-based approach to investigating long-term clogging in green stormwater infrastructure. *Journal of Sustainable Water in the Built Environment*, *5*(1). https://doi.org/10.1061/jswbay.0000875

Yamaguchi, M., Chan, J. C. L., Moon, I. J., Yoshida, K., & Mizuta, R. (2020). Global warming changes tropical cyclone translation speed. *Nature Communications*, *11*(1). https://doi.org/10.1038/s41467-019-13902-y

Yang, L., Smith, J. A., Baeck, M. L., & Zhang, Y. (2016). Flash flooding in small urban watersheds: Storm event hydrologic response. *Water Resources Research*, *52*(6), 4571–4589. https://doi.org/10.1002/2015WR018326

Zhang, W., Villarini, G., Vecchi, G. A., & Smith, J. A. (2018). Urbanization exacerbated the rainfall and flooding caused by Hurricane Harvey in Houston. *Nature*, *563*(7731), 384–388. https://doi.org/10.1038/s41586-018-0676-z

Zhang, Y. J., Ye, F., Yu, H. *et al.* (2020). Simulating compound flooding events in a hurricane. *Ocean Dynamics*, *70*, 621–640. https://doi.org/10.1007/s10236-020-01351-x

Zhang, Z., & Colle, B. A. (2017). Changes in extratropical cyclone precipitation and associated processes during the twenty-first century over eastern North America and the western Atlantic using a cyclone-relative approach. *Journal of Climate*, *30*(21), 8633–8656. https://doi.org/10.1175/JCLI-D-16-0906.1

Zhao, G., Gao, H., & Cuo, L. (2016). Effects of urbanization and climate change on peak flows over the San Antonio River basin, Texas. *Journal of Hydrometeorology*, *17*(9), 2371–2389. https://doi.org/10.1175/JHM-D-15-0216.1

Zhu, L., & Quiring, S. M. (2013). Variations in tropical cyclone precipitation in Texas (1950 to 2009). *Journal of Geophysical Research Atmospheres*, *118*(8), 3085–3096). Blackwell Publishing Ltd. https://doi.org/10.1029/2012JD018554

Zscheischler, J., Martius, O., Westra, S., Bevacqua, E., Raymond, C., Horton, R. M., van den Hurk, B., AghaKouchak, A., Jézéquel, A., Mahecha, M. D., Maraun, D., Ramos, A. M., Ridder, N. N., Thiery, W., & Vignotto, E. (2020). A typology of compound weather and climate events. *Nature Reviews Earth & Environment*, *1*(7), 333–347. https://doi.org/10.1038/s43017-020-0060-z

Section II

Impacts of Climate Change on Natural Ecosystems

II.1 Extending the Planetary Boundaries Framework to Texas

Are We Beyond the Safe Operating Space?

Ryan A. McManamay

II.1.1 INTRODUCTION

Addressing climate change and other anthropogenic pressures on Earth's system requires not only appropriate scientific understanding, but also accurate communication of scientific findings to the public. This is especially true for communicating the potential impacts of climate change on Earth's ecosystems and the services derived from those ecosystems – both notoriously complex to study and understand. In his keynote address at the Austrian World Summit in 2018, United Nations Secretary-General António Guterres said: "Climate change is, quite simply, an existential threat for most life on the planet – including, and especially, the life of humankind" (UN, 2018). Since that time, climate activists have used the terms 'existential,' 'crisis,' and 'urgency' to describe the threats of climate change to modern society to mobilize government action and climate mitigation policies (Han and Ahn, 2020). There are, however, discrepancies in the adoption of this terminology between the scientific community and the public (Huggel et al., 2022). The language used to communicate climate change is critical, as it has not only led to confusion but also increasing skepticism by the public (Pasquini et al., 2023). The assertion that climate change is an existential risk, presumably to the continuation of current society and natural systems, has not been adequately defined or analyzed within the scientific literature (Huggel et al., 2022). Existential risks are synonymous with global catastrophic risks (GCRs), natural or human-caused, such as an asteroid impact or nuclear war (Bostrom, 2013) – these threaten the entirety of human civilization (Baum and Barrett, 2018). Avin et al. (2018) provides a classification for GCRs based on three components: (1) a critical system whose safety boundaries have been breached; (2) the mechanisms by which the threat spreads globally and affects most of the human population; and (3) the failure to prevent or mitigate the risk. Hence, to characterize climate change as a GCR appropriately would suggest that all three components have been adequately studied and evaluated. Alternatively, the potential pitfalls for iterative and endless clarification over the threats of climate change could lead to continued 'pluralistic ignorance' (Andre et al., 2024) or further 'spirals of inaction' (Bouman and Steg, 2022).

To clarify, I am not suggesting that climate change fails to meet the criteria of a GCR, but rather, there is a need to rapidly evaluate, synthesize, and consistently communicate such criteria and how it relates to human livelihoods. As one example, the World Health Organization reports that from 2030 to 2050 an additional 250,000 deaths each year will stem from climate-change related illness or stress (WHO, 2025). While this estimate seems large, climate-related deaths are projected to be only 0.39% of the current global mortality rate (64 million deaths yr^{-1}). In comparison, environmental pollution is estimated to cause 9 million deaths yr^{-1}, the vast majority (6.7 million) from

DOI: 10.1201/9781003406600-8

air pollution, and, to a lesser extent, water pollution (1.3 million) (Fuller et al., 2022; WHO, 2024). Addressing air pollution alone could reduce global mortality by 10%, an additional justification for reducing the byproducts of fossil combustion, particularly particulate matter and aerosols. By comparison, human displacement from climate change appears to be a much larger issue. By 2050, as many as 216 million people could be displaced, exhibiting bilateral (domestic) migration away from areas affected by sea level rise or storm surge (Clement et al., 2021). Even so, this displacement represents 2% of the world's potential inhabitants by 2050 (9 billion people).

There are a couple of challenges that arise with synonymizing climate change with GCRs, at least without appropriate characterization. The first lies in the misinterpretation of time and spatial scales of climate change. An asteroid strike, for instance, operates over immediate time frames and comprehensive spatial scales, whereas climate change impacts are locally heterogenous and operate on far longer time scales (Huggel et al., 2022). Another challenge is that this seemingly endorses a myopic view of climate change as a singular and independent risk, inadvertently undermining the necessity of understanding Earth as an inherently complex and interdependent system of systems. Unlike an asteroid strike, the impacts of climate change on humans are interdependent with the anthropogenic impacts to the biosphere, geosphere, and hydrosphere. Unfortunately, many global environmental problems (e.g., pollution, climate change) are treated as independent issues rather than simultaneously considering the aggregate and interactive effects of these perturbations on the overall state on the Earth system (Richardson et al., 2023). Indeed, Earth's bio-, geo-, and hydrospheric processes are coupled and co-dynamic. Hence, without an appropriate understanding and diagnosis of climate change, attempts to fix the problem will not only be misguided but, in fact, could exacerbate the problem. For instance, Caro et al. (2022) note a common misconception among scientists and the media, i.e., that climate change is the leading driver of biodiversity loss. Not only is this unsupported by compilations of the global threats to fauna by conservation scientists, but this also detracts from efforts needed to address the primary threats to biodiversity, specifically habitat loss (from landscape alteration) and overexploitation (Caro et al., 2022). Indeed, a myopic view of climate change could lead to uninformed climate mitigation policies, such as unconstrained renewable energy development, only continuing to expand humanity's land footprint, leading to accelerated habitat destruction and increasing biodiversity loss.

Ultimately, to solve the climate problem, we cannot separate Earth's climate system, ideologically or scientifically, from other spheres of the planet. A potential solution to this issue lies in the communication strategy underlying the Planetary Boundaries Framework (PBF), which defines a safe operating space for humanity within Earth's biophysical systems (Rockström et al., 2009a) (Figure II.1.1). The Earth's safe operating space is represented by nine boundaries, each representing biophysical components of Earth's system affected by human activities and pressures. Rather than climate change being siloed as 'the' existential threat, the PBF jointly considers the aggregate influence of all nine planetary systems, including the climate system, on Earth's overall state (Richardson et al., 2023). Specifically, the PBF portrays Earth as a complex geosphere-biosphere system, characterized by non-linear interactions within and among subsystems where thresholds, or tipping points, if exceeded, lead to dramatic reductions in resilience and stability (Rockström et al., 2024). While interactions among subsystems are not formally evaluated by the PBF, the purposeful selection of each biophysical system and boundaries are aimed at encompassing processes and interactions that lead to a holistic view of interdependence and non-linear interactions that maintain Earth's entire systems stability (Richardson et al., 2023).

As a case in point, the integrity of one of the nine planetary processes characterized by the PBF, the biosphere, is measured using two control variables: genetic diversity and functional integrity (Figure II.1.1). Human appropriation of net primary production (HANPP) is used as a surrogate of functional biospheric integrity since net primary production (NPP) serves as the basis for energy and material flows to support living organisms (Richardson et al., 2023). Currently, 30% of Earth's NPP has been appropriated by humans, primarily for agricultural purposes, whereas Richardson et al. propose 10% HANPP as a safe planetary boundary, suggesting that

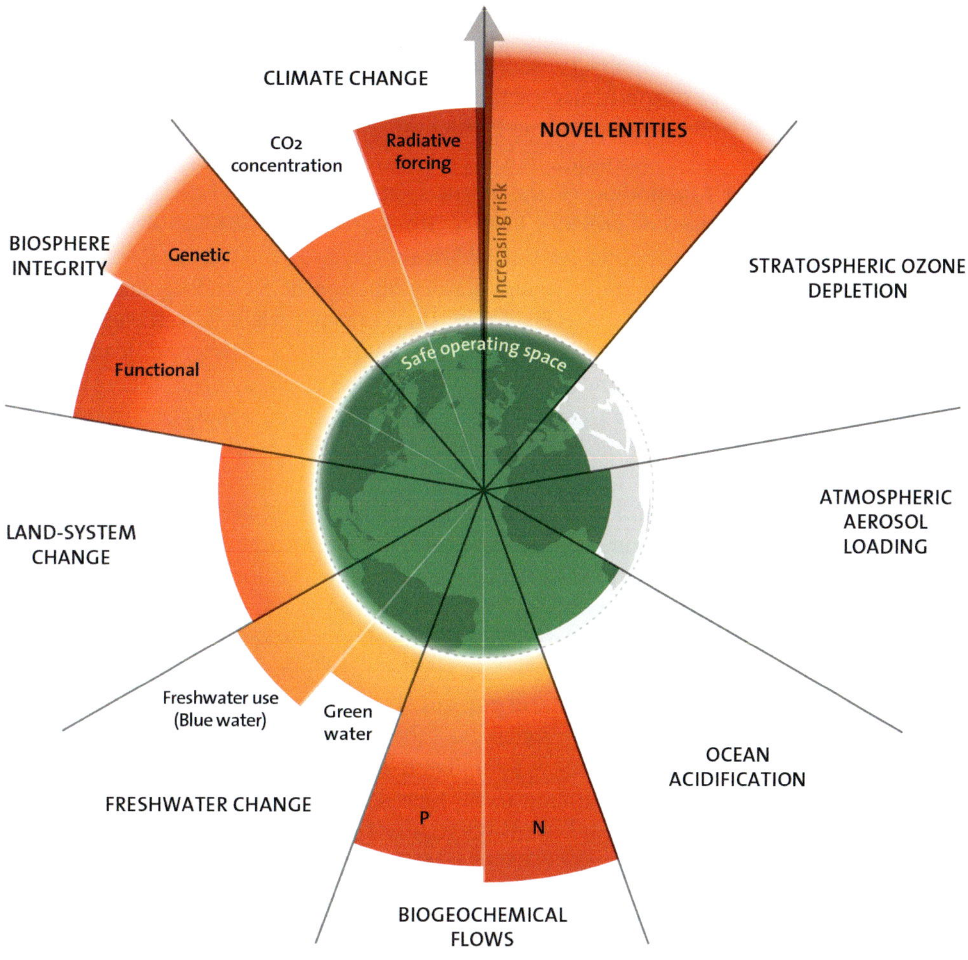

FIGURE II.1.1 The planetary boundaries framework with most recent assessment in 2023 including six of the nine earth system processes. Licensed under CC BY-NC-ND 3.0. (Credit: Azote for Stockholm Resilience Centre, based on analysis in Richardson et al. (2023).)

human society transgressed this boundary well over a century ago. Earth's NPP has been directly manipulated by humans through harvesting or landscape alteration, completely independent from human-induced climate change; therefore, intentional mitigation of impaired biospheric functioning, particularly losses in biodiversity from HANPP, are most acutely achieved through reforestation and land management, rather than fossil fuel reduction. Nonetheless, climate change and NPP are tightly coupled. For instance, loss of NPP has interactive consequences on climate change by reducing Earth's CO_2 sequestration capacity and altering global hydrologic cycles, via altered evapotranspiration, important in mediating global temperatures. Subsequently, climate alteration further accelerates functional losses through feedbacks on remaining intact and disrupted ecosystems.

In this chapter, the PBF framework is presented and briefly discussed, first because it provides a valuable communication tool to expand a myopic view of climate change, and second, it introduces the subsequent chapters in this section that cover some of the impacts to natural systems in Texas. Through some justified assumptions and calculations, the PBF framework is extended to quantify boundaries in Texas.

II.1.2 THE PLANETARY BOUNDARY FRAMEWORK

Rockström et al. (2009a, b) originally presented planetary boundaries as a concept to promote the maintenance (or return) to 'safe limits' of the influence of human activities on nine processes that regulate the stability and resilience of the Earth system (Stockholm Resilience Center, 2025). The boundaries are not just conceptual, but quantitative, and represent the appropriate pre-industrialization state and functioning of the Earth system within the Holocene, a 10,000-year period of unusual stability based on previous history (Rockström et al., 2009a). Boundaries are represented by thresholds or tipping points, beyond which abrupt, and even irreversible, environmental changes can take place, leading to new system states with significant consequences for humans. Each of the thresholds is defined by critical values for 'control variables,' selected because they represent Earth-system processes and have related thresholds that, if surpassed, lead to unacceptable environmental change (Rockström et al., 2009a). The nine Earth-system processes of the PBF include: climate change, novel entities, stratospheric ozone depletion, atmospheric aerosol loading, ocean acidification, modification of biogeochemical flows, freshwater change, land system change, and biosphere integrity (Figure II.1.1).

In the original conceptualization of the PBF, seven of the nine planetary boundaries had proposed approaches for quantification, i.e., control variables, and, based on evidence, only three of the boundaries (climate change, biodiversity loss, and nitrogen flows) were considered transgressed by human society (Rockström et al., 2009a, b) (Figure II.1.1). Subsequently, the PBF was holistically revisited twice more in 2015 and 2024, each time re-assessing representative control variables and boundaries with new datasets and methods (Figure II.1.1). Steffen et al. (2015) documented that four boundaries were transgressed, including land change. The authors also reported updates to the quantification approach used for control variables, i.e., a tiered approach, for each boundary. Biospheric integrity was updated to include both genetic and functional diversity, where genetic diversity continued to be represented by the extinction rate and where functional diversity was represented by the Biodiversity Intactness Index (Sanchez-Ortiz et al., 2019a). Additionally, sufficient evidence existed to show that human alteration of global phosphorus cycles, in addition to nitrogen, had exceed biogeochemical flow boundaries. New global assessments documented the remaining planetary processes, including novel entities (Persson et al., 2022), but also filled gaps in freshwater availability, particularly green water (Porkka et al., 2024; Wang-Erlandsson et al., 2022). Through a synthesis of literature and global data products, Persson et al. (2022) concluded that the boundary for novel entities had been transgressed, specifically that the production rate of new chemical products exceeded society's ability to conduct monitoring and safety assessments. In contrast to previous assessments relying on water withdrawal impacts to freshwater resources, Porkka et al. (2024) conducted a novel global and spatially explicit assessment estimating non-temporary departures in streamflow (blue water) and soil moisture (green water) from pre-industrial conditions, concluding that the freshwater change boundary had also been transgressed. Richardson et al. (2023) summarize these latest findings, suggesting that six of nine planetary boundaries have now been transgressed by human society (Figure II.1.1).

Obviously, climate change is only one of nine planetary systems evaluated by the PBF. Climate change is portrayed as having a role equal with other processes, rather than serving a governing force on other systems. As Rockström et al. (2009a) indicate: "Although the planetary boundaries are described in terms of individual quantities and separate processes, the boundaries are tightly coupled. We do not have the luxury of concentrating our efforts on any one of them in isolation from the others." Crossing one boundary could affect others, even pushing them others out of safe operating space, suggesting that planetary boundaries cannot be addressed separately (Stockholm Resilience Center, 2025). This, of course, challenges a parochial vision of prioritizing some Earth processes over others, particularly the notion that addressing climate change will resolve other planetary boundary exceedances. Existential threats exceeding the boundaries of Earth's system are complex, multifaceted, and rarely singular in nature. Ensuring a resilient Earth system suggests that the safe operating space requires maintenance of all processes within their respective boundaries.

One example of the importance of co-maintenance among boundaries is the preservation of freshwater resources, which are divided into blue and green sources (Porkka et al., 2024). Blue water refers to water in streams, lakes, groundwater, and glaciers, whereas green water refers to the water falling as precipitation, evaporating or contributing to soil moisture, eventually being used by plants (Wang-Erlandsson et al., 2022). Major alterations to Earth's green water can lead to direct or indirect cascading impacts to at least six other planetary processes, including the climate system, which subsequently creates feedbacks to other cycles. For example, deforestation and land cover alterations can result in soil drying and erosion, reducing soil moisture, leading to plant physiological stress and the release of biogenic volatile organic compounds (aerosols). Degraded soil conditions reduce primary productivity, which, in turn, reduces climate regulation by dampening CO_2 sequestration and evaporative cooling. Climate stress in conjunction with losses in primary productivity also impact the biosphere, by reducing biomass and energy flows for supporting biodiversity. Reductions in precipitation can further reduce primary production, whereas extreme rainfall can lead to impairments to biogeochemical cycles via loss of nutrients through erosion or leaching. Shifts in vegetation and soil conditions alter hydrologic cycles, influencing subsequent blue water availability for human use.

The selection of control variables is equally important to defining the boundaries; this requires careful consideration of parameters (or close surrogates) that govern the state of planetary systems and subsystems that, if sufficiently altered, can potentially cause irreversible changes to Earth's system stability (Richardson et al., 2023; Steffen et al., 2015). The identification of these variables also reflects the need to quantify boundaries and thresholds associated with each variable, where sufficient evidence is collected to understand how much perturbations place planetary systems at risk (Steffen et al., 2015). To suggest numerical boundaries, especially in comparison to Holocene conditions, requires considerable effort, as documented in the papers cited above and their associated supplements. More accurately, these numerical boundaries represent the safe operating space, where values for control variables are a safe distance from actual thresholds, which are completely uncertain or non-existent if planetary systems exhibit less dramatic, non-linear responses (Rockström et al., 2009b; Steffen et al., 2015).

11.1.3 ARE PLANETARY BOUNDARIES BOGUS, REAL, OR JUST ELUSIVE?

Of course, with any new provocative framework, there will be criticism and caution regarding universal application. Montoya et al. (2018) suggest that the PBF is vague, treating Earth's processes as a 'black box,' but also suggests the PBF promotes 'harmful policies' to biodiversity protection, primarily because the notion of planetary boundaries adds little insight into measures for sustaining biodiversity or ecosystem functioning. The Editorial (Nature, 2009), published in the same issue as the induction of the PBF (Rockström et al., 2009a), presents a good overview of the limitations, summarizing commentaries by seven leading experts. The noted limitations of the PBF are centralized, as expected, around the boundaries or thresholds:

- Boundaries are arbitrary relative to uncertainty among scientific evidence.
- Boundaries do not always apply globally, even for processes that regulate the planet.
- Setting 'acceptable' limits to support continued human existence could undermine the resources needed for supporting humanity (e.g., advent of the green revolution by way of industrial N fixation).
- Boundaries should not be interpreted as targets, as environmental rules will be broken.

Part of the criticism arises from the uncertain evidence for environmental thresholds in scientific literature. In a global meta-analysis of over 4,600 environmental impacts on natural communities,

Hillebrand et al. (2020) found that thresholds were rarely detectable, suggesting that thresholds, if present, could be too elusive to prove useful to management (Dudney and Suding, 2020). Likewise, Capon et al. (2015) found little evidence of consistent regime-shifts or abrupt non-linear changes in aquatic ecosystems under anthropogenic pressures. Others, however, have clearly documented thresholds for ecosystems. For example, Berdugo et al. (2020) provide evidence for multiple thresholds indicative of abrupt global ecosystem shifts arising from increasing aridity. The authors further suggest that 20% of Earth's land surface may cross these boundaries leading to land degradation and desertification. Specific to Texas, Lueders et al. (2025) show that thresholds to hydrologic alteration in river systems, although elusive, are present but have been previously undetected in many instances.

While there are clear limitations of the PBF, the importance of the framework, also noted in the Editorial, is an attempt to quantify and communicate limitations to planet Earth. Attempts to document the Earth's limits to sustaining the ever-growing demands of human society are not unprecedented. For instance, the 'Carrying Capacity' concept was developed as early as the late 19th century and was extended to game populations in the early 20th century (Sayre, 2008). The 'law of population growth' and carrying capacities (Lotka, 1925) have been a staple components of ecology texts for some time. Verhulst (1838) originally proposed what is now known as the 'logistic growth curve' to explain human population changes in France, Belgium, and England. The growth curve is a differential equation to model logistic population growth where the rate of changes in populations slow as they approach a given capacity limit (eq 1):

$$\frac{dP}{dt} = rP\left(1 - \frac{P}{K}\right)$$

Changes in population (dP) over time (dt) are dependent upon population growth rate (r) and a carrying capacity (K), which serves as a limit to a population's size based on existing space or resources. The equation can be rearranged to provide an instantaneous or analytic estimate of population at time t (eq 2):

$$P(t) = \frac{P_0 K e^{rt}}{(K - P_0) + P_0 e^{rt}}$$

Pearl and Reed (1920) infamously rediscovered the logistic growth curve, applying it to the U.S. human population to project a value for K, which, at that time, was roughly estimated at 200 million people based on existing agriculture production rates and per capita caloric needs. The curve very accurately modeled U.S. human population until 1950, afterward grossly underestimating continued growth due to their estimate of K. To be fair, Pearl and Reed noted that the estimates could change if radical shifts occurred in either Americans' food habitats or agriculture production. Of course, that is exactly what happened. Technological innovations (e.g., industrial fertilizer) within the Green Revolution dramatically increased the production-per-unit-land area for agriculture and led to exponential growth in worldwide commodity exchange, significant global population growth, and mortality decline (Gollin et al., 2021). Even so, the physics of a carrying capacity, albeit subject to resource efficiencies, are still as relevant today as they were 100 years ago and are occasionally revisited (e.g., Cohen, 1995).

Regardless of the peculiar notions of single 'control variables' governing the planet's safe operating boundaries, human society still benefits from an understanding, or at least a reminder, of the limitations of our planet. Just because we cannot easily detect thresholds, should we assume none have been crossed? Does Earth not have limits? Should we continue the status quo of ever-increasing extraction, manipulation, and alteration of natural processes on Earth until thresholds become fully realized? And if so, should we expect the continued provision of ecological services from those processes – clean water, food production, and mental well-being? Even if we cannot accurately measure thresholds, the PBF suggests that a precautionary approach is best. For instance, according to the PBF, some boundaries have been crossed for a long time, specifically early post-industrialization. Could it be that we just haven't seen the full consequences yet?

II.1.4 EXTENDING THE PLANETARY BOUNDARIES FRAMEWORK TO TEXAS

Although the Planetary Boundaries is an attractive concept, how to operationalize the 'safe operating space' is not immediately evident. Steffen et al. (2015) suggest that the PBF was not designed to be 'downscaled' or 'disaggregated' to smaller units, such as nations or local communities, although the authors endorse localized 'PB thinking.' It seems reasonable that nations and even local entities would desire to localize the PBF framework to adopt policies to reduce their contribution to the global environmental footprint, since policies are developed at the regional scale. Indeed, there are numerous examples of downscaling the PBF to regions (Häyhä et al., 2018) or extending the PBF to include social dimensions (Ferretto et al., 2022; Meyer and Newman, 2018; O'Neill et al., 2018). O'Neill et al. (2018) downscale the PBF to the national level and to demonstrate disconnects between quality of life and planetary boundary exceedances. Dearing et al. (2014) merge the PBF with the social doughnut framework (Raworth, 2012) to extend planetary boundaries to environmental 'ceilings' for two rural communities in China. The authors use sediment cores representing decades of information as proxies of long-term air, water, and sediment quality. By constructing their own regionally explicit baseline of safe operating space, they demonstrate that air, water, and sediment quality have exceeded environmental ceilings in at least one of the provinces, despite not all social targets being met. Kronenberg et al. (2024) suggest cities should adopt degrowth polices by showing that planetary boundaries are directly connected to at least one critical component of urban areas, including production/consumption, environmental disconnections, and novel environments and entities.

Here, an overall assessment of the state of environmental conditions in Texas is presented as a localized extension of the PBF. Ideologically, the extension is approached from the standpoint of Texas playing a respective role in reducing its own environmental footprint collectively to move the planet back into safe operating space. There are noted limitations, the largest of which include the arbitrary extension of universal control variables to specific regions and making no consideration of the social implications of meeting or exceeding thresholds. Additionally, Rockström et al. (2009a) and Steffen et al. (2015) both suggest that boundaries will vary regionally. Even so, the exercise is valuable as extending the PBF to Texas provides a holistic accounting of environmental challenges and the cumulative state of transgression – including but in addition to climate change. This is important in communicating that Texas, like the remainder of planet Earth, has its limits.

The approaches used to estimate current environmental conditions and extending planetary boundaries to Texas are described in Table II.1.1. Three of the nine Earth system processes were not assessed based on available data or inability to isolate the role of individual regions on a global process: aerosol loading, stratospheric ozone depletion, and novel entities. For the processes that were evaluated, all control variables presented in Richardson et al. (2023) were directly used or minimally modified, except the climate change variables, i.e., global CO_2 concentration and radiative forcing. Current environmental conditions were obtained by surveying the literature or by obtaining regional or global datasets (e.g., the Biodiversity Intactness Index) for spatial summarization within Texas's boundaries (Figure II.1.2). Planetary boundaries were extended to Texas using numerous approaches, depending on the control variable. Generally, control variables that represented universal values and thresholds (e.g., HANPP, Ocean carbonate ion concentration) were directly applied as boundaries in Texas. In other cases, planetary boundaries provided more relevant regional estimates (e.g., temperate biomes) that were used over global estimates. Any control variables presented as global rates (e.g., Phosphorus flows to oceans Tg yr^{-1}) were converted to a per-unit-area estimate and then applied to Texas based on applicable area.

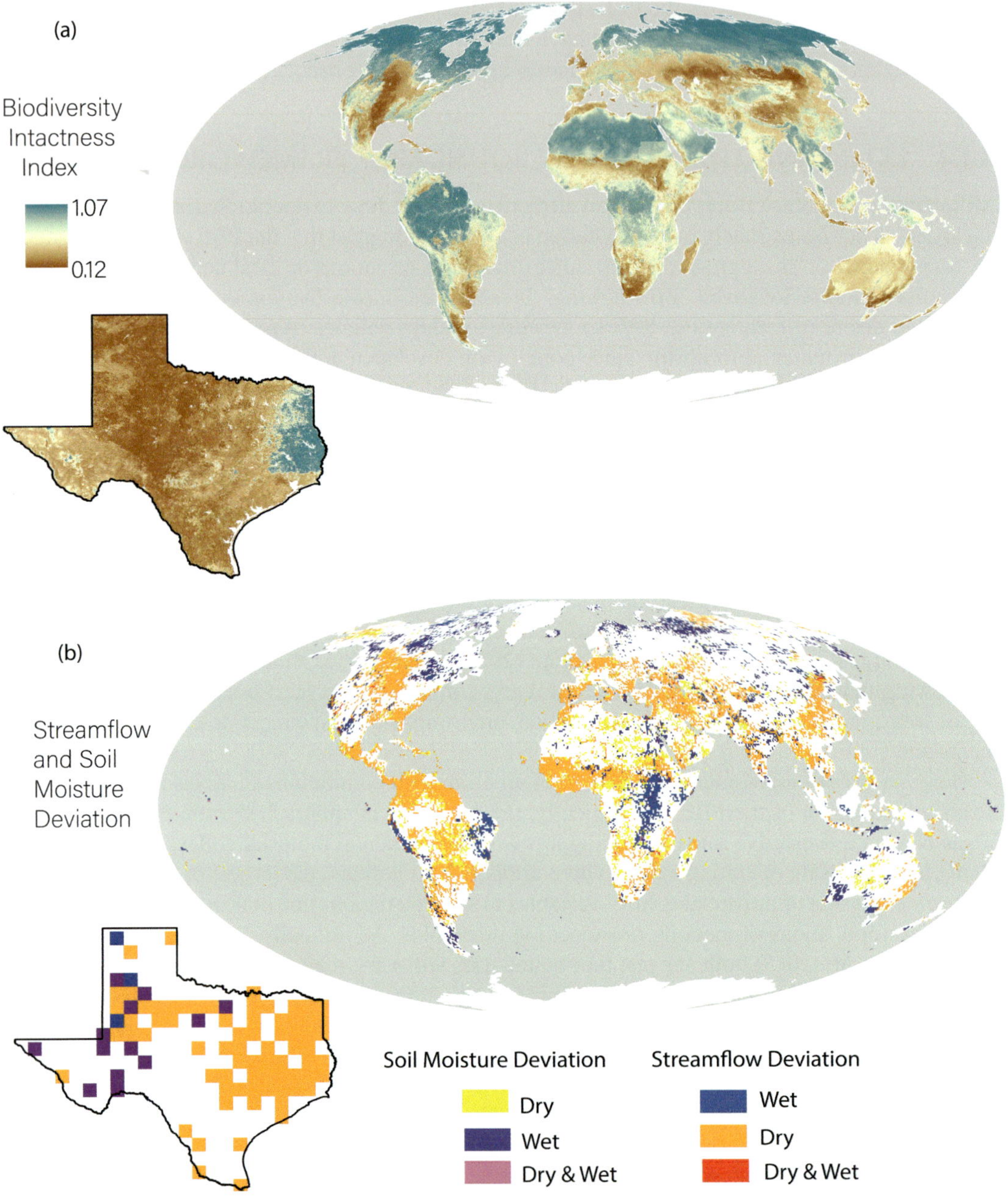

FIGURE II.1.2 Examples of summarizing global control variables in Texas including (a) Biodiversity Intactness Index (BII), a measure of the average abundance and richness of organisms in a given area relative to their reference populations (data taken and remapped from Sanchez-Ortiz et al., 2019b), and (b) global deviations in blue water, i.e., persistent streamflow deviations from pre-industrial conditions, at a fine spatial scale (data obtained and remapped from Porkka et al., 2024).

The current environmental conditions relative to the downscaled planetary boundaries are reported in Table II.1.2. Ultimately, the results suggest that, assuming direct extensions of the PBF, Texas has exceeded five of the six planetary boundaries that were assessed (Figure II.1.3). The assessment for Texas is generally congruent with O'Neill et al.'s (2018) extension of the PBF for the remainder of the U.S. Ocean acidification is the only boundary not transgressed based on

TABLE II.1.1
Methods Used to Extend Planetary Boundaries to Texas, Including Documenting Current Environmental Conditions and Estimating Planetary Boundaries for Control Variables

Earth System Process	Control Variable	Description for Extension and Estimation in Texas
Climate change	CO_2 emissions ($GtCO_2$ yr^{-1})	To achieve a 350-ppm atmospheric CO_2 condition (pre-industrialization), global CO_2 emissions (currently 37 Gt CO_2 yr^{-1}) are estimated to need to reach 1-2 Gt CO_2 yr^{-1} by 2050 (Riahi et al., 2017). Considering the global human inhabited/modified area (107 million km^2) contributing to CO_2 emissions, a per-unit-area emission rate was applied to Texas's human modified area (172,000 km^2). Texas CO_2 emissions documented by TCEQ (2024). Human modified areas estimated by Winkler et al. (2021) and Theobald et al. (2025).
Change in biosphere integrity	Genetic diversity: E/MSY	Extinction rate in Texas includes global extinctions and local extirpations. Based on available records, timeline considered 1800 to present (225 years). Eight mammals (Schmidly and Bradley, 2016), 11 fish (Birdsong et al., 2021), 9 plants (Knapp et al., 2021), and 3 birds (Casto, 2002) are extirpated or globally extinct in Texas.
	Functional integrity: %HANPP	Paudel et al. (2023) provides product-specific human appropriation of NPP in US counties. Average percentages were calculated for Texas. As a comparison, Graff (2010) estimates are very close to Paudel et al. (2023).
	Biodiversity Intactness	Global maps of Biodiversity Intactness Index for richness and abundance were obtained from Sanchez-Ortiz et al. (2019b) and averaged for Texas.
Ocean acidification	Carbonate ion conc: Ωarag	Values for aragonite saturation states for Texas estuaries were obtained from Hu et al. (2022).
Biogeo-chemical flows: P and N cycles	P flow from freshwater systems into the ocean (Tg P yr^{-1})	P loading estimates for all major Texas estuaries (Marshall and Montagna, 2025) were summed to estimate total loading (yr^{-1}). Assuming 83% of world's entire terrestrial surface (123 million km^2) is a contributing drainage area to the oceans, the planetary boundary condition of 11 Tg yr^{-1} was calculated on a per-unit-area basis (km^{-2}). This rate was applied to the drainage area for the Texas-Gulf region (471,000 km^2) to determine P flow boundary for Texas.
	P fertilizers applied to soils (Tg P yr^{-1})	Fertilizer application rates (N and P) for farm and urban areas in US counties (Falcone, 2021) were totaled for Texas agriculture and urban lands (192,000 km^2). Planetary boundary estimate of 6.2 Tg yr^{-1} was converted to per-unit area, assuming 48 million km^2 of global agricultural lands. Boundary applied to Texas agriculture land (151,000 km^2) to determine TX P fertilizer boundary.
	Nitrogen industrial fixation of N (Tg N yr^{-1})	It is assumed that N fixation approximates the application rate of N fertilizer. Same methods are used for N as with P fertilizer estimates, except total global land and total land in Texas was used because boundary estimates are for N fixation.
Land system change	% forests	Estimate is based on % of current forest area relative to historical (1873) forested area, estimated by Liknes et al. (2013). Current deciduous, evergreen, and mixed forest coverage in Texas from National Land Cover Database.
Freshwater change	Human disturbance of blue water	Land areas with persistent deviations in streamflow from preindustrial periods (Porkka et al., 2024) were summarized as % of total area in Texas.
	Human disturbance of green water	Land areas with persistent deviations in soil moisture from preindustrial periods (Porkka et al., 2024) were summarized as % of total area in Texas.

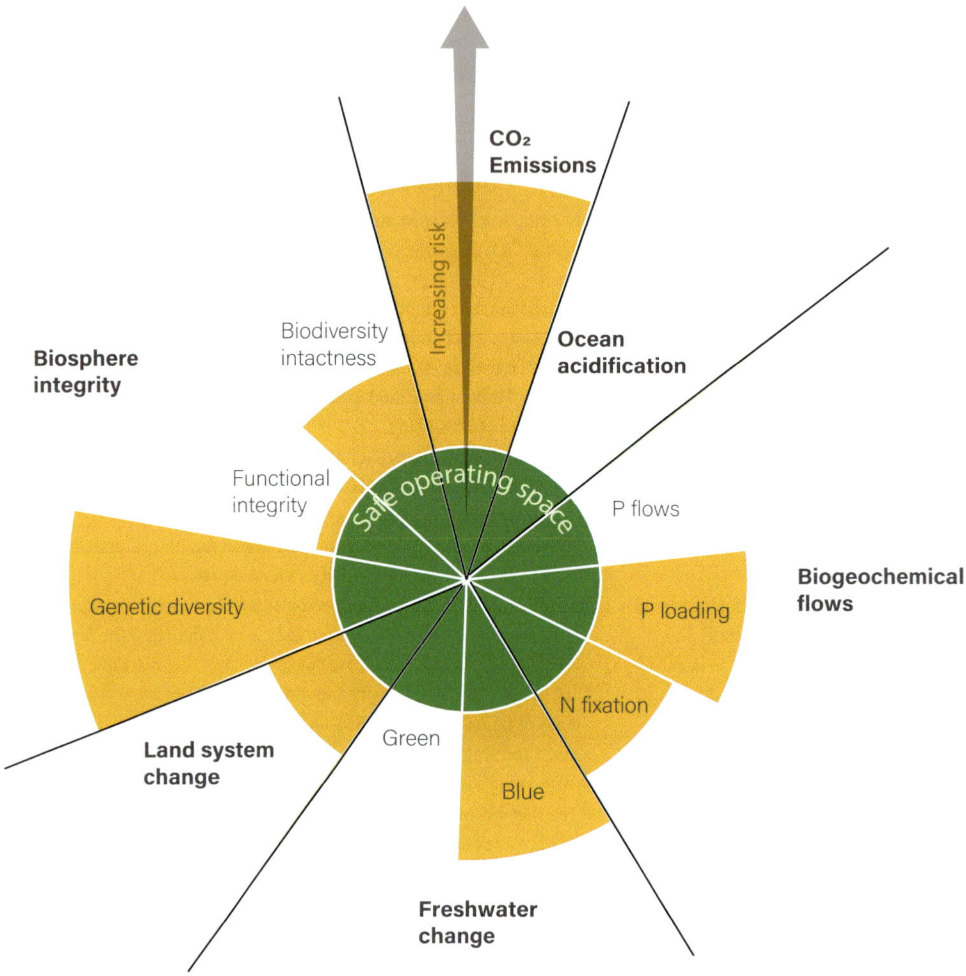

FIGURE II.1.3 Planetary boundaries extended to Texas. Transgressions in each control variable are shown as increasing beyond green safe operating space.

monitoring by Hu et al. (2022). Genetic diversity, represented by extinction rate per million-species-years (E/MSY), shows among the highest exceedances of safe operating space. Taxonomic-specific E/MSY were calculated for species originally found in Texas that were either globally extinct or extirpated from the state based on the total number of resident taxa. Sources estimating the total number of taxa in Texas are reported in Chapter I.1, "Why Texas?" Species reported as globally extinct or extirpated in Texas include 8 mammals (145 total), 11 fish (191 total), 3 birds (672 total), and 9 plants (~ 5,000 total) from 1800 to the present, meaning E/MSY ranged from 8 to 256, far exceeding the background extinction rate of one species per MSY. The Biodiversity Intactness Index (BII) (Sanchez-Ortiz et al., 2019a) suggests that biodiversity, measured as richness and abundance, is jeopardized in Texas, with BII values roughly half of the value reported as a planetary boundary (Figure II.1.2a). Based on assessments by Graff (2010) and Paudel et al. (2023), HANPP averages 10.1% for Texas, just cresting the boundary condition for biospheric functional integrity; however, values can locally exceed 80% HANPP for various counties. Land change was assessed based on the current area of forests compared to original forested areas. According to historical forest surveys reconstructed by Liknes et al. (2013), Texas harbored almost 217,000 km² of forested

TABLE II.1.2
Six of Nine Planetary Boundaries Extended to Texas Based on Control Variables, Including Values Reported in Richardson et al. (2023) and Boundaries Extended to Texas Relative to Current Observed Values

Earth System Process	Control Variable(s)	Planetary Boundary	Texas Boundary	Texas Boundary of Increasing Risk	Current Value
Climate change	CO_2 emissions ($GtCO_2\ yr^{-1}$) to achieve 350 ppm atmospheric CO_2 concentration	1 $GtCO_2\ yr^{-1}$ (2 $GtCO_2\ yr^{-1}$)	0.001 $GtCO_2\ yr^{-1}$	0.003 $GtCO_2\ yr^{-1}$	0.873 $GtCO_2\ yr^{-1}$
Change in biosphere integrity	Genetic diversity: E/MSY	1 E/MSY (10 E/MSY)	1 E/MSY	10 E/MSY	8 - 255 E/MSY
	Functional integrity: measured as energy - HANPP (Gt C/yr) % of preindustrial NPP	<10% HANPP (20% HANPP)	10%	20%	10.08%
	Biodiversity Intactness Index	>= 90%(30%)	>= 90%	>= 30%	40 - 41%
Ocean acidification	Carbonate ion concentration, average global surface ocean saturation state with respect to aragonite (Ωarag)	4.44 Ωarag (2.75 Ωarag)	3.44 Ωarag	2.75 Ωarag	3.25 - 4.08 Ωarag
Biogeochemical flows: P and N cycles	Phosphate flow from freshwater systems into the ocean (Tg of P yr^{-1})	11 Tg P yr^{-1} (100 Tg P yr^{-1})	42,000 t yr^{-1}	383,000 t yr^{-1}	13,000 t yr^{-1}
	P from fertilizers mined and applied to erodible (agricultural) soils (Tg of P yr^{-1})	6.2 Tg P yr^{-1} (11.2 Tg P yr^{-1})	19,500 t yr^{-1}	35,200 t yr^{-1}	90,224 t yr^{-1}
	Nitrogen fixation from industrial process (Tg of N yr^{-1})	62 Tg N yr^{-1} (82 Tg N yr^{-1})	291,000 t yr^{-1}	385000 t yr^{-1}	738,000 t yr^{-1}
Land system change	Percentage of original forest cover (%).	75% global; 50% temperate (30%)	50%	30%	26%
Freshwater change	Blue water: human induced disturbance of blue water (% of land area with deviations in streamflow)	10.2% (50%)	10.20%	50%	35%
	Green water: human induced disturbance of green water (% of land area with deviations in soil moisture)	11.1% (50%)	11.1	50%	10%

land, predominantly in the eastern portion of the state, whereas, based on current land cover, there are only 56,500 km^2 of forested lands, roughly one quarter of the original areas. Fertilizer application rates, taken from Falcone (2021), suggest that Texas far exceeds phosphorus application rates and industrialized nitrogen fixation rates, transgressing biogeochemical flows. Eutrophic conditions have been commonly reported for Texas estuaries due to nutrient loading (Bugica et al., 2020; Thronson and Quigg, 2008; Wetz et al., 2025). Therefore, it is surprising that freshwater loading of phosphorus to Texas estuaries, estimated by Marshall and Montagna (2025), was below the downscaled boundary for Texas. This demonstrates limitations to the university applicability of the PBF where boundary conditions are best regionally determined. Additionally, there are numerous assumptions made in extending the boundaries. For example, if phosphorus loading had been scaled by global freshwater discharge totals rather than total contributing drainage area, the boundary might have been far more restrictive for Texas estuaries. Global deviations in blue and green water from Porkka et al. (2024) were spatially summarized in Texas and suggest that, although green water alterations have not transgressed boundaries, blue water alterations are outside of the boundary (Figure II.1.2b).

For Earth's climate system boundary, all three iterations of the PBF assessment maintain that 350-ppm atmospheric CO_2 concentration and +1 W m^{-2} radiative forcing are the safe operating limits for Earth's climate system (Richardson et al., 2023; Rockström et al., 2009a; Steffen et al., 2015). Atmospheric concentrations are challenging to proportionalize directly to sub-global levels, whereas emissions can be assigned to localities, i.e., emitters. Emissions are eventually manifested within atmospheric CO_2 levels and contribute to radiative forcing following feedbacks via global carbon cycles. Based on alternative projections of future socioeconomic conditions (Riahi et al., 2017), achieving +1 to +1.5 W m^{-2} radiative forcing by 2100 would, at a minimum, require reducing global CO_2 emissions from current levels of 38 Gt CO_2 yr^{-1} to 1 Gt CO_2 yr^{-1} by 2050. By converting this rate to a per-unit-area estimate based on the global human modified area (Theobald et al., 2025; Winkler et al., 2021), a boundary estimate for Texas can be obtained (0.001 Gt CO_2 yr^{-1} or 1 MMt yr^{-1}). The Texas Commission on Environmental Quality (TCEQ) estimated current CO_2 emissions in Texas as 873 MMt CO_2 yr^{-1} (TCEQ, 2024), over two orders of magnitude above the current boundary. Texas currently leads the nation in CO_2 emissions, contributing roughly 13–14% of the nation's total; however, TCEQ reports that through multiple aggressive measures in industry and transportation (TCEQ, 2024), emissions could be reduced by 592 MMt yr^{-1}, a 68% reduction, by 2050. Even still, 280 MMt yr^{-1} far exceeds the proposed boundary.

II.1.5 IS TEXAS BEYOND THE SAFE OPERATING SPACE?

Of course, one cannot answer this question. There are too many uncontrolled variables and undefined parameters, and undelineated scope. For instance, "Under what temporal scale is being considered?" Presumably, most of the public is skeptical that safe boundaries have been transgressed given their day-to-day existence is seemingly unaffected by current environmental conditions. Planetary systems, however, operate over much longer time frames. The question should be restated, "Is Texas Beyond the Safe Operating Space for my great-grandchildren?" We also cannot overlook social incognizance with respect to surpassing safe operating space. For example, most of the public is unaware that their drinking water is sourced, at least partially, from wastewater (Turner et al., 2021). Hence, most people are incognizant of their own exposure, via numerous pathways, to countless novel entities, the majority of which are unregulated.

It is also equally probable that the boundaries are real but socially unrealistic. For example, O'Neill et al. (2018) found 100% of the 150 nations assessed in their study failed to meet the basic needs for citizens, while also maintaining resource use within safe operating boundaries. The authors found that physical needs, such as nutrition, sanitation, access to electricity, and elimination of poverty could technically be achieved without transgressing planetary boundaries; however, if goals representing high quality of life were achieved, this would require resources between two and six times the current safe operating limits of Earth (O'Neill et al., 2018). There is, of course, some room for efficiency in maintaining high quality of life while reducing environmental footprint. For example, an average person in the U.S. consumes 77,000 kwh of energy, almost three times that of a person in the UK (U.S. EIA, 2025). If planetary limits have already been exceeded, then we will have increasing challenges as social development continues. This also means we have yet to experience the full consequences. Even if boundaries are too restrictive, i.e., 'too safe,' there are limits, albeit uncertain, to the biophysical functioning of the planet. Uncertainty in boundaries, however, is not a good excuse to keep marching forward in the same direction and at the same pace.

We know many of the solutions. Yes, they are daunting, possibly challenging our ideologies, but they are not out of reach: a blue-green economy, alternative fuels, reducing resource (energy, water, and land) demands, increasing resource efficiency, lifestyle shifts, dietary shifts, transportation reform, urban planning, health reform, sustainable material sourcing, nature-based designs, regenerative farming, reforestation, and increasing protected areas, to name a few.

The most critical argument presented in this chapter is the plurality of the planetary boundaries, and cautionary advice to not fixate on one planetary challenge as 'the' existential threat or global catastrophic risk (GCR) in need of solving, at the expense of other Earth systems. Beard et al. (2021) present an analytical framework to understand climate change's contribution to GCRs by extending the GCR concept to planetary boundaries, reinforcing the need to understand the impacts of climate change within the Earth system appropriately, specifically that complex feedbacks between societal and ecological systems can contribute to increasing risks. Reductions in the resilience of planetary boundaries increase the risk of climate-induced tipping points - these are further amplified by global systems death spiral, illustrating social system failures (Beard et al., 2021). It could be that none of the planetary boundaries are necessarily existential, in and of themselves. Perhaps the true existential threat is our inability to see our own fragility to planetary boundaries.

REFERENCES

Andre P, Boneva T, Chopra F et al., 2024. Globally representative evidence on the actual and perceived support for climate action. Nature Climate Change 14, 253–259. https://doi.org/10.1038/s41558-024-01925-3

Avin S, Wintle BC, Weitzdörfer J et al., 2018. Classifying global catastrophic risks. Futures 102, 20–26. https://doi.org/10.1016/j.futures.2018.02.001

Baum SD, Barrett AM. 2018. The most extreme risks: Global catastrophes. In: Bier V, (eds) Risk in Extreme Environments: Preparing, Avoiding, Mitigating, and Managing (pp. 174–184). New York, New York: Routledge.

Beard SJ, Holt L, Tzachor A, Kemp L, Avin S, Torres P, Belfield H, 2021. Assessing climate change's contribution to global catastrophic risk. Futures 127, 102673.

Berdugo M, Delgado-Baquerizo M, Soliveres S, Hernández-Clemente R, Zhao Y, Gaitán JJ, Gross N, Saiz H, Maire V, Lehmann A, Rillig MC, 2020. Global ecosystem thresholds driven by aridity. Science 367, 787–790. https://doi.org/10.1126/science.aay5958

Birdsong T, Bean M, Mayes K, Robertson S, Curtis S, 2021. Conservation status of Texas freshwater fishes and protected species recommendations. Journal of Southeastern Association of Fish and Wildlife Agencies 8, 40–52.

Bostrom N. 2013. Existential risk prevention as global priority. Glob Policy 4, 15–31. https://doi.org/10.1111/1758-5899.12002

Bouman T, Steg L, 2022. A spiral of (in) action: Empowering people to translate their values in climate action. One Earth 5(9), 975–978.

Bugica K, Sterba-Boatwright B, Wetz MS, 2020. Water quality trends in Texas estuaries. Marine Pollution Bulletin 152, 110903.

Capon SJ, Lynch AJ, Bond N, Chessman BC, Davis J, Davidson N, Finlayson M, Gell PA, Hohnberg D, Humphrey C, Kingsford RT, 2015. Regime shifts, thresholds and multiple stable states in freshwater ecosystems; a critical appraisal of the evidence. Science of the Total Environment 15, 122–130.

Caro T, Rowe Z, Berger J, Wholey P, Dobson A, 2022. An inconvenient misconception: Climate change is not the principal driver of biodiversity loss. Conservation Letters 15, e12868. https://doi.org/10.1111/conl.12868

Casto SD, 2002. Extinct and extirpated birds of Texas. Bulletin of the Texas Ornithological Society 35(2), 17–32.

Clement V, Rigaud KK, de Sherbinin A, Jones B, Adamo S, Schewe J, Sadiq N, Shabahat E, 2021. Groundswell Part 2: Acting on internal climate migration. World Bank. http://hdl.handle.net/10986/36248. Accessed June 2, 2025.

Cohen JE, 1995. How many people can the earth support? The Sciences 35(6), 18–23.

Dearing JA, Wang R, Zhang K, Dyke JG, Haberl H, Hossain MS, Langdon PG, Lenton TM, Raworth K, Brown S, Carstensen J, 2014. Safe and just operating spaces for regional social-ecological systems. Global Environmental Change 28, 227–238. https://doi.org/10.1016/j.gloenvcha.2014.06.012

Dudney J, Suding KN, 2020. The elusive search for tipping points. Nature Ecology and Evolution 4(11), 1449–1450. https://doi.org/10.1038/s41559-020-1273-8

Falcone JA, 2021. Tabular county-level nitrogen and phosphorus estimates from fertilizer and manure for approximately 5-year periods from 1950 to 2017: U.S. Geological Survey data release. https://doi.org/10.5066/P9VSQN3C. Accessed June 2, 2025.

Ferretto A, Matthews R, Brooker R, Smith P, 2022. Planetary boundaries and the doughnut frameworks: A review of their local operability. Anthropocene 39, 100347.

Fuller R, Landrigan PJ, Balakrishnan K, Bathan G, Bose-O'Reilly S, Brauer M et al., 2022. Pollution and health: A progress update. The Lancet Planetary Health 6(6), e535–e547.

Gollin D, Hansen CW, Wingender AM, 2021. Two blades of grass: The impact of the green revolution. Journal of Political Economy 129(8), 2344–2384.

Graff CP, 2010. Human Appropriation of Net Primary Productivity (HANPP) in Texas: A statewide analysis of sustainability in the agricultural and timber sectors. Thesis. Texas A&M University.

Han H, Ahn SW, 2020. Youth mobilization to stop global climate change: Narratives and impact. Sustainability 12(10), 4127. https://doi.org/10.3390/su12104127

Häyhä T, Cornell SE, Hoff H, Lucas P, van Vuuren D, 2018. Operationalizing the concept of a safe operating space at the EU level—first steps and explorations. Stockholm: Stockholm Resilience Centre.

Hillebrand H, Donohue I, Harpole WS et al., 2020. Thresholds for ecological responses to global change do not emerge from empirical data. Nature Ecology and Evolution 4, 1502–1509. https://doi.org/10.1038/s41559-020-1256-9

Hu X, Yao H, McCutcheon MR, Dias L, Staryk CJ, Wetz MS, Montagna PA, 2022. Aragonite saturation states in estuaries along a climate gradient in the northwestern Gulf of Mexico. Frontiers in Environmental Science 10, 951256.

Huggel C, Bouwer LM, Juhola S et al., 2022. The existential risk space of climate change. Climatic Change 174, 8. https://doi.org/10.1007/s10584-022-03430-y

Knapp WM, Frances A, Noss R, Naczi RF, Weakley A, Gann GD, Baldwin BG, Miller J, McIntyre P, Mishler BD, Moore G, 2021. Vascular plant extinction in the continental United States and Canada. Conservation Biology 35(1), 360–368.

Kronenberg J, Andersson E, Elmqvist T, Łaszkiewicz E, Xue J, Khmara Y, 2024. Cities, planetary boundaries, and degrowth. The Lancet Planetary Health 8(4), e234–e241.

Liknes GC, Nelson MD, Kaisershot DJ, 2013. Net Change in Forest Density, 1873–2001. Using Historical Maps to Monitor Long-Term Forest Trends. USDA Forest Service RMAP-NRS-4 (12 p). Newtown Square, PA: U.S. Department of Agriculture, Forest Service, Northern Research Station. https://doi.org/10.2737/NRS-RMAP-4

Lotka AJ, 1925. Elements of Physical Biology. Baltimore, MD: Williams & Wilkins.

Lueders MB, McManamay RA, Le Bouille D, Garrett K, Roush L, Powers S, Young D, Bean P, Mayes K, Smith R, Trungale JF, 2025. BBEST practices: Intentional efforts to identify elusive flow-ecology thresholds. Ecological Indicators 172, 113311.

Marshall DA, Montagna PA, 2025. Nitrogen and phosphorus budgets for Texas estuaries. In: Montagna, P.A., Douglas, A.R. (eds) Freshwater Inflows to Texas Bays and Estuaries. Estuaries of the World. Cham: Springer. https://doi.org/10.1007/978-3-031-70882-4_15

Meyer K, Newman P, 2018. The planetary accounting framework: A novel, quota-based approach to understanding the impacts of any scale of human activity in the context of the planetary boundaries. Sustainable Earth 1, 4.

Montoya JM, Donohue I, Pimm SL, 2018. Planetary boundaries for biodiversity: Implausible science, pernicious policies. Trends in Ecology & Evolution 33(2), 71–73.

Nature E, 2009. Earth's boundaries? Nature 461, 447–448. https://doi.org/10.1038/461447b

O'Neill DW, Fanning AL, Lamb WF, Steinberger JK, 2018. A good life for all within planetary boundaries. Nature Sustainability 1(2), 88–95. https://doi.org/10.1038/s41893-018-0021-4

Pasquini G, Spencer A, Tyson A, Funk C, 2023. Why some Americans do not see urgency on climate change. Pew Research Center 32. https://www.pewresearch.org/science/2023/08/09/why-some-americans-do-not-see-urgency-on-climate-change/. Access June 2, 2025.

Paudel S, Mueller K, Ovando-Montejo G, Rushforth R, Tango L, Lant C, 2023. Product-specific human appropriation of net primary production in US counties. Ecological Indicators 150, 110241.

Pearl R, Reed LJ, 1920. On the rate of growth of the population of the United States since 1790 and its mathematical representation. Proceedings of the National Academy of Sciences of the United States of America 6(6), 275–288. https://doi.org/10.1073/pnas.6.6.275

Persson L, Carney Almroth BM, Collins CD, Cornell S, De Wit CA, Diamond ML, Fantke P, Hassellov M, MacLeod M, Ryberg MW, Søgaard Jørgensen P, 2022. Outside the safe operating space of the planetary boundary for novel entities. Environmental Science & Technology 56(3), 1510–1521.

Porkka M, Virkki V, Wang-Erlandsson L, Gerten D, Gleeson T, Mohan C, Fetzer I, Jaramillo F, Staal A, Te Wierik S, Tobian A, 2024. Notable shifts beyond pre-industrial streamflow and soil moisture conditions transgress the planetary boundary for freshwater change. Nature Water 2(3), 262–273.

Raworth K, 2012. A safe and just space for humanity: Can we live within the doughnut? Oxfam Discussion Paper. Oxfam, Oxford, UK.

Riahi K, Van Vuuren DP, Kriegler E, Edmonds J, O'Neill BC, Fujimori S, Bauer N, Calvin K, Dellink R, Fricko O, Lutz W, 2017. The shared socioeconomic pathways and their energy, land use, and greenhouse gas emissions implications: An overview. Global Environmental Change 42, 153–168.

Richardson K, Steffen W, Lucht W, Bendtsen J, Cornell SE, Donges JF, Drüke M, Fetzer I, Bala G, Von Bloh W, Feulner G, 2023. Earth beyond six of nine planetary boundaries. Science Advances 9(37), eadh2458.

Rockström J, Donges JF, Fetzer I, Martin MA, Wang-Erlandsson L, Richardson K, 2024. Planetary boundaries guide humanity's future on earth. Nature Reviews Earth & Environment 5(11), 773–788.

Rockström J, Steffen W, Noone K, Persson A, Chapin FS III, 2009a. A safe operating space for humanity. Nature 461, 472–475. https://doi.org/10.1038/461472a

Rockström J, Steffen W, Noone K, Persson Å, Chapin FS III, Lambin E, Lenton TM, Scheffer M, Folke C, Schellnhuber HJ, Nykvist B, 2009b. Planetary boundaries: Exploring the safe operating space for humanity. Ecology and Society 14(2), 32. http://www.ecologyandsociety.org/vol14/iss2/art32/

Sanchez-Ortiz K, Gonzalez RE, De Palma A, Newbold T, Hill SL, Tylianakis JM, Börger L, Lysenko I, Purvis A. 2019a, Land-use and related pressures have reduced biotic integrity more on islands than on mainlands. BioRxiv:576546.

Sanchez-Ortiz K, Newbold T, Purvis A, De Palma A, 2019b. Global maps of biodiversity intactness index. Figshare. https://doi.org/10.6084/m9.figshare.7951415.v1. Accessed June 2, 2025.

Sayre NF, 2008. The genesis, history, and limits of carrying capacity. Annals of the Association of American Geographers 98(1), 120–134. https://doi.org/10.1080/00045600701734356

Schmidly DJ, Bradley RD, 2016. The Mammals of Texas, 7th ed. Austin, TX: University of Texas Press.

Steffen W, Richardson K, Rockström J, Cornell SE, Fetzer I, Bennett EM, Biggs R, Carpenter SR, De Vries W, De Wit CA, Folke C, 2015. Planetary boundaries: Guiding human development on a changing planet. Science 347(6223), 1259855.

Stockholm Resilience Center, 2025. Planetary boundaries. https://www.stockholmresilience.org/research/planetary-boundaries.html. Accessed June 2, 2025.

TCEQ (Texas Commission on Environmental Quality), 2024. Climate pollution reduction grants priority action plan for the State of Texas. State and Local Climate Energy Program U.S. Environmental Protection Agency, Austin, TX. EPA Grant Number 02F35501. March 1, 2024. https://www.tceq.texas.gov/agency/climate-pollution-reduction-grants

Theobald DM, Oakleaf JR, Moncrieff G et al., 2025. Global extent and change in human modification of terrestrial ecosystems from 1990 to 2022. Science Data 12, 606. https://doi.org/10.1038/s41597-025-04892-2

Thronson A, Quigg A, 2008. Fifty-five years of fish kills in coastal Texas. Estuaries and Coasts 31(4), 802–813. http://www.jstor.org/stable/40663477

Turner SWD, Rice JS, Nelson KD et al., 2021. Comparison of potential drinking water source contamination across one hundred U.S. cities. Nature Communications 12, 7254. https://doi.org/10.1038/s41467-021-27509-9

UN (United Nations). 2018. Climate change: An 'existential threat' to humanity, UN chief warns global summit. https://news.un.org/en/story/2018/05/1009782. Accessed June 2, 2025.

US EIA (U.S. Energy Information Administration), 2025. International. https://www.eia.gov/international/overview/world. Access June 3, 2025.

Verhulst, P-F, 1838. Notice sur la loi que la population poursuit dans son accroissement. Correspondence Math and Physics 10, 113–121.

Wang-Erlandsson L, Tobian A, Van der Ent RJ, Fetzer I, te Wierik S, Porkka M, Staal A, Jaramillo F, Dahlmann H, Singh C, Greve P, 2022. A planetary boundary for green water. Nature Reviews Earth & Environment 3(6), 380–392.

Wetz MS, Beecraft L, McBride M, Steichen JL, Quigg A, 2025. Nutrient-phytoplankton dynamics in Texas estuaries, In: Montagna, P.A., Douglas, A.R. (eds) Freshwater Inflows to Texas Bays and Estuaries: A Regional-Scale Review, Synthesis, and Recommendations (pp. 191–204). Cham: Springer Nature Switzerland.

WHO (World Health Organization), September 23, 2024. Sustainable development goal indicator 3.9.1: Mortality attributed to air pollution. 80pg. https://www.who.int/publications/i/item/9789240099142. Accessed June 2, 2025.

WHO (World Health Organization), 2025. Climate change. https://www.who.int/news-room/fact-sheets/detail/climate-change-and-health. Accessed June 2, 2025.

Winkler K, Fuchs R, Rounsevell M, 2021. Global land use changes are four times greater than previously estimated. Nature Communications 12, 2501. https://doi.org/10.1038/s41467-021-22702-2

II.2 Climate Change and the Living Planet

Impacts on the Biosphere and Consequences to Ecosystem Services

Jordan T. Jatko, Levi T. Sweet-Breu, and Ryan A. McManamay

II.2.1 INTRODUCTION

Earth's biosphere, i.e., the living organisms and the ecosystems they habitat, is undoubtedly the most complex of the planetary subsystems to understand, much less to project the future implications of climate change with certainty. Apart from climate change, humans have already sufficiently manipulated Earth's systems to the point that the current geological epoch of the Quaternary Period is no longer called the Holocene, but the Anthropocene – an era of rapid global-scale change brought about by human domination of the planet. For the past 10,000 years, humans have brought about unfathomable changes to the biosphere, including inducing the Earth's most rapid extinction of species, primarily through the transformation of terrestrial land cover, complete manipulation of terrestrial freshwater cycles, changes in the acidity and temperature of the ocean, alteration of biogeochemical flows and cycles, and shifts in the atmospheric concertation of constituents (Chure et al., 2022).

Approximately 8.7 million eukaryotic species are estimated to exist on Earth (Mora et al., 2011), whereas the International Union of the Conservation of Nature (IUCN) estimates that 2.1 million have been described by humans (IUCN, 2025). Furthermore, only 169,500 have been assessed for risk of extinction through the IUCN Red List (IUCN, 2025). Of the assessed species, at least 1,900 animals and plants have gone extinct since 1500 (Andermann et al., 2020; IUCN, 2025), which suggests a 1% species extinction rate per 500 years, or 3.8 species yr^{-1}. Although seemingly small, this rate is orders of magnitude larger than the background extinction rates from fossil records. Even so, there are a couple reasons to suspect this could be an underestimate based on the fossil record in the previous 5,000 years of human occupation on Earth (Andermann et al., 2020), and the fact that modern extinction estimates are solely based on species that have been inventoried.

Not only are actual extinction rates higher than estimated but they are projected to dramatically increase in the future (Kaiho, 2022; Monroe et al., 2019). Of the species assessed by the Red List, 28% are currently threatened with risk of extinction (IUCN, 2025). Applying this rate to global biodiversity estimates would suggest that roughly 2.5 million of the total species on Earth are threatened with extinction, and at least 87,000 are likely to go extinct in at least 500 years, perhaps 8,700 within our lifetime. Unfortunately, in each subsequent Red List re-assessment, there is an increasing trend in the proportion of species classified as threatened, suggesting a continued path toward extinction (Johnson et al., 2017; Monroe et al., 2019). Based on increasing extinction risk in IUCN records, Monroe et al. (2019) project that 200 to 600 of assessed bird species (2–5% of total

DOI: 10.1201/9781003406600-9

bird diversity) will go extinct in the next 500 years. If we apply this to all un-inventoried species across all taxonomic groups, this suggests 435,000 species could go extinct in 500 years, 43,500 in our lifetime. That estimate, however, assumes the condition of Earth's systems remains relatively constant; however, that is unlikely the case (Diaz et al., 2019; Ostberg et al., 2018). Even under stabilizing environmental conditions, Kaiho (2022) projects a 10%–15% animal loss rate in the next 40 years (already 10 times greater than current); however, under elevated rates of deforestation, environmental pollution, greenhouse gas (GHG) emissions, and even potential nuclear war, extinction rates could increase to 20% to 50%.

As with species extinctions, we have also underestimated the magnitude of our collective transformations of Earth's ecosystems (Winkler et al., 2021). At least 75% of the Earth's land surface is experiencing human pressures (Venter et al., 2016), with two-thirds of the world's biomes and half of the ecoregions moderately modified by humans (Theobald et al., 2025). Only one-third of the planet's rivers remain free-flowing to the ocean (Grill et al., 2019). This raises an important point – the global biodiversity loss, to date, has been primarily driven by habitat loss and overexploitation, not climate change (Maxwell et al., 2016).

Indeed, it is upon this current stage of Earth's extremely modified biospheric condition that climate change enters the play. Fundamental to understanding the potential biospheric impacts of climate change is the appreciation of the current state of ecosystem resilience, specifically that we have reduced ecosystem functional integrity to the point of extreme vulnerability to climate shifts and extremes. Climate shifts will inexplicably induce behavioral responses to adapt to extremes brought about by warming temperatures. Thermal tolerance variation within colonies of social insects could reduce resiliency of the colony by imposing costs of colony performance due to "weak links", or divisions of insect workers that adapt poorly to temperature increases (Baudier and O'Donnell, 2017). Fricke et al. (2022) estimate a 59.7% average reduction in climate-tracking seed dispersal due to historical defaunation and current threatened birds and mammals, as these animals are vital to plant seed dispersal. This reduces the resilience of plants to adapt to climate change due to their inability to seek suitable habitats on their own. Morphological adaptation is shown not to be a viable response to climate change, likely due to the macroevolutionary time scale, although phenological changes have shown increased adaptive responses compared to maladaptive responses to climate change, although overall adaptation strategies may be insufficient (Radchuk et al., 2019).

Shifting our attention to Texas, the situation is an extension of the global crisis. Texas currently leads the U.S. in population growth, gaining over half a million new residents in just the last four years (U.S. Census Bureau, 2024). Although much of Texas's landscape does not occur in its natural, pre-disturbed state, 81% of Texas is classified as "working lands", supporting crop production, livestock grazing and farm operations – these naturalized landscapes are important for species conservation in the state (Martinez et al., 2025). Due to population growth, economic expansion, and urbanization over the last 25 years, at least 15,000 km^2 of working lands, predominantly low intensity grazing lands, have been lost to fragmentation (Smith et al., 2025). Streamflows in at least 75% of rivers in Texas are declining (Lueders and McManamay, 2023), and with projected water demands and new infrastructures, pressures on surface and groundwaters will only increase (Texas Water Development Board (TWDB), 2025). Ultimately, these trends are expected to continue, meaning that ecosystems will be more vulnerable, i.e., less resilient, to the impacts of climate stressors.

As reported in the previous chapter (Chapter II.1), at least 11 species have been extirpated, globally or locally, from Texas based on existing records. We estimate there are 35,000 animal and plant species within the state, and if we can assume the extinction rates above can be extended locally, then anywhere from 350 to 1,750 species will go extinct or extirpated from Texas in the next five decades, 35 to 175 in our lifetime. These estimates are not unfathomable, given that there are

1,053 rare, endangered, or threatened species in the state. A total of 1,551 are considered 'species of greatest conservation need' (TPWD, 2025e).

So why should we care? As is customary in any discussion of human impacts on the environment, including but not limited to climate change and biodiversity loss, questions of ethics arise. In modern times, particularly the era of post-environmentalism, Western society rarely asks, "Why should we care?" but rather asks, "What should we really do about it?". This latter question is motivated from a socially synthetic juncture of moral crossroads, forcing one to make an arbitrary choice between prioritizing human needs, presumably exogenous to the natural world, and the intrinsic value of non-human nature. Based on this utilitarian and anthropocentric view, any protection of nature then must be justified on the basis of nature's instrumental use by humans (Batavia and Nelson, 2017). Of course, if nature is valuable to humans, then it can be monetized. Indeed, there are many benefits of nature to humans, often called ecosystem services, and these services have been used as a justification for conservation (Batavia and Nelson, 2017). *Co-production* of ecosystem services, or the contributions from both human and natural assets, has capitalized on monetization through production of food, clean water, and services from the biosphere, be it intentional or unintentional (Woodhead et al., 2025). We must recognize, however, that nature is valuable beyond its services to humans. Conservationists, for some long time, have advocated for the projection of non-human nature purely on the grounds of intrinsic value (e.g., John Muir). In other words, the Earth's biosphere has inherent significance and importance, beyond human value, where organisms and ecosystems also have an inalienable right to exist and function. Even so, the biosphere does provide many benefits, some critical to society, and it is important to understand the potential loss of those services due to degraded biospheric integrity from climate change. The positive relationship of biodiversity and ecosystem resilience is well documented, thus showing that maintaining high integrity of the biosphere, i.e., biodiversity, will be necessary to maintain ecosystem services (Bone et al., 2016; Millar et al., 2007). Here, we do not evaluate or question either moral standpoint but rather acknowledge that both exist and provide grounds for human behavioral adaptation.

In this chapter, we cover the effects of climate change through the lens of biodiversity and ecosystem services in Texas. Climate change has significant implications regarding the biosphere through the loss of habitat, poor adaptation to climate shifts, and reduced ecosystem quality. Anthropogenic interactions with the environment have incurred significant loss of biodiversity, although climate change and human reaction to climate change will likely compound the continued loss of biodiversity. The downstream implications to this are the loss of ecosystem services that humans rely upon for sustainability and resiliency (Perry et al., 2024). This chapter will delve into details and specific examples of climate change impacts on the biosphere and resulting perturbations to ecosystem services.

II.2.2 TEXAS ECOSYSTEMS

Due to its large size, Texas contains many diverse ecosystems with unique flora and fauna composition and climatic variables. As a whole, Texas can be included with Kansas and Oklahoma as the Southern Great Plains in the U.S. This region has largely flat terrain at a low latitude with large swaths of grassland, scrubland, rangeland, and cropland (Crimmins et al., 2023). Within Texas, however, at least 10 separate ecoregions (areas where ecosystems and environmental resources are similar) are represented (Omernik, 1987). These include the Pineywoods, Gulf Prairies and Marshes, Post Oak Savanah, Blackland Prairie, Crosstimbers, South Texas Plains, Edwards Plateau, Rolling Plains, High Plains, and the Trans-Pecos (TPWD, 2022) (Figure II.2.1). These distinct regions have their own ecological structures and communities that will be variously affected by climate change, and we describe each ecoregion below (Table II.2.1).

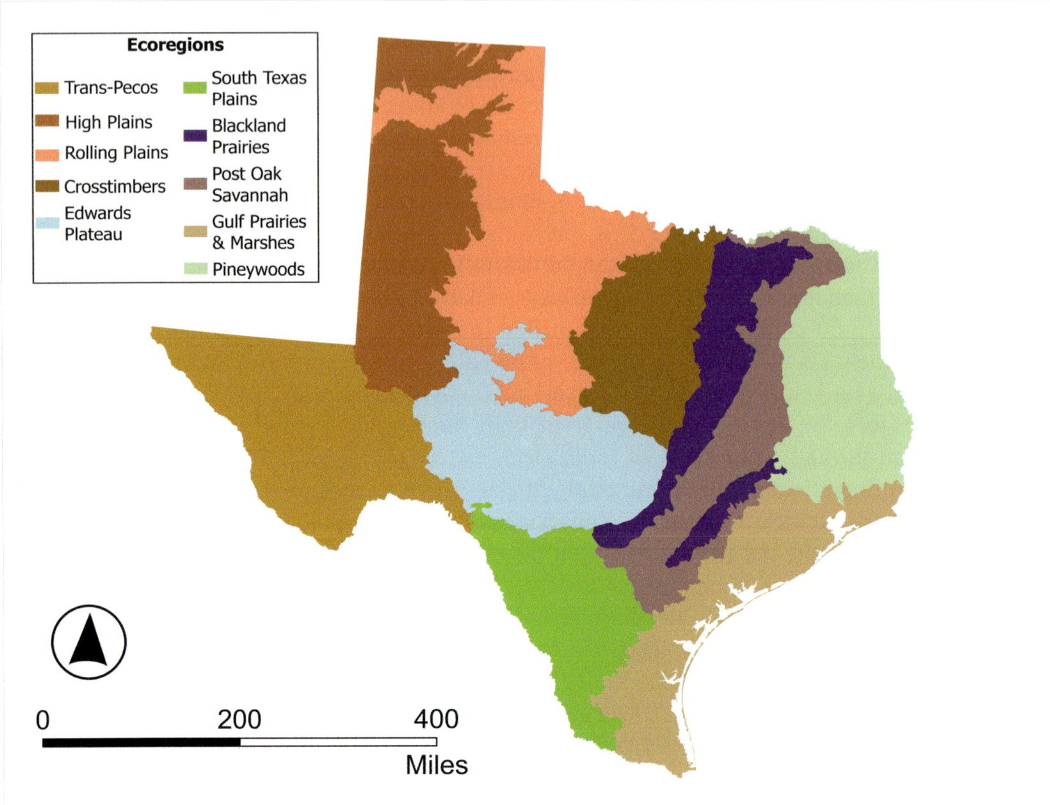

FIGURE II.2.1 The 10 ecoregions of Texas. Artwork courtesy of Texas Parks and Wildlife Department © 2022.

II.2.3 CLIMATE CHANGE AND THE BIOSPHERE

Quantifying climate change effects on the biosphere is often a multifaceted problem that is challenging to study. In particular, climate change does not operate as a manipulated variable in a highly controlled experiment – we only have one Earth and one climate system, and hence, only one observational unit to draw inferences. Additionally, there are other uncontrolled variables at play that influence the integrity of the biosphere. Aside from climate change, other anthropogenic activities, such as land alteration, water use, overexploitation of resources, and pollution, are primarily responsible for global biodiversity decline (Caro et al., 2022). Climate change acts upon this stage, further stressing the integrity of global ecosystems. The impact of climate stability additionally affects biodiversity by changing the timing, intensity, and duration of weather patterns and natural phenomena, such as droughts, wildfires, and floods. These changes induce phenological changes in organisms, or timings of recurring biological events in plants and animals, such as bird migrations, flower blooming, and hibernation. Phenology is not the only facet of the biosphere affected, but physiology and habitat distributions are also affected.

Many organisms exhibit some flexibility in their physiological tolerances and phenology. The concern then becomes how, if at all possible, can organisms adapt when their physiological tolerance is exceeded by shifting environmental conditions, induced by changes in temperature or timing of climate events. Organisms have numerous adaptive mechanisms, including dispersal, phenotypic plasticity, behavioral shifts, or morphological changes (Weiskopf et al., 2020); however,

TABLE II.2.1
Texas Ecoregion Physical and Climate Descriptions

Ecoregion	Description	Climate
Pineywoods	Pine forests throughout with marshes and hardwood trees in the south	Temperatures peaks at 100 °F in the summer with lows in the 30–40 °F range. Wettest region in Texas with 40–52 in/yr
Gulf Coast Prairies & Marshes	Estuaries, marshes, and tallgrass prairies	High humidity, warm temperatures, and 30–50 in/yr of rain
Post Oak Savannah	Hilly mixture of oak woodland and grassland	Mild average annual temperatures at 65–70 °F with 35–45 in/yr precipitation
Blackland Prairie	Largely grasslands with fertile soil	Similar temperatures as the Post Oak Savannah with precipitation at 30–40 in/yr
Crosstimbers	Forested to the west and east and grasslands through the center	Mild temperatures throughout the region. Precipitation averages 25 in/yr in the west and 35 in/yr in the east
South Texas Plains	Flat to gently rolling landscapes of grassland and dense thorn/scrub forests	A mix of subtropical and semi-arid climates with hot summers and mild winters. Receives 20–32 in/yr of rainfall
Edwards Plateau	"Hill Country" with a mix of savannas, forests, grasslands, and steep limestone hills	Generally mild climate with hot summers and cool winters. Averages 15–34 in/yr of precipitation
Rolling Plains	Mix of prairie grasslands, small woodlands, and scattered rocky outcrops	Semi-arid climate that receives 20–28 in/yr of precipitation
High Plains	Shortgrass prairies and occasional mesas that are a part of the larger Great Plains	Hot summers, cold winters, and low annual rainfall (15–22 in/yr) that makes the region prone to drought
Trans-Pecos	Part of the larger Chihuahuan Desert, features open plains, canyons, and high mountain ranges	Arid desert environment that is hot and dry with extreme temperature fluctuations. Precipitation averages < 12 in/yr

Source: **Adapted from the Texas Parks and Wildlife Department.**

these may induce bioenergetic and physiological tradeoffs that lead to lower survival, abundance of populations, and eventual extinction. More importantly, there are limits to adaptation. Radchuk et al. (2019) suggest that adaptive phenological responses for a large number of bird species were insufficient to cope with seasonal shifts due to climate change, increasing the likelihood of extinction risk.

As one example of a behavioral adaptation, the National Audubon Society has shown that between the years 1966 and 2013 the average northward distance birds migrate in North American has increased by more than 40 miles (Figure II.2.2) (Audubon Society, 2025). Some species have moved further – the United States Environmental Protection Agency reports that 48 species have moved more north by nearly 200 miles (U.S. EPA, 2025). That equates to roughly the drive from Houston to Dallas. As humans continue to fragment habitats, we must expect that compounded disturbances, i.e., human-induced localized impacts in conjunction with climate change, will continue to jeopardize species existence. To understand the potential impacts for the future, we must know what the future will look like through the lens of climate change. This section will overview climate projections from multiple perspectives and the current and potential impacts on Texas's species of greatest conservation need (SGCN).

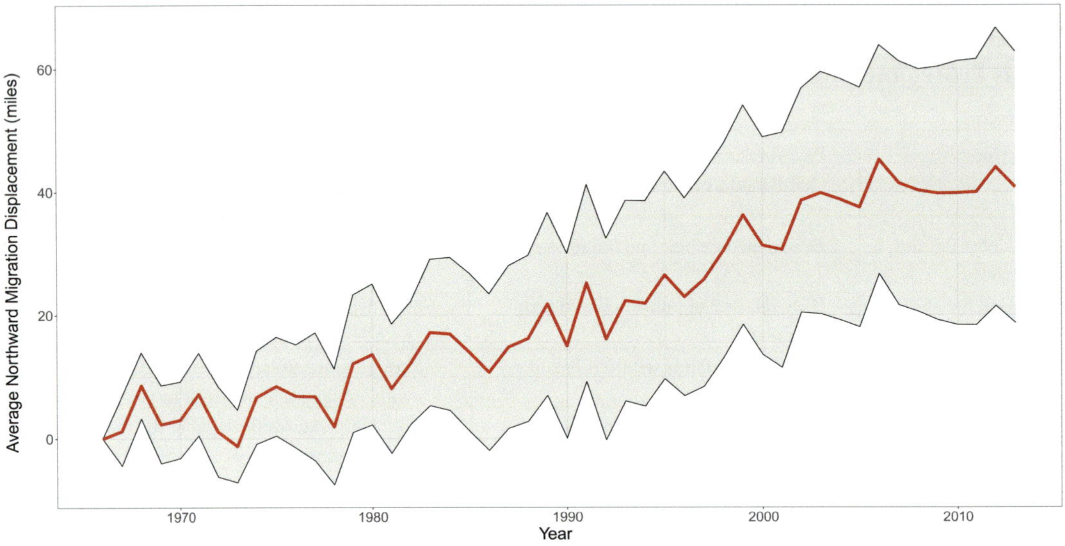

FIGURE II.2.2 Visualization of the National Audubon Society bird center distributions for 305 unique bird species from 1966 to 2014, adopted from the United States Environmental Protection Agency. Average bird center distributions have moved northward by more than 40 miles since 1966, likely due to increasing winter temperatures.

II.2.4 TEMPERATURES ARE INCREASING ACROSS THE BOARD – TEXAS IS NO EXCEPTION

The Office of the Texas State Climatologist at Texas A&M University has reported on Texas climate projections through 2036 based on historical trends, current observations, and larger climate projections (Nielsen-Gammon et al., 2024). The authors predict an increase in average Texas temperatures of 1.8° F compared to the 1991–2020 average and three degrees higher than the 1950–1999 average. Historical temperature changes have increased, on average, 1° F per decade in southeastern and western Texas. Additionally, days of greater than 100° F are expected to occur four times as often by 2036 compared to 1970–1990, and the minimum winter temperatures are also expected to continue to rise (Nielsen-Gammon et al., 2024). These shifts in temperatures will correspond with a shift in growing season for crops and natural vegetation. Projections for the Southern Great Plains region predict increases in annual average temperatures by 2.5° and 5° F by 2050 and 2100, respectively, for an intermediate scenario and by 5° and 10° F in a worst-case scenario (Crimmins et al., 2023).

The National Climate Assessment (NCA) summarizes climate projections from global circulation models under numerous scenarios, depicting different future socioeconomic conditions, outlined by the Shared Socioeconomic Pathways (SSP), and future GHG emissions, characterized by the Representative Concentration Pathways (RCP) – meaning scenarios on how humans will embrace the future and the resulting impacts on global temperatures. The figures are based on SSP scenario 5 (SSP5), a future of rapid economic growth fueled by fossil fuels, and RCP scenario 8.5, the highest concentration of GHGs as solar irradiance (W/m^2). This combination can be viewed as a "worst case scenario" for climate change. In the top left of Figure II.2.3, we see changes in average annual temperature per Texas ecoregion compared to the 1991-2020 average under the scenario that global temperatures have increased by 5.4° F (3° C) (Figure II.2.3). For reference, global average temperatures have already increased by about 2° F from pre-industrial levels (Lindsey and Dahlman, 2025). Most ecoregions sit around a 5.2° F temperature increase. Gulf Prairies and Marshes lower distribution reach around 4.4° F average temperature increase compared to the

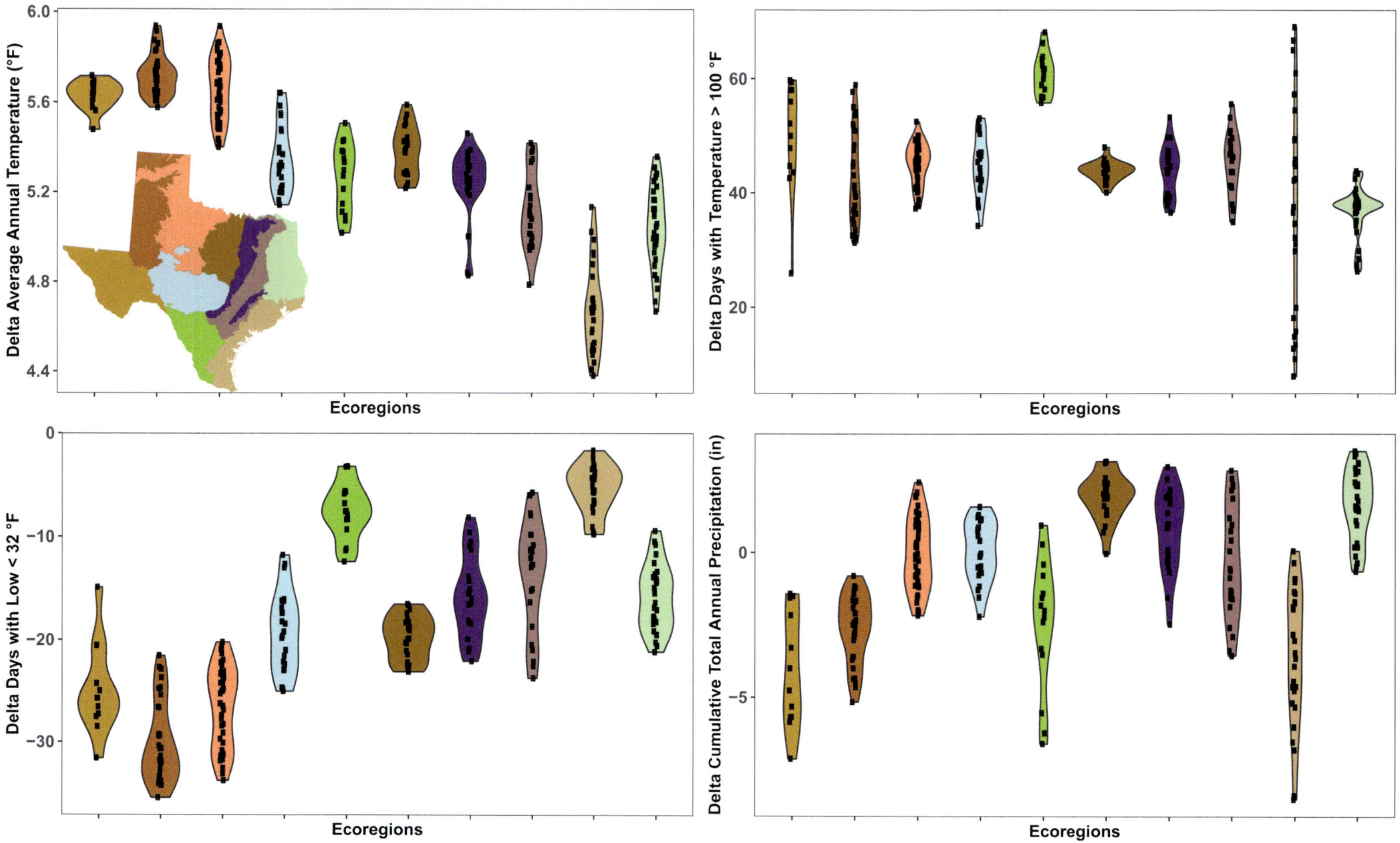

FIGURE II.2.3 Violin plots summarizing National Climate Assessment (NCA) climate outcomes under the SSP585 future scenarios. Top left: change in average annual temperature. Top right: change in number of days with temperatures above 100° F. Bottom right: change in number of days with low temperatures below 32° F. Top right: change in cumulative total annual precipitation in inches.

highest estimate in the High Plains showing about a 5.9° F average temperature increase. Increased average annual temperatures will certainly have a significant impact on the biological timings and behaviors of Texas's organisms.

II.2.5 FRESHWATER IS AT RISK, OCEANS ARE ON THE RISE

Other climate change trends in Texas are more variable than temperature. Precipitation has and is expected to continue to increase in eastern Texas, while it will continue to decrease in western Texas (Nielsen-Gammon et al., 2024). Rainfall intensity is projected to increase more than 10% and the probability of extreme precipitation could increase by over 50% by 2036 compared to the last two decades, thus increasing flood risk. While many of the contributing factors of drought are expected to increase or worsen, drought itself is difficult or impossible to project (Nielsen-Gammon et al., 2024). Likewise, other extreme weather events, such as thunderstorms and tornadoes, are similarly challenging to project. Wildfire susceptibility, however, is expected to increase and spread further into eastern Texas due to rising temperatures at earlier times in the year leading to drier spring and summer conditions.

The National Climate Assessment has also provided annual precipitation projections using the same scenarios previously mentioned, SSP585 (5.4° F average global temperature increase). They are summarized by Texas ecoregions at the bottom right of Figure II.2.3. Remember, the current average annual precipitation for each ecoregion is summarized above in the beginning of this introduction. Most scenarios in every ecoregion indicate a loss of precipitation, although all scenarios from the Gulf Prairies and Marshes, High Plains, and Trans-Pecos are exclusively negative. This is particularly impactful in the Trans-Pecos ecoregion because this region receives less than 12 inches of precipitation on average annually. The lower distribution of scenarios indicates a >5 inch reduction in annual average precipitation, nearly half of the current average. Not all ecoregions will be affected equally and the ecological hotspot that is the Trans-Pecos ecoregion is at particular risk with its abundance of SGCNs. Because climate change effects are not ubiquitous, droughts are difficult to project, as previously mentioned. Reduced annual precipitation, however, is a powerful contributing factor.

Although the magnitude of future sea level rise is uncertain, sea level rise is an observed global phenomenon, including along the Texas Gulf Coast. Sea levels at Galveston Bay have risen 0.67 m (2.18 ft) over the past 100 years, increasing by 8 inches (20.3 cm) in the last 14 years (Pathak, 2024). Rising global temperatures have expanded ocean waters, aided by contributions from melting glaciers. In addition, groundwater pumping in Texas is causing the land to sink (Meadows Center, 2024) and relative sea levels to rise. Current tidal gauge records along the Gulf Coast show a rise more rapidly than at any time over the last three millennia. As a result, climate projections show an additional increase in sea level by 10–12 inches by 2050 and potentially greater than four feet by 2100 (Meadows Center, 2024). These trends are linear projections and are considered conservative, especially if sea levels rise non-linearly in the future.

II.2.6 THE HIGHS AND LOWS ARE INCREASING – PHENOLOGICAL AND PHYSIOLOGICAL CHANGES IN THE BIOSPHERE

Increasingly, record-breaking temperatures are being documented globally, as well as increasing trends in heat wave frequency and magnitude (Coumou and Robinson, 2013; Perkins-Kirkpatrick and Lewis, 2020). How this specifically has and will impact the biosphere, i.e., the plants, animals, and their environment, is still uncertain and varies locally. Generally speaking, shifts in temperature – more warm days and fewer cold days – can significantly disrupt an organism's phenology, i.e., the timing and mechanisms of life history phases based on daily and seasonal changes. The National Climate Association, under the same SSP585 (5.4° F average increase in global temperatures), shows an average of 40–50 increase in days over 100° F for all Texas ecoregions, as seen in the top right of Figure II.2.3. Some scenarios in the Gulf Prairies and Marshes and the South Texas Plains project nearly 70 more days of these extreme temperatures. Projections like these indicate a longer summer period, cutting into the spring, autumn, and winter seasons when many significant biological events occur, such as flowering, reproduction, and dispersal. This is evident in the NCA projections at the bottom left of Figure II.2.3 that shows the change in number of days below 32° F. The High Plains and Rolling Plains are projected to have lower scenario distributions of nearly 40 fewer days below freezing temperatures. Important phenological events are characterized by the first freeze, including dormancy of insects (some of which become vectors of disease), bird migrations, and leaf shedding.

Although a few degrees of temperature might seem insignificant, these small changes could induce major disruptions to ecological systems and species that cannot adapt in time. Organisms often rely on a narrow set of habitat and physiological conditions in order to thrive; hence, seemingly miniscule shifts in climate can cause altered timings for growth, reproduction, and other important functions. The following section will discuss the effects of changing climate on ecological systems, particularly for their timing and cyclical nature.

II.2.7 GROWING SEASONS, POLLEN PRODUCTION, AND SOIL HEALTH – THE IMPORTANCE OF PLANTS

Phenological changes for plants often mean altering when plant functions occur, such as leafing, flowering, and fruiting. Some of the most noticeable changes arising from seasonal climate shifts include events occurring earlier in the year (like those triggered by the arrival of spring) and lasting longer (such as an extended growing season). For example, climate instability can affect the flowering of plants with late spring frosts (Qiu et al., 2024). These phenological changes can affect growing seasons for farmers but also affect pollen producing plants that commonly cause allergies. Texas cities already rank among the most challenging "allergy capitals" in the U.S. (AAFA, 2025). Warming temperatures and delayed onset of the first frost increased the length of the pollen season for ragweed (*Ambrosia* spp., a common allergen) by as much as 16 days between 1995 and 2009 (Ziska et al., 2011). Apart from climate stress, the abundance of CO_2 concentrations in the atmosphere can directly influence plants. In Texas, woody shrubs are invading prairie grasslands due to favorable CO_2 concentrations (Acharya et al., 2017). Increased CO_2 concentrations also increase primary productivity, further increasing pollen production by plants, thus compounding climate related effects on allergen production (U.S. EPA, 2016). As climate stability becomes more variable, other temperature-dependent ecosystem processes will be affected, such as decomposition rates, nutrient cycles, and growth rates (Li et al., 2022); these processes are complicated by species interactions and functional changes to ecosystems. For example, woody shrub encroachment on grasslands can alter soil bacterial and fungal communities (Hollister et al., 2010), creating second-order alterations to decomposition and nutrient cycles. These impacts are not just isolated to natural systems, but can have major economic consequences, such as leading to degraded grazing land, alterations to nutrient availability for agricultural purposes, and water quality degradation (The Nature Conservancy, 2025).

II.2.8 FUNCTIONAL AND GENETIC INTEGRITY OF ANIMALS UNDER CLIMATE STRESS

Analogous to plant life histories, climate change also affects phenological responses in animals. Animals rely on seasonal and temperature queues for important life history events, such as hibernation or migration. Shifts in seasonal queues for hibernation and migration lead to adaptive behaviors in mammals, such as earlier emergence in spring (Chmura et al., 2023). With amphibians and reptiles, however, increased temperatures can have physiological implications, such as reductions in hatching and growth rates (Mendoza et al., 2021). Although most mammals in Texas do not exhibit true hibernation due to the milder climate, some organisms undergo periods of torpor during winter months, such as ground squirrels, bats, and reptiles. The little brown bat (*Myotis lucifugus*) lowers its body temperature to just above air temperature in the winter. This reduces energy consumption, although if the cave temperature increases, the bat's body temperature also increases, thus raising energy consumption. If these bioenergetic adjustments occur too often, the bat could starve before the end of winter (Hiller, 1983). Interestingly, one study shows that temperature had little impact on the long-term migration changes of the Brazilian (Mexican) free-tailed bat (*Tadarida brasiliensis*), but rather wind and precipitation pattern changes had the most significant effect (Haest et al., 2021). One of the largest bat colonies in the world occurs at the Braken Cave in San Antonio, Texas, dominated by the Brazilian (Mexican) free-tailed bat.

The American alligator (*Alligator mississippiensis*), one of the most charismatic fauna of the southern American coast, occurs in marshes and swamps within 11 U.S. states with their range generally extending as far north as North Carolina. As temperatures have warmed, however, alligator

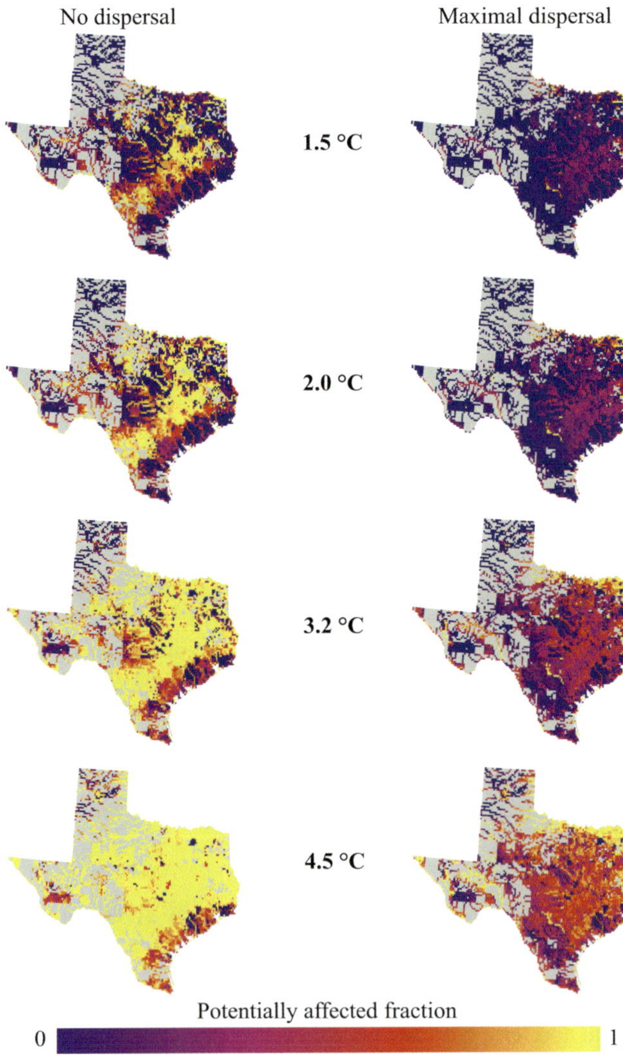

FIGURE II.2.4 Proportion of fish populations potentially affected by the consequences of global warming scenarios and climate responses, 0 being no species being affected and 1 being 100% of species being affected. This figure is adapted from Barbarossa et al. (2021).

sightings have increased in Virginia (Hagan, 2022), indicating dispersal and range expansion. Indeed, this is in congruence with very long observational records. Over the course of its evolutionary history, the American alligator has largely remained unchanged in its present form as it did over the course of 30 million years. The fossil record suggests that distributional shifts, as opposed to evolving climate tolerances, were how the American alligator adapted to major climate fluctuations of glacial and interglacial cycles (Ryberg and Lawing, 2018). Unfortunately, this does not necessarily suggest future success for alligators or other crocodilians – modern climate change is occurring at a much faster rate than it did prehistorically. Rapid sea level rise could compromise existing habitat, narrowing areas of suitable habitat. Furthermore, alligator egg development has interesting interactions with the air temperature, such as the temperature-dependent sex determination. Incubation at 33° C produces all male offspring, while incubation at 30° C produces all female offspring (Ferguson and Joanen, 1982). Recent studies also show that relative humidity has significant control on egg-hatching success, with lower humidity levels (~80%) having nearly zero percent hatching success (Reigh and Williams, 2020). Additionally, many species, including crocodilians, display temperature-dependent sex determination during nesting. Projected increases in American alligator nest temperatures (1.6 to 3.7° C) by 2100 are estimated to have severe consequences on offspring sex ratios, leading to subsequent population recruitment challenges (Bock et al., 2020). Increased temperature and drought intensity, duration, and frequency could have significant impacts on the American alligator success and genetic variability.

Similar to the American alligator dispersion along the Gulf of Mexico, many organisms will disperse to find habitats with more favorable conditions. This is naturally apparent with fish, such as those with niche spawning habitats. However, human infrastructures, such as dams, have long disrupted riverine organisms' abilities to disperse (Chan et al., 2025). As climate change continually affects organisms, the ability to disperse will be paramount for survival. Fish are restricted to channels within connected streams. Restricting a dispersal ability will increase its risk to consequences of climate change. Barbarossa et al. (2021) assessed the proportion of fish populations at risk to the consequences of climate change, assuming no dispersal capabilities. Consequently, not all fish will be affected equally, as one would expect, since survival strategies and adaptive capacities vary considerably among fish. As humans adapt to the ongoing climate crisis, further fragmentation of river systems for drinking water purposes will significantly impact fish populations at a large scale, as seen by the yellow portions of streams under the "no dispersal" scenario (Figure II.2.4).

II.2.9 CLIMATE CHANGE RISK TO SPECIES

Generally, it is difficult to quantify the direction and magnitude of impacts to biodiversity that are caused by climate change. Often perturbations on biodiversity are an unforeseen consequence of human activities, evaluated retrospectively rather than prospectively. Many times, adaptive responses by organisms are difficult to predict, as they also can be influenced by numerous interactions between other species. Insufficient adaptive responses of animals to climate and other anthropogenic pressures (Radchuk et al., 2019) constitute a significant risk to organisms, especially those already endangered or threatened with risk of extinction. Due to uncertainties in species and ecosystem responses, scientists generally quantify potential impact as "risk", either to individual species or ecosystems.

Risk is often described as the combined product of two factors: (1) the hazard, such as increased temperatures or lower precipitation, and (2) exposure. Quantifying risk across a region, such as Texas, also requires accounting for the relative susceptibility of the organisms found in that region. Bateman et al. (2020) assessed the relative risk of birds across the U.S., which was adapted for Texas (Figure II.2.5). The study shows that birds are at risk of losing suitable ranges in Texas, particularly during the non-breeding season, under two different climate scenarios. In Texas, at-risk organisms are monitored regularly and documented thoroughly. Organisms that are listed as threatened or endangered at the global, national, and subnational levels are all documented by the Texas Parks and

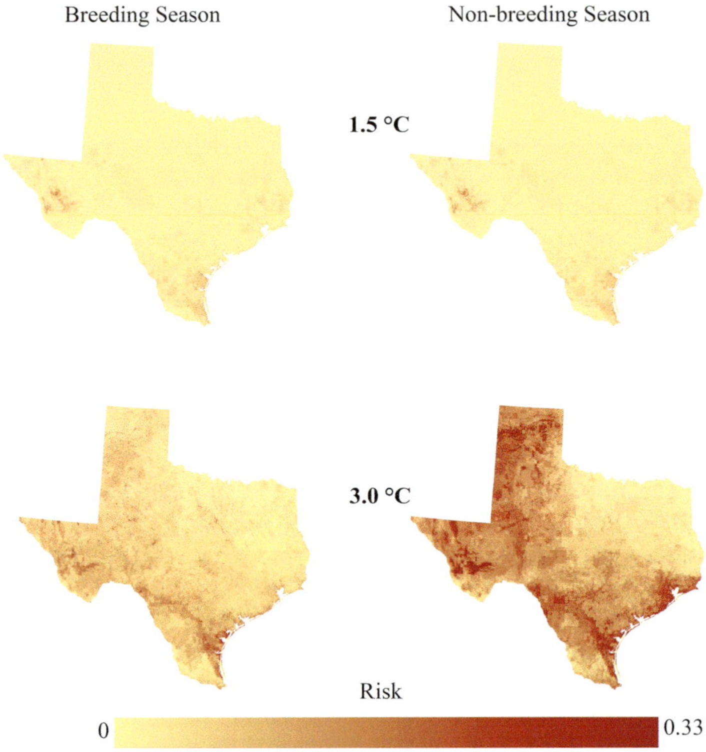

FIGURE II.2.5 Risks for Texas birds under two global warming scenarios during breeding and non-breeding seasons. Risk is a product of hazard and exposure, which accounts for total species and total sensitive species in this figure adapted from Bateman et al. (2020).

Wildlife Department (TPWD, 2025e). Currently, there are 2,269 organisms listed on the TPWD Rare, Threatened, and Endangered Species list found through the TPWD's website (Figure II.2.6). There are 779 endemic species found only in Texas that are listed as threatened or endangered at a global, national, or subnational level (Figure II.2.6). The specific details of the organisms described below, and more, can be found through the TPWD Rare, Threatened, and Endangered Species list.

Uncertainties of the impacts of climate change on biodiversity are compounded by uncertainties in human society evolution in the future, particularly how humans respond to climate stress. For example, in Texas, most municipalities have adapted to increasing drought frequencies by securing additional potential drinking water sources – an imminent need in Texas. The population is projected to reach over 50 million by the year 2080 in Texas, while municipal water demand is projected to reach over 8 million acre-ft (TWDB, 2025). Texas utilizes both ground and surface water sources for municipal needs, the latter primarily derived through reservoir development. Many at-risk aquatic organisms, such as the Texas fawnsfoot (*Truncilla macrodon*) and Trinity pigtoe (*Fusconaia chunii*), are adapted to lotic environments and hence are intolerant to reservoirs (U.S. FWS, 2024). These freshwater mussels are found in large rivers with low to moderate velocities. Large rivers are often targets for reservoir development, and thus would impair these critical habitats. Groundwater extraction can also significantly affect endangered and threatened species, like those that occupy groundwater-fed headwaters, springs, and aquifers. Peck's Cave amphipod (*Stygobromus pecki*) is an aquatic crustacean that lives in the Edwards Aquifer, one of the largest aquifers in Texas that is at risk of overconsumption (U.S. FWS, 2025). The Comanche Springs pupfish (*Cyprinodon elegans*) is restricted to a small series of springs in the Trans-Pecos ecoregion and the Clear Creek gambusia (*Gambusia heterochir*) is restricted to the headwaters of Clear Creek

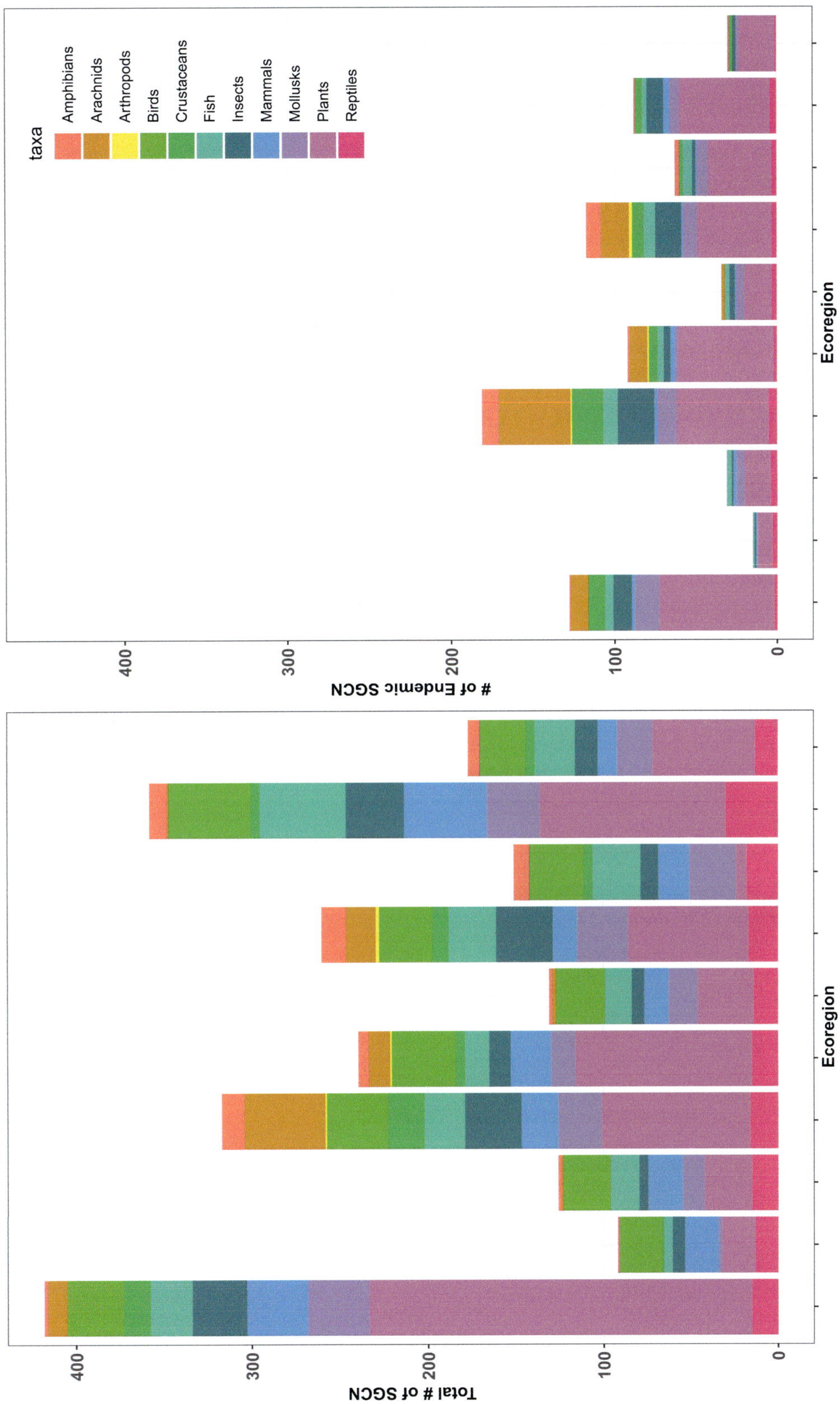

FIGURE II.2.6 Summary of the total (left) and endemic (right) species classified as species of greatest conservation need at a global, national, or subnational level. Colors are representative of each taxa level as described by the Texas Parks and Wildlife Department.

in the Edwards Plateau (TPWD, 2025a; TPWD, 2025b). Interestingly, the Clear Creek gambusia was also threatened initially because of a series of dams that altered the range of water temperature suitable for the small fish (TPWD, 2025b).

Precipitation also plays significant roles in maintaining habits and conditions for these threatened and endangered species. Drought conditions can impact already widely fluctuating flowering populations of the Navasota ladies'-tresses (*Spiranthes parksii*) in the Post Oak Savannah ecoregion. These plants rely on favorable hydrologic conditions, such as a perched water table, to flourish. Because an individual plant is not likely to flower every year, stable climate conditions are necessary for the survival of an individual until it flowers again (TPWD, 2025c). The Texas Poppy-Mallow (*Callirhoe scabriuscula*) of the Rolling Plains has an interesting lifestyle dependent on precipitation in the Great Plains. Flowering occurs from May to June. By mid-summer, the above-ground portion of the plant dies back and a rosette forms at the end of summer (TPWD, 2025d). In times of drought, the above-ground parts will be absent while the plant suffocates in the deep sandy soils.

II.2.10 IMPACTS ON ECOSYSTEM SERVICES – WHY SHOULD WE DO ANYTHING?

Functional ecosystems provide benefits, often called ecosystem services, to humans living in or near those ecosystems. These services range from provision of food to improved water quality to cultural values (IPBES, 2019). As ecosystems are degraded through land use change, pollution, or climate change, the services they offer also degrade which can harm their monetary and cultural values. A report on the ecosystem services provided in Texas found four general categories of services: provisioning, regulating, cultural, and supporting services (Putman et al., 2022). Based on market estimations and consumer willingness to pay for services, this report valued Texas's ecosystem services at $89 billion per year (Table II.2.2). As discussed in Chapter II.1, climate change is only one of many planetary boundaries being exceeded. Other human activities (e.g., urbanization, pollution, water use, land change) are having a large effect on ecosystem services, but here, we focus on discussing how climate change will alter ecosystems and the services they provide.

TABLE II.2.2
Ecosystem Services Provided in Texas

Ecosystem Service Type	Ecosystem Service Description	Texas Ecosystems Offering Services	Estimated Monetary Value of Service (billion USD)
Provisioning	Services that provide tangible resources including freshwater supply, food, fiber, and timber	Pineywoods, Crosstimbers, plains/prairies	15
Regulating	Services that regulate quality of water and air as well as mitigate erosion and floods	Forested areas, wetlands, playas	11
Cultural	Services that offer recreational, cultural, or spiritual value	All Texas ecosystems	3
Supporting	Services that support all other ecosystem services and ensure ecological function such as nutrient cycling and water capture	All Texas ecosystems	60

Source: Adapted from **Putman et al. (2022).**

II.2.11 PROVISIONING SERVICES UNDER NEW CLIMATE REGIMES

Provisioning services are those that provide tangible materials to people, such as timber and food (Putman et al., 2022). Regions such as the Pineywoods and Crosstimbers offer trees for timbering and paper production, and rangeland in the plains and prairies of Texas supply food for cattle and other grazing animals. Additionally, the Gulf waters are prime sources of seafood for Texas citizens. Much of the agricultural industry in Texas relies on the food and fiber provided by its ecosystems, evidenced by the estimated $14.6 billion in service valuation (Putman et al., 2022).

Climate change, specifically increases in temperature and lower precipitation, could jeopardize this provisioning of resources. Texas is predicted to grow warmer and drier, which could reduce surface water supplies in reservoirs (Nielsen-Gammon et al., 2020), as well as crop productivity and livestock products (Bakure et al., 2022; Challinor et al., 2009). Corn and sorghum yields, in particular, are estimated to decline by 35% and 13%, respectively, in dryland regions of Texas by end of the century (see Chapter III.1 by Kothari et al., 2025). However, in some cases, positive responses in services are projected; for instance, wheat yields could increase up to 18% under climate projections (III.1). Warming oceans additionally could push fish populations to new locations, reducing fisheries' yield in the Texas gulf. Accompanied by land use change and population growth, climate change threatens the ecosystems and the direct resources they supply to Texans.

II.2.12 REGULATING SERVICES – ENVIRONMENTAL, CULTURAL, AND SUPPORTING SERVICES

Services that contribute to improving water and air quality or erosion and flood control are considering regulating services (MEA, 2005). These services are delivered mostly by vegetation filtering out pollutants from the atmosphere, water, or land surface, as well as trapping water and soil. Pollination is another regulating service that has significant economic value for agriculture. A diversity of plant species aids in pollution filtering, and intact forest ecosystems retain soil and water to reduce flood risks and erosion. Land use change threatens these services more directly than climate change but, as discussed previously, changing base conditions from climate change negatively impacts ecosystem resilience. Shifting phenology such as the timing of blooming and growing seasons will likely significantly affect pollinators (Marshman et al., 2019). Although difficult to evaluate, regulating ecosystem services is estimated to provide approximately $11 billion to Texas (Putman et al., 2022). This significant monetary value signifies the importance of regulating services and the necessity to preserve them in Texas.

Ecosystems also provide aesthetic values, specifically tourism, recreation, or cultural or religious values. These cultural services are typically valued indirectly through hunting permits or park entrance fees, but they are also important for their impact on human well-being (MEA, 2005). Value from hunting alone is estimated at $3 billion annually (Putman et al., 2022), but Texas also receives a large amount of tourism for parks such as Big Bend National Park, as well as other locales throughout the state that is difficult to quantify. Warming temperatures due to climate change can discourage tourism and hunting, and any disruption of ecosystem integrity will reduce the aesthetic or cultural values from Texas environments. Despite the difficulty in assigning a dollar amount to cultural ecosystem services, they are nonetheless an important component when considering the impact of climate change and human activity on the environment.

Not often consumed directly by humans, supporting ecosystem services ensure the proper cycling of resources throughout the environment, allowing for the provision of other ecosystem services. This includes the production of biomass needed for the species providing the provisioning, regulating, and cultural services, as well as the nutrient cycling required for vegetation growth (Bakure et al., 2022). A large supporting service is the replacement and enhancement of water quantity. The price of replacing water falling on developed land that would have been captured on undeveloped land was estimated to average almost $350/acre across Texas and $49 billion per year for the state

(Putnam et al., 2022). Supporting services account for over half of the ecosystem service value in Texas, highlighting their importance.

As with predicting impacts on the biosphere, it is also difficult to estimate precisely how climate change will affect supporting ecosystem services, especially in comparison to other human activities. For water services, higher temperatures, a decrease in rain in western Texas, and the removal of wetlands will dry out reservoirs and limit the water recharging into aquifers (Veettil et al., 2024). Accompanied by a continuing increase in the population, climate change will place an even greater stress on Texas water resources. Changes in phenology will disrupt the growth of biomass and therefore the cycling of nutrients, on which all ecosystems depend. The important takeaway for supporting services is that, of all ecosystem services, climate change endangers the resilience of ecosystems. Ecosystems will become more vulnerable to change and disruption, and in conjunction with the other planetary boundaries being crossed, the monetary and cultural benefits Texans receive from their ecosystems will be threatened.

REFERENCES

AAFA (Asthma and Allergy Foundation of America), 2025. Allergy capitals. https://aafa.org/wp-content/uploads/2025/03/aafa-2025-allergy-capitals-report.pdf

Acharya BS, Hao Y, Ochsner TE et al., 2017. Woody plant encroachment alters soil hydrological properties and reduces downward flux of water in tallgrass prairie. Plant Soil. 414:379–391. https://doi.org/10.1007/s11104-016-3138-0

Andermann T, Faurby S, Turvey ST, et al., 2020. The past and future human impact on mammalian diversity. Sci Adv. 6:eabb2313. https://doi.org/10.1126/sciadv.abb2313

Audubon Society, 2025. Audubon Christmas bird count. Audubon. https://www.audubon.org/community-science/christmas-bird-count

Bakure BZ, Hundera K, Abara M, 2022. Review on the effect of climate change on ecosystem services. IOP Conf. Ser.: Earth Environ. Sci. 1016:012055. https://doi.org/10.1088/1755-1315/1016/1/012055

Barbarossa V, Bosmans J, Wanders N et al., 2021. Threats of global warming to the world's freshwater fishes. Nat Commun. 12:1701. https://doi.org/10.1038/s41467-021-21655-w

Batavia C, Nelson MP, 2017. For goodness sake! What is intrinsic value and why should we care? Biol. Conserv. 209:366–376. https://doi.org/10.1016/j.biocon.2017.03.003

Bateman BL, Wilsey C, Taylor L, et al., 2020. North American birds require mitigation and adaptation to reduce vulnerability to climate change. Conserv. Sci. Pract. 2:e242. https://doi.org/10.1111/csp2.242

Baudier KM, O'Donnell S, 2017. Weak links: How colonies counter the social costs of individual variation in thermal physiology. Curr. Opin. Insect Sci. 22:85–91. https://doi.org/10.1016/j.cois.2017.06.004

Bock SL, Lowers RH, Rainwater TR, et al., 2020. Spatial and temporal variation in nest temperatures forecasts sex ratio skews in a crocodilian with environmental sex determination. Proc. R. Soc. B. 287(1926):20200210. https://doi.org/10.1098/rspb.2020.0210

Bone C, Moseley C, Vinyeta K, et al., 2016. Employing resilience in the United States forest service. Land Use Policy. 52:430–438. https://doi.org/10.1016/j.landusepol.2016.01.003

Challinor AJ, Ewert F, Arnold S et al. 2009. Crops and climate change: progress, trends, and challenges in simulating impacts and informing adaptation. J. Exp. Bot. 60:2775-2789. https://doi.org/10.1093/jxb/erp062

Caro T, Rowe Z, Berger J, et al., 2022. An inconvenient misconception: Climate change is not the principal driver of biodiversity loss. Conserv. Lett. 15:e12868. https://doi.org/10.1111/conl.12868

Chan JC, Lam BY, Dudgeon D, et al., 2025. Global consequences of dam-induced river fragmentation on diadromous migrants: A systematic review and meta-analysis. Biol. Rev. https://doi.org/10.1111/brv.70032

Chmura HE, Duncan C, Burrell G, et al., 2023. Climate change is altering the physiology and phenology of an arctic hibernator. Science. 380:846–849. https://doi.org/10.1126/science.adf5341

Chure G, Banks RA, Flamholz AI, et al., 2022. Anthroponumbers.org: A quantitative database of human impacts on planet earth. Patterns. 3(9):100552. https://doi.org/10.1016/j.patter.2022.100552

Coumou D, Robinson A, 2013. Historic and future increases in the global land area affected by monthly heat extremes. Environ. Res. Let. 8:034018. https://doi.org/10.1088/1748-9326/8/3/034018

Crimmins AR, Avery CW, Easterling DR, et al., 2023. Fifth National Climate Assessment. U.S. Global Change Research Program, Washington, DC, USA. https://doi.org/10.7930/NCA5.2023.CH1

Díaz S, Settele J, Brondízio ES, et al., 2019. Pervasive human-driven decline of life on earth points to the need for transformative change. Science. 366:eaax3100. https://doi.org/10.1126/science.aax3100

Ferguson M, Joanen T, 1982. Temperature of egg incubation determines sex in *alligator mississippiensis*. Nature. 296:850–853. https://doi.org/10.1038/296850a0

Fricke EC, Ordonez A, Rogers HS, et al., 2022. The effects of defaunation on plants' capacity to track climate change. Science. 375(6577):210–214. https://doi.org/10.1126/science.abk3510

Grill G, Lehner B, Thieme M et al., 2019. Mapping the world's free-flowing rivers. Nature. 569:215–221. https://doi.org/10.1038/s41586-019-1111-9

Haest B, Stepanian PM, Wainwright CE, et al., 2021. Climatic drivers of (changes in) bat migration phenology at bracken cave (USA). Glob. Chang Biol. 27(4):768–780. https://doi.org/10.1111/gcb.15433

Hagan E, 2022. As global temperatures rise, alligators are moving North. https://www.greenmatters.com/news/how-far-north-do-alligators-live

Hiller I, 1983. Young Naturalist. The Louise Lindsey Merrick Texas Environment Series, No. 6, Texas A&M University Press, College Station.

Hollister EB, Schadt CW, Palumbo AV, et al., 2010. Structural and functional diversity of soil bacterial and fungal communities following woody plant encroachment in the southern great plains. Soil Biol. Biochem. 42(10):1816–1824. https://doi.org/10.1016/j.soilbio.2010.06.022.

IPBES (Intergovernmental Science-Policy Platform on Biodiversity and Ecosystem Services), 2019. Global assessment report on biodiversity and ecosystem services of the Intergovernmental Science-Policy Platform on Biodiversity and Ecosystem Services (Version 1). Zenodo. https://doi.org/10.5281/zenodo.6417333other

IUCN (International Union for Conservation of Nature), 2025. IUCN red list of threatened species. https://www.iucnredlist.org/

Johnson CN, Balmford A, Brook BW, et al., 2017. Biodiversity losses and conservation responses in the Anthropocene. Science. 356(6335):270. https://doi.org/10.1126/science.aam9317

Kaiho K, 2022. Extinction magnitude of animals in the near future. Sci Rep. 12:19593. https://doi.org/10.1038/s41598-022-23369-5

Kothari K, Singh B, Samanta S, et al., 2025. Impacts of climate change on human-environmental systems. In Climate Change Impacts in Texas: Integrated Analysis and Adaptation, Wells M, McManamay RA, eds. CRC Press, Taylor & Francis Group, Boca Raton, Florida.

Li J, Ravi S, Wang G, et al., 2022. Woody plant encroachment of grassland and the reversibility of shrub dominance: Erosion, fire, and feedback processes. Ecosphere. 13(3):e3949. https://doi.org/10.1002/ecs2.3949

Lindsey R, Dahlman L, 2025. Climate change: Global temperature. https://www.climate.gov/news-features/understanding-climate/climate-change-global-temperature

Lueders MB, McManamay RA, 2023. Species depletion profiles as an alternative to streamflow alteration thresholds in a hydroecological risk assessment. Ecol. Indic. 147:109989. https://doi.org/10.1016/j.ecolind.2023.109989

Marshman J, Blay-Palmer A, Landman K, 2019. Anthropocene crisis: Climate change, pollinators, and food security. Environments. 6(2):22. https://doi.org/10.3390/environments6020022

Martinez LA, Campbell TA, Lopez RR, 2025. Enabling endangered species conservation on private land: A case study of the ocelot in Texas. Wildl. Soc. Bull. 49:e1594. https://doi.org/10.1002/wsb.1594

Maxwell S, Fuller R, Brooks T et al., 2016. Biodiversity: The ravages of guns, nets and bulldozers. Nature. 536:143–145. https://doi.org/10.1038/536143a

MEA (Millennium Ecosystem Assessment), 2005. Ecosystems and Human Well-Being: Synthesis. Island Press, Washington, DC., USA. https://www.millenniumassessment.org/documents/document.356.aspx.pdf

Mendoza P, Cerdan I, Garcia B, et al., 2021. Influence of incubation temperature on embryo development, hatchling morphology and early growth rate in red-footed tortoise (*Chelonoidis carbonaria*). Comp. Biochem. Physiol. A Mol. Integr. Physiol. 259:110999. https://doi.org/10.1016/j.cbpa.2021.110999

Millar CI, Stephenson NL, Stephens SL, 2007. Climate change and forests of the future: Managing in the face of uncertainty. Ecol. Appli. 17(8):2145–2151. https://doi.org/10.1890/06-1715.1

Monroe MJ, Butchart SHM, Mooers AO, et al., 2019. The dynamics underlying avian extinction trajectories forecast a wave of extinctions. Biol. Lett. 1520190633. http://doi.org/10.1098/rsbl.2019.0633

Mora C, Tittensor DP, Adl S, et al., 2011. How many species are there on earth and in the ocean? PLoS Biol. 9(8):e1001127. https://doi.org/10.1371/journal.pbio.1001127

Nielsen-Gammon J, Holman S, Buley A, et al., 2024. Assessment of historic and future trends of extreme weather in Texas, 1900–2036: 2024 update. Document OSC-202401, 40 pp, Office of the State Climatologist, Texas A&M University, College Station

Nielsen-Gammon JW, Banner JL, Cook BI, et al., 2020. Unprecedented drought challenges for Texas water resources in a changing climate: What do researchers and stakeholders need to know? Earth's Future. 8:e2020EF001552. https://doi.org/10.1029/2020EF001552

Omernik JM, 1987. Ecoregions of the conterminous United States. Ann. Am. Assoc. Geogr. 77(1):118–125. https://doi.org/10.1111/j.1467-8306.1987.tb00149.x

Ostberg S, Boysen LR, Schaphoff S, et al., 2018. The biosphere under potential Paris outcomes. Earth's Future. 6:23–39. https://doi.org/10.1002/2017EF000628

Pathak A. 2024. Climate Resilient Galveston: Understanding climate vulnerabilities and adaptation strategies to build resilience. Austin, TX: National Wildlife Federation. https://texaslivingwaters.org/wp-content/uploads/Galveston-Assessment-Final-5.29-Web.pdf

Perkins-Kirkpatrick SE, Lewis SC, 2020. Increasing trends in regional heatwaves. Nat. Comm. 11:3357. https://doi.org/10.1038/s41467-020-16970-7

Perry D, Praskievicz S, McManamay RA, et al., 2024. Resilient riverine social-ecological systems: A new paradigm to meet global conservation targets. WIREs Water. 11(6):e1753. https://doi.org/10.1002/wat2.1753

Putman A, Lopez R, Smith L, et al., 2022. Texas ecosystem services: A statewide assessment. Texas A&M Natural Resources Institute, Research Report Number 2022-1. College Station, Texas, USA.

Qiu H, Yan Q, Yang Y et al., 2024. Flowering in the Northern hemisphere is delayed by frost after leaf-out. Nat. Commun. 15:9123. https://doi.org/10.1038/s41467-024-53382-3

Radchuk V, Reed T, Teplitsky C et al., 2019. Adaptive responses of animals to climate change are most likely insufficient. Nat Commun. 10:3109. https://doi.org/10.1038/s41467-019-10924-4

Reigh RC, Williams MB, 2020. Effects of relative humidity and elevated oxygen concentrations on egg-hatching success of American alligator, *Alligator mississippiensis*. J. World Aquacult. Soc. 51:982–989. https://doi.org/10.1111/jwas.12719

Ryberg WA, Lawing AM, 2018. Genetic consequences and management implications of climate change for the American Alligator (Alligator mississippiensis). In American Alligators: Habitats, Behaviors, and Threats, Hanke SE, Eversole CB, eds. Nova Science Publishers, INC, Hauppauge, New York.

Smith LA, Lopez RR, Lund AA, et al., 2025. Status update and trends of Texas working lands. Texas A&M Natural Resources Institute, College Station, TX, USA.

The Meadows Center for Water and the Environment, 2024. Sea-level rise. Texas State University. https://www.meadowscenter.txst.edu/climatechange/climatedashboard/sealevelrise.html

The Nature Conservancy, 2025. Seeding the future. https://www.nature.org/en-us/about-us/where-we-work/united-states/texas/stories-in-texas/seeding-the-future/

Theobald DM, Oakleaf JR, Moncrieff G et al., 2025. Global extent and change in human modification of terrestrial ecosystems from 1990 to 2022. Sci Data. 12:606. https://doi.org/10.1038/s41597-025-04892-2

TPWD (Texas Parks and Wildlife Department), 2022. Texas ecoregions. https://tpwd.texas.gov/education/hunter-education/online-course/wildlife-conservation/texas-ecoregions

TPWD (Texas Parks and Wildlife Department), 2025a. Comanche Springs Pupfish (Cyprinodon elegans). https://tpwd.texas.gov/huntwild/wild/species/comanchespringspupfish/

TPWD (Texas Parks and Wildlife Department), 2025b. Clear Creek Gambusia (Gambusia heterochir). https://tpwd.texas.gov/huntwild/wild/species/clearcreekgambusia/

TPWD (Texas Parks and Wildlife Department), 2025c. Navasota Ladies'-Tresses (Spiranthes parksiii). https://tpwd.texas.gov/huntwild/wild/wildlife_diversity/nongame/listed-species/plants/navasota_ladies_tresses.phtml

TPWD (Texas Parks and Wildlife Department), 2025d. Texas Poppy-Mallow (Callirhoe scabriuscula). https://tpwd.texas.gov/huntwild/wild/wildlife_diversity/nongame/listed-species/plants/texas_poppy_mallow.phtml

TPWD (Texas Parks and Wildlife Department), 2025e. Rare, threatened, and Endangered Species of Texas. https://tpwd.texas.gov/gis/rtest/

TWDB (Texas Water Development Board), 2025. 2026 regional water plan board-adopted population and municipal demand. https://www.twdb.texas.gov/waterplanning/data/projections/2027/municipal.asp

U.S. Census Bureau, 2024. 2024 census survey. https://www.census.gov/programs-surveys/decennial-census/decade/2030/planning-management/plan/research-and-testing/2024-census-survey.html

U.S. EPA (U.S. Environmental Protection Agency), 2016. Climate change indicators: Ragweed Pollen season. https://www.epa.gov/climate-indicators/climate-change-indicators-ragweed-pollen-season#:~:text=Ragweed%20plants%20mature%20in%20mid,pollen%20until%20the%20first%20frost.

U.S. EPA (U.S. Environmental Protection Agency), 2025. Climate change indicators: Bird wintering ranges. https://www.epa.gov/climate-indicators/climate-change-indicators-bird-wintering-ranges#:~:text=A%20total%20of%2048%20species,27%20percent)%20have%20shifted%20southward

U.S. FWS (United States Fish and Wildlife Service), 2024. Service approves conservation agreement for six aquatic species in the Trinity River basin. https://www.fws.gov/press-release/2024-03/conservation-agreement-six-aquatic-species-trinity-river-basin

U.S. FWS (United States Fish and Wildlife Service), 2025. Peck's Cave Amphipod. https://www.fws.gov/species/pecks-cave-amphipod-stygobromus-pecki

Veettil AV, Fares A, Awal R, et al., 2024. Assessing future water scarcity in Texas under climate change and growing water demands. J. Hydrol. Eng. 29(6):05024022. https://doi.org/10.1061/JHYEFF.HEENG-6253

Venter O, Sanderson E, Magrach A et al., 2016. Sixteen years of change in the global terrestrial human footprint and implications for biodiversity conservation. Nat. Commun. 7:12558. https://doi.org/10.1038/ncomms12558

Weiskopf SR, Rubenstein MA, Crozier LG, et al., 2020. Climate change effects on biodiversity, ecosystems, ecosystem services, and natural resource management in the United States. Sci. Total Environ. 733, 137782. https://doi.org/10.1016/j.scitotenv.2020.137782

Winkler K, Fuchs R, Rounsevell M et al., 2021. Global land use changes are four times greater than previously estimated. Nat. Commun. 12:2501. https://doi.org/10.1038/s41467-021-22702-2

Woodhead AJ, Kenter JO, Thomas CD et al., 2025. How ecosystem services are co-produced: a critical review identifying multiple research framings. Ecosyst. Serv. 71: 101694. https://doi.org/10.1016/j.ecoser.2024.101694

Ziska L, Knowlton K, Rogers C, et al., 2011. Recent warming by latitude associated with increased length of ragweed pollen season in central North America. Proc. Natl. Acad. Sci. U.S.A. 108(10):4248–4251. https://doi.org/10.1073/pnas.1014107108

II.3 Climate Change Impacts on Freshwaters

Recent Trends and Future Predictions for Biodiversity, Ecosystem Function, and Human Wellbeing

Matthew J. Troia

II.3.1 A GENERAL OVERVIEW OF FRESHWATER BIODIVERSITY, ECOSYSTEMS, AND SERVICES

Globally, freshwater environments are diverse in their physical, chemical, and biological characteristics (Figure II.3.1). These environments include lentic (i.e., non-flowing) waterbodies like lakes, ponds, oxbows, wetlands, resacas, and ciénegas. Lotic (i.e., flowing) waterbodies range from clear and turbulent streams flowing off steep mountain slopes to large muddy rivers draining vast regions of continents. Freshwater scientists and natural resource practitioners have long focused their research and management efforts on perennial waterbodies because they are the most conspicuous aquatic features on the landscape, yet ephemeral and intermittent waterbodies are increasingly recognized for their key roles in harboring biodiversity and transporting nutrients and materials (Datry et al., 2014). Over half of the world's streams are non-perennial and this proportion is increasing as climate change increases the frequency and severity of droughts and human populations increasingly pump groundwater and divert surface water for societal uses. Streams and rivers are intimately connected to terrestrial landscapes (Allan, 2004) and therefore play a significant role in the storage and transport of sediments (Li et al., 2020; Milliman and Meade, 1983) and key nutrients like carbon (Battin et al., 2009; Butman et al., 2016; Liu et al., 2022), nitrogen (Fowler et al., 2013; Vitousek et al., 1997), and phosphorus (Filippelli, 2008; Withers and Jarvie, 2008).

The physicochemical diversity of freshwater bodies and their inherent isolation among one another has generated incredible biodiversity that is unmatched in terrestrial and marine ecosystems. Freshwater environments cover only 2.3% of the Earth's surface, yet these ecosystems harbor at least 9.5% of described animal species (Reid et al., 2019). Nevertheless, this biodiversity is threatened by multiple anthropogenic stressors including habitat degradation, water pollution, hydrologic alteration, invasive species, and overexploitation of organisms (Dudgeon et al., 2006). Freshwater is essential to human societies. Streams, rivers, lakes, reservoirs, and wetlands provide numerous supporting, provisioning, regulating, and cultural services (Keeler et al., 2012; Postel and Carpenter, 1997).

II.3.2 AN OVERVIEW OF CLIMATE CHANGE IMPACTS ON FRESHWATERS

Direct effects of anthropogenic climate change—increasing temperature and changing precipitation regimes—affect the quantity and quality of water in flowing and non-flowing environments, which then impact numerous physical, chemical, and biological processes (Figure II.3.2A). Prior

DOI: 10.1201/9781003406600-10

FIGURE II.3.1 Examples of freshwater environments highlighting the geographic diversity beyond the borders of Texas. (a) A cold-water stream draining the Blue Ridge Mountains of Tennessee, USA. (b) Laguna Toreadora, A high-elevation natural lake in the Ecuadorean Andes. (c) Norris Lake, a reservoir on the Clinch River in Tennessee, USA, resulting from the first dam completed by the Tennessee Valley Authority in 1936. (d) Peat swamp forest in Sumatra, Indonesia. (Photo credits: Matthew Troia and wikicommons.org.)

investigators have reviewed the impacts of climate change on freshwater ecosystems, focusing on particular environmental responses (e.g., biodiversity or human health) and different spatial extents (e.g., southeastern United States or global) (Covich, 2009; Dawson et al., 2003; Knouft and Ficklin, 2017; Kundzewicz et al., 2008; Kundzewicz et al., 2014; Meyer et al., 1999; Mulholland et al., 1997; Poff et al., 2002; Woodward et al., 2010). Reduced water quality manifests as increasing water temperature and (in some geographic contexts) salinity and increasing frequency, duration, and spatial coverage of hypoxia (Kaushal et al., 2010; Olson 2019; Zhi et al., 2023). Water quantity generally declines as precipitation decreases and evapotranspiration increases for streams, wetlands, and reservoirs (Van Vliet et al., 2023). At the same time, and for streams in particular, shifts in the seasonality of precipitation and the increase in frequency of extreme precipitation events (including

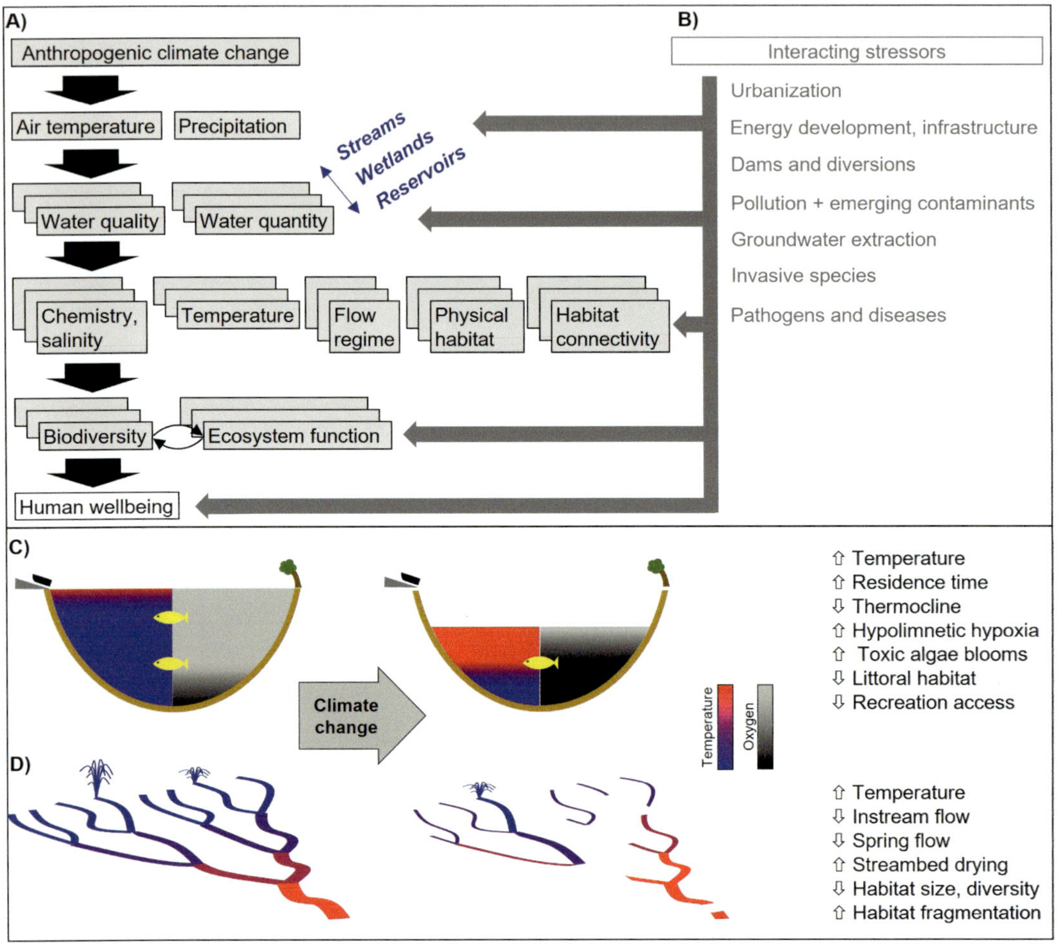

FIGURE II.3.2 Chapter outline and conceptual overview of (A) components of climate change, their impacts on three primary freshwater ecosystem types (blue text) and human wellbeing, and (B) interactions with other anthropogenic stressors. Visual depiction of impacts of climate change on (C) lentic and (D) lotic waterbodies.

tropical storms that dump large volumes of rainfall on coastal landscapes) and heatwaves alter the more nuanced hydrologic regime of streams (Stewart et al., 2005; Tonkin, 2022)—defined as the magnitude, frequency, timing, duration, and rate of change of low- and high-flow events in streams (Poff et al., 1997). Lower flows in lotic systems reduce the amount and diversity of in-stream habitat (Schlosser, 1982; Vannote et al., 1980) and, in extreme scenarios when stream channels dry, habitat is lost completely and remaining habitat patches become fragmented (Malish et al., 2024; Perkin et al., 2015). Similarly, lower water levels in lakes and reservoirs erode access to the littoral zone where important habitat features like woody debris and emergent vegetation are most prevalent (Gaeta et al., 2014; Miranda et al., 2020). Wetlands, by their very definition, have both 'wet' and 'land' characteristics which are maintained by the natural hydroperiod. Hence, any changes in temperature, precipitation, runoff, and evapotranspiration disrupt this natural hydroperiod and subsequently the structure and function of wetland ecosystems with consequences for biodiversity and services to human societies.

Changes in water quality and quantity both affect biodiversity. In streams, increasing temperatures shift suitable thermal habitat upstream (Isaak et al., 2016; Troia et al., 2019) and these shifts often cause erosion of the downstream limits of cold-adapted species and expansion of the upstream limits of warm-adapted species (Comte et al., 2013). Increasing temperatures and hypoxia in lakes and reservoirs are expected to cause extirpations of recreationally-desirable cold-adapted species (e.g., Smallmouth Bass, *Micropterus dolomieu*) and persistence and establishment of undesirable non-native species that are physicochemically tolerant (e.g., Common Carp, *Cyprinus carpio*) (Jones et al., 2013). Increasing salinity in inland freshwaters also is causing the replacement of stenohaline species with euryhaline species (Hoagstrom, 2009; Kefford et al., 2016; Mahan et al., 2022). A positive relationship between stream discharge (i.e., a correlate of habitat size and diversity) and lotic species richness is a universal pattern (Tedesco et al., 2013). Thus, declining flows caused by climate change are forecasted to erode species richness in streams and rivers (Atkinson et al., 2014; Xenopoulos et al., 2005), although strong empirical evidence documenting the start of such biodiversity erosion is lacking. Similarly, declining water surface levels in lakes and reservoirs impact growth, recruitment, and reproduction of fishes and other aquatic taxa (Sass et al., 2006; Yuan et al., 2013).

Ecosystem function is the totality of ecosystem processes that maintain biodiversity and provide services to human societies. These functions include but are not limited to the production of biomass (including that of microbes, plants, and animals), decomposition of organic matter, cycling of nutrients, and transport of sediments and nutrients. Field observations and modeled forecasts indicate that increasing temperatures increase biomass production at all trophic levels (Brylinsky and Mann, 1973; Mooij et al., 2005; Tranvik et al., 2009; Woolway et al., 2020; but see O'Reilly et al., 2003) and also accelerate rates of organic matter decomposition (Dang et al., 2009; Woodward et al., 2010; but see Meyer and Pulliam, 1992). Whole ecosystem metabolism is the balance between the rates of autotrophy (i.e., inorganic carbon converted to organic carbon) and heterotrophy (i.e., organic carbon used for metabolism or growth) and therefore determines whether a river, for example, sequesters carbon or exports carbon downstream (Bernhardt et al., 2018; Dodds, 2007). Because respiration is more sensitive to temperature than photosynthesis, climate warming generally is shifting freshwater ecosystems toward heterotrophy and therefore reduced carbon retention (Battin et al., 2023; Busch and Fisher, 1981). Rates of autotrophy and heterotrophy set limits on the cycling of key macronutrients like nitrogen and phosphorus (Dodds, 2007; Tank and Dodds, 2003). In streams, bulk export of organic carbon such as algae mats and leaf litter during floods affect ecosystem metabolism (Ruffing et al., 2022; Uehlinger, 2000). Thus, changes in the frequency and magnitude of flood and drought events caused by climate change affect these ecosystem processes (Gudmundsson et al., 2021). Lastly, hypoxia in lentic waterbodies is increasing with climate change with cascading effects on nutrient cycling, biodiversity, and the quality of water used by human societies (Jane et al., 2021; Jansen et al., 2024).

Other pervasive and emerging anthropogenic stressors interact with climate change to impact freshwater ecosystems typically in additive or multiplicative ways (Figure II.3.2B). For example, nutrient pollution and increasing temperature combine to increase the frequency and intensity of hypoxic conditions and harmful algae blooms (Jane et al., 2021; Paerl et al., 2016). Increasing frequency of flood events that disrupt the natural flow regimes of streams will become more intense as urbanization increases impervious surfaces that transmit runoff to streams more rapidly (Jha et al., 2012; Shao et al., 2020). Alternatively, climate change-induced stream drying causes extirpations of obligate aquatic species and dams prevent recolonization following the return of flow (Perkin et al., 2015). In karst landscapes like the Edwards Plateau of Central Texas, stable thermal regimes are buffered from rising air temperatures but this buffering capacity of springs is eroded when groundwater extraction for irrigation and municipal use reduces spring flow (Caldwell et al., 2020; Craig and Bonner, 2021). Native cold-adapted fishes are losing suitable thermal habitat, while also facing competition and predation by warm-adapted non-native fishes (Wenger et al., 2011). Similarly, climate change is facilitating the expansion of invasive crayfish in Europe that transmit the crayfish plague (*Aphanomyces astaci*) to native crayfishes (Capinha et al., 2013)—such emerging infection diseases will likely continue with climate change (Reid et al., 2019).

Thresholds, tipping points, and regime shifts are increasingly recognized in the freshwater climate change literature (Carrier-Belleau et al., 2022; Hernández Martínez de la Riva et al., 2023). For example, biogeochemical studies of Arctic lake sediment cores have shown that gradual warming has led to abrupt changes in sediment organic matter content and isotopic composition (Luoto et al., 2019). Similarly, small changes in drought intensity cause large changes (i.e., threshold effects) in abundances of macroinvertebrates in spring-fed headwater streams (Aspin et al., 2019). Although many knowledge gaps still remain, it is clear that freshwater ecosystem characteristics representing all levels of biological organization are likely to change in such non-linear ways in response to climate change.

II.3.3 TEXAS FRESHWATERS: THE NATURAL HISTORY BACKDROP

Texas is big and therefore spans a diverse suite of climatic and geological backdrops on which waterbodies are situated. It is no surprise, then, that freshwater environments in Texas are also very diverse (Figure II.3.3). Thirteen drainage basins larger than 5,000 km^2 and numerous smaller basins drain the Texas landscape generally from a northwest to a southeasterly direction (Figure II.3.4A). The prominent east to west precipitation gradient drives the transition of lotic waterbodies from perennial to intermittent or ephemeral (Figure II.3.4B). From north to south, waterbodies experience fewer and lower magnitude freezing events in the winter, which affect biodiversity, ecosystem functions, and human interactions (Lopez et al., 2023). Wetlands are also more prominent in the east and are more likely to be forested in the east as opposed to herbaceous (Figure II.3.4C–D). While large natural lakes are common in other areas of North America—think Minnesota's Land of 10,000 Lakes nickname—these features are rare in the southern United States, including in Texas. Caddo Lake, located on Texas's northeastern border with Louisiana, is the only natural lake in the state and uniquely formed in the late 18th century when a colossal logjam impounded the flow of Twelve Mile Bayou in the Red River basin (Figures II.3.3B, II.3.4F). At least 8,278 springs have been mapped across the state (Brune, 2002), but the highest density is along the Balcones Escarpment that defines the southeastern boundary of the Edwards Plateau (Figure II.3.4). Two of these springs—Comal Springs and San Marcos Springs—are first magnitude springs, meaning that their average discharge exceeds 100 cubic feet per second. San Felipe Springs regularly exceeds 10 cubic feet per second and is a second magnitude spring (Figure II.3.4A). The diversity of geological formations over and through which freshwaters flow drives substantial spatial variation temperature regimes and chemical signatures of streams and rivers. For example, surface water emanating from the abundant springs on the Edwards Plateau is clear, thermally moderated, and rich in dissolved minerals like bicarbonate, sulfate, and chloride taken up from the limestone aquifer. In contrast, streams draining the High Plains of the Texas Panhandle such as the Red and Canadian Rivers are turbid, thermally variable, and particularly high in salinity owing to naturally salty springs and soils.

The diversity of freshwater habitats has produced and maintained an equally impressive diversity of life (Figure II.3.5). Biodiversity can be measured in many ways (Magurran, 2013), but species richness (i.e., the total number of species occupying a habitat) and endemism (i.e., the proportion of species that occur nowhere else on Earth) are common and useful measures—freshwater habitats in Texas are impressive by both measures. Spatial variation in biodiversity is also a natural and important component of biodiversity, but anthropogenic activities contribute to the homogenization of biodiversity by facilitating geographic expansion of generalist species and extirpation of specialist species (Olden et al., 2004). The biodiversity of freshwater environments in Texas is no exception to this threat (Miyazono et al., 2015; Perkin and Bonner, 2011).

Riparian and aquatic plants form the base of aquatic food webs and provide essential habitat for aquatic animals. Of the hundreds of species associated with Texas freshwaters, 31 are classified as Species of Greatest Conservation Need (TPWD, 2023). Texas Wild Rice (*Zizania texana*) is a federally endangered species that is restricted to the upper two miles of the San Marcos River in Hays County. Several non-native aquatic plants have established in Texas and have substantial impacts on freshwater ecosystems and are economically costly. For example, Hydrilla (*Hydrilla verticillata*) is

FIGURE II.3.3 Examples of freshwater environments highlighting the geographic diversity in Texas. (A) San Felipe Creek, the third largest spring-fed stream in Texas, Val Verde County. (B) Caddo lake, the only natural lake in Texas formed by a natural logjam during the 18th century, Harrison/Marion Counties. (C) McKittrick Creek, an intermittent stream in the Chihuahuan Desert, Culberson County. (D) The tailwater reach of the Guadalupe River downstream from Canyon Lake Dam, Comal County, supports the southern-most cold-water trout fishery in the United States. (Photo credits: Matthew Troia and wikicommons.org.)

a submerged vascular plant that is common in Texas reservoirs and forms dense stands that impact, both positively and negatively, sportfish production, angling success, and recreationist satisfaction (Bellinger and De Jesús, 2024). Rivercane (*Arundo donax*), in contrast, is an emergent bamboo-like plant that outcompetes native riparian vegetation along streams and rivers, draws down water tables, and alters natural channel morphology (Jain et al., 2015; Seawright et al., 2009).

At least 50 species of native freshwater mussels of the family *Unionidae* occur in Texas freshwaters and approximately 15 of these species occur nowhere else but in Texas (Dascher et al., 2018; Howells et al., 1996). These mussels are important components of freshwater ecosystems given their potential to reach high densities on riverbeds combined with their filtering and burrowing behaviors that mobilize and cycle nutrients (Figure II.3.5A; Vaughn, 2010). Several non-native mussels have

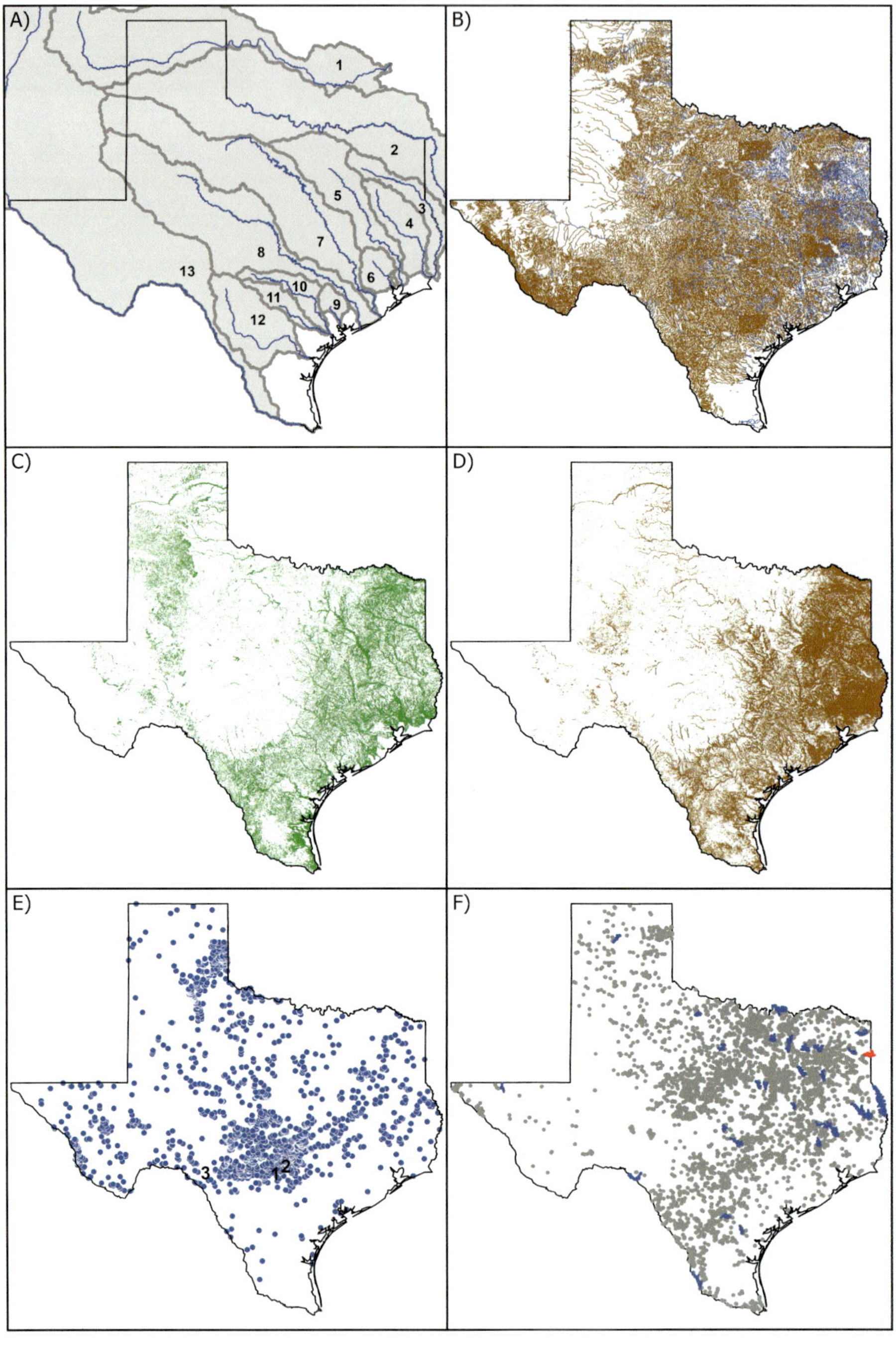

FIGURE II.3.4 Distribution of freshwater environments in Texas. (A) River basins >5,000 km^2 and their mainstems (1-Canadian, 2-Red, 3-Sabine, 4-Neches, 5-Trinity, 6-San Jacinto, 7-Brazos, 8-Colorado, 9-Lavaca, 10-Guadalupe, 11-San Antonio, 12-Nueces, 13-Rio Grande). (B) Distribution of perennial (blue) and intermittent/ephemeral (brown) streams following the National Hydrography Dataset FCodes. Wetlands representing (C) emergent herbaceous and (D) woody wetlands following the National Land Cover Dataset. (E) Documented springs following Brune (2002) including the three largest springs in Texas (1-Comal springs, 2-San Marcos springs, 3-San Felipe springs). (F) Human-made reservoirs >50 km^2 (blue) plus Caddo lake (red), the sole natural lake in Texas. Gray symbols show the 4,020 documented dams from the National Anthropogenic Barrier Dataset.

FIGURE II.3.5 Examples of invertebrates, fishes, and herptiles highlighting native and non-native components of freshwater biodiversity in Texas. (A) Texas Fawnsfoot (*Lampsilis bracteata*), a native freshwater mussel. (B) Australian Redclaw Crayfish (*Cherax quadricarinatus*), a recently-established non-native crustacean capable of overland dispersal. (C) Freshwater shrimp (genus *Macrobrachium*), a migratory crustacean that reproduce in the ocean and migrate into Texas rivers. (D) Spotted Sucker (*Minytrema melanops*), a broadly distributed native fish that reaches its southwesterly geographic range limit in East Texas. (E) Rio Grande Darter (*Etheostoma grahami*), the most southwesterly distributed species of the highly diverse *Etheostomatinae* subfamily. (F) Asian Carp (genus *Hypophthalmichthys*), an invasive fish that is established in the Red river and may spread to other Texas river basins. (G) San Marcos Salamander (*Eurycea nana*), an amphibian endemic to San Marcos Springs. (H) Concho Watersnake (*Nerodia paucimaculata*), an endemic species restricted to Central Texas streams and reservoirs. (Photo credit: Garrett Tucker and wikicommons.org.)

established, including the Asian fingernail clam (*Corbicula fluminea*) and Zebra mussel (*Dreissena polymorpha*). The former mussel is widespread throughout Texas (Howells et al., 1996) and, when abundant, may have stronger effects on nutrient cycling than native mussels (Hopper et al., 2022). The latter mussel has wreaked havoc on lakes and reservoirs elsewhere in North America (Nalepa and Schloesser, 1992) and potentially is an emerging threat to reservoirs and rivers in Texas (Beason and Schwalb, 2022; Robertson et al., 2020). The diversity of crayfish (Infraorder *Astacidea*) in Texas freshwaters is modest and comprehensive knowledge is limited relative to that of other freshwater taxa (Troia and McManamay, 2016), although some investigators have expanded knowledge on particular river basins and species (Birnbaum et al., 2007; Hays et al., 2023). Several species of crayfish are globally invasive and new species continue to emerge. In Texas, the Australian Redclaw Crayfish (*Cherax quadricarinatus*) was detected in Brownsville (Cameron County) in 2022 and represents an emerging threat to native species if it continues to spread to other areas of Texas (Figure II.3.5B; Sanjar et al., 2023). Interestingly, several species of freshwater prawn of the genus *Macrobrachium* (Infraorder *Caridea*) migrate between freshwater and marine environments in Texas but, like crayfish, also remain poorly studied (Figure II.3.5C; Horne and Beisser, 1977). Springs in West Texas harbor at least 12 species of hyperendemic springsnails (Perez et al., 2023). The iconic Comal and San Marcos Springs in Central Texas are home to hyperendemic springsnails, amphipods, and beetles (Longley, 1986). Many of these spring-associated invertebrates are listed endangered under the U.S. Endangered Species Act.

One hundred and ninety-six native fish species and another 21 non-native species have been documented in Texas freshwaters (Thomas et al., 2007) and this number continues to grow as taxonomists continue to identify unique populations and describe new species (e.g., Dye et al., 2024). Non-native fishes began establishing in the 19th century and—for better or worse—new species continue to establish (e.g., Best et al., 2022). Fish diversity increases from west to east along the precipitation gradient ostensibly because the amount of freshwater habitat increases, which invokes diversity-maintaining mechanisms of the species-area relationship (Figure II.3.5D; Hoeinghaus et al., 2007; Muneepeerakul et al., 2008). Although there are more fish species in the east, central and western Texas harbor more endemic species. Indeed, Texas has more endemic freshwater fishes than any other U.S. state (Figure II.3.5E; Troia, 2017). These species tend to occupy isolated springs and include various roundnose minnows (genus *Dionda*), livebearers (genus *Gambusia*), and pupfishes (genus *Cyprinodon*) (Hoagstrom et al., 2011). Texas freshwater fish face numerous threats to their long-term persistence—including but not limited to climate change! Many of these species are imperiled because they occupy small geographic ranges—some like the Leon Springs pupfish (*Cyprinodon bovinus*) are limited to a single spring in arid West Texas (Thomas et al., 2007). Other species including the Blue Sucker (*Cycleptus elongatus*), Plains Minnow (*Hybognathus placidus*), and River Darter (*Percina shumardi*) have broader geographic distributions, but reach their southwestern limits in Texas and are experiencing population declines and consequent range erosion (Birdsong et al., 2020).

All of these Texas freshwater fishes play essential roles in the ecosystems they occupy (Franssen et al., 2011; Hargrave, 2009) and have intrinsic and economic value (Monroe et al., 2009; Thomas, 2015; Wilson, 1986). A subset of these species possesses direct economic value. For example, tilapias (genus *Oreochromis*) and Channel Catfish (*Ictalurus punctatus*) are produced at aquaculture facilities and generated above $42 million in annual revenue in Texas in 2022 (USDA, 2024). Other species including the iconic State Fish of Texas—the Guadalupe Bass (*Micropterus treculii*)—as well as hybrid striped bass (genus *Morone*), crappie (genus *Pomoxis*), Largemouth Bass (*Micropterus salmoides*), and the massive Alligator Gar (*Atractosteus spatula*) are sought after by subsistence and recreational anglers and generate economic revenue for angling guides and local communities that support these fisheries (Thomas, 2015). For example, bass anglers at just one popular reservoir, Lake Fork in northeastern Texas, contributed over $18 million per year directly into that local economy during the 2014–2015 period of study (Bennett et al., 2024). Still other species such as invasive suckermouth armored catfishes (genera *Hypostomus* and *Pterygoplichthys*) and Asian carps (genus *Hypophthalmichthys*) degrade freshwater ecosystems, interfere with native species, and cost economies (Havel et al., 2015; McGarrity, 2019) (Figure II.3.5F). In 2023, for example, more than $260,000 were allocated to the removal of non-native plants and fishes in the Comal and San Marcos Rivers (Edwards Aquifer Authority, 2023).

Texas is home to 44 frog and toad species and 28 salamander species (Hibbitts et al., 2012). Like fishes, these amphibian groups also generally increase in richness and abundance from west to east as the amount of fresh surface water increases (Rogers, 1976). Springs emanating from the Edwards-Trinity Aquifer harbor numerous species of spring-associated salamanders of the genus *Eurycea* and the cryptic diversity of these species has only recently been recognized with the application of molecular ecology techniques (Figure II.3.5G; Devitt et al., 2019). Interestingly, one of these species—the Texas Blind Salamander (*Eurycea rathbuni*)—has colonized subterranean habitats within the Edwards-Trinity Aquifer and lacks eyes and external pigmentation. The Houston Toad (*Anaxyrus houstonensis*) is a noteworthy endangered amphibian that is geographically restricted to southeastern Texas and reproduces in ephemeral ponds (Duarte et al., 2011). Aquatic reptiles also increase in richness and abundance from west to east. Noteworthy species include the Concho Watersnake (*Nerodia paucimaculata*) from central Texas (Figure II.3.5H; Greene et al., 1994; Greene et al., 1999) and the Alligator Snapping Turtle (*Macrochlys temminckii*) in East Texas, both of which are endangered. Wetlands (Figure II.3.4C–D) are also essential habitats for waterfowl such as ducks, geese, and cormorants, as well as wading birds like herons, spoonbills, and sandpipers (Shackelford et al., 2005). Waterfowl hunters injected over $200 million to the Texas economy in 2006 (Carver, 2006), while travel and equipment expenditures by birdwatchers contribute additional millions of U.S. dollars each year (Lopez-Hoffman et al., 2017).

II.3.4 TEXAS FRESHWATERS: THE HUMAN HISTORY BACKDROP

The most recent archeological knowledge indicates that prehistoric humans arrived in Texas no later than 16,000 years ago (Williams et al., 2018). Since then, human societies have relied on perennial freshwater environments, particularly in arid central and western regions. For example, prehistoric sites abound near Comal, San Marcos, San Antonio, and San Felipe Springs, as well as numerous other smaller springs in central and western Texas (Collins, 2004; Nickels and Bousman, 2010). Prehistoric sites are also common along East Texas rivers including the Neches and Sabine Rivers (Abernethy, 2008; Speir, 2017). Indigenous peoples living in Texas prior to the arrival of Europeans include but are not limited to the Coahuiltecans in South Texas, Jumano in West Texas, and Caddo in East Texas (Hickerson, 1994; Jensen et al., 2006; Speir, 2017). At the time of European arrival, and likely long before, indigenous peoples had developed irrigation practices to cultivate beans and other crops in West Texas, including at San Solomon Springs (Jensen et al., 2006). Indigenous peoples in Texas also harvested fish and mussels from streams and rivers (Abernethy, 2008; Haag, 2012; Patterson, 2009).

With the arrival of Europeans came more widespread alteration of freshwater bodies for agricultural irrigation. Dams and canals were constructed to divert river and spring flow to agricultural fields. For example, Spanish missionaries arriving in the early 18th century to the area that is now the municipality of San Antonio relied on diversion dams and an elaborate network of canals to distribute flow from San Antonio and San Pedro Springs to the missions they constructed nearby and to surrounding fields in which they raised cattle and grew crops (Kemp et al., 2024). In addition to diversions, early Europeans recognized the bountiful water resources available below ground in the Edwards Aquifer, and quickly excavated wells to exploit this source of water for agriculture (Brune, 2002; Craig and Bonner, 2021).

The mid-twentieth century brought the 'Golden Age' of dam building to the United States and to Texas (Reisner, 1993). At least 4,020 dams are documented in Texas with construction years spanning from 1849 to 2007 (Ostroff et al., 2013). While most of these dams are small—66% are less than 10 meters tall—there are 30 dams that impound reservoirs that cover 50 or more square kilometers (Figure II.3.4F). These dams provide key services to the people of Texas, including water supply during droughts, protection from floods, hydroelectricity, and recreation (Brismar, 2002; Schorr et al., 1995). At the same time, dams and reservoirs eliminate and fragment natural lotic habitats, alter natural flow and sediment regimes, and harbor non-native species that negatively impact native freshwater species and ecosystem services (Graf, 1999; Kondolf, 1997; Liew et al., 2016; Perkin et al., 2013; Perkin et al., 2017). The later 20th and early 21st centuries have seen exponential growth of the human population in Texas—there are nearly double the number of residents in the early 2020s compared to 1990. With this growth, both groundwater and surface water resources are increasingly exploited for industrial, agricultural, and municipal consumption. At the same time, large metropolitan populations and industry produce substantial volumes of wastewater that are discharged into streams and rivers (Luthy et al., 2015). Some streams that were historically spring fed like the San Antonio River (Bexar County) are now maintained exclusively by treated wastewater.

II.3.5 CLIMATE CHANGE IMPACTS ON TEXAS FRESHWATER ECOSYSTEMS

Texas is not unique in that the impacts of climate change and interacting anthropogenic stressors observed and predicted globally also threaten freshwaters in the Lone Star State (Figures II.3.1, II.3.6). However, because Texas is big and encompasses so many different climates, physiographic provinces, and floral and faunal zones, the nature and severity of climate change impacts vary across the state and among freshwater environments. Research on the impacts of climate change in Texas has accumulated since the early 1990s (Figure II.3.7). This research is unevenly distributed among different freshwater habitats (Figure II.3.7A–B) and the impacts on those habitats (Figure II.3.7C–D). The following subsections detail this current state of knowledge organized by type of freshwater environment: streams and rivers, wetlands, and reservoirs.

II.3.5.1 Streams and Rivers

Flowing ecosystems in Texas may be the most vulnerable to climate change for several reasons. First, given the lack of natural lakes in Texas (Figure II.3.4F), streams and rivers are the predominant *natural* freshwater environments and harbor more species than wetlands and the many human-made reservoirs. Second, humans in Texas have already modified stream ecosystems by removing water for irrigation, municipal, and industrial use; returning wastewater to streams; and constructing large reservoirs that destroy and fragment stream habitat (Zhang and Wurbs, 2018). Third, and most importantly, most streams are naturally intermittent (Figure II.3.4B), meaning that climate change is pushing these streams toward ephemeral flow and the few perennial streams are transitioning to intermittent flow regimes. Gaging stations maintained by the United States Geological Survey (www.waterdata.usgs.gov) are documenting a decline in streamflow in West Texas (Figure II.3.6A). Groundwater extraction, particularly in the Ogallala Aquifer in northwestern Texas and in the Edwards Aquifer of Central Texas, is also exacerbating climate change impacts (Loáiciga et al., 2000; Wolaver et al., 2024). In West Texas, groundwater and surface water extraction is largely a consequence of crop irrigation that ultimately is fed to cattle for beef and dairy production (Awal et al., 2018; Richter et al., 2020). Modeled forecasts for central Texas suggest that flows will continue declining as a consequence of reduced precipitation and evapotranspiration brought on by climate change (Wootten et al., 2023).

The loss of perennial flow negatively affects fishes that are fluvial specialists (e.g., pelagic spawning minnows of the genus *Macrhybopsis*) and positively affect generalist lentic-adapted species (e.g., *Lepomis* sunfishes) (Perkin et al., 2015). These changes have been documented in the aforementioned Guadalupe River (Perkin and Bonner, 2011) and the Brazos River (Nguyen et al., 2023). Spring-associated and headwater fishes that disperse poorly, including darters (genus *Etheostoma*) and roundnose minnows (genus *Dionda*) are also particularly vulnerable to drying as documented, respectively, in Pinto Creek in South Texas (Schlechte and Fleming, 2015) and in Hickory and Clear

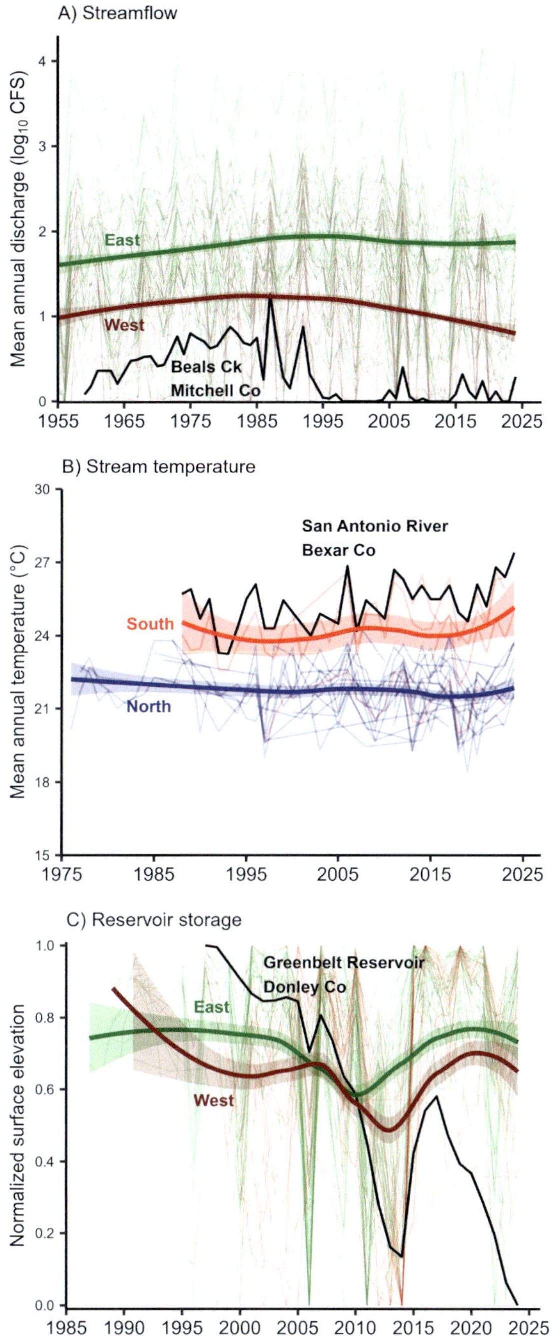

FIGURE II.3.6 Long-term trends in (A) stream discharge, (B) stream temperature, and (C) reservoir levels. Data from USGS gaging stations (www.waterdata.usgs.gov). Thin lines represent annual medians of individual stream discharge gages (n = 103), stream temperature monitoring stations (n = 50), and reservoir surface elevation monitoring stations (n = 82). Thick lines represent fits (±95% confidence intervals) for all gages or reservoirs using a generalized additive model. Colors distinguish gages or reservoirs east (blue) or west (red) of Austin (97.7°W). Black lines and labels represent select streams and reservoirs showing strong impacts of climate change (and potentially other interacting anthropogenic stressors): (A) USGS Station ID 08123800, 32.966°N, 101.830°W; (B) USGS Station ID 08181800, 29.222°N, −98.356°W, and (C) USGS Station ID 07299840, 35.001°N, 100.895°W.

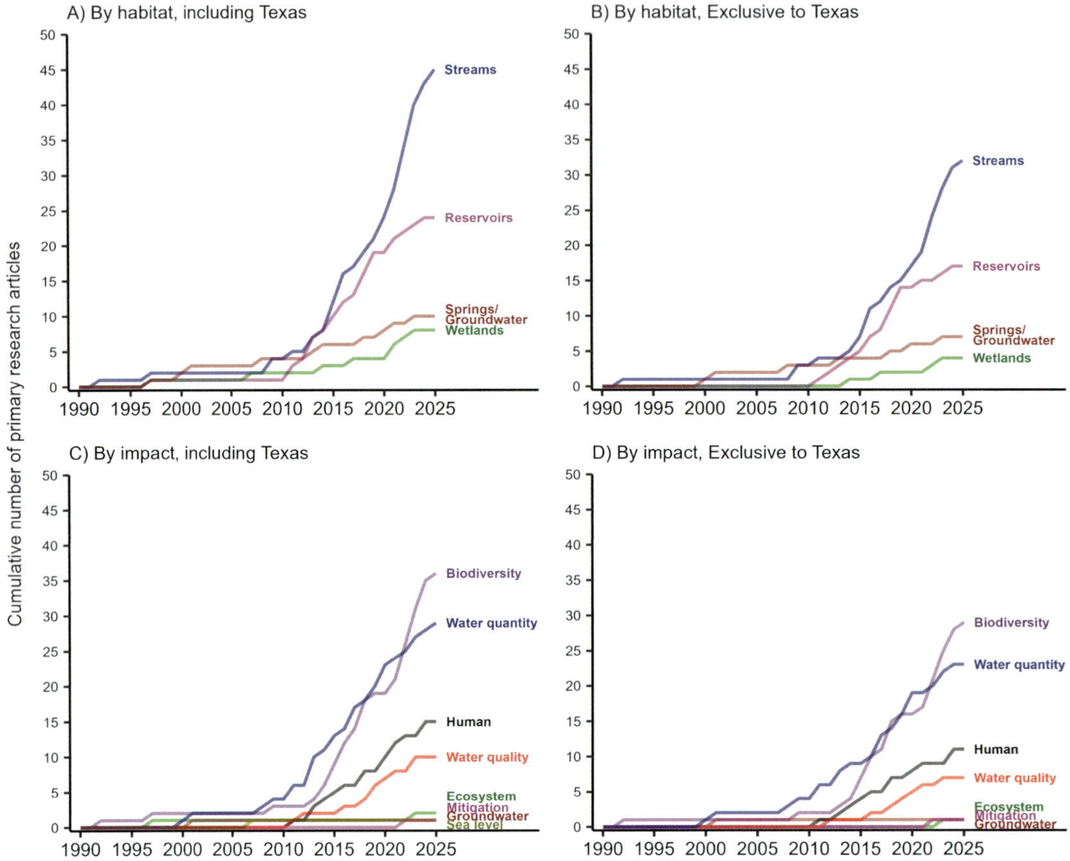

FIGURE II.3.7 Accumulation of peer-reviewed primary research articles ($n = 95$) published on the impacts of climate change on freshwater ecosystems in Texas. Articles were compiled by M. Troia in February of 2025 using standard keyword searches in Google Scholar (e.g., Texas AND climate change AND stream*). Articles were screened for relevance and classified (A-B) by climate change impact (C-D) by freshwater habitat type. Articles describing research within Texas and beyond (including up to global extent) are shown in panels A and C ($n = 95$), whereas articles describing research only in Texas are shown in panels B and D ($n = 73$). Note that colors in top and bottom panels are not related.

Creeks in North Texas (Driver and Hoeinghaus, 2015). Benthic macroinvertebrates provide the base of stream food webs and are important indicators of water quality and ecosystem health. Studies in Texas have documented declines in abundance and diversity of stream macroinvertebrates from perennial to intermittent streams and that this effect is exacerbated by conversion of watersheds to urban land use (Carvallo et al., 2022; King et al., 2016). The transition from perennial to intermittent flow regimes is particularly impactful on freshwater mussels because they are less able to retreat to permanent pools when flows cease and recolonize rewetted reaches when flows return (Cushway and Schwalb, 2024; Mitchell et al., 2018). In the San Saba River, a tributary of the Middle Colorado River, mussel abundance and diversity are lower in intermittent compared to perennial reaches (Randklev et al., 2018b). Erosion of mussel abundance and diversity disrupts ecosystem function in other regions (Vaughn, 2010) and drought can exacerbate these disruptions (Atkinson et al., 2014), but such studies have not been carried out in Texas.

For streams that remain perennial, maintaining environmental flows (i.e., eFlows) is important to sustain biodiversity and ecosystem functions. In Texas, a series of laws (Senate Bills 1, 2, and 3) passed by the state legislature between 1997 and 2007 formulated statewide water planning aimed at balancing the growing human use of surface waters with the need for environmental flows that maintain biodiversity and ecosystem functions (Roach, 2013). This latter objective spurred numerous studies that provided key information on the flow needs of fishes, mussels, and riparian trees (Kiser et al., 2022; Lueders et al., 2025; Lueders and McManamay, 2023; Perkin et al., 2023; Winemiller et al., 2024).

Texas streams and rivers are also vulnerable to temperature increases. Long-term records of stream temperatures are sparsely available across the United States and show that temperatures have increased over the last century (Kaushal et al., 2010). Such long-term records are not available in Texas, although monitoring in the last several decades indicates a warming trend for some streams (Figure II.3.6B). Predictive models also forecast upward trends in stream temperatures (Wootten et al., 2023). Several studies of spring temperature regimes indicate that groundwater has some capacity to buffer warming trends, but declining spring flows are likely to temper this effect (Caldwell et al., 2020). How these temperature increases affect aquatic biodiversity depends on species-specific physiological sensitivity which can be measured in a variety of ways (Angilletta, 2009). Upper thermal tolerances are known for many common Texas freshwater fishes (Beitinger et al., 2000), as well as several endemic and imperiled fishes like the Devils River Minnow (*Dionda diaboli*) (Fries and Gibson, 2010). Thermal tolerances of freshwater mussels are comparatively poorly understood, but recent studies in Texas have quantified tolerances for four species (Mildenberger et al., 2025; Rangaswami et al., 2023). While upper thermal tolerances can be used to elucidate vulnerability to extreme heat events (Troia, 2023), understanding sublethal thermal performance is also important to predict impacts of warming on growth, recruitment, and reproduction. Comparatively few studies have quantified these sublethal metrics for Texas species. Emadi et al. (2024) quantified swimming performance of four fish species of conservation need. Growth of Guadalupe Bass has also been measured directly in lab studies (Sullivan et al., 2013) and projected to field conditions using bioenergetics modeling (Troia and Perkin, 2022). Warming may have either positive or negative impacts on invasive species. For example, Zebra mussels (*Dreissena polymorpha*) appear to be at their warm-edge geographic range limit in Central Texas and so warming may push this invasive species northward and out of reservoirs in South and Central Texas (Schwalb et al., 2023). Alternatively, tropical invasive fishes including various tilapia species (genus *Oreochromis*) and Suckermouth Armored Catfish (genera *Hypostomus* and *Pterygoplichthys*) are currently limited to South Texas but may expand their ranges northward as temperatures rise (Osland et al., 2021). Five-plus decades of fish kill reporting by the TPWD Kills and Spills Team has shown that winter cold snaps frequently kill tilapias (Lopez et al., 2023). As warming progresses, these kills will likely occur less frequently. Also, the popular and infamous aquarium fish, the Red-bellied Piranha (*Pygocentrus nattereri*), is increasingly likely to persist in Texas freshwaters as temperatures continue to warm (Bennett et al., 1997).

Texas streams and rivers are also vulnerable to salinization. Along the Gulf Coast, declining river flow and rising sea levels are contributing to saltwater intrusion into the lower reaches of coastal rivers. For example, the Lower Neches River Authority installed a permanent saltwater barrier on the Neches River 25 kilometers upstream from its mouth at Sabine Lake in 2003. This barrier has proven effective at reducing saltwater intrusion and has measurable positive effects on freshwater fish species that are most sensitive to high salinity (Pizano-Torres et al., 2017). This barrier is particularly effective during low flow periods and thus will be increasingly important to limit saltwater intrusion as the frequency and magnitude of droughts continue to increase. Saltwater intrusion also has notable effects on freshwater mussels. For example, Tarter et al. (2023) documented a decline in the abundance and diversity of unionid mussels in the lower Neches River with the exception of *Glebula rotunda*, a uniquely euryhaline species. In West Texas, streams are naturally high in salinity, but decreasing flows, increasing evapotranspiration, and irrigation are causing salinity to increase (Hoagstrom, 2009). Model projections forecast future increases in salinity (Olson, 2019). Studies in the Pecos and Rio Grande Rivers have demonstrated that increasing salinization is increasing the relative abundances of euryhaline fish species such as the Red Shiner

(*Cyprinella lutrensis*) and western Mosquitofish (*Gambusia affinis*) and decreasing relative abundances of stenohaline species such as Rio Grande Shiner (*Notropis jemezanus*) and Longnose Dace (*Rhinichthys cataractae*) (Cheek and Taylor, 2016; Miyazono et al., 2015; Pease and Delaune, 2021). Other studies have shown that increasing salinity is limiting the distributions of the endemic and endangered Texas Hornshell mussel (*Popenaias popeii*), Rio Grande Cooter turtle (*Pseudomys gorzugi*) (Mahan et al., 2022; Randklev et al., 2018a), and amphibians (Hua and Pierce, 2013).

Streams and rivers in Texas are impacted by nutrient loading, hypoxia, harmful algae blooms, and other pollutants. In 2024, over 260 stream segments in Texas were designated by the United States Environmental Protection Agency as impaired (Clean Water Act Section 303(d)). Much of this impairment is driven by urban centers—Texas has four of the 11 largest cities in the U.S.—that generate nutrient-rich wastewater that is discharged into streams (Brooks et al., 2006; Luthy et al., 2015). As the urban population in Texas grows and as climate change drives stream flows down and water temperatures up, these existing water quality issues likely will become more severe and geographically widespread in Texas. Field measurements of nitrogen, phosphorus, dissolved oxygen, and chlorophyll-a in Texas streams reveal that values of these water quality parameters have not improved over the last three decades following a period of improvement after the Clean Water Act of 1972 (Kuwayama et al., 2020; Perkin and Bonner, 2016; Sears et al., 2025). Rates of oxygen decline are projected to accelerate through the remainder of the 21st century, particularly in agricultural and urbanized watersheds (Zhi et al., 2023). Eutrophication of streams in Texas drive changes in the species composition of algae, macroinvertebrate, and fish assemblages (Taylor et al., 2014) and harmful algae blooms have caused massive fish kills (Rhodes and Hubbs, 1992). The aforementioned urbanization in Texas is also forecasted to interact with climate change to impact flow regimes drastically. For example, in San Antonio—the seventh largest municipality in the U.S. and fastest growing of the 47 largest cities—peak flows in the San Antonio River are projected to increase with urbanization, but climate change will increase this urbanization effect by three-fold (Shao et al., 2020; Zhao et al., 2016).

II.3.5.2 Wetlands

The diverse array of wetlands in Texas—herbaceous freshwater wetlands along the Gulf Coast, wooded inland wetlands in East Texas, and playas in Northwest Texas—are vulnerable to different climate change-related stressors (Doyle et al., 2007; Fern et al., 2023). For example, precipitation is increasing in East Texas and decreasing in West Texas, meaning that the hydroperiods (and even persistence) of wetlands in these two regions is moving in opposite directions (Vose et al., 2014). In the Galveston Bay area, sea level rise is projected to reduce coastal freshwater wetlands by nearly 30% between 2025 and 2100 (Thompson et al., 2014). Playa wetlands in West Texas naturally vary in hydroperiod (i.e., the duration and timing of inundation) and these hydroperiods are altered by urban and agricultural land uses (Collins et al., 2014). Coastal wetlands in Texas play a crucial role in carbon sequestration and changes in temperature, hydroperiods, and salinity are likely to affect carbon reservoirs and rates of sequestration (Hansen and Nestlerode, 2014). Playas also vary in salinity and affects the plant and invertebrate communities that use these wetlands (Piña and Lougheed, 2022). While increasing temperature and altered precipitation patterns are likely changing the hydroperiod and salinity of playas, there is a dearth of forecasts quantifying the magnitude of these effects, the geographic distribution of these effects, and subsequent impacts on biodiversity, ecosystem function, and human activities (e.g., access to agricultural lands).

II.3.5.3 Reservoirs

Human-built reservoirs—while they have only appeared in the last 70 or so years and contribute little to freshwater biodiversity in Texas—are immensely important to the present and future status of human societies and will also play a key role in the persistence of the aforementioned biodiversity that requires flowing streams and rivers. Humans rely on reservoirs and their dams for water supply, flood control, hydropower, and recreation. All of these ecosystem services will be affected by climate change in Texas and will interact with a myriad of other anthropogenic stressors. As the human population grows and water demand increases through the remainder of the 21st century, the many reservoirs strategically placed throughout Texas will serve their intended purpose of storing water during the increasingly frequent and intense droughts that are anticipated and dampening downstream flooding following more intense rainfall events (Longyang and Zeng, 2024). Reservoirs also supply cooling water to coal-fired power plants that increase demand for this cooling water when electrical demand is high during heat waves, which is precisely when droughts cause reduction in reservoir storage. Nevertheless, the ongoing transition from coal to natural gas and renewable electricity production will continue to reduce the demand for cooling water (Scanlon et al., 2013; Turner et al., 2021).

Water quality in reservoirs is an ongoing concern as urban land cover, agricultural practices, and wastewater effluent contribute nutrients and other pollutants that make their way into reservoirs. Chronic hypoxia is increasingly common across reservoirs in Texas (Bellinger et al., 2018; Brooks et al., 2011; Dawson et al., 2015; Gelca et al., 2016). For example, Eagle Mountain Lake near Fort Worth experiences hypoxia throughout the entire water column in August and September (Wagner et al., 2023). Blooms of harmful cyanobacteria and golden algae (*Prymnesium parvum*) are increasingly common in Texas reservoirs and their detection is associated with temperature and dissolved oxygen (among other physicochemical variables), suggesting that climate change will exacerbate these harmful algae blooms (Gámez et al., 2019; Patiño et al., 2014; Roelke 2012; VanLandeghem et al., 2015). Hypoxic events and blooms of golden algae cause mass mortality events of freshwater fishes, mussels, and amphibians, which have direct impacts on recreational fisheries and biodiversity conservation (e.g., TPWD News, 2017). Increasing concentrations of nutrients and ions in Texas reservoirs are generally increasing the number of reservoirs that fail to meet total maximum daily loads, meaning that water is increasingly unsuitable for drinking, swimming, and boating (TCEQ, 2024).

Water levels in Texas reservoirs have declined in recent decades (Figure II.3.6C; Dawson et al., 2015; Gelca et al., 2016) and will continue to decline as temperatures increase and precipitation decreases (Turner et al., 2021). Declining water level is directly related to loss of boat ramp access and, to a lesser extent, shoreline access for reservoir recreationists in central Texas (Daugherty et al., 2011). Water level declines also reduce the quantity and quality of littoral zone habitat, which affect populations of sportfish and angler satisfaction (Daugherty et al., 2015; Smith et al., 2024). Reservoir storage also constrains the availability of water for downstream environmental flows. Recent studies of instream flow and ecological responses demonstrate that reservoir releases at critical times of the year are important for lotic-adapted species (Kiser et al., 2022; Lueders et al., 2025; Lueders and McManamay, 2023; Perkin et al., 2023; Winemiller et al., 2024). As climate change drives water levels down in reservoirs, there will be fewer opportunities to allocate stored water for environmental flows and more conflict with human users over such allocations.

II.3.6 OPPORTUNITIES FOR MITIGATION AND ADAPTATION

The preceding section paints a not-so-rosy outlook for freshwater ecosystems in Texas as the climate continues to change through the remainder of the 21st century. Nevertheless, there are opportunities to *mitigate* climate change impacts and *adapt* to them. Mitigation mostly is a global challenge that requires international political cooperation and action to curb greenhouse gas emissions and transition to renewable energy production. While political entities in Texas are not directly involved in such international affairs, many of the elected officials representing Texans *are* and therefore residents of Texas *can and should* make their voices heard by exercising their rights to vote and by lobbying their representatives in Washington D.C. to address issues of climate change, globally. Indeed, the large and growing human population in Texas and its congressional delegation place Texans in the fortunate position to affect climate change policies at the federal level that are then reverberated internationally. Within the borders of Texas, climate mitigation is already happening with the installation of solar and wind farms. As of 2023, the installed electricity production capacity was 41 gigawatts from wind turbines and 18.5 gigawatts from solar panels, which represents 16% of all renewable energy capacity in the United States (eia.gov, accessed 21 March 2025). Carbon sequestration represents the other side of the mitigation coin and there are diverse opportunities related to management of agricultural and natural vegetation across the state (Bell et al., 2023; Johnson and Mitchell-McCallister, 2025). For example, no-till practices in agricultural fields can increase soil organic carbon, as can regular burning of native grasslands (Potter et al., 2009). Municipally owned natural areas can also be managed to sequester carbon (e.g., via prescribed burning), but these relatively small areas have minimal impacts on the statewide carbon budget (McCaw, 2012).

There are also numerous opportunities for climate change adaptation in Texas (Table II.3.1). These opportunities include actions implemented by governmental entities operating within the state (e.g., municipalities, state agencies) and include changes in dam operations, installation of infrastructure, and management of native and invasive species. Some municipalities use treated wastewater to maintain flows in streams that historically were fed by springs (Craig and Bonner, 2021; Hamdhani et al., 2020). Other opportunities are aimed at improving per-person water use. For example, enacting lawn watering restrictions, consuming foods that are produced locally and with efficient water use, and augmenting in-stream flows with recycled wastewater. As the human

TABLE II.3.1
Actions to Adapt to the Impacts of Anthropogenic Climate Change on Freshwater Ecosystems

Category	Impact	Actual and Potential Adaptations in Texas	References
Water quality	Temperature	Riparian shade restoration; maintain instream flows; maintain spring flow	Caldwell et al., 2020; Knouft et al., 2021; Van Vliet et al., 2013
	Hypoxia	Curtail nutrient loading; maintain flows; mechanical aeration of lentic waterbodies	Baker et al., 2007; Bellinger et al., 2018; Visser et al., 1996
	Salinity	Maintain flows in inland streams; install saltwater intrusion barriers on coastal streams	Miyazono et al., 2015; Pizano-Torres et al., 2017
Water quantity	Droughts	Municipal conservation; dam releases, maintain aquifer levels and spring flows	Longyang and Zeng, 2024
	Floods*	Capture floodwaters to fill surface reservoirs; recharge aquifers	Longyang and Zeng, 2024
	Instream habitat	Maintain flows to curtail habitat loss and fragmentation	Perkin et al., 2023; Winemiller et al., 2024
Biodiversity	Native extirpations	Maintain, restore riparian vegetation; establish, maintain arc populations; managed relocation	Capon et al., 2013; Olden et al., 2011
	Exotic introductions	Strengthen plant, animal trade regulations; public education	Bowles and Bowles, 2001; Kemp et al., 2017
Ecosystem function	Nutrient, sediment transport, cycling	Maintain, restore riparian vegetation; maintain instream flows; maintain functional biodiversity	Knouft et al., 2021; Renfrow et al., 2021

* Including more frequent and intense rainfall from hurricanes.

population continues to grow in Texas, these latter water-use efficiency opportunities will be increasingly important. Some good news is that some mitigation actions have already been implemented in Texas. Many municipalities in Central and West Texas restrict non-essential water use. For example, the San Antonio Water System advances through four stages of water use restrictions as drought conditions increase (i.e., measured as the water level in Edwards Aquifer at monitoring wells). These restrictions limit residential landscape irrigation, car washes, and washing impervious surfaces. Comal Springs and San Marcos Springs are highly susceptible to the establishment of invasive tropical fishes given their proximity to highly-populated municipalities and their buffered thermal regimes. Managers of these unique habitats regularly conduct public outreach and education to curtail new introductions. As climate change progresses, other freshwater bodies will become increasingly suitable for tropical invasive fishes and similar outreach and education actions may become effective at a wider range of waterbodies throughout Texas. During recent droughts, the United States Fish and Wildlife Service and Texas Parks and Wildlife have partnered to salvage endangered stream fishes trapped in isolated and drying pools (Hennessy-Fisk, 2011; Mayes et al., 2019). Droughts and stream drying have also provided unique opportunities for the restoration of aridland fish communities. For example, during the historic drought of the summer of 2011, a significant length of the Blanco River in Central Texas dried, which facilitated the removal of invasive Smallmouth Bass (*Micropterus treculii*) that hybridize with native Guadalupe Bass (*Micropterus treculii*). When flows returned in the spring of 2012, hatchery-produced Guadalupe Bass were repatriated in the Guadalupe River and have persisted as a genetically-pure population since (Magnelia et al., 2019). While many perennial streams are increasingly susceptible to drying, effluent from municipal wastewater can sustain flows where natural spring flow has diminished. For example, the San Antonio River is sustained by effluent released by the San Antonio Water System. In 2015, the Texas Parks and Wildlife Department reestablished a population of Guadalupe Bass in the San Antonio River. While this population is susceptible to stressors related to urbanization (e.g., flashy hydrologic regime, nutrient enrichment, etc.), effluent augmentation ensures perennial flow (Sears et al., 2025). Finally, the Texas Water Development Board manages the Texas Water Trust, which was established in 1997 and facilitates the voluntary release of privately held water rights so that water remains in the stream channel to support biodiversity and ecosystem services that depend on sustained environmental flows (Garmany, 2020, twdb.texas.gov, accessed 21 March 2025).

II.3.7 KNOWLEDGE GAPS AND RESEARCH NEEDS FOR TEXAS

The previous sections demonstrate a wealth of knowledge that researchers and managers have accumulated over the past several decades. Nevertheless, gaps in our understanding of ongoing climate change impacts and our ability to forecast future impacts remain (Figure II.3.7).

To date, most research has focused on impacts of climate change on freshwater biota and water quantity, whereas relatively few studies have assessed impacts on ecosystem-level responses, mitigation actions, groundwater, and sea level rise (Figure II.3.7A–B). From a taxonomic perspective, much knowledge has accumulated for fishes and to a lesser extent for mussels, whereas crayfishes remain poorly studied in Texas (Troia and McManamay, 2016, but see Birnbaum et al., 2007; Hays et al., 2023). Basic habitat association and life history information are needed for native crayfishes because species differ in their use of perennial, intermittent, and wetland habitats, which are highly sensitive to climate change (Bloomer et al., 2021). From an ecosystem perspective, intermittent streams and wetlands remain poorly studied compared to perennial streams and reservoirs in Texas (Levick et al., 2008; Figure II.3.7C–D). Improved understanding of the geomorphology and hydrology of these ecosystems are needed as well as biological responses at all levels of organization (i.e., population, community, ecosystem). Also, with much of Texas situated in an historically subtropical climate, there is a need to understand and predict the impacts of tropicalization—that is, the comprehensive transformation of temperate ecosystems by northward expanding tropical organisms (Osland et al., 2021). Indeed, several tropical fishes (e.g., tilapias and suckermouth armored catfishes) are firmly established in South Texas and will continue expanding northward as temperatures increase. However, the full complement of tropical freshwater microbes, plants, invertebrates, and vertebrates moving northward into Texas remains poorly known. Lastly, and from the perspective of human wellbeing, there is a need to understand the changing landscape of environmental justice. There is a history of disadvantaged communities having greater exposure to floods and lower quality municipal water supplies (Cook et al., 2024). As climate change exacerbates flood risk and water supply for all communities in Texas, it will be important to allocate climate adaptation resources equitably.

REFERENCES

Abernethy FE, 2008. The flow of the Neches. East Texas Historical Journal, 46(1), 6.

Allan JD, 2004. Landscapes and riverscapes: The influence of land use on stream ecosystems. Annual Review of Ecology, Evolution, and Systematics, 35(1), 257–284.

Angilletta MJ Jr, 2009. Thermal adaptation: A theoretical and empirical synthesis. Oxford University Press.

Aspin TW, Hart K, Khamis K, Milner AM, O'Callaghan MJ, Trimmer M, Wang Z, Williams GMD, Woodward G, Ledger ME, 2019. Drought intensification alters the composition, body size, and trophic structure of invertebrate assemblages in a stream mesocosm experiment. Freshwater Biology, 64(4), 750–760.

Atkinson CL, Julian JP, Vaughn CC, 2014. Species and function lost: Role of drought in structuring stream communities. Biological Conservation, 176, 30–38.

Awal R, Fares A, Bayabil H, 2018. Assessing potential climate change impacts on irrigation requirements of major crops in the Brazos Headwaters Basin, Texas. Water, 10(11), 1610.

Baker ME, Weller DE, Jordan TE, 2007. Effects of stream map resolution on measures of riparian buffer distribution and nutrient retention potential. Landscape Ecology, 22(7), 973–992.

Battin TJ, Lauerwald R, Bernhardt ES, Bertuzzo E, Gener LG, Hall RO Jr, Hotchkiss ER, Maavara T, Pavelsky TM, Ran L, Raymond P, Rosentreter JA, Regnier P, 2023. River Ecosystem metabolism and carbon biogeochemistry in a changing world. Nature, 613(7944), 449–459.

Battin TJ, Luyssaert S, Kaplan LA, Aufdenkampe AK, Richter A, Tranvik LJ, 2009. The boundless carbon cycle. Nature Geoscience, 2(9), 598–600.

Beason E, Schwalb AN, 2022. Impact of zebra mussels on physiological conditions of unionid mussels in Texas. Aquatic Sciences, 84(2), 21.

Beitinger TL, Bennett WA, McCauley RW, 2000. Temperature tolerances of North American freshwater fishes exposed to dynamic changes in temperature. Environmental Biology of Fishes, 58, 237–275.

Bell J, DeLaune PB, Fischer BL, Foster JL, Lewis KL, McCarl BA, Outlaw JL, 2023. Carbon sequestration and water management in Texas–One size does not fit all. Agrosystems, Geosciences and Environment, 6(2), e20372.

Bellinger BJ, De Jesús MJ, 2024. Hydrilla management impacts on a largemouth bass fishery: A case for a balanced management approach. Journal of the Southeastern Association of Fisheries and Wildlife Agencies, 11, 14–22.

Bellinger BJ, Richter A, Porras A, Davis SL, 2018. Drought and management effects on biophysicochemistry in a rapidly-flushed reservoir. Lake and Reservoir Management, 34(2), 182–198.

Bennett DL, Driscoll MT, Norman JD, 2024. Dynamics and economic contribution of large bass tournaments at Lake fork reservoir, Texas. Journal of the Southeastern Association of Fish and Wildlife Agencies, 11, 23–35.

Bennett WA, Currie RJ, Wagner PF, Beitinger TL, 1997. Cold tolerance and potential overwintering of the red-bellied piranha *pygocentrus nattereri* in the United States. Transactions of the American Fisheries Society, 126(5), 841–849.

Bernhardt ES, Heffernan JB, Grimm NB, Stanley EH, Harvey JW, Arroita M, Appling AP, Cohen MJ, McDowell WH, Hall RO Jr, Read JS, Roberts BJ, Stets EG, Yackulic CB, 2018. The metabolic regimes of flowing waters. Limnology and Oceanography, 63(S1), S99–S118.

Best A, Perkin JS, Pinion AK, Binkley H, Conway KW, 2022. First record of the Gangetic swamp eel, *ophichthys cuchia* (Hamilton, 1822) (Teleostei: Synbranchidae), from Texas (USA) based on Museum vouchered material, and confirmation of a second established non-native population in the USA. Check List, 18(3), 475–482.

Birdsong T, Garrett G, Bean M, Bean P, Curtis S, Fleming P, Grubh A, Lutz-Carillo D, Mayes KB, Roberson C, Robertson S, Schlechte JW, Smith N, 2020. Conservation of Texas freshwater fish diversity: Selection of Species of Greatest Conservation Need. Texas Parks and Wildlife Department. PWD RP T3200-2780 (10/20). Austin, Texas.

Birnbaum JS, Winemiller KO, Shen L, Munster CL, Wilcox BP, Wilkins RN, 2007. Associations of watershed vegetation and environmental variables with fish and crayfish assemblages in headwater streams of the Pedernales River, Texas. River Research and Applications, 23(9), 979–996.
Bloomer CC, DiStefano RJ, Taylor CA, 2021. A global review of life history studies on burrowing crayfish. Crustaceana, 94(3), 357–379.
Bowles DE, Bowles BD, 2001. A Review of the Exotic Species Inhabiting the Upper San Marcos River, Texas, USA. Austin, TX: Texas Parks Wildlife Department.
Brismar A, 2002. River Systems as providers of goods and services: A basis for comparing desired and undesired effects of large dam projects. Environmental Management, 29, 598–609.
Brooks BW, Riley TM, Taylor RD, 2006. Water quality of effluent-dominated ecosystems: Ecotoxicological, hydrological, and management considerations. Hydrobiologia, 556, 365–379.
Brooks BW, Valenti TW, Cook-Lindsay BA, Forbes MG, Doyle RD, Scott JT, Stanley JK, 2011. Influence of climate change on reservoir water quality assessment and management. Effects of reduced inflows on Diel pH and site-specific contaminant Hazards. In I Linkov, TS Bridges (Eds.), Climate: Global Change and Adaptation (pp. 491–522). Springer Netherlands.
Brune GM. 2002. Springs of Texas (Vol. 1). Texas A&M University Press.
Brylinsky M, Mann K, 1973. An analysis of factors governing productivity in lakes and reservoirs. Limnology and Oceanography, 18(1), 1–14.
Busch DE, Fisher SG, 1981. Metabolism of a desert stream. Freshwater Biology, 11(4), 301–307.
Butman D, Stackpoole S, Stets E, McDonald CP, Clow DW, Striegl RG, 2016. Aquatic carbon cycling in the conterminous United States and implications for terrestrial carbon accounting. Proceedings of the National Academy of Sciences, 113(1), 58–63.
Caldwell TG, Wolaver BD, Bongiovanni T, Pierre JP, Robertson S, Abolt C, Scanlon BR. 2020. Spring discharge and thermal regime of a groundwater dependent ecosystem in an arid karst environment. Journal of Hydrology, 587, 124947.
Capinha C, Larson ER, Tricarico E, Olden JD, Gherardi F, 2013. Effects of climate change, invasive species, and disease on the distribution of native European crayfishes. Conservation Biology, 27(4), 731–740.
Capon SJ, Chambers LE, Mac Nally R, Naiman RJ, Davies P, Marshall N, Pittock J, Reid M, Capon T, Douglas M, Catford J, Baldwin DS, Stewardson M, Roberts J, Parsons M, Williams SE. 2013. Riparian ecosystems in the 21st century: Hotspots for climate change adaptation? Ecosystems, 16(3), 359–381.
Carrier-Belleau C, Pascal L, Nozais C, Archambault P, 2022. Tipping points and multiple drivers in changing aquatic ecosystems: A review of experimental studies. Limnology and Oceanography, 67, S312–S330.
Carvallo FR, Strickland BA, Kinard SK, Reese BK, Hogan JD, Patrick CJ, 2022. Structure and functional composition of macroinvertebrate communities in coastal plain streams across a precipitation gradient. Freshwater Biology, 67(10), 1725–1738.
Carver E. 2006. Economic Impact of Waterfowl Hunting in the United States. Addendum to the 2006 National Survey of Fishing, Hunting, and Wildlife Associated Recreation. United States Fish and Wildlife Report 2006-2. https://www.fws.gov/sites/default/files/documents/2024-04/1153.pdf
Cheek CA, Taylor CM, 2016. Salinity and geomorphology drive long-term changes to local and regional fish assemblage attributes in the lower Pecos River, Texas. Ecology of Freshwater Fish, 25(3), 340–351.
Collins MB, 2004. Archeology in central Texas. In TK Perttula (Ed.), The Prehistory of Texas, Texas A&M University Press, College Station, Texas, USA. (pp. 101–126).
Collins SD, Heintzman LJ, Starr SM, Wright CK, Henebry GM, McIntyre NE. 2014. Hydrological dynamics of temporary wetlands in the southern Great Plains as a function of surrounding land use. Journal of Arid Environments, 109, 6–14.
Comte L, Buisson L, Daufresne M, Grenouillet G, 2013. Climate-induced changes in the distribution of freshwater fish: Observed and predicted trends. Freshwater Biology, 58(4), 625–639.
Cook M, Tremaine D, Wyatt B, Banner J, Charles J, Berg M, Bruno T, Glazer YR, Callison C, Mace ME, Miller V, Bare R, Sanchez Flores R, Seefeldt J, Fuller A, Niyogi D. 2024. Addressing challenges to ensuring justice and sustainability in policy and infrastructure for Texas water resources in the 21st century. Texas Water Journal, 15(1), 104–139.
Covich AP, 2009. Emerging climate change impacts on freshwater resources. A perspective on transformed watersheds, RFF Report, Resources for the Future, Washington DC, USA.
Craig CA, Bonner TH, 2021. Spring flow lost: A historical and contemporary perspective of an urban fish community. Urban Ecosystems, 24(3), 417–427.
Cushway KC, Schwalb AN, 2024. When rivers run dry: Perennial pools as ecological refuges for freshwater mussels during drought. Freshwater Biology, 69(2), 226–239.
Dang CK, Schindler M, Chauvet E, Gessner MO. 2009. Temperature oscillation coupled with fungal community shifts can modulate warming effects on litter decomposition. Ecology, 90(1), 122–131.

Dascher ED, Burlakova LE, Karatayev AY, Ford DF, Schwalb AN, 2018. Distribution of unionid freshwater mussels and host fishes in Texas. A study of broad-scale spatial patterns across basins and a strong climate gradient. Hydrobiologia, 810, 315–331.
Datry T, Larned ST, Tockner K, 2014. Intermittent rivers: A challenge for freshwater ecology. BioScience, 64(3), 229–235.
Daugherty DJ, Bennett DL, VanZee B, Morgan T, Tibbs J, 2015. Effects of water-level reductions on littoral habitat and recreational access in Brazos River reservoirs, Texas. Journal of the Southeastern Association of Fish and Wildlife Agencies, 2, 1–7.
Daugherty DJ, Buckmeier DL, Kokkanti PK, 2011. Sensitivity of recreational access to reservoir water level variation: An approach to identify future access needs in reservoirs. North American Journal of Fisheries Management, 31(1), 63–69.
Dawson D, VanLandeghem MM, Asquith WH, Patiño R, 2015. Long-term trends in reservoir water quality and quantity in two major river basins of the southern great plains. Lake and Reservoir Management, 31(3), 254–279.
Dawson TP, Berry PM, Kampa E, 2003. Climate change impacts on freshwater wetland habitats. Journal for Nature Conservation, 11(1), 25–30.
Devitt TJ, Wright AM, Cannatella DC, Hillis DM, 2019. Species delimitation in endangered groundwater salamanders: Implications for aquifer management and biodiversity conservation. Proceedings of the National Academy of Sciences, 116(7), 2624–2633.
Dodds WK, 2007. Trophic state, eutrophication and nutrient criteria in streams. Trends in Ecology and Evolution, 22(12), 669–676.
Doyle TW, O'Neil CP, Melder MPV, From AS, Palta MM, 2007. Tidal freshwater swamps of the southeastern United States: Effects of land use, hurricanes, sea-level rise, and climate change. In WJ Conner, TW Doyl, KW Krauss (Eds.), Ecology of Tidal Freshwater Forested Wetlands of the Southeastern United States (pp. 1–28). Springer. DOI: 10.1007/978-1-4020-5095-4_1.
Driver LJ, Hoeinghaus DJ, 2015. Spatiotemporal dynamics of intermittent stream fish metacommunities in response to prolonged drought and reconnectivity. Marine and Freshwater Research, 67(11), 1667–1679.
Duarte A, Brown DJ, Forstner MR, 2011. Estimating abundance of the endangered Houston toad on a primary recovery site. Journal of Fish and Wildlife Management, 2(2), 207–215.
Dudgeon D, Arthington AH, Gessner MO, Kawabata ZI, Knowler DJ, Lévêque C, Naiman RJ, Prieur-Richard A-H, Soto D, Stiassny MLJ, Sullivan CA. 2006. Freshwater biodiversity: importance, threats, status and conservation challenges. Biological Reviews, 81(2), 163–182.
Dye KR, Fields AT, Bean MG, Robertson SM, Hollenbeck CM, Conway KW, Portnoy DS, 2024. Assessment of genomic diversity within and between two cryptic shiners, the West Texas shiner (*Notropis megalops*) and the Texas shiner (*Notropis amabilis*). Journal of Fish Biology, 106, 836–845.
Edwards Aquifer Authority, 2023. Edwards Aquifer Habitat Conservation Plan 2023 Annual Report.
Emadi CM, Bean PT, Mager EM, 2024. Swimming performance assessments of fish species of greatest conservation need to inform future stream crossing designs in Texas. Fishes, 9(6), 234.
Fern RR, Baron MD, England AE, Giese JC, Kraai KJ, Lancaster JD, Oldenburger SL, Shipes JC, Wilson BC, Wyckoff SR, 2023. The state of Texas wetlands: A review of current and future challenges. Texas Water Journal, 14(1), 136–174.
Filippelli GM, 2008. The global phosphorus cycle: Past, present, and future. Elements, 4(2), 89–95.
Fowler D, Coyle M, Skiba U, Sutton MA, Cape JN, Reis S, Sheppard LJ, Jenkins A, Grizzetti B, Galloway JN, Vitousek P, Leach A, Bouwman AF, Butterbach-Bahl K, Dentener F, Stevenson D, Amann M, Voss M. 2013. The global nitrogen cycle in the twenty-first century. Philosophical Transactions of the Royal Society B: Biological Sciences, 368(1621), 20130164.
Franssen NR, Tobler M, Gido KB, 2011. Annual variation of community biomass is lower in more diverse stream fish communities. Oikos, 120(4), 582–590.
Fries JN, Gibson JR, 2010. Critical thermal maxima of captive-bred devils River minnows (*Dionda diaboli*). The Southwestern Naturalist, 55(4), 544–550.
Gaeta JW, Sass GG, Carpenter SR, 2014. Drought-driven lake level decline: Effects on coarse woody habitat and fishes. Canadian Journal of Fisheries and Aquatic Sciences, 71(2), 315–325.
Gámez TE, Benton L, Manning SR. 2019. Observations of two reservoirs during a drought in central Texas, USA: Strategies for detecting harmful algal blooms. Ecological Indicators, 104, 588–593.
Garmany K, 2020. Market strategies for addressing water scarcity: An introduction to environmental water transactions in Texas. Texas State University (masters thesis)
Gelca R, Hayhoe K, Scott-Fleming I, Crow C, Dawson D, Patiño R, 2016. Climate–water quality relationships in Texas reservoirs. Hydrological Processes, 30(1), 12–29.

Graf WL, 1999. Dam nation: A geographic census of American dams and their large-scale hydrologic impacts. Water Resources Research, 35(4), 1305–1311.

Greene BD, Dixon JR, Mueller JM, Whiting MJ, Thornton OW Jr, 1994. Feeding ecology of the concho water snake, *nerodia harteri paucimaculata*. Journal of Herpetology, 28(2), 165–172.

Greene BD, Dixon JR, Whiting MJ, Mueller JM, 1999. Reproductive ecology of the concho water snake, *nerodia harteri paucimaculata*. Copeia, 1999(3), 701–709.

Gudmundsson L, Boulange J, Do HX, Gosling SN, Grillakis MG, Koutroulis AG, Leonard M, Liu J, Muller Schmied H, Papadimitriou L, Pokhrel Y, Seniratne SI, Satoh Y, Thiery W, Westra S, Zhang X, Zhao F, 2021. Globally observed trends in mean and extreme river flow attributed to climate change. Science, 371(6534), 1159–1162.

Haag WR, 2012. North American Freshwater Mussels: Natural History, Ecology, and Conservation. Cambridge University Press.

Hamdhani H, Eppehimer DE, Bogan MT, 2020. Release of treated effluent into streams: A global review of ecological impacts with a consideration of its potential use for environmental flows. Freshwater Biology, 65(9), 1657–1670.

Hansen VD, Nestlerode JA, 2014. Carbon sequestration in wetland soils of the Northern Gulf of Mexico coastal region. Wetlands Ecology and Management, 22, 289–303.

Hargrave CW, 2009. Effects of fish species richness and assemblage composition on stream ecosystem function. Ecology of Freshwater Fish, 18(1), 24–32.

Havel JE, Kovalenko KE, Thomaz SM, Amalfitano S, Kats LB, 2015. Aquatic invasive species: Challenges for the future. Hydrobiologia, 750, 147–170.

Hays HC, Pease AA, Fleming P, Barnes MA, 2023. Distribution and habitat use of a rare native crayfish: Implications for conserving data deficient species. Aquatic Conservation: Marine and Freshwater Ecosystems, 33(7), 751–760.

Hennessy-Fisk M, 2011. Texas scientists try to rescue drought-threatened fish. Los Angeles Times. Accessed 24 October 2024.

Hernández Martínez de la Riva A, Harper M, Rytwinski T, Sahdra A, Taylor JJ, Bard B, Bennett JR, Burton D, Creed IF, Haniford LSE, Hanna DE, Harmsen EJ, Robichaud CD, Smol JP, Thapar M, Cooke SJ, 2023. Tipping points in freshwater ecosystems: An evidence map. Frontiers in Freshwater Science, 1, 1264427.

Hibbitts TJ, LaDuc TJV, Hibbitts TL, Hibbitts TD. 2012. Texas amphibians: A field guide. University of Texas Press.

Hickerson NP, 1994. The Jumanos: Hunters and Traders of the South Plains. University of Texas Press.

Hoagstrom CW, 2009. Causes and Impacts of Salinization in the Lower Pecos River. Great Plains Research, 27–44.

Hoagstrom CW, Brooks JE, Davenport SR, 2011. A large-scale conservation perspective considering endemic fishes of the North American plains. Biological Conservation, 144(1), 21–34.

Hoeinghaus DJ, Winemiller KO, Birnbaum JS, 2007. Local and regional determinants of stream fish assemblage structure: Inferences based on taxonomic vs. functional groups. Journal of Biogeography, 34(2), 324–338.

Hopper GW, Buchanan JK, Sánchez González I, Kubala ME, Bucholz JR, Lodato MB, Lozier JD, Atkinson CL, 2022. Little clams with big potential: Nutrient release by invasive *corbicula fluminea* can exceed co-occurring freshwater mussel (*Unionidae*) assemblages. Biological Invasions, 24(8), 2529–2545.

Horne F, Beisser S, 1977. Distribution of river shrimp in the Guadalupe and San marcos rivers of Central Texas, USA (Decapoda, *Caridea*). Crustaceana, 33(1), 56–60.

Howells RG, Neck RW, Murray HD, 1996. Freshwater Mussels of Texas. University of Texas Press.

Hua J, Pierce BA, 2013. Lethal and sublethal effects of salinity on three common Texas amphibians. Copeia, 2013(3), 562–566.

Isaak DJ, Young MK, Luce CH, Hostetler SW, Wenger SJ, Peterson EE, Ver Hoef JM, Groce MC, Horan DL, Nagel DE, 2016. Slow climate velocities of mountain streams portend their role as refugia for cold-water biodiversity. Proceedings of the National Academy of Sciences, 113(16), 4374–4379.

Jain S, Ale S, Munster CL, Ansley RJ, Kiniry JR, 2015. Simulating the hydrologic impact of *Arundo donax* invasion on the headwaters of the Nueces River in Texas. Hydrology, 2(3), 134–147.

Jane SF, Hansen GJ, Kraemer BM, Leavitt PR, Mincer JL, North RL, North RL, Pilla RM, Stetler JT, Williamson CE, Woolway RI, Arvola L, Chandra S, DeGasperi CL, Diemer L, Dunalska J, Erina O, Flaim G, Grossart H, Hambright KD, Hein C, Hejzlar J, Janus LL, Jenny J, Jones JR, Knoll LB, Leoni B, Mackay E, Matsuzaki SS, McBride C, Müller-Navarra DC, Paterson AM, Pierson D, Rogora M, Rusak JA, Sadro S, Saulnier-Talbot E, Schmid M, Sommaruga R, Thiery W, Verburg P, Weathers KC,

Weyhenmeyer GA, Yokota K, Rose KC, 2021. Widespread deoxygenation of temperate lakes. Nature, 594(7861), 66–70.

Jansen J, Simpson GL, Weyhenmeyer GA, Härkönen LH, Paterson AM, del Giorgio PA, Prairie YT, 2024. Climate-driven deoxygenation of northern lakes. Nature Climate Change, 14(8), 832–838.

Jensen R, Hatler W, Mecke M, Hart C, 2006. The influences of human activities on the waters of the Pecos Basin of Texas: A brief overview. Texas Water Resources Institute. SR- 2006-03.

Jha AK, Bloch R, Lamond J, 2012. Cities and Flooding: A Guide to Integrated Urban Flood Risk Management for the 21st Century. World Bank Publications.

Johnson TD, Mitchell-McCallister D, 2025. Agricultural carbon sequestration: Incentives, contracts, and impacts on water resources in the Texas high plains. Journal of Natural Resources and Agricultural Ecosystems. doi: 10.13031/jnrae.16207

Jones R, Travers C, Rodgers C, Lazar B, English E, Lipton J, Vogel J, Strzepek K, Martinich J, 2013. Climate change impacts on freshwater recreational fishing in the United States. Mitigation and Adaptation Strategies for Global Change, 18, 731–758.

Kaushal SS, Likens GE, Jaworski NA, Pace ML, Sides AM, Seekell D, Belt KT, Secor DH, Wingate RL, 2010. Rising stream and river temperatures in the United States. Frontiers in Ecology and the Environment, 8(9), 461–466.

Keeler BL, Polasky S, Brauman KA, Johnson KA, Finlay JC, O'Neill A, Kovacs K, Dalzell B, 2012. Linking water quality and well-being for improved assessment and valuation of ecosystem services. Proceedings of the National Academy of Sciences, 109(45), 18619–18624.

Kefford BJ, Buchwalter D, Cañedo-Argüelles M, Davis J, Duncan RP, Hoffmann A, Thompson R, 2016. Salinized rivers: Degraded systems or new habitats for salt-tolerant faunas? Biology Letters, 12(3), 20151072.

Kemp C, van Riper CJ, BouFajreldin L, Stewart WP, Scheunemann J, van den Born RJ, 2017. Connecting human–nature relationships to environmental behaviors that minimize the spread of aquatic invasive species. Biological Invasions, 19, 2059–2074.

Kemp L, Mauldin RP, Munoz C, Smith S. 2024. Archaeology Along the San Antonio River: The Mission Reach Project, San Antonio, Bexar County, Texas. Volume 1: Introduction, Background and Summary. Center for Archaeological Research, The University of Texas at San Antonio.

King RS, Scoggins M, Porras A, 2016. Stream biodiversity is disproportionately lost to urbanization when flow permanence declines: Evidence from southwestern North America. Freshwater Science, 35(1), 340–352.

Kiser AH, Khan JM, Robertson CR, Lopez R, Randklev CR, 2022. The effect of flow and mussel species traits on the occurrence of rare mussels: A case study within selected rivers of the West gulf coastal plain. Aquatic Conservation: Marine and Freshwater Ecosystems, 32(1), 98–111.

Knouft JH, Botero-Acosta A, Wu CL, Charry B, Chu ML, Dell AI, Hall DM, Herrington SJ, 2021. Forested riparian buffers as climate adaptation tools for management of riverine flow and thermal regimes: A case study in the Meramec River basin. Sustainability, 13(4), 1877.

Knouft JH, Ficklin DL, 2017. The potential impacts of climate change on biodiversity in flowing freshwater systems. Annual Review of Ecology, Evolution, and Systematics, 48(1), 111–133.

Kondolf GM, 1997. PROFILE: Hungry water: Effects of dams and gravel mining on River channels. Environmental Management, 21(4), 533–551.

Kundzewicz ZW, Gerten D, 2015. Grand challenges related to the assessment of climate change impacts on freshwater resources. Journal of Hydrologic Engineering, 20(1), A4014011.

Kundzewicz ZW, Mata LJ, Arnell NW, Döll P, Jimenez B, Miller K, Oki T, Sen Z, Shiklomanov I, 2008. The implications of projected climate change for freshwater resources and their management. Hydrological Sciences Journal, 53(1), 3–10.

Kundzewicz ZW, Kanae S, Seneviratne SI, Handmer J, Nicholls N, Peduzzi P, Mechler R, Bouwer LM, Arnell N, Mach K, Muir-Wood R, Brakenridge GR, Kron W, Benito G, Honda Y, Takahashi K, Sherstyukov B. 2014. Flood risk and climate change: Global and regional perspectives. Hydrological Sciences Journal, 59(1), 1–28.

Kuwayama Y, Olmstead SM, Wietelman DC, Zheng J, 2020. Trends in nutrient-related pollution as a source of potential water quality damages: A case study of Texas, USA. Science of the Total Environment, 724, 137962.

Levick L, Fonseca J, Goodrich D, Hernandez M, Semmens D, Stromberg J, Leidy R, Scianni M, Guertin DP, Tluczek M, Kepner W, 2008. The ecological and hydrological significance of ephemeral and intermittent streams in the arid and semi-arid American Southwest. U.S. Environmental Protection Agency and USDA/ARS Southwest Watershed Research Center, EPA/600/R-08/134, ARS/233046, 116 pp.

Li L, Ni J, Chang F, Yue Y, Frolova N, Magritsky D, Borthwick AGL, Ciais P, Wang Y, Zheng C, Walling DE, 2020. Global trends in water and sediment fluxes of the world's large rivers. Science Bulletin, 65(1), 62–69.

Liew JH, Tan HH, Yeo DC, 2016. Dammed rivers: Impoundments facilitate fish invasions. Freshwater Biology, 61(9), 1421–1429.

Liu S, Kuhn C, Amatulli G, Aho K, Butman DE, Allen GH, Lin P, Pan M, Yamazaki D, Brinkerhoff C, Gleason C, Xia X, Raymond PA, 2022. The importance of hydrology in routing terrestrial carbon to the atmosphere via global streams and rivers. Proceedings of the National Academy of Sciences, 119(11), e2106322119.

Loáiciga HA, Maidment DR, Valdes JB, 2000. Climate-change impacts in a regional karst aquifer, Texas, USA. Journal of Hydrology, 227(1–4), 173–194.

Longley G, 1986. The biota of the Edwards aquifer and the implications for paleozoogeography. In The Balcones Escarpment-Geology, Hydrology, Ecology and Social Development in Central Texas (pp. 51–54). Geological Society of America.

Longyang Q, Zeng R, 2024. Assessing the effectiveness of reservoir operation and identifying reallocation opportunities under climate change. Environmental Research Letters, 19(7), 074010.

Lopez SG, Aspbury AS, Fritts SR, Tidwell T, Bonner TH, 2023. Long-term patterns in inland fish kills associated with cold-shock and winter stress: A regional case study from Texas. Journal of Fish Biology, 103(3), 472–480.

Lopez-Hoffman L, Chester CC, Semmens DJ, Thogmartin WE, Rodríguez-McGoffin MS, Merideth R, Diffendorfer JE, 2017. Ecosystem services from transborder migratory species: Implications for conservation governance. Annual Review of Environment and Resources, 42(1), 509–539.

Lueders MB, McManamay RA, 2023. Species depletion profiles as an alternative to streamflow alteration thresholds in a hydroecological risk assessment. Ecological Indicators, 147, 109989.

Lueders MB, McManamay RA, Le Bouille D, Garrett K, Roush L, Powers S, Young D, Bean P, Mayes K, Smith R, Trungale JF, 2025. BBEST practices: Intentional efforts to identify elusive flow-ecology thresholds. Ecological Indicators, 172, 113311.

Luoto TP, Rantala MV, Kivilä EH, Nevalainen L, Ojala AE, 2019. Biogeochemical cycling and ecological thresholds in a High Arctic Lake (Svalbard). Aquatic Sciences, 81, 1–16.

Luthy RG, Sedlak DL, Plumlee MH, Austin D, Resh VH, 2015. Wastewater-effluent-dominated streams as ecosystem-management tools in a drier climate. Frontiers in Ecology and the Environment, 13(9), 477–485.

Magnelia S, Linam G, McGillicuddy R, Saunders K, Parker M, Birdsong T, Lutz-Carrillo D, Williamson J, Ranft RL, Bonner T, 2019. Repatriation of Guadalupe bass in the Blanco River, Texas: A case study in the opportunistic use of drought as A fisheries management tool. In MJ Siepker, JW Quinn (Eds.), Managing Centrarchid Fisheries in Rivers and Streams (Vol. 87, No. 270, pp. 213–230). American Fisheries Society Symposium.

Magurran AE, 2013. Measuring Biological Diversity. John Wiley and Sons.

Mahan LB, Bassett LG, Duarte A, Forstner MR, Mali I, 2022. Effects of salinization on the occurrence of a long-lived vertebrate in a desert river. Scientific Reports, 12(1), 15907.

Malish MC, Gao S, Allen DC, Neeson TM, 2024. Impacts of stream drying depend on stream network size and location of drying. Ecological Applications, 34(6), e3015.

Mayes KB, Wilde GR, McGarrity ME, Wolaver BD, Caldwell TG, 2019. Watershed-scale conservation of native fishes in the Brazos River basin, Texas. In Multispecies and Watershed Approaches to Freshwater Fish Conservation (pp. 315–343). Bethesda, MD: American Fisheries Society Symposium 91.

McCaw WM, 2012. Sequestration vs. emissions: On the carbon sequestration potential of the natural areas of the City of Austin, Texas. Natural Areas Journal, 32(1), 86–95.

McGarrity ME, 2019. Spatial conservation assessment for balancing avoidance of impacts of tilapia introduction on imperiled fish biodiversity with economic impacts to the aquaculture industry. In Multispecies and Watershed Approaches to Freshwater Fish Conservation: Science, Planning, and Implementation (pp. 161–182). Bethesda, MD: American Fisheries Society Symposium 91.

Meyer JL, Pulliam WM, 1992. Modification of terrestrial–aquatic interactions by a changing climate. In Global Climate Change and Freshwater Ecosystems (pp. 177–191). New York, NY: Springer New York.

Meyer JL, Sale MJ, Mulholland PJ, Poff NL, 1999. Impacts of climate change on aquatic ecosystem functioning and health 1. JAWRA Journal of the American Water Resources Association, 35(6), 1373–1386.

Mildenberger DP, Bonner TH, Corbett T, Robertson CR, Lopez RR, Randklev CR, 2025. Evaluating lethal thermal tolerances and vulnerability of three east Texas freshwater mussels. Freshwater Science, 44(1), 45–60.

Milliman JD, Meade RH, 1983. World-wide delivery of river sediment to the oceans. The Journal of Geology, 91(1), 1–21.
Miranda LE, Coppola G, Boxrucker J, 2020. Reservoir fish habitats: A perspective on coping with climate change. Reviews in Fisheries Science and Aquaculture, 28(4), 478–498.
Mitchell ZA, McGuire J, Abel J, Hernandez BA, Schwalb AN, 2018. Move on or take the heat: Can life history strategies of freshwater mussels predict their physiological and behavioural responses to drought and dewatering? Freshwater Biology, 63(12), 1579–1591.
Miyazono S, Patiño R, Taylor CM, 2015. Desertification, salinization, and biotic homogenization in a dryland river ecosystem. Science of the Total Environment, 511, 444–453.
Monroe JB, Baxter CV, Olden JD, Angermeier PL, 2009. Freshwaters in the public eye: Understanding the role of images and media in aquatic conservation. Fisheries, 34(12), 581–585.
Mooij WM, Hülsmann S, De Senerpont Domis LN, Nolet BA, Bodelier PL, Boers PC, Dionisio Pires LM, Gons HJ, Ibelings BW, Noordhis R, Portielje R, Wolfstein K, Lammens EH, 2005. The impact of climate change on lakes in the Netherlands: A review. Aquatic Ecology, 39, 381–400.
Mulholland PJ, Best GR, Coutant CC, Hornberger GM, Meyer JL, Robinson PJ, Stenberg JR, Turner RE, Vera-Herrera F, Wetzel RG, 1997. Effects of climate change on freshwater ecosystems of the south-eastern United States and the Gulf Coast of Mexico. Hydrological Processes, 11(8), 949–970.
Muneepeerakul R, Bertuzzo E, Lynch HJ, Fagan WF, Rinaldo A, Rodriguez-Iturbe I, 2008. Neutral metacommunity models predict fish diversity patterns in Mississippi–Missouri basin. Nature, 453(7192), 220–222.
Nalepa TF, Schloesser DW, eds., 1992. Zebra Mussels Biology, Impacts, and Control. CRC Press.
Nguyen E, Mayes KB, Smith R, Trungale J, Perkin JS, 2023. The duality of drought: Pelagic-and benthic-spawning stream fishes show opposing responses to drought in the southern Great Plains. North American Journal of Fisheries Management, 43(5), 1276–1293.
Nickels DL, Bousman CB, 2010. Archaeological Testing at San Marcos Springs (41HY160) for the Texas Rivers Center, Hays County, Texas. Index of Texas Archaeology: Open Access Gray Literature from the Lone Star State, 2010(1), 8.
O'Reilly CM, Alin SR, Plisnier PD, Cohen AS, McKee BA, 2003. Climate change decreases aquatic ecosystem productivity of Lake Tanganyika, Africa. Nature, 424(6950), 766–768.
Olden JD, Poff NL, Douglas MR, Douglas ME, Fausch KD, 2004. Ecological and evolutionary consequences of biotic homogenization. Trends in Ecology and Evolution, 19(1), 18–24.
Olden JD, Kennard MJ, Lawler JJ, Poff NL. 2011. Challenges and opportunities in implementing managed relocation for conservation of freshwater species. Conservation Biology, 25(1), 40–47.
Olson JR, 2019. Predicting combined effects of land use and climate change on river and stream salinity. Philosophical Transactions of the Royal Society B, 374(1764), 20180005.
Osland MJ, Stevens PW, Lamont MM, Brusca RC, Hart KM, Waddle JH, Langtimm CA, Williams CM, Keim BD, Terando AJ, Reyier EA, Marshall KE, Loik ME, Boucek RE, Lewis AB, Seminoff JA, 2021. Tropicalization of temperate ecosystems in North America: The northward range expansion of tropical organisms in response to warming winter temperatures. Global Change Biology, 27(13), 3009–3034.
Ostroff A, Wieferich D, Cooper A, Infante D, 2013. 2012 National Anthropogenic Barrier Dataset (NABD). U.S. Geological Survey - Aquatic GAP Program: Denver, CO.
Paerl HW, Gardner WS, Havens KE, Joyner AR, McCarthy MJ, Newell SE, Qin B, Scott JT, 2016. Mitigating cyanobacterial harmful algal blooms in aquatic ecosystems impacted by climate change and anthropogenic nutrients. Harmful Algae, 54, 213–222.
Patiño R, Dawson D, VanLandeghem MM. 2014. Retrospective analysis of associations between water quality and toxic blooms of golden alga (*Prymnesium parvum*) in Texas reservoirs: Implications for understanding dispersal mechanisms and impacts of climate change. Harmful Algae, 33, 1–11.
Patterson LW, 2009. Characteristics of Shell Middens on the Coastal Margins of Southeast Texas. Houston Archeological Society, 5, 5–10.
Pease A, Delaune K, 2021. Dried and Salted: Cumulative Impacts of Diminished Flows and Salinization on Lower Pecos River Food Webs. Desert Fishes Council.
Perez KE, Saenz V, Salazar-Lozano N, Schwartz BF, Hutchins BT, 2023. DNA barcoding of springsnails (Mollusca, Gastropoda, Caenogastropoda) endemic to the Trans-Pecos region of Texas (USA). Check List.
Perkin JS, Acre MR, Ellard JK, Rodger AW, Trungale JF, Winemiller KO, Yancy LE, 2023. Flow–recruitment relationships for Shoal Chub and implications for managing environmental flows. North American Journal of Fisheries Management, 43(5), 1260–1275.

Perkin JS, Bonner TH, 2011. Long-term changes in flow regime and fish assemblage composition in the Guadalupe and San Marcos rivers of Texas. River Research and Applications, 27(5), 566–579.
Perkin JS, Bonner TH, 2016. Historical changes in fish assemblage composition following water quality improvement in the mainstem Trinity River of Texas. River Research and Applications, 32(1), 85–99.
Perkin JS, Gido KB, Cooper AR, Turner TF, Osborne MJ, Johnson ER, Mayes KB, 2015. Fragmentation and dewatering transform Great Plains stream fish communities. Ecological Monographs, 85(1), 73–92.
Perkin JS, Knorp NE, Boersig TC, Gebhard AE, Hix LA, Johnson TC, 2017. Life history theory predicts long-term fish assemblage response to stream impoundment. Canadian Journal of Fisheries and Aquatic Sciences, 74(2), 228–239.
Perkin JS, Shattuck ZR, Gerken JE, Bonner TH, 2013. Fragmentation and drought legacy correlate with distribution of Burrhead chub in subtropical streams of North America. Transactions of the American Fisheries Society, 142(5), 1287–1298.
Piña AE, Lougheed VL, 2022. Macroinvertebrate community composition in wetlands of the desert southwest is driven by wastewater-associated nutrient loading despite differences in salinity. Wetlands, 42(8), 128.
Pizano-Torres RI, Roach KA, Winemiller KO. 2017. Response of the fish assemblage to a saltwater barrier and paper mill effluent in the Lower Neches River (Texas) during drought. Journal of Freshwater Ecology, 32(1), 147–162.
Poff NL, Allan JD, Bain MB, Karr JR, Prestegaard KL, Richter BD, Sparks RE, Stromberg JC, 1997. The natural flow regime. BioScience, 47(11), 769–784.
Poff NL, Brinson ME, John WD, 2002. Potential impacts on inland freshwater and coastal wetland ecosystems in the United States. Aquatic Ecosystems and Global Climate Change, 1–44.
Postel S, Carpenter S, 1997. Freshwater ecosystem services. In Nature's Services: Societal Dependence on Natural Ecosystems (p. 195). Island Press.
Potter KN, Unger PW, Allen Torbert H, 2009. Management effects on soil organic carbon in Texas soils. In R Lal, RF Follett (Eds.), Soil Carbon Sequestration and the Greenhouse Effect (pp. 115–122, 2nd ed.). American Society of Agronomy and Soil Science
Randklev CR, Miller T, Hart M, Morton J, Johnson NA, Skow K, Inoue K, Tsakiris ET, Oetker S, Smith R, Robertson C, Lopez R, 2018a. A semi-arid river in distress: Contributing factors and recovery solutions for three imperiled freshwater mussels (Family Unionidae) endemic to the Rio Grande basin in North America. Science of the Total Environment, 631, 733–744.
Randklev CR, Tsakris ET, Johnson MS, Popejoy T, Hart MA, Khan J, Geeslin D, Robertson CR, 2018b. The effect of dewatering on freshwater mussel (Unionidae) community structure and the implications for conservation and water policy: A case study from a spring-fed stream in the southwestern United States. Global Ecology and Conservation, 16, e00456.
Rangaswami XL, Goldsmith AM, Khan JM, Robertson CR, Lopez RR, Randklev CR, 2023. Thermal tolerances of *Popenaias popeii* (Texas hornshell) and its host fish from the Rio Grande Basin, Texas. Scientific Reports, 13(1), 4603.
Reid AJ, Carlson AK, Creed IF, Eliason EJ, Gell PA, Johnson PT, Kidd KA, MacCormack TJ, Olden JD, Ormerod SJ, Smol JP, Taylor WW, Tockner K, Vermaire JC, Dudgeon D, Cooke SJ, 2019. Emerging threats and persistent conservation challenges for freshwater biodiversity. Biological Reviews, 94(3), 849–873.
Reisner M, 1993. Cadillac Desert: The American West and Its Disappearing Water. Penguin.
Renfrow J, Harvison S, Briggs MK, 2021. Assessing the vulnerability of reestablishing obligate riparian trees along a desert stream to climate change: An example from Terlingua creek, West Texas. In Renewing Our Rivers: Stream Corridor Restoration in Dryland Regions (p. 146). The University of Arizona Press.
Rhodes K, Hubbs C, 1992. Recovery of Pecos River fishes from a red tide fish kill. The Southwestern Naturalist, 178–187.
Richter BD, Bartak D, Caldwell P, Davis KF, Debaere P, Hoekstra AY, Li T, Marston L, McManamay R, Mekonnen MM, Ruddell BL, Rushforth RR, Troy TJ, 2020. Water scarcity and fish imperilment driven by beef production. Nature Sustainability, 3(4), 319–328.
Robertson JJ, Swannack TM, McGarrity M, Schwalb AN, 2020. Zebra mussel invasion of Texas lakes: Estimating dispersal potential via boats. Biological Invasions, 22(11), 3425–3455.
Roach KA. 2013. Texas water wars: How politics and scientific uncertainty influence environmental flow decision-making in the Lone Star state. Biodiversity and Conservation, 22(3), 545–565.
Roelke DL, Brooks BW, Grover JP, Gable GM, Schwierzke-Wade L, Hewitt NC. 2012. Anticipated human population and climate change effects on algal blooms of a toxic haptophyte in the south-central USA. Canadian Journal of Fisheries and Aquatic Sciences, 69(8), 1389–1404.

Rogers JS. 1976. Species density and taxonomic diversity of Texas amphibians and reptiles. Systematic Zoology, 25(1), 26–40.

Ruffing CM, Veach AM, Schechner A, Rüegg J, Trentman MT, Dodds WK, 2022. Prairie stream metabolism recovery varies based on antecedent hydrology across a stream network after a bank-full flood. Limnology and Oceanography, 67(9), 1986–1999.

Sanjar A, Davis DR, Kline RJ, 2023. Evidence of an established population of *Cherax quadricarinatus* (von Martens, 1868) in south Texas, USA. BioInvasions Records.

Sass GG, Gille CM, Hinke JT, Kitchell JF, 2006. Whole-lake influences of littoral structural complexity and prey body morphology on fish predator–prey interactions. Ecology of Freshwater Fish, 15(3), 301–308.

Scanlon BR, Duncan I, Reedy RC, 2013. Drought and the water–energy nexus in Texas. Environmental Research Letters, 8(4), 045033.

Schlechte JW, Fleming BP, 2015. Response of Devils River minnow and other fish in Pinto Creek, Kinney County, Texas during a severe drought. The Southwestern Naturalist, 60(1), 45–55.

Schlosser IJ, 1982. Fish community structure and function along two habitat gradients in a headwater stream. Ecological Monographs, 52(4), 395–414.

Schorr MS, Sah J, Schreiner DF, Meador MR, Hill LG, 1995. Regional economic impact of the Lake Texoma (Oklahoma-Texas) striped bass fishery. Fisheries, 20(5), 14–18.

Schwalb AN, Swearingen D, Robertson JJ, Locklin JL, Moore JS, McGarrity M, 2023. Living on the edge: Thermal limitations of zebra mussels (*Dreissena polymorpha*) in Central Texas. Biological Invasions, 25(3), 847–861.

Sears MM, Myers RA, Nisbet MT, Tidwell MT, Troia MJ, 2025. Assessing short-and long-term anthropogenic threats to a reintroduced fish in a restored urban riverscape. Freshwater Biology, 70(1), e14377.

Seawright EK, Rister ME, Lacewell RD, McCorkle DA, Sturdivant AW, Yang C, Goolsby JA, 2009. Economic implications for the biological control of *Arundo donax*: Rio Grande Basin. Southwestern Entomologist, 34(4), 377–394.

Shackelford CE, Rozenburg ER, Hunter WC, Lockwood MW, 2005. Migration and of Texas. Austin, TX, USA: TPWD.

Shao M, Zhao G, Kao SC, Cuo L, Rankin C, Gao H, 2020. Quantifying the effects of urbanization on floods in a changing environment to promote water security—A case study of two adjacent basins in Texas. Journal of Hydrology, 589, 125154.

Smith DR, Schlechte JW, Myers RA, Dance MA, Norman JD, Nisbet MT, 2024. Seasonal, spatial, and water level predictors of angler catch and effort within Texas black bass reservoir fisheries. North American Journal of Fisheries Management, 44(1), 79–92.

Speir T, 2017. Historic survey of Waskom, Harrison County, Texas. East Texas Historical Journal, 55(2), 6.

Stewart IT, Cayan DR, Dettinger MD, 2005. Changes toward earlier streamflow timing across western North America. Journal of Climate, 18(8), 1136–1155.

Sullivan ML, Zhang Y, Bonner TH, Tomasso JR, 2013. Temperature modulation of growth and physiology of juvenile Guadalupe Bass. North American Journal of Aquaculture, 75(3), 373–376.

Tank JL, Dodds WK, 2003. Nutrient limitation of epilithic and epixylic biofilms in ten North American streams. Freshwater Biology, 48(6), 1031–1049.

Tarter AA, Ford DF, Symonds DE, Ford NB, Schwalb AN, 2023. Impact of extreme climatic events on unionid mussels in a subtropical river basin. Hydrobiologia, 850(6), 1427–1442.

Taylor JM, King RS, Pease AA, Winemiller KO, 2014. Nonlinear response of stream ecosystem structure to low-level phosphorus enrichment. Freshwater Biology, 59(5), 969–984.

Tedesco PA, Oberdorff T, Cornu JF, Beauchard O, Brosse S, Dürr HH, Grenouillet G, Leprieur F, Tisseuil C, Zaiss R, Hugueny B, 2013. A scenario for impacts of water availability loss due to climate change on riverine fish extinction rates. Journal of Applied Ecology, 50(5), 1105–1115.

Texas Commission on Environmental Quality (TCEQ). 2024. 2024 Texas integrated report of surface water quality for clean water act sections 305(b) and 303(d).

Texas Parks and Wildlife Department (TPWD) News. 2017. Golden alga returns to several West Texas reservoirs. Released 20 December 2017. https://tpwd.texas.gov/newsmedia/releases/?req=20171220a

Thomas C, Bonner TH, Whiteside BG, 2007. Freshwater Fishes of Texas: A Field Guide. Texas A&M University Press.

Thomas ZA. 2015. Fishing warmwater streams with limited public access: Angling behavior, economic impact, and the role of Guadalupe Bass in a twenty-four county region of Texas (Doctoral dissertation).

Thompson M, Brenner J, Gilmer B, 2014. Informing conservation planning using future sea-level rise and storm surge modeling impact scenarios in the Northern Gulf of Mexico. Ocean and Coastal Management, 100, 51–62.

Tonkin JD, 2022. Climate change and extreme events in shaping river ecosystems. Encyclopedia of Inland Waters, 2, 653–664.

Tranvik LJ, Downing JA, Cotner JB, Loiselle SA, Striegl RG, Ballatore TJ, Dillon P, Finlay K, Fortino K, Knoll LB, Kortelainen PL, Kutser T, Larsen S, Laurion I, Leech DM, McCallister SL, McKnight DM, Melack JM, Overholt E, Porter JA, Prairie Y, Renwick WH, Roland F, Sherman BS, Schindler DW, Sobek S, Tremblay A, Vanni MJ, Verschoor AM, von Wachenfeldt E, Weyhenmeyer GA, 2009. Lakes and reservoirs as regulators of carbon cycling and climate. Limnology and Oceanography, 54(6part2), 2298–2314.

Troia MJ, 2017. What should your state fish be? Improving public awareness of endemic fishes in the United States. Fisheries, 42(1), 9–15.

Troia MJ, 2023. Magnitude–duration relationships of physiological sensitivity and environmental exposure improve climate change vulnerability assessments. Ecography, 2023(1), e06217.

Troia MJ, Kaz AL, Niemeyer JC, Giam X, 2019. Species traits and reduced habitat suitability limit efficacy of climate change refugia in streams. Nature Ecology and Evolution, 3(9), 1321–1330.

Troia MJ, McManamay RA, 2016. Filling in the GAPS: Evaluating completeness and coverage of open-access biodiversity databases in the United States. Ecology and Evolution, 6(14), 4654–4669.

Troia MJ, Perkin JS, 2022. Can fisheries bioenergetics modelling refine spatially explicit assessments of climate change vulnerability? Conservation physiology, 10(1), coac035.

Turner SW, Nelson K, Voisin N, Tidwell V, Miara A, Dyreson A, Cohen S, Mantena D, Jin J, Warnken P, Kao SC, 2021. A multi-reservoir model for projecting drought impacts on thermoelectric disruption risk across the Texas power grid. Energy, 231, 120892.

Uehlinger U, 2000. Resistance and resilience of ecosystem metabolism in a flood-prone river system. Freshwater Biology, 45(3), 319–332.

United States Department of Agriculture (USDA) 2024. 2023 Census of Aquaculture. Volume 3, Special Studies, Part 2, AC-22-SS-2. Issued December 2024. https://www.nass.usda.gov/Publications/AgCensus/2022/Online_Resources/Aquaculture/Aqua.pdf

Vannote RL, Minshall GW, Cummins KW, Sedell JR, Cushing CE, 1980. The river continuum concept. Canadian Journal of Fisheries and Aquatic Sciences, 37(1), 130–137.

VanLandeghem MM, Farooqi M, Southard GM, Patiño R. 2015. Spatiotemporal associations of reservoir nutrient characteristics and the invasive, harmful alga *Prymnesium parvum* in West Texas. JAWRA Journal of the American Water Resources Association, 51(2), 487–501.

Van Vliet MT, Franssen WH, Yearsley JR, Ludwig F, Haddeland I, Lettenmaier DP, Kabat P. 2013. Global river discharge and water temperature under climate change. Global Environmental Change, 23(2), 450–464.

Van Vliet MT, Thorslund J, Strokal M, Hofstra N, Flörke M, Ehalt Macedo H, Nkwasa A, Tang T, Kaushal SS, Kumar R, van Griensven A, Bouwman L, Mosley LM, 2023. Global river water quality under climate change and hydroclimatic extremes. Nature Reviews Earth & Environment, 4(10), 687–702.

Vaughn CC, 2010. Biodiversity losses and ecosystem function in freshwaters: Emerging conclusions and research directions. BioScience, 60(1), 25–35.

Visser P, Ibelings BAS, Van Der Veer B, Koedood JAN, Mur R. 1996. Artificial mixing prevents nuisance blooms of the cyanobacterium Microcystis in Lake Nieuwe Meer, the Netherlands. Freshwater Biology, 36(2), 435–450.

Vitousek PM, Aber JD, Howarth RW, Likens GE, Matson PA, Schindler DW, Tilman DG, 1997. Human alteration of the global nitrogen cycle: Sources and consequences. Ecological Applications, 7(3), 737–750.

Vose RS, Applequist S, Squires M, Durre I, Menne MJ, Williams CN Jr, Fenimore C, Gleason K, Arndt D, 2014. Improved historical temperature and precipitation time series for US climate divisions. Journal of Applied Meteorology and Climatology, 53(5), 1232–1251.

Wagner ND, Osburn FS, Robbins CJ, Ernst MR, Owens J, Powers SM, Scott JT, 2023. Lake stability and anoxia dynamics revealed from high frequency vertical profiling in a eutrophic polymictic reservoir. Inland Waters, 13(2), 167–181.

Wenger SJ, Isaak DJ, Luce CH, Neville HM, Fausch KD, Dunham JB, Dauwalter DC, Young MK, Elsner MM, Rieman BE, Hamlet AF, Williams JE, 2011. Flow regime, temperature, and biotic interactions drive differential declines of trout species under climate change. Proceedings of the National Academy of Sciences, 108(34), 14175–14180.

Williams TJ, Collins MB, Rodrigues K, Rink WJ, Velchoff N, Keen-Zebert A, Gilmer A, Frederick CD, Ayala SJ, Prewitt ER. 2018. Evidence of an early projectile point technology in North America at the Gault Site, Texas, USA. Science Advances, 4(7), eaar5954.

Wilson EO, 1986. Biophilia. Harvard University Press.

Winemiller KO, Perkin JS, Trungale JF, Hoeinghaus DJ, Moore GW, Schwalb AN, Mitchell ZA, Trimble A, Reeves C, Hardy TB, Buzan D, 2024. Advancing environmental flows science: Hindcasting and forecasting flow–ecology relationships. Fisheries, 49(8), 353–368.

Withers PJA, Jarvie HP, 2008. Delivery and cycling of phosphorus in rivers: A review. Science of the Total Environment, 400(1–3), 379–395.

Wolaver BD, Reynolds LV, Caldwell TG, Bongiovanni T, Pierre JP, Breton C, Mayes KB, 2024. Hydrologic changes in the Brazos River Basin and implications for Great Plains fishes. Journal of Hydrology, 629, 130351.

Woodward G, Perkins DM, Brown LE, 2010. Climate change and freshwater ecosystems: Impacts across multiple levels of organization. Philosophical Transactions of the Royal Society B: Biological Sciences, 365(1549), 2093–2106.

Woolway RI, Kraemer BM, Lenters JD, Merchant CJ, O'Reilly CM, Sharma S, 2020. Global lake responses to climate change. Nature Reviews Earth and Environment, 1(8), 388–403.

Wootten AM, Martin E, Randklev CR, Smith R, 2023. Projected changes to streamflow and stream temperature in central Texas: How much will the river flow? Earth Interactions, 27(1), 220021.

Xenopoulos MA, Lodge DM, Alcamo J, Märker M, Schulze K, Van Vuuren DP, 2005. Scenarios of freshwater fish extinctions from climate change and water withdrawal. Global Change Biology, 11(10), 1557–1564.

Yuan XZ, Zhang YW, Liu H, Xiong S, Li B, Deng W, 2013. The littoral zone in the Three Gorges Reservoir, China: Challenges and opportunities. Environmental Science and Pollution Research, 20, 7092–7102.

Zhang Y, Wurbs R, 2018. Long-term changes in river system hydrology in Texas. Proceedings of the International Association of Hydrological Sciences, 379, 255–261.

Zhao G, Gao H, Cuo L, 2016. Effects of urbanization and climate change on peak flows over the San Antonio River Basin, Texas. Journal of Hydrometeorology, 17(9), 2371–2389.

Zhi W, Klingler C, Liu J, Li L, 2023. Widespread deoxygenation in warming rivers. Nature Climate Change, 13(10), 1105–1113.

II.4 Texas Gulf Coast Ecosystems and Climate Change

Coral Reefs, Barrier Islands, Bays, and Blue Carbon

Susan Power Bratton

II.4.1 INTRODUCTION: THE UNSEEN

Public discussions of climate change can overlook the Gulf of Mexico (Gulf of America) and Texas coastal ecosystems because the popular media focuses on the Arctic and the Antarctic, and the plankton-rich southern inshore waters appear immutable. Planning conversations often center on vulnerability to hurricanes and do not look any further, as Texas seaside communities suffer frequent, costly storm damage. Yet, the atmosphere and the oceans are intimately connected by the exchange of heat and gases like oxygen and carbon dioxide. Differences in the absorption and reflection of solar radiation across the planet's surface drive ocean currents critical to transferring oxygen to the depths, circulating nutrients toward the surface, and carrying warm water toward the poles. Weather-related phenomena, like high- and low-pressure cells and the winds, drive surface currents and wave formation. Atmospheric and oceanic geophysical and chemical changes progress in tandem.

In marine ecosystems, these oft-invisible shifts are already stressing the foundation species, such as reef-building corals and seagrasses, on which other species depend for photosynthesis, and as structures to attach to, hide in, or forage around. Changes originating with atmospheric gases exacerbate stresses from other coastal environmental threats, such as sewage effluent, nutrient enrichment, petroleum spills, intensive fish harvest, and channel dredging. Rather than begin with a complex summary of marine chemistry, this chapter will first introduce two maritime ecosystems already suffering ecological disruption from anthropogenic disturbances – coral reefs and the bottom communities subject to "dead zones" – and will explore the impacts of atmospheric change and the interactions of climate with other physical and chemical environmental factors serving as ecosystemic and physiological stressors. The chapter then selects examples of vulnerable Gulf and coastal ecosystems and documents the biophysical drivers causing loss of habitat area, declines in productivity, and modification of species assemblages. Among the affected ecological communities, coral reefs and mature coastal live oak stands are rare in Texas; thus, conservationists have targeted them for protection in national marine sanctuaries, national wildlife refuges, state parks, and nature reserves. Organisms of exceptional interest or economic value to humans, including sea turtles, red snapper, and migrant birds, exemplify the impact of atmospheric change on the geographic ranges and population health of individual Texas species. The last quarter of the chapter presents potential means of mitigating the ecological consequences of climate change, including public education concerning marine ecology and conserving and expanding the role of blue carbon. The chapter uses standard abbreviations for metric measurements, including millimeter (mm), centimeter (cm), meter (m), kilometer (km), hectare (ha), degrees Celsius (°C), milligrams (mg), liter (l or L) second (s), and year (y); and for the elements hydrogen (H), oxygen (O), carbon (C), and calcium (Ca).

DOI: 10.1201/9781003406600-11

II.4.2 CORAL REEF BLEACHING AND BIODIVERSITY LOSS

II.4.2.1 Elevated Water Temperatures Stress Foundation Species

To begin with an exceptionally temperature-sensitive ecosystem, many Texans are unaware that small coral reefs have formed on Texas offshore salt domes, separated from the coral reefs of Mexico and western Florida by 400 miles. More than 100 miles by boat from Freeport, East and West Flower Garden Banks are now a National Marine Sanctuary, popular with experienced scuba divers. With the shallowest points just under 18 m (60 feet) from the surface, their depth and strong currents discourage novice divers and casual snorkelers. A few other small, scattered submarine pinnacles in the northern Gulf, including Stetson and Geyer Banks, also provide substrates and water clarity suitable for reef-building by hard or scleractinian corals. Huge manta rays, colorful parrot and angel fish, and stoic loggerhead sea turtles glide above these treasure troves of biodiversity (Cancelmo, 2008). As relatively deep reefs, the Texas coral ecosystems lack the large branching corals, like elkhorn, and the showy gorgonian octocorals that a snorkeler observes in a healthy, shallow Florida reef. At Flower Garden Banks, the upper *Diploria-Monatastrea-Porites* zone extends from the crest, circa 18 m to 36 m, where dome-like star corals (*Montastrea* and *Orbicella* spp.) and brain corals (*Colpophyllia natans* and *Pseudodiploria strigosa*) form heads that can exceed two meters in diameter. As depth increases, the community structure changes, and biodiversity declines. At 28 to 46 m on East Flower Garden Bank, a smaller, cream-colored branching coral, yellow pencil coral (*Madracis mirabilis*), and patches of leafy macroalgae are common. From 36 to 50 m, in the *Stephanocoeniz-Millepora* zone, a bright yellow hydrozoan fire coral, sea ginger (*Millepora alcornis*), increases in dominance in a mix of scleractinian corrals, including brain, star, and plate corals, able to tolerate the attenuated illumination. Although the flanks of the salt domes extend below 50 m, there is not enough light penetration to support the larger coral heads at greater depths; thus, sponges, sea whips, encrusting organisms, echinoderms like sea stars, and scattered soft corals take over (Britton and Morton, 1989; Cancelmo, 2008).

During the 1970s and 1980s, multiple reefs in the Gulf of Mexico and the Caribbean began to display widespread symptoms of environmental stress, including diminished coral area, disease outbreaks, coral bleaching, algal overgrowth, loss of predators, and the disappearance of rare species. Sewage effluent and terrestrial runoff bearing bacteria, pesticides, petroleum products, and the nutrients stimulating algal blooms easily reach inshore coral formations, as does sediment from dredging and the construction of vacation homes, airports, and marinas. The pollutants are toxic to coral polyps or cover their delicate mouths and tentacles. Overfishing and invasive species, like Asian lionfish (*Pterois* spp.), add to the burden by disrupting food webs. Tourists break off and step on corals and kick sediment over their surfaces (Griggs, 2017; Sheppard, 2021).

Initially in the 1980s, when bleaching incidents began to occur at shallow and heavily utilized hermatypic (reef-building) coral colonies worldwide, marine scientists attributed the speedy death of corals, leaving ghostly white calcium carbonate skeletons, to diverse local factors like extreme low tides, freshwater runoff, road building, and toxic algal blooms. Corals have symbiotic algae, called zooxanthellae, that use the light-rich environment for photosynthesis, producing glucose and amino acids, and greatly enhancing the energy available to the coral polyps, which, as smaller relatives of sea anemones (*Cnidaria*), feed on plankton. The algae benefit from the nitrogen and other nutrients the polyps acquire as carnivores. When bleaching occurs, the corals lose their algal sugar producers. The polyp can survive for a short period without the zooxanthellae but will, in essence, quickly die from starvation (Griggs, 2017; Kaiser et al., 2020). Although global warming was a suspect, some scientists argued against it as a significant cause and pointed out that, during the 1980s, many reefs, like the Australian Great Barrier Reef, remained little affected. Further, it was unclear why the algae would respond to minor shifts in ocean temperature. Astrophysicists Baliunas and Soon (2005) protested incorrectly that 250 years ago, before the industrial revolution, temperatures in the South Pacific had been 2 °C higher than today, and reefs had flourished. They noted that reefs subject to bleaching in 1995 had "almost all" recovered by 1996. Indeed, at a time when the Florida Keys National Marine Sanctuary was already losing larger branching corals, such as staghorn (*Acropora cervicornis*) and elkhorn (*Acropora palmata*), at an alarming rate, the Flower Garden Banks' massive formations of star and brain corals remained relatively free of coral diseases and bleaching. In reviewing environmental concerns for the Southern Gulf of Mexico coral reefs, Horta-Puga (2007) reported that a single mass bleaching event occurred in Mexico during the 1990s, probably caused by a severe temperature drop accompanying a cold front. However, he confirmed that low-level bleaching was "not uncommon."

Water has a high specific heat, which means the oceans absorb and retain much of the additional energy from the atmosphere caused by the greenhouse effect. Oceanographers have installed monitoring stations at Flower Garden Banks and multiple other coral reefs worldwide to collect data on changes in water temperature throughout the year at different depths. Remote sensing via satellite provides a detailed portrait of global sea surface temperatures. Over the past forty years, all the broad-scale international bleaching events have correlated with elevated sea surface temperatures, including exceptionally strong El Nino-generated ocean warming occurrences (Griggs, 2017). Ultimately, in 1995, 2005, and 2010, Flower Garden Banks experienced bleaching episodes, with about 45% of corals displaying early-stage bleaching. In the 2005 event, the United States' (U.S.) Caribbean coral reefs lost roughly half their corals overall. In 2016, a major bleaching incident with significantly more days of water temperatures above 30°C than the previous summer average caused mass mortality in Texas. Analyzing photographs taken from permanent photo stations before and after the 2016 event, Johnston et al. (2019) found the percentage of affected corals was about twice as great on East as on West Flower Garden Banks, and the East Banks had more days of temperatures exceeding 30°C. The bleaching impact was more severe at shallower depths at 18 to 24 m (60–80 feet), where the water was warmer than around the cooler base of the reefs at 32 to 40 m (100 to 130 feet). Due to currents and the topography of the reefs, ocean water temperature does not decline at a constant rate with depth among the Texas coral communities. In October 2016, the shallow stations on the East Bank had 21% bleaching and 46% paling for 67% or about two-thirds of the coral area. In comparison, the deep stations at East Bank had 0.5% bleaching and 29% paling. At the same time, the comparable shallow stations at the West Bank had 9% bleaching and 16% paling for 25% of the shallow coral area, while the deep stations had 5% bleaching and 11% paling. Like the Australian Great Barrier Reef and the Florida Keys, Texas is experiencing an increasing frequency and intensity of bleaching events, with the most recent, as of this writing, beginning in August 2023 and continuing into 2024 (Office of National Marine Sanctuaries, 2024).

As hermatypic corals are adapted to the tropics and live in some of the naturally warmest coastal regions, the question arises as to why corals would be so sensitive to temperature shifts of a degree or two. As Morgan et al. (2016) summarize: "Warming increases performance at the lower end of an organism's thermal range but has the reserve effect above a thermal optimum … Species from warmer (e.g. upper intertidal) or more stable (e.g. tropical) environments live close to the upper acute thermal tolerance limits and are therefore particularly susceptible to warming." Coral reef ecosystems have relatively stable physical environments, with only minor temperature fluctuations throughout the year. Hermatypic corals are generally confined to waters between 20 and 30°C. In very hot lagoons and those with elevated levels of salinity, such as Shark Bay, Australia, corals do not thrive, and cyanobacteria in doomed formations of calcium carbonate called stromatolites replace them. Individual coral species have a limited range of temperature tolerance. Worldwide, short-term exposures of 3–4°C greater temperatures or longer-term events with temperatures elevated by 1–2°C will cause bleaching. Biologists have developed the concept of degree heating weeks (DHW), which is based on the degrees Celsius above normal temperatures times the number of weeks of elevated temperatures. Many species will pale at approximately four DHW, and widespread bleaching emerges at about eight DHW. The light and temperature raise the photosynthetic activity of the zooxanthellae, which, aside from sugars, generates additional oxygen, part of it as free radicals. Rather than being beneficial to the coral polyp, the excess oxygen is toxic, and the polyp ejects its symbionts (Kaiser et al., 2020; Sheppard, 2021; Sheppard et al., 2018). Johnston et al. (2019)

concluded that, in addition to exposures of over 30°C, longer exposures to temperatures of 29°C at Flower Garden Banks stressed the corals sufficiently to initiate loss of zooxanthellae.

Individual coral species are adapted to a specific range of water depths, with some growth forms, such as branching corals, growing primarily in shallower habitats, and others, such as many of the plate corals, found at deeper locations with less light and less wave action. Bleaching causes greater mortality among shallow-water corals and thus changes the species composition of the affected reefs. Nonetheless, bleaching also kills soft or ahermatypic (non-reef building) corals, which are present at Flower Garden Banks, with their area increasing with depth. As they do not generate hard skeletons, the loss can be underestimated (Sheppard et al., 2018). Through the middle of the 20th century, most of the world's coral reefs had been keeping up with sea level rise and maintaining their location in the high light zone via accretion of calcium carbonate, typically at growth rates of 1–10 mm per year. Due to heavy shallow coral mortality, many reefs are now eroding rather than accreting and are, therefore, slowly sinking relative to the water's surface (Sheppard, 2006). Scleractinian corals can recover from bleaching if the mortality is limited, and a significant loss of associated species, like fish, may not follow a moderate bleaching event. However, corals are foundation species that maintain the complex architecture of the reef. Sheppard (2006) concludes that: "where reefs have now been reduced to planar, two-dimensional surfaces instead of rugose, three-dimensional surfaces, fish quantity has recently become considerably reduced." Further, as major coral die-offs occur, filamentous green algae speedily cover the ghostly coral heads, blocking light to any remaining living polyps and making it difficult for coral planula (the larval dispersal phase) to settle in the competing algal mat and establish new coral colonies. This process also displaces fish and invertebrates, such as shrimp and sponges (Kaiser et al., 2020). Due to their depth, the Flower Garden Banks have thus far escaped mass mortalities, causing radical biodiversity loss. A National Oceanic and Atmospheric Administration (NOAA)-sponsored team has conducted a rapid climate vulnerability survey for Texas's National Marine Sanctuary, which predicts that following global trends, the Flower Gardens and other Texas coral communities will suffer increases in the frequency and severity of bleaching events (Dias et al., 2023). As bleaching events become more frequent, even occurring annually or biannually, the corals will be unable to recover metabolically (Sheppard et al., 2018).

11.4.2.2 Ocean Acidification and Dissolution of Calcium Carbonate

In addition to heat from the greenhouse effect, the world's oceans, which are undersaturated with carbon dioxide, absorb massive amounts of CO_2 from burning fossil fuels. The uptake is approximately 30% of the anthropogenic CO_2 released in the past two centuries. Aside from shifting the composition of gases in seawater, the ever-increasing concentration of CO_2 is modifying the natural buffering capacity of ocean water by disassociating hydrogen ions (H^+), bicarbonate (HCO^-), and carbonate ions (CO_3^{2-}). The increase in H^+ ions is making the oceans more acidic by increasing the concentration of carbonic acid. Although the impact on ocean chemistry is universal, coral reef health is a particular concern due to the importance of marine calcifying organisms, including corals, mollusks, and crustaceans, in these ecosystems. Numerous studies have documented declines in pH from a projected pre-industrial average of 8.16 to levels of 8.05 and lower. This might appear to be a minor change. However, it seriously affects animals manufacturing shells or skeletons from calcium carbonate via biomineralization. This process depends on the adequate availability of carbonate ions to maintain the saturation state optimal for producing aragonite for shells. More acidic waters "steal" the CO_3^{2-} necessary to form HCO_3 with the more available H^+ ions, thereby inhibiting the secretion of reef-building calcium carbonate, $CaCO_3$. Changes in the natural conditions, including pressure and temperature with depth in the world's oceans, establish a saturation horizon at varying distances from the surface. Below the saturation horizon, species utilizing $CaCO_3$ for biogenic structures are vulnerable to dissolution. This saturation horizon has been moving upward over the past two centuries by 50 to 200 m. It takes about 30 years for CO_2 and other atmospheric gases to mix through the water column and fully dissolve in the oceans. If we were to stop releasing CO_2 today, the process of acidification would continue for another three decades, regardless of our best efforts toward sustainability (Iglesias-Rodríguez et al., 2016; Kaiser et al., 2020; Sheppard, 2021).

The Gulf of Mexico has warm ocean temperatures and a relatively high total alkalinity to inorganic carbon ratio, which helps sustain relatively high aragonite saturation (Ω_{ar}) values relative to regions such as the North Atlantic. Nonetheless, multiple studies have found that acidification of Gulf of Mexico estuaries, open surface waters, and subsurface waters has progressed in the past four decades. Some Gulf ecosystems, such as mangrove-dominated wetlands, tolerate moderate pH changes. Others, such as oyster reefs, are more vulnerable. Experimental studies of oysters, particularly the commercially important native species, *Crassostrea virginica*, indicate that lower pH can cause shell weakening, delays in metamorphosis from larval stages, greater mortality, and decreased tolerance to extreme temperatures. Site-specific research has documented the ongoing dissolution of calcium carbonate for coral reefs in the upper Florida Keys and phase shifts from hard corals to macroalgae and soft corals, which are less dependent on calcium biomineralization (Osborne et al., 2022). Lunden et al. (2013) concluded that the aragonite saturation (Ω_{ar}) for Gulf waters at greater than 300 m depth was between .98 and 1.69 and that the scleractinian cold-water coral *Lophelia pertusa* was present at sites between 1.25 and 1.69 but absent from more acidic locales. Ongoing ocean acidification could, therefore, also limit the sites suitable for the formation of deeper cold-water coral structures off the Texas coast.

II.4.2.3 Coral Recovery from Hurricanes

Hurricanes disturb coral reefs via the effects of winds, waves, and surges that can topple even the biggest heads and batter branching corals into fragments. Storm surge can cover reefs with sediment, as can the terrestrial run-off generated by heavy rains. Massive discharges of freshwater reduce salinity, temporarily creating unfavorable conditions for corals. Although very shallow and near-shore reefs are the most vulnerable to this natural process, in September 2005, Hurricane Rita, which produced widespread flooding and wind damage in Texas, moved over the Flower Garden Banks as a Category 3 storm. The Banks were 35 miles from the eyewall and sustained physical damage to corals as deep as 72 m. Rita generated "a huge slug of brown water," reaching 115 miles offshore to the sanctuary. Monitoring at the National Marine Sanctuary recorded the salinity and temperature changes, but the instrumentation was insufficient to detect potential chemical pollutants. Despite damage to living coral surfaces, NOAA managers have concluded that Rita cooled the reef waters, which helped to relieve the elevated temperature causing the 2005 bleaching event (Hickerson and Schmahl, 2007). Nonetheless, Lugo-Fernandez and Gravois (2010) found an 11% reduction in coral cover following Hurricane Rita. Despite its Category Two status, Hurricane Ike caused similar impacts due to its vast area in October 2008. NOAA staffers found coral boulders up to 4 meters across ripped from the reef and blocks of coral that had landed as far as 7.5 m (25 feet) away from their original location. Ike had scoured sand from some areas and deposited it in others. In August 2017, the destructive freshwater flooding generated by Hurricane Harvey did not quite reach the National Marine Sanctuary. Nonetheless, the water turnover resulting from Harvey lowered the water temperature at the Flower Garden Banks to a safe 27.2°C, inhibiting renewed bleaching (Flower Garden Banks National Marine Sanctuary, 2024).

Corals can reestablish from hurricane-caused physical disturbances, although it may require a decade or more in cases where the architecture of a reef and the larger corals have been severely damaged. Detecting hurricane frequency and intensity increases is difficult due to the wide differences from year to year for specific geographic regions. Nonetheless, predictive modeling based on forcings from Earth system models has found that intensities will increase, and changes in the seasons' lengths and tropical storm frequency are also likely. Hurricanes already interact with anthropogenic disturbances to coral reef ecosystems – often negatively. Corals, for example, can eject sediment particles deposited by storms and clear their surfaces, but can manage thin sprinklings of coral sand more easily than thick terrigenous grit from road building or particles coated with petroleum products. An increase in hurricane intensity could increase the number of disturbance events toppling and damaging corals at deeper, offshore reefs, like the Flower Garden Banks, and inhibit recovery from recent bleaching events. Kossin (2018) found that the translation speed of tropical cyclones slowed by 10% between 1949 and 2016. On the one hand, with the expected increase of tropical storm rainfall due to higher atmospheric temperature, these slower-moving weather events will deliver more rainfall to individual locales, causing more intrusion of fresh water onto coral reefs. On the other hand, as NOAA has documented, the water temperature changes associated with late summer hurricanes, especially those that do not directly strike the reefs, can alleviate bleaching. The continuation of the permanent monitoring of the Gulf's coral reefs is critical to understanding future coral ecosystem dynamics and ensuring reef health.

II.4.3 WARMING WATERS AND OXYGEN MINIMUM ZONES

II.4.3.1 Mass Mortality of Benthic Faunas at the Mouth of the Mississippi River

An equally dramatic example of anthropogenic stressors causing mass mortality of marine organisms is the appearance of dead zones in soft and hard sea bottom environments in more than 400 locations worldwide. One of the largest and best-studied is off the Mississippi River delta, where it extends across the border from Louisiana into Texas waters. Consisting of patches of greater and lesser impact, a marine dead zone does not have set boundaries but expands and contracts from year to year and season to season, depending on environmental conditions. Originating from phytoplankton blooms fed by high levels of nutrients, at its worst, the dead zone at the mouth of the Mississippi has been larger than the state of Connecticut. Bacteria decompose the dying algae near the surface as they sink toward the ocean bottom. This creates a greater biological demand (BOD) for the dissolved oxygen in the water and results in oxygen-minimum zones (OMZs) with oxygen levels below 2 mg L^{-1}, where most multicellular organisms cannot survive. OMZs can occur naturally. In Oregon, for instance, intrusions of cold and lower oxygen waters have coincided with natural nutrient upwelling events, producing high levels of algal production. This has then led to hypoxia or low oxygen conditions, which kill invertebrates like crabs and fish. However, the most common cause is excess anthropogenic input of nutrients, particularly nitrogen, from sewage, burning of fossil fuels, industrial emissions, or agricultural pollution. Fertilizer-laden run-off from farms in Iowa, Nebraska, and other states in the Mississippi River watershed fuels the massive dead zone creeping into Texas (Griggs, 2017; Kaiser et al., 2020).

II.4.3.2 Vulnerability of Shallow Estuaries to Hypoxia

Although climate change is ordinarily not the primary cause of dead zones, cold water has a greater capacity to retain dissolved oxygen than warm water, and heat inputs from the greenhouse effect can exacerbate oxygen depletion. Warmer water accelerates the metabolic processes of many marine organisms, raising their need for oxygen. Texas continues the eastern U.S.'s magnificent sequence of barrier islands – the longest in the world. Fed by multiple rivers forming brackish estuaries, shallow bays fringed by salt marshes rim the entire Texas coast. The larger tidal bodies of water immediately behind the islands are primary bays with salinity levels and water chemistry similar to the open Gulf. The smaller lagoons at the mouths of rivers are the secondary bays and have greater inputs of freshwater and lower salinity. These enclosed bodies of water exchange better-oxygenated water with the Gulf of Mexico through channels and passes between the islands, which is a slower process than waters mixing in the open seas. In addition, shallower bodies of water, such as Texas's bays, are prone to developing an upper layer of warm water, causing stratification by trapping cooler, denser water underneath it. This reduces the movement of better-oxygenated water from the surface to the bottom of a bay. Locales like microtidal Corpus Christi Bay are prone to hypoxia or severely reduced oxygen availability, especially in the summer. Montagna and Froeschke (2009) examined both the epifauna, or the animals living on top of the sediments, and the infauna, or animals living within the sediments, from hypoxic and non-hypoxic areas in Corpus Christi Bay. They found that the hypoxic zone had lower populations, less biomass, and lower diversity of bottom-dwelling fauna than their better-oxygenated control sites. The infauna in the hypoxic area had lower accumulation levels. The infauna thus did not recover as quickly from hypoxic events as the epibenthos, even though the infaunal invertebrates are generally more tolerant of low oxygen levels.

Texas is already experiencing periodic fish kills associated with nutrient pollution and elevated water temperatures of 70°F and greater in estuaries and inshore waters. For example, dead Gulf menhaden (*Brevoortia patronus*) covered multiple Texas beaches in June 2023. Hypoxic zones deplete biomass, unravel marine food webs, and reduce estuarine nursery areas, negatively impacting fish and shellfish populations. The primary dead zone in the Gulf of Mexico costs commercial fisheries and tourism more than $80 million in annual losses (Griggs, 2017). As Kaiser et al. (2020) conclude: "ongoing warming together with rising atmospheric CO_2 will see an expansion of oxygen minima zones, perhaps by more than 50 percent by volume, by the end of the 21st century." However, the interactions with climate are complex. If temperatures and rainfall increase, it will result in more nutrient-rich runoff and more stratification of Texas bays, expanding oxygen minimum zones. Conversely, lower rainfall and more intense tropical storms reduce stratification, while intense rainfall events can flush large amounts of soil and fertilizer into rivers. As McKinney and Allred (2021) of the Harte Research Institute, Texas A&M, Corpus Christi have pointed out, conditions precipitating low oxygen clear quickly in some Texas bays, while others, including the upper reaches of Galveston Bay and Baffin Bay in the Laguna Madre, are prone to lingering impacts, due to elevated nutrients. Currently, more Texas bays are experiencing decreases in dissolved oxygen than increases, with the Laguna Madre especially subject to low values. Coral reefs are not exempt from low-oxygen waters, including those resulting from tropical storms producing intrusions of freshwater. The current predictions for the Texas coastal region of higher temperatures and more extended dry periods punctuated by intense rainfall events suggest mixed outcomes. However, temperature increases and periodic flushing of nutrients are likely to promote hypoxia. Overall, the warming ocean with lower dissolved oxygen levels will contribute to expanding Gulf dead zones for the foreseeable future.

II.4.4 RISING SEA LEVELS AND WETLAND RECESSION IN TEXAS BAYS

II.4.4.1 The Barrier Islands and Bays

The West Coast of the U.S. is a geologically active, leading-edge coast, where the movement of tectonic plates generates volcanic activity and slowly lifts the coast. Rocky shores, cliffs, a narrow continental shelf, and bays formed on active fault lines (like San Francisco Bay), characterize states like California and Washington. In contrast, the East Coast is a trailing-edge coast with a broad coastal plain grading gently into a wide continental shelf. Rivers build large deltas and deposit the sand that feeds the Gulf's wide, flat beaches, sand bars, and barrier islands. Aside from being easily eroded and reconfigured by storms, barrier islands and natural sand beaches are notoriously sensitive to changes in currents, sediment input, sea level, and vegetation cover. Hurricanes can transport materials, fill an inlet with sediments, bury a road, and dump sand on a marsh. They can also erode a shoreline back under a row of homes, remove entire dune lines, and cut a new inlet lacking a bridge. Throughout the Pleistocene, the coastline of the Gulf of Mexico moved seaward during glacial expansion as sea levels fell and then retreated inland as glaciers melted and sea levels rose. The barrier island systems moved in tandem with the coastline. Today, Texas has "fossil" dune lines and sand plains from previous eras of higher sea levels, forming bluffs or elevated shorelines along Texas bays. Although the barrier islands can slowly realign with sea levels, a significantly accelerated rate of sea level rise, intensification of storm surges, or declines in sediment transport by the Mississippi or Texas rivers will cause them to erode faster than they accrete sand. Consequently, they will lose area and their shorelines will recede (Anderson, 2007; Davis, 2011; Pilkey and Pilkey, 2019).

11.4.4.2 Rising Sea Levels and Reduction of Seagrass Meadows

Landward of the barrier islands, the Texas bays, and their wetland margins support highly productive plant communities, fixing hundreds of thousands of tons of carbon annually. Historically, much of their submerged area supported seagrasses, which, like corals, are foundation species. They contribute food and structure to the bays, which, in turn, foster highly productive nurseries and foraging areas for fish, shellfish, and, of course, juvenile and adult shrimp. Unfortunately, recent decades have seen significant declines in seagrasses in Texas and nationwide. Seagrasses are vascular plants rather than algae, yet they are sensitive to sedimentation, chemical spills, and sewage effluents. A combination of coastal development and modern agricultural chemical use has caused eutrophication in all the Texas bays, although inshore waters adjoining cities, like Corpus Christi Bay and Galveston Bay, have sustained greater impacts. The excess nutrients, such as nitrogen and phosphorus, that cause hypoxia also stimulate algal blooms that coat the seagrasses and block their light reception and photosynthesis. Relatively small concentrations of commercial herbicides in run-off water can kill seagrasses directly. In the early 1960s, for example, what had been lush seagrass beds disappeared almost entirely from the bay shoreline of Galveston Island – probably due to multiple human-generated stressors (Blackburn, 2004).

The recession of the Pleistocene glaciers toward the poles initiated a period of global sea level rise. Barrier island and bay ecosystems have shifted as the coastline of Texas has slowly moved inland. Where storm surge and submergence disturbed vegetation, plant succession kept up with changing water depth and reestablished salt marshes and sea grass flats. Currently, however, global warming is accelerating the rate of sea level rise. Due to the internal wind circulation patterns and warmer waters, the rate of rise in the Gulf of Mexico is two to three times the typical global rates and has climbed to >10 mmy^{-1} (or about a meter a century) since 2010 (Dangendorf et al., 2023). To complicate matters, the Texas coast has experienced widespread subsidence due to hydrocarbon and water extraction – reaching 9.5 mmy^{-1} in the region of the Coastal Bend around Corpus Christi and roughly doubling changes in water depth. Storms displacing large amounts of water, like Hurricane Harvey, cause weeks of terrain depression, temporarily contributing to the phenomenon (Capistrant-Fossa and Dunton, 2024).

Capistrant-Fossa and Dunton (2024) studied trends in seagrass coverage at permanent monitoring stations in the Laguna Madre, Texas's most pristine bay. Since 2014, water depth has increased by 1.4 to 2.5 cm per year. They found that at a deep fixed-edge station, two keystone seagrasses, shoal grass (*Halodule wrightii*) and manatee grass (*Syringodium filifrome*) had entirely disappeared in a mere five years. Seagrasses were gone from 23% of the 144 stations at different depths spaced throughout the Upper Laguna behind South Padre Island and south into Mexico. While algal epiphytes growing on the seagrasses were at relatively low levels compared to monitoring stations in other Texas bays, the water chemistry had been changing significantly, including a switch from hyposaline (lower salinity) to hypersaline (very salty). The study organized the stations into clusters, averaging from .45 m to 1.66 m in water depth. The seagrass biomass responded to temperature changes and the amount of solar radiation reaching the bottom, while water depth and light penetration determined whether seagrasses would be present or not. For example, the optimal conditions for manatee grass were water temperatures between 25 to 30°C and 370 μmol (micromoles) photons of light per $m^{-2}s^{-1}$. Irradiance levels below the optimal ranges inhibited seagrass from re-establishing. Increasing water depth by only .25 m can decrease the light necessary for photosynthesis by 25 to 50%. Like the loss of zooxanthellae for corals, the resulting loss of production will "starve" the seagrasses, extirpating them and greatly reducing the fish and invertebrate populations dependent on their presence.

While government agencies, such as the National Marine Fisheries Service and Texas Parks and Wildlife Department, and conservation non-profits, such as the Galveston Bay Foundation, have been cooperating on experimental projects restoring seagrass beds and struggling to protect them from the barrage of stressors, losses of the submerged meadows have increased in some of the bays' least developed zones. As Capistrant-Fossa and Dunton (2024) report, seagrass communities once occupied an estimated 200 km^2 of the Upper Laguna Madre, but as of 2018, habitat for seagrass had been depleted by 41 km^2 or about 20%. Continuing sea-level rise could expand the unsuitable area to 113 km^2, gobbling away more than half the historically available habitat. However, rising water levels are also flooding current shorelines and could provide new submerged "flats" for seagrasses, adding a potential 90 to 135 km^2 to the seagrass meadows in Upper Laguna Madre. The best-case scenarios could result in gains for seagrass ecosystems. Unfortunately, further development of the coastal fringe and expanded shoreline hardening to prevent erosion and flooding of buildings is likely to constrain the expansion of seagrasses into sites currently occupied by tidal marshes, sand flats, and even backyards.

As coastal conservationist B.C. Robinson (2020) discusses in his readable account of Texas bird life at risk, Laguna Madre has sheltered memorable populations of migratory redhead ducks (*Aythya americana*). Seagrasses are necessary for their support and that of other waterfowl enjoying mild border winters. However, in the case of the redheads, it is not just the area of seagrass meadows that matters, but also the specific species present. Redheads forage preferentially on shoal grass (about 85% of their diet) rather than on the other four species of seagrass native to the Laguna Madre. Tipping their heads and beaks down, they excavate shoal grass rhizomes from the sediments, along with lesser amounts of widgeon grass (*Ruppia cirrhosa*) and invertebrates like clams. Higher salinity levels favor shoal grass and widgeon grass, while lower levels favor manatee grass and turtle grass (*Thalassia testudinum*), which lack the accessible, nutritious rhizomes. Characterized by its limited water exchange with the Gulf, high evaporation rates, and a lack of rivers diluting its salt concentration, the Laguna Madre has historically been the only negative estuary on the U.S. east coast. In a negative estuary, evaporation exceeds water inflow from rivers, and thus more saline water flows in from the ocean. Before industrial engineering arrived, the Laguna Madre's salinity was so high, reaching 100 ppt (parts per thousand), or nearly three times the average salinity of the Gulf, that it inhibited seagrass establishment.

The construction and dredging of the Intracoastal Waterway through the Laguna to the Mexican border, the opening of Mansfield Pass, and other projects to support ship traffic altered the hydrology and enhanced the exchange of water with the Gulf and circulation within the bay. These developments modified the water chemistry, reduced the salinity, and, by the 1950s, encouraged seagrasses. Initially, the salinity levels favored shoal grass and redhead ducks, and then, as salinity stabilized and declined, manatee grass and turtle grass proportionately increased. This led to less welcoming conditions for redheads, who have sought out alternative sanctuaries (Robinson, 2020). In 2011 and 2012, drought-induced hypersaline conditions in Laguna Madre caused losses of manatee grass. However, as salinity declined after the event, the manatee grass and the depleted shoal grass populations did not recover. Sea level rise has further inhibited seagrass recovery and thus indirectly impacted waterfowl populations (Capistrant-Fossa and Dunton, 2024). Climate change will continue to affect the Laguna Madre's salinity via factors elevating salt concentrations, including higher temperatures that increase evaporation and more prolonged droughts that reduce freshwater flows. Conversely, intense storm events with greater freshwater inputs via rainfall can reduce salinity temporarily. Although these variations might seem minor chemical footnotes, they will determine the fate of the shoal grass, determining whether redhead ducks will continue to migrate to the still relatively isolated Laguna Madre.

11.4.4.3 Saltmarsh Migration, Replacement by Mangroves, and Loss of Low Marsh

Although Texas's salt marshes are more resistant to eutrophication than the seagrass communities, thousands of acres have disappeared due to shoreline development, drainage, channelization, and dredging. Salt marsh area naturally decreases from Sabine Lake in the north, with over

300,000 wetland acres, including freshwater wetlands, to Corpus Christi Bay, with about 25,000 acres of estuarine marsh. The north also has the greatest freshwater inflow; thus, tree-dominated swamps and flooded bottomlands are more expansive, as the canoe-worthy bayous of the Houston and Beaumont regions attest (Blackburn, 2004). The species composition of Texas's estuarine wetlands and their dominant emergent vegetation change naturally from north to south due to physical factors such as freshwater input, salinity, and seasonal temperature minima. For the Upper Coast, rainfall for cities like Beaumont and Houston often exceeds 50 inches (127 cm) in a year and is greater than evapotranspiration, producing more runoff and greater river flow. In the Middle Coast, from Matagorda Bay to Corpus Christi, the climate shifts from temperate to subtropical, and precipitation and evaporation are close to balanced. On the Lower Coast, the rainfall drops to 16 inches (41 cm) per year, and freshwater flow becomes sporadic, with rivers only filling their channels in response to tropical storms (Chapman and Bolen, 2016).

Eastern U.S. coastal salt marshes change in species composition along a gradient from sites flooded much of the day, called low marshes, to sites that are only flooded at the highest tides, called high marshes. From the Louisiana border to Rockport, the salt marsh cord grasses, *Spartina patens* and *Spartina alterniflora*, dominate the tidally flooded zones, as they do in the sounds of North Carolina and Georgia. They begin to thin out around Aransas Pass and Corpus Christi. Although *Spartina alterniflora*'s range extends to the Laguna Madre, it is not adapted to hypersaline conditions and is no longer a dominant wetland species around Padre Island. Mangroves dominate many similar Gulf shores in Central America and Mexico. The black mangrove (*Avicennia germinans*), which grows to about 10 m in height, forms scattered stands in Texas about as far north as Aransas Pass. Other species, like sea lavender (*Limonium nashii*), take advantage of the reduction in cord grass dominance, and their cover increases on the lower coast. Although the Aransas Pass marks the gradual shift from grass-dominated marshes to tree-dominated swamps in the intertidal zone, the rich *Spartina* flats of Aransas National Wildlife Refuge continue to succor the migrant whooping cranes (*Grus americana*) and enrich the diets of the invertebrates on which the endangered cranes and other wading and shore birds depend. Heading south, the aridity reduces the productivity and cover of the vegetation above the high tide line. The dense stands of shrubs, like *Baccharis halimifolia*, one observes at Galveston Island, become open and then, on the lower coast, give way to more drought-adapted species (Britton and Morton, 1989).

Global warming is facilitating a northward expansion of black mangroves in Texas. While black mangroves can tolerate temperatures falling briefly below freezing, they are sensitive to prolonged cold. The three other types of mangrove building Mesoamerican mangals – red mangroves, white mangroves, and buttonwood – are even more vulnerable to a deep or multi-day freeze. For example, red mangroves (*Rhizophora mangle*) occasionally germinate north of the Mexican border, but they have not historically formed mature stands due to frost kill. Over the past nine decades, small populations of black mangroves have established footholds in Nueces, Matagorda, and Galveston Bays (Chapman and Bolen, 2016). Mangals are rich in bird life and support thriving fisheries. Logging, conversion to agriculture, and urbanization have decimated them globally, making mangrove preservation a priority for action among conservation organizations. However, their potential slow displacement of Texas salt marshes is a mixed blessing. Smee et al. (2017) compared the faunas of Texas *Spartina* marshes with and without mangroves present. They found that the abundance of some keystone invertebrates like polychaetes (worms) and infaunal communities, in general, decreased with abundant mangroves. While crabs and fish increased with mangrove invasion, the species changed. In their study areas, blue crabs (*Callinectes sapidus*) and brown shrimp (*Penaeus aztecus*) declined with more mangroves present. Mud crabs (*Xanthidae/Panopeidae*), which are of less commercial value, increased. These explorations are site-specific, and other researchers, such as Caudill (2005) investigating mangroves in Louisiana, have come to different conclusions concerning mangrove impacts on specific taxa, such as the fate of blue crab and shrimp populations. Nonetheless, there is a consensus that mangroves form hunting corridors and perches for predators, increase the rate of sediment deposition, elevate wetlands, and modify food webs.

The rate of eustatic sea level rise differs around the rim of the Gulf. For Texas, the highest rate of 12 mm/y is at the Louisiana border, while 4 to 5 mm/y is typical for the middle and lower coasts. Sabine Pass and Galveston, with their extensive wetlands, have experienced greater increases than Corpus Christi and Padre Island, with their more limited estuarine wetlands. The highest level of shoreline erosion, from 1 to 2.9 m per year, is on the stretch of coast from the Louisiana border to east Galveston Bay due to the impacts of the petroleum industry and management of the Mississippi River for navigation and flood control. Pumping of hydrocarbons and water has caused subsidence, and the channel, levee, and spoil disposal systems have constricted sediment deposition on the Mississippi River delta. Houston has a significant problem with subsidence from fluid extraction, while the weight of the buildings contributes to the compaction of the thick layers of sediment underlying the metroplex. Baytown sank about three meters during the 20th century (Davis, 2011). Climate and subsidence-driven sea level rise and climate change-driven shifts in precipitation patterns that cause a reduction in sediment transport combine with other factors, such as wave action and the wakes from ships, to erode marshes. The ecological dynamic is complex; however, rising sea levels initiate marsh migration and can create new wetlands.

Jung et al. (2024) used historic topographic sheets and recent satellite imagery to compare changes in Texas coastal marshes between the 1850s and 2019. They determined that, during the past century, 58% of historic natural marshes were lost, primarily due to erosion, but 72% of salt marshes were recently generated, mainly due to the submergence of lands on the coastal fringe. Although there was a net gain of 14% in the area of salt marsh over the past century, the ecological dynamics have changed the type of marsh present. The flooding has favored the upland migration of the high marsh. At the same time, the erosion has eaten away the low marsh, reducing its overall area. While both contribute to marsh productivity and form excellent wildlife habitats, the low marsh makes the greater contribution to fisheries and storm protection. The low marsh's rich invertebrate communities join with those of the seagrass meadows in supporting Texas's spectacular flocks of wading birds. Thus, the proportional loss of low marsh has a negative impact on wildlife populations and fisheries. Jung et al. (2024) estimate the loss of property value and other fiscal damages of Texas marsh migration at $0.7–1 trillion. They conclude that "the loss of ecosystem services underscores the complexity and importance of considering not only the quantity but the quality in marshland conservation efforts."

II.4.4.4 Saltwater Intrusion into Bottomland Forests, Freshwater Wetlands, Agricultural Lands, and Groundwater

Rising seas raise salinity in estuaries, which modifies species composition. As brackish tides reach further into the interior, they change the water chemistry. Salt intrusion into aquifers displaces freshwater swales on the barrier islands and slowly kills the less salt-tolerant tree species in the bayous. In some cases, mangroves may ultimately replace bottomland forests. As the degradation of what had been harvestable stands of bald cypress (*Taxodium distichum*) in Louisiana demonstrates, the initial impact of saltwater intrusion is the loss of the bald cypress canopy and decay of the root systems, which hold wetland soils in place. The result is further erosion and a reduction in the capacity of the wetlands to buffer the coast from storms. Canals built to facilitate boat traffic, the petroleum industry, and logging have accelerated the effects of sea level rise by allowing the salt water to penetrate further inland (Gomez, 2008). In addition, the rise in ionic strength, in turn, increases ion exchange, modifying the release of nutrients. An outcome of particular concern is the liberation of ammonium from soil particles. This process removes nitrogen from bottomland forest and wetland soils and contributes to the eutrophication of estuaries and bays (Tully et al., 2019). Mangrove species are better morphologically and physiologically adapted to anaerobic soil conditions and high salt concentrations than most freshwater floodplain and bayou tree species. For example, the mangroves' aerial roots enhance respiration. Mangroves partially exclude root absorption of salt, tolerate high salt concentrations in tissue, and remove salt by secretion (Hogarth,

2015). Although mangroves could slowly replace other wetland forested communities, other factors, including the continuation of winter freezes, elevated erosion rates, and the frequency and intensity of hurricanes, will influence their colonizing ability.

Saltwater is not just higher in sodium chloride than freshwater; it also has greater concentrations of basic cations and sulfate than most interior waters, except for those that are naturally alkaline or "bitter." Of particular concern to farmers, saltwater intrusion delivers magnesium and calcium carbonates that can elevate soil pH and negatively affect phosphorus availability to plants by binding phosphorus in insoluble compounds. Saltwater can increase bonding between cations and dissolved organic matter (DOM), which encourages flocculation or the formation of particles that are less available to microbes. This has the ironic advantage of increasing carbon sequestration and burial in wetland soils, removing the carbon from atmospheric circulation. However, in agricultural ecosystems, it can interfere with nutrient cycling. Meanwhile, the additional sodium can cause clay dispersion. This clogs soil pores, thereby slowing water infiltration. These changes in soil structure can impede drainage, interfere with irrigation, and delay planting – all of which decrease crop yields. Both saltwater flooding from phenomena like storm surges and the slow intrusion of brackish water deposit sulfate in soils. Microbes driving the organic matter mineralization pathways are very sensitive to sulfate availability. Additionally, microbial community changes can affect key processes such as carbon sequestration. The steady sulfate deposition caused by saltwater intrusion is more likely than briefer flood events to use up the available iron to form precipitates and accumulate in the free sulfides that are toxic to plants. As Tully et al. (2019) conclude: "Although plants vary widely in their tolerances to high sulfide conditions concentrations and salinity, few nonadapted plant species can tolerate waterlogged and saline environmental conditions Therefore, saltwater intrusion can drive inland retreat of upland plant communities and landward migration of coastal plant communities … far in advance of regular tidal inundation with seawater …."

Sea level rise interacts with other climate impacts to cause the loss or abandonment of coastal agricultural lands. For example, lower rainfall levels and higher evapotranspiration increase the demand for irrigation water. They also increase the risk of mismanaged irrigation and accelerated drying, accumulating salts in the plant rooting zone. Semi-arid regions, like southeast Texas, are particularly prone (Corwin, 2020). A study of 250,000 coastal U.S. water wells found that landward hydraulic gradients, indicating the movement of saltwater into fresh groundwater, already characterized more than 17% of Gulf Coast wells, providing water for drinking and irrigation (Jasechko et al., 2020). Barrier islands are especially vulnerable to slowly losing all their freshwater wells, and the problem is exacerbated where nearby major metropolitan areas are creating cones of depletion due to freshwater demand (Pilkey and Pilkey, 2019). In addition to threatening urban water supplies, saltwater intrusion raises the cations in irrigation water that will be applied to soils already subject to salinization.

II.4.5 BARRIER ISLAND AND COASTAL FRINGE ECOLOGICAL COMMUNITIES

II.4.5.1 Dunes and Island Interiors

The unique ecological communities occupying barrier island beaches, dunes, and swales are adapted to disruption by hurricanes, which may both erode them away and bury them in sediment. A cadre of drought and salt-tolerant dune grasses and herbaceous species disperses quickly to new swaths of open sand. However, the complete process of ecological succession on dunes, from scattered pioneers to hardwood forests, requires several decades. Species with tough, long rhizomes, like sea oats (*Uniola paniculata*), facilitate the establishment of other plants by stabilizing the wind-blown sand and adding soil-building organic matter to the nutrient-poor substrate. Among the pioneers are legumes like partridge pea (*Cassia fasciculata*), which facilitate species diversification by fixing nitrogen, increasing its availability. Some of the shrubs and small trees, like the mesquite (*Prosopis glandulosa*) and huisache (*Acacia farnesiana*) occupying the dunes' sheltered lee side or elevated shell mounds, are also legumes with microbial symbionts that manufacture fertilizer. As organic matter and nutrients accumulate and the soil profile matures, trees begin to form protective groves and shade out the grasses. Taking Matagorda Island as an example, above the intertidal zone is a sparsely vegetated primary dune backed by a better vegetated secondary dune, with wetter swales between them. Behind the dunes is a barrier flat with a thick carpet of grasses, such as marsh hay cordgrass (*Spartina patens*) and seacoast bluestem (*Schizachyrium scoparium*). Freshwater ponds and wetlands dot the interior. On the bay side are tidal flats, subject to daily flooding, fronted by a rim of higher shell ridges at the bay margins (McAlister and McAlister, 1993). Intact dune lines inhibit the intrusion of storm surge, and the entire configuration buffers the mainland coast behind it from wind and wave action. Like salt marshes, dunes can move and recover their plant and animal communities as barrier islands respond to sea level rise by shifting landward.

Rows of beach homes, access roads, and other developments for tourism and recreation have stripped vegetation from many barrier island dunes. In addition, industrial and navigational projects, like the Houston Ship Channel, interfere with the sediment deposition necessary to maintain the shorelines. Direct anthropogenic removal of dunes and their natural plant cover combines with factors like subsidence and climate-driven sea level rise to accelerate the erosion of the islands. Paine et al. (2017) estimated an average annual retreat rate of 1.3 m/year from 1930 to 2007 along the Texas coast. Shahtakhtinskiy et al. (2023) utilized remote sensing to measure Hurricane Harvey's immediate impacts on the barrier islands of the central Texas coast, including North Padre, Mustang, and San Jose. They detected an average shoreline retreat of 16.2 m (53 feet) from the single event, with maximum losses reaching 59 m (193 feet) at locations near landfall. Storms of this magnitude can obliterate an entire dune line in a single day.

Although storm-eroded dunes initially migrate landward by rear slope accretion of sediment, they can recover their position by seaward accretion and vertical aggradation of sand. Dune vegetation is resilient to storm disturbance and can recover given enough time. Ajedegaba et al. (2019) examined dune vegetation recovery following Hurricane Dolly in 2008 on an undeveloped area of South Padre Island. Using remote sensing, they quantified vegetation cover changes from 1996 to 2016. They found that the sparse plants of the foredune were more vulnerable to hurricane winds and waves than the more aggregated plant communities of the back dune. Despite the storm, vegetation cover in their dune study area increased by 38% from 1996 to 2016. There was a steady increase in the productivity of the resilient dune ecosystem during this period. However, the resistance and resilience of the dune ecosystems are contingent on adequate recovery times between storms, moderate erosion rates, and adequate sediment supplies to replenish the beaches. Climate-driven changes like an increase in hurricane frequency or intensity, or declines in precipitation that reduce river flow, will make it more difficult for dunes to rebuild and their biota to re-establish. A study by the Texas Bureau of Economic Geology (2021) examined the region from Freeport to East Matagorda Bay and found that, from the 1950s to 2002, there was an increase in the total marsh area but a systemic decrease in tidal flats from 899 to 363 ha. Gulf beaches declined from 460 ha (1,135 acres) to 227 ha. Meanwhile, open Gulf water and open estuarine water increased by more than 500 ha combined. If the islands slowly decline in circumference, dune and tidal flat habitats will also be reduced. The same study found an increase in mangals from 12 ha in the 1950s to 93 ha in 2002.

II.4.5.2 Coastal Woodlands

Superficially, coastal woodlands might appear to have a low vulnerability to accelerated sea level rise and the intensification of tropical storms because driving along the barrier islands or the interior edge of the bays, one does not see much evidence of undisturbed forests. Further, those forests that are present are on higher ground. However, since European colonization, logging, conversion to grazing and farmlands, and development for housing and industry have removed much of what had been relatively continuous bands of hardwood forests and oak savannahs, stationed like beads on a two-strand necklace on the leeward side of barrier islands and the sand ridges and salt domes along the mainland edge of the bays (Anderson, 2003). For example, the Aransas National Wildlife Refuge's mainland area is on the Blackjack Peninsula, part of the Ingleside Barrier, a geological formation that originated during an interglacial interval of higher seas circa 135,000 years ago. Geologists have argued over the origins of the broad Ingleside sand ridge extending south of Corpus Christi to the Mississippi River delta. One hypothesis has proposed that the Ingleside Formation was once a well-developed barrier island system, and another has proposed that it was a strand plain generated by the deposition of river-borne sediments on an open shoreline. Regardless of its origins, the Ingleside Formation offers nutrient-rich, well-drained soils favorable to forest communities. The escarpment live oak (*Quercus fusiformis*) and laurel oak (*Quercus laurifolia*) dominated old-growth stands of coastal fringe forest growing on the sand ridges of the Ingleside Barrier and Terrace at Aransas National Wildlife Refuge and Goose Island State Park, which harbor exceptionally large trees, some 400 to 500 years old. The oaks mix with other canopy-sized species, like red bay (*Persea borbonia*) and net-leaf hackberry (*Celtis reticulata*), and shade an understory dominated by yaupon holly (*Ilex vomitoria*). Although the elevations at the Aransas Refuge average around 20 feet (6 m), the woodlands rule the high points and the bluffs, reaching about 45 feet (14 m) (McAlister and McAlister, 2006). Further north, the southern live oak (*Quercus virginiana*) dominated woods of High Island, on the Bolivar Peninsula north of Galveston, stand atop salt domes where the elevation is about 12 m (38 feet) above mean sea level (Robinson, 2020). As is true of eastern deciduous and oak forests in general, fragmentation and interference with successional processes deplete biodiversity and inhibit recovery from natural disturbances like hurricanes (McShea and Healy, 2002; Yahner, 2000).

These scattered intact coastal woodlands have multiple natural values, including their record-sized trees. They are prime habitat for perching birds and provide breeding and foraging territories for rare or more southerly bird species, such as the pauraque (*Nyctidromus albicollis*), a large nighthawk native to Mexico (McAlister and McAlister, 2006). As Robinson (2020) records, these unassuming woodlands play a crucial role in the lives of mid-continental migratory species. Diminutive songbirds fly straight across the Gulf's daunting breadth and, aside from expending precious energy reserves, are unable to eat or drink freshwater during the strenuous journey. The coastal woodlands are especially important to migrants, arriving exhausted and dehydrated on their spring return from the tropics. The transients must stop in sheltered woody vegetation for rest and refreshment as soon as possible. High Island is situated right on the Central Flyway, and freshwater is available. Thus, it receives remarkable numbers of temporary feathered visitors each spring. So many may descend at once that bird watchers call the colorful, twittering hordes of orioles, tanagers, grosbeaks, warblers, buntings, and other songbirds a "fall out." The four bird sanctuaries on High Island draw thousands of committed bird watchers annually, many traveling from out of state or other countries.

As the woodlands are less extensive than historically, hurricane damage is already a concern for the remaining stands. In 2007, Hurricane Humberto, a mere Category 1, uprooted large trees on High Island, opening the canopy while not leveling the woods completely. Hurricane Ike's strong storm surge poured around the salt dome but did not inundate the elevated woodlands (Robinson, 2020). An increase in higher-intensity storms will cause more storm events with hurricane-force winds. The coastal woodlands can recover via forest succession, and toppled oaks can regenerate via sprouting from their stumps and roots. However, regrowing the splendid oaks is a multi-decade process, which the next storm can disrupt. A more terminal problem is evident at the Aransas National Wildlife Refuge, where coastal erosion is cutting into the sand bluffs and the coastal woodlands at the edge of San Antonio Bay (Figure II.4.1). Like the wetlands, the terrestrial shorelines of the Texas bays are retreating. The Aransas Refuge experienced a 3 m (10 foot) storm surge during Hurricane Carla in 1961. The event inundated the barrier islands and pushed a mass of water across San Antonio Bay that battered the mainland shoreline. In 1988, Hurricane Gilbert's robust waves caused easily observable erosion between the Refuge headquarters and Dagger Point, causing oaks to pitch over from the bluff's crest. While many daunting storms, like Hurricanes Katrina and Rita, have done little to accelerate erosion at Aransas, others make landfall with a

FIGURE II.4.1 The live oak and hardwood forest of Aransas National Wildlife Refuge stands on bluffs undercut by erosion at the edge of San Antonio Bay in 2024. Shoreline hardening at Dagger Point has been ineffective, and the sea has dislocated or removed protective concrete slabs. Erosion has created the steep bank, truncating the sand dunes in the foreground. The woodland in the background has merged with the beach, where the trees will eventually end up in San Antonio Bay. Despite the height of the forested bluff, surge and waves produced by tropical storms have been cutting away both the forest and the mature soils that coastal woodland requires. (Photo: Susan Bratton.)

wallop and take a nasty bite out of the precious oaks. Although both the barrier islands and the associated vegetation bands rimming the bays have relocated back and forth as sea levels have risen and fallen through the Pleistocene, the current generation of coastal woodlands cannot simply shift landward. Saltwater swales, shrubby sand plains, and freshwater prairies are in the way (McAlister and McAlister, 2006).

II.4.6 VULNERABLE, EXPLOITED, DECLINING, AND ENDANGERED SPECIES

II.4.6.1 Vulnerable Species in Gulf of Mexico Fisheries

NOAA has developed a fisheries climate strategy and analytic tools for appraising the vulnerability of individual Gulf of Mexico fish and invertebrate species exploited commercially or recreationally to atmospheric change. Of the multiple stressors the NOAA research teams have investigated, changes in temperature, acidity, salinity, and dissolved oxygen are likely to have the most significant impacts. Among the species characteristics that increase sensitivity are small populations, habitat and prey specificity, diadromous life histories (migrating between salt and fresh water), complex reproductive cycles, and a limited range of tolerances for salinity, pH, temperature, or available oxygen. Although all the commercial species in the Gulf of Mexico will experience impacts, those with more specialized ecological niches, low propensity to distributional change, and populations already diminished by overharvest or other anthropogenic disturbances are more likely to suffer further population declines. Quinlan et al. (2023) found that the anadromous Gulf sturgeon (*Acipenser oxyrinchus desotoi*) has very high vulnerability, along with black grouper (*Mycteroperca bonaci*), yellowmouth grouper (M. *interstitialis*), and Nassau grouper (*Epinephelus striatus*). Eleven of the 75 species they studied have high vulnerability. All of these are elasmobranchs (sharks and rays) and bottom-dwelling groupers or hinds, except for snook (*Centropomus undecimalis*) and eastern oysters. Of the 75 species investigated, 21 are moderately vulnerable, and 31 have low vulnerability. Elasmobranchs and groupers are also prevalent in the moderate sensitivity category, which incorporates the commercially important red and mutton snappers (*Lutjanus campechanus*, *Lutjanus analis*), dolphin or dorado (*Coryphana hippunus*), and spiny lobster (*Panulinus argus*), and the recreationally charismatic tarpon (*Megalops atlanticus*). On the positive side, four shrimp species and the blue crabs, important to commercial fisheries, are in the low vulnerability category. Nonetheless, nationally threatened or endangered species, like the Gulf sturgeon and the Nassau grouper, are in the high vulnerability group.

Among the challenges the National Marine Fisheries Service and Texas Parks and Wildlife face in protecting and regulating Texas marine fisheries are the range of habitats utilized by species with high to moderate vulnerability to climate change and the degree to which the health of individual fish stocks originates in multiple, interdependent habitats. For example, snook and oysters are heavily dependent on the health of estuaries. The habitat preferences of grouper depend on the age class and the species. While juvenile black and Nassau grouper may prefer seagrass beds, oyster rubble, and shallow inshore waters, adult grouper often congregate around reefs and hard structures offshore. Similarly, young snappers prefer sand or muddy bottom shallow habitats, while mature red snappers share reefs and high relief structures with grouper in deeper waters. The eastern oyster is a keystone species that reduces hypoxia in the bays by filtering phytoplankton and forms reefs that slow marsh erosion and add critical three-dimensional structure to the estuaries. The presence of oysters thus enhances environments for juvenile groupers and snappers. An experienced scuba explorer visiting Flower Garden Banks can encounter several high-vulnerability fish species on a single dive. Yet, far fewer groupers and snappers will be cruising among the coral heads if the Texas estuaries are more severely compromised.

II.4.6.2 Threatened and Endangered Sea Turtles Utilizing Diverse Habitats

Sea turtles are representative of the complex of problems that atmospheric change presents to endangered and threatened Texas marine and coastal species populations. Five of the seven sea turtle species are found in the Gulf of Mexico, and all five at least occasionally forage in Texas waters. Loggerhead (*Caretta caretta*), green (*Chelonia mydas*), and Kemp's Ridley (*Lepidochelys kempii*) sea turtles all nest annually on Texas beaches, while nests of leatherbacks (*Dermochelys coriacea)* and hawksbills (*Eretmochelys imbricata*) are sporadic at best, with zero in most years. Ironically, the Kemp's Ridley's primary nesting area was historically in Mexico, but today this species is the most common Texas nester. This range extension is due to a National Park Service and Texas Parks and Wildlife-sponsored program to transfer eggs from Mexico to Padre Island and establish a reproductively viable population in Texas. Among the climate-related concerns for sea turtle nesting and hatching success are elevated beach sand temperatures, sea level rise, and hurricane frequency and intensity increases. For sea turtles, nest temperature determines the gender of the hatchlings, with higher temperatures producing more females. A 1–4°C elevation in sand temperature can eliminate the males. Although sea turtles do not form pair bonds, a biased sex ratio with few males available reduces genetic diversity in the next generation of already minimal populations. Female turtles need to be able to crawl well above the high tide line to locate nests safely where they will not flood. Thus, the truncation or steepening of beach profiles due to accelerated erosion is problematic. Historically, many sea turtle nests hatched before the core hurricane season, avoiding heavy rains and storm surges that drown or wash away their eggs. The potential earlier onset of the hurricane season reduces the available window for safe hatching. Female response to warming waters can, however, move nesting dates forward (Patricio et al., 2021; Poloczanska et al., 2009; Tomillo and Spotila, 2020).

In addition, changing conditions can reduce food availability or modify the species available. For example, green turtles often forage on seagrasses, including the aptly named turtle grass, which are vulnerable to water depth and salinity changes. Temperature change, hypoxia, and acidification reduce the biomass of edible invertebrates. They also degrade foraging sites, like coral reefs. Sea turtles migrate between feeding areas and nesting beaches, in some cases, by hundreds of kilometers. Changes in temperatures, winds, and currents can modify the distributions of sea turtles and their preferred prey organisms at sea. The recent northward range extension of the Kemp's Ridley is evidence that climate change may prompt or accommodate geographic shifts. Nonetheless, atmospheric change affects all the life stages of marine turtles, which, like many other long-lived vertebrates, may be slow to adapt despite their motility (Patricio et al., 2021; Poloczanska et al., 2009; Tomillo and Spotila, 2020).

11.4.7 MITIGATION AND RESTORATION

11.4.7.1 Public Education

One of the most powerful tools for addressing atmospheric change is public education, and the Texas coast is blessed with an array of outreach-oriented institutions and nature reserves. Among the keys to successful educational programs concerning topics such as coral reef conservation are encouraging providers, identifying audiences, selecting appropriate learning approaches, and developing effective tools (Browning et al., 2006). However, much of the scientific analysis of climate change and Texas marine ecosystems is relatively recent and has not yet been summarized and edited for educational venues. For example, the Texas State Aquarium, Corpus Christi, offers attractive, attention-grabbing exhibits committed to Gulf of Mexico habitats, including a "room" and a large tank committed to Flower Garden Banks National Marine Sanctuary. As visitors enter, they encounter signs and exhibits that assist in identifying Texas species and understanding the processes and geology underpinning the Gulf's biodiversity. Colorful, conservation-oriented displays reinforce the concept that the ecological health of the Gulf depends on all of us, and the waters are integrally connected to what is happening on land. In 2024, the Flower Garden exhibit messaging did not include the threat of coral bleaching. However, it did incorporate a panel describing hurricane disturbance and a sequence covering establishing a permanent monitoring system for the banks. This approach notes that the Flower Garden Banks represent the northern extent of coral habitat and "the Sanctuary is a unique place to address the impacts of climate change." It personalizes by asking the visitor, "What do you think?" concerning the fate of the corals and whether the reef species will adapt or shift. These displays were already in place when Johnston et al. (2019) published the analysis of the monitoring data from the 2016 bleaching event, and interpreting the bleaching trends will require a carefully designed revision of the exhibit. Additionally, as of 2024, the aquarium's multi-tank Caribbean reef exhibit did not interpret bleaching or climate, even though the impacts of ocean warming are just as critical to the much larger reefs off the Yucatan Peninsula.

The Texas State Aquarium maintains a multifaceted website to link teachers, parents, and the public to their many educational opportunities, such as class visits, remotely delivered lectures, and sea camps. The site also informs the public of current research and conservation activities. Web information is easier to expand and update than the stationary storyboards or lighted panels adjoining the tanks. In 2024, the Aquarium's "Conservation" webpage featured a Texas A&M Corpus Christi research project led by Dr. Keisha Bahr on coral responses to changing physical conditions. The educational web pages accessed videos of Dr. Bahr and the aquarium staff conducting research on corals at the aquarium and in Hawaii. They describe the development of a color comparison tool for monitoring coral health for use by citizen scientists, as well as field and experimental research on coral health changes related to changing water temperatures, coral bleaching, and acidity. Intended for school children exploring careers in science and members of the public already interested in conservation, Dr. Bahr has posted videos and updates as the project has progressed. Thus, the aquarium's online posts provide more detail and up-to-date information than the aquarium displays. The current challenge is to convert the results of such research endeavors into "soundbite" or quick-messaging forms that expose the less-informed public to the core concepts and raise awareness of the immediacy of the threats to reef health and integrity.

Although most public educational nexuses in Texas have surprisingly little interpretation of climate impacts such as sea level rise and changes in seagrass and salt marsh communities, the Patton Marine Science Bay Education Center at Port Aransas integrates panels on the impact of climate change on Texas's bays and coast studied into the greater framework of stewardship to curtail water pollution and overfishing. The displays feature research by University of Texas at Austin scientists, including the impacts of climate change in other parts of the globe. As of 2024, the child-friendly exhibits designed for class visits offer hands-on activities like touching screens, dropping tokens, handling stuffed animals, and observing live ones. Among the themes are potential drying of the coastal Texas climate, changes in estuarine salinity, loss of beach sand (for the Arctic), and changes in populations of economically important species such as blue crabs. Because updating the exhibits requires a significant investment in design and installation, the University of Texas scientists report outcomes from recent investigations, such as Dr. Kyle Capistrant-Fossa's verification that sea level rise is reducing the area of seagrass beds, on their educational websites or via press releases. Despite outstanding efforts by organizations like the Texas State Aquarium and the Patton Marine Science Bay Education Center and the many non-profits active on the Texas coast, the speedy development of scientific knowledge concerning the impacts of atmospheric change on the coastal and marine ecosystems is making it difficult for educators to stay up to date and to disseminate the pertinent conclusions, not just to children and families, but to other stakeholders, such as city and county governments, farm managers, commercial fishers, and business owners.

II.4.7.2 Shoreline Hardening and Beach Sand Replenishment

Two strategies implemented to slow shoreline erosion are hardening shorelines and replenishing beaches with sand from other locales. Both are financially expensive, vulnerable to damage or removal by high-intensity storms, and disruptive of natural geological and ecosystemic processes. Although the Galveston seawall, accompanied by the construction of sediment-trapping groins and elevation of much of the city, has inhibited storm surge intrusion, it has also truncated the beach profile, eliminating native dune biota (Anderson, 2007; Wright-Gidley, 2019). Galveston's shoreline hardening has caused sand starvation and recession of the shoreline and native beach vegetation south (down-drift) of the seawall. The city has transported sand to the beaches, but only to rebuild an adequate strand for sunbathers and swimmers. The sea wall has obliterated the dunes and inhibited natural processes like sea turtle and shorebird nesting (Figure II.4.2).

In the case of recreational beaches, a core issue is where to obtain sand for beach nourishment without disturbing other habitats or depleting sediment necessary to rebuild beaches or deltas elsewhere. As Pilkey and Pilkey (2019, 70–78) summarize, beach nourishment has two sites where it has high ecological impacts – the beach itself, where it initially kills almost all the fauna, and the sea floor where dredges or pumps remove thousands of cubic meters of sediment. Although birds will congregate on newly distributed beach sand and pick up the dying animals transported to the shoreline from offshore sand bars, the avifauna will soon disappear from the replenished strand which is devoid of life. In locales where sediment transport will replace the extracted sand, the offshore benthic organisms will eventually recover. However, in the meantime, the fish and other carnivores dependent on them will have to find another foraging area. Further, some offshore banks

FIGURE II.4.2 Storm waves and severe erosion have removed the sand beach in front of the Galveston seawall in 2003. The sea wall, development, and erosion have eliminated the natural dunes and vegetation, and degraded shoreline habitat for animals like shorebirds. The shoreline cannot naturally shift landward with sea level rise. Heavy equipment and beach nourishment with transported sand are employed to maintain a beach for recreation.

formed of sediment deposited in earlier geologic eras do not regain their structure following dredging, because the processes that previously provided the sediment, such as transport by glaciers, are no longer operating.

Aransas National Wildlife Refuge (NWR) has utilized shoreline armoring via revetments to protect the Dagger Point area, but storm surge has battered the concrete slabs and partially destabilized the system. The concrete has not been effective in preventing the collapse of the higher bluffs rimming the oak-hardwood forest (Figure II.4.1). Beginning in 2022, the State of Texas and the U.S. Fish and Wildlife Service initiated a program budgeted at $26,705,000 to construct a three-mile-long rock breakwater, extend and repair the armoring at the base of the bluffs, regrade collapsing slopes, and plant upland vegetation to hold soil and sand in place. The U.S. Fish and Wildlife Service contracted Anchor QEA (2021) to prepare an environmental assessment that proposes mitigation measures for protecting rare and endangered species, soil conditions, and archeological resources, particularly during construction. The project should help to reduce shoreline erosion for the near future. A more difficult question is how effective it will be over multiple decades. The armor will protect over a mile of eroding bluffs, and the breakwater should reduce the impact of wave action well beyond. Forwarding a campaign for "a new coast," Peterson (2019) has pointed out that the federal and state agencies involved in such projects are "overcommitted and underfunded." As sea level rises and coastal hydrology changes, immobile structures such as revetments and breakwaters are notoriously subject to undercutting and overtopping by storm surge and wave action. Erosive processes will cause shoreline recession around and behind concrete slabs. The issue becomes not just how to save Daggar Point and the adjoining tidal flats, but how to slow shoreline recession along the entire rim of San Antonio Bay.

Where possible, maintaining a substantial building setback from the high tide and natural dune lines is superior to seawalls for reducing storm surge in planning zones adjoining recreational beaches. Natural areas also need the most expansive buffer zones that are practical to set aside from development. The degree to which shoreline hardening can eliminate or degrade shoreline dune and wetland ecosystems is evident in the struggle to maintain the Galveston beaches (Figure II.4.2). On the margins of bays, restoring oyster reefs and marshes can reduce the need for bulkheads, as can limiting shoreline disturbance by developments. As armoring can be necessary for protecting harbors, bridges, and other structures, coastal planners and engineers have been experimenting with designing seawalls and breakwaters as marine habitats. Among the novel features are habitat panels that attract mussels and other species that attach to hard surfaces, artificial tide pools integrated into stone walls, and fish habitat enhancement devices (FishHEDs) mounted on bulkheads (Hindle, 2019).

II.4.7.3 Alternative Energy Development

Texas is in the process of diversifying its economic dependence on fossil fuels and expanding its investment in alternative energy production. The infrastructure for fossil fuel extraction and the depletion of petroleum deposits have exacerbated environmental problems, including coastal subsidence, shoreline erosion, and changes in estuarine hydrology. However, Texas remains the largest petroleum producer of any U.S. state, accounting for 43% of U.S. crude oil pumped in 2023 (U.S. EIA, 2024). Replacing hydrocarbons with alternatives will reduce major environmental risks such as oil spills, leaks, and well blowouts. Taking Texas's already profitable investments in wind energy generation as an example, the state is expanding plans for turbines along the coast and offshore. Compared to wind farms in the Texas Panhandle, coastal sites for wind farms reduce transmission distance to major urban centers like Houston. Among the impacts of windmills is the additional mortality of migratory birds. The Texas coast is a critical link on flyways and an indispensable wintering habitat for species such as whooping cranes. On the one hand, although this problem is difficult to eliminate completely, and killing an endangered species is a form of "taking" that violates the U.S. Endangered Species Act, possible mitigation measures include installing turbines away from migratory routes, shutting turbines down during seasons and times when migrants are likely to be present, and installing detection devices that shut down the turbines when migrants are present. On the other hand, stopping turbines is a fiscal loss for the developers and needs to be minimized to maintain the availability of low-cost, sustainable energy (Burkholder, 2016).

Regional residents and the tourist industry can protest and attempt to politically block offshore wind development as visually intrusive, destructive of scenery, or posing risks to navigation, as has happened repeatedly concerning leases for wind generation around Cape Cod, Massachusetts. In contrast, advocates for recreational and commercial fisheries often favor building platforms, particularly if the designs forward the establishment of invertebrate colonies and attract fish like grouper and snapper. Artificial structures such as drilling platforms form vertical reefs that are alternative habitats for Texas's coral reef species and other organisms requiring hard or rigid substrates (Wicksten, 2015). Texas legislators have attempted to inhibit the installation of offshore wind farms, and a 2023 wind lease auction in federal waters off Texas remarkably drew no bidders despite some of the most favorable wind conditions in the Gulf. The Bureau of Ocean Energy Management does intend to put sites in Texas waters up for auction again, although Donald Trump has taken a position discouraging further development of wind energy resources at the beginning of his second presidential term. Nonetheless, the energy generated from a single lease could power 1.2 million homes. Meanwhile, Louisiana has passed legislation to fast-track wind farm leases in state waters (Baurick, 2024). The Texas coast has the potential to serve as a major locale for alternative energy development, thereby slowing climate change and contributing to the state's already vigorous energy economy. However, the state needs more sophisticated and less conflicted planning to deal with ecological, legal, and social barriers in order to optimize the opportunities.

II.4.7.4 Conserving Blue Carbon and Ecosystem Restoration

Texas's marine and coastal ecosystems already contribute to maintaining an optimal balance of gases in the Earth's atmosphere by fixing many kilotons of carbon annually. Early models of the Earth's carbon cycles often underestimated the importance of the "blue carbon" sequestered by photosynthesis and carbon accumulation in aquatic and wetland ecosystems. Salt marshes can sequester 2 metric tons and more of carbon per hectare per year. In his plan for the Texas coast, Blackburn (2017) distinguished three categories of assets: green or ecosystemic, grey or human developments, and blue or the water itself. Blackburn has proposed protecting and improving the resilience of ranch and farmlands and supporting ecosystem services via economic means, such as wetland loss mitigation via restoration and long-term commodity contracts (rather than easements in perpetuity) to restrict property development in areas critical to maintaining ecosystem services, such as buffering the coast from storm surge. Blackburn (2017) distinguishes four keystone systems as essential to coastal ecosystem services and resilience: salt and freshwater wetlands, coastal prairies, bottomland forests, and oyster reefs. He backs a voluntary carbon and eco services market, selling carbon credits to offset carbon releases from the combustion of hydrocarbon reserves. As Blackburn (2017) points out, federal and state agencies have continued to purchase coastal properties, and conservation organizations, such as the Nature Conservancy and the Audubon Society, have established multiple reserves. However, much of the coast remains in private hands. Successful climate change planning must synthesize ecological futures with economic pressures and business interests to generate successful means of conserving blue carbon, fisheries, living storm buffers, agricultural production, and healthy environments for human residence. Among the highest priorities for environmental action are enhanced regulation and reduction of pollutants, such as nitrogen, which cause ecological degradation and interact synergistically with climate change.

As noted, Texas conservation agencies and organizations have been protecting and restoring salt marshes with some notable successes. Not all ecosystems are equally easy to heal or reestablish. The most challenging are those with the longest natural recovery times and the greatest sensitivity to water chemistry and temperature changes. Marine biologists have been experimentally restoring coral reefs by methods such as restabilizing corals that have been knocked over, seeding corals and coral larvae, and growing corals in tanks or moving them from a less disturbed area and "planting" them on a suitable substrate. Restoration can speed recovery from disturbances like ship groundings and hurricanes (Japp et al., 2006). Biologists have found that due to natural genetic and physiological variation, some populations of corals within a species are more tolerant of elevated temperatures than others, and these can be artificially propagated in hopes of reintroducing them to heat-stressed reefs. Howells et al. (2021) have asserted that "corals from the hottest reefs in the world transfer sufficient heat tolerance to a naïve population to withstand end-of-century warming projections." Nonetheless, starting a reef from scratch or reestablishing one where algae have taken over is difficult and expensive, as is any extensive ecosystemic repair. To date, genetic studies of Texas corals have been insufficient for conservation planning. Some marine species are shifting their ranges north, but corals need relatively stable conditions, substrates free of thick sediments, and clear water with high light penetration. As Texas's outlying coral reefs demonstrate, suitable spaces for corals in the northern Gulf of Mexico are naturally extremely limited, and, thus, there is no place for reef species to move, other than the various types of offshore platforms like oil rigs. However, the most significant barrier to reef restoration remains the increasing frequency and intensity of bleaching events, which can kill newly established corals faster than they can rebuild a reef.

II.4.8 CONCLUSION

Modifying the composition of atmospheric gases and the global climate already has well-documented impacts on ecosystem structure, species populations, and even the location and dynamics of the shorelines. Key ecological stressors include elevated water temperatures, ocean acidification, local hypoxia, higher and lower salinities, water and soil chemistry changes, and sediment erosion and deposition dynamics modification. Atmospheric change interacts with other sources of environmental degradation, including the release of pollutants like fertilizers, withdrawal of river and groundwater, and removal of vegetation from barrier island and bay shorelines. Among the economic sectors affected are fisheries, agriculture, forestry, recreation, construction, and energy production. Although government agencies are generating plans responding to climate change relative to their missions or individual locales like Aransas NWR, Texas lacks comprehensive planning to address the foundational causes of atmospheric change and its most significant consequences, such as sea level rise. Texas's unique marine ecosystems, rich fisheries, energy economy, and the quality of life of Texas's human coastal residents are at stake.

REFERENCES

Ajedegba JO, Perotto-Baldivieso HL, Jones KD, 2019.Coastal dune vegetation resilience on South Padre Island, Texas: A spatiotemporal evaluation of the landscape structure. J Coast Res 35, 3: 534–544, doi:10.2112/jcoastres-d-18-00034.1.

Anchor QEA, 2021. Draft Environmental Assessment for the Aransas National Wildlife Refuge Dagger Point Coastal and Marine Habitat Protection and Restoration Project, Anchor QEA, LLC, Dauphin, AL, htpps://www.fws.gov/sites/default/files/documents/dagger-point-ea-oct22-aransasnwr.pdf.

Anderson JB, 2007. The Formation and Future of the Upper Texas Coast, Texas A&M University Press, College Station.

Anderson K, 2003. Nature Culture and Big Old Trees: Live Oaks and Ceibas in the Landscapes of Louisiana and Guatemala, University of Texas Press, Austin.

Baliunas S, Soon W, 2005. Global warming does not cause coral bleaching. In Are the World's Coral Reefs Threatened? Ferguson C, ed. Greenhaven Press, New York.

Baurick T, 2024. Gulf's Next Wind Auction Puts Focus on Louisiana after Texas Shuns Renewables. Louisiana Illuminator, Baton Rouge, LA, https://lailluminator.com/2024/03/31/gulf-wind/.

Blackburn J, 2004. The Book of Texas Bays, Texas A&M University Press, College Station.

Blackburn J, 2017. A Texan Plan for the Texas Coast, Texas A&M University Press, College Station.

Britton JC, Morton B, 1989. Shore Ecology of the Gulf of Mexico, University of Texas Press, Austin.

Browning LJ, Finlay RAO, Fox LRE, 2006. Education as a tool for coral reef conservation: Lessons from marine protected areas. In Coral Reef Conservation, Cote IM, Reynolds JD, eds., Cambridge University Press, Cambridge, UK.

Burkholder O, 2016. A shift in the wind: Siting more wind projects along Texas's 367-mile coast of Gulf winds and mitigating potential risk to migratory bird populations. Seattle J Env Law 6: 1, https://digitalcommons.law.seattleu.edu/sjel/vol6/iss1/6

Cancelmo J, 2008. Texas Coral Reefs, Texas A&M Press, College Station.

Capistrant-Fossa KA, Dunton K, 2024. Rapid sea level rise causes loss of seagrass meadows. Comm Earth Env 5: 87, doi:10.1038/s43274-024-01236-7.

Caudill MC, 2005. Nekton utilization of black mangrove (*Avicennia germinans*) and smooth cordgrass (*Spartina alterniflora*) sites in southwestern Caminada Bay, Louisiana. MS Thesis, Louisiana State University, No. 273, https://repository.lsu.edu/gradschool_theses.

Chapman BR, Bolen EG, 2016. The Natural History of Texas, Texas A&M University Press, College Station.

Corwin DL, 2020. Climate change impacts on soil salinity in agricultural areas. Euro J Soil Sci 72, 2: 842–862, doi:10.111/ejss.13010.

Dangendorf S, Hendricks N, Tornqvist T, 2023. Acceleration of US Southeast and Gulf coast sea-level rise amplified by internal climate variability. Nat Comm 14: 1935, doi:10.1038/s41467-023-37649-9.

Davis RA Jr., 2011. Sea Level Change in the Gulf of Mexico, Texas A&M Press, College Station.

Dias LM, Johnston MA, O'Connell K, Clift LW, Eisenbach O, Hannum R, Williams K, French D, Cannizzo ZJ, Hutto S, 2023. Rapid climate vulnerability assessment for Flower Garden Banks National Marine Sanctuary, National Marine Sanctuaries Conservation Science Series ONMS-23-12, https://nmssanctuaries.blob.core.windows.net/sanctuaries-prod/media/docs/20231201-fgnms-cva-final-report.pdf

Flower Garden Banks National Marine Sanctuary, 2024. Webpage: Hurricanes, https://flowergarden.noaa.gov/science/hurricanes.html.

Gomez GM, 2008. The Louisiana Coast: Guide to an American Wetland, Texas A&M Press, College Station.

Griggs G, 2017. Coasts in Crisis: A Global Challenge, University of California Press, Oakland.

Hickerson EL, Schmahl GP, 2007. Hurricane Rita impacts at the flower garden banks National Marine Sanctuary, NOAA: Flower Garden Banks National Marine Sanctuary, https://nmsflowergarden.blob.core.windows.net/flowergarden-prod/media/archive/doc/reports/hurricaneritareport.pdf.

Hindle RL, 2019. The hard habitats of coastal armoring. In Sustainable Coastal Design and Planning, Mossop E, ed. The CRC Press, Boca Raton.

Hogarth PJ, 2015. The Biology of Mangroves and Seagrasses, Oxford University Press, New York.

Horta-Puga G, 2007. Environmental impacts. In Coral Reefs of the Southern Gulf of Mexico, Tunnel JW, Chavez E, Withers E, eds., Texas A&M University Press, College Station.

Howells EJ, Abrego D, Liew YJ, Burt JA, Meter E, Aranda M, 2021. Enhancing the heat tolerance of reef-building corals to future warming. Sci Adv 7, 34: eabg6070, doi:10.1126/sciadv.abg6070.

Iglesias-Rodríguez MD, Fabricius E, McElhany P, 2016. Ecological effects of ocean acidification. In Stressors in the Marine Environment, Physiological and Ecological Response, Societal Implications, Solan M, Whiteley NM, eds., Oxford University Press, Oxford, UK.

Japp WC, Hudson JH, Dodge RE, Gilliam D, Shaul R, 2006. Coral reef restoration with case studies from Florida. In Coral Reef Conservation, Cote IM, Reynolds JD, eds., Cambridge University Press, Cambridge, UK.

Jasechko S, Perrone D, Seybold H, Fan Y, Kirchner JW, 2020. Groundwater level observations in 250,000 coastal US wells reveal the scope of potential seawater intrusion. Nat Comm 11: 3229, doi:10.1038/s41467-020-17038-2.

Johnston M, Hickerson EL, Nuttall MF, Blakeway RD, Sterne TK, Eckert RJ, Schmahl GP, 2019. Coral bleaching and recovery from 2016 to 2017 at East and West Flower Garden Banks, Gulf of Mexico. Coral Reefs 38: 787–799, doi:10.1007/s00338-019-01788-7.

Jung NW, Doe TA, Jung Y, Dellapenna TM, 2024. Massive sea-level-driven marsh migration and implications for coastal resilience along the Texas coast. Remote Sens 16, 13: 2269, doi:10.3390/rs16132268.

Kaiser MJ, Attrill MJ, Jennings S, Thomas DN, Barnes DKA, Brierley AS, Graham NAJ, Hiddink JG, Howell K, Kaartokallio H, 2020. Marine Ecology: Process, Systems and Impacts, Oxford University Press, Oxford, UK.

Kossin JP, 2018. A global slowdown of tropical cyclone translation speed. Nat 588: 104–107, doi:10.1038/s41586-018-0158-3.

Lugo-Fernandez A, Gravois M, 2010.Understanding the impacts of tropical storms and hurricanes on submerged bank reefs and coral communities in the northwestern Gulf of Mexico. Con Shelf Res 30, 10–11: 1226–1240, doi:10.1016/j.csr.2010.03.014.

Lunden JL, Georgian SE, Cordes EE, 2013. Aragonite saturation states at cold-water coral reefs structured by *Lophelia pertusa* in the northern Gulf of Mexico. Limnol Ocean 58, 1: 354–362, doi:10.4319/lo.2013.58.1.0354.

McAlister WH, McAlister MK, 1993. A Naturalist's Guide: Matagorda Island, University of Texas Press, Austin.

McAlister WH, McAlister MK, 2006. Guidebook to the Aransas National Wildlife Refuge, Mince County Press, Victoria.

McKinney L, Allred L, 2021. Texas dead zones: trapped layers of oxygen-starved water threaten the state's coastal wildlife and coral reefs, Texas Parks & Wildlife Magazine (July): 41–45.

McShea WJ, Healy WM, 2002. Oak Forest Ecosystems: Ecology and Management for Wildlife, Johns Hopkins University Press, Baltimore, MD.

Montagna PA, Froeschke J, 2009. Long-term biological effects of coastal hypoxia in Corpus Christi Bay, Texas, USA. J Exp Marine Biol & Ecol 381, Suppl. 1: S21–S30, doi:10.1016/j.jembe.2009.07.006.

Morgan EA, Brown A, Ciotti BJ, Panton A, 2016. Effects of temperature stress on ecological processes. In Stressors in the Marine Environment: Physiological and Ecological Responses; Societal Implications, Solan M, Whiteley NM, eds. Oxford University Press, Oxford, UK.

Office of National Marine Sanctuaries, 2024. Flower Garden Banks National Marine Sanctuary 2023 research, monitoring, and resource protection report, https://nmsflowergarden.blob.core.windows.net/flowergarden-prod/media/archive/doc/reports/2023researchreport.pdf.

Osborne E, Hu X, Holl ER, Yotes K, Vreeland-Dawson J, Shamberger K, Barbero L, Hernandez-Ayon JM, Gomex FA, Hicks T, Xu Y, McCutcheon MR, Acquafredda M, Chapa-Balcorta C, Norzagaray O, Pierrot D, Munoz-Caravaca A, Dobson KL, Williams N, Rabalais N, Dash P, 2022. Ocean acidification in the Gulf of Mexico: Drivers, impacts, and unknowns. Prog Oceanog 209: 102882, doi:10.1016/j.pocean.2022.102882.

Paine JC, Caudle TL, Andrews JR, 2017. Shoreline and sand storage dynamics from annual airborne LIDAR surveys, Texas Gulf Coast. J Coast Res 33, 3: 487–506, doi:10.212/jcoastres-d-15-00241.1.

Patricio AR, Hawkes LA, Monsinjon JR, Godley BJ, Fuentes MMPB, 2021. Climate change and marine turtles: Recent advances and future directions. Endang Species Res 44: 363–385, doi:10.3354/ers01110.

Peterson J, 2019. A New Coast: Strategies for Responding to Devastating Storms and Rising Seas, Island Press, Washington, DC.

Pilkey OH, Pilkey KC, 2019. Sea Level Rise: A Slow Tsunami on America's shores, Duke University Press, Durham, NC.

Poloczanska ES, Limpus CJ, Hays GC, 2009. Vulnerability of marine turtles to climate change. Adv Marine Bio 50: 151–211, Burlington, VT, Academic Press.

Quinlan JA, Nelson M, Savoia C, Scott JD, Ailloud L, Ainsworth C, Alvarez D, Bacheler NM, Burton M, Calay S, Cummings N, Doerr JC, Driggers W, Erisman B, Gandy R, Grove LJ, Hanisko D, Heublein J, Hoffmayer E, Isely J, Johnson M, Jones C, Karnauskas M, Kelble C, Kirkland T, Langwiser C, Leo J, Lombardi L, McCarthy K, Nylander-Asplin H, O'Boyle M, Orbesen E, Orhun R, Patterson W III, Pollack AG, Powers S, Potts J, Rios A, Sagarese S, Schueller A, Serafy J, Snodgrass D, Switzer T, Zink I, Griffis R, 2023. Results from the Gulf of Mexico climate vulnerability analysis for fishes and invertebrates, NOAA Technical Memorandum, NMFS-SEFSC-767, doi:10.25923/5svf-se47.

Robinson BC, 2020. A Haven in the Sun: five Stories of Bird Life and Its Future on the Texas Coast, Texas Tech University Press, Lubbock.

Shahtakhtinskiy A, Khan DS, Rojas SS, 2023. Quantifying the impact of hurricane Harvey on Beach–Dune systems of the Central Texas Coast and monitoring their changes using UAV photogrammetry. Remote Sens 15, 24: 5779, doi:10.3390/rs15245779.

Sheppard C, 2021. Coral Reefs: A Natural History, Princeton University Press, Princeton, NJ.

Sheppard CR, 2006. Longer term impacts of climate change on coral reefs. In Coral Reef Conservation, Cote M, Reynolds JD, eds. Cambridge University Press, Cambridge, UK.

Sheppard CRC, Davy SK, Pilling GM, 2018. The Biology of Coral Reefs, Oxford University Press, Oxford, UK.

Smee DL, Sanchez JA, Diskin M, 2017. Mangrove expansion into salt marshes alters associated faunal communities. Estuar Coast Shelf Sci 187: 306–313, doi:10.1016/j.ecss.2017.02.005.

Texas Bureau of Economic Geology, 2021. Barrier Islands, Freeport to East Matagorda Bay and South Padre Island: Status and trends of inland wetland and aquatic habitats, University of Texas, Austin, TX, http://beg.texas.edu.research/programs/coastal/wetlands/sargent-quintana-spi.

Tomillo PS, Spotila JR, 2020. Temperature-dependent sex determination in sea turtles in the context of climate change: Uncovering the adaptive significance. BioEssays 42, 11: 2000146, doi:10.1002/bies.202000146.

Tully K, Gedan K, Epanchin-Niell R, Strong A, Bernhardt ES, BenDor T, Mitchell M, Kominoski J, Jordan TE, Neubauer SE, Weston NB, 2019. The invisible flood: The chemistry, ecology, and social implications of coastal saltwater intrusion. BioSci 69, 5: 368–378, doi:10.1093/biosci/biz027.

US EIA, 2024. Texas state profile and energy estimates, U.S. Energy Administration, https://eia.gov/state/index.php?sid=TX,referenced9/28/2024.

Wicksten MK, 2015. Vertical Reefs: Life on Oil and Gas Platforms in the Gulf of Mexico, Texas A&M University, College Station.

Wright-Gidley J, 2019. Reshaping Galveston Island after the 1900 hurricane. In Protecting Historic Coastal Cities: Case Studies in Resilience, Pelz M, ed., Texas A&M University Press, College Station.

Yahner RH, 2000, Eastern Deciduous Forest: Ecology and Wildlife Conservation, University of Minnesota Press, Minneapolis.

II.5 Climate Change Impacts on the Terrestrial Ecosystems of Texas

Joseph D. White

II.5.1 INTRODUCTION

Climate affects every terrestrial organism by setting the main constraints of most biological activity, including (1) temperature and (2) moisture availability. With continued climate change, individual species within a community will respond slightly differently, reflecting a species' evolution over time within that community. This evolution is driven by competition with co-occurring species within the community to establish an apportionment of resource space, termed ecological niche, which allows for the species' survival and propagation within the community. Because evolution is a slow process, a community with numerous species and well-defined niches reflects a past stable environment and climate that has allowed these niches to develop, often over centuries or millennia. With rapid climate change (< century) currently occurring, species present within a community will be quickly filtered out as niches disappear due to a changing climate. What species will survive the current climate change event is a source of concern in Texas, given the large area's unique natural plant and animal population. This chapter will assess the connection between climate and biological response as a guide for predicting changes in Texas's terrestrial ecosystems.

II.5.1.1 Ecosystems and Climate Stability

A species' ecological niche encompasses more than biological organisms, as terrestrial systems encompass climate, soil conditions, and a disturbance regime. The "system" of an ecosystem refers to the interactions between the biotic and abiotic components that, over time, stabilize regarding species, vegetation production, and soil type. Within an ecosystem, characteristic plant and animal species may emerge as those with the most adapted niches for the given ecosystem environment.

II.5.1.1.1 Lessons from the Recent Past – The Wisconsin Ice Age

To better appreciate the impact of climate on Texas ecosystems, it is helpful to examine Earth's past climate changes and how ecosystems responded. The most recent event is the Wisconsin Ice Age, which was named after the glacial geologic deposits first described in Wisconsin. Approximately 20,000 years ago, this period was freezing. Ice sheets, now only present in the Arctic Ocean, Greenland, and Antarctica, covered land as far south as the states of Idaho, Montana, Wisconsin, and New York in North America. Imagine traveling to upper Wisconsin 20,000 years ago and looking at ice formations a mile thick! Such a large ice sheet affected the global climate by slowing the hydrologic cycle, lowering sea levels, and creating a high-pressure dome over much of the northern latitudes.

In Texas, cooler and wetter conditions prevailed compared to today. This included milder summers (6°C lower than present; Toomey et al., 1993) and cold winters. In West Texas, precipitation was estimated to be 50% greater than today, with the landscape dotted with small lakes supported by frequent, intense rainfall events (Reeves, 1973). From these lakes and others across the state, pollen has been analyzed from preserved sediments, indicating the surprising presence of northern

DOI: 10.1201/9781003406600-12

tree species, including spruce, birch, and maples (Bryant and Holloway, 1985). For the Panhandle Plains, a mix of sagebrush (*Artemisia* spp.) grasslands with pockets of spruce-pine forest covered much of the landscape (Hall and Valastro, 1995).

More rainfall in Texas during the Ice Age resulted in higher vegetation cover, which supported forest growth over much of the state. For the expansive grassland ecosystem present today, climate conditions supported more widespread woodlands and trees forming extensive savannas. The Chihuahuan Desert was less arid, with vegetation dominated by shrubby plants and grasses rather than succulents such as cacti and agave, which are currently found there. For the pine and oak forests now found in East Texas, cooler, wet conditions allowed the boundary of these forests to extend farther west.

II.5.1.1.1.1 Relic Pleistocene Texas Ecosystems

Examples of the Wisconsin Ice Age ecosystems are still found in Texas that include the Lost Maples (bigtooth maple; *Acer grandidentatum)* found in Bandera and Real counties in southern Texas, the Lost Pines (loblolly pine; *Pinus taeda*) near Bastrop, Texas, and the isolated forests on the highest elevations of the Chisos Mountains in Big Bend National Park that include common Rocky Mountain species Douglas fir (*Pseudotsuga menziesii*) and quaking aspen (*Populus tremuloides*). All these species are relics from the cooler and wetter climate conditions prevalent in Texas during the last Ice Age.

With the end of the Ice Age, the climate warmed and dried approximately 12,000 years ago; these relic plant communities have survived in areas where combinations of elevation, shade, and consistent water sources have created cooler and wetter microhabitats, similar to previous climate conditions. For example, the big-tooth maples survive today in the narrow, shaded canyons of the Edwards Plateau, where the canyon bottoms retain persistent moisture due to the surrounding topography (Van Auken et al., 2022).

The cool and wet climate of the past enabled forests, including loblolly pines, to expand across broader regions. Today, the Lost Pines are an isolated population of loblolly pine forests within the post oak savannah. The sandy soils and the elevated topography of the Bastrop region provide microclimate and soil conditions ideal for this species. The Bastrop area is part of the Western Gulf Coastal Plain, characterized by deep, sandy soils ideal for loblolly pine growth. These soils, derived from ancient rivers and marine deposits, help the trees retain moisture even in drier conditions. Historically, natural fires and indigenous land management helped maintain forests, such as the Lost Pines ecosystem, by controlling competing vegetation and promoting pine regeneration (Miller and Corby, 2022). This balance of fire and growth allowed the forest to thrive in an otherwise unsuitable region. As the climate warmed up and became drier at the end of the Ice Age, most loblolly pine populations retreated to the eastern portion of the state, where humidity levels are higher, and now dominate as extensive forests.

Ice Age temperatures across southwest Texas were also significantly cooler and wetter. This climate supported forests of Douglas fir, aspen, and other montane species at much lower elevations and farther south than they are typically found today. During cooler climatic periods, seeds from Douglas firs and aspens are dispersed across broader ranges via wind, water, or animal vectors. As the Ice Age ended, these species retreated to higher elevations, where the montane environment persisted. In the Chisos Mountains, these trees found suitable refuge at elevations between 5,000 and 7,800 feet, particularly on north-facing slopes where temperatures remain cooler and moisture levels are higher. Today's Chisos Mountains are considered "sky island" ecosystems (Gehlbach, 1993), representing the southernmost range of these species in the United States, surrounded by the vast expanse of the Chihuahuan Desert.

These examples demonstrate that climate change has led to dramatic ecosystem changes over 12,000 years. Warmer, drier conditions since the Ice Age have transformed Texas ecosystems, creating the diverse biogeographic regions seen today that include deserts, grasslands, and forests. The climate is changing at a rate of less than 100 years, raising questions about the future of our familiar Texas landscapes. If the ecosystem comprises characteristic species and soils emerge as a stable entity after 1000s of years with constant climate, what happens when the climate changes at a historically unprecedented rate? Recent events may provide clues as to what to expect. For example, during the extreme La Niña event of 2011, record-breaking temperatures and drought were reported across the state. At the same time, multiple large fires burned across all regions of Texas, including half of the area of the Lost Pines Forest (Nielsen-Gammon, 2012). In this case, it is easy to see that energy in equals energy out; as the climate warms and provides heat energy to the forests, fires within these forests balance that energy in addition to the warming climate. This also highlights that ecosystems subjected to rapid climate change are more vulnerable to increased disturbance.

Our modern climate is beyond the evolutionary adaptation rate of many modern plants, with migration unable to keep pace. Plant community turnovers within Texas's physiographic regions will likely result from selecting the most heat and drought-adapted plants and local extirpation of those that are not.

II.5.1.2 Present and Future Texas Climate

As a backdrop to the discussion of climate change impacts on present ecosystems in Texas, the worst-case scenario is considered, where a lack of action to reduce greenhouse gas emissions and continued land-use changes result in the highest expected warming by 2100. As a framework, the temperature and precipitation values used here are based on mean global circulation model results, referred to as ensemble mean values, for the recent Coupled Model Intercomparison Project Phase 6 (CMIP6) simulations for scenarios 5–8.5 times the current radiative forcing, referred to as the high-emission scenario. Under these conditions, the state's average annual temperatures are projected to be 4–5°C warmer by the end of this century (Figure II.5.1; data source: https://psl.noaa.gov/ipcc/cmip6/ccwp6.html). The current mean temperature is approximately 15.5°C, thus a future mean temperature of 20°C. Also, the frequency of days with higher temperatures is expected to rise. For example, Nielsen-Gammon et al. (2020) projected that the number of days with temperatures

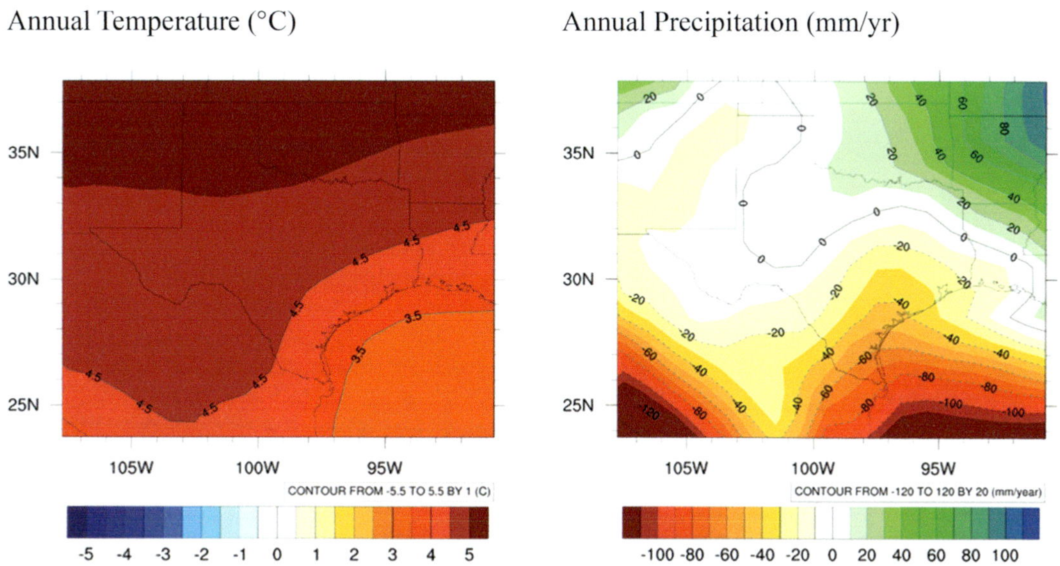

FIGURE II.5.1 Differences in the annual temperature (°C, left panel) and annual precipitation (mm/yr, right panel) calculate from the CMIP6 ensemble model for the radiative forcing mean for the future period (ssp585: 2070–2099) minus the values from the historical reference period (1985–2014). (Source: https://psl.noaa.gov/ipcc/cmip6/)

exceeding 40°C will double by 2036. For precipitation, annual rainfall will be 20 to 60 mm/year less by 2100 based on this model in the south and southeast parts of Texas, compared to west Texas, where more rain is anticipated.

Seasonal differences in projected precipitation are also apparent across different regions (Figure II.5.2). For the western half of the state, most rainfall deficits may occur during winter months, with the South Texas Plains having the most significant deficit of 24 mm during those months (Table II.5.1). Spring may be the driest for East Texas with up to 10 mm less precipitation. During the summer, much of the central and western regions are expected to experience increased rainfall, up to 15 mm in the dry Big Bend Country. Autumn rains are variable across the state,

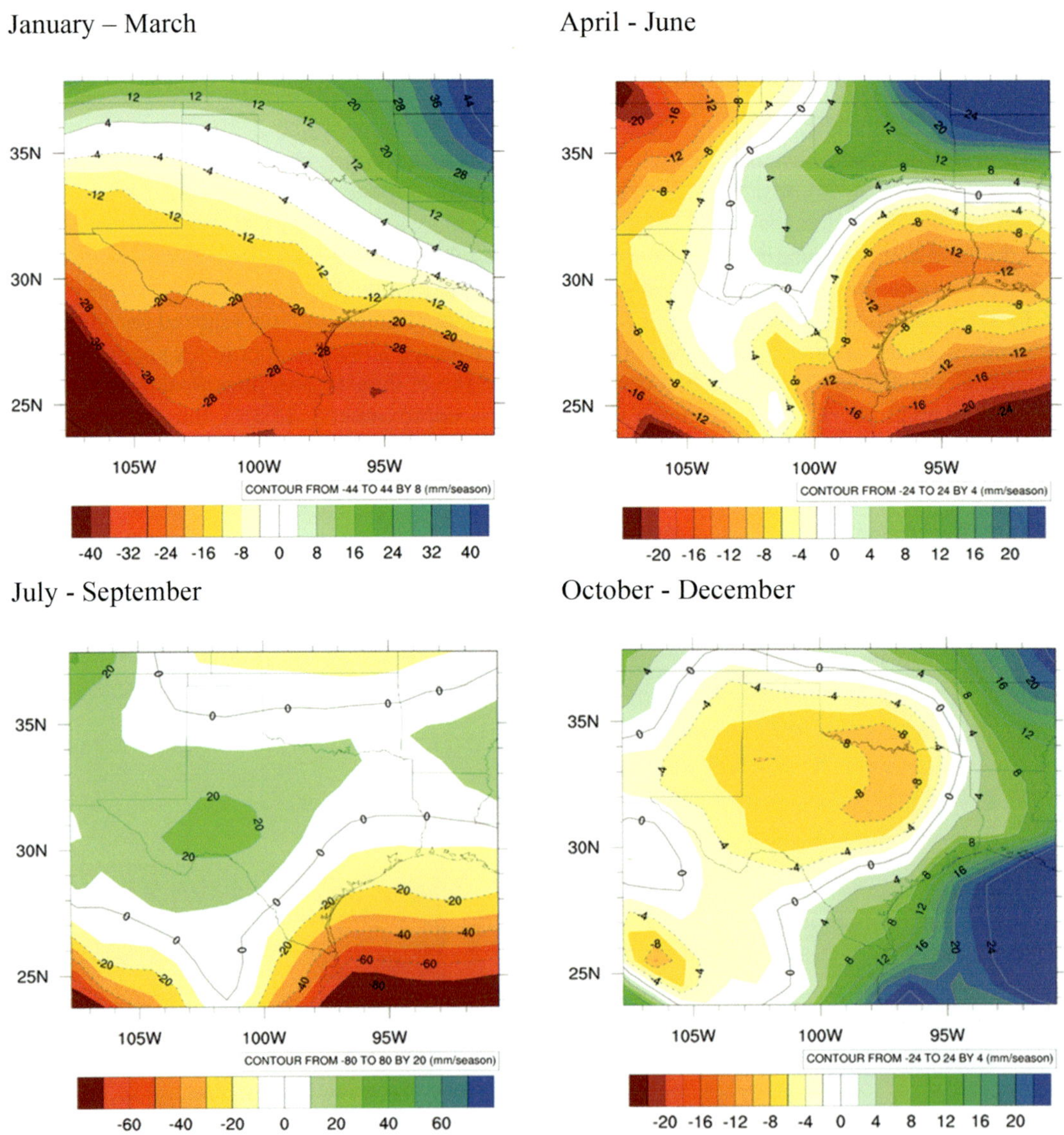

FIGURE II.5.2 Differences in the seasonal precipitation (mm/season, right panel) calculated from the CMIP6 ensemble model for the radiative forcing mean for the future period (ssp585: 2070–2099) minus the values from the historical reference period (1985–2014). (Source: https://psl.noaa.gov/ipcc/cmip6/)

TABLE II.5.1
Differences in the Seasonal Precipitation (mm/season; JFM = January through March; AMJ = April through June; JAS = July through September; OND = October through December) Calculated from the CMIP6 Ensemble Model for the Radiative Forcing Mean for the Future Period (ssp585: 2070–2099) Minus the Values from the Historical Reference Period (1985–2014)

Physiographic Region	JFM	AMJ	JAS	OND
Pineywoods	*equiv.*	−10	*equiv.*	+2
Prairies and Lakes	−4	−10	*equiv.*	−4
Gulf Coast	−16	−10	−10	+4
Panhandle Plains	−4	+2	+10	−6
Hill Country	−12	−4	+10	−6
South Texas Plains	−24	−8	−10	+2
Big Bend Country	−16	−2	+15	−4

Source: https://psl.noaa.gov/ipcc/cmip6/

Estimated mean values by comparing seasonal differences (Figure II.5.2) with physiographic regions (Figure II.5.3). Where no differences occurred, cells are marked as equivalent (Equiv.)

ranging from 6 mm deficits to a 4 mm surplus. These seasonal precipitation differences across Texas regions are essential in affecting plant reproduction and peak growing periods and influencing disturbances such as disease and wildfire.

In addition to average temperature and precipitation changes, mesoscale climatic events, such as the El Niño–Southern Oscillation (ENSO), are also expected to be modified. An ENSO event is caused by shifts in the Southern Pacific Ocean's current strength that control the prominence of cold or warm water off the coast of northern South America. Due to its proximity, the ENSO affects weather globally, particularly in the southern United States. The ENSO has two major phases that change every two to three years. When warm water is present near South America, it is referred to as El Niño; in contrast, cold water is referred to as La Niña. The phase of ENSO affects seasonal and extreme weather events in Texas. For example, Wolter et al. (1999) analyzed past climate records for Texas. They found that extreme temperatures, both cold and warm, were affected by the phase of ENSO mainly during the winter and spring seasons. Colder temperatures can affect plants by causing direct damage from freezing and delaying seed germination. Warm temperatures during spring promote early and vigorous growth, facilitating the germination of seeds and the emergence of winter-dormant plants. However, warm temperatures during La Niña in South Texas may exacerbate expected drier conditions in winter and spring (Figure II.5.2), leading to higher mortality rates of seedlings and existing plants.

The vapor pressure deficit (VPD) is also expected to change as temperatures increase. Vapor pressure deficit is the difference between the potential pressure of water vapor that the air can maintain at a given temperature (saturated) and the actual pressure of water vapor in the ambient atmosphere. It is related to relative humidity, the ratio of the actual to the potential water vapor, expressed as a percentage. For terrestrial organisms, liquid water in their tissues is lost through various evaporation forms, proportional to the vapor pressure deficit (VPD), due to the pressure imbalance, resulting in energy loss from saturated to unsaturated conditions. Soil water evaporation is also influenced by VPD, particularly near the soil surface. For example, a typical summer day today, with a temperature of 35°C and a relative humidity of 50%, would have a VPD of 2.8 kPa. For a summer day in 2100 with the same humidity of 50% a 5°C increase in temperature in 2100 to 40°C

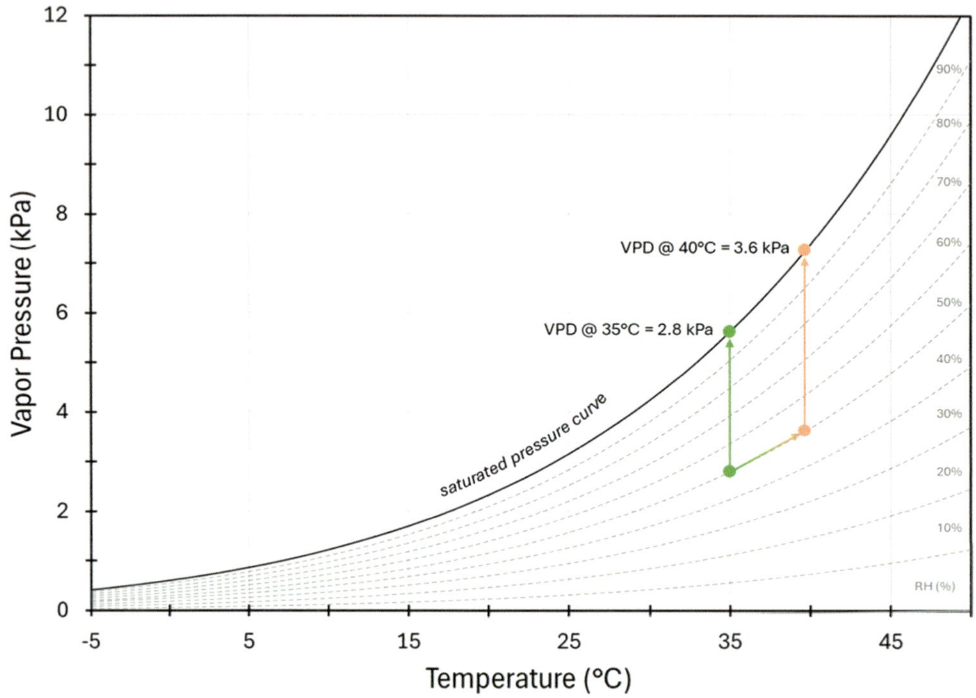

FIGURE II.5.3 Saturated vapor pressure curve showing vapor pressure by relative humidity. The effect of temperature on the vapor pressure deficit (VPD) is shown.

would have a VPD value of 3.6 kPa (Figure II.5.3). This increased VPD would increase evaporation rates by 10–25%, depending on other conditions. Therefore, despite similar precipitation for 2100, increased temperatures will result in more evaporated water from plants, animals, and soil simply due to the physics of the atmosphere. The expected higher temperatures in Texas suggest a future arid climate, even with rainfall equivalent to the present.

II.5.1.3 The Physiographic Regions of Texas

For simplicity, the physiographic regions of Texas (Van Auken, 2018; Figure II.5.4), provide organization for discussing climate change across the large expanse of Texas. This classification defines seven regions using climate, topography, geology, and dominant vegetation types (Landers, 1987). The regions include the (1) Pineywoods, (2) Prairies and Lakes, (3) Gulf Coast, (4) Panhandle Plains, (5) Hill Country, (6) South Texas Plains, and (7) Big Bend Country (Figure II.5.3). The Pineywoods are in easternmost Texas, characterized by pine forests on gentle hill terrain with sandy soils (e.g., *Pinus taeda*). They also feature hardwoods such as bur oak (*Quercus macrocarpa*) and post oak (*Q. stellata*), as well as in moister habitats, sweetgum (*Liquidambar styraciflua*) and black hickory (*Carya texana*). The region receives over 150 cm of rainfall per year. The Prairies and Lakes region receives 75 cm/yr in precipitation, including the Blackland and the Grand Prairies and the forested Cross Timbers area. Grass dominates the prairies growing on deep clay soils, including little bluestem (*Schizachyrium scoparium*) and side-oats grama (*Bouteloua curtipendula*). In the Crosstimbers, tree species such as cedar elm (*Ulmus crassifolia*), hackberry (*Celtis laevigata*), and blackjack oak (*Quercus marilandica*) are common. The Gulf Coast is a broad, flat area bordering the Gulf, characterized by sandy soils with clay inclusions, which support various large species, including big bluestem (*Andropogon gerardii*) and switchgrass (*Panicum virgatum*). The Panhandle Plains is a drier region in the northernmost part of Texas, with an average annual precipitation of 50 cm. Rugged topography with deep canyons is typical in the northern portion, which becomes broad and flat toward the south. Common grasses of this region include buffalograss (*Bouteloua dactyloides*) and blue grama grass (*Bouteloua gracilis*). The Hill Country region is situated on the 250,000 km^2 limestone called the Edwards Plateau. Thin soils and varying topography support grasslands and patches of woodland that include oaks such as plateau live oak (*Q. fusiformis*), Spanish oak (*Q. buckleyi*), and shin oak (*Q. sinuata* var. *brevifolia*) in addition to Ashe's juniper (*Juniperus ashei*). Rainfall varies across this large region, with an average annual rainfall of 50 cm. The South Texas Plain has flat to undulating topography in the southernmost region with approximately 60 cm/yr of precipitation. The area is a mixture of grasslands and thick shrublands, containing leguminous species, such as honey mesquite (*Neltuma glandulosa*) and palo verde (*Parkinsonia texana*). The Big Bend Country is the westernmost region, encompassing the northern portion of the Chihuahuan Desert and small mountain belts that support montane forest communities. The average annual rainfall over the region is 25 cm. Chihuahuan Desert vegetation includes succulents such as Lechugilla (*Agave lechugilla*), prickly-pear cacti (*Opuntia* spp.), and the highly specialized ocotillo (*Fouquieria splendens*). In the mountains, piñon pine (e.g., *P. cembroides*), ponderosa pine (*P. ponderosa*), Douglas fir (*Pseudotsuga menziesii*), and quaking aspen (*Populus tremuloides*) are found on climatically protected slopes.

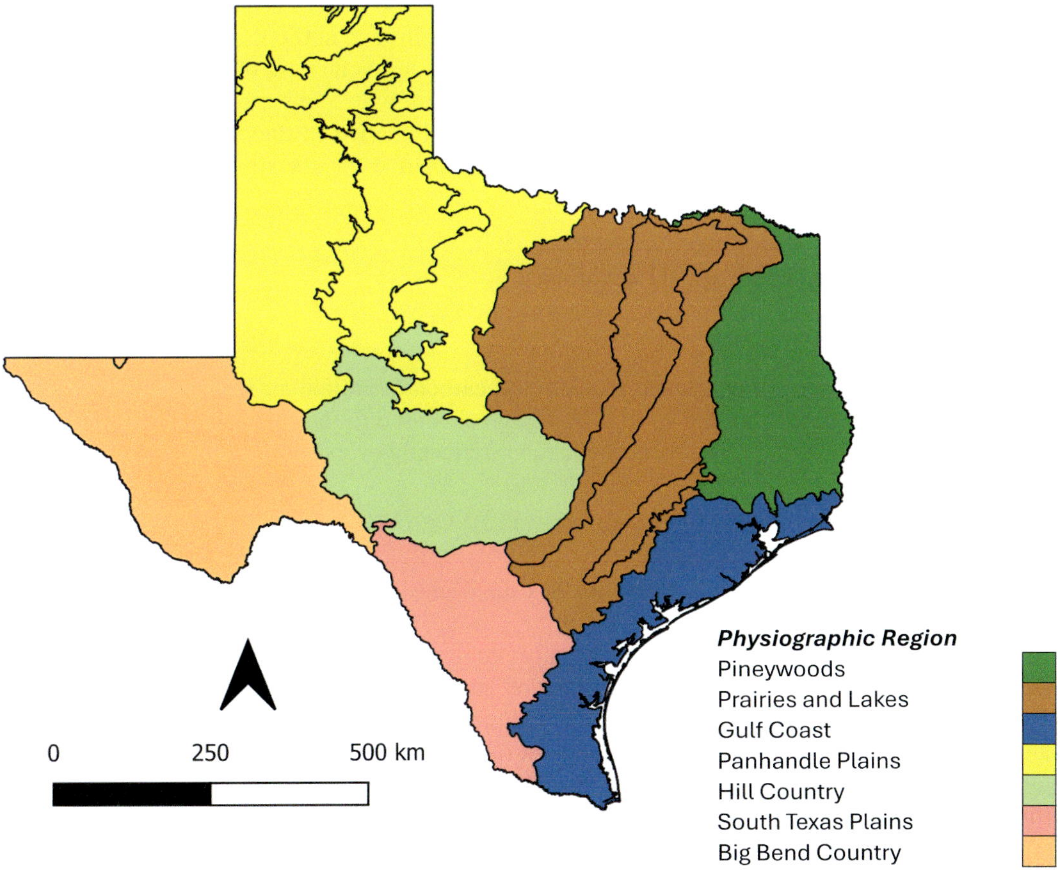

FIGURE II.5.4 Physiographic regions of Texas proposed by Van Auken (2018).

II.5.1.4 General Ecosystem Responses to Climate Change

The study of terrestrial ecosystems involves measuring the transfer of energy and mass among their biotic and abiotic components, which can be quantified over time as fluxes. For example, the energy flux from sunlight into the ecosystem involves the absorption (and reflection) of sunlight by plant leaves over a day, month, or year. This energy is used to drive the biochemical reactions of fixing carbon dioxide from the atmosphere (photosynthesis) and/or result in the serendipitous evaporation of water within the leaves (transpiration). As the leaves assimilate carbon, carbohydrates from this process may be redistributed throughout the rest of the plant to the stem, roots, and reproductive organs. Animals may consume this matter, releasing waste containing carbon and other nutrients into the soil. Microbes in the soil then utilize these inputs for biological activity, after which they may die, releasing carbon back into the atmosphere and making nutrients available in the soil for plant uptake. The energy inputs into the system drive the rates of these cycles and fluxes within the ecosystem. Heat energy, for example, increases the rate of most biological activities by increasing the activity of enzymes as part of the metabolic pathways of organisms. Beyond a certain level, heat can reduce or stop metabolism by altering the structure of enzymes as protein molecules, a process known as denaturation. Thus, the mechanism for ecosystem changes, such as those caused by increased temperature, may be controlled by the most basic components of the ecosystem elements.

This chapter focuses on the vegetation as plants are the primary producers, and the energy is available to consumers at higher trophic levels because whatever impacts plants will also affect animals. To simplify, ecosystem processes to be discussed include: (1) how plants fix atmospheric carbon in their leaves, losing water as a consequence; (2) the factors affecting the utilization of assimilated carbon for growth and metabolism; (3) the control of nutrient cycling associated with the climate-vegetation-soil interaction; (4) reproduction as a special case of carbon allocation and essential for the sustainability of the ecosystem through time; and (5) disturbance related in internal and external factors influencing ecosystem perturbation. Reliance on the basic principles of biology, chemistry, and physics provides the most reliable method for assessing future climate stress expected for Texas ecosystems.

II.5.2 PLANT BIOLOGY BASICS

II.5.2.1 Plant Production

Photosynthesis, also known as carbon fixation and carbon assimilation, is the process by which plants absorb light energy through special photopigments used in the biochemical bonding of carbon from atmospheric CO_2. The chemical equation frequently oversimplifies this:

$$6CO_2 + 6H_2O \rightarrow C_6H_{12}O_6 + 12O_2$$

shown in its balanced form to emphasize that six molecules are necessary for the synthesis of one molecule of glucose ($C_6H_{12}O_6$).

Within this basic chemical equation are subcellular structures, substrates, and enzymes that facilitate the conversion of absorbed solar radiation, resulting in carbon assimilation into the plant. Two key components include phytopigments, with chlorophyll being the most abundant, and the enzyme that "fixes" CO_2 gas within the leaf into a three-carbon sugar called ribulose *bisphosphate* carboxylase/oxygenase, or Rubisco for short. One should note that, as the name implies, this enzyme can bind to both carbon dioxide and oxygen. Additionally, since the first product of this reaction contains three carbon atoms, it is called C_3 photosynthesis. What is significant about these enzymes is that both require nitrogen. With nitrogen essential for plant growth, there is a link between the plant and the decomposition process of soil as the primary source of nitrogen and other nutrients for the plant ecosystem.

Under the most ideal conditions, including an ample water and nitrogen supply, plants can achieve the highest intrinsic rate of photosynthesis, A_{max}, which refers to maximum assimilation. This rate varies among plant species due to leaf shape affecting light absorption, nitrogen limitation, root density and absorption, and different rates of chlorophyll and Rubisco biosynthesis related to leaf nitrogen concentration. A_{max} and leaf nitrogen concentration (N_L) correlate globally with other plant characteristics, including leaf shape and lifespan (Reich, 2012). The relationships among all modern plant species are so strong that a change in one of these four variables inherently implies a shift in the others, representing a change in species. Because these variables are affected by climate change, they also provide a framework for assessing ecosystem shifts driven by plant characteristics, which are the fruition of associated functional process changes.

II.5.2.1.1 Temperature Impacts on Photosynthesis

Temperature plays a crucial role in the rate of photosynthesis due to its impact on the activity of key photosynthetic enzymes and its influence on membrane stability. All biological enzymes have a range of temperatures for maximization of the activity rate. This is due to the kinetic energy associated with different temperatures, which makes the enzyme more likely to interact with its corresponding substrates. Below these optimum temperatures, enzymes reduce binding to substrates simply due to reduced kinetic energy. Above optimum temperatures, the enzyme's shape and stability may degrade, referred to as denaturation. The structure of the enzyme is essential as the physical binding of the substrate and enzyme is conferred by the shape of both and the location of specific chemical binding points. With hot temperatures, the shape of the primary, secondary, and tertiary structures of enzymatic proteins is disrupted by changes in bonding among amino acids. At a certain temperature specific to different plant species, unique proteins within the cell, known as heat shock proteins (HSPs), can help a heat-damaged protein recover its shape and enzymatic function. However, HSPs also have thermal limits, after which all protein and enzyme functions are disabled. For Rubisco, the active temperature ranges broadly from 15 to 55°C (Galmes et al. 2015; Sage et al., 2008).

In addition to enzymes, photosynthesis requires special plasma membranes found in the chloroplasts of leaf cells. These organelles contain chlorophyll and other photopigments crucial for harvesting sunlight, which is then utilized to drive biochemical reactions, including Rubisco's enzymatic assimilation of CO_2. Like all plasma membranes within the cell, the chloroplast membranes are formed by individual molecules called phospholipids. The chemical structure of the membrane is formed by the phospholipid molecules that do not bond to each other, allowing the membrane to exhibit a fluid-like behavior. In the case of the chloroplast, the fluid membrane serves as an attachment surface for the chlorophyll molecules and the associated proteins crucial for light harvesting. The fluid-like nature of the membrane is key to its proper function: (1) if it is less fluid, the membrane is rigid and stiff; and (2) if it is too fluid, the membrane may dissolve. The fluidity of the membranes is controlled across varying temperatures by whether the lipid tails of the phospholipid contain saturated or unsaturated fatty acids. As an unsaturated fatty acid has structural "kinks," a phospholipid containing more unsaturated fatty acids will be looser to membrane fluidity at lower temperatures. In contrast, saturated fatty acids are straight; thus, compactness and reduced fluidity are essential at high temperatures. The thermal stability of the plasma membrane is related to its permeability, which is crucial for efficient light harvesting within the chloroplast. At temperatures of around 50 to 52°C, light use by the plant is significantly reduced (Quinn and Williams, 1985), with complete failure occurring at temperatures above 55°C (Berry and Björkman, 1980). On August 12, 1936, in Seymour, located in North Texas, and on June 28, 1994, in Monahans, West Texas, recorded maximum daily temperatures reached 49°C (https://www.ncei.noaa.gov/access/monitoring/scec/records/all/tmax). With predicted temperatures in 2100 under the worst-case scenario increasing by 4–5°C, there is the real possibility of irreparably disrupting the essential photosynthetic capacity of many current Texas plants simply due to heat damage to plant cell enzymes and structures.

II.5.2.1.2 Autotrophic Respiration

Like most living organisms, plants utilize organic compounds, primarily carbohydrates, to generate chemical energy for metabolism and tissue biosynthesis, a process referred to as autotrophic respiration. For plants, specifically, their carbohydrate nutrition relies solely on photosynthesis, with all the previously described constraints. Net primary productivity refers to the gross amount of photosynthesis minus the autotrophic respiration of fixed carbon. Whatever carbon is left becomes the substrate for all other plant carbon-based compounds.

Temperature affects the rate of carbohydrate utilization via enzyme activity rates; higher temperatures increase carbohydrate use. For the entire plant, the budget of carbohydrate demand is based on the amount of living tissue. This is also related to the overall partitioning of photosynthetic and non-photosynthetic biomass, referred to as source and sink of carbohydrates, respectively, of different plant types. Grasses and herbaceous plants have the highest photosynthetic biomass per total biomass. Only the roots and smaller vascular and reproductive tissues are carbohydrate sinks. In contrast, shrubs and trees, having evolved extensive woody stem tissues to better compete for light resources in denser vegetation canopies, must maintain a large amount of biomass in their stems and more extensive roots for mechanical support and the vascular transport of water and nutrients. As such, woody plants are more likely to be stressed by higher temperatures, as the larger biomass, combined with high enzyme rates, leads to a greater use of photosynthetically derived carbohydrates.

Woody plants face a carbohydrate shortage at higher temperatures due to higher respiratory rates that outpace their ability to fix carbon through photosynthesis. The consequence of this imbalance is less carbon to allocate to new or replacement growth of various plant tissues, including leaves, roots, and reproductive tissue. With drought, higher autotrophic respiration, particularly of stems, is inversely proportional to growth (Rowland et al. 2021). Also, fixed carbon is essential for synthesizing hormones and defensive compounds. This carbon starvation affects the growth and development of the plant, as well as its ability to withstand attacks from disease-causing pathogens and herbivory (Kolb et al., 2016).

II.5.2.2 Plant Water

Any discussion of climate change and plant water availability is intrinsically connected to plants, as CO_2 uptake is related to water loss through the small openings on the leaves' epidermis, called stomata. It is not, however, an equivalent exchange. This is due to the molecular mass differences between CO_2 and H_2O and the effect these differences have on the diffusion rates of these two molecular species. A glance at the periodic table reveals that carbon has an atomic mass of 12 and oxygen has an atomic mass of 16. Therefore, the molecular mass of CO_2 is approximately 44 [12 (C) + 2 • 16 (O) = 44 (CO_2)]. By comparison, water is lighter, having a molecular mass of 18. Based on the ratio of the CO_2 and H_2O masses, the relative gas diffusion predicts that approximately four water molecules are lost for a single molecule of CO_2 entering the stomata ($44/12 \approx 3.6$). Simply from the physics of gas diffusion, plant carbon uptake is expensive in terms of water loss. The necessity of photosynthesis by plants inevitably results in water loss via transpiration.

Land plants continuously absorb sunlight energy through their leaves. Some of this energy is harvested to drive photosynthesis, with the remainder (a) converted into heat energy, raising the temperature of the leaves, and (b) transforming liquid water in the leaves into a vapor form. When the leaf stomata open to facilitate CO_2 conductance, water vapor from light absorption diffuses into the atmosphere, a process called transpiration. However, as different land plant species have evolved through time, so have the variety of adaptations for balancing the rates of photosynthesis relative to transpiration, termed water use efficiency (WUE). As one travels from the tall pine forests of East Texas across the grassy savannas of the Edwards Plateau to the Chihuahuan Desert of the west, it is easy to see how plant water-use efficiency (WUE) and climate are closely related. Higher temperatures result in higher VPD and lower WUE (Stoy et al., 2019) per species. However, plants with inherently higher WUE are more likely to survive and persist in future plant communities within Texas ecosystems.

Soil depth and texture also influence water availability for plants. Soil depth develops regionally due to past weathering rates, which are influenced by climate, vegetation, and the geologic substrate, as well as topography, which affects erosion or deposition of weathered material. However, soil texture determines the water-holding capacity, as soil particles of different sizes will affect the amount of water in the soil that can drain. Coarse textures, such as sand, drain quickly, whereas clay, the finest soil texture, retains water in the soil as charged minerals bind to water molecules. In between, silty soils, often best suited for agriculture, have an intermediate grain size, balancing drainage and molecular adhesion. Across Texas, soil textures vary as a function of geologic parent material across physiographic regions (Table II.5.2). These soil characteristics set the limits for water uptake by plants derived from precipitation.

As seen in sandy soils, plant water limitations can manifest as low water-holding capacity. Looking across Texas, evergreen Loblolly pines (*P. taeda*) grow prolifically in east Texas's sandy soil with abundant rainfall and warm temperatures. By comparison, trees in far west Texas are reduced to small, specialized Havard shin oaks (*Q. havrardii*) characterized by woody sprigs poking out of sandy soil with a substantial root network in the hot, dry climate there. The sandy soil in East Texas reduces nutrient availability and retention, favoring evergreen species like loblolly pine. However, the sandy soil in west Texas allows the roots of the shin oak to grow deep enough to access groundwater.

Plant water availability (PWA; mm) is a helpful metric for integrating the limits of soil and climate. It is calculated as the amount of water left from precipitation (P; mm) in the soil following extraction by transpiration (T; mm) from plants and direct evaporation (E; mm) [$PWA = P - T - E$], assuming no runoff from saturated soils. Plant transpiration and soil evaporation rates are influenced by soil texture, which controls the amount and availability of water in the soil. Higher VPD inherently results in higher evaporation rates.

TABLE II.5.2
General Soil Characteristics of the Physiographic Regions of Texas Include a Qualitative Description of Water Holding Capacity, Soil Textural Classification, and Related Moisture Retention (Landers, 1987)

Physiographic Region	Water Holding Capacity	Key Soil Characteristics	PWA (mm/yr)
Pineywoods	Moderate to High	Sandy loam drains fast; clay loam retains moisture	1100
Prairies and Lakes	Moderate to Very High	Loamy soil retains some water; limestone limits its depth; expansive clay holds water but dries out hard.	683
Gulf Coast	Moderate to High	Clay holds water; sandy areas drain faster.	378
Panhandle Plains	Moderate to High	Clay loam retains water well; deep loess holds moisture, but high evaporation rates reduce availability.	780
Hill Country	Low	Thin, rocky soil drains quickly.	273
South Texas Plains	Low to Moderate	Varies from clay loam that holds water but dries hard to sandy loam that drains fast.	364
Big Bend Country	Very Low	Desert soil drains quickly and retains little moisture.	368

Modeled plant water availability (PWA; mm/yr) for 2065 to 2074 Derived from Li et al. (2024), who utilized CMIP6 Shared Socioeconomic Pathway For Radiative Forcings 5–8.5 (SSP 585; High Emission Scenario). Data from Thrasher et al. (2022). Values were calculated for regions broadly overlapping with the Van Auken (2018) Physiographic Areas of Texas.

Because of autotrophic respiration costs for plants supporting non-photosynthetic tissue (e.g., woody stems), forests and woodlands require more water to support tree productivity for extended periods of the growing season, the PWA limit of an area to support trees is approximately 350–400 mm/yr (Thrasher et al., 2022). In a study using GCM data for the SSP 585 scenario, future PWA was calculated for the different physiographic regions of Texas (Table II.5.2). Surprisingly, this research showed that the Hill Country region had an average PWA of 273 mm/yr, lower than the 350 mm/yr necessary for supporting trees. This area currently supports woodland patches within a grass matrix. This low PWA reflects this region's predicted future rainfall deficit in 2011, with fragile soil and low water-holding capacity.

II.5.2.2.1 Stomatal Conductance

Stomata limit water loss by opening and closing, primarily due to tissue water concentration. Stomata pores, typically found on plant leaves, are formed by two specialized epidermal cells called guard cells that regulate the amount of water inside to open (with full turgor) or close (in a wilted state). The rate at which carbon dioxide and water vapor move through these stomatal openings is called stomatal conductance (g_s), which regulates the balance between photosynthesis and transpiration.

Across geological time, the size and density of stomata on leaves have also changed, considered as a response to past climate and atmospheric CO_2 levels that affect stomatal conductance. Generally, plants with smaller stomata also have more stomata per unit area of leaf and vice versa. Thus, the combination of guard cell regulation of stomatal opening, the maximum size of the opening, and density of the stomata on leaves combine to constrain how fast water is lost from the leaf, while supplying the internal cells with ample CO_2.

Stomatal closure mechanisms reduce transpiration losses under high VPD, reducing carbon uptake (Urban et al., 2017). Plant species vary in stomatal number and size, and guard cells' sensitivity to internal water and CO_2 concentrations. As higher temperatures contribute to higher VPD, species with lower WUE due to stomatal conductance characteristics will be most stressed within a plant ecosystem. However, plants with inherently higher WUE are more likely to survive and dominate future vegetation communities.

Texas plant species have stomatal conductance values that vary with PWA and VPD. Loblolly pines in east Texas have higher stomatal conductance ($\approx$160 mmol m^{-2} s^{-1}; Murthy et al., 1997) as well as co-occurring oak species, including Shumard oak (*Q. shumardii*; gs 400 mmol m⁻2 s⁻1) and bur oak (*Q. macrocarpa*; g_s = 200 mmol m⁻2 s⁻1), in this moist climate. Drier, warmer conditions prevail in the Gulf Coast, where coastal live oak (*Q. virginiana*) has a lower conductance of $\approx$ 80 (Cooper et al., 2018). Thorny shrubs of the South Texas Plains exhibit lower conductance values, consistent with the region's lower rainfall, sub-humid climate, and soil with moderate drainage. These include sweet acacia (*Vacellia farnesiana*), desert hackberry (*Celtis ehrenbergiana*), narrowleaf forestiera (*Forestiera angustifolia*), palo verde (*Parkinsonia aculeata*), and honey mesquite (*N. glandulosa*) with a mean g_s of 80 mmol m^{-2} s^{-1} (Adhikari and White, 2014). Some trees in the Prairies and Lake region exhibit variable stomatal conductance, consistent with this region serving as a transition zone between the moist east and the drier west. For example, eastern redcedar (*J. virginiana*) has a low g_s of 20 mmol m^{-2} s^{-1} compared to post oak (*Q. stellata*) with an average g_s of 100 mmol m^{-2} s^{-1} (Torquato Rettondini, 2019) and blackjack oak (*Q. marilandica*; g_s = 120 mmol m^{-2} s; Choinski and Wise, 1999).

In the Hill Country, woodlands are comprised of oaks and juniper, with conductance values that vary despite differences in their leaf phenology. The deciduous Spanish oak (*Q. buckleyi*) has a g_s of 110 mmol m^{-2} s (Balok and Hilaire, 2002). This is like the semi-deciduous plateau live oak (*Q. fusiformis*; g_s = 200 mmol m^{-2} s) and coniferous evergreen Ashe's juniper (*J. ashei*; g_s = 100 mmol m^{-2} s; Bendevis et al., 2010). However, the lower stomatal conductance of the juniper may be an adaptation by this species to reduce water loss as a conservation strategy and maintain longer-term PWA. In this region, the Texas climate is conducive to growth throughout the year.

In the Big Bend Country, plants have very low stomatal conductance to minimize water loss in the extreme environment. For the woody shrub creosote (*Larrea tridentata*), g_s is approximately 80 mmol m^{-2} s (Pataki et al., 2000) compared to the succulent prickly pear cacti (*Opuntia* spp.) with only four mmol m^{-2} s (Graham and Nobel, 1999). Native grasses have remarkably high stomatal conductance values. These include blue grama grass (*Bouteloua gracilis*; g_s = 200 mmol m^{-2} s; Read et al., 1997), little bluestem (*Schizachyrium scoparium*; g_s = 300 mmol m^{-2} s, silver bluestem (*Bothriochloa saccharoides*; g_s = 350 mmol m^{-2} s), and Texas wintergrass (*Stipa leucotricha*; g_s = 300 mmol m^{-2} s; Anderson and Briske, 1990). These grasses exhibit a pulse-driven phenology, where starch storage tissues are activated to initiate leaf growth once specific precipitation and temperature thresholds are met.

II.5.2.2.2 Alternate Photosynthetic Pathways for Plant Water Conservation

The higher stomatal conductance enables grass to assimilate rapidly CO_2 while conditions are favorable, followed by a return to dormancy, a strategy referred to as "live fast, die young." In addition, some grasses are adapted to grow even when soil moisture and VPD are less than optimal and stomatal conductance is reduced. These grass species employ a slightly different form of carbon capture that does not utilize the Rubisco enzyme; instead, they use phosphoenopyruvate carboxylase, also known as PEP-case. This enzyme does not bind to oxygen but rather solely to CO_2. In addition, the first product of PEP-case fixation of CO_2 is a four-carbon sugar, hence called C_4 photosynthesis. For plants exhibiting C_3-type photosynthesis (e.g., *S. leucotricha*), assimilation of O_2 over CO_2 is energy inefficient. However, as plants produce O_2 by splitting water due to sunlight harvesting, C_3 plants are challenged during drought periods when stomatal conductance may be reduced. For C_4 plants, such as many Texas grass species (e.g., *B. gracilis*, *S. scoparium*, *B. saccharoides*), this type of photosynthesis enables plants to continue assimilating carbon even when stomatal conductance is minimized and internal leaf concentrations of oxygen are high, without incurring an extra energetic cost.

As warmer weather results in less plant water, this favors grasses that exhibit C4-type photosynthesis. Interestingly, these grasses are referred to as "warm-season" grasses. Cooler weather favors plants with C_3-type photosynthesis and is called "cool-season" grasses. Given the current warming projections for Texas, warm-season grasses are likely to expand further and dominate grassland communities. However, many current invasive grass species in Texas (e.g., *B. ischaemum*, *Eragrostis lehmanniana*, *Cynodon dactylon*) are warm-season grasses originating from the warm climates of Africa, Asia, and other regions. Thus, as Texas temperatures continue to rise, these non-native species may expand their ranges, resulting in ecological changes that include reduced faunal forage, decreased nutrient cycling, and increased fire frequency (Wied et al., 2020).

Finally, desert plants, including succulents like cacti, have the lowest stomatal conductance values, as expected. In its most specialized form, a succulent's thick, waxy coat minimizes water loss and protects its leaves from intense solar radiation. However, like warm-season grasses, another photosynthetic adaptation helps plants grow in the driest climates. Succulents also fix carbon using the PEP-case pathway, which avoids oxygen as a by-product of waste during assimilation. However, this occurs at night, allowing stomata to open to limit water loss. When first discovered, scientists studying succulents from the Crassulaceae plant family noted a change in the acidic taste of the plants from day to night, hence the name crassulacean acid metabolism (CAM) for short. During the day, CAM plants will produce CO_2 from the compounds produced at night, increasing internal CO_2 concentrations fixed internally by Rubisco. Separating photosynthetic activity between day and night allows these plants to survive with minimal water requirements. This strategy successfully copes with drought; however, as previously described, temperature damage to key enzymes associated with photosynthesis under extremely high temperatures can be detrimental, even to some highly specialized plant types. However, the Chihuahuan Desert native Lechuguilla (*A. lechuguilla*) can endure temperatures greater than 60°C, indicating that expected plant community changes may involve the extirpation of species lacking extreme tolerance to heat and drought, especially in the desert.

Finally, while most plants transpire most of their water from their leaves during the daytime, places with high temperatures, such as Texas, store heat in plant canopies and soil throughout the day. This added heat energy may cause plants to transpire water at night. Based on field measurements, transpiration from various plant types and climates ranges from 1% to 25% of daytime transpiration (Dawson et al., 2007). Increased temperature will increase nocturnal evapotranspiration as more heat is retained overnight by vegetation and soil. As nighttime temperatures also increase, higher VPD will occur, accompanied by higher evapotranspiration rates. This additional water loss will further reduce plant water availability for Texas plants and exacerbate drought conditions.

II.5.2.2.3 Stem Hydraulics

While stomatal pores are the final point of soil water moving through the plant into the atmosphere, plants with woody stems, such as trees, have varying tissue forms that affect the water flux rate supplied to the leaves via transpiration. Under moderate VPD, the rate of water transport may be equivalent to the loss rate through the stomata in the plant canopy. This occurs because liquid water molecules are electrostatically bound to each other in a continuous stream from the root to the leaf through the xylem. This connection results in an internal tensional force that is partially related to the VPD of the air outside the stomata.

However, as evaporation rates increase with higher VPD, the stem's structural hydraulic limits may not sufficiently transport enough water to resupply the canopy leaves after water loss. In such a situation, the continuous water support by the stem vascular system may be broken, allowing air from local tissue to enter the vascular tissue, resulting in embolism. Embolic xylem shuts down the water supply from roots to leaves of the plants. Rising temperatures with high VPD exacerbate drought conditions, increasing tree mortality due to carbon starvation resulting from stem xylem embolism (McDowell, 2011). In the Hill Country, Ashe's juniper (*J. ashei*) exhibits embolism-resistant xylem compared to the co-occurring live oak (*Q. fusiformis*), highlighting species differences that influence competition, particularly in response to drought (Northup et al., 2021).

II.5.2.2.4 Roots

Regarding water supply, roots serve as the interface with the soil's moisture and vary significantly among plant types. For trees, the lateral extension of roots is approximately four times the diameter of the tree's crown. For grasses, roots may extend to depths of more than 8 m, particularly in well-developed prairie soils. Desert plants such as cacti and agave have smaller roots that can absorb water quickly from heavy, infrequent rainfall. Each plant type is adapted to differences in environmental water supply, particularly precipitation, which affects rooting depth and lateral root development (Schenk and Jackson, 2002). Forests have deep roots to access persistent ground and more lateral roots to compete effectively with neighboring trees. Grasses develop new leaves in response to episodic precipitation events, utilizing deep roots to extend the duration of active development. For cacti and agave, smaller roots reduce the metabolic cost of maintaining non-photosynthetic tissue in an extreme water-limited environment.

Cumulative environmental stress, particularly from drought and nutrients, increases net primary production belowground to form roots (Runyon et al., 1994). While this strategy supports the continued growth of plants over the short term, chronic stress results in less photosynthetic material to provide carbon for maintenance and development. As Texas plants experience increased heat and drought, more carbon may be allocated to root growth to offset moisture and nutrient deficits, albeit at reduced productivity. This may favor phreatophyte plants, such as honey mesquite, that produce deep roots as part of their normal development.

II.5.2.2.4.1 Mesquite – A Climate Change Resilient Plant

Mesquite (*Neltuma* spp.) has one of the deepest root systems, extending 20 m into the soil. This allows groundwater access during prolonged periods of drought. Consequently, it can grow in areas that survive with as little as 15 cm of annual precipitation. As a legume, mesquites thrive in nutrient-poor, sandy, or rocky soils due to supporting symbiotic nitrogen-fixing bacteria in their roots.

The mesquites' root system allows them to outcompete grass in areas that experience long-term drought and disturbance. Once established, mesquites will dominate, outcompeting other plants for soil moisture. In addition, mesquite trees' small, pinnate leaves reduce surface area and water loss in areas with intense sunlight, such as grasslands and deserts with consistently clear skies. Lastly, the mesquites can enter a semi-dormant metabolic state during extreme drought, shed leaves, or limit flowering and seed production. Honey mesquite (*N. glandulosa*) is also tolerant of wildfire, resprouting quickly from roots if the above-ground portion of the plant is destroyed. The combination of these characteristics supports the finding of Yang et al. (2024) that honey mesquite (*N. glandulosa*) will expand its range across two-thirds of the area of Texas by 2100 based on the SSP585 scenario. The potential future expansion of this nitrogen-fixing woody species into areas historically dominated by grasses will have complex outcomes for soil carbon and nutrient cycling.

11.5.2.3 Nutrient Availability

For an ecosystem, decomposition, or heterotrophic respiration, occurs when microorganisms, known as heterotrophs, in the soil utilize detrital carbon compounds for metabolism and growth. Extreme temperatures, such as those experienced in 2011 in Texas, reduce the activity of enzymes in soil microorganisms, resulting in slower decomposition rates (Acosta-Martinez et al., 2014). Where water is limited, organic matter in the soil may not decompose as hotter temperatures increase VPD in the upper soil horizons (Dirks et al., 2010).

Soil organic matter can vary in terms of decomposability depending on the source. Leaves shed from the vegetation canopy have the highest decomposability of plant matter. On the one hand, the amounts of simple sugars, such as sucrose, and structural carbohydrates, including cellulose and lignin, influence the decomposability of plants. Simple sugars stored within plant cells are readily accessible for microbial degradation. Cellulose is less available due to enzymatic limitations. On the other hand, Lignin is highly resistant to decomposition due to the abundance of phenolic ring structures that comprise this complicated molecule. Plants use lignin to stiffen structures, such as leaves, and to form secondary cell walls in stems and roots, which help vascular tissue resist collapse under the tensional pressure of water transport in the xylem. Because of this, the amount of lignin embedded in plant tissue is related to the aridity of the climate in which plants live (Meentemeyer, 1978).

Biomass shed from plants also contains nitrogen, which becomes part of the soil's nutrients through decomposition. Nitrogen in leaves can be reabsorbed into stem tissue before leaf abscission. This is especially important for plants that lose all their leaves in their canopy seasonally (deciduous) to help foster the growth of new leaves for the next growing season. Alternatively, in environments where decomposition rates are slow, the resupply of nutrients, such as nitrogen, from soil decomposition processes is reduced. Consequently, plants retain their leaves for more than one season to maintain adequate tissue nutrient concentrations, a trait referred to as evergreen. For evergreen plant species, lignin is embedded in the leaves through sclerification, reducing herbivory, and is particularly prevalent in low-nutrient-available ecosystems (Wright and Cannon, 2001).

Ecosystem feedback occurs in drier climates, where plants with woody tissues are supported, increasing evergreen dominance. As organic matter from these plants is added to the soil, the amount of lignin from the detritus inhibits rapid decomposition, thereby slowing the recycling of nutrients, such as nitrogen, back into the soil for plant uptake (Pei et al., 2019). It also promotes soil carbon accumulation. For example, in the South Texas Plains, the conversion of grasslands to thorny woodlands has increased soil carbon by 25 times, as lignin-embedded woody tissue from the shrubs dies and is added to the soil detritus (Hibbard et al., 2003). With nitrogen being key to producing chlorophyll and the key photosynthetic enzyme (i.e., Rubisco), the low availability of nitrogen reduces photosynthetic rates and growth. However, nitrogen cycling limited by lignin is overcome where nitrogen-fixing plant species are present, such as many of the thorny shrub species in the South Texas Plains (McCulley et al., 2004).

II.5.2.4 Reproduction

Plant reproduction is a key component of vegetation changes, although it is understudied, especially in the context of climate change across such a large spatial scale as an entire state. First, at a community level, different plant reproductive strategies will influence the composition of a plant community over time. Second, reproduction is costly for the plant's carbon budget, as inflorescences, fruits, and seeds often contain specialized carbohydrate and lipid compounds that utilize metabolic energy, carbon, and other plant nutrients during growth. Climate change is, therefore, likely to impact the reproductive success of plants due to temperature and precipitation limits that influence pollination, germination, dispersal, and seedling survival, as well as changes in fauna that are also affected by climate change stress.

Regarding species shifts in communities affected by climate change, plants are more likely to thrive under hotter, drier conditions than those with inherently broader physiological tolerances for germination. The germination of many plant seeds requires preconditioning in either cooler or warmer temperatures. Plants with cool temperature requirements may be eliminated from communities if the necessary duration of exposure to cold is not met. In contrast, plants that prefer warm temperature conditions will be more likely to germinate but may do so earlier than previously due to seasonal shifts in the onset of warm weather. For Texas, annual temperatures at the end of the century are projected to rise the most in the northern Panhandle region by approximately 5.0°C compared to the south, with slightly lower temperature increases of 4.0 to 4.5°C. This geographical difference may be more significant for plants in the north, which are more adapted to cooler temperatures and shorter growing seasons. This is likely to affect cool-season grasses more in that region, leading to shifts in the community toward warm-season grasses.

II.5.2.5 Disturbance

Changes in plant growth and survivorship are due to disturbances. Natural disturbances in Texas take various forms for terrestrial plants, including wind, ice damage, pathogens, and fire. Each of these results in widespread mortality and may increase the amount of considerable standing and downed plant material.

II.5.2.5.1 Wind

Wind gusts, or instantaneous pulses of wind most likely to affect standing vegetation, decrease from west to east, with gusts of 34 m/s in the far west to 26 m/s in central Texas (Letson et al., 2018). Due to hurricanes, wind gusts in southern coastal Texas may approach 70 m/s during intense storms, causing considerable and widespread wind damage to forests. In the Gulf Coast, hurricanes down trees, contributing to large amounts of fuel, and increasing forest fire occurrence in affected stands (Myers and Van Lear, 1998). For the most extreme hurricanes in this region, over 100 kg/ha of large fuels may be added following these events, dramatically increasing the intensity of any subsequent wildfire if not managed (Vihnanek et al., 2009).

II.5.2.5.2 Ice Storms

Ice storms are infrequent in Texas. However, the damage caused by the excess weight on leaves and branches is substantial, resulting in the total collapse of the canopy. Most Texas trees are poorly adapted to this, and as branches break, a significant portion of the canopy may be lost. For taller trees, such as loblolly pine, more than 70% of crown loss results in a 25% chance of tree death (Bragg et al. 2004), contributing to the accumulation of downed, dead woody material. In 1944, one of the most significant ice storms affected over a million hectares of eastern forested areas, primarily damaging loblolly pines (White, 1944). With continuing climate change, less freezing precipitation is likely (Nielsen-Gammon et al., 2020). However, the past century of warming affecting ocean circulation has been implicated in the conditions that gave rise to the extreme cold of February 2021

(Hsu et al. 2022). This storm produced extensive heavy snow and ice areas, affecting the entire state and sparking speculation about the potential for similar future events driven by continued climate change.

II.5.2.5.3 Disease

Pathogens also affect vegetation health by causing changes in tissue function as a disease syndrome. Large areas can be affected if the disease-causing pathogen is exceptionally virulent and has an effective transmission capability. For example, since the 1930s the fungal pathogen *Ceratocystis fagacearum* has infected various species of oak trees in Texas. This fungus is spread from an infected tree by a beetle attracted to the scent of fungal mats that proliferate in the sapwood under the tree bark. If beetles carrying spores acquired from the mat land on a neighboring tree with limb or trunk damage, the spores may be transferred to complete the fungal life cycle within the infected tree's vascular system (Appel and Lewis, 1985). This results in vascular system failure and a general loss of water transport from roots to leaves. In 2000, the area of trees infected with *C. fagacearum* was 58.1 km^2, primarily in the Hill Country region. Since then, oak wilt has spread to 30 counties across the state, expanding to the Panhandle Plains, Lake and Prairies, and Gulf Coast regions (https://texasoakwilt.org/).

In the Piney Woods of East Texas, infestations of southern pine beetles (e.g., *Ips avulsus, I. calligraphus, I. grandicollis, and Dendroctonus frontalis*) have been damaging trees by boring into tree stems to feed on cambial tissue, a process referred to as engraving. As cambium is destroyed, the tree's ability to grow from xylem and phloem production is compromised, often leading to tree death (Thatcher, 1960). Under moist growing conditions, trees can redirect carbohydrates derived from photosynthesis into resin, a complex compound that inhibits the excavation and feeding of beetles. During drought, site water limitations reduce photosynthetic gains by restricting stomatal opening and reducing plant carbohydrate production. As less resin is produced, more trees become infected, increasing tree mortality during dry periods. Extreme temperatures also affect loblolly pine beetle infestations in tandem with rates of beetle attacks and stand tree age (Friedenberg et al., 2008). The last factor is particularly significant, given that the Piney Woods are regrowing from large-scale timber felling, making the trees at their most susceptible to beetle attacks.

II.5.2.5.4 Wildfire

A common outcome of all previously described disturbances is the widespread mortality and production of dead biomass, which can be available to fuel future wildfires. For the past, estimated natural fire intervals for all non-desert Texas ecosystems have ranged from 3 to 7 years, increasing to 8 to 23 years for deserts (Stambaugh et al., 2014). Specific meteorological conditions in Texas increase wildfire potential through wind speed and aridity changes. Fires in Texas can occur every month in the state (Jurney et al., 2000) and are most prevalent during the La Niña phase of the El Niño-Southern Oscillation (ENSO) (Lindley et al., 2014). This is because the weather during La Niña years typically features more dry, low-pressure systems, resulting in increased southerly winds from northern Mexico, unstable air masses aloft, and an increased frequency of lightning. All these conditions are perfect for wildfire ignition and spread.

Drought accentuates this potential by decreasing the moisture content of live and dead fuels in an ecosystem. Long-term (>10 years) extreme temperature and precipitation events are highly correlated with the ENSO phase (Bhatia et al., 2019). For example, in 2011, during an extreme La Niña event, the Bastrop Complex Fire in the Lost Pines burned approximately 137 km^2 of the 303 km^2 of the remnant loblolly forest, highlighting its vulnerability to extreme weather events. Historically, loblolly pine forests have a 10-year fire return frequency of low-intensity surface burns from natural ignition sources (Oliver and Larson, 1996). However, the accumulation of fuels due to urbanization and subsequent fire exclusion near the Lost Pines may have contributed to the size and severity of this fire (Stafford, 2011). In the juniper-oak woodlands of the eastern Edwards Plateau in the Hill Country, near Austin, fire frequency was approximately eight years before 1965 (Murray and

White, 2013). After that, fire frequency increased to an average of 40 years, attributed to urbanization and the need for fire suppression. Suppression does not alter the potential for fire; instead, it simply delays its occurrence. Thus, the role of humans in changing Texas ecosystems beyond land-use conversion cannot be ignored.

In grass-dominated ecosystems in Texas, the vigor and productivity of the prior growing season influence wildfire potential by altering the mass and composition of grass and herbaceous cover. Invasive species in grasslands, such as guinea grass (*Urochloa maxima*) and Lehmann lovegrass (*Eragrostis lehmanniana*), have altered fire regimes in the South Texas Plains. Guinea grass is drought and shade-tolerant (Wied et al., 2020) and grows within thorn shrub thickets. The South Texas Plains, originally open grasslands, are now dominated by various thorny shrubs, including mesquite and acacia, with little fine fuel for carrying a surface fire. As a result, fire return frequencies have been estimated to be variable, ranging from 5 to 30 years in this region. These thorn shrubs are fire-intolerant, with thin bark that allows the burning of cambial tissue.

An extensive review of Texas's natural vegetation fire history has been produced by Stambaugh et al. (2014). Much of the study assesses the fire return intervals for different physiographic regions. For the entire state, historical estimates range from 1 to 12 years. This indicates that fire played a significant role in shaping plant communities and ecosystem processes. Past fires were most frequent in the east and least in the west. In the Prairies and Lakes, areas of intermixing between grass and woodlands increased fire intensity and spread as fires transitioned from fine fuel beds to woodlands. Along the Gulf Coast, grassed plains supported semiannual burning. Past fires were generally extensive in the Panhandle Plains due to the continuity of fine fuels, which allowed fires to spread rapidly. Recently, fires in the Panhandle Plains regions during the 2005–2006 drought cumulatively burned over 4,000 km^2 of grassland, pastures, and small woodlands at a rate of 1.5 m/s (Gray et al., 2007). Again in 2024, fires in the upper panhandle of Texas occurred during February and March, with a single fire, the Smokehouse Creek Fire, burning 4283 km^2 through grass and small shrublands. This fire is the largest in Texas's recorded history, fueled by warm, dry conditions and high wind speeds approaching 30 m/s (King et al., 2024). Given the future projections of increased temperatures, higher vapor pressure deficit (VPD), and seasonal shifts in precipitation, wildfires are likely to occur even more frequently than in the past.

II.5.3 TEXAS ECOSYSTEM RESPONSE TO FUTURE CLIMATE CHANGE

Rising temperatures will push Texas ecosystems closer to their thermal limits. This could lead to reduced biodiversity as species unable to adapt to heat stress decline or migrate. Warmer temperatures will also alter plant phenology (seasonal growth patterns) due to changes in rainfall predicted across Texas's physiographic regions (Figure II-5-4), affecting food availability for herbivores and pollinators. Hotter and drier conditions will lead to more frequent and intense wildfires, especially in grasslands, savannas, and wooded areas. While some ecosystems are already fire-adapted, the increased frequency could prevent the regrowth of current plant species, leading to long-term transformation.

All the factors and processes described provide a framework for projecting future climate-induced changes in Texas ecosystems through the feedback of interrelations (Figure II.5.5). This represents the temperature increases and drought of the worst-case scenario with projected ecosystem responses. Beginning with A_{max}, climate influences the productivity of current vegetation across all physiographic regions (Figure II-5-4) through warmer and extreme temperatures and drier precipitation with altered seasonality. With hotter climate conditions, membrane stability and enzyme denaturation may deteriorate, leading to lower A_{max} for all species. Reduced precipitation, aridity, and subsequent lower plant water availability will lower stomatal conductance as plants conserve tissue water. This will further reduce the CO_2 supply, reducing photosynthetic assimilation and biomass accumulation. With higher temperatures, VPD will also increase, favoring the survivorship of plants with lignin embedded in their organs. As lignin-laden dead biomass from

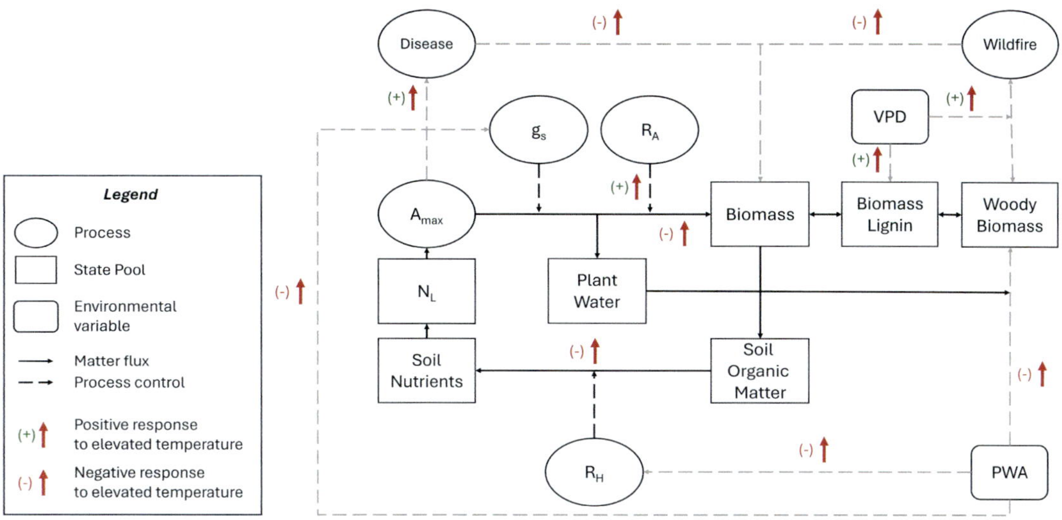

FIGURE II.5.5 Diagram of some interrelated components within an ecosystem. Red arrows indicate central effects of higher temperatures on ecosystem interactions. Positive [(+)] indicates an increase to higher temperatures and negative [(–)] indicates lower activity. (A_{max} = maximum assimilation; g_s = stomatal conductance; R_A = autotrophic respiration; R_H = heterotrophic respiration/decomposition; VPD = vapor pressure deficit; PWA = plant water availability; N_L = leaf nitrogen)

vegetation is added to drier soils, less nitrogen will be released, resulting in lower nutrient availability for plant uptake. This will also support the dominance of plants with an evergreen habit. With a further nutrient deficit, less nitrogen may be allocated to plant leaves, leading to another factor that reduces A_{max}.

With lower plant water availability from higher temperatures, there will be less metabolic support for maintaining trees. Coupled with lower photosynthetic gains, reduced carbohydrate reserves in existing woodlands and forests will result in mortality and considerable fuel accumulation. Higher temperatures and VPD will support intense fires, primarily in dead canopies, resulting in large-scale biomass consumption. Wildfires may result in a community's conversion from wooded to non-wooded areas, given that the reproductive capacity of trees is compromised, contingent upon PWA and further warming.

From west to east, smaller formed plants, reduction in survival of trees, and expansion of plants with high WUE, such as grasses and succulents, may occur. This is attributed to lower PWA than is expected in the West. Seasonal rainfall will also affect plant survivorship. For example, higher summer precipitation is predicted for the Big Bend Country and western Hill Country regions, favoring the growth of rhizomatous grasses. Pathogenic mortality, especially from fungal pathogens, may be enhanced early on for diseases, as reducing defense compounds limits the plant's carbon budget. However, later, hot, dry temperatures would be less favorable for microbial pathogens' growth. However, successful infections from insect pests, such as bark beetles, may increase as climate stress compromises the production of defense compounds in susceptible trees.

With mortality, a pulse of downed and dead woody biomass may occur, leading to large, intense fires with a higher potential for canopy fires, which can eliminate woodlands and forests from their current locations. A combination of PWA limitations to sustaining trees and a higher frequency of intense fires and disease outbreaks may shift plant dominance to shrubs and grass/herbaceous vegetation. Over time, the prominence of finer fuels from these smaller plant types will promote frequent surface fires, further reducing the establishment of fire-intolerant tree species.

II.5.3.1 Pineywoods

Changes in productivity are expected in the Pineywoods due to the combined effects of warmer temperatures, higher vapor pressure deficit (VPD), and seasonal drought. Texas loblolly forest productivity is estimated to fall by 50% over the next century (McNulty et al., 1996). The dark canopies of *P. taeda* stands are more likely to absorb and store heat due to their lower albedos (Betts and Ball, 1997), which leads to plant cell damage, lower photosynthetic capacity, and increased nighttime transpiration water loss. This reduces PWA, which will already be lower due to this region's predicted reduced annual precipitation for 2100. In 2011, the Texas drought led to substantial tree mortality in the Pineywoods and Gulf Coast regions, primarily due to heat and water stress (Klockow et al., 2018). With heat and drought becoming the norm for this region in the future, forest canopies will become more open, characterized by less canopy cover and lower tree density.

Warmer climates also expand pests' range and reproductive cycles, such as the southern pine beetle (*D. frontalis*), leading to increased infestations and tree mortality (Lesk et al., 2017). Reduced growth and increased pest outbreaks may lead to further forest fragmentation, loss of canopy cover, and decreased carbon storage in the soil. As more woody fuels accumulate due to forest dieback, higher temperatures, and prolonged droughts, the likelihood of wildfires will increase. While some pine species, like longleaf pine, are adapted to periodic fires, the intensity and frequency of wildfires may exceed their resilience thresholds. Changes in temperature and precipitation patterns may render current habitats unsuitable for some pine species.

II.5.3.2 Prairies and Lakes

In this region, short warm-season grasses (e.g., *S. scoparium* and *B. saccharoides*) are the most heat and drought-tolerant and, therefore, most likely to persist in this region's warmer, drier climate. Tallgrass species (e.g., *A. gerardii*) will be extirpated, as higher leaf area absorbing radiation and warmer leaf temperatures will likely exceed their photosynthetic limits. Invasive grasses such as Bermuda grass (*C. dactylon*) and Johnsongrass (*Sorghum halepense*) may fill the niche left by the loss of the tallgrass species.

Due to less precipitation during certain seasons, which can lead to prolonged droughts, species with deep roots will be more successful in withstanding environmental stress. Clay soils under prolonged drying will crack and shrink more often, reducing soil water availability as precipitation infiltration is directed to deeper groundwater from bypass flow conduits formed by the cracked soils. Trees in woodlands, such as blackjack oak (*Q. marilandica*) and post oak (*Q. stellata*), will have reduced growth due to lower PWA and are likely to be replaced by juniper (*J. ashei*) with its lower stomatal conductance and embolism-resistant xylem. With the loss of native fire-adapted prairie species, fire effects may reduce grass cover and lead to encroachment by woody species, including juniper and mesquite (*N. glandulosa*).

II.5.3.3 Gulf Coast

This region is expected to experience significantly reduced precipitation by 2100, except for autumn, which is attributed to higher hurricane intensity (Table II.5.1). This drought stress will impact less drought-tolerant prairie species such as Gulfdune paspalum (*Paspalum monostachyum*), sedges (e.g., *Carex* spp., *Rhyncospora* spp., *Fimbistylis* spp.), and Spotted Beebalm (*Monarda fruticulosa*) (Diamond and Fulbright, 1990). With changes in grass species, lower productivity is expected, given the heat and precipitation predictions. With most precipitation occurring in the autumn, the growing season will be truncated. Heavy rains and reduced grass cover can lead to soil erosion. Increases in grass invasive coverage, such as from Kelberg bluestem (*Dichanthium annulatum*), are likely to occur due to these species having higher adaptation to hot, dry climates. Like other Texas grasslands, woody plant encroachment is expected due to the invasion of mesquite and huisache (*V. farnesiana*) into these ecosystems.

II.5.3.4 Panhandle Plains

The Panhandle Plains are projected to have the highest change in mean temperature compared to all other areas, approximately 5°C. This heat level will reduce plant diversity in the small remnant grass and shrub vegetation. As much of the land is used for agriculture, these changes could affect grazing animals and disrupt production, as pasture forage is replaced by drought- and heat-tolerant species, such as Mexican feathergrass (*Nassella tenuissima*) (García-Cancel, 2023). With warmer temperatures, there will also be fewer freezes. This will allow more plants, both native and non-native, from warmer climates.

With higher temperatures, this region is projected to have more precipitation each season, with the highest amount for the summer season (10mm/season; Table II.5.1), likely associated with increased North American monsoon activity. This increased rainfall will favor the warm-season shortgrass species present there. With higher precipitation, plant water availability is also predicted to grow to more than 700 mm, well above the 350 mm threshold required to support trees. While afforestation is not expected, this does support the further expansion of woody shrubs such as mesquite (*N. glandulosa*) and redberry juniper (*J. pinchottii*). Adding these plants in grasslands, characterized by vegetation with darker leaves and canopies, can have a feedback effect on the climate, increasing precipitation and temperature in this area, associated with changes in surface-atmosphere interactions (Ge and Zou, 2013).

II.5.3.5 Hill Country

The most significant future threat to the Hill Country region is the potential loss of woodlands, with a projected reduction of PWA that falls below a threshold level that supports woody vegetation (Table II.5.2). This low future PWA value is attributed to thin soils and increased evaporation under the future climate projection. In a recent Elkington et al. (2014) study, dominant tree species of the Edwards Plateau water budgets are restricted by soil depth. Woodlands are predicted to be replaced by existing grasslands, which serve as the landscape matrix. As a preview of this conversion, during the prolonged drought of the 1930s to 1950s in Texas, 10% of woodlands were replaced by grasslands, primarily attributed to fire (Murray et al., 2013). Analysis of fire weather factors for this area supports the conclusion that fire occurrence is more attributed to drought than fuel loading. Additionally, fires associated with extreme droughts can lead to rapid community change in this region (Murray and White, 2013). A future climate with increased drought and aridity will likely eliminate nighttime humidity recovery, leading to fire spread between successive days and more extensive areas being burned in single fires.

In addition to abiotic factors, deciduous oaks and remnant maples in deeper canyons of the Hill Country are not regenerating (Van Auken et al., 2022). This is attributed to increased deer browsing. This is similar to the lack of recruitment for upland oaks, such as *Q. buckleyi*, as Russell and Fowler (2004) postulated occurred earlier in the 20th century, following the eradication of screwworm (*Cochliomyia hominivorax*). The role of wildlife in future land cover conversion is less understood.

Loss of woodlands from fire can result in rapid soil erosion, transporting nutrients such as nitrogen off-site, particularly on steeper slopes (Wright et al., 1976). Open areas within woodland patches and grasslands, typically found on limestone plateaus, are likely remnants of soil loss associated with past droughts and overgrazing (Van Auken, 2000). Will little soil development and further vegetation decline occur around these patches, with these areas acting as a nucleus for the exposure of outcrops? Small, warm-season grasses will fill in vegetation gaps, including little bluestem (*Schizachyrium scoparium*), sideoats grama (*B. curtipendula*), and buffalograss (*B. dactyloides*). Cool-season grasses, such as Texas wintergrass (*S. leucotricha*) and tallgrass species (e.g., *Bouteloua. pectinata*), are likely to be extirpated across the region due to the predicted harsher climate and soil deterioration. Like other Texas grasslands, invasive species will expand. The ubiquitous King Ranch bluestem (*B. ischaemum*) is already a dominant member of grass communities in the state's eastern half, with other species, such as rescuegrass (*Bromus catharticus*), poised to expand their range.

11.5.3.6 South Texas Plains

The southeastern part of Texas is projected to experience hotter and drier conditions than all other regions. This will further stress the current vegetation, which already inhabits a harsh climate adjacent to the Chihuahuan Desert to the west. However, not all vegetation is expected to respond similarly to this change. Adhikari and White (2014) determined from their study of five common woody thorn shrubs from the South Texas Plains, including sweet acacia (*V. farnesiana*), desert hackberry (*C. ehrenbergiana*), narrowleaf forestiera (*F. angustifolia*), palo verde (*P. aculeata*), and honey mesquite (*N. glandulosa*), that only mesquite and palo verde could withstand extreme water stress. In a climate change modeling study for this region, the productivity of shrublands along the Rio Grande was reduced by only 10% by 2050 under the most severe scenario (Adhikari and White, 2016). The most significant impact of climate change identified in this study was in the region closest to the coast, where a 25% decline in shrub production was observed. This area has the highest production today due to slightly more current rainfall. With decreased production, this area may facilitate the growth of more invasive grasses, such as guinea grass (*U. maxima*) and Lehmann's lovegrass (*E. lehmanniana*), potentially increasing fire frequency. In the West, the South Plains, with further heating and drying, will be climatically subsumed into the Chihuahuan Desert. This has been accelerated by human factors, including the diversion of river water in the upper catchment of the Rio Grande, land clearing for agricultural purposes, and soil salination resulting from groundwater irrigation (Rappole et al., 1986). Additionally, increased minimum temperatures will enable freeze-susceptible succulents to expand their ranges northward and eastward.

11.5.3.7 Big Bend Country

The Chihuahuan Desert is projected to experience significant temperature increases by the end of the 21st century, as climate change is expected to become hotter and drier, further stressing water resources. Like other western regions, shrubland encroachment in this region will occur with a reduction in desert grasslands (White et al., 2008). However, like the Panhandle Plains, more rainfall from the summer monsoon could have the opposite impact. Peters et al. (2012) demonstrated that increased annual precipitation (greater than 6 cm) would lead to greater grass cover in the northern Chihuahuan Desert. Pulse precipitation, as predicted for this area during summer, is favorable for the growth of desert grasses, such as black grama (*B. eriopoda*), with increases in soil respiration and nutrient cycling (Thomey et al., 2011). However, VPD, not precipitation, may limit grass growth as an environmental variable that increases with temperature (Kidron and Gutschick, 2017). The encroachment of shrubs and grass replacement will limit fire occurrence, as the fine fuels necessary for surface burning are eliminated.

It is uncertain how succulents such as cacti and agave will respond. Lechuguilla (*A. lechuguilla*) is a characteristic species of the Chihuahuan Desert and is tolerant of extreme heat and arid conditions. Its thermal tolerance reflects its adaptation to some of the harshest environments in North America (Nobel and Quero, 1986). Lechuguilla thrives in rocky, well-drained soils of arid and semi-arid regions, often on slopes or limestone outcrops that help stabilize the soil. Its presence usually indicates well-drained, nutrient-poor soils where few other plants can survive. It is also crucial for providing pollinators with nectar, such as bats and moths (Silva-Montellano and Eguiarte, 2003). For prickly pear (Opuntia spp.), the survivorship of plants is halved with temperatures > 60°C (Didden-Zopfy and Nobel, 1982).

Prolonged droughts, more intense heat, and reduced snowpack could make the mountain regions in the Big Bend Country less hospitable for forest species over time, reducing their survival ability. Warming temperatures and changing precipitation patterns may further stress stands of Douglas firs (*Pseudotsuga menziesii*) and aspens (*Populus tremuloides*) as ongoing climate change increases vapor pressure deficit (VPD), particularly at higher elevations, where precipitation additions are offset, affecting plant water availability (PWA). Changes in montane soil nutrient cycling are also expected, as soil bacterial and fungal populations are affected differentially due to reduced nitrogen availability in the future (Bell et al., 2008).

11.5.4 CONCLUDING REMARKS

The future of Texas ecosystems depends on global climate policies and local conservation efforts. While ecosystems will face significant challenges, initiative-taking measures to protect biodiversity, manage resources sustainably, and adapt to a changing climate can help mitigate some of the most severe impacts. Heavy rainfall may erode soil, reduce habitat quality, and disrupt ecosystems, particularly in East Texas forests and coastal wetlands. Coastal ecosystems, including salt marshes, mangroves, and estuaries, along the Texas coast are threatened by rising sea levels. Saltwater intrusion will alter soil composition, potentially harming freshwater plant species and disrupting fish, birds, and other wildlife habitats. Warmer climates can favor the spread of invasive species and pests, which may outcompete native flora and fauna. Many species may be unable to adapt quickly to the rapid pace of the current climate change. With its diverse ecosystems, Texas risks losing species in sensitive environments, including the alpine regions of the Big Bend Country area, freshwater springs in the Hill Country, and Gulf Coast wetlands.

While human activities contribute to these challenges, mitigation and adaptation strategies can help reduce their impact. These include designating migration corridors already blocked by human infrastructure to allow species to relocate to more suitable habitats, thereby reducing migratory limits. Additionally, it is essential to recognize that water is a limited resource, both now and in the future, and its conservation is crucial for both ecosystems and humans. Lastly, in a future climate, fires will be more prevalent, posing a challenge to human safety at urban and wildland interfaces. As such, prescribed burns and firebreaks will be needed to reduce the severity and extent of future wildfires.

REFERENCES

Acosta-Martinez V, Moore-Kucera J, Cotton J, Gardner T, Wester D. 2014. Soil enzyme activities during the 2011 Texas record drought/heat wave and implications to biogeochemical cycling and organic matter dynamics. Applied Soil Ecology. 75:43–51.

Adhikari A, White JD. 2014. Plant water use characteristics of five dominant shrub species of the Lower Rio Grande Valley, Texas, USA: Shrubland restoration and conservation implications. Conservation Physiology. 2(1):cou005.

Adhikari A, White JD. 2016. Climate change impacts on regenerating shrubland productivity. Ecological Modelling. 337:211–220.

Anderson VJ, Briske DD. 1990. Stomatal distribution, density, and conductance of three perennial grasses native to the southern true prairie of Texas. American Midland Naturalist. 1:152–159.

Appel DN, Lewis RJ Jr. 1985. Prospects for oak wilt control in Texas. In: Insects and Diseases of Southern Forests. Proc. 34th Annu. For. Symp., La. Agric. Exp. Stn., Louisiana State Univ., Baton Rouge (pp. 60–68).

Balok CA, Hilaire RS. 2002. Drought responses among seven southwestern landscape tree taxa. Journal of the American Society for Horticultural Science. 127(2):211–218.

Bell C, McIntyre N, Cox S, Tissue D, Zak J. 2008. Soil microbial responses to temporal variations of moisture and temperature in a Chihuahuan Desert grassland. Microbial Ecology. 56:153–167.

Bendevis MA, Owens MK, Heilman JL, McInnes KJ. 2010. Carbon exchange and water loss from two evergreen trees in a semiarid woodland. Ecohydrology: Ecosystems, Land and Water Process Interactions, Ecohydrogeomorphology. 3(1):107–115.

Berry J, Björkman O. 1980. Photosynthetic response and adaptation to temperature in higher plants. Annual Review of Plant Physiology. 31(1):491–543.

Betts AK, Ball JH. 1997. Albedo over the boreal forest. Journal of Geophysical Research: Atmospheres. 102(D24):28901–28909.

Bhatia N, Singh VP, Lee K. 2019. Variability of extreme precipitation over Texas and its relation with climatic cycles. Theoretical and Applied Climatology. 138(1):449–467.

Bragg DC, Shelton MG, Heitzman E. 2004. Relative impacts of ice storms on loblolly pine plantations in central Arkansas. Gen. Tech. Rep. SRS–71. Asheville, NC: US Department of Agriculture, Forest Service, Southern Research Station. pp. 132-137.

Bryant VM Jr, Holloway RG. 1985. A late-quaternary paleoenvironmental record of Texas: An overview of the pollen evidence. In: Bryant, Vaughn M., Holloway, Richard G., eds. Pollen Records of Late-Quaternary North American Sediments (pp. 39–70). American Association of Stratigraphic Palynologists Foundation, Dallas, TX.. 436 p.

Choinski JS Jr, Wise RR. 1999. Leaf growth development about gas exchange in *Quercus marilandica* Muenchh. Journal of Plant Physiology. 154(3):302–309.

Cooper CE, Vogel JG, Muir JP, Moore GW. 2018. Leaf functional trait responses to changes in water status differ among three oak (*Quercus*) species. Plant Ecology. 219:1463–1479.

Dawson TE, Burgess SS, Tu KP, Oliveira RS, Santiago LS, Fisher JB, Simonin KA, Ambrose AR. 2007. Nighttime transpiration in woody plants from contrasting ecosystems. Tree Physiology. 27(4):561–575.

Diamond DD, Fulbright TE. 1990. Contemporary plant communities of upland grasslands of the Coastal Sand Plain, Texas. The Southwestern Naturalist. 385–392.

Didden-Zopfy B, Nobel PS. 1982. High temperature tolerance and heat acclimation of *Opuntia bigelovii*. Oecologia. 52:176–180.

Dirks I, Navon Y, Kanas D, Dumbur R, Gruenzweig JM. 2010. Atmospheric water vapor as driver of litter decomposition in Mediterranean shrubland and grassland during rainless seasons. Global Change Biology. 16(10):2799–2812.

Elkington RJ, Rebel KT, Heilman JL, Litvak ME, Dekker SC, Moore GW. 2014. Species-specific water use by woody plants on the Edwards Plateau, Texas. Ecohydrology. 7(2):278–90.

Friedenberg NA, Sarkar S, Kouchoukos N, Billings RF, Ayres MP. 2008 Temperature extremes, density dependence, and southern pine beetle (Coleoptera: *Curculionidae*) population dynamics in East Texas. Environmental Entomology. 37(3):650–659. https://doi.org/10.1093/ee/37.3.650

Galmes J, Kapralov MV, Copolovici LO, Hermida-Carrera C, Niinemets Ü. 2015. Temperature responses of the Rubisco maximum carboxylase activity across domains of life: Phylogenetic signals, trade-offs, and importance for carbon gain. Photosynthesis Research. 123:183–201.

García-Cancel JG. 2023. Modeling future invasion of three non-native grasses and their effects on native species in semiarid lands (Doctoral dissertation, Texas Tech University).

Ge J, Zou C. 2013. Impacts of woody plant encroachment on regional climate in the Southern Great Plains of the United States. Journal of Geophysical Research: Atmospheres. 118(16):9093–9104.

Gehlbach FR. 1993. Mountain Islands and Desert Seas: A Natural History of the U.S.-Mexican Borderlands (326 pp). Texas A&M University Press, College Station, TX.

Graham EA, Nobel PS. 1999. Root water uptake, leaf water storage and gas exchange of a desert cacti: Implication for root system redundancy. Annals of Botany. 84:213–223.

Gray R, Dunivan M, Jones J, Ridenour K, Leathers M, Stafford K. 2007. Cross Plains, Texas Wildland Fire Case Study. Texas Forest Service, Lufkin.

Hall SA, Valastro S Jr. 1995. Grassland vegetation in the southern great plains during the last glacial maximum. Quaternary Research. 44(2):237–245.

Hibbard KA, Schimel DS, Archer S, Ojima DS, Parton W. 2003. Grassland to woodland transitions: Integrating changes in landscape structure and biogeochemistry. Ecological Applications. 13(4):911–926.

Hsu PC, Hsu HH, Hong HJ, Chen YT, Chen YL, Tseng WL. 2022. Texas cold snap: Manifestation of natural variability and a recent warming trend. Weather and Climate Extremes. 37:100476.

Jurney DH, Ippolito J, Bergstrom V. 2000. The frequency of fire in east Texas forests. Journal of Northeast Texas Archaeology. 13:40–49.

Kidron GJ, Gutschick VP. 2017. Temperature rise may explain grass depletion in the Chihuahuan desert. Ecohydrology. 10(4):e1849.

King K, Burrows D, Hunter T, Abraham J, Henderson J. 2024. Investigative Committee Report on the Panhandle Wildfires Presented to the Eighty-eighth Texas Legislature. 48 p. (https://www.house.texas.gov/pdfs/committees/reports/interim/88interim/House-Interim-Committee-on-The-Panhandle-Wildfires-Report.pdf)

Klockow PA, Vogel JG, Edgar CB, Moore GW. 2018. Lagged mortality among tree species four years after an exceptional drought in east Texas. Ecosphere. 9(10):e02455.

Kolb TE, Fettig CJ, Bentz BJ, Stewart JE, Weed AS, Hicke JA, Ayres MP. 2016. Forest insect and fungal pathogen responses to drought [Chapter 6]. In: Vose, James M.; Clark, James S., Luce, Charles H., Patel-Weynard, Toral, eds. Effects of Drought on Forests and Rangelands in the United States: A Comprehensive Science Synthesis (pp. 113–133, Vol. 93). Gen. Tech. Rep. WO-93b. Washington, DC: US Department of Agriculture, Forest Service, Washington Office.

Landers RQ. 1987. Native vegetation of Texas. Rangelands Archives. 9(5):203–207.

Lesk C, Coffel E, D'Amato AW, Dodds K, Horton R. 2017. Threats to North American forests from southern pine beetle with warming winters. Nature Climate Change. 7(10):713–717.

Letson F, Pryor SC, Barthelmie RJ, Hu W. 2018. Observed gust wind speeds in the coterminous United States, and their relationship to local and regional drivers. Journal of Wind Engineering and Industrial Aerodynamics. 173:199–209.

Li W, Li D, El-Askary H, Fisher JB, Fang ZN. 2024. Evaluating future water availability in Texas through the lens of a data-driven approach leveraged with CMIP6 general circulation models. Science of The Total Environment. 921:171136.

Lindley TT, Murdoch GP, Guyer JL, Skwira GD, Schneider KJ, Nagle SR, Van Speybroeck KM, Smith BR, Beierle MJ. 2014. Southern Great Plains Wildfire Outbreaks. E-Journal of Severe Storms Meteorology. 9(2):1–43.

McCulley RL, Archer SR, Boutton TW, Hons FM, Zuberer DA. 2004. Soil respiration and nutrient cycling in wooded communities developing in grassland. Ecology. 85(10):2804–2817.

McDowell NG. 2011. Mechanisms linking drought, hydraulics, carbon metabolism, and vegetation mortality. Plant Physiology. 155:1051–1059. doi: 10.1104/pp.110.170704

McNulty SG, Vose JM, Swank WT. 1996. Potential climate change effects on loblolly pine forest productivity and drainage across the southern United States. Ambio. 449–453.

Meentemeyer V. 1978. Macroclimate and lignin control of litter decomposition rates. Ecology. 59(3):465–472.

Miller SR, Corby MK. 2022. Fire in pines of the Southeastern US. In: John R. Weir and J. Derek Scasta, eds. Global Application of Prescribed Fire (304 p). CSIRO Publishing., Clayton, Victoria, Australia.

Murray DB, White JD. 2013. Loss of neighbors, fire, and climate effects on Texas red oak growth in a juniper-dominated woodland ecosystem. American Midland Naturalist. 170:348–369.

Murray DB, White JD, Swint P. 2013. Woody vegetation persistence and disturbance in central Texas grasslands inferred from multidecadal historical aerial photographs. Rangeland Ecology & Management. 66(3):297–304.

Murthy R, Zarnoch SJ, Dougherty PM. 1997. Seasonal trends of light-saturated net photosynthesis and stomatal conductance of loblolly pine trees grown in contrasting environments of nutrition, water, and carbon dioxide. Plant, Cell & Environment. 20(5):558–568.

Myers RK, Van Lear DH. 1998. Hurricane-fire interactions in coastal forests of the South: A review and hypothesis. Forest Ecology and Management. 103(2–3):265–276.

Nielsen-Gammon, J. 2012. The 2011 Texas drought. Texas Water Journal. 3(1):59–95. DOI: 10.21423/twj.v3i1.6463

Nielsen-Gammon J, Escobedo J, Ott C, Dedrick J, Van Fleet A. 2020. Assessment of historic and future trends of extreme weather in Texas, 1900–2036. Officer of the Texas State Climatologist, Texas A&M University. Document OSC-202001.

Nobel PS, Quero E. 1986. Environmental productivity indices for a Chihuahuan desert CAM plant, *Agave lechuguilla*. Ecology. 67(1):1–1.

Northup AP, Keitt TH, Farrior CE. 2021. Cavitation-resistant junipers cease transpiration earlier than cavitation-vulnerable oaks under summer dry conditions. Ecohydrology. 15(5):e2337.

Oliver CD, Larson BC 1996. Forest Stand Dynamics. John Wiley and Sons, New York.

Pataki DE, Huxman TE, Jordan DN, Zitzer SF, Coleman JS, Smith SD, Nowak RS, Seeman JR. 2000. Water use of two Mojave Desert shrubs under elevated CO_2. Global Change Biology. 6:889–897.

Pei G, Liu J, Peng BO, Gao D, Wang C, Dai W, Jiang P, Bai E. 2019. Nitrogen, lignin, C/N as important regulators of gross nitrogen release and immobilization during litter decomposition in a temperate forest ecosystem. Forest Ecology and Management. 440:61–69.

Peters DP, Yao J, Sala OE, Anderson JP. 2012. Directional climate change and potential reversal of desertification in arid and semiarid ecosystems. Global Change Biology. 18(1):151–163.

Quinn PJ, Williams WP. 1985. Environmentally induced changes in chloroplast membranes and their effects on photosynthetic function. Photosynthetic Mechanisms and the Environment. 6:1–47.

Rappole JH, Russell CE, Norwine JR, Fulbright TE. 1986. Anthropogenic pressures and impacts on marginal, neotropical, semiarid ecosystems: the case of south Texas. Science of the Total Environment. 55:91–99.
Read JJ, Morgan JA, Chatterson NJ, Harrison PA. 1997. Gas exchange and carbohydrate and nitrogen concentrations in leaves of *Pascopyrun smithii* (C3) and *Bouteloua gracilis* (C4) at different carbon dioxide concentrations and temperatures. Annals of Botany. 79:197–206.
Reeves CC Jr. 1973. The full-glacial climate of the southern High Plains, west Texas. The Journal of Geology. 81(6):693–704.
Reich PB. 2012. Key canopy traits drive forest productivity. Proceedings of the Royal Society B: Biological Sciences. 279(1736):2128–2134.
Rowland L, Oliveira RS, Bittencourt PR, Giles AL, Coughlin I, Costa PD, Domingues T, Ferreira LV, Vasconcelos SS, Junior JA, Oliveira AA. 2021. Plant traits controlling growth change in response to a drier climate. New Phytologist. 229(3):1363–1374.
Runyon J, Waring RH, Goward SN, Welles JM. 1994. Environmental limits on net primary production and light-use efficiency across the Oregon transect. Ecological. Applications. 4:226–237.
Russell FL, Fowler NL. 2004. Effects of white-tailed deer on the population dynamics of acorns, seedlings, and small saplings of *Quercus buckleyi*. Plant Ecology. 173:59–72.
Sage RF, Way DA, Rubien DS. 2008. Rubisco, Rubisco activase, and global climate change. Journal of Experimental Botany. 59(7):1581–1595. https://doi.org/10.1093/jxb/ern053
Schenk HJ, Jackson RB. 2002. Rooting depths, lateral root spreads and below-ground/above-ground allometries of plants in water-limited ecosystems. Journal of Ecology. 480–494.
Silva-Montellano A, Eguiarte LE. 2003. Geographic patterns in the reproductive ecology of *Agave lechuguilla* (Agavaceae) in the Chihuahuan desert. I. Floral characteristics, visitors, and fecundity. American Journal of Botany. 90(3):377–387.
Stafford KL. 2011. Effects of fuel loading on fire behavior and intensity in the wildland urban interface in Bastrop County, Texas (Master's thesis, Stephen F. Austin State University).
Stambaugh MC, Sparks JC, Abadir ER. 2014. Historical pyrogeography of Texas, USA. Fire Ecology. 10:72–89.
Stoy PC, El-Madany TS, Fisher JB, Gentine P, Gerken T, Good SP, Klosterhalfen A, Liu S, Miralles DG, Perez-Priego O, Rigden AJ. 2019. Reviews and syntheses: Turning the challenges of partitioning ecosystem evaporation and transpiration into opportunities. Biogeosciences. 16(19):3747–3775.
Thatcher RC. 1960, Bark beetles affecting southern pines: A review of current knowledge, U.S.D.A. Forest Service. Occas. Pap. 180. Southwestern Forestry Experiment Station.
Thomey ML, Collins SL, Vargas R, Johnson JE, Brown RF, Natvig DO, Friggens MT. 2011. Effect of precipitation variability on net primary production and soil respiration in a Chihuahuan Desert grassland. Global Change Biology. 17(4):1505–1515.
Thrasher B, Wang W, Michaelis A, Melton F, Lee T, Nemani R. 2022. NASA global daily downscaled projections, CMIP6. Scientific Data. 9:262. https://doi.org/10.1038/s41597-022-01393-4
Toomey RS, Blum MD, Valastro S. 1993. Late Quaternary climates and environments of the Edwards Plateau, Texas. Global and Planetary Change. 7(4):299–320. https://doi.org/10.1016/0921-8181(93)90003-7.
Torquato Rettondini P. 2019. Water relation and photosynthetic performance of eastern redcedar (Juniperus virginiana) and post oak (*Quercus stellata*) in the Cross Timbers Forest. MS Thesis Oklahoma State University.
Urban J, Ingwers M, McGuire MA, Teskey RO. 2017. Stomatal conductance increases with rising temperature. Plant Signaling & Behavior. 12(8):e1356534.
Van Auken OW. 2000. Characteristics of intercanopy bare patches in Juniperus Woodlands of the southern Edwards Plateau, Texas. The Southwestern Naturalist. 45(2):95–110.
Van Auken OW. 2018. Ecology of Plant Communities of South-Central Texas. Scientific Research Publishing, Inc. USA.
Van Auken OW, Taylor DL, Bush JK, Singhurst JR. 2022. Isolated deciduous woodland in central Texas. Phytologia. 104:13–23.
Vihnanek RE, Balog CS, Wright CS, Ottmar RD, Kelly JW. 2009. Stereo photo series for quantifying natural fuels. Volume XII: Post-hurricane fuels in forests of the Southeast United States. Gen. Tech. Rep. PNW-GTR-803. Portland, OR: US Department of Agriculture, Forest Service, Pacific Northwest Research Station. 53 p.
White JD, Gutzwiller KJ, Barrow WC, Randall LJ, Swint P. 2008. Modeling mechanisms of vegetation change due to fire in a semi-arid ecosystem. Ecological Modelling. 214(2–4):181–200.
White WE. 1944. Texas ice storm. American Forests. 50(1):108–109.

Wied JP, Perotto-Baldivieso HL, Conkey AA, Brennan LA, Mata JM. 2020. Invasive grasses in South Texas rangelands: Historical perspectives and future directions. Invasive Plant Science and Management. 13(2):41–58.

Wolter K, Dole RM, Smith CA. 1999. Short-term climate extremes over the continental United States and ENSO. Part I: Seasonal temperatures. Journal of Climate. 12(11):3255–3272.

Wright HA, Churchill FM, Stevens WC. 1976. Effect of prescribed burning on sediment, water yield, and water quality from dozed juniper lands in central Texas. Rangeland Ecology & Management/Journal of Range Management Archives. 29(4):294–298.

Wright IJ, Cannon K. 2001. Relationships between leaf lifespan and structural defences in a low-nutrient, sclerophyll flora. Functional ecology. 15(3):351–359.

Yang J, Will R, Zhai L, Zou C. 2024. Future climate change shifts the ranges of major encroaching woody plant species in the Southern Great Plains. USA. Earth's Future. 12(7):e2024EF004520.

Section III

Impacts of Climate Change on Human-Environmental Systems

III.1 Impacts of a Warming and Drying Climate on Agriculture and Food Production

Kritika Kothari, Bhupinder Singh, Sayantan Samanta, and Srinivasulu Ale

III.1.1 OVERVIEW OF TEXAS AGRICULTURE

Agriculture is an important part of Texas's economy. According to the 2022 census of agriculture, Texas has 230,662 farms, which account for 12% of the total farms in the United States (U.S.) (USDA, 2022). The total farmland in Texas spans 50,776,991 hectares, out of which 73% is pastureland, 22% is cropland, and 5% is woodland (USDA, 2022). Out of the total cropland, nearly 13% is irrigated. Texan farmers sold $32.2 billion worth of agricultural products in 2022, out of which the share of livestock, poultry, and associated products was 77%, and the share of crops was 23%. Specifically, cattle, calves, broilers, dairy products – particularly milk, cotton, and corn – are the top agricultural commodities. Researchers at Texas A&M AgriLife Research estimated that the food and fiber systems accounted for nearly 9.1% of Texas's gross domestic product (GDP) in 2021 (TAMU, 2023). In 2022, Texas's GDP was $2.4 trillion, which ranked second among the states in the U.S., and contributed 9.3% to the U.S. GDP (Guci and Wetzler, 2023). The food and fiber systems encompass not only the livestock and crop production, but also other activities linked to agriculture, such as machinery, fertilizer, food processing, wholesale and retail sales, transportation, warehousing, and food services. In this chapter, principal and minor crops in Texas are covered. The spatial distribution of crops across Texas (Figure III.1.1) and the long-term trends in their area and demand are summarized in the following sections.

III.1.1.1 COTTON

Texas is the leading producer of cotton (*Gossypium hirsutum* L.) in the U.S., contributing approximately 40% of the total cotton production. In 2024, cotton was harvested from 1.4 million ha, yielding 893 million kg of cotton lint (USDA-NASS, 2024). The top cotton-producing counties in the state are Lubbock, Nueces, Floyd, Hockley, San Patricio, Hale, and Crosby. Most of the Texas cotton is grown in the Southern High Plains and Northern High Plains regions, which account for 51% and 16% of the state's cotton production, respectively. About 65% of Texas cotton is grown under dryland conditions and 35% is either fully or partially irrigated. The drought-tolerant nature of cotton makes it a significantly less water intensive crop, requiring 305–610 mm of water, compared to corn, which requires 610–762 mm. However, supplemental irrigation during critical growth stages, such as the peak bloom, may significantly boost cotton yield (Himanshu et al., 2023).

Cotton acreage in Texas has mostly remained stable since the late 19th century (Figure III.1.2a) except in the Northen High Plains, where the planted acres have significantly increased in area between 2013 and 2019 (Ledbetter, 2020). The rapid decline of the saturated thickness of the Ogallala Aquifer resulted in inadequate irrigation water supply for corn production in the Northern High Plains. This caused producers in the area to ramp up cotton and sorghum production, which are drought-tolerant crops to meet their (corn-based) economic expectations while conserving water. Cotton contributes $2.4 billion to the state's GDP, generates more than 40,000 jobs, and

DOI: 10.1201/9781003406600-14

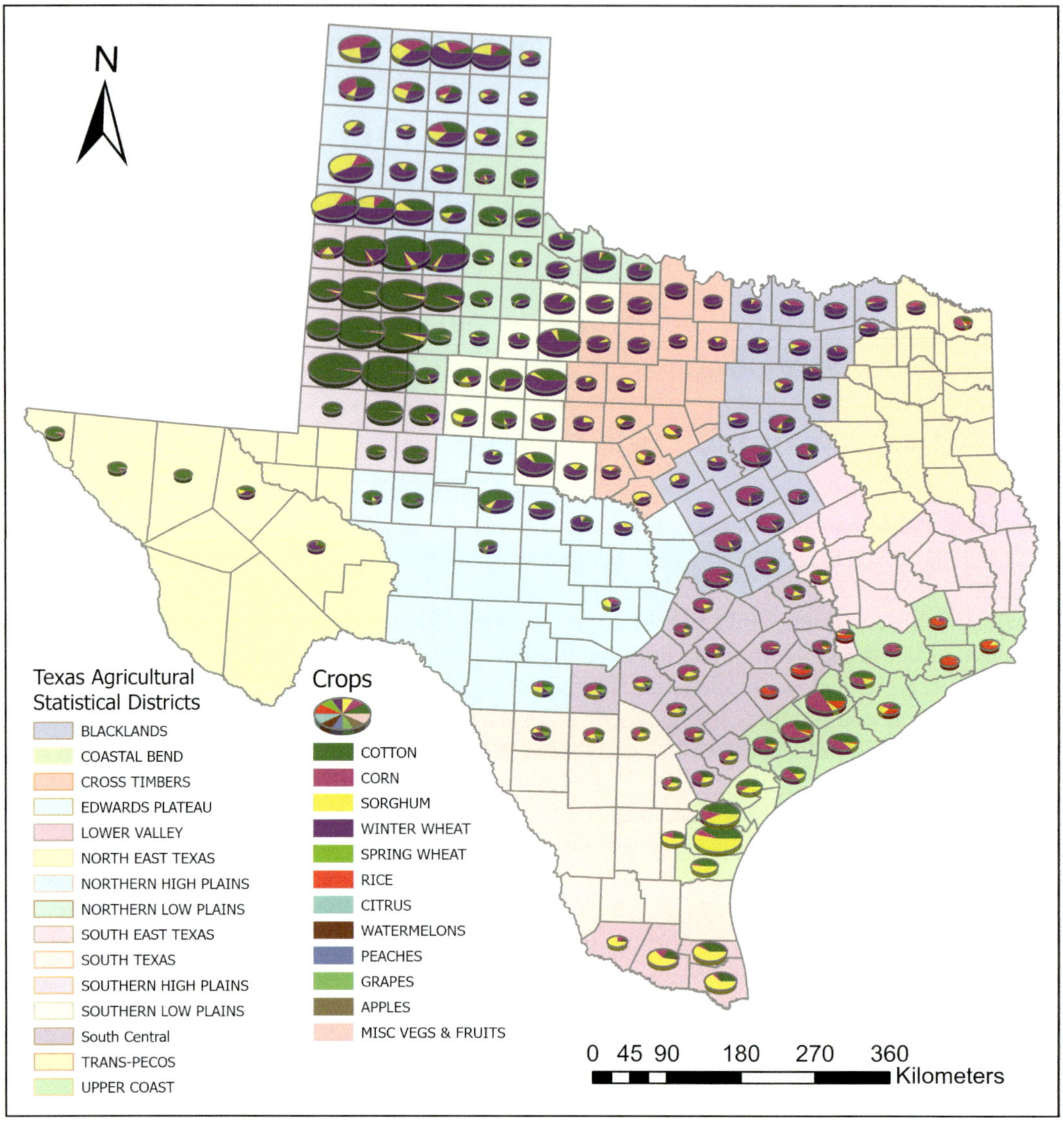

FIGURE III.1.1 Crop distribution in Texas based on the 2024 United States Department of Agriculture (USDA) National Agricultural Statistics Service (NASS) Cropland Data Layer.

$1.6 billion in hired labor expense annually. Although cotton production plays a key role in Texas's economy, it faces multifaceted challenges. The increasing shift to dryland cultivation to conserve water has increased the risk of cotton production related to climate variability and simultaneously increased the abandonment rates.

III.1.1.2 Corn

Corn (*Zea mays* L.) is the second largest agricultural commodity among field crops in Texas after cotton (TAMU, 2023). In 2024, corn was planted in 870,084 ha across Texas, of which 87% was harvested for grain, resulting in 5.3 billion kg of grain production and 7% was harvested for silage, producing 3.9 million kg of silage. Among the counties, the top grain corn producers are Sherman,

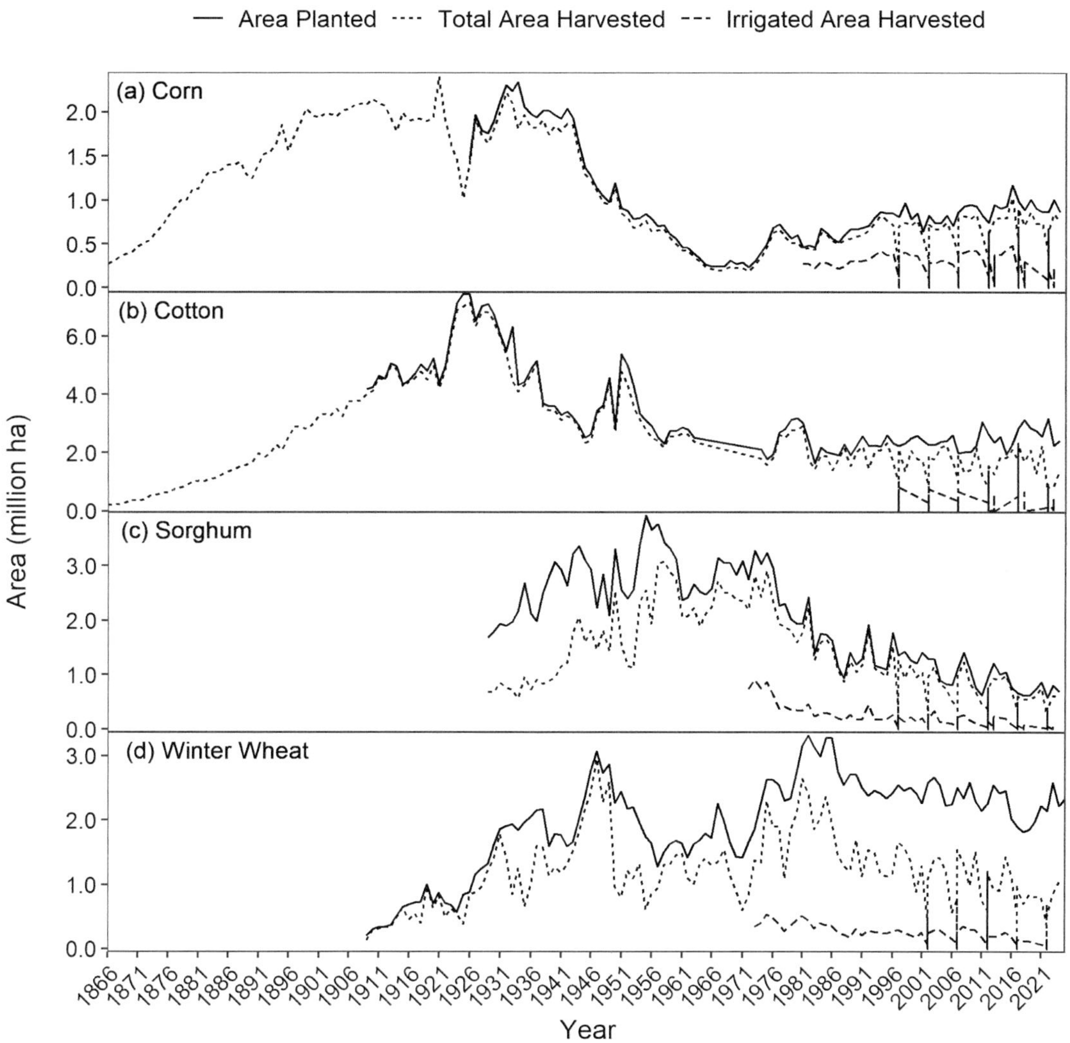

FIGURE III.1.2 Trends of crops acreages in Texas: (a) corn, (b) cotton, (c) sorghum, and (d) wheat.

Dallam, Hartley, Castro, Hansford, Moore, Wharton, Jackson, Williamson, Parmer, and Lamb. Among the Texas agricultural districts, the highest planted corn acres in 2024 were in the Northern High Plains followed by Blacklands, Upper Coast, South Central, and Southern High Plains (Figure III.1.1). About half of the total planted corn acres in Texas are irrigated, and the majority of these irrigated areas (>75%) are concentrated in the Northern High Plains.

Grain corn has a variety of uses, namely livestock feed, ethanol production, food and beverages, and exports. The demand for corn in the U.S. has been influenced by a combination of factors, including increasing global demand for livestock feed and biofuel (Saavoss et al., 2021). In Texas, the corn area peaked in the 1930s and then declined until the 1960s. The corn acreage picked up again and has been rising since then, from 404,825 ha in the 1960s to 936,857 ha in the 2010s (Figure III.1.2b). The increased area of corn can be attributed to the advances in seed technologies and commercial availability of insect-resistant, and drought- and herbicide-tolerant seeds (Saavoss et al., 2021). The importance of corn can be understood from its significant contribution to Texas's economy, with annual cash receipts of nearly $1 billion (TAMU, 2023).

III.1.1.3 Sorghum

Texas is the second largest producer of sorghum (*Sorghum bicolor* L. Moench) in the U.S. after Kansas. These two states account for 75% of U.S. sorghum production (Macke, 2020). In 2024, sorghum was planted on 687,967 ha in Texas and resulted in the production of 2.1 billion kg of grain. Grain sorghum is used for multiple purposes including feed, food, and fuel. Most of the planted sorghum is used for grain production and a small fraction is used for silage (typically 3–5%) and syrup production. Among the counties, top sorghum producers (with > 50 million kg grain production) include Nueces, San Patricio, Willacy, Cameron, Hidalgo, Sherman, Ochiltree, Hale, Moore, Refugio, Hockley, Deaf Smith, Carson, Kleberg, Wharton, Lamb, Matagorda, and Lubbock (Figure III.1.1). Among the agricultural districts, the sorghum acreage is the highest in the Northern High Plains followed by the Coastal Bend, Southern High Plains, and Lower Valley (Figure III.1.1). About 85% of Texas sorghum is produced under rain-fed conditions and 15% is irrigated.

Long-term trends reveal that the area in Texas under sorghum declined over the past few decades from 2.8 million ha in the 1970s to 811,646 ha in the 2010s (Figure III.1.2c). The decline in the acreage of sorghum in Texas and across the U.S. was due to market competition with other crops such as corn and cotton (Laingen, 2015). Nonetheless, sorghum remains popular among Texan farmers due to its ability to thrive in a warm and arid climate, and its better drought tolerance and lower water requirement than corn (Assefa et al., 2010). The renewed interest in sorghum is also due to the interest in alternative gluten-free cereals for food, and because of the increased foreign demand for grain sorghum as a livestock feed. In Texas, grain sorghum annual cash receipts are valued at $0.376 billion (TAMU, 2023).

III.1.1.4 Wheat

Texas is the ninth-largest producer of wheat (*Triticum aestivum* L.) in the U.S., with around 2.3 million ha planted annually between 2010 and 2024 (USDA-NASS, 2024). Around 50% of planted wheat is used for grain production and the remainder for forage production. In 2024, wheat was planted on 2.2 million ha in Texas and produced 2.2 billion kg of grain valued at $431 million (USDA-NASS, 2024). Wheat production is concentrated in the central and northwest portions of the state with leading acreages in the High Plains, Rolling Plains, and Blackland Prairies. The top wheat producing counties are Hansford, Ochiltree, Deaf Smith, Sherman, Haskell, and Wilbarger. In Texas, wheat is produced under both dryland and irrigated conditions, with around one-third acreage under irrigation (Texas-Almanac, 2021).

In the early 21st century, Texas was the third largest wheat producer in the country, harvesting around 1.4 million ha annually and generating $973 million for the economy. However, the harvested acres have dropped significantly in the recent years (Figure III.1.2d), down to 1.1 million ha in 2024. The primary reasons for the loss of wheat acreage vary across different regions and they include intense rainfall, injuries from hail and strong winds, inadequate moisture during the grain filling stage, and low grain prices (Ledbetter, 2024). An important aspect of wheat production is making it available for the livestock industry. Low wheat grain prices can be offset by the price of cattle gain, making it more favorable for producers to graze wheat biomass or produce silage (Ledbetter, 2024). Winter wheat is also largely used as a cover crop in Texas. It is relatively cheaper to produce wheat, and it provides a good amount of residue to protect the soil from erosion and enhance water conservation by improving infiltration rate, water holding capacity, and soil structure. When grown as a cover crop, wheat is not harvested for grain and it is typically terminated before the cash crop growing season and left in the field to decompose, enhancing nutrient cycling.

III.1.1.5 Rice

Texas ranks fifth in rice (*Oryza sativa* L.) production in the country behind Arkansas, California, Louisiana, and Missouri. In 2024, rice was planted on 59,900 ha in Texas, which produced

644 million kg of grain valued at $193 million (USDA-NASS, 2024). Rice farming is extremely water intensive, requiring around 650–830 mm under aerobic conditions and around 1,350 mm under flooded conditions between transplanting and harvesting (Castaneda et al., 2002). An additional 300 to 450 mm of water is required to produce the second (ratoon) crop. Historically, the primary rice-growing areas in Texas are along the Gulf coast, where the terrain is relatively flat with abundant water resources, which favor rice cultivation. The top rice producing counties are Wharton, Colorado, Jefferson, and Brazoria.

In recent years, the declining water availability and extreme weather events have threatened rice cultivation in Texas. Groundwater in the Gulf Coast is mostly saline, requiring producers in Wharton, Matagorda, and Colorado Counties to depend primarily on the water from the Colorado River. However, during the periods of drought, the Lower Colorado River Authority prioritizes the water distribution depending on urban water demand and reduces water allocation for agricultural use. Also, the repeated occurrence of extreme rainfall events leads to flooding for long periods of time that eventually damages rice crops. Despite these challenges, rice generates around $145 million for producers and contributes more than $480 million to the state's economy annually (Smith and Anciso, 2000).

III.1.1.6 Fruits

A wide variety of fruit is grown across the diverse landscapes and climate regions of Texas. Texas ranks third in citrus production in the U.S., with 5,423 bearing ha in 2024 (Figure III.1.3) yielding $85 million. Citrus is grown mostly in the South Texas Plains region (Lower Rio Grande Valley) with 85% and 14% of the acreage in Hidalgo and Cameron Counties, respectively. The subtropical climate in this region aided with irrigation and cold protection enables the growth of citrus fruits such as grapefruits (*Citrus × paradisi*), oranges (*Citrus × sinensis*), lemons (*Citrus × limon*), and limes (*Citrus × aurantiifolia*). Texas is also the third largest producer of watermelons (*Citrullus lanatus*), which is grown in over 100 counties across the state. However, watermelon production is largely concentrated in the southcentral and eastern parts of Texas, where the climate is hot with ample sunshine and soils are typically well-drained and hold consistent moisture. Central Texas

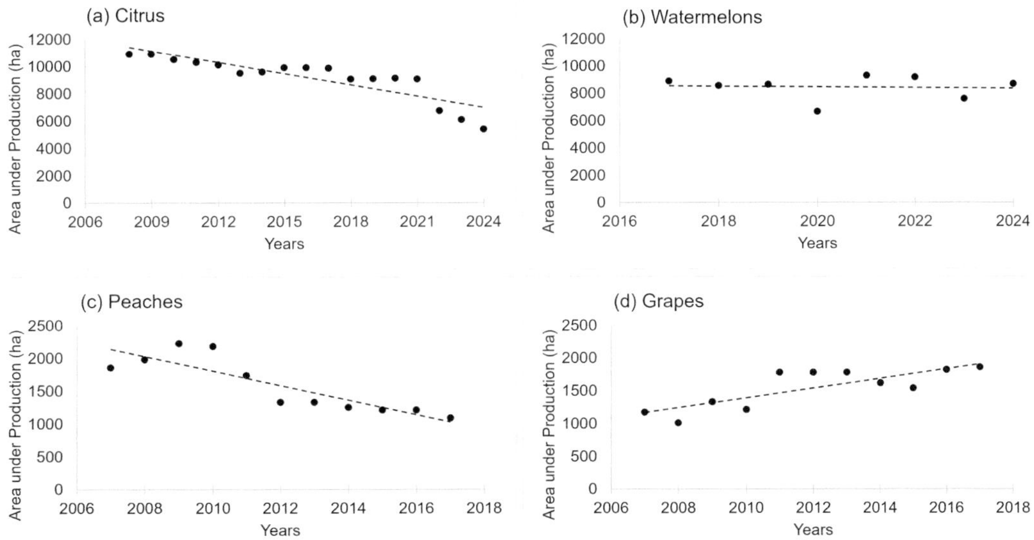

FIGURE III.1.3 Trends of fruit acreages in Texas: (a) citrus, (b) watermelons, (c) peaches, and (d) grapes.

provides favorable growing conditions for stone fruits such as peach (*Prunus persica*), plum (*Prunus domestica*), and nectarine (*Prunus persica var nectarina*). Grapes (*Vitis vinifera*) are commonly grown in the Panhandle region and the Hill Country. In 2021, there were 9,300 acres of vineyards with 70 varieties of grapes in Texas, the majority of which were in the Panhandle (2,430 ha) followed by the Hill Country (915 ha), North Texas (240 ha), and the Gulf Coast (200 ha) regions.

Long-term trends (Figure III.1.3) reveal a wide variability in fruit acreages. A loss of acreage was noted for citrus and peaches production. A series of events consisting of hurricanes, winter storms and droughts, and water disputes with Mexico led to a decline in citrus production in the Rio Grande valley. On the one hand, urban encroachment is a leading cause for the drop in peach farms. High real-estate values in the central Texas counties of Parker, Palo Pinto. Wise, Tarrant, Hood, Somervell, and Johnson, combined with frequent droughts have led to a substantial loss of peach acreages in Texas (Hardin, 2022). Watermelon acreage has mostly remained steady, with portions of the Rio Grande Valley losing watermelon acreage, while South Texas gained acreage. On the other hand, the grape acreage has been increasing in Texas, following the demands of the growing wine industry. Grape, although a minor crop in Texas, plays a major role in the state's economy, as the wine industry generates nearly $20.35 billion in total economic activity in the state.

III.1.1.7 Forage Crops

Forage crops are typically classified as either perennials or annuals. Cool-season perennials grown in Texas include alfalfa (*Medicago sativa*), tall fescue (*Festuca arundinacea*), and Texas bluegrass (*Poa arachnifera*), while the warm-season perennials include bermudagrass (*Cynodon dactylon*), buffalograss (*Bouteloua dactyloides*), buffelgrass (*Cenchrus ciliaris*), dallisgrass (*Paspalum dilatatum*), and Johnsongrass (*Sorghum halepense*). Cool-season annuals grown in Texas include wheat and oats (*Avena sativa*), while corn, sorghum, sorghum-sudan grass (*Sorghum × drummondii*), and pearl-millet (*Pennisetum glaucum*) are warm-season annuals. Texas currently ranks third in milk production after California and Wisconsin. The dairy industry has been booming in Texas, replacing New York for the fourth rank in 2021 and Idaho for the third rank in 2024 (USDA-NASS, 2024). More than 80% of milk in the state comes from the High Plains region (Whitlock, 2021). Favorable climate, land availability, and processing capacity have enhanced the growth of the dairy industry in the High Plains. To meet the growing demand for the dairy industry, the pressure to ensure supply of reliable high-quality forage has been growing. Corn is a high-quality forage crop, popular in the High Plains, due to its good digestibility and high energy. However, rapid decline of groundwater levels in the Ogallala Aquifer drastically impacted corn production in the High Plains. Alternatively, sorghum, pearl millet, and wheat silage utilize limited water compared to corn and ensure a steady supply of forage crops (Bhattarai et al., 2019). The introduction of the Brown midrib (BMR) trait in sorghum and pearl millet significantly increased the forage quality and fiber digestibility.

III.1.2 CLIMATE VARIABLES AFFECTING CROP GROWTH AND DEVELOPMENT

Understanding the physiological and morphological responses to projected elevated carbon dioxide (CO_2) concentrations and high temperatures is critical for understanding the climate change impacts on crop production and developing effective adaptation strategies, including the selection of crops, the creation of climate-resilient crop varieties, and the implementation of sustainable agricultural practices. These strategies are essential for coping with the challenges posed by a changing climate. The impact of climate variations on crop production may differ significantly among crop species. The C_3 and C_4 crops exhibit distinct stomatal responses, reflecting their mechanisms to optimize photosynthesis while minimizing transpiration (Wang et al., 2012). The C_3 plants are characterized by the initial formation of a three-carbon compound during photosynthesis. However, under conditions of elevated temperature and light intensity, oxygen has a strong affinity for the photosynthetic enzyme Ribulose-1,5-bisphosphate carboxylase/oxygenase (Rubisco; Wang et al., 2012). This leads to oxygen binding to Rubisco instead of CO_2, resulting in photorespiration, which decreases both the photosynthesis and water use efficiency of C_3 plants. Consequently, C_3 crops do not possess adaptations to mitigate the effects of photorespiration. Approximately 85% of plant species are categorized as C_3 crops. The predominant C_3 crops grown in Texas include crops such as cotton, rice, soybean, wheat, citrus, and peach.

In contrast to C_3 plants, C_4 plants exhibit a unique leaf anatomy and biochemical pathway that allows them to efficiently capture CO_2 upon entering the leaf, forming a four-carbon compound that is then utilized to enhance CO_2 concentration around the Rubisco enzyme (Wang et al., 2012). This CO_2 concentration mechanism significantly improves both photosynthetic activity and water use efficiency. The C_4 pathway offers distinct advantages by minimizing oxidative reactions, thereby reducing the likelihood of photorespiration and supporting high productivity and adaptability in conditions characterized by low CO_2 concentrations, elevated temperatures, high solar radiation, drought, and salinization—factors that can exacerbate carbon loss in C_3 plants due to photorespiration (Srinivasarao et al., 2016). In Texas, the predominant C_4 crops include corn, sorghum, Indiangrass, big bluestem, grapes, and watermelon. Research indicates that, at CO_2 levels of 330 parts per million (ppm), the threshold temperature for optimal performance in C_3 plants is approximately is approximately between 15°C and 25°C, which is notably lower than that for C_4 plants (20–35°C) (Allen, 2000). Consequently, in environments with high light intensity and elevated temperatures, C_4 plants generally demonstrate superior productivity in comparison to C_3 plants.

Photosynthesis constitutes a biological process through which plants convert light, CO_2, and water into sugars that are essential for their growth, primarily utilizing the photosynthetic enzyme Rubisco. The optimal temperature for photosynthesis varies depending on physiological adaptations between C_3 and C_4 crop species. In C_4 species such as sorghum and corn, the carboxylation mechanism is effectively saturated by ambient CO_2 concentrations (400 ppm), whereas C_3 species including rice, soybean (*Glycine max* L.), and wheat (*Triticum aestivum* L.) require significantly higher CO_2 levels (700 ppm) to reach saturation. Importantly, C_4 species are projected to demonstrate minimal responses to increasing CO_2 concentrations in the future, with the potential exception of drought-stressed conditions (Leakey, 2009). Elevated CO_2 may enhance photosynthesis in C_4 species solely under drought stress; however, the reduction in stomatal conductance due to elevated CO_2 is unlikely to be compensated for by canopy area, as observed in C_3 species. The C_3 crops exhibit a heightened response to increased CO_2 concentrations, which also aids in alleviating drought stress. This response is manifested through improved water use efficiency and enhanced photosynthesis and yield resulting from a higher transpiration (ET) rate (van der Kooi et al., 2016).

The C_4 crops exhibit a unique leaf anatomy known as "Kranz" (German for "wreath"), in which specialized cells called bundle sheath cells surround the vascular bundles. This structure is crucial for the C_4 photosynthetic pathway because it concentrates CO_2 around the enzyme RuBisCO, thereby minimizing photorespiration and enhancing photosynthetic efficiency. This anatomical adaptation allows C_4 plants to thrive in dry environments, where CO_2 concentrations are lower, and the likelihood of photorespiration increases. Additionally, this CO_2-concentrating mechanism in C_4 species could make them comparatively less sensitive than C_3 species to changes in transpiration in response to projected increases in CO_2 concentrations in the future. For example, Hussain et al. (2013) reported that the reduction in the evapotranspiration (ET) in corn under elevated CO_2 was half of that observed in soybean. Their findings also indicated that transpiration accounted for a lesser proportion of total ET in corn compared to soybean, suggesting that stomatal function is less critical in mediating ET responses to elevated CO_2 in C_4 species. Additionally, Jarvis and McNaughton (1986) projected a 24.7% reduction in transpiration for corn as opposed to 15.1% for soybean when CO_2 concentrations doubled to 700 ppm. Similarly, Wilson et al. (1999) simulated a decrease in ET of 5.4% for soybeans and 8.6% for corn when CO_2 concentration was increased from 350 to 700 ppm. The anticipated reductions in ET for corn and wheat were 23.3 mm and 8.9 mm, respectively, corresponding to 4.3% and 1.8% of the annual ET. Under a scenario of doubled CO_2

concentration (660 ppm), the annual ET for corn and wheat is predicted to decrease by 4.3% and 1.8%, respectively.

III.1.3 CLIMATE CHANGE IMPACTS ON CROP PRODUCTION

Studies assessing the climate change impacts on crop production in Texas used climate data projected by the General Circulation Models (GCMs) and analyzed under phases 3, 5, and 6 of the Coupled Model Intercomparison Project (CMIP). This section summarizes those studies and presents results on irrigated and dryland crop yields and irrigation water use projections (Table III.1.1 and Figure III.1.4).

III.1.3.1 Cotton

Cotton is the leading cash crop in Texas, which ranks first in cotton production in the U.S. Since 1980, Texas has typically left an average of 21% of its planted acres abandoned, and this trend has been increasing over time due to a combination of factors, including drought, rising input costs, and the potential increase in acreage for other crops, such as corn, soybeans, and sorghum. The average yield of upland cotton over the past 10 years (2015–2024) was 757 kg/ha, resulting in a production of 1.4 million tons (USDA-NASS, 2024). Different regions across Texas experience varying growing season lengths and cultivation areas. The Southern High Plains is recognized as the world's largest cotton-producing area, where acreage exceeds 1.21 million ha in some years. This region has an average temperature of approximately 22°C and receives about 457 mm of annual rainfall. Approximately 60% of the cotton fields in the Southern High Plains are irrigated, with the growing season extending from June to October. The reported optimal temperature range for most

TABLE III.1.1
Simulated Average Changes in Future Irrigated and Dryland Crop Yields, and Projected Irrigation Water Use Compared to the Historical Period

Crop	Period	Irrigation Water Use	Dryland Yield	Irrigated Yield	References
Cotton	Mid-century	9 (57)	4 (57)	16 (57)	(Kothari et al., 2021; Wen et al., 2024)
	Late-century	11 (57)	−1 (57)	15 (57)	
Corn	Mid-century#	−25* (101)†	−21 (54)	−20 (101)	(Chen et al., 2019; Chen et al., 2021; Kothari et al., 2022; Wen et al., 2024)
	Late-century	−29 (101)	−35 (54)	−28 (101)	
Sorghum	Mid-century	−31 (101)	−7 (57)	−11 (101)	(Chen et al., 2019; Chen et al., 2021; Kothari et al., 2020b; Wen et al., 2024)
	Late-century	−36 (101)	−13 (57)	−17 (101)	
Wheat	Mid-century	−14 (101)	13 (116)	6 (101)	(Chen et al., 2021; Kothari et al., 2019; Salinas et al., 2024; Wen et al., 2024)
	Late-century	−19 (101)	18 (116)	8 (101)	

Averages calculated from multiple future scenarios that represent different locations, general circulation models (GCMs), representative concentration pathways (RCPs) or shared socioeconomic pathways (SSPs) from climate change impact assessment studies conducted in Texas.

The periods used for specifying mid- and late-century were different in the cited studies: Wen: 2041–2070 and 2071–2100; Chen: 2040–2069 and 2070–2099; Kothari: 2036–2065, and 2066–2095; Salinas: 2046–2070 and 2071–2095

* Numbers show percent change in irrigation water use/yield in the future compared to baseline.

† Numbers in parenthesis show the number of data points averaged to get the percent change value.

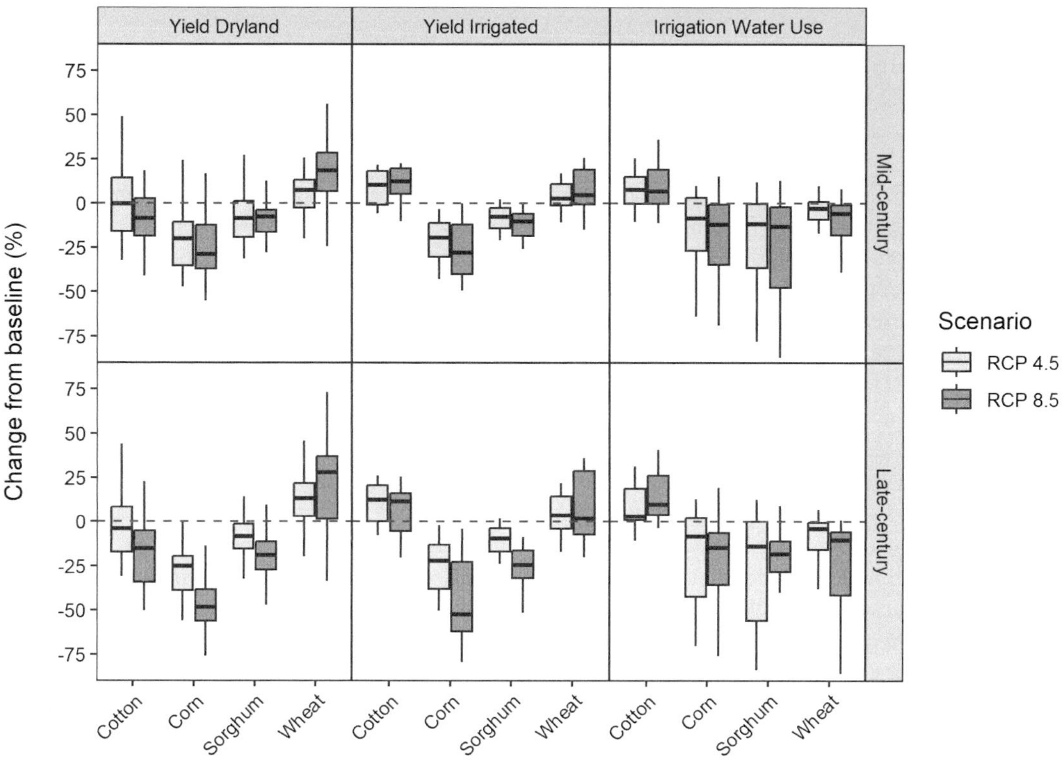

FIGURE III.1.4 Variability in projected changes in simulated crop yield and irrigation water use as compared to the historical/baseline period [based on the results from multiple future scenarios that represent different locations in Texas High Plains, General Circulation Models (GCMs) and Representative Concentration Pathways (RCPs)]

cotton physiological processes is 23–32°C, with 30°C being considered as ideal (Saini et al., 2023). Temperatures outside this range can negatively affect cotton yields (Reddy et al., 1999). The water requirement for cotton in Texas ranges from 353 to 625 mm (Koudahe et al., 2021). Daily average water usage by cotton increases from the initial growth phase (around 2.3 mm/day) to mid-season (7.1 mm/day), and the period from the first square to the first bloom is critical in preventing severe water deficiency stress (Koudahe et al., 2021).

The growth and development of cotton are notably influenced by three primary ecological factors: air temperature, water availability, and atmospheric CO_2 concentration (Reddy et al., 1999; Saini et al., 2023; Singh et al., 2018; Snider and Oosterhuis, 2012). Climate change studies project that average daily temperatures in Texas could rise by 2°C to 5°C by the end of the 21st century (Kothari et al., 2021; Modala et al., 2017; Singh et al., 2023), along with an expected CO_2 concentration of 690 ppm in the Northern Hemisphere compared to the current levels (Meinshausen et al., 2020). An increase in temperature could potentially reduce seed cotton yield by as much as 15% if CO_2 levels do not rise in the future (Adhikari et al., 2016). A projected increase in CO_2 could boost seed cotton yields by up to 53%, with a lesser increase anticipated in the southern region compared to the northern region of the state (Kothari et al., 2021) (Figure III.1.4). Seasonal irrigation water usage is projected to increase by 9–11% (Kothari et al., 2021; Singh et al., 2023) (Table III.1.1), primarily due to increasing ET demands and CO_2 concentration. A projected rise of 5°C in temperature could shorten the cotton growing season (from germination to maturity) by up to 35 days (Ahmad et al., 2020; Singh et al., 2023).

III.1.3.2 Corn

Corn is typically planted in Texas between March and May and harvested between July and November (USDA-NASS, 2010). The planting dates and growing season length depend on the location and crop variety. A typical long-season corn variety takes about 111–115 days to reach maturity in the High Plains (Jiang et al., 2020). The grain corn yield in Texas was estimated as 7,030 kg ha^{-1} in 2024 (USDA-NASS, 2024). The average water requirement of corn is 745 mm, and it ranges between 670 and 970 mm (Domínguez et al., 2022; Kapanigowda et al., 2010). The optimal temperature for reproductive growth and failure temperature threshold for corn are reported as 18–25°C and 35°C, respectively (Hatfield et al., 2011). Corn acreage and yield are susceptible to year-to-year weather variability and extreme weather events, particularly drought (Ray et al., 2018).

Under climate change scenarios, researchers have simulated a decline in corn yields in Texas (Chen et al., 2021; Kothari et al., 2022; Wen et al., 2024). The future grain corn yields are simulated to decline, on average, by 20% in the mid-century and 28% in the late-century under irrigated conditions (averaged from n = 101 number of data points from published studies), and by 21% in the mid-century and 35% in the late-century under rainfed condition (n = 54) (Table III.1.1 and Figure III.1.4). The future irrigation water use is also simulated to reduce by 25% in the mid-century and 29% in the late-century (n = 101). The decline in the future grain corn yield and irrigation water use could be attributed to the shorter growing season, reduced above-ground biomass, grain unit weight, and grain number reduction under warmer temperatures (Kothari et al., 2022). The simulated future grain corn yield trends for Texas are consistent with those obtained from gridded modeling studies, which simulated a decline in yield by 22% globally and by 35% in the U.S. for late-century under RCP 8.5 (Jägermeyr et al., 2021; Rezaei et al., 2023). The yield losses under climate change have been attributed to increasing temperature and a shorter growing season (Kothari et al., 2022; Yang et al., 2021).

III.1.3.3 Sorghum

In Texas, sorghum is grown under diverse climatic and management conditions. It can be planted as early as February in the Rio Grande Valley and as late as June in the High Plains (Ohadi et al., 2018; O'Shaughnessy et al., 2014). The planting date advances into the summer as we move from south to north due to the temperature gradient. The optimal temperatures for sorghum have been reported as 26–34°C for vegetative growth and 25–28°C for reproductive growth (Prasad et al., 2008). The typical water requirement of a medium- to late-maturing sorghum (110–130 days) is 450–650 mm (Assefa et al., 2010). The grain sorghum yield in Texas was estimated as 3891 kg ha^{-1} in 2024 (USDA-NASS, 2024). On the one hand, although sorghum is a drought tolerant crop, moisture stress during critical stages, namely booting and flowering, can significantly reduce the grain yield (Samanta et al., 2024), by up to 85% (Thapa et al., 2020). On the other hand, extremely high rainfall events can also reduce sorghum yield. Sharma et al. (2019) analyzed yield data from 1973–2000 in 25 counties across Texas and reported that excess precipitation reduced grain sorghum yield by 18–38%. Barring the extreme rainfall events, sorghum yield typically shows a positive correlation with the rainfall and a negative correlation with air temperatures (Assefa et al., 2024).

Climate change impact assessments conducted in Texas using the process-based models suggest that sorghum yield could decline in the future (Chen et al., 2021; Kothari et al., 2020b; Wen et al., 2024). On average, future sorghum yields are simulated to decline by 11% in the mid-century and 17% in the late-century under irrigated conditions (n = 101), mainly due to shortening of growing season because of heat stress. Dryland sorghum yields are expected to decline by 7% in the mid- and 13% in the late-century under rainfed conditions (n = 57), primarily due to an increase in growing season temperature and reduction in rainfall (Table III.1.1). Additionally, the irrigation water demand is also expected to decline by 31% in the mid-century and 36% in the late-century (n = 101). The reduced irrigation water demand could be attributed to the shorter growing season, lower biomass and leaf area, and improved crop water use efficiency under elevated CO_2 concentration (Kothari et al., 2020b).

III.1.3.4 Wheat

Winter wheat is typically planted in Texas during September–November and harvested during June–July (USDA-NASS, 2010). The water requirement of wheat has been reported to vary between 491 mm in Uvalde to 877 mm in Bushland (Howell et al., 1997; Ko et al., 2009). The differences in water requirement in those studies were due to the differences in growing season length (ranging from 182 to 283 days), local environment, and frequency of irrigation. Substantial yield difference can be observed between rain-fed (1,000–3,000 kg ha^{-1}) and irrigated (3,000–8,000 kg ha^{-1}) wheat yields in the Texas High Plains (Thapa et al., 2017). In addition to drought stress, wheat yield losses in the U.S. Great Plains, which includes the Texas High Plains, are attributed to compound hot, dry, and windy conditions (Zhao et al., 2022), rising temperatures and days with maximum temperature greater than 31°C (Shrestha et al., 2020).

On the one hand, under climate change scenarios, researchers projected an increase in winter wheat yield and a reduction in irrigation water use in Texas (Chen et al., 2021; Kothari et al., 2019; Salinas et al., 2024; Wen et al., 2024). On average, future wheat yields are simulated to increase by 6% in the mid-century and by 8% in the late-century under irrigated conditions (n = 101), and by 13% in the mid-century and 18% in the late-century under rainfed conditions (n = 116) (Table III.1.1 and Figure III.1.4). The irrigation water use, on the other hand, is projected to reduce on average by 14% in the mid-century and 19% in the late-century (n = 101). The increase in wheat yields and irrigation water use was attributed to CO_2 fertilization effect. While wheat yield is projected to increase under majority of the future scenarios, researchers have also projected negative yield response to climate change, particularly for the sites with coarser soils and the locations that are currently close to the optimal temperatures and are negatively affected by rising temperatures (Kothari et al., 2019). Overall, the yield gains projected for Texas align with the yield gains projected on average globally (10%) and for the U.S. (21%) under RCP 8.5 by the late-century (Jägermeyr et al., 2021; Rezaei et al., 2023). The wheat yield gains under climate change are attributed to the CO_2 fertilization effect (Kothari et al., 2019).

III.1.3.5 Rice

Rice is typically planted in Texas between March and June and harvested between August and October, when producers start managing the crop as a ratoon crop. Planting dates vary with the variety and cropping system. The best planting window for the main (first) crop production is in April, whereas that for the total (main + ratoon) crop production would be late March (Dou et al., 2016). In Texas, due to the short growing season, early maturing rice varieties are typically used. This also provides an opportunity for rice ratooning before the onset of cooler temperatures. The declining water resources threaten the production of this water-intensive crop, and it is critical to develop management practices that could potentially improve water use efficiency to ensure sustainable rice production. Practices such as alternate wetting and drying, direct seeding, saturated soil culture, and utilizing aerobic rice varieties have shown promising impacts on rice yield by reducing unproductive percolation losses and evaporation (Bouman et al., 2007).

The projections from the Coupled Model Intercomparison Project Phase 6 (CMIP6) indicate that the Gulf Coast region could experience an increase in temperature between 1°C and 3.5°C. Changes in average rainfall are expected to reduce between 1 to 12% mid-century (2040s) and between 7 to 9% in the late century (2070s) relative to the historical period (2000s; Li et al., 2024). Rice, being a C_3 crop, is expected to benefit from the projected increase in CO_2 concentrations. Li et al. (2024) found that across both SSP2-4.5 (moderate emission scenario) and SSP5-8.5 (high emission scenario), rice yield in the Gulf Coast is expected to increase by up to 20% in the 2040s and 30% in the

2070s. Projected variations in temperatures are low in the region, compared to historical temperatures. Consequently, the increase in heat stress would not be severe enough to impact the spikelet fertility and hence would not negatively impact yields.

III.1.3.6 Fruits and Trees

Citrus is an important fruit crop, mainly grown in the Rio Grande Valley. and its production has been reported to benefit greatly from the increasing temperatures, which would help reduce crop losses due to freezing. Citrus trees are not fully dormant during the winter months in the northern hemisphere, thereby exposing them to freezing events that can reduce the yield (De Leon et al., 2024). The Winter Storm Uri in 2021 significantly damaged the citrus production in subsequent years, killing about 10% of grapefruit trees. Tubiello et al. (2002) estimated a 20–30% yield increase by 2030 and a 50–70% yield increase by 2090 at Corpus Christi, TX. They also reported a significant reduction in irrigation water use, mainly due to reduction in ET by up to 12% and 11% in the 2055s and 2090s, respectively. Similar reductions in irrigation requirements were reported by Fares et al. (2017).

Unlike citrus, peach production is expected to suffer in the future in Texas, and the effects have already started to surface. One of the challenges faced by Northeast Texas, where peach is largely grown, is the lack of chill hours (between 0°C and 7°C) for peach trees to break dormancy. This is attributed to the warmer than normal winter temperatures (Madson and Montague, 2024; Watson, 2017). With projected rising temperatures, this situation is likely to aggravate in the future. Watermelon has also been reported to be negatively impacted by rising temperatures (de Melo et al., 2020). However, Hong et al. (2022) found that CO_2 enrichment could dramatically improve yields. The impact of changing climate can also drastically impact wine production. Changing temperature and precipitation patterns may affect the quality of grapes, but the Texas wine industry is more reliant on heat-tolerant grape varieties (Alarcón, 2022) such as the Black Spanish (Lenoir), Blanc du Bois, and Tempranillo. However, increasing water shortages in the Texas Panhandle may negatively impact future grape production.

III.1.3.7 Forage Crops

Increasing climate variability may adversely affect the livestock industry as it may disrupt the forage supply (Briske et al., 2021). Increasing frequency of droughts may drastically reduce forage yields, especially in the Texas High Plains, where water availability is already a major challenge. The impact of changing climate on pasture and rangeland is largely dependent on the forage species (C_3 or C_4). Legumes and C_3 species may be favored by CO_2 enrichment, whereas warmer environments may favor C_4 species' metabolism (Izaurralde et al., 2011). In addition, the quality of forage may also deteriorate (Briske et al., 2021; Havstad et al., 2018; Izaurralde et al., 2011), posing severe challenges to dairy production. Both harvested forages and rangeland forages have been estimated to have reduced digestibility in nitrogen-limited systems on account of increased cellulosic and lignin contents. In pastures without nitrogen limitation, the forage quality may improve (Izaurralde et al., 2011). Integrating crop–livestock systems, such as combining grain crop production with perennial pastures and livestock operations, may help in improving sustainability of forage production under projected future climates.

III.1.4 ADAPTATION STRATEGIES

Climate adaptation can be defined as an activity to ameliorate the negative impacts or enhance the positive impacts of climate change on crop production (Rezaei et al., 2023). The number of studies on climate change adaptation in Texas is limited. This section provides an overview of the research on climate adaptation under three categories, focusing mainly on Texas, and few studies from other parts of the U.S.

III.1.4.1 Management-based Adaptations

Farmers can adapt to the changing climate by altering crop management practices, such as planting, irrigation, and fertilizer management (Hasegawa et al., 2022). The rationale behind adjusting the planting date is to move the crop cycle into favorable temperature and rainfall regimes and thereby escape periods of heat and drought stresses. Research shows that farmers in the U.S. have already been adjusting planting dates in response to warming temperatures. For example, Sacks and Kucharik (2011) found that farmers planted corn, on an average, 10 days earlier in 2005 compared to 1981. Similarly, Baum et al. (2020) reported advancing corn planting date in Iowa by 13 d yr^{-1}, based on long-term (1980–2015) data. Miller et al. (2021) reported that optimizing planting and harvest dates could offset the negative effects of warming on sorghum yields in Kansas. The benefits of shifting planting date, however, depend on the crop variety (long-season versus short-season) and the local environment such as air temperature and frost-free period (Baum et al., 2020; Jiang et al., 2020). Warming temperatures result in longer growing seasons and are therefore particularly beneficial for cooler regions where growth is currently temperature-limited. In addition to planting date, changing row spacing can serve as a climate adaptation strategy. For example, mid-season (May 15) cotton planting in narrow (75 cm or 30 in) row spacing was identified as a promising climate adaptation strategy to sustain cotton production at Halfway in the Texas High Plains and Chillicothe in the Texas Rolling Plains in the future (Singh et al., 2024).

Irrigation and fertilizer management are among the most prominent agronomic adaptations reported in global studies (Rezaei et al., 2023). These adaptations refer to the changes in the timing and amount of application (Hasegawa et al., 2022). Evaluating critical crop growth stages for water stress and ensuring adequate irrigation water application during those stages can enable crops to better adapt to changing climate. For instance, losses in sorghum yield could be offset by timing irrigation during panicle initiation to early grain filling (Araya et al., 2018; Kothari et al., 2020a). Likewise, losses in seed cotton yield can be minimized by allocating more irrigation water for the peak bloom growth stage (Himanshu et al., 2021).

Similarly, incorporating organic fertilizers, and adopting precision and variable rate irrigation and nutrient application methods can help offset the negative impacts of climate change (McCarl, 2025). The nutrient management not only serves as an adaptation strategy but also as a strategy to mitigate climate change effects by reducing greenhouse gas (GHG) emissions (Fargione et al., 2018). Overall, the management related adaptations are location-specific and depend on several factors such as the availability of irrigation water, and the disturbances caused to existing crop rotation practices due to shifting of the crop cycle.

III.1.4.2 Cultivar-based Adaptations

The negative impacts of climate change on crop yields could be offset by using cultivars that are better suited to warmer conditions, such as longer maturing, and drought- and heat-tolerant cultivars (Hasegawa et al., 2022). Researchers have used process-based crop models to identify cultivars with improved genotype characteristics that could perform better than the conventional cultivars under future climate. For example, based on the simulation studies conducted in the Texas High Plains, Kothari et al. (2019, 2020b, 2021, 2022) reported that the cultivars with high yielding traits could provide 1–21% yield advantage for corn, cotton, sorghum, and winter wheat than the conventional cultivars under climate change (Table III.1.2). These cultivars were designed to have bigger leaf size, more assimilate partitioning to reproductive growth, and higher radiation use efficiency than the traditional cultivar. Consistent with these results, Asseng et al. (2019) suggested increasing radiation use efficiency as one of the promising strategies to increase wheat yield under climate change.

Drought tolerant cultivars with more vigorous root system could benefit sorghum and wheat with yield advantage of 1–14% over the conventional cultivars (Table III.1.2). However, the efficacy of drought tolerant cultivar was found to vary depending on the soil type and irrigation status (irrigated versus dryland). The ability of cultivars to extract water from deeper soils was identified

TABLE III.1.2
Summary of Literature from the Texas High Plains on Genotype–Based Climate Adaptation Strategies (Change in Irrigated Crop Yield and Seasonal Irrigation with and without Adaptation – Averaged across Nine Global Climate Models (GCMs), 2 Representative Concentration Pathways RCPs (RCP 4.5 and 8.5), and 2 Future Periods (2050s and 2080s)

		Percent Change from Reference Cultivar		
Crop	**Adaptative Cultivar (Modified Characteristics)**	**Irrigated Yield**	**Seasonal Irrigation**	**Reference**
Cotton	Drought tolerant (greater root water uptake)	2–7%	−1 to +8%	(Kothari et al., 2021)
	Heat tolerant (+2°C higher thresholds for failure temperature for boll addition and partitioning to boll)	3–13%	<1%	
	High yielding (increased area of full leaf and enhanced partitioning of assimilates to reproductive growth)	5–8%	0–2%	
	Long maturity (longer grain filling period)	5–11%	7–9%	
Corn	Heat tolerant (+2°C higher thresholds for failure temperature)	8–28%	<2%	(Kothari et al., 2022)
	High yield potential (more kernels per head, higher radiation use efficiency)	8–21%	<1%	
	Long maturity (longer grain filling period)	15–40%	4–11%	
Sorghum	Drought tolerant (higher root density)	4–14%	−6 to −2%	(Kothari et al., 2020b)
	Heat tolerant (+2°C higher thresholds for failure temperature)	1–9%	−1 to 2%	
	High yielding (bigger size leaf, greater partitioning to reproductive organ, and higher radiation use efficiency)	7–19%	−2 to 2%	
	Long maturity (longer emergence to juvenile-end phase, and longer grain filling period)	−3 to +17%	9–14%	
Winter wheat	Drought tolerant (denser roots)	1–10%	−4 to −2%	(Kothari et al., 2019)
	High yielding (more kernels per head)	2–11%	0%	
	Long maturity (longer grain filling period)	1–3%	1–4%	

as an important drought tolerant trait in the field experiments conducted in Texas (Thapa et al., 2017). Heat tolerant cultivars with higher (+2°C) threshold of failure temperature than conventional cultivar resulted in yield advantage of 3–13% for cotton, 8–28% for corn, and 1–9% for sorghum (Table III.1.2). Longer maturity cultivar could also provide yield advantage over conventional cultivar by 2–40%, however, it could also result in greater seasonal irrigation water use on average by 1–14% (Table III.1.2). It is worth noting that the simulation models allow for evaluating a wide range of genotyping adaptations rapidly, but incorporating the desirable traits into crop varieties may take several breeding cycles. Development of cultivar-based adaptation strategies, therefore, requires collaborative efforts between crop modelers, agronomists, and plant breeders.

III.1.4.3 Alternative Crops and Cropping Systems

While the management- and cultivar-based adaptations focus on improving the yield of existing crops, farmers may adapt to climate change by introducing new cropping practices, such as planting new crop mixes, adopting double-cropping and cover crops, switching to short-season forages and agroforestry (IPCC, 2022; McCarl, 2025). Research based on econometric models showed that the locations of crop mixes in the U.S. have shifted between 1970 and 2010 due to climate change. For instance, areas where winter wheat, hay, soybean, and corn are grown have shifted northward (Cho and McCarl, 2017). The crop mixes are projected to shift further northward and to higher elevations in the future. With increasing temperatures, farmers may choose to grow cotton, rice, sorghum, and winter wheat, while the share of corn, soybeans, barley, and spring wheat crops could decline (Cho and McCarl, 2017). As the climate change intensifies, farmers in the southern region of the U.S. would likely find a shift in crop mixes, and even conversion of cropland into pasture and grassland (Cho and McCarl, 2021).

Globally, many researchers have been advocating for agroecological and nature-based solutions to improve climate resilience and sustainability of cropping systems (IPCC, 2022). These approaches focus on conserving biodiversity, improving soil health, and reducing GHG emissions. Research shows that crop diversification measures including agroforestry, cover crops, intercropping, and crop rotations provide a variety of benefits, including enhanced crop production, better yield stability, and improved soil and water quality (Beillouin et al., 2021). Research on agroecological measures as climate adaptation strategies is limited in Texas. Considering the geographic diversity of Texas, a variety of adaptation measures would be required to maintain the profitability of agriculture under a changing climate (Bell et al., 2023).

III.1.5 SUMMARY AND CONCLUSIONS

This chapter discusses the impact of climate change on agriculture and food production in Texas. Due to Texas's diverse geography and ecology, a range of C_3 and C_4 crops are cultivated, including fiber crops like cotton, grain crops like corn and rice, dual-purpose grain and forage crops such as wheat and sorghum, and fruit trees such as citrus and peaches. Among all crops, hay occupies the largest area in Texas, followed by cotton, corn, sorghum, wheat, rice, and fruit trees. However, climate change studies project that by the end of the 21st century, average daily temperatures in Texas may increase by 2°C to 5°C, which could significantly influence the crop selection and management practices across Texas. Researchers employed crop-specific models to predict how anticipated climate changes may affect growth, yield, and irrigation water demand of crops. On one side, crop models have indicated a decrease in yield for crops like cotton (dryland), corn, peaches, and sorghum in Texas, mainly due to heat stress and a decline in rainfall under projected future conditions. Conversely, yield of crops such as wheat and citrus is simulated to increase in the future due to CO_2 fertilization. The irrigation water needed for some of the crops such as cotton is predicted to increase in the future due to various climatic influences such as increased evapotranspiration demand, and CO_2 fertilization, while the water use for crops like sorghum and corn is expected to decrease due to lower biomass and a shortened growing season. The crop models have also been used to evaluate adaptation strategies to address specific challenges related to crop production in the future. Among these strategies, management practices like changes in row spacing and planting time could help crops mitigate projected heat stress during the season, thereby lessening the impact of future heat stress on yields. Traits such as a robust root system and extensive leaf area may play a crucial role in developing germplasm for enhanced resilience to severe heat stress in the future. Moreover, integrating livestock production systems with cropping systems could contribute to the sustainability of forage production given projected future climate conditions.

REFERENCES

Adhikari P, Ale S, Bordovsky JP, Thorp KR, Modala NR, Rajan N, Barnes EM, 2016. Simulating future climate change impacts on seed cotton yield in the Texas high plains using the CSM-CROPGRO-cotton model. Agricultural Water Management, 164:317–330. doi:10.1016/j.agwat.2015.10.011

Ahmad F, Perveen A, Mohammad N, Ali MA, Akhtar MN, Shahzad K, Danish S, Ahmed N, 2020. Heat stress in cotton: Responses and adaptive mechanisms. In S. Ahmad & M. Hasanuzzaman (Eds.), Cotton Production and Uses: Agronomy, Crop Protection, and Postharvest Technologies (pp. 393–428). Singapore: Springer Singapore.

Alarcón C, 2022. How climate change is impacting winemakers in Texas. Forbes. Retrieved from https://www.forbes.com/sites/claudiaalarcon/2022/10/03/how-climate-change-is-impacting-winemakers-in-texas/ (accessed March 10, 2025).

Allen LH Jr, 2000. Evapotranspiration responses of plants and crops to carbon dioxide and temperature. Journal of Crop Production, 2(2):37–70. doi:10.1300/J144v02n02_02

Araya A, Kisekka I, Gowda PH, Prasad PVV, 2018. Grain sorghum production functions under different irrigation capacities. Agricultural Water Management, 203:261–271. doi:10.1016/j.agwat.2018.03.010

Assefa Y, Holman JD, Obour AK, O'Brien D, Prasad PVV, 2024. Historic grain sorghum production, value, yield gap, and weather relation trends. Agronomy, 14(11). doi:10.3390/agronomy14112582

Assefa Y, Staggenborg SA, Prasad VPV, 2010. Grain sorghum water requirement and responses to drought stress: A review. Crop Management, 9(1):1–11. doi:10.1094/CM-2010-1109-01-RV

Asseng S, Martre P, Ewert F, Dreccer MF, Beres BL, Reynolds M, Braun H-J, Langridge P, Le Gouis J, Salse J, 2019. Model-driven multidisciplinary global research to meet future needs: The case for "improving radiation use efficiency to increase yield. Crop Science, 59(3):843–849.

Baum ME, Licht MA, Huber I, Archontoulis SV, 2020. Impacts of climate change on the optimum planting date of different maize cultivars in the central US Corn Belt. European Journal of Agronomy, 119:126101. doi:10.1016/j.eja.2020.126101

Beillouin D, Ben-Ari T, Malézieux E, Seufert V, Makowski D, 2021. Positive but variable effects of crop diversification on biodiversity and ecosystem services. Global Change Biology, 27(19):4697–4710. doi:10.1111/gcb.15747

Bell J, DeLaune PB, Fischer BL, Foster JL, Lewis KL, McCarl BA, Outlaw JL, 2023. Carbon sequestration and water management in Texas—One size does not fit all. Agrosystems, Geosciences & Environment, 6(2):e20372. doi:10.1002/agg2.20372

Bhattarai B, Singh S, West CP, Saini R, 2019. Forage potential of pearl Millet and Forage Sorghum alternatives to corn under the water-limiting conditions of the Texas High Plains: A review. Crop, Forage & Turfgrass Management, 5(1):190058. doi:10.2134/cftm2019.08.0058

Bouman BA, Humphreys E, Tuong TP, Barker R, 2007. Rice and water. Advances in Agronomy, 92:187–237.

Briske DD, Ritten JP, Campbell AR, Klemm T, King AEH, 2021. Future climate variability will challenge rangeland beef cattle production in the Great Plains. Rangelands, 43(1):29–36. doi:10.1016/j.rala.2020.11.001

Castaneda A, Bouman B, Peng S, Visperas R, 2002. The potential of aerobic rice to reduce water use in water-scarce irrigated lowlands in the tropics. Proceedings of the International Work-shop on Water-wise Rice Production:165–176.

Chen Y, Marek GW, Marek TH, Moorhead JE, Heflin KR, Brauer DK, Gowda PH, Srinivasan R, 2019. Simulating the impacts of climate change on hydrology and crop production in the Northern High Plains of Texas using an improved SWAT model. Agricultural Water Management, 221:13–24. doi:10.1016/j.agwat.2019.04.021

Chen Y, Marek GW, Marek TH, Porter DO, Brauer DK, Srinivasan R, 2021. Modeling climate change impacts on blue, green, and grey water footprints and crop yields in the Texas High Plains, USA. Agricultural and Forest Meteorology, 310:108649. doi:10.1016/j.agrformet.2021.108649

Cho SJ, McCarl B, 2021. Major United States land use as influenced by an altering climate: A spatial econometric approach. Land, 10(5):546.

Cho SJ, McCarl BA, 2017. Climate change influences on crop mix shifts in the United States. Scientific Reports, 7(1):40845. doi:10.1038/srep40845

De Leon D, Nelson S, Sétamou M, Bhandari A, Zamora E, Zaragoza A, Donato M, Reyes-Cabrera J, 2024. Enhanced nutrition programs to rehabilitate freeze-stressed citrus trees in subtropical regions. HortScience, 59(12):1763–1771.

de Melo TK, Espínola J, de Medeiros JF, Figueiredo VB, da Silva JS, da Silva Sá FVS, 2020. Impacts of climate change scenarios in the Brazilian semiarid region on watermelon cultivars. Revista Caatinga, 33(3):794–802.

Domínguez A, Schwartz RC, Pardo JJ, Guerrero B, Bell JM, Colaizzi PD, Louis Baumhardt R, 2022. Center pivot irrigation capacity effects on maize yield and profitability in the Texas High Plains. Agricultural Water Management, 261:107335. doi:10.1016/j.agwat.2021.107335

Dou F, Lee T, Chen K, Wright AL, Mohammed AR, 2016. Planting date and variety effects on rice main and ratoon crop production in South Texas. Communications in Soil Science and Plant Analysis, 47(21):2414–2420. doi:10.1080/00103624.2016.1243705

Fares A, Bayabil HK, Zekri M, Mattos D Jr, Awal R, 2017. Potential climate change impacts on citrus water requirement across major producing areas in the world. Journal of Water and Climate Change, 8(4): 576–592. doi:10.2166/wcc.2017.182

Fargione JE, Bassett S, Boucher T, Bridgham SD, Conant RT, Cook-Patton SC, Ellis PW, Falcucci A, Fourqurean JW, Gopalakrishna T, 2018. Natural climate solutions for the United States. Science Advances, 4(11):eaat1869.

Guci L, Wetzler N, 2023. New and Updated Estimates of the Regional Economic Accounts, Results of the 2023 Comprehensive Update. Survey of Current Business, Bureau of Economic Analysis, 1–26.

Hardin J. 2022. Why Parker County Texas peach farms are disappearing. Fort Worth Star-Telegram. Retrieved from https://www.star-telegram.com/news/business/growth/article262462682.html (accessed March 10, 2025).

Hasegawa T, Wakatsuki H, Ju H, Vyas S, Nelson GC, Farrell A, Deryng D, Meza F, Makowski D, 2022. A global dataset for the projected impacts of climate change on four major crops. Scientific Data, 9(1):58.

Hatfield JL, Boote KJ, Kimball BA, Ziska LH, Izaurralde RC, Ort D, Thomson AM, Wolfe D, 2011. Climate impacts on agriculture: Implications for crop production. Agronomy Journal, 103(2):351–370. doi:10.2134/agronj2010.0303

Havstad KM, Brown JR, Estell R, Elias E, Rango A, Steele C, 2018. Vulnerabilities of Southwestern U.S. Rangeland-based animal agriculture to climate change. Climatic Change, 148(3):371–386. doi:10.1007/s10584-016-1834-7

Himanshu SK, Ale S, Bell J, Fan Y, Samanta S, Bordovsky JP, Gitz Iii DC, Lascano RJ, Brauer DK, 2023. Evaluation of growth-stage-based variable deficit irrigation strategies for cotton production in the Texas High Plains. Agricultural Water Management, 280:108222. doi:10.1016/j.agwat.2023.108222

Himanshu SK, Fan Y, Ale S, Bordovsky J, 2021. Simulated efficient growth-stage-based deficit irrigation strategies for maximizing cotton yield, crop water productivity and net returns. Agricultural Water Management, 250:106840. doi:10.1016/j.agwat.2021.106840

Hong T, Cai Z, Li R, Liu J, Li J, Wang Z, Zhang Z, 2022. Effects of water and nitrogen coupling on watermelon growth, photosynthesis and yield under CO_2 enrichment. Agricultural Water Management, 259:107229. doi:10.1016/j.agwat.2021.107229

Howell T, Steiner J, Schneider A, Evett S, Tolk J, 1997. Seasonal and maximum daily evapotranspiration of irrigated winter wheat, sorghum, and corn—Southern High Plains. Transactions of the ASAE, 40(3):623–634.

Hussain MZ, VanLoocke A, Siebers MH, Ruiz-Vera UM, Cody Markelz RJ, Leakey ADB, Ort DR, Bernacchi CJ, 2013. Future carbon dioxide concentration decreases canopy evapotranspiration and soil water depletion by field-grown maize. Global Change Biology, 19(5):1572–1584. doi:10.1111/gcb.12155

IPCC, 2022. Chapter 5 - Food, Fibre and other ecosystem products. In Climate Change 2022: Impacts, Adaptation and Vulnerability. Cambridge University Press, UK and New York, NY, USA, Cambridge.

Izaurralde RC, Thomson AM, Morgan J, Fay P, Polley H, Hatfield JL, 2011. Climate impacts on agriculture: Implications for forage and rangeland production. Agronomy Journal, 103(2):371–381.

Jägermeyr J, Müller C, Ruane AC, et al., 2021. Climate impacts on global agriculture emerge earlier in new generation of climate and crop models. Nature Food, 2(11):873–885. doi:10.1038/s43016-021-00400-y

Jarvis PG, McNaughton KG, 1986. Stomatal control of transpiration: Scaling up from leaf to region. In A. MacFadyen & E. D. Ford (Eds.), Advances in Ecological Research (Vol. 15, pp. 1–49). Academic Press.

Jiang W, Thapa S, Jessup KE, Hao B, Hou X, Marek T, Becker J, Bell J, Xue Q, 2020. Corn response to later than traditional planting dates in the Texas High Plains. Crop Science, 60(2):1004–1020. doi:10.1002/csc2.20042

Kapanigowda M, Stewart BA, Howell TA, Kadasrivenkata H, Baumhardt RL, 2010. Growing maize in clumps as a strategy for marginal climatic conditions. Field Crops Research, 118(2):115–125. doi:10.1016/j.fcr.2010.04.012

Ko J, Piccinni G, Marek T, Howell T, 2009. Determination of growth-stage-specific crop coefficients (Kc) of cotton and wheat. Agricultural Water Management, 96(12):1691–1697. doi:10.1016/j.agwat.2009.06.023

Kothari K, Ale S, Attia A, Rajan N, Xue Q, Munster CL, 2019. Potential climate change adaptation strategies for winter wheat production in the Texas High Plains. Agricultural Water Management, 225:105764. doi:10.1016/j.agwat.2019.105764

Kothari K, Ale S, Bordovsky JP, Munster CL, 2020a. Assessing the climate change impacts on grain sorghum yield and irrigation water use under full and deficit irrigation strategies. Transactions of the ASABE, 63(1):81–94.

Kothari K, Ale S, Bordovsky JP, Munster CL, Singh VP, Nielsen-Gammon J, Hoogenboom G, 2021. Potential genotype-based climate change adaptation strategies for sustaining cotton production in the Texas High Plains: A simulation study. Field Crops Research, 271:108261. doi:10.1016/j.fcr.2021.108261

Kothari K, Ale S, Bordovsky JP, Porter DO, Munster CL, Hoogenboom G, 2020b. Potential benefits of genotype-based adaptation strategies for grain sorghum production in the Texas High Plains under climate change. European Journal of Agronomy, 117:126037. doi:10.1016/j.eja.2020.126037

Kothari K, Ale S, Marek GW, Munster CL, Singh VP, Chen Y, Marek TH, Xue Q, 2022. Simulating the climate change impacts and evaluating potential adaptation strategies for irrigated corn production in Northern High Plains of Texas. Climate Risk Management, 37:100446. doi:10.1016/j.crm.2022.100446

Koudahe K, Sheshukov AY, Aguilar J, Djaman K, 2021. Irrigation-water management and productivity of cotton: A review. Sustainability, 13(18). doi:10.3390/su131810070

Laingen C, 2015. A spatiotemporal analysis of sorghum in the United States. Papers in Applied Geography, 1(4):307–311.

Leakey AD, 2009. Rising atmospheric carbon dioxide concentration and the future of C4 crops for food and fuel. Proceedings of the Royal Society B: Biological Sciences, 276(1666):2333–2343.

Ledbetter K, 2020. Northern High Plains cotton acres more than double in last 5 years. Retrieved from https://agrilifetoday.tamu.edu/2020/03/26/northern-high-plains-cotton-acres-more-than-double-in-last-5-years/ (accessed 21 March, 2025): Texas A&M AgriLife Today.

Ledbetter K, 2024. Wheat harvest struggling across the state. Texas A&M AgriLife Today. Retrieved from https://agrilifetoday.tamu.edu/2024/06/25/wheat-harvest-struggling-across-the-state/ (accessed March 10, 2025).

Li S, Fleisher DH, Barnaby JY, 2024. Quantifying the impact of climate change and extreme heat on rice in the United States. Agricultural and Forest Meteorology, 355:110145. doi:10.1016/j.agrformet.2024.110145

Macke Y, 2020. International agricultural trade report. Foreign Agricultural Service. Retrieved from https://www.fas.usda.gov/sites/default/files/2020-07/2020-07_sorghum_iatr.pdf (accessed January 23, 2025).

Madson D, Montague C. 2024. Texas peach farmer faces climate challenges. Yale Climate Connections. Retrieved from https://yaleclimateconnections.org/2018/07/texas-peach-farmer-faces-climate-challenges/ (accessed March 10, 2025).

McCarl BA, 2025. Climate change: What do we do about it? Economic issues regarding agricultural adaptation and mitigation. American Journal of Agricultural Economics, 107(2):368–389. doi: 10.1111/ajae.12517

Meinshausen M, Nicholls ZRJ, Lewis J, Gidden MJ, Vogel E, Freund M, Beyerle U, Gessner C, Nauels A, Bauer N, Canadell JG, Daniel JS, John A, Krummel PB, Luderer G, Meinshausen N, Montzka SA, Rayner PJ, Reimann S, Smith SJ, van den Berg M, Velders GJM, Vollmer MK, Wang RHJ, 2020. The shared socio-economic pathway (SSP) greenhouse gas concentrations and their extensions to 2500. Geoscientific Model Development, 13(8):3571–3605. doi:10.5194/gmd-13-3571-2020

Miller N, Tack J, Bergtold J, 2021. The impacts of warming temperatures on US Sorghum yields and the potential for adaptation. American Journal of Agricultural Economics, 103(5):1742–1758. doi:10.1111/ajae.12223

Modala NR, Ale S, Goldberg DW, Olivares M, Munster CL, Rajan N, Feagin RA, 2017. Climate change projections for the Texas High Plains and Rolling Plains. Theoretical and Applied Climatology, 129(1):263–280. doi:10.1007/s00704-016-1773-2

O'Shaughnessy SA, Evett SR, Colaizzi PD, Tolk JA, Howell TA, 2014. Early and late maturing grain sorghum under variable climatic conditions in the Texas High Plains. Transactions of the ASABE, 57(6):1583–1594.

Ohadi S, Littlejohn M, Mesgaran M, Rooney W, Bagavathiannan M, 2018. Surveying the spatial distribution of feral sorghum (*Sorghum bicolor* L.) and its sympatry with johnsongrass (S. halepense) in South Texas. PLoS One, 13(4):e0195511. doi:10.1371/journal.pone.0195511

Prasad PV, Pisipati S, Mutava R, Tuinstra M, 2008. Sensitivity of grain sorghum to high temperature stress during reproductive development. Crop Science, 48(5):1911–1917.

Ray RL, Fares A, Risch E, 2018. Effects of drought on crop production and cropping areas in Texas. Agricultural & Environmental Letters, 3(1):170037.

Reddy KR, Davidonis GH, Johnson AS, Vinyard BT, 1999. Temperature regime and carbon dioxide enrichment alter cotton boll development and fiber properties. Agronomy Journal, 91(5):851–858. doi:10.2134/agronj1999.915851x

Rezaei EE, Webber H, Asseng S, Boote K, Durand JL, Ewert F, Martre P, MacCarthy DS, 2023. Climate change impacts on crop yields. Nature Reviews Earth & Environment, 4(12):831–846. doi:10.1038/s43017-023-00491-0

Saavoss M, Capehart T, McBride W, Effland A, 2021. ERR-294. Trends in production practices and costs of the US corn sector. U.S. Department of Agriculture, Economic Research Service.

Sacks WJ, Kucharik CJ, 2011. Crop management and phenology trends in the U.S. Corn Belt: Impacts on yields, evapotranspiration and energy balance. Agricultural and Forest Meteorology, 151(7):882–894. doi:10.1016/j.agrformet.2011.02.010

Saini DK, Impa SM, McCallister D, Patil GB, Abidi N, Ritchie G, Jaconis SY, Jagadish KSV, 2023. High day and night temperatures impact on cotton yield and quality—current status and future research direction. Journal of Cotton Research, 6(1):16. doi:10.1186/s42397-023-00154-x

Salinas C, Osei E, Yu M, Guney S, Lovell A, Kan E, 2024. Climate change effects on Texas dryland winter wheat yields. Agriculture, 14(2). doi:10.3390/agriculture14020232

Samanta S, Ale S, Himanshu SK, Singh B, Kothari K, 2024. Identification of priority-based variable deficit irrigation strategies for grain sorghum production in the Texas High Plains under increasing climate variability. Journal of Agricultural Engineering (India), 61(3): 311–323. doi: 10.52151/jae2024613.1850

Sharma OP, Kannan N, Cook S, Pokhrel BK, McKenzie C, 2019. Analysis of the effects of high precipitation in Texas on rainfed sorghum yields. Water, 11(9):1920.

Shrestha R, Thapa S, Xue Q, Stewart BA, Blaser BC, Ashiadey EK, Rudd JC, Devkota RN, 2020. Winter wheat response to climate change under irrigated and dryland conditions in the US southern High Plains. Journal of Soil and Water Conservation, 75(1):112–122. doi:10.2489/jswc.75.1.112

Singh B, Norvell E, Wijewardana C, Wallace T, Chastain D, Reddy KR, 2018. Assessing morphological characteristics of elite cotton lines from different breeding programmes for low temperature and drought tolerance. Journal of Agronomy and Crop Science, 204(5):467–476. doi:10.1111/jac.12276

Singh B, Samanta S, Ale S, Barnes EM, Thorp KR, 2024. *Projected Climate Change Impacts on Irrigated Cotton Production in Two Distinct Regions of the United States Cotton Belt.* Paper presented at the ASA, CSSA, SSSA International Annual Meeting.

Singh B, Samanta S, Kothari K, Singh H, Ale S, Barnes EM, 2023. *Projected Climate Change Impacts on Cotton Production in the Texas High Plains.* Paper presented at the ASA, CSSA, SSSA International Annual Meeting.

Smith D, Anciso J, 2000. Rice in Texas. Crop Briefs. Retrieved from https://aggie-horticulture.tamu.edu/vegetable/guides/crop-briefs/rice-in-texas/ (accessed 03/10/2025).

Snider JL, Oosterhuis DM, 2012. Heat Stress and Pollen-Pistil Interactions. Flowering and Fruiting in Cotton. The Cotton Foundation, Cordova, TN:59–78.

Srinivasarao C, Kundu S, Shanker AK, Naik RP, Vanaja M, Venkanna K, Maruthi Sankar GR, Rao VUM, 2016. Continuous cropping under elevated CO_2: Differential effects on C4 and C3 crops, soil properties and carbon dynamics in semi-arid alfisols. Agriculture, Ecosystems & Environment, 218:73–86. doi:10.1016/j.agee.2015.11.016

Tamu A, 2023. The Food and Fiber System and Production Agriculture's Contributions to the Texas Economy. Texas A&M AgriLife Extension.

Texas-Almanac, 2021. Principal crops in Texas. Principal Crops in Texas | TX Almanac. Retrieved from https://www.texasalmanac.com/articles/principal-crops-in-texas (accessed March 10, 2025).

Thapa S, Xue Q, Jessup KE, Rudd JC, Liu S, Pradhan GP, Devkota RN, Baker J, 2017. More recent wheat cultivars extract more water from greater soil profile depths to increase yield in the Texas High Plains. Agronomy Journal, 109(6):2771–2780.

Thapa S, Xue Q, Stewart BA, 2020. Alternative planting geometries reduce production risk in corn and sorghum in water-limited environments. Agronomy Journal, 112(5):3322–3334. doi:10.1002/agj2.20347

Tubiello F, Rosenzweig C, Goldberg R, Jagtap S, Jones J, 2002. Effects of climate change on US crop production: simulation results using two different GCM scenarios. Part I: wheat, potato, maize, and citrus. Climate research. 20(3):259–270.

USDA, 2022. Census of agriculture. Retrieved from https://www.nass.usda.gov/Publications/AgCensus/2022/index.php (accessed January 19, 2025)

USDA-NASS. 2010. Field crops usual planting and harvesting dates. US Department of Agriculture Washington, DC.

USDA-NASS, 2024. US department of agriculture, national agricultural statistics service, quick stats. Retrieved from https://quickstats.Nass.Usda.Gov/ (accessed January 20, 2025)

van der Kooi CJ, Reich M, Löw M, De Kok LJ, Tausz M, 2016. Growth and yield stimulation under elevated CO2 and drought: A meta-analysis on crops. Environmental and Experimental Botany, 122:150–157. doi:10.1016/j.envexpbot.2015.10.004

Wang C, Guo L, Li Y, Wang Z, 2012. Systematic comparison of C3 and C4 plants based on metabolic network analysis. BMC Systems Biology, 6(2):S9. doi:10.1186/1752-0509-6-S2-S9

Watson S. 2017. Texas peach crop may be in short supply. Texas Agriculture Daily. Retrieved from https://texasfarmbureau.org/texas-peach-crop-may-short-supply/ (accessed March 10, 2025).

Wen N, Marek GW, Srinivasan R, Brauer DK, Qi J, Wang N, Han Y, Zhang X, Feng P, Liu DL, Chen Y, 2024. Assessing the impacts of long-term climate change on hydrology and yields of diversified crops in the Texas High Plains. Agricultural Water Management, 302:108985. doi:10.1016/j.agwat.2024.108985

Whitlock J. 2021. Texas dairy producing steady stream despite challenges. Texas Agriculture Daily. https://texasfarmbureau.org/texas-dairy-producing-steady-stream-despite-challenges/ (accessed March 10, 2025).

Wilson KB, Carlson TN, Bunce JA, 1999. Feedback significantly influences the simulated effect of CO2 on seasonal evapotranspiration from two agricultural species. Global Change Biology, 5(8):903–917. doi:10.1046/j.1365-2486.1999.00280.x

Yang X, Menefee D, Cui S, Rajan N, 2021. Assessing the impacts of projected climate changes on maize (Zea mays) productivity using crop models and climate scenario simulation. Crop and Pasture Science, 72(12):969–984.

Zhao H, Zhang L, Kirkham MB, Welch SM, Nielsen-Gammon JW, Bai G, Luo J, Andresen DA, Rice CW, Wan N, Lollato RP, Zheng D, Gowda PH, Lin X, 2022. U.S. winter wheat yield loss attributed to compound hot-dry-windy events. Nature Communications, 13(1):7233. doi:10.1038/s41467-022-34947-6

III.2 Extreme Temperatures, Air Quality, and Public Health

Melissa R. Allen-Dumas and TC Chakraborty

III.2.1 EXTREME HEAT

Extreme heat poses an escalating threat to Texas, driven by a confluence of factors, including background climate change with significant public health implications. Of note, urban areas also tend to have higher local temperatures, the urban heat island effect, with disproportionate impacts on rapidly urbanizing populations. This chapter examines the multifaceted nature of extreme heat in Texas, beginning with an analysis of long-term warming trends. Reanalysis datasets reveal general increases in statewide average maximum and minimum air temperatures, with minimum temperatures, in particular, rising at a faster rate, thereby reducing the diurnal temperature range. Spatially, recent warming ([2020–2024] vs. [1950–1954]) is pronounced, especially for minimum temperature, which has increased by over 1°C across large portions of the state. Urban areas exhibit distinct localized warming or heat islands, with mean daytime surface urban heat island (SUHI; based on land surface temperature (LST)) intensities of around 0.82°C (i.e. 0.82°C higher LST than their rural backgrounds) and mean air temperature or canopy urban heat island (CUHI) intensities near 0.09°C. The SUHI can exceed 2.5°C in some major urban cores. Critically, intra-urban heat distribution is often inequitable, with evidence suggesting that lower-income neighborhoods frequently experience higher land surface and air temperatures compared to wealthier areas, indicating disparities in heat hazard and exposure. Finally, a comparison of surface temperature and air temperature over urbanized areas in Texas establishes the importance of using physiologically relevant heat metrics for understanding public health implications. Of note, while LST is often used to estimate local urban warming impacts, it is often a weak proxy for heat hazard. Despite decreasing relative humidity, the heat index (HI) in Texas shows an increasing trend (approx. 0.027°C/year), primarily driven by rising air temperatures (approx. 0.026°C/year), indicating an intensification of overall moist heat experienced by the population. This chapter underscores the complex and worsening nature of extreme heat in Texas, necessitating targeted strategies for mitigation and adaptation.

III.2.1.1 A Warming Climate

Texas has experienced long-term warming (Nielsen-Gammon, 2011), a trend evident in multiple meteorological datasets (Figure III.2.1). Long-term records of summer average maximum and minimum 2-m air temperatures (T2) reveal consistent warming trajectories for most datasets. For instance, the European Centre for Medium-Range Weather Forecasts (ECMWF) Re-analysis v5 (ERA5)-Land re-analysis dataset (Muñoz-Sabater et al., 2021) (spanning 1950–2023) indicates an increase in statewide T2 at a rate of approximately 0.009°C per year and minimum T2 at 0.016°C per year (Figure III.2.1). The Gridded Surface Meteorological (GRIDMET) dataset (Abatzoglou, 2013) (1979–2023) shows even steeper increases, with maximum temperatures rising at 0.028°C per year and minimums at 0.031°C per year. Notably, average minimum temperatures (nighttime lows) are increasing at a faster rate than average maximum temperatures (daytime highs) for all the datasets. This reduction in diurnal temperature range is expected for various factors that influence long-term warming, including greenhouse gases (GHGs), aerosols, cloud cover feedback, and land use change (Chakraborty and Lee, 2019; Kalnay and Cai, 2003; Karl et al., 1991). This diminishing diurnal temperature range has significant public health implications, reducing the potential for nighttime

DOI: 10.1201/9781003406600-15

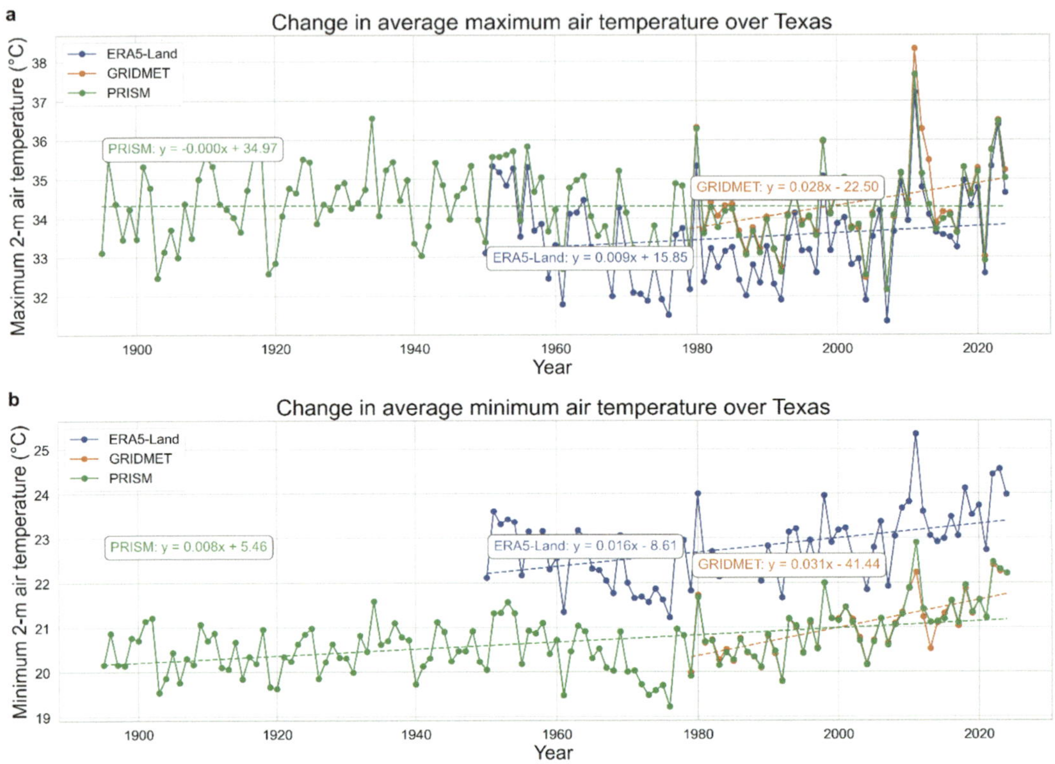

FIGURE III.2.1 Long-term trends in average a maximum and b minimum 2-m air temperature from ERA5-Land, GRIDMET, and PRISM over Texas for northern hemisphere summer (June, July, August). The lines of best fit regressed against the years are shown.

cooling and recovery from daytime heat exposure (He et al., 2022; Murage et al., 2017; Obradovich et al., 2017). The Parameter-elevation Regressions on Independent Slopes Model (PRISM) dataset (Daly et al., 2008) (1895–2023) shows a flatter trend for maximum temperatures but a positive trend for minimum temperatures. The flat trend for PRISM is for a much longer period, and it should be noted this includes the 1930s Dust Bowl, which was an extreme event that led to some of the highest temperatures in Texas and other states in the Great Plains (Donat et al., 2016).

Spatially, the warming in Texas has not been uniform across the state. A comparison of average T2 between the recent period 2020–2024 and a mid-20th century baseline (1950–1954) highlights distinct regional patterns (Figure III.2.2). Average maximum temperatures show pronounced warming, exceeding 2.0°C in extensive areas of West and North Texas, while some parts of East and Southeast Texas exhibit slight cooling or less significant warming. Conversely, average minimum temperatures demonstrate widespread and substantial increases across virtually the entire state, with many regions, particularly in West, Central, and South Texas, experiencing rises of 1°C to over 1.5°C. This robust increase in minimum temperatures further emphasizes the trend of warmer nights, which can exacerbate heat stress, particularly for vulnerable populations.

III.2.1.2 Urban Heat Across Scales

III.2.1.2.1 The Urban Heat Island Effect

Urban areas tend to be warmer than their surroundings, also known as the urban heat island (UHI) effect (Oke, 1982). With the majority of U.S. population, and around 84% of Texas population,

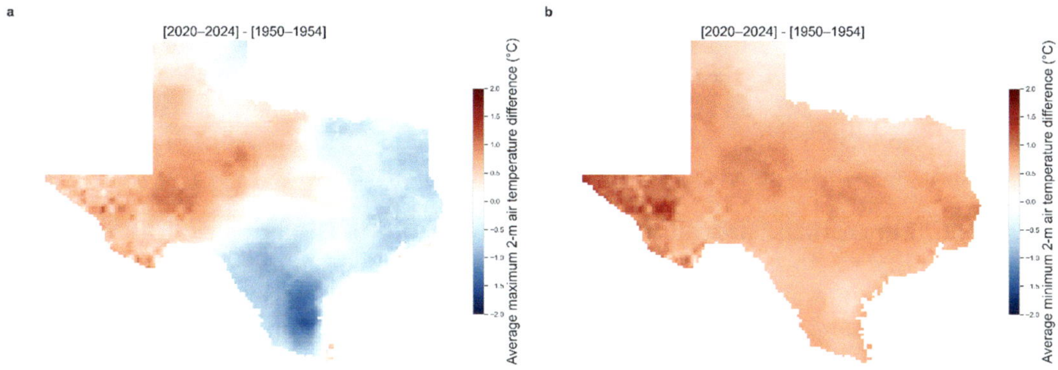

FIGURE III.2.2 Spatial variability of the change in average a maximum and b minimum 2-m air temperature from ERA5-Land between 1950–1954 to 2020–2024 over Texas for Northern Hemisphere summer (June, July, August).

living in urban areas, the UHI has the potential to impact human health disproportionately. The UHI intensity can be quantified using both land surface temperature or LST (namely the SUHI) and near-surface T2 (canopy urban heat island, CUHI) with different implications (Venter et al., 2021) (see next section). However, note that most of the gridded datasets considered above do not have this urban signal, or have weaker than expected urban signals (Chen et al., 2024; Nogueira et al., 2022). This is because the weather station data assimilated into these datasets are rarely available within cities (Muller et al., 2013). As such, a majority of studies on urban thermal environment have focused on LST, which we explain in more detail below. We also leverage a recent urban-resolving 1 km dataset (Zhang et al., 2022) to also discuss the CUHI and contrast the two signals.

Urban environments within Texas, based on U.S. census urbanized areas (2010 Census Urban and Rural Classification and Urban Area Criteria, n.d.), exhibit clear UHI signals. During the day, the mean SUHI reaches approximately 0.82°C (Figure III.2.3), while the mean daytime CUHI is about 0.09°C (Figure III.2.5). These are based on an updated version of a publicly available nationwide UHI dataset (Chakraborty et al., 2020). At night, the mean SUHI is 0.32°C, and the mean nighttime CUHI is also around 0.09°C. The UHI intensity varies by city. Maps of summer daytime SUHI show intensities of close to 2.5°C or even more in the cores of major metropolitan areas like

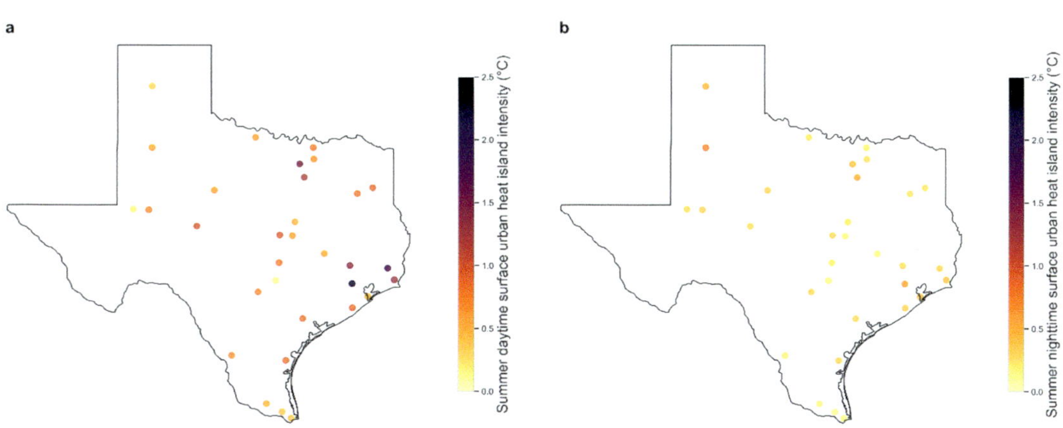

FIGURE III.2.3 Five-year (2013–2017) average summer a daytime and b nighttime surface urban heat island intensities for urbanized areas in Texas based on MODIS Aqua daily land surface temperature product.

Houston, Dallas-Fort Worth, and San Antonio. Nighttime SUHI, while generally of lesser magnitude, persists across these urban areas. Canopy UHI intensities are typically lower than SUHI but still present a consistent local warming effect. Daytime CUHI and nighttime CUHI across Texas cities generally show values up to approximately 0.5°C. While these CUHI values are smaller than SUHI, they represent a direct increase in the T2 experienced by urban dwellers, contributing to increased energy demand for cooling and potential health impacts.

III.2.1.2.2 Intra-Urban Variability in Heat and Relevance to Disparities in Hazard

Heat hazard is not uniform even within a single city. Significant intra-urban temperature variations exist, mainly driven by the heterogeneity of urban environments Buyantuyev and Wu, 2010). In the U.S., heat hazards tend often to covary with socioeconomic factors (Voelkel et al., 2018), thereby raising concerns about disparities in urban environmental hazards. Multiple previous studies have shown the pervasiveness of these disparities (Benz and Burney, 2021; Chakraborty et al., 2020; Hsu et al., 2021; McDonald et al., 2021). Here, we look at a subset of the nationwide dataset to explore the relationship between temperature and income in Texas cities. For instance, analysis of summer daytime LST sensitivity to median household income shows that in several urban locations, higher income levels are associated with significantly lower LSTs, with sensitivities of up to −1°C per $10,000 higher median income (red markers in Figure III.2.4a). Conversely, some areas

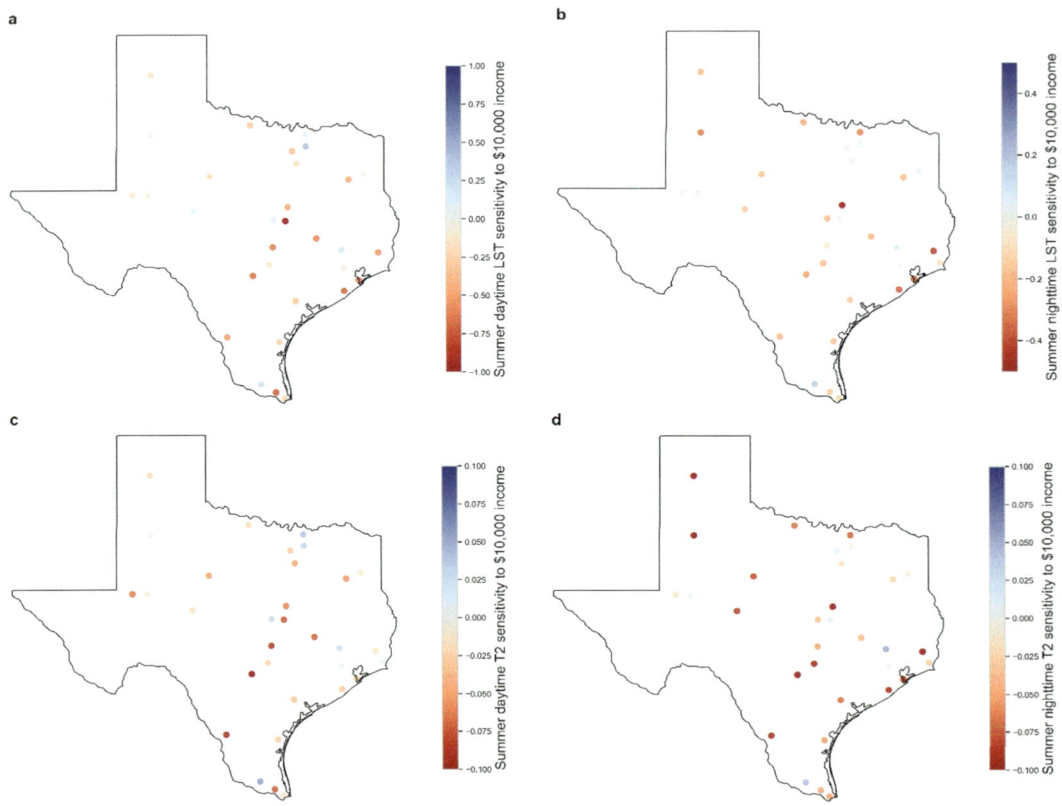

FIGURE III.2.4 Sensitivity of average summer a daytime and b nighttime LST and c average maximum and d average minimum 2-m air temperature (T2) to a $10,000 higher median income for urbanized areas in Texas. The LST and T2 are for 2013–2017 and the income data is from the American Community Survey. The LST data are based on MODIS Aqua daily land surface temperature product while T2 is from Zhang et al. (2022).

show a positive correlation, where higher income is linked to higher LSTs (blue markers), although these instances appear to be less frequent (only seen for around 20% of the urbanized areas) or of smaller magnitude. Similar trends, albeit with generally smaller sensitivity magnitudes (slopes between approximately −0.4 and +0.4°C per $10,000 higher income), are observed for nighttime LST (Figure III.2.4b).

These disparities are also evident in near-surface T2. Data on summer daytime T2 sensitivity to income indicates that many urban areas exhibit a negative correlation, with sensitivities reaching −0.100°C per $10,000 higher income, signifying that wealthier neighborhoods tend to experience lower T2 (Figure III.2.4c). This pattern generally persists for nighttime T2 as well (Figure III.2.4d), where higher-income areas are often associated with cooler conditions. These findings suggest that lower-income communities in Texas may bear a disproportionate burden of heat exposure, a critical consideration for public health interventions and urban planning aimed at mitigating heat-related risks.

III.2.1.3 Physiologically Relevant Heat Stress

III.2.1.3.1 Land Surface Temperature Versus Air Temperature

From a human health perspective, it is important to distinguish between LST and T2, as they have different implications for health (Ho et al., 2016). In general, radiative skin temperature directly influences the amount of longwave radiation emitted from the ground and other surfaces, which means higher mean radiant temperature, a key component of human thermal comfort and heat stress. However, satellites only provide bulk estimates of the temperature of the surfaces they see, which we refer to as LST (Du et al., 2023). This includes a combination of surfaces, including roads, roofs, and walls, along with other urban features (Stewart et al., 2021). As such, it is generally difficult to match satellite-derived LST with estimates of heat hazard relevant for human exposure at the pedestrian scale (Li et al., 2023). Quantifying heat hazards using LST is particularly problematic as, even when examining the LST of a flat surface, we would rarely experience that temperature unless we were walking barefoot on the ground. T2, in comparison, is the more commonly relevant variable modulating direct heat exchange between the human body and the surrounding air. The urban impact on LST and T2, or the SUHI and CUHI, respectively, are also substantially different in magnitude and diurnality (Venter et al., 2021). Overall, the mean daytime SUHI (0.82°C) is substantially greater than the mean daytime CUHI (0.09°C). Even at night, the mean SUHI (0.32°C) remains higher than the mean CUHI (0.09°C). Generally, SUHI shows higher daytime values while CUHI shows higher nighttime values, although the dataset we use here does not sufficiently account for this latter signal. The different estimates in SUHI and CUHI make sense since local land cover changes, such as urbanization, would more strongly impact LST than it would impact T2. The latter is strongly moderated through advection.

The differences between LST and T2 also matter when examining disparities in heat hazard (Chakraborty et al., 2023). As seen in Figure III.2.4, the magnitude of disparity is much higher for LST compared to T2, which follows the higher spatial variability expected of LST. As such, while LST is useful for large-scale analyses such as these, one must be cautious about quantifying the strength of unequal exposures using LST alone (Figure III.2.5).

III.2.1.3.2 Moist Heat Stress

Beyond dry-bulb T2, the human experience of heat is profoundly modified by atmospheric moisture (Raymond et al., 2020). High humidity levels impede the body's primary cooling mechanism—evaporation of sweat—thus exacerbating heat stress (Sherwood and Huber, 2010). Measures for moist heat incorporate both temperature and humidity to provide a more physiologically relevant assessment of how hot it feels. Across scales, humidity tends to have negative feedback to moist heat, since many climate-related perturbations that increase temperature (or vice versa) tends to

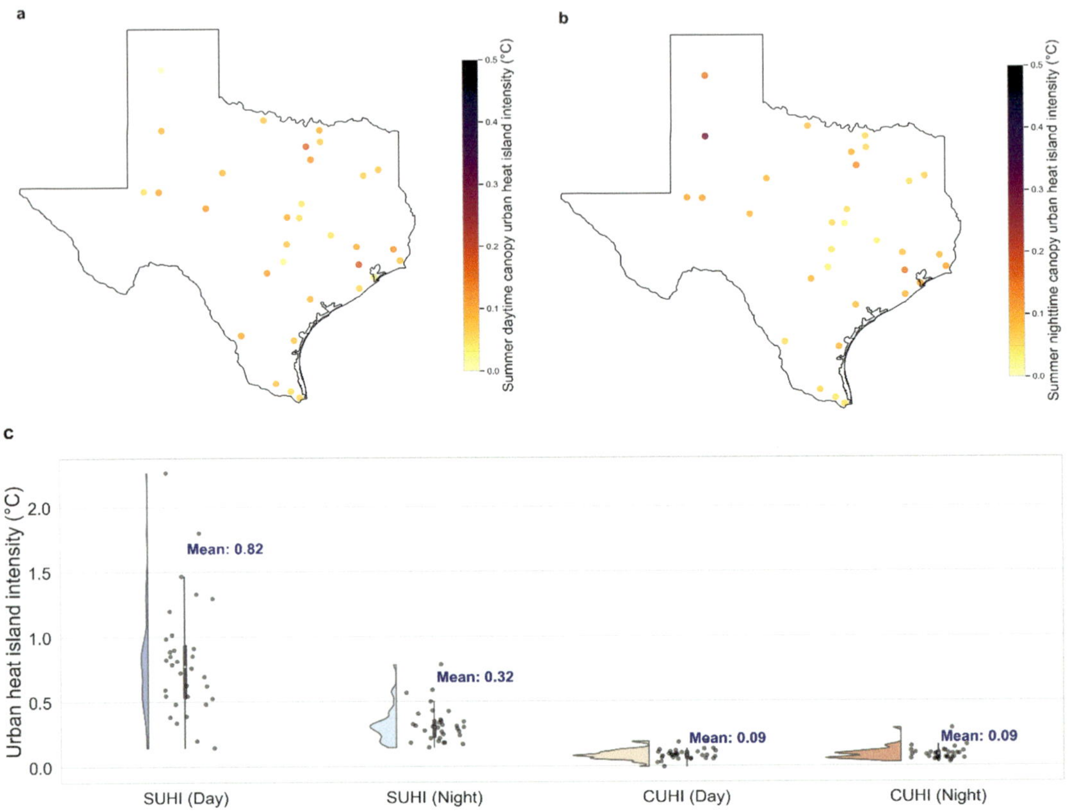

FIGURE III.2.5 Five-year (2013–2017) summer average (a) maximum and (b) minimum canopy urban heat island intensities for urbanized areas in Texas based on air temperature estimates from Zhang et al. (2022). Sub-figure (c) shows distribution of daytime and nighttime surface urban heat islands and corresponding canopy urban islands for urbanized areas in Texas.

decrease relative humidity (RH2) (Chakraborty et al., 2022; Chakraborty et al., 2025). Over Texas, the average minimum RH2 from ERA5-Land decreased between 1980 to 2023 (−0.071% per year) (Smith and Chang, 2020), while, as already seen previously, the average maximum T2 increased (0.026°C per year). Since RH2 has the opposite diurnal cycle as T2 (it is generally minimum when T2 is maximum), we can combine these values to get a reasonable estimate of average maximum moist heat (Chakraborty et al., 2023). Here, we calculate two metrics for moist heat stress (see Rothfusz (1990) and Lee (2018) for methods), the HI, used operationally by the U.S. National Weather Service, and the wet-bulb temperature (WBT), often used as proxy for moist heat in the geoscience literature. These two metrics have very different responses to humidity, with HI being weakly influenced by humidity and WBT being strongly modulated by humidity (Sherwood, 2018). Both average maximum HI and WBT show increases over Texas, suggesting an increase in moist heat, and thus the ability to survive under prolonged exposure, in Texas over the decades. Note that while WBT is generally a poor proxy of heat stress on a continuous scale under most realistic conditions (Chakraborty et al., 2025), it is shown here since it represents the upper bound of how much the change in RH2 can moderate moist heat. Basically, the significant decrease in RH2 is not sufficient to negate the impact of rising temperatures on moist heat stress. These trends collectively underscore an escalating risk of physiologically significant heat stress across Texas (Figure III.2.6).

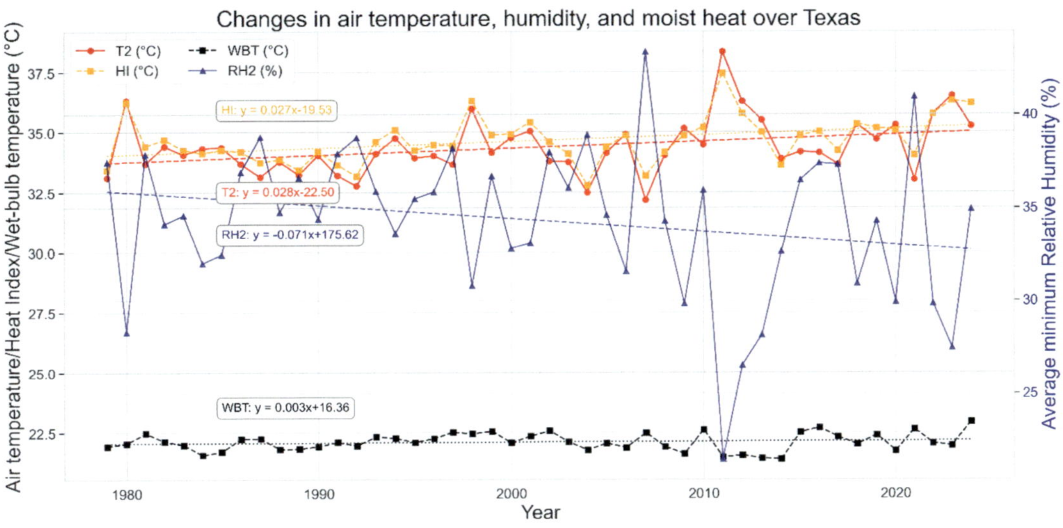

FIGURE III.2.6 Long-term trends in average maximum 2-m air temperature, Heat Index, and wet-bulb temperature and average minimum relative humidity from ERA5-Land for northern hemisphere summer (June, July, August). The lines of best fit regressed against the years are shown.

III.2.2 EXTREME COLD

While most publicity around climate change focuses on overall warming, some of the changes in local climate are coming in the form of anomalous extreme cold events. Many scientific studies are showing that changes to the polar vortex due to Arctic warming are stretching the waves of the polar jet stream and weakening it enough to pull Arctic air as far south as southern Texas (e.g., Cohen et al., 2021; McPherson et al., 2023). These exaggerations of the polar jet can sometimes stall the jet stream in its rotation around the Earth allowing large and lasting winter storms to develop. We describe the climate mechanisms behind this type of storm and identify three recent Texas winter storms that exemplify this phenomenon. Winter Storm Uri (February, 2021) was the most publicized storm of this type (e.g., Bolinger et al., 2022), but other storms such as Winter Storm Landon (February, 2022) and Winter Storm Mara (February, 2023) followed similar patterns.

III.2.2.1 Climate Variability: Teleconnections

Although the overall average temperature of Earth is warming, the regional impacts of climate change manifest as different meteorological phenomena in different parts of the Earth. These differing geographical meteorological realizations can be connected to processes occurring in parts of the Earth that are seemingly disconnected from experienced weather. Climate teleconnections are significant relationships between weather phenomena at widely separated locations on earth (Zavadoff and Arcodia, 2022). These weather phenomena are tied to climate patterns that span thousands of miles. Some of these teleconnections are related simply to the Earth's rotation and resulting atmospheric circulation. Some of them are related to atmospheric waves that result from the Earth's rotation (Section 1) including Rossby Waves (Section 2), and others are due to decadal cycles of ocean heating and cooling (Section 3).

III.2.2.1.1 Global Atmospheric Circulation

The global atmosphere can be characterized as a continuous fluid medium encircling planet Earth. This continuum is governed by the fundamental physical laws of conservation of mass, momentum and energy and is subject to body and surface forces such as gravity, inertia, and pressure (Holton, 2004). Gravity affects mass elements (molecular constituents) of the atmosphere as a force related to the square of the distance between the center of mass of the earth and the center of mass of the atmospheric mass element. Fluids are subject to friction, or viscosity, acting as a force resisting fluid flow. The pressure gradient force on a mass element (for this discussion, in the atmosphere) is the change in pressure with the density and volume of the mass element (Holton, 2004). As features on Earth are heated by the sun and then cooled throughout the Earth's rotation, changes in atmospheric pressure occurs contributing to the high- and low-pressure systems we experience as weather.

III.2.2.1.2 Rossby Waves

Atmospheric Rossby waves, or planetary waves, are large-scale, high-altitude waves. These waves are generated in wind due to the Earth's rotation and variations in the Earth's curvature with latitude (Holton, 2004). Changes in these wind waves are associated with weather-related high- and low-pressure systems, the jet stream and the polar vortices (Figure III.2.7). Deviations in the generally sinusoidal configuration of these waves can occur due to sudden stratospheric warming (SSW) events, that can weaken or reverse the stratospheric polar vortex (SPV) causing the polar vortex to stretch southward pulling Arctic air into lower latitudes (Baldwin et al., 2021; Butler et al., 2017). This thermally driven stratospheric wind system usually develops in winter, and its highest winds typically occur near 60°N (Waugh et al., 2017).

III.2.2.1.3 Decadal Cycles

Some of the large changes seen in various regions of the global climate are a result of natural climate variability. This variability occurs in different global regions on scales of years to decades

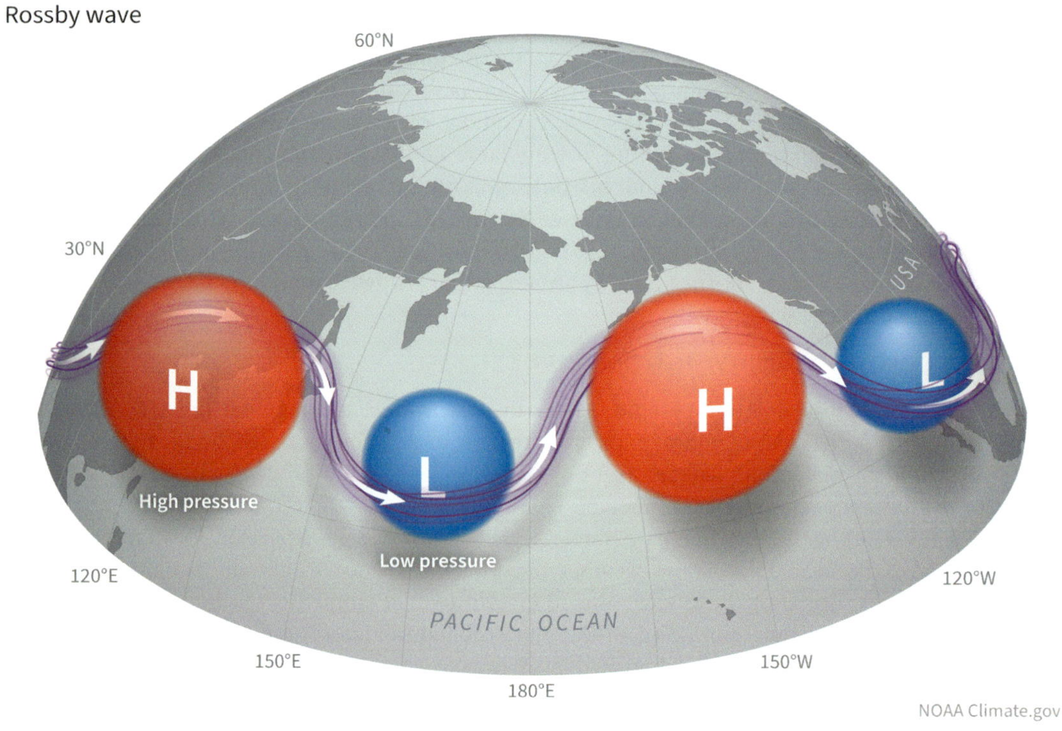

FIGURE III.2.7 Rossby wave perturbation of the Northern Hemisphere mid-latitude jet stream resulting in warmer, high-pressure air pushing poleward and cooler, lower-pressure air pulling equatorward. NOAA Climate.gov.

and is often represented as indices associated with those regions. Five of the most well-known of these *decadal cycles* are the El Niño-Southern Oscillation (ENSO), the Pacific Decadal Oscillation (PDO), the East-Pacific-North Pacific teleconnection pattern (EP-NP), the Atlantic Multidecadal Oscillation (AMO), and the North Atlantic Oscillation (NAO). Although each of these cycles are defined within specific ocean regions and are measured across different time scales, all can impact environments around the world. Some of these impacts include severe storms or mild weather, drought or flooding (NOAA, 2024). For example, during the El Nino period of the ENSO cycle, Texas usually experiences lower temperatures and higher rainfall. During the La Nina period, the weather is hotter and drier (CPC, 1997).

Hsu et al. (2022) demonstrated that a large contributor to the February 2021 Winter Storm Uri was due to the confluence of a few of these decadal cycles. They cited a negative Pacific–North American pattern (PNA)—that is, a weaker jet stream across the Central Pacific Ocean along with a high-pressure blocking of atmospheric flow in the northern North Pacific; a lower than average ENSO sea surface temperature anomaly in the equatorial eastern Pacific; and a sea surface temperature warming trend in the subtropical North Atlantic (related to the NAO).

III.2.2.2 Climate Change Contributors to Recent Winter Storms in Texas

The geneses of three Texas winter storms that severely affected Texas residents in three recent subsequent years were due in part to the climate variability discussed in Section III.2.2.1, along with observed Arctic warming.[1] These three storms were given the names used here by the Weather Channel because of their severity. Because large winter storms are rare historically in Texas, the impacts of these storms were not foreseen by city planners or system managers, although studies had shown that days with two or more inches of precipitation had become more frequent across the Southern Great Plains, with larger increases in the eastern half of the region than the western half (Kirchmeier-Young and Zhang, 2020; McPherson et al., 2023). Each storm caused record damage to people and infrastructure resulting in high restoration costs and intensification of planning for future extreme weather events. While these events taken in isolation may not specifically be indicators of large-scale climate change, together they may represent the implications of part of that predicted trend.

III.2.2.2.1 Winter Storm Uri (February 13–17, 2021)

In early January 2021, the upper stratosphere in the Northern Hemisphere rapidly warmed in response to planetary-scale waves that disrupted the normal circulation. This disruption was a precursor to the February 2021 Texas extreme winter storm named Winter Storm Uri (Bolinger et al., 2022). Cohen et al. (2021) attribute the storm development to an atmospheric response to SPV stretching. Referred to also as the "Valentine Storm," the U.S. National Weather Service (NWS 2021) at the time alerted the public to "a strong Arctic cold front moving across the country," which was bringing snow, sleet, and freezing rain to the South-Central region and would affect 1,854 counties in all.

III.2.2.2.2 Winter Storm Landon (February 1–9, 2022)

Unlike Winter Storm Uri, Winter Storm Landon was caused not only by perturbations to the jet stream. It was also the result of the meeting of that cold, dry air mass from the jet stream moving south from the northern great plains and warm, wet air from the Gulf of Mexico moving inland. The meeting of the two air masses resulted in a band of weather spanning the eastern U.S. diagonally from Dallas to Maine, throughout which snow, sleet, and rain fell for nine days. Along with the precipitation were strong winds from tornadoes that formed because of the added atmospheric energy on the warm side of the storm. Ahead of this storm, the National Oceanic and Atmospheric Administration (NOAA) had measured the Gulf temperature two degrees Fahrenheit warmer than normal, and water near the Alabama and Louisiana coasts eight to ten degrees Fahrenheit warmer than normal (Kiefer, 2022).

III.2.2.2.3 Winter Storm Mara (January 31–February 2, 2023)

Winter Storm Mara moved across the South-Central Plains as primarily an ice storm which started on January 29 as an arctic cold front dipping southward through the Southern Plains (NWS, 2023). Framed by a low-pressure system to the south and a high-pressure system to the north, the storm included rain, freezing rain and sleet, and extended from Texas through Oklahoma and Arkansas and northeast through West Virginia (Hernandez, 2023).

III.2.2.3 Texas Impacts from the Winter Storms

Multiple impacts were observed in Texas after each of the three extreme winter storms. Of the three storms, Winter Storm Uri caused the most deaths in the state at 246 people (Svitek, 2022). Eight people in Texas died in during Winter Storm Mara (Cabral and Yousif, 2023), while only three deaths in Texas occurred as a result of Winter Storm Landon. These deaths occurred due to vehicle accidents on the icy roads. However, many additional impacts to infrastructure resulted from the storm, and some of these impacts affected human health and the economy as well.

Winter Storm Uri caused outages in power, water, communication, and transportation infrastructure. As a result of these outages, emergency and medical services were also affected. The storm caused the largest blackout (over 9.9 million customers, including 4.5 million Texans) in the U.S. since the Northeast blackout of 2003 (Sullivan and Malick, 2021). More than 40% of customers in Austin, including hospitals, were without power by the fifth day of the storm. The power outages caused water treatment plants to become inoperable, and boil water notices were issued in cities such as Houston, San Antonio, Fort Worth, Abilene, Austin, Kileen, and Arlington. However, communication strategies to increase public awareness of City and County services during disasters were not in place during the storm. For the services that were available, icy and impassable roads reduced access to these services. Additionally, medical resources were difficult to reach, and transport of patients from hospitals without water to those with water was treacherous. Fire and other emergency responders lacked necessary equipment suitable for the road conditions, and numerous accidents were reported on interstate highways (HSEM, 2021; WFAA, 2021).

By the third day of Winter Storm Landon, 70,000 people in Texas had lost power. The storm's icing conditions across the remaining states caused an additional 300,000 customers to lose power (Kiefer, 2022). In Texas, a research and analytics team named "CrisisReady" found that the north-central and northeastern counties of Lamar, Fannin, Eastland, Noland, Runnels, and Delta counties contained high numbers of electricity-dependent medical equipment users who were significantly affected by the outages (Nallen, 2022).

Winter Storm Mara downed significant numbers of tree limbs, many of which struck power lines leading to some fraction of the overall 500,000 power outages reported (Childs and Brackett, 2023). Ice accumulation due to the storm caused several hundreds of traffic accidents across the region (NWS, 2023).

Of the three storms, Winter Storm Uri was the costliest. It was also the costliest natural disaster ever to occur in both Texas and the United States with over $195 billion in damage in Texas alone (HSEM, 2021). For the first quarter (January, February, March) of the year 2022, the U.S. recorded $6 billion in economic losses. Much of this total is attributed to winter storms and severe convective storms occurring during that time (AON, 2022). Damage associated with Winter Storm Mara was around $86 million (NCEI, 2023).

III.2.3 AIR QUALITY DEGRADATION

Not only do emissions (e.g., carbon dioxide and GHGs) contribute to climate change, but impacts of emissions from both natural and human sources on human health can also be exacerbated by climate change. We discuss here how emissions of various substances to the atmosphere and the atmosphere itself are not unrelated, but they are part of what we can think of as a chemistry-climate system. Left unmanaged, this system can contribute to and be affected by both local consequences and imbalances in the Earth's global energy budget.

III.2.3.1 Climate Influence

Changes in climate are triggered by changes in the Earth's energy balance due to changes in one of the components of that balance, such as increases or decreases in carbon dioxide or other heat-absorbing atmospheric gas concentrations or in solar radiation reaching the Earth. These changes are characterized by a quantity called radiative forcing, which represents the change in the energy available to the climate system caused by a change in one or more of those components (North et al., 2015). The chemistry of the atmosphere can both affect and be affected by changes in the climate. The combination of these effects comprises the chemistry-climate system (Figure III.2.8).

Radiative forcing can trigger interactive changes within the chemistry-climate system resulting in changes to surface air quality. These changes can manifest as perturbations in the rising of air from the surface, local wind speed and larger-scale frontal passages, changes in precipitation patterns, chemical production and loss rates, and emissions from natural sources, resulting in changes in local and global atmospheric composition and background concentrations (Jacob and Winner, 2009). Changes in precipitation patterns can lead to more occurrences of drought and flooding (Trenberth, 2005). Wildfire risk increases with long drought periods during which vegetation dries becoming more susceptible to disease and ultimately more flammable (An et al., 2015). Airborne particulate matter from these events can spread raising pollution in nearby areas. Alternatively, lingering moisture due to flooding can lead to mold inside buildings (West et al., 2023). Finally, potential increases in volatile organic compound (VOC) emissions from vegetation and soil nitrous oxide (N_2O) emissions can occur with atmospheric warming (Forkel and Knoche, 2006). These compounds are precursors for the generation of ground-level ozone, which, in large quantities, is

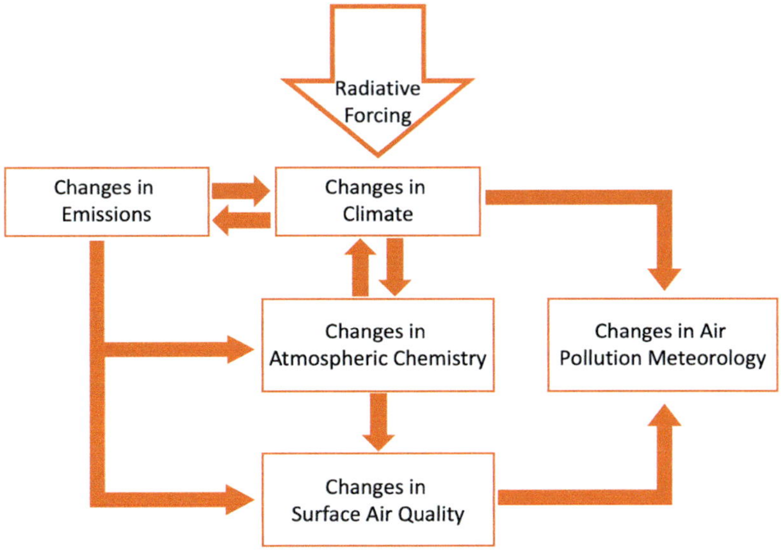

FIGURE III.2.8 The Chemistry-Climate System (after Jacob and Winner, 2009). Radiative forcing affects changes in climate, which are also affected by changes in emissions. Changes in climate affect and are affected by changes in atmospheric chemistry. These changes in atmospheric chemistry are also affected by changes in emissions as they affect changes in surface air quality. Changes in surface air quality are also affected by changes in emissions. Changes in both climate and surface air quality in turn, produce changes in air pollution meteorology.

harmful to human health (Ebi and McGregor, 2008). Increased abundance of these compounds from natural sources can amplify quantities produced by industrial sources.

III.2.3.1.1 Impacts of Atmospheric Warming

Annual average temperatures have increased from 1900 to 2020: 1.5°F for Texas (Runkle et al., 2022). While rising warm air can lift pollutants in the atmosphere out of an area, it can also lead to conditions more favorable to ozone production. Additionally, when a warm layer of air caps a cold layer of air, a temperature inversion occurs that can trap particulate matter and other toxic air substances in low-lying areas (Hou and Wu, 2016). Also, because a warmer atmosphere can hold more moisture before becoming fully saturated and producing rain, precipitation patterns are changing from smaller more frequent rain events to periods of drought followed by large rain events (Adam, 2023). This relationship, articulated mathematically, by among others, Rudolf Clausius, a German physicist (1850), and French engineer and physicist Émile Clapeyron (1834), can affect the ambient relative humidity available to condense on aerosols and particulate matter to produce rain to remove these pollutants from the atmosphere. It can also lead to drought conditions that are conducive to wildfires or, alternatively, produce the type of flooding that soaks buildings and other infrastructure and leads to mold production (Zingales et al., 2022).

III.2.3.1.2 Changes in Air Pollution Due to Regional Warming and Local Sources

Heightened air pollution due to heat waves and temperature inversions have been shown to be worsening in a variety of North American locations. Many of these locations are in Texas, where heat waves are predicted to increase further by up to 10% and temperature inversions are predicted to increase by up to 20%. More of these increases are in eastern Texas than in western Texas (Hou and Wu, 2016). While it is known that atmospheric warming is due primarily to atmospheric gas absorption of heat radiating from the ground due to its capture of energy from the sun (Trenberth et al., 2009), these atmospheric gases themselves can be categorized as pollutants as well (Jacobson, 2012). While many gases contribute to air pollution, we describe only those defined as criteria air pollutants according to the Clean Air Act (1970) National Ambient Air Quality Standards (NAAQS): Carbon Dioxide (CO_2), Carbon Monoxide (CO), Ozone (O_3), Nitrogen Dioxide (NO_2), and Sulfur Dioxide (SO_2). We also describe NAAQS particulate matter pollutants in the atmosphere as these particles (aerosols) cause even more severe health problems than the gases (USC, 2013).

III.2.3.1.2.1 Carbon Dioxide

Both CO_2 and CO are colorless odorless gases. While CO_2 does not directly impact health at typical outdoor air concentrations, it can produce city-scale carbon dioxide domes that increase local temperatures and water vapor, both of which increase concentrations of other pollutants such as ozone and particulate matter that damage human and vegetative health (Jacobson, 2010). Local sources of CO_2 include evaporation from the Pacific Ocean and the Gulf of Mexico, bacterial decomposition of dead organic matter, biomass burning (including wildfire), fossil fuel and biofuel combustion, bacterial fermentation for food and beverage production, and atmospheric chemical reaction (Jacobson, 2012).

III.2.3.1.2.2 Carbon Monoxide

CO is an important component of urban and indoor air pollution because it has harmful short-term health effects. It is the most abundantly emitted gas after water vapor and CO_2, and it absorbs some wavelengths of radiant heat leading to higher local atmospheric temperatures. Local sources of CO include plant and other biological activity in the oceans, fossil fuel and biofuel combustion, biomass burning, and atmospheric chemical reaction (Jacobson, 2012).

III.2.3.1.2.3 Ozone

Ozone (O_3) is a short-lived GHG produced by chemical reactions of precursor species in the atmosphere (Solomon et al., 2007). Precursor species can include oxides of Nitrogen (NOx), VOCs,

Methane (CH4), and CO, among others (Jacobson, 2012). Nitrous oxide, emitted from soils, can react with oxygen to produce the O_3 precursor, nitric oxide (NO). Peak O_3 concentrations are typically observed during periods of high temperature and increased sunlight because photons from sunlight are required to begin the chemical reactions that produce O_3. At temperatures above 90°F (32°C), a strong positive association has been found between temperature and ground-level O_3 production (Knowlton, 2004; Patz, 2000). Additionally, high temperatures increase ozone production more in polluted air than in clean air. Enhancements above 100% in seasonal averages of O_3 concentration due to fire events in northern, western, eastern, and southern Texas from 2001–2010 have also been shown (Hou and Wu, 2016).

III.2.3.1.2.4 Nitrogen Dioxide

A brown gas with a strong odor, NO_2 is a precursor for O_3 formation and the production of nitric acid, a component of damaging acid rain to both vegetation and built infrastructure on which it deposits. Sources of nitrogen dioxide are biomass burning, fossil fuel and biofuel combustion, and atmospheric chemical reaction. Many atmospheric reactions that produce NO_2 start with NO emissions, which occur in part due to soil denitrification and lightning (Jacobson, 2012).

III.2.3.1.2.5 Sulfur Dioxide

SO_2 is a precursor to the aerosol particle component, sulfuric acid (H_2SO_4) that can deposit on to surfaces in acid rain. Although it is a colorless gas, at high enough concentrations in the atmosphere, it can be perceived as taste or odor. Sources of SO_2 include atmospheric chemical reaction of biological emissions, fossil fuel combustion, chemical manufacturing, and mineral ore processing (Jacobson, 2012). Ore mining and processing has waned recently in Texas, but it was once a strong economic enterprise in the Trans Pecos region (Figure III.2.9) of west Texas (Becker and Kyle, 2011).

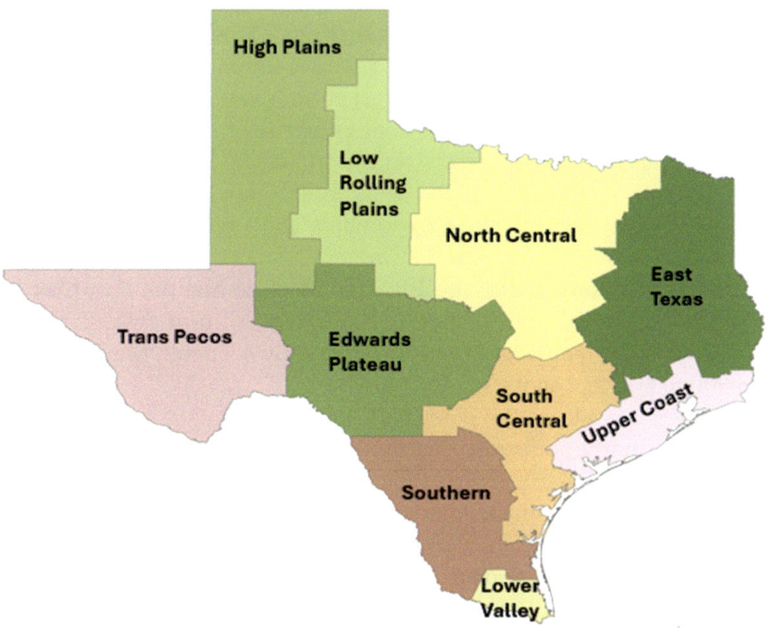

FIGURE III.2.9 Texas Climate Zones. Texas consists of 10 climate zones. Zones are defined by geographical and topographical features.

III.2.3.1.2.6 Particulate Matter

Particulate matter, or particles in the atmosphere, are often referred to by size such as PM10 or PM2.5. The number associated with the size is the number of microns in diameter of the individual particles. However, the composition of particulate matter can be quite varied. The particles can be emitted directly from natural or human made sources, or they can be converted from gases to solids (e.g., H_2SO_4 above). Natural sources of particulate matter in Texas include sea spray, soil dust, and wildfires. Sources from human activity include dust from road paving, vehicle emissions, erosion of tires, fossil fuel burning, and industrial processes (Winkler et al., 2018). Fire events have produced enhancements of up to 140% PM2.5 over seasonal average concentrations due to fire events across Texas from 2001–2010 (Hou and Wu, 2016). A particularly toxic solid substance found in air pollution is lead, a solid heavy metal with a low melting point. Sources of lead today are primarily from lead ore crushing and smelting, and battery manufacturing. Additional emissions are associated with solid waste disposal (Jacobson, 2012). Some companies in Texas are developing "next generation" batteries, more powerful than their predecessors, that use chemicals other than lead, eliminating this emission hazard (e.g., Noah, 2025).

III.2.3.1.3 Application to Texas Climate Zones

Differences in climate and weather across the state of Texas have to do with the geography and the topography of the state. The National Oceanic and Atmospheric Administration (NOAA, 2025) defines 10 climate zone divisions in the state of Texas (Figure III.2.9). General topographical and meteorological characteristics of these zones are defined as follows by Texas Parks and Wildlife (TPW, 2025). In locations where terrain elevation is above 4,500 feet, high levels of CO can cause elevated levels of carbon in the bloodstream because of reduced oxygen concentration at high altitudes (US EPA, 2006). In locations of minimal rainfall, fewer opportunities for pollutants to wash out of an area are available.

High Plains: Elevations range from 3,000 to 4,500 feet above sea level. Average annual rainfall is 15 to 22 inches. Rainfall is lowest in winter and mid-summer and highest in April or May and September or October. Extended droughts have occurred here several times this century.

Low Rolling Plains: Average annual rainfall is 20 to 28 inches, with peaks in May and September. A summer dry period with high temperatures and high evaporation rates is typical. Elevations vary from 800 to 3,000 feet above sea level.

North Central: The North Central climate zone comprises the Blackland Prairie (east) and the Crosstimbers and Prairies (west). Average annual rainfall in the Blackland Prairie ranges from 28 to 40 inches. May is the peak rainfall month. Elevations range from 300 to 800 feet above sea level. For the Crosstimbers and Prairies, rainfall is moderate but somewhat erratic.

South Central: The southern part of the Blackland Prairie and the Post Oak Savanah are included in this region. Rainfall distribution is fairly uniform throughout the year and averages 28 to 40 inches per year. Peak rainfall usually occurs in May or June. Elevations range from 300 to 800 feet above sea level.

East Texas: East Texas contains the pinewoods area with annual average rainfall of 36 to 50 inches per year, fairly uniformly distributed throughout the year. Humidity and temperatures are typically high. Elevations range from 200 to 500 feet above sea level.

Upper Coast: The upper coast climate region includes gulf prairies and marshes. Topography is fairly level with elevation less than 150 feet above sea level. Average annual rainfall varies from 30 to 50 inches per year distributed fairly uniformly throughout the year.

Southern (including Lower Valley): Average annual rainfall of 20 to 32 inches increases from west to east. Average monthly rainfall is lowest during winter and highest during spring (May or June) and fall (September). Summer temperatures are very high with very high evaporation rates.

Edwards Plateau: Average annual rainfall ranges from 15 to 34 inches. Rainfall is highest in May or June and September. Elevations range from slightly less than 100 feet to over 3,000 feet above sea level.

Trans Pecos: Elevations range from 2500 feet to more than 8,749 feet above sea level at Guadalupe Peak. Average annual rainfall is less than 12 inches but it varies greatly from year to year and from lower to higher elevations. July and August are usually the higher rainfall months.

III.2.3.2 Impacts of Reduced Air Quality on Texas Life

III.2.3.2.1 Human Health

Overall rising temperatures can exacerbate air pollution by increasing near-surface O_3. Measurements of O_3 taken in the Dallas–Fort Worth and Houston–Galveston metropolitan areas in 2023 showed that these areas were in exceedance of national O_3 standards (CDC, 2024). Total population in these areas is approximately 12 million. Human health symptoms related to elevated O_3 levels can start with headaches, chest pain and shortness of breath. If exposure to high levels of O_3 is sustained over the long term, O_3 can cause respiratory illness, aging of lung tissue, and lung disease. All of these conditions can be fatal (Jacobson, 2012).

Both indoor and outdoor atmospheric dust particulate matter in Texas, especially in areas affected by drought and human activities associated with land use, can also increase health problems such as respiratory and cardiovascular issues. The Cook Children's Medical Center (2025) found that 7–13% of children between the ages of 0 and 14 in the counties of Collin, Denton, Grayson, Hood, Johnson, Parker, Tarrant, and Wise were reported to have had an asthma diagnosis at some point in their lifetime. In a study by Yue et al. (2021) it was found that nearly half of pneumonia-related deaths in children under five are attributable to inhalation of indoor particulate matter pollution. Additionally, the small particulate matter, such as PM2.5, cause more respiratory illness and premature death than do larger particulate matter (Jacobson, 2012).

III.2.3.2.2 Infrastructure Health

Air pollution damages not only human health, but also the health of the infrastructure needed to sustain human life in society. Acid rain accelerates corrosion of limestone, sandstone, mortar, and many metals, including aluminum siding. It dissolves paints and breaks apart stonework. Additionally, acid rain reduces fish stocks and causes loss of soil nutrients for tree growth and available lumber. Ground level O_3 also harms forests and agricultural crops, restricting uptake of CO_2 by growing trees and limiting food yield from farming (Canada, 2010). Dust particles can deposit on electricity transmission line insulators and cause power transmission failures. Adhesion of these particles to these electric grid components is exacerbated by high humidity, which becomes more common with higher temperatures (Jiang et al., 2020).

NOTE

1. Although many researchers have found connections between Arctic warming and unprecedented cold snaps occurring in southern regions of the Northern Hemisphere, complete consensus of the scientific community has not been reached.

REFERENCES

2010 Census Urban and Rural Classification and Urban Area Criteria., n.d. https://www.census.gov/programs-surveys/geography/guidance/geo-areas/urban-rural/2010-urban-rural.html

Abatzoglou JT, 2013. Development of gridded surface meteorological data for ecological applications and modelling. International Journal of Climatology, 33(1):121–131.

Adam D, 2023. What a 190-year-old equation says about rainstorms in a changing climate. Proceedings of the National Academy of Sciences, 120(14):e2304077120.
An H, Gan J, Cho SJ, 2015. Assessing climate change impacts on wildfire risk in the United States. Forests, 6(9):3197–3211.
AON, 2022. Q1 Global Catastrophe Recap. https://www.aon.com/reinsurance/getmedia/af1248d6-9332-4878-8c92-572c1bf3c19d/20221204-q1-2022-catastrophe-recap.pdf (accessed on May 7, 2022).
Baldwin MP, Ayarzagüena B, Birner T, Butchart N, Butler AH, Charlton-Perez AJ, Domeisen DI, Garfinkel CI, Garny H, Gerber EP, Hegglin MI, 2021. Sudden stratospheric warmings. Reviews of Geophysics, 59(1):e2020RG000708.
Becker GR, Kyle JR, 2011. Geographic distribution of metal and industrial mineral occurrences in Trans-Pecos Texas and border regions: A Google Earth database and compilation: The University of Texas at Austin, Bureau of Economic Geology.
Benz SA, Burney JA, 2021. Widespread race and class disparities in surface urban heat extremes across the United States. Earth's Future, 9(7). https://doi.org/10.1029/2021EF002016
Bolinger RA, Brown VM, Fuhrmann CM et al., 2022. An assessment of the extremes and impacts of the February 2021 South-Central US Arctic outbreak, and how climate services can help. Weather and Climate Extremes, 36:100461. https://doi.org/10.1016/j.wace.2022.100461
Butler AH, Sjoberg JP, Seidel DJ, Rosenlof KH, 2017. A sudden stratospheric warming compendium. Earth System Science Data, 9(1):63–76.
Buyantuyev A, Wu J, 2010. Urban heat islands and landscape heterogeneity: Linking spatiotemporal variations in surface temperatures to land-cover and socioeconomic patterns. Landscape Ecology, 25(1): 17–33. https://doi.org/10.1007/S10980-009-9402-4
Cabral S, Yousif N, 2023. Coldest wind chill ever recorded in continental US, say forecasters. British Broadcasting Corporation (BBC). https://www.bbc.com/news/world-us-canada-64485092 (accessed May 7, 2023).
Canada, 2010. Air pollution damage to infrastructure and industry. Environment and Climate Change Canada (ECCC). Government of Canada. https://www.canada.ca/en/environment-climate-change/services/air-pollution/quality-environment-economy/economic-issues/damage-infrastructure-industry.html (accessed May 8, 2025).
CDC (Centers for Disease Control), 2024. Regional health effects – Southern great plains. Climate and health. Centers for Disease Control. https://www.cdc.gov/climate-health/php/regions/southerngreatplains.html (accessed May 8, 2025).
Chakraborty T, Hsu A, Manya D, Sheriff G, 2020. A spatially explicit surface urban heat island database for the United States: Characterization, uncertainties, and possible applications. ISPRS Journal of Photogrammetry and Remote Sensing, 168:74–88.
Chakraborty T, Lee X, 2019. Land cover regulates the spatial variability of temperature response to the direct radiative effect of aerosols. Geophysical Research Letters, 46:8995–9003.
Chakraborty TC, Newman AJ, Qian Y, Hsu A, Sheriff G, 2023. Residential segregation and outdoor urban moist heat stress disparities in the United States. One Earth, 6(6):738–750.
Chakraborty TC, Qian Y, Li J, Leung LR, Sarangi C, 2025. Daytime urban heat stress in North America reduced by irrigation. Nature Geoscience, 18(1), 57–64.
Chakraborty T, Venter Z, Qian Y, Lee X, 2022. Lower urban humidity moderates heat stress [Preprint]. Meteorology. https://doi.org/10.1002/essoar.10511293.1
Chen J, Qian Y, Chakraborty TC, Yang Z, 2024. Complexities of urban impacts on long-term seasonal trends in a mid-sized arid city. Environmental Research Communications, 6(2), 021004.
Childs JW, Brackett R, 2023. Ice Storm Blamed In At Least 10 Deaths; Power Outages Top 500,000. https://weather.com/news/news/2023-02-01-winter-storm-mara-ice-storm-live-updates (accessed on March 2, 2025).
Clapeyron BPE, 1834. Memoir sur la puissance motrice de la chaleur. Journal De l'École Royale Polytechnique. 0:153–190.
Clausius R, 1850. On the moving force of heat and the laws of heat which May be deduced therefrom. Poggendorf's Annalen, 79:368–369.
Clean Air Act, 1970. CAAA70, PL 91-604. https://www.govinfo.gov/content/pkg/STATUTE-84/pdf/STATUTE-84-Pg1676.pdf (accessed May 8, 2025)
Cohen J, Agel L, Barlow M, Garfinkel CI, White I, 2021. Linking Arctic variability and change with extreme winter weather in the United States. Science, 373(6559):1116–1121.
Cook Children's Medical Center, 2025. Community health needs assessment. https://www.cookchildrens.org/siteassets/documents/about/2024-chna_full-report_4-29-25.pdf (accessed May 8, 2025).

CPC (Climate Prediction Center), 1997. The 1997 El Nino: Potential effects in Texas – November 1997 to April 1998. Climate Prediction Center (CPC), NWS/NOAA. https://www.cpc.ncep.noaa.gov/products/analysis_monitoring/ensostuff/states/tx.disc.html (accessed May 8, 2025).
Daly C, Halbleib M, Smith JI et al., 2008. Physiographically sensitive mapping of climatological temperature and precipitation across the conterminous United States. International Journal of Climatology: A Journal of the Royal Meteorological Society, 28(15):2031–2064.
Donat MG, King AD, Overpeck JT, Alexander LV, Durre I, Karoly DJ, 2016. Extraordinary heat during the 1930s US Dust Bowl and associated large-scale conditions. Climate Dynamics, 46:413–426.
Du H, Zhan W, Liu Z, Krayenhoff ES, Chakraborty TC, Zhao L, Jiang L, Dong P, Li L, Huang F, 2023. Global mapping of urban thermal anisotropy reveals substantial potential biases for remotely sensed urban climates. Science Bulletin, 68(16):1809–1818.
Ebi KL, McGregor G, 2008. Climate change, tropospheric ozone and particulate matter, and health impacts. Environmental Health Perspectives, 116(11):1449–1455.
Forkel R, Knoche R, 2006. Regional climate change and its impact on photooxidant concentrations in southern Germany: Simulations with a coupled regional climate-chemistry model. Journal of Geophysical Research: Atmospheres, 111:D12.
He C, Kim H, Hashizume M, Lee W, Honda Y, Kim SE, Kinney PL, Schneider A, Zhang Y, Zhu Y, 2022. The effects of night-time warming on mortality burden under future climate change scenarios: A modelling study. The Lancet Planetary Health, 6(8):e648–e657.
Hernandez J, 2023. An ice storm is unleashing treacherous conditions across parts of the Southern US. https://www.npr.org/2023/01/31/1152781168/ice-storm-texas-south-flights-travel. (accessed March 3, 2025)
Ho HC, Knudby A, Xu Y, Hodul M, Aminipouri M, 2016. A comparison of urban heat islands mapped using skin temperature, air temperature, and apparent temperature (Humidex), for the greater Vancouver area. Science of The Total Environment, 544:929–938. https://doi.org/10.1016/j.scitotenv.2015.12.021
Holton JR, 2004. Dynamic Meteorology. Elsevier.
Hou P, Wu S, 2016. Long-term changes in extreme air pollution meteorology and the implications for air quality. Scientific Reports, 6(1):23792.
HSEM, 2021. 2021 Winter Storm Uri after-action review findings report. City of Austin and Travis County. https://www.austintexas.gov/sites/default/files/files/HSEM/2021-Winter-Storm-Uri-AAR-Findings-Report.pdf (accessed May 6, 2025).
Hsu A, Sheriff G, Chakraborty T, Manya D, 2021. Disproportionate exposure to urban heat island intensity across major US cities. Nature Communications, 12(1):2721. https://doi.org/10.1038/s41467-021-22799-5
Hsu PC, Hsu HH, Hong HJ, Chen YT, Chen YL, Tseng WL, 2022. 2021 Texas cold snap: Manifestation of natural variability and a recent warming trend. Weather and Climate Extremes, 37:100476.
Jacob DJ, Winner DA, 2009. Effect of climate change on air quality. Atmospheric Environment, 43(1):51–63.
Jacobson MZ, 2010. Enhancement of local air pollution by urban CO_2 domes. Environmental Science & Technology, 44(7):2497–2502.
Jacobson MZ, 2012. Air Pollution and Global Warming: History, Science, and Solutions. Cambridge University Press.
Jiang M, Guo J, Jiang Y, Li L, Lu M, 2020. Dust contamination on surface of transmission line insulators in air-polluted regions in China: Statistical characteristics, adhesion mechanism, and environmental impact factors. Environmental Science and Pollution Research, 27:23643–23654.
Kalnay E, Cai M, 2003. Impact of urbanization and land-use change on climate. Nature, 423(6939):528–531.
Karl TR, Kukla G, Razuvayev VN, Changery MJ, Quayle RG, Heim RR Jr, Easterling DR, Fu CB, 1991. Global warming: Evidence for asymmetric diurnal temperature change. Geophysical Research Letters, 18(12):2253–2256.
Kiefer P, 2022. Ice, tornadoes, and cracking trees: Why winter storm Landon dragged a melee of weather across the US. Popular Science. https://www.popsci.com/environment/winter-storm-landon-weather-effects/ (accessed May 7, 2025).
Kirchmeier-Young MC, Zhang X, 2020. Human influence has intensified extreme precipitation in North America. Proceedings of the National Academy of Sciences of the United States of America, 117(24):13308–13313. https://doi.org/10.1073/pnas.1921628117
Knowlton K, Rosenthal JE, Hogrefe C, Lynn B, Gaffin S, Goldberg R, Rosenzweig C, Civerolo K, Ku JY, Kinney PL, 2004. Assessing ozone-related health impacts under a changing climate. Environmental Health Perspectives, 112(15):1557–1563.
Lee X., 2018. Fundamentals of Boundary-Layer Meteorology (Vol. 256). Springer.
Li X, Chakraborty TC, Wang G, 2023. Comparing land surface temperature and mean radiant temperature for urban heat mapping in Philadelphia. Urban Climate, 51:101615.

McDonald RI, Biswas T, Sachar C, Housman I, Boucher TM, Balk D, Nowak D, Spotswood E, Stanley CK, Leyk S, 2021. The tree cover and temperature disparity in US urbanized areas: Quantifying the association with income across 5,723 communities. PLoS One, 16(4):1–27.

McPherson RA, Fay PA, Alvarez SG, Bertrand D, Broadbent TL, Bruno T, Fares A, McCullough B, Moore GW, Moorhead B, Patiño L, Petersen A, Smith NG, Steiner JL, Taylor A, Warziniack T, 2023. Ch. 26. Southern Great plains. In: Fifth National Climate Assessment. Crimmins AR, Avery CW, Easterling DR, Kunkel KE, Stewart BC, Maycock TK, Eds. Washington, DC, USA: US Global Change Research Program. https://doi.org/10.7930/NCA5.2023.CH26

Muller CL, Chapman L, Grimmond CSB, Young DT, Cai X, 2013. Sensors and the city: A review of urban meteorological networks: Sensors and the city. International Journal of Climatology, 33(7):1585–1600. https://doi.org/10.1002/joc.3678

Muñoz-Sabater J, Dutra E, Agustí-Panareda et al., 2021. ERA5-land: A state-of-the-art global reanalysis dataset for Land applications [Preprint]. Data, Algorithms, and Models. https://doi.org/10.5194/essd-2021-82

Murage P, Hajat S, Kovats RS, 2017. Effect of night-time temperatures on cause and age-specific mortality in London. Environmental Epidemiology, 1. https://doi.org/10.1097/EE9.0000000000000005

Nallen J, 2022. Winter storm Landon leaves thousands of Texas residents without power, putting medical device users at risk. CrisisReady. https://www.crisisready.io/winter-storm-landon-leaves-thousands-without-power/ (accessed May 7, 2025).

NCEI (National Centers for Environmental Information), 2023. Storm events database, event types: Winter storm. National Centers for Environmental Information. https://www.ncdc.noaa.gov/stormevents/listevents.jsp?eventType=%28Z%29+Winter+Storm&beginDate_mm=01&beginDate_dd=30&beginDate_yyyy=2023&endDate_mm=02&endDate_dd=02&endDate_yyyy=2023&hailfilter=0.00&tornfilter=0&windfilter=000&sort=DT&submitbutton=Search&statefips=-999%2CALL (accessed May 7, 2025).

Nielsen-Gammon JW, 2011. The changing climate of Texas. The Impact of Global Warming on Texas, 39:86.

NOAA, 2024. Climate. National marine ecosystem status website. https://ecowatch.noaa.gov/ (accessed May 8, 2025).

NOAA (National Oceanic and Atmospheric Administration), 2025. NOAA/NCEI climate division data. National Oceanic and Atmospheric Administration. https://psl.noaa.gov/data/usclimdivs/ (accessed May 8, 2025).

Noah Chemicals, 2025. Working with customers to develop the next generation of batteries. https://www.noahchemicals.com/batteries (accessed May 6, 2025).

Nogueira M, Hurduc A, Ermida S, Lima DC, Soares PM, Johannsen F, Dutra E, 2022. Assessment of the Paris urban heat island in ERA5 and offline SURFEX-TEB (v8. 1) simulations using METEOSAT land surface temperature product. Geoscientific Model Development Discussions, 2022:1–29.

North GR, Pyle JA, Zhang F, eds., 2015. Encyclopedia of Atmospheric Sciences. Elsevier.

NWS, 2021. Valentine's week winter outbreak 2021: Snow, ice, record cold. https://www.weather.gov/hgx/2021ValentineStorm. (accessed March 3, 2025).

NWS (National Weather Service), 2023. Late January – Early February 2023 sleet and ice event. https://www.weather.gov/fwd/jan-feb2023 (accessed May 7, 2025).

Obradovich N, Migliorini R, Mednick SC, Fowler JH, 2017. Nighttime temperature and human sleep loss in a changing climate. Science Advances, 3(5):e1601555.

Oke TR, 1982. The energetic basis of the urban heat island. Quarterly Journal of the Royal Meteorological Society, 108(455):1–24.

Patz JA, 2000. Climate change and health: New research challenges. Ecosystem Health, 6(1):52–58.

Raymond C, Matthews T, Horton RM, 2020. The emergence of heat and humidity too severe for human tolerance. Science Advances, 6(19):eaaw1838. https://doi.org/10.1126/sciadv.aaw1838

Rothfusz LP., 1990. The Heat Index Equation (or, more than You Ever Wanted to Know About Heat Index) (pp. 23–90). Fort Worth, Texas: National Oceanic and Atmospheric Administration, National Weather Service, Office of Meteorology.

Runkle J, Kunkel KE, Nielson-Gammon J, Frankson R, Champion SM, Stewart BC, Romolo L, Sweet W, 2022. Texas State Climate Summary 2022. NOAA Technical Report NESDIS 150-TX. National Oceanic and Atmospheric Administration, National Environmental Satellite, Data, and Information Service, Silver Spring, MD. https://statesummaries.ncics.org/chapter/tx (accessed May 8, 2025).

Sherwood SC, 2018. How important is humidity in heat stress? Journal of Geophysical Research: Atmospheres, 123(21). https://doi.org/10.1029/2018JD028969

Sherwood SC, Huber M, 2010. An adaptability limit to climate change due to heat stress. Proceedings of the National Academy of Sciences, 107(21):9552–9555. https://doi.org/10.1073/pnas.0913352107

Smith RK, Chang D-C, 2020. Utilizing recent climate data in eastern Texas to calculate trends in measures of aridity and estimate changes in watering demand for landscape preservation. Journal of Applied Meteorology and Climatology, 59(1):143–152.

Solomon S, Qin D, Manning M, Alley RB, Berntsen T, Bindoff NL, Chen Z, Chidthaisong A, Gregory JM, Hegerl GC, Heimann M, 2007. Technical summary. *Climate Change 2007: The Physical Science Basis. Contribution of Working Group I to the Fourth Assessment Report of the Intergovernmental Panel on Climate Change [Solomon S, Qin D, Manning M, Chen Z, Marquis M, Averyt KB, Tignor M, Miller HL (eds.)].*

Stewart ID, Krayenhoff ES, Voogt JA, Lachapelle JA, Allen MA, Broadbent AM, 2021. Time evolution of the surface urban heat island. Earth's Future, 9(10). https://doi.org/10.1029/2021EF002178

Sullivan BK, Malick NS, 2021. Five million Americans have lost power from Texas to North Dakota after devastating winter storm. Time, New York. https://time.com/5939633/texas-power-outage-blackouts/ (accessed May, 6, 2025).

Svitek P, 2022. Texas puts final estimate of winter storm death toll at 246. The Texas Tribune. https://www.texastribune.org/2022/01/02/texas-winter-storm-final-death-toll-246/ (accessed May 6, 2025).

TPW (Texas Parks and Wildlife), 2025. Texas Ecoregions. https://tpwd.texas.gov/education/hunter-education/online-course/wildlife-conservation/texas-ecoregions. (accessed May 8, 2025).

Trenberth KE, 2005. The impact of climate change and variability on heavy precipitation, floods, and droughts. Encyclopedia of Hydrological Sciences, 17:1–11.

Trenberth KE, Fasullo JT, Kiehl J, 2009. Earth's global energy budget. Bulletin of the American Meteorological Society, 90(3):311–324.

US EPA (United States Environmental Protection Agency), 1978, revised 2006. Altitude as a factor in air pollution. US Environmental Protection Agency, Washington, D.C., EPA/600/9-78/015 (NTIS PB285645).

USC (United States Code), 2013. United States Code, 2013 Edition Title 42 – The public health and welfare chapter 85 – Air pollution prevention and control, Subchapter I – Programs and activities, Part A - Air quality and emission limitations, Sec. 7409 - National primary and secondary ambient air quality standards. From the US Government Publishing Office. https://www.govinfo.gov/content/pkg/USCODE-2013-title42/html/USCODE-2013-title42-chap85-subchapI-partA-sec7409.htm (accessed May 5, 2025).

Venter ZS, Chakraborty T, Lee X, 2021. Crowdsourced air temperatures contrast satellite measures of the urban heat island and its mechanisms. Science Advances, 7(22):eabb9569. https://doi.org/10.1126/sciadv.abb9569

Voelkel J, Hellman D, Sakuma R, Shandas V, 2018. Assessing vulnerability to urban heat: A study of disproportionate heat exposure and access to refuge by socio-demographic status in Portland, Oregon. International Journal of Environmental Research and Public Health, 15(4):640.

Waugh DW, Sobel AH, Polvani LM, 2017. What is the polar vortex and how does it influence weather? Bulletin of the American Meteorological Society, 98(1):37–44.

West JJ, Nolte CG, Bell ML, Fiore AM, Georgopoulos PG, Hess JJ, Mickley LJ, O'Neill SM, Pierce JR, Pinder RW, Pusede S, 2023. Fifth national climate assessment: Chapter 14-air quality. In Fifth National Climate Assessment, Crimmins AR, Avery CW, Easterling DR, Kunkel KE, Stewart BC, Maycock TK, Eds. US Global Change Research Program, Washington, DC, USA. 14.1–14.49. Chapter 14, 5:1–49.

WFAA, 2021. Multiple cities in North Texas under a boil water order as power outages continue. https://www.wfaa.com/article/news/local/north-texas-cities-under-boil-water-notices/287-45044bc8-2d6a-4610-89b6-cdc37d6ad4b1 (accessed on May 6, 2025).

Winkler SL, Anderson JE, Garza L, Ruona WC, Vogt R, Wallington TJ, 2018. Vehicle criteria pollutant (PM, NO x, CO, HCs) emissions: how low should we go? NPJ Climate and Atmospheric Science, 1(1):26.

Yue X, Ma NL, Sonne C, Guan R, Lam SS, Van Le Q, Chen X, Yang Y, Gu H, Rinklebe J, Peng W, 2021. Mitigation of indoor air pollution: A review of recent advances in adsorption materials and catalytic oxidation. Journal of Hazardous Materials, 405:124138.

Zavadoff B, Arcodia M, 2022. What are teleconnections? Connecting Earth's climate patterns via global information superhighways. https://www.climate.gov/news-features/blogs/enso/what-are-teleconnections-connecting-earths-climate-patterns-global (accessed May 5, 2025).

Zhang T, Zhou Y, Zhao K, Zhu Z, Chen G, Hu J, Wang L, 2022. A global dataset of daily maximum and minimum near-surface air temperature at 1 km resolution over land (2003–2020). Earth System Science Data, 14(12):5637–5649.

Zingales V, Taroncher M, Martino PA, Ruiz MJ, Caloni F, 2022. Climate change and effects on molds and mycotoxins. Toxins, 14(7):445.

III.3 Water Supply and Wastewater Infrastructure System Challenges Under Climate Uncertainty

Perpetua I. Okoye, Jordan T. Jatko, and Ryan A. McManamay

III.3.1 INTRODUCTION

Managed water systems are the deliberate control and regulation of water resources to maximize use, such as with drinking water, or minimize damage, as with wastewater. Managed water systems provide sustainable options for populations, although they have negative impacts on the environment. Disruption of run-of-river with dam construction fragments aquatic systems and encroaches on critical terrestrial habitats that riparian organisms occupy. Wastewater effluent mobilizes contaminants into the environment, because removal of contaminants, such as nutrients, pharmaceuticals, and industrial waste, is rarely 100% effective. There is a critical need to understand the impacts of climate change on managed water systems as populations increase, particularly in Texas, one of the fastest growing populations in the United States as of 2025.

Reservoirs are an old technological advancement to provide populations with drinking water supplies, while providing other services, such as flood protection, hydropower generation, and recreation (Brazos River Authority, 2025). Increasing frequencies and intensities of drought conditions and floods pose risks for reservoir volumes and water quality. Population increases result in higher demand for water, thus compounding the effects of worsening drought frequencies and intensities in water availability (United Nations, 2025). Reservoir stratification results in anoxic conditions in the deepest parts of the reservoir, or the hypolimnion (LaBrie and Maranger, 2023). This is where contaminants accumulate over time, thus as reservoir volumes decrease in drought conditions and with increased demand, the exposure risk to these contaminants increases while also decreasing the dilution factors of contaminants from upstream (Medeiros et al., 2025; Preece et al., 2019).

Wastewater treatment plants remediate grey and black water, and often stormwater, sewerage systems to minimize impact of municipal and industrial waste on the environment. Drought conditions can result in reduced availability of surface waters to dilute effluent, increasing risk of downstream populations who experience de facto re-use, or unintentional use of treated wastewater in drinking water supplies (Marino et al., 2024). Conversely, intense flooding can stress combined stormwater sewerage systems that often results in untreated wastewater being released into the environment. These challenges can be compounded by population increases where wastewater generation increases (Aziz et al., 2024). This positive feedback loop of demand and scarcity can pose additional risks for public and environmental health.

This chapter explores the history and future predictions of reservoir storage supplies and wastewater management in Texas through the lens of climate change and population growth. Texas is a large state with varying distributions of climate and environmental influences that prevent the use of blanket approaches to managed water systems to accommodate population growth and climate

 DOI: 10.1201/9781003406600-16

changes. These critical analyses of managed water systems will be paramount for informing regulatory bodies when counseling legislative decision-makers.

III.3.2 RESERVOIRS

Reservoirs are a critical infrastructure for society, providing stable water supplies and reducing flood risks (Randle et al., 2021). The ability of this infrastructure to provide these services, particularly for water supplies, depends on storage of incoming flows to maintain conservation capacity. Conservation capacity facilitates the strategic use of temporarily variable river inflows for diverse water needs such as municipal, agricultural, and industrial uses, water quality regulation, hydropower generation, and recreation (Wurbs, 2021). Due to climate change impacts, such as droughts and irregular precipitation patterns, in conjunction with population growth, the pressures of water demand on finite resources can intensify and create critical water scarcity situations, such as rapidly declining reservoir storage (Apurv and Cai, 2021). For instance, Lake Lanier, which supplies 70% of the demand in Atlanta, experienced a severe decline in storage due to an extreme drought event in 2007, leaving about 35 days of usable water for more than 5 million people (Missimer et al., 2014). Lake Mead, located on the Colorado River, witnessed a similar drop in storage due to prolonged drought and overallocation of resources in Las Vegas by 2022 (Hannoun and Tietjen, 2023). Continued declines in Lake Mead storage levels over the past 22 years have led to iterative placement of new water intakes at increasing depths in the reservoir (Forsythe et al., 2012). Developing long-term water management strategies is crucial to mitigating water supply challenges and meeting varied water demands (Apurv and Cai, 2021). Hence, water supply planning increasingly relies on proactively considering reservoir storage in relation to both immediate and future climate variabilities and water demands (Rathnayaka et al., 2016).

Reservoir water storage is a critical strategy for mitigating exposure to extreme climate variabilities in Texas. Texas hydrology varies dramatically, ranging from extremely intense and destructive floods to persistent and detrimental multi-year droughts (Ferencz et al., 2024). Reservoir construction has substantially diminished streamflow irregularity (Mix et al., 2016) while improving water supply availability and reliability (Wurbs, 2005). Reservoirs stationed partially or wholly in the state have conservation capacities equal to or above 500,000 acre-feet (Wurbs, 2021); in addition, there are over 50,000 free-flowing streams and roughly 200 strategically positioned surface water supply intakes in Texas (Figure III.3.1a). The largest reservoirs have more than 44 million acre-feet total in capacity, with over 29 million acre-feet of storage, and about 882,000 acres of conservation

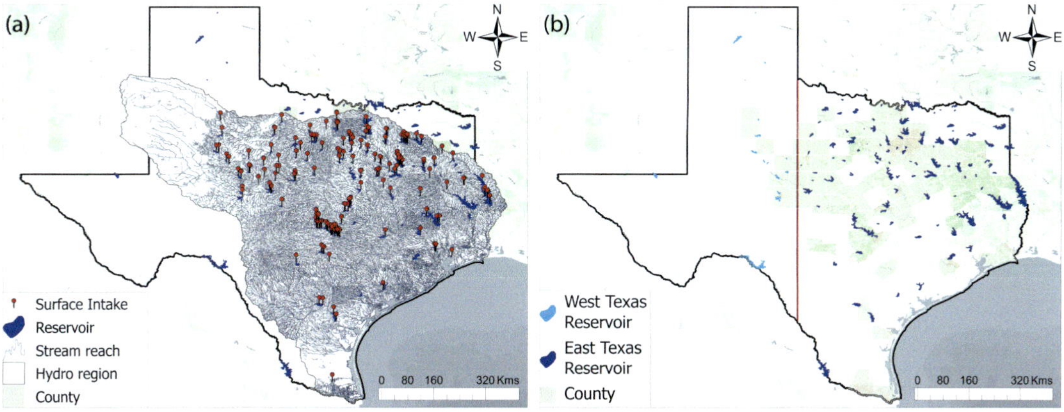

FIGURE III.3.1 Map of (a) Texas Intakes located on streams and reservoirs, and (b) Reservoirs in West and East Texas, divided by the 100th meridian (shown as a red line).

pool surface areas (Wurbs, 2021). Municipal water use accounts for 62% of surface waters, 33% of groundwater, and 4% of reclaimed water (TWDB, 2022). Reservoirs supply 90% of the municipal surface water supplies compared to 10% from smaller waterbodies, run-of-river intakes in free-flowing stream reaches. Texas has made significant investments in constructing reservoirs and surface intakes to ensure water security in the face of a rapidly changing climate and a growing population over the past century. By the early 2000s, Texas had constructed over 200 major reservoirs with conservation capacity greater than 5,000 acre-feet. The Texas Water Development Board (TWDB) works with regional water districts and utilities to develop a future water plan for the state every five years. In the latest 2022 plan, future surface water supply projects, including new reservoirs, will require an additional $ 8.2 billion financial investment for construction (TWDB, 2017). For example, Lake Ralph Hall reservoir, currently under development, has been projected to have a construction cost of approximately $490 million (UTRWD, 2020). These investments may address immediate, localized issues (Greer, 2020), but could fail to mitigate future-scale water supply challenges (McDonald et al., 2014; Vörösmarty et al., 2000) and could lead to poor water management planning if not considered proactively.

Reservoir management strategies in Texas centers on recognizing the hydroclimatic asymmetry between East and West Texas (Nielsen-Gammon et al., 2020), which influences reservoir construction, operability, and reliability (Figure III.3.1b). While East Texas reservoirs are larger and provide a relatively stable backbone for statewide supply, West Texas reservoirs highlight the fragility of water supply systems under climate stress, necessitating flexible, region-specific strategies to ensure future reliability (Zhao et al., 2018). Exploring pathways of regional hydrological disparities that consider the dynamic interplay between resource management and future climate scenarios is essential for ensuring sustainability under uncertain future conditions (Weinthal and Sowers, 2020). These pathways must consider reservoir storage in the light of progressing climate conditions, ensuring alignment with broader environmental and policy objectives. Furthermore, it should account for both historical and future climate changes, as well as changes in system water availability and demand, ensuring that management strategies are robust and comprehensive. However, a site-specific approach is essential for promoting water supply resilience against climate stressors and navigating local sociopolitical challenges (Randle et al., 2021; Uhlmann et al., 2011), which complicate resource management planning (US Water, 2021). Only a few regional instances (e.g., Harris-Lovett et al., 2019; McManamay et al., 2019) have been successfully implemented in local water infrastructure planning.

Previous climate change impacts and population growth have raised concerns about the sustainability of reservoir storage and supply in Texas (Brown et al., 2019; Chen et al., 2001). Declining reservoir storage decreases water supply reliability if sufficient alternatives are not available (Zhao et al., 2018). Without sufficient adaptive strategies, extended drought has resulted in heavy dependence on static surface water systems, thereby increasing regional susceptibility to severe water shortages (Nielsen-Gammon et al., 2020). Intensifying drought extremes are characterized by seasonal shifts and uneven fluctuations in rainfall and streamflow, both within and between years (Winters, 2013). Highly variable streamflow under climatic fluctuations (Mishra et al., 2011) can be related to past reservoir storage patterns to reveal the sensitivities and vulnerabilities to water supply shortages. Subsequently, a comparative assessment can help effectively connect temporally variant reservoir storage with population-driven demand and climate variability, providing a holistic approach to contextualizing future impacts at the regional and site-specific levels (Dash et al., 2023).

III.3.2.1 Historical Reservoir Storage in Relation to Drought

The 1950s Drought of Record marked the most notable drought in Texas, which lasted nearly a decade, drying up countless smaller reservoirs (Winters, 2013). Earlier droughts, prior to the 1950s, also impacted significant portions of the state, while more recent droughts were more economically damaging, primarily due to population growth and economic expansion. For example, Lake Travis and Lake Buchanan reservoirs reached critically low storage levels, declining to less than 30% from 2011 to 2015. The severity of the more recent droughts corresponds with that of the earlier droughts in several regions of Texas. The driest year was notably recorded in 2011, and at the opposite end, 2015 was one of the wettest years in Texas history, characterized by multiple flood events. Since then, Texas hydrologic extremes have become increasingly volatile, alternating between periods of severe drought and high-impact flooding, underscoring the need for improved drought prediction. Refer to Chapter I.4 for more comprehensive details on the historical drought events in Texas.

The U.S. Drought Monitor produces weekly drought conditions by combining climatic, hydrologic, soil, and field reports – this resource can be used to assess drought severity associated with reservoir storage dynamics in West and East Texas (Figure III.3.2). In Texas, drought conditions are recurring phenomena that occur approximately every two to four years, exhibiting varying degrees of severity, with both regions displaying cyclical fluctuations in reservoir storage levels. These cycles showed up spikes in increased spatial coverage of low-to-extremely-high (D1–D4) drought intensity from 2011 to 2014. The most intense drought occurred from 2011 to 2013, during which nearly 100% of Texas was in drought, and large portions were in extremely high D3 and D4 levels, confirming the statewide crisis that year (Ferencz et al., 2024). Periods such as between 2004 and 2005 and mid-2007, with low or no drought, represent recovery when reservoir levels were replenished, especially after heavy rainfall or wetter years. The post-2010 period showed sharp fluctuations between drought and non-drought years, indicating increased climate instability that challenges long-term reservoir storage and supply planning. For example, such notable swings include a small drought from 2010 to 2011, followed by an extreme drought from 2014 (high drought) to 2015–2017 (recovery). The short non-drought period from 2021 to 2022 was followed by a sharp resurgence of drought conditions across the state from 2022 to 2023, with severity levels reaching D4 again, starting in mid-2022. As of early 2025, conditions rose again, suggesting a trend toward prolonged dry periods, especially in the West Texas region.

III.3.2.2 Major Patterns and Trends in Reservoir Storage

Real-time and historical storage data for major reservoirs in Texas are publicly available and widely used for drought monitoring and early warning (TWDB, 2024). Storage levels of existing reservoirs in East and West Texas from 2008 to 2024 display highly variable patterns and fluctuations that mirror drought conditions. Over this time period, long-term trends suggest that storage is decreasing, on average, over time and this decline is more severe for western than Eastern reservoirs. Reservoir storage, expressed in million gallons per day (MGD), refers to the total amount of water held in a reservoir at a specific point in time. Conservation storage and capacity are the portion of storage allocated for planned human use and the maximum water volume that can be stored for conservation purposes, respectively. Mean average change in storage per reservoir represents the slope of the change in reservoir storage over the time period. The interannual changes in mean storage categorizes the magnitude of changes based on five main classes – major loss (≤ –2,000 MGD), moderate loss (–2,000 to –0.5), stable (–0.5 to 0.5), moderate gain (0.5 to 2,000), and major gain (≥ 2,000 MGD), illustrating trends of storage declines or recovery.

Figure III.3.3 illustrates spatial and temporal contrasts in West and East Texas reservoirs during the period from 2004 to 2024. Temporary declines in storage follow drought conditions (Figure III.3.3a, b). The annual averages showed a consistent decline in reservoir and conservation storage during extended drought in 2011 to 2014, with partial recovery in subsequent years in West Texas. The storage slope and delta validate these patterns, with most reservoirs experiencing major losses, particularly in 2011 and 2013, followed by recovery starting from 2014 (Figure III.3.3c, d). However, these storage metrics consistently remained well below conservation capacity, reflecting sustained underutilization. This trend aligned with the increased D0–D4 land area coverage from 2011 to 2014, during which nearly 100% of Texas experienced drought conditions (Figure III.3.2).

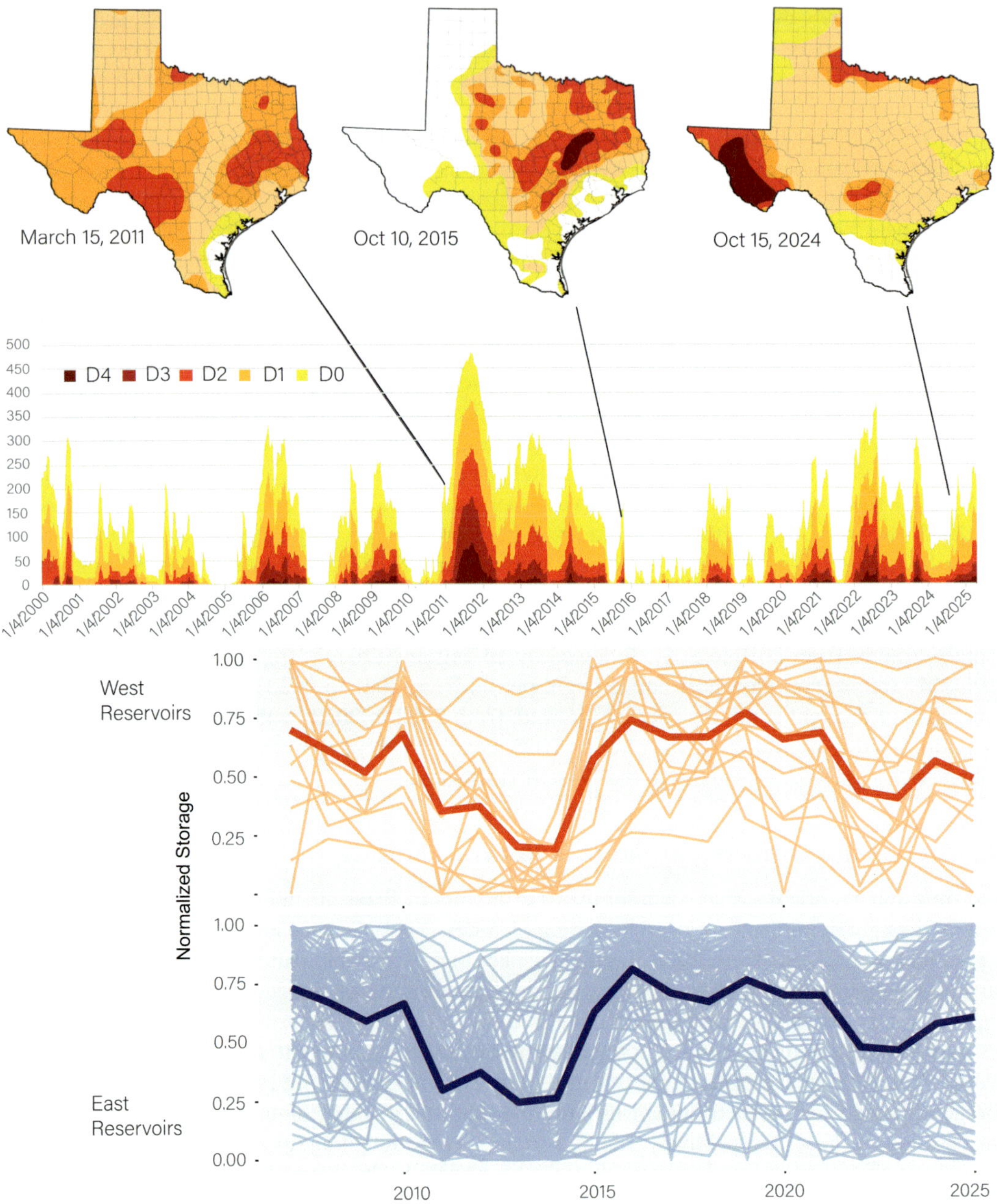

FIGURE III.3.2 Temporal distribution of drought severity across Texas based on the U.S. Drought Monitor classifications (D0–D4) and reservoir storage level fluctuations in West and East Texas. *Note:* D0 – abnormally dry, D1 – moderate drought, D2 – severe drought, D3 – extreme drought, D4 – exceptional drought. Data from TWDB (2025).

Despite some interannual fluctuations, reservoir storage rarely reached capacity, indicating a chronic storage deficit. Unlike West Texas, East Texas reservoirs showed greater temporal stability, with storage closely tracking capacity, suggesting that these systems are buffered against drought and more consistently replenished in wetter years. Following a decline in 2011, coinciding with the extreme drought, reservoirs alternately replenished until hitting a pronounced storage deficit in

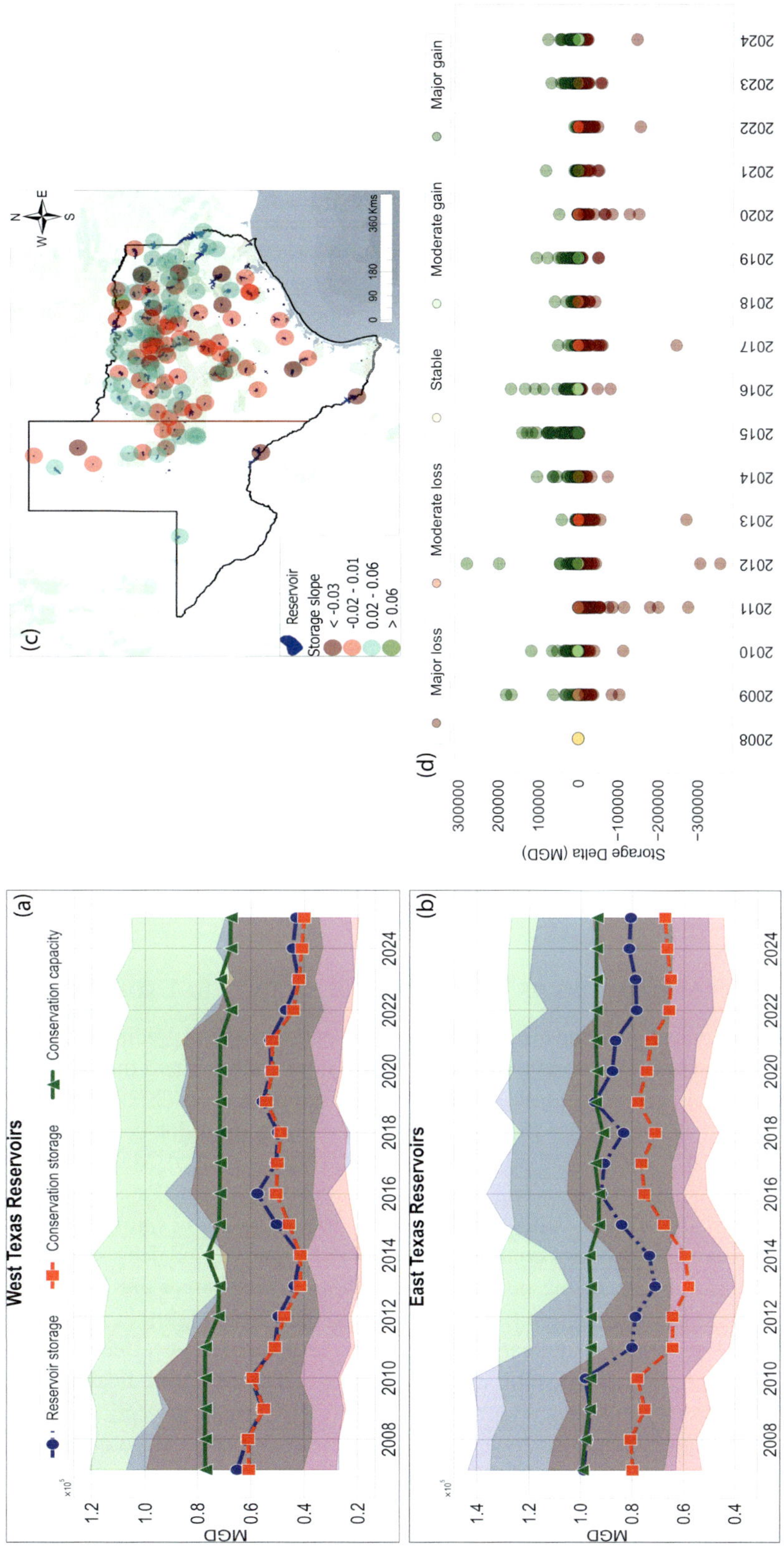

FIGURE III.3.3 Reservoir storage trends (a, b) and long-term storage slope and interannual change detection (c, d) in East and West Texas (2008–2024). The shaded area in (a) and (b) represents the 95% confidence interval around the mean line.

2022. The resurgence in drought conditions in 2022 were characterized by many areas falling under D4 conditions, the most severe level on the U.S. Drought Monitor (TWDB, 2025) (Figure III.3.3e). Recent years (2023–2024) showed signs of storage recovery, but drought resurgence in early 2025 (Figure III.3.2) signaled the onset of prolonged dry periods that could once again strain reservoir storage in Texas.

III.3.2.3 Interannual Variability in Reservoir Storage

Another approach to exploring reservoir sensitivity to climate conditions is by relating storage to streamflow at intra-annual scales. Seasonal patterns in storage gains and losses reflect reservoir management strategies, generally guided by reservoir operation curves, relative to anticipated streamflow conditions. For example, for the U.S. Army Corps of Engineers (USACE), reservoir operations follow guidelines that vary based on wet vs. drought conditions, ensuring conservation storage while maintaining downstream releases for the lower river environment (Patterson and Doyle, 2018).

Figure III.3.4 shows the historical (1970s–1990s) and projected (2040, 2100) changes in streamflow and seasonal storage in East and West Texas, illustrating the inverse relationship between basin runoff and storage trends. Historically, streamflow was highest from February to June, peaking in May, and lowest during the late summer to early fall (August–October) in both regions, a typical pattern of rainfall-runoff dynamics in Texas, driven by spring precipitation and summer evapotranspiration losses (Figure III.3.4a, b). Most reservoirs in West Texas store substantial water from March to July (excluding April), aligning with the peak flow months. In contrast, significant storage losses occur in late summer and fall (August–October). Generally, West Texas reservoirs display sharper seasonal fluctuations especially in the late winter to early spring (February–March), while in East Texas, storage changes are moderate, with smaller gains and losses compared to those in West Texas. Storage gains in winter months (e.g., December) suggest greater hydrological buffering and potentially more regulated inflow-outflow regimes. Under future climate scenarios, projected reductions in streamflow, particularly in summer and fall, could exacerbate current storage challenges, especially in West Texas, where seasonal recharge is crucial for maintaining water supplies (Figure III.3.4c, d).

III.3.2.4 Future Considerations of Water Supply in Relation to Drought

Climate model projections align in the direction and spatial pattern of total precipitation change, with a general inclination toward drier conditions, particularly in summer, where there is strong agreement on declining rainfall (Schaller et al., 2011). Nielsen-Gammon et al. (2020) projected drought severity and soil moisture availability through 2100 under a high-emission scenario (RCP 8.5) in West and East Texas under multimodel ensembles (MME) (Figure III.3.5). MME results showed a clear drift toward drier conditions across all drought indicators by the late 21st century. Shallow soil moisture (SM-30cm) also exhibited values comparable to moderate 20th century droughts. Palmer Drought Severity Index (PDSI) values consistently remained negative, and deep soil moisture (SM-2m) showed particularly severe drying in West Texas. These drying trends are more extreme than those of historical megadroughts. The differences between indicators highlight the metric-specific sensitivity of drought projections, indicating that no single measure can effectively capture the complexity of future drought risk. Despite differences across models and indicators, summer soil moisture drying consistently emerges as a robust indicator in climate projections in Texas. This agreement indicates a fundamental regional drift toward a drier baseline, potentially surpassing conditions observed during the driest periods of the last millennium.

The most critical interacting factors affecting reservoir storage are compounding stressors of simultaneous increases in future demand and drought intensity, alongside their associated

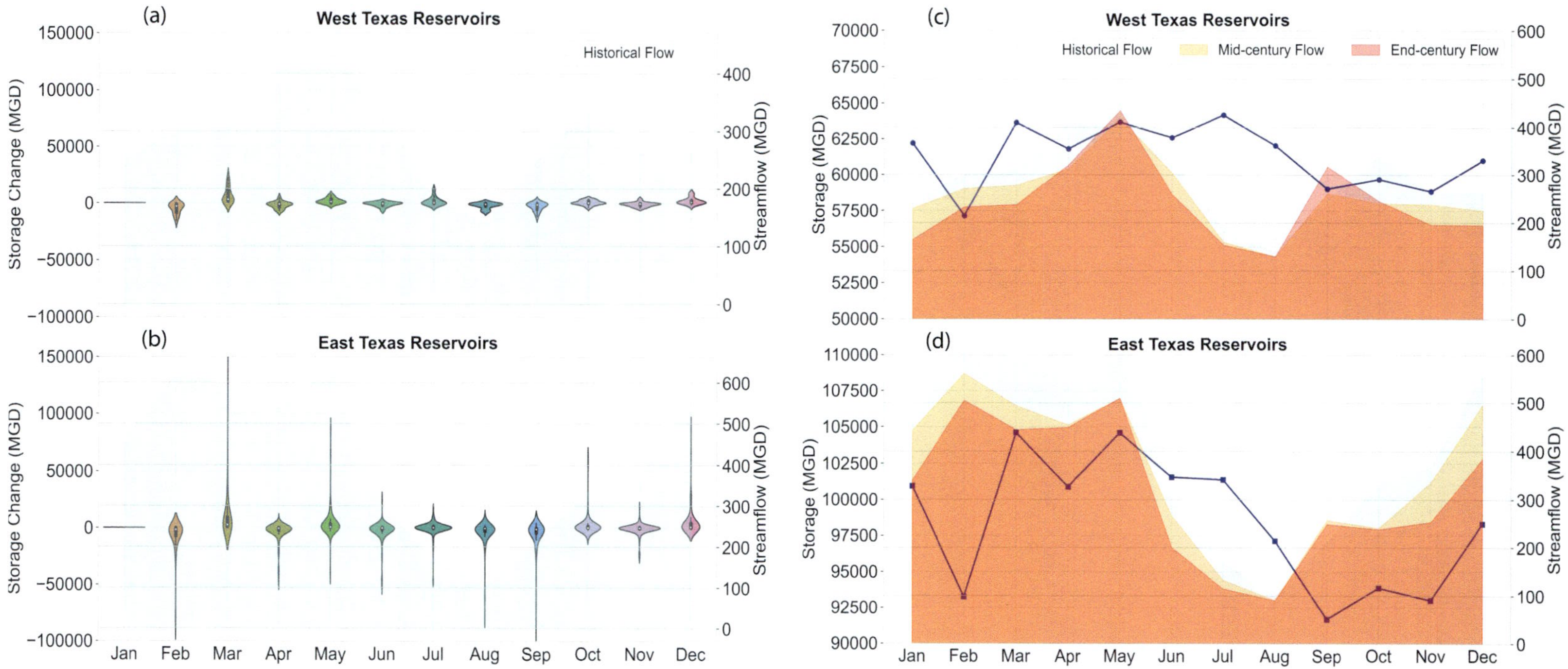

FIGURE III.3.4 Cumulative monthly historical streamflow and reservoir storage changes for West Texas (a) and East Texas (b), with statewide reservoir storage overlaid against historical, mid-century, and end-of-century streamflow projections (c, d) (USDA, 2022). The U.S. Department of Agriculture has modeled streamflow changes by comparing historical (1990s) and projected mid-century (2040s) and end-of-century (2080s) runoff, using updated climate scenarios and gridded simulations (USDA, 2022). A comparative analysis of reservoir storage and historical flow data (1970s–1990s) was conducted to assess intra-annual dynamics and identify directional trends in reservoir behavior in East and West Texas. Storage changes were classified as either gains or losses, highlighting short-term fluctuations and seasonal draw-down patterns. Reservoir storage overlaid with mid and late century streamflow illustrates the potential amplification of current storage challenges under future climate scenarios.

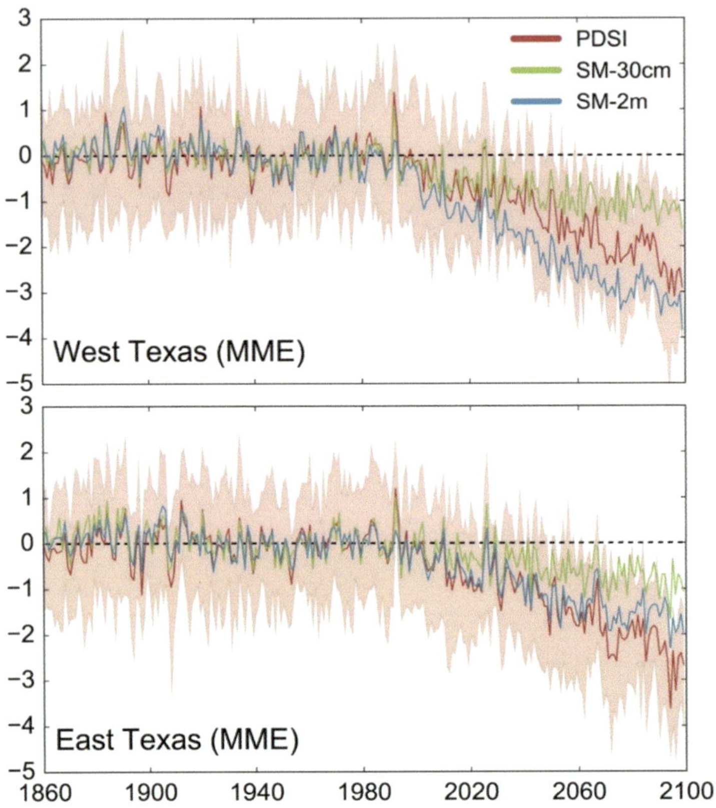

FIGURE III.3.5 Multimodel projections of PDSI and soil moisture at 30 cm and 2 m depths for West Texas and East Texas, showing long-term trends toward increased drought drying. Multimodel Ensemble (MME) model taken directly from Nielsen-Gammon et al. (2020).

uncertainties. Accordingly, Zhao et al. (2018) simulated relative reservoir storage and supply reliability for seven reservoirs near Dallas, TX from the ensemble of droughts for the 1950s historic drought and future droughts, including near-future (2000–2049) and late-century (2050–2099) periods, in conjunction with observed (2010) and projected (2050 and 2090) water demands (Figure III.3.6). Droughts in the late century are projected to be more protracted (about 36 vs. 21 months) and severe, causing a decline in average storage from about 70% to 54%, a 16% reduction compared to the near-future century (Figure III.3.6a). During the near-future century, reservoirs are anticipated to maintain 100% reliability even under severe droughts (Figure III.3.6b). In the late century, however, reliability is projected to decline sharply due to extended droughts and rising demand, declining as low as 26% after 50 months in one case. On average, reliability is projected to drop from roughly 96% to 82% between periods. The initial failure point, where reliability falls below 100%, occurs earlier, with the median decreasing from 24 to 14 months. Under the 1950s drought scenario, this threshold drops from 29 months (2010 demand) to 13 months (2090 demand), underscoring how future demand intensifies vulnerability.

On the demand side, uncertainty stems from population growth and changes in per capita water use, both of which significantly alter future demand projections (Dawadi and Ahmad, 2013). As population and water use per capita increase, municipal demand intensifies, resulting in higher withdrawals, particularly during peak usage months in summer. Figure III.3.7 shows how uncertainty in projected demand (minimum, median, maximum) affects simulated reservoir storage and water supply reliability under future drought scenarios. Relative storage showed substantial seasonal

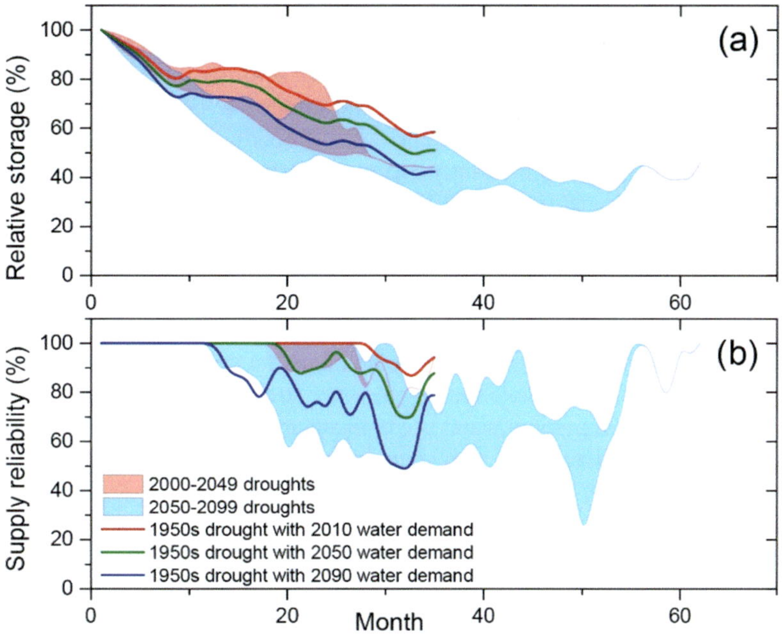

FIGURE III.3.6 Simulated responses of (a) relative reservoir storage and (b) water supply reliability to projected drought events (paired with corresponding future water demands) and the 1950s drought scenario under 2010, 2050, and 2090 demand levels for seven reservoirs near Dallas, TX. Taken directly from Zhao et al. (2018).

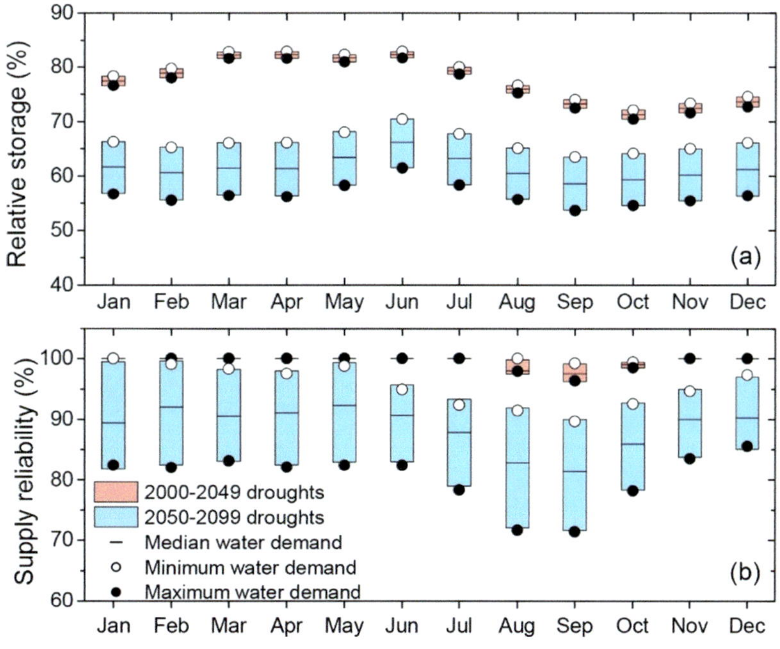

FIGURE III.3.7 Monthly impacts of water demand scenarios – minimum, median, and maximum – on (a) relative reservoir storage and (b) water supply reliability, highlighting seasonal sensitivity to demand fluctuations for seven reservoirs near Dallas, TX. Taken directly from Zhao et al. (2018).

variation, with the lowest levels occurring in September, mainly due to peak summer water use and high evaporation rates. This seasonal decline in storage resulted in the lowest reliability in the same month, highlighting the system's vulnerability during late summer months when demand is high and inflows are limited. The impacts are more evident under late-century (2050–2099) droughts, indicating growing stress under future climate and demand conditions. Compared to the near-future century, the late century displays significantly greater uncertainty in both storage and supply reliability. In September, the storage uncertainty range increases from 2% in the near-future century to 19% in the late century. Similarly, reliability is expected to remain near 100% in the near future; however, under future conditions, demand variability leads to a 19% uncertainty range (71–90%), highlighting the growing influence of demand-side uncertainty on reservoir system performance.

Another approach to evaluating potential risk to reservoir storage is directly evaluating water stress based on current and projected stream flows, based on the USDA modeled streamflow assessment described earlier. Here, we evaluate municipal water stress on reservoir systems[1], where stress risk is calculated as a ratio of municipal water use to mean annual streamflow. Aside from changing streamflow conditions, water demand is also changing over time. Water demand projections were taken from Texas Water Development Board's 2022 Water Plan (TWDB, 2022). Stress is classified based on severity – low stress (<10%), moderate stress (10–30%), high stress (30–60%), severe stress (60–100%), and extreme stress (>100%).

Figure III.3.8 presents municipal water use stress risk for historical and future climate scenarios, categorizing stress levels at intake points in the Texas hydrologic region. Across all scenarios,

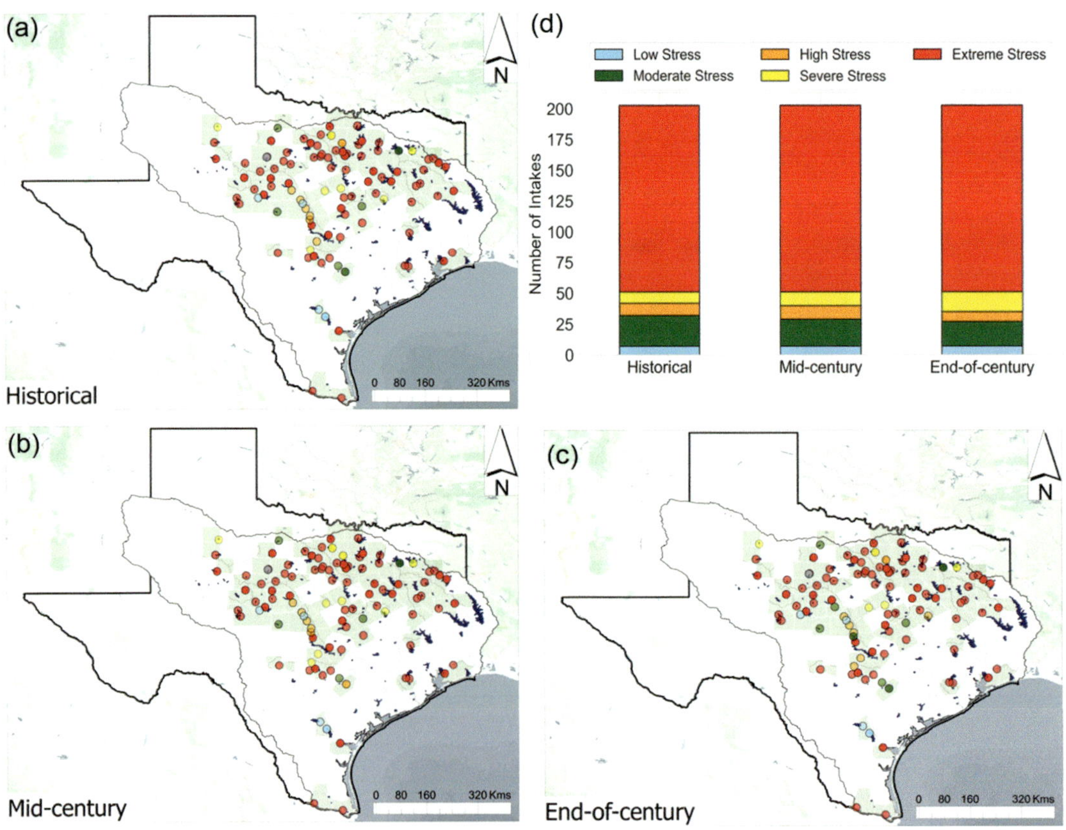

FIGURE III.3.8 Spatial and temporal patterns of municipal water use stress risk across Texas under (a) historical and future (b) mid-century and (c) late-century climate and water demand scenarios. (d) The plot summarizes the number of municipal intakes falling into each stress category for the three time periods.

extreme stress became the dominant condition, with many intakes classified at the highest level (Figure III.3.8a, c). This reflects the complex effects of declining streamflow and increasing or sustained municipal water use, with East Texas emerging as a persistent hotspot of stress. Few intakes worsened in classification from historical to mid-century and from historical to late century, suggesting localized vulnerabilities where stress is projected to intensify beyond their historical condition (Figure III.3.8d). Most intakes stayed the same over time, reflecting localized resilient hydrological system, where water availability and demand remain balanced through stable catchment-level streamflow distribution, or adaptive management. Without improvement in the intake stress condition, future surface water demand is expected to decline, necessitating that the demand be met through different sources or conservation. The majority of intakes are already operating near or above sustainable levels, with high usage relative to flow, under both historical and future conditions. This consistent trend, together with localized shifts, stresses the importance of adaptive, site-specific strategies, including demand management, diversified supply portfolios, and infrastructure redesign.

III.3.3 PUBLIC WASTEWATER SYSTEMS

Wastewater is a significant contributor to contaminants found in surface waters and can escalate environmental health challenges during low flow conditions. In most cases, wastewater in the United States is treated by permitted infrastructures called wastewater treatment plants, or WWTP. In the United States, to release effluent into any surface waters, one must be permitted by the National Pollution Discharge and Elimination System (NPDES) under the Clean Water Act of 1972. In Texas, the Texas Commission on Environmental Quality (TCEQ) takes on the permitting role and reports permits to the United States Environmental Protection Agency (U.S. EPA) for compliance. Currently, there are approximately 1,215 municipal wastewater treatment plants that discharge into surface waters in Texas. These permits enforce discharging requirements into surface waters so as not to jeopardize water quality. Many inland surface waterbodies in the U.S. are assigned a designated use, thus have varying degrees of water quality requirements to maintain (U.S. EPA, 2012). For example, waterbodies used primarily as a drinking water source will have strict water quality standards to not overburden municipal drinking water facilities. Other designated uses are recreational use, biological integrity, and aquifer recharge protection, to name but a few.

Wastewater can also be used to augment aquatic systems that have been reduced or depleted due to anthropogenic activities, such as the case with the Trinity River here in Texas where wastewater discharge from Dallas-Fort Worth sustains base flows (Luthy et al., 2015). This is not without drawbacks, such as input of wastewater-derived organic compounds and nutrients in surface waters (Plumlee et al., 2012). Some studies estimate that global nutrient contamination into surface waters will increase about 70% by mid-century under different development scenarios (Beusen et al., 2022; van Puijenbroek et al., 2015). Nutrient contamination is often linked to harmful algal blooms, surface water hypoxia, and fish kills, and although agriculture is imperative for supporting growing populations, fertilizer and soil runoff compound nutrient contamination in surface waters. Public opinions and concerns about wastewater present in surface waters also play a significant role in the use or re-use of wastewater. Modern technologies have made re-use of water a viable strategy to combat water supply limitations, although public acceptance has limited the use of this technology (Rice et al., 2016). Interestingly, de facto re-use, or unintentional re-use of treated wastewater from upstream WWTPs, accounts for 7 to 100% of the top 25 most impacted drinking water treatment plant (DWTP) supplies across the United States under low flow conditions (Rice et al., 2013). Wastewater is shown to be significant for both maintaining baseflow conditions and for drinking water supply, thus the concern is for public health and safety. NPDES primary goal is to maintain water quality standards for a system's designated use, often through the use of dilution. Dilution using water already present in the aquatic system is secondary only to primary treatment of wastewater. An aquatic system has an innate assimilative capacity, or self-rejuvenation, to accommodate moderate insult from wastewater discharge (Hashemi Monfared et al., 2022), which takes dilution into account. As populations rise and water become scarcer in Texas, the assimilative capacity of surface waters, including dilution, becomes reduced, thus posing potential public health and safety concerns.

III.3.3.1 Wastewater Dilution Factors Related to Surface Waters

The U.S. EPA has experimental requirements for whole effluent toxicity (WET) analysis for freshwater organisms. For acute and chronic WET toxicology studies, the EPA recommends a dilution factor of effluent at ≥0.5, or 50% (U.S. EPA, 2000). This means that wastewater should be diluted by at least half in freshwater for toxicology experiments concerning freshwater organisms. For this purpose, we can assume that a 50% dilution of wastewater effluent could be interpreted as a threshold of wastewater exposure in surface waters. In-stream dilution factors for regulatory purposes are often calculated using the flow statistic 7Q10, or the lowest seven-day average flow that has a probability of occurring once every 10 years (USGS, 2014). This flow statistic provides a probable low flow scenario where dilution factors could be at their worst. Dilution factors can then be calculated using effluent concentrations with the 7Q10 statistic to generate a worst-case hydrologic scenario for dilution factors (U.S. EPA, 2010) Figure III.3.9 shows outputs for dilution factors calculated under historical, modeled mid-century, and modeled late-century 7Q10 flows. Effluent loads for dilution factors calculations were conducted under 25%, 50%, and 75% of WWTP maximum design flows to simulate realistic scenarios and future wastewater inputs as the population in Texas increases.

In Figure III.3.9, for all flow scenarios, nearly 50% of all surface waters receiving effluent discharge are never in non-attainment of that 0.5 dilution factor. In fact, non-attainment status across all wastewater conditions and across all flow scenarios show no significant difference, with only 10 WWTPs reaching a non-attainment status at 25% WWTP maximum design flow between historical flow and modeled late-century flow. The primary driver of differences within each flow scenario is the percentages of WWTP maximum design flow, which can be seen as a simulation of population growth. As the population of Texas increases, chemical loads and strains on WWTPs will also increase (O'Brien et al., 2014). Although these are estimated dilution factors under static conditions, these analyses provide insight into what could occur under global warming scenarios in Texas. These dilution factors also do not account for upstream effluents or intensification/extensification of WWTPs as populations in Texas increase. It is simple to identify the compounding problems with wastewater as populations grow and water is lost through drinking water intakes and climate instability. Conservation of surface water is imperative for the availability of drinking water, but also for ecosystem and public health concerns with wastewater dilution.

III.3.3.2 Wastewater Infrastructure Future Considerations

Historically, wastewater infrastructure was developed and built to site conditions – topography, proximity, and cost, for example. The American Society of Civil Engineers (ASCE) state that WWTPs are built for a lifespan of 40–50 years, whereas the average age for WWTPs in the U.S. is 45 years (ASCE, 2020). According to a 2022 survey by the U.S. EPA, the U.S. will need to invest at least $630 billion over the next 20 years to maintain wastewater, and other clean water, infrastructure to maintain protection of the nation's inland waters (U.S. EPA, 2024). Wastewater treatment extensification could provide opportunities to augment baseflows for depleted aquatic systems in order to restore ecosystem services. Arid and semi-arid regions in the southwestern U.S. are becoming dominated by municipal effluent discharges (Brooks et al., 2006). Focuses on use of wastewater effluent in stream restoration could provide habitat and restore abiotic and biotic systems, although only for more robust species as water quality is likely to be another hurdle to conquer (Brooks et al., 2006; Luthy et al., 2015). Focusing on sustainable development, such as augmenting depleted surface water systems, will generally be less cost efficient as site selection will be based on the need of the aquatic system rather than topography and proximity. WWTPs, known as centralized systems, are the most common type of treatment in the United States, although decentralized systems, or

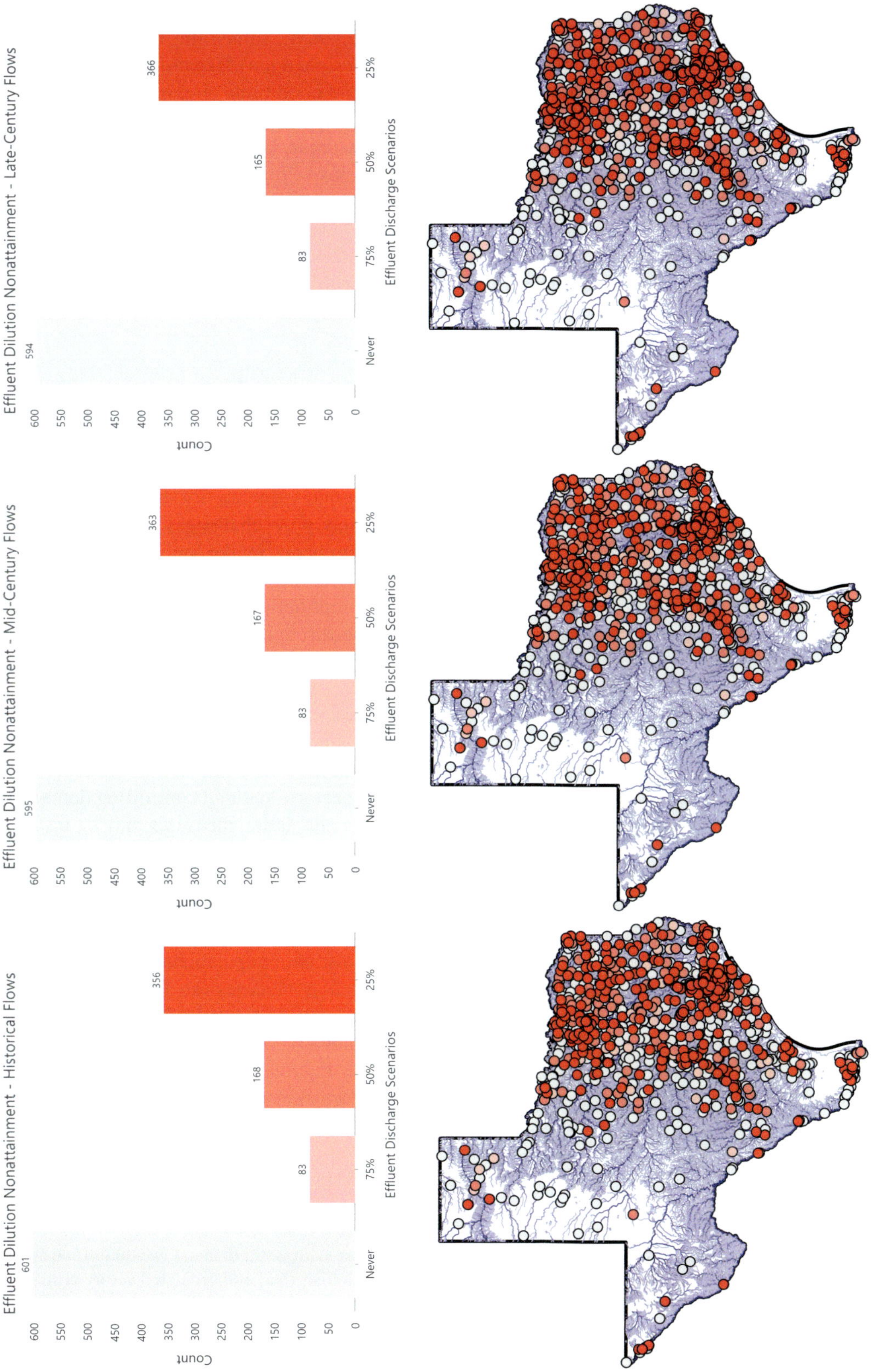

FIGURE III.3.9 Non-attainment status of wastewater treatment plants in Texas based on three streamflow scenarios based on historical, mid-century, and late-century magnitudes of 7Q10 with three wastewater effluent scenarios (25%, 50%, and 75% of maximum treatment plant capacity).

on-site sewage facilities (OSSFs), are the second most common wastewater treatment infrastructure. OSSFs, commonly referred to as septic systems, in essence, do the same thing as WWTPs, but scaled for a single building or small communities. OSSFs are designed for lifespans of 20–30 years, although due to lack of census data, the remaining lifespans and conditions of existing OSSFs are difficult to ascertain (ASCE, 2020). OSSFs are regulated at a state level, rather than at the federal level like WWTPs. TCEQ does the permitting for WWTPs in Texas, but reports to the U.S. EPA, although this is not the case for OSSFs, where TCEQ is the exclusive regulatory body. OSSF popularity is increasing and the use of OSSFs will be a facet of the response to managing population increases in Texas, according to Dr. June Wolf of Texas A&M AgriLife Research (TWRI, 2022). Research in the use of decentralized approaches is a facet of the Texas A&M AgriLife Research and can be significant in offsetting the burdens on aging wastewater infrastructure in the United States.

III.3.4 CONCLUSIONS

Demand is as critical, if not more critical, than climate change for managed water systems in Texas. As Texas faces more frequent and intense droughts, reservoir storage is increasingly under pressure. The magnitude and pattern of demand, driven by population growth, have an equal influence on storage outcomes (Dawadi and Ahmad, 2013). The severity of municipal water use stress in the states, especially under future scenarios, further signals a systemic vulnerability that is worsening over time. Extreme stress conditions at most intake points may pose challenges in maintaining reliable water supplies for communities that rely on a limited number of intake sources. Even as climate models project more frequent and extreme droughts (Nielsen-Gammon et al., 2020), the rapid increase in municipal water demand emerges as a significant amplifier of system stress. For Texas, this requires a dual strategy, strengthening climate resilience through infrastructure and policy adaptation, while assertively managing demand through efficiency upgrades, conservation incentives, and diversified supply portfolios. Integrated water resource planning at the river basin level, supported by real-time monitoring and adaptive reservoir operations, is already being implemented as a critical tool for advancing practical and resilient adaptation and mitigation efforts (TWDB, 2024). In sum, improving water resilience in Texas requires climate adaptation and a fundamental shift in how demand is forecasted and managed across various sectors. Without controlling demand increase, even the most advanced reservoir systems and climate adaptation measures may be insufficient in preventing severe water shortages across the state.

Wastewater management is critical for maintaining public and environmental health. Current centralized infrastructure is reliant on the presence of surface waters for dilution of wastewater effluent. As climate conditions change, such as increased intensity and frequencies of droughts and floods, the stress on the aging infrastructure can result in risks to downstream populations that experience de facto re-use of wastewater for their drinking supplies. Population increases can also intensify the risks associated with wastewater by decreasing availability of surface water through drinking water demands. Increasing wastewater production is positively associated with worsening dilution factors of whole effluent. These conservative estimations described in Figure III.3.9 show that increased wastewater effluent discharge can result in worse dilution factors, even under historical flows. Alternative wastewater treatment, such as decentralized approaches can help reduce the burden on our nation's aging WWTPs, although more information is needed on their lifespans and efficacies.

NOTE

1. To understand how climate-driven changes in streamflow interact with water demand, we quantified and spatially resolved municipal surface water use stress under both historical and projected future climate conditions (Okoye and McManamay, 2025). We evaluated municipal stress risk, often spatially concentrated and tied to fixed intake infrastructure, by integrating streamflow projections, estimated water availability as an index (WAI) for each intake, and municipal surface water demand within each county. WAI, or available water for withdrawal, was estimated for reservoirs and free-flowing streams at each

surface intake location, using the inflow rate and storage capacity of respective systems (McKay et al., 2012). WAI was estimated in streams based on a 10% withdrawal fraction of the minimum monthly flow. WAI in reservoirs were subject to environmental and operational constraints, ensuring that environmental flows and storage are sustained. Total water use in each county was proportionally allocated to each intake based on weighting intakes using WAI values. The stress risk for each water system was computed as the ratio of adjusted municipal water use to historical and future mean annual streamflow. Note that municipal water use was also dynamic over time, varying for mid- and late-century values.

REFERENCES

Apurv T, Cai X, 2021. Regional drought risk in the contiguous United States. Geophys. Res. Lett. 48(5), e2020GL092200.

ASCE (American Society of Civil Engineers), 2020. 2020 Report Card for America's Infrastructure. https://infrastructurereportcard.org/wp-content/uploads/2020/12/Wastewater-2021.pdf

Aziz F, Wang X, Mahmood MQ, Guild R, 2024. Wastewater flooding risk assessment for coastal communities: Compound impacts of climate change and population growth. J. Hydrol. 645(A). https://doi.org/10.1016/j.jhydrol.2024.132136

Beusen AHW, Doelman JC, van Beek LPH, van Puijenbroek PJTM, Mogollón JM, van Grinsven HJM, Stehfest E, van Vuuren DP, Bouwman AF, 2022. Exploring river nitrogen and phosphorus loading and export to global coastal waters in the Shared Socio-economic pathways. Glob. Environ. Change. 72:102426. https://doi.org/10.1016/j.gloenvcha.2021.102426

Brazos River Authority, 2025. Flood control vs. water supply vs. recreation. https://brazos.org/about-us/news/news-room/resource-library/flood-control-vs-water-supply-vs-recreation

Brooks BW, Riley TM, Taylor RD, 2006. Water quality of effluent-dominated ecosystems: Ecotoxicological, hydrological, and management considerations. Hydrobiologia. 556:365–379. https://doi.org/10.1007/s10750-004-0189-7

Brown TC, Mahat V, Ramirez JA, 2019. Adaptation to future water shortages in the United States caused by population growth and climate change. Earth's Future. 7:219–234.

Chen CC, Gillig D, McCarl BA, 2001. Effects of climatic change on a water dependent regional economy: A study of the Texas Edwards Aquifer. Clim. Change. 49:397–409.

Dash SS, Sahoo B, Raghuwanshi NS, 2023. An integrated reservoir operation framework for enhanced water resources planning. Sci. Rep. 13:21720.

Dawadi S, Ahmad S, 2013. Evaluating the impact of demand-side management on water resources under changing climatic conditions and increasing population. J. Environ. Manage. 114:261–275.

Ferencz SB, Sun N, Turner SWD, Smith BA, Rice JS, 2024. Multisectoral analysis of drought impacts and management responses to the 2008-2015 record drought in the Colorado basin, Texas. Nat. Hazards Earth Syst. Sci. 24:1871–1896.

Forsythe KW, Schatz B, Swales SJ, Ferrato LJ, Atkinson DM, 2012. Visualization of lake mead surface area changes from 1972 to 2009. ISPRS Int. J. Geo-Information. 1:108–119.

Greer RA, 2020. A review of public water infrastructure financing in the United States. Wiley Interdiscip. Rev. Water. 7(5):e1475.

Hannoun D, Tietjen T, 2023. Lake Management under severe drought: Lake Mead, Nevada/Arizona. JAWRA J. Am. Water Resour. Assoc. 59:416–428.

Harris-Lovett S, Lienert J, Sedlak D, 2019. A mixed-methods approach to strategic planning for multi-benefit regional water infrastructure. J. Environ. Manage. 233:218–237.

Hashemi Monfared SA, Dehghani Darmian M, Snyder SA et al., 2022. Water quality planning in rivers: Assimilative capacity and dilution flow. Bull. Environ. Contam. Toxicol. 99:531–541. https://doi.org/10.1007/s00128-017-2182-7

LaBrie R, Maranger R, 2023. Predicting the presence of hypoxic hypolimnia in lakes at large spatial scales. Limnol. Oceanogr. 69(2):355–366. https://doi.org/10.1002/lno.12488

Luthy R, Sedlak D, Plumlee M, Austin D, Resh V, 2015. Wastewater-effluent-dominated streams as ecosystem-management tools in a drier climate. Front. Ecol. Environ. 13:477–485. https://doi.org/10.1890/150038

Marino A, Bertolotti S, Macrì M, Bona F, Bonetta S, Falasco E, Minella M, Fenoglio S, 2024. Impact of wastewater treatment and drought in an alpine region: A multidisciplinary case study. Heliyon. 10(15). https://doi.org/10.1016/j.heliyon.2024.e35290

McDonald RI, Weber K, Padowski J, Flörke M, Schneider C, Green PA, Gleeson T, Eckman S, Lehner B, Balk D, Boucher T, Grill G, Montgomery M, 2014. Water on an urban planet: Urbanization and the reach of urban water infrastructure. Glob. Environ. Chang. 27:96–105.

McKay L, Bondelid T, Dewald T, Johnston J, Moore R, Rea A, 2012. NHDPlus Version 2: User guide.
McManamay RA, DeRolph CR, Surendran-Nair S, Allen-Dumas M, 2019. Spatially explicit land-energy-water future scenarios for cities: Guiding infrastructure transitions for urban sustainability. Renew. Sustain. Energy Rev. 112:880–900.
Medeiros CR, Paiva FF, Ferreira LMR, Costa LDA, Nóbrega RLB, Cunha J, Molozzi J, Galvão CO, 2025. How does the storage volume of semi-arid reservoirs change water quality and modulate the diversity of benthic macroinvertebrates? Sci. Total Environ. 958:177881. https://doi.org/10.1016/j.scitotenv.2024.177881
Mishra AK, Singh VP, Özger M, 2011. Seasonal streamflow extremes in Texas river basins: Uncertainty, trends, and teleconnections. J. Geophys. Res. Atmos. 116:8108.
Missimer TM, Danser PA, Amy G, Pankratz T, 2014. Water crisis: The metropolitan Atlanta, Georgia, regional water supply conflict. Water Policy. 16:669–689.
Mix K, Groeger AW, Lopes VL, 2016. Impacts of dam construction on streamflows during drought periods in the upper Colorado River basin, Texas. Lakes Reserv. Sci. Policy Manag. Sustain. Use. 21:329–337.
Nielsen-Gammon JW, Banner JL, Cook BI, Tremaine DM, Wong CI, Mace RE, Gao H, Yang ZL, Gonzalez MF, Hoffpauir R, Gooch T, Kloesel K, 2020. Unprecedented drought challenges for Texas water resources in a changing climate: What do researchers and stakeholders need to know? Earth's Futur. 8:e2020EF001552.
O'Brien JW, Thai PK, Eaglesham G, Ort C, Scheidegger A, Carter S, Lai FY, Mueller JF, 2014. A model to estimate the population contributing to the wastewater using samples collected on census day. Environ. Sci. Technol. 48(1):517–525. https://doi.org/10.1021/es403251g
Okoye PI, McManamay RA, 2025. Extending shared socioeconomic pathways to future water supply infrastructure scenarios: A case study of San Antonio, TX (in review). Environ. Res. Communication.
Patterson LA, Doyle MW, 2018. A nationwide analysis of U.S. Army corps of Engineers reservoir performance in meeting operational targets. J. Am. Water Resour. Assoc. 54:543–564.
Plumlee MH, Gurr CJ, Reinhard M, 2012. Recycled water for stream flow augmentation: Benefits, challenges, and the presence of wastewater-derived organic compounds. Sci. Total Environ. 438:541–548. https://doi.org/10.1016/j.scitotenv.2012.08.062
Preece EP, Moore BC, Skinner MM, Child A, Dent S, 2019. A review of the biological and chemical effects of hypolimnetic oxygenation. Lake Reserv. Manag. 35(3):229–246. https://doi.org/10.1080/10402381.2019.1580325
Randle TJ, Morris GL, Tullos DD, Weirich FH, Kondolf GM, Moriasi DN, Annandale GW, Fripp J, Minear JT, Wegner DL, 2021. Sustaining United States reservoir storage capacity: Need for a new paradigm. J. Hydrol. 602:126686.
Rathnayaka K, Malano H, Arora M, 2016. Assessment of sustainability of urban water supply and demand management options: A comprehensive approach. Water 8, 595.
Rice J, Wutich A, Westerhoff P, 2013. Assessment of De Factor Wastewater Reuse across the U.S.: Trends between 1980 and 2008. Environ. Sci. Technol. 47(19):11099–11105. https://doi.org/10.1021/es402792s
Rice J, Wutich A, White DD, Westerhoff P, 2016. Comparing actual de facto wastewater reuse and its public acceptability: A three city case study Sustain. Cities Soc. 27:467–474. https://doi.org/10.1021/es402792s
Schaller N, Mahlstein I, Cermak J, Knutti R, 2011. Analyzing precipitation projections: A comparison of different approaches to climate model evaluation. J. Geophys. Res. Atmos. 116:10118.
TWDB (Texas Water Development Board), 2017. 2017 State Water Plan [WWW Document]. https://www.twdb.texas.gov/waterplanning/swp/2017/index.asp (accessed 4.30.25).
TWDB (Texas Water Development Board), 2022. 2022 State Water Plan [WWW Document]. https://www.twdb.texas.gov/waterplanning/swp/2022/index.asp (accessed 4.30.25).
TWDB (Texas Water Development Board), 2024. Water data for Texas: Texas reservoirs - Monitored water supply reservoirs are 77.8% full on 2025-05-12 [WWW Document]. https://waterdatafortexas.org/drought/drought-monitor?period=2021-08-24&areaType=state&areaName=tx (accessed 5.5.25).
TWDB (Texas Water Development Board), 2025. Water data for Texas: Drought monitor [WWW Document]. https://waterdatafortexas.org/drought/drought-monitor?period=2019-12-17&areaType=state&areaName=tx (accessed 5.14.25).
TWRI (Texas Water Resources Institute), 2022. Bigger is not always better: Decentralizing Texas' wastewater infrastructure. https://twri.tamu.edu/publications/txh2o/2022/winter-2022/bigger-is-not-always-better-decentralizing-texas-wastewater-infrastructure/
Uhlmann, D., Paul, L., Hupfer, M., Fischer, R., 2011. Lakes and Reservoirs. Treatise Water Sci. 2:157–213.
U.S. EPA (U.S. Environmental Protection Agency), 2000. Method guidance and recommendation for whole effluent toxicity (WET) testing (40 CFR Part 136). Office of Water. https://www.epa.gov/sites/default/files/2016-02/documents/method-guidance-recommendations-wet-testing_2000.pdf

U.S. EPA (U.S. Environmental Protection Agency), 2010. NPDES Permit Writer's Manual. (EPA-833-K-10-001). Office of Water. https://www.epa.gov/system/files/documents/2025-09/pwm_2010_edits_2025_06.pdf

U.S. EPA (U.S. Environmental Protection Agency), 2012. Water quality standards handbook – Chapter 2: Designation of uses. (40 CFR 131.10). Office of Water. https://www.epa.gov/sites/default/files/2014-10/documents/handbook-chapter2.pdf

U.S. EPA (U.S. Environmental Protection Agency), 2024. New EPA survey highlights wastewater infrastructure needs to protect waterbodies in communities across the country. EPA Press Office. https://www.epa.gov/newsreleases/new-epa-survey-highlights-wastewater-infrastructure-needs-protect-waterbodies

United Nations, 2025. Water – At the end of the climate crisis. https://www.un.org/en/climatechange/science/climate-issues/water

US Water, 2021. Recovering stronger : A federal policy blueprint. texaswater.org.

USDA, F.S.G.T. and A.C. (GTAC), 2022. Hydro flow metrics – Modeled streamflow and hydrologic metrics for the contiguous U.S. enterprise data warehouse [WWW Document]. https://data.fs.usda.gov/geodata/edw/datasets.php?xmlKeyword=hydro flow metrics (accessed 5.12.25).

USGS (United States Geological Survey), 2014. Low-flow frequency and flow duration of selected South Carolina streams in the Catawba-Wateree and Santee River Basins through March 2012. https://pubs.usgs.gov/of/2014/1113/pdf/ofr2014-1113.pdf

UTRWD (Upper Trinity Regional Water District), 2020. UTRWD breaks ground on North Texas' newest lake - Lake Ralph Hall [WWW Document]. https://lakeralphhall.com/news/utrwd-breaks-ground-on-north-texas-newest-lake/ (accessed 4.30.25).

van Puijenbroek PJTM, Bouwman AF, Beusen AHW, Lucas PL, 2015. Global implementation of two shared socioeconomic pathways for future sanitation and wastewater flows. Water Sci. Technol. 71(2):227–233. https://doi.org/10.2166/wst.2014.498

Vörösmarty CJ, Green P, Salisbury J, Lammers RB, 2000. Global water resources: Vulnerability from climate change and population growth. Science (80-.). 289:284–288.

Weinthal E, Sowers J, 2020. The water-energy nexus in the middle East: Infrastructure, development, and conflict. Wiley Interdiscip. Rev. Water. 7:e1437.

Winters KE, 2013. A historical perspective on precipitation, drought severity, and streamflow in Texas during 1951–56 and 2011. U.S. Geol. Surv. Sci. Investig. Rep. No. 2013–5113, 24.

Wurbs RA, 2005. Texas water availability modeling system. J. Water Resour. Plan. Manag. 131:270–279.

Wurbs RA, 2021. Storage and regulation of river flows by dams and reservoirs. Texas Water J. 12:10–39.

Zhao G, Gao H, Kao SC, Voisin N, Naz BS, 2018. A modeling framework for evaluating the drought resilience of a surface water supply system under non-stationarity. J. Hydrol. 563:22–32.

III.4 Impact of Climate Change on Texas Traffic

Opportunities to Leverage Crowdsourced Data to Understand Impacts on Traffic

Francisco Contreras and Cristina Torres-Machi

III.4.1 INTRODUCTION

III.4.1.1 The Critical Role of Transportation Infrastructure

Civil infrastructure, such as transportation networks, water distribution systems, and power lines, are critical infrastructure systems that directly contribute to essential societal functions, quality of living, and the economy. Within civil infrastructure, transportation is a critical lifeline for economic activity, access to opportunities and healthcare, and evacuation and emergency response during and after extreme events. Transportation is used to commute to work, access goods and services, visit family and friends, and travel for leisure and work. In 2023, the U.S. transportation system served 335 million U.S. residents, including people who may not own a vehicle or rarely travel (U.S. Department of Transportation 2024). Transportation also plays a crucial role in sustaining the American economy by enabling and supporting economic activity. In 2022, transportation accounted for 9.0% of the national gross domestic product (GDP). Within the different modes of transportation, the U.S. road system, which includes highways and bridges, is critically important. There are over 4 million miles of public roadways in the U.S. that move 72% (or nearly $17 trillion) of the nation's goods (ASCE 2021).

Texas has the largest highway network of any state in the country, with more than 324,000 centerline-miles and more than 703,000 lane-miles[1] of public roads. Every day, Texas's road network supports more than 825 million vehicle miles traveled (VMT), with this metric representing the total sum of miles traveled by every vehicle (TxDOT 2024c). Texas's highway system is crucial for travelers' mobility and is also an important contributor to the state economy. In 2022, Texas's freight system moved 3.4 billion tons of freight, valued at $3.1 trillion – the largest value of freight moved of all states (TRIP 2023). From 2022 to 2050, freight moved annually in Texas by trucks is expected to increase 92% in value (inflation-adjusted dollars) and 52% by weight (TRIP 2023).

III.4.1.2 Investment Needs Derived from Transportation Deterioration and Increased Demand

Despite the importance of transportation infrastructure in the economy and well-being, its condition is inadequate. Approximately 39% of major roads in the U.S. are in poor or mediocre condition (ASCE 2025a). This deterioration, paired with congestion, results in estimated average costs of $1,400 per driver in additional repairs, accelerated vehicle depreciation, and increased fuel consumption and tire wear, as well as time lost to congestion (ASCE 2025a).

DOI: 10.1201/9781003406600-17

Over the last decades, Texas has developed a rapidly growing road network to accommodate its expanding population, which collectively drove 51.2 billion miles in 2023 (ASCE 2025b). While pavement conditions in Texas are good (nearly 90% of pavements are rated as having good condition), traffic congestion, particularly for trucks, is a growing concern (ASCE 2025b). Texas tackles this by dedicating significant resources to highways – more than $37 billion budgeted in 2024–2025 – and exploring innovative solutions, such as traffic flow technology (ASCE 2025b). Despite these efforts, automobile commuters in Texas spend more than 30 hours annually stuck in peak-hour traffic congestion (7:00 to 9:00 AM in the morning and 4:00 to 7:00 PM in the evening on weekdays) (ASCE 2025b). Currently, 67% of freeway travel in urban metro counties in Texas experiences heavy congestion, and this is expected to grow more than 80% by 2035 because traffic volume in Texas's highway system is projected to increase by 66% for automobiles and 123% for truck traffic from 2008 to 2035 (ASCE 2025b).

The pressing needs derived from a deteriorated transportation system needing to accommodate an increasing number of trips induced from a growing population requires a systematic approach to managing this infrastructure – a process known as *transportation asset management*. Transportation asset management is defined by the American Association of State Highway Officials (AASHTO) as a "strategic and systematic process of operating, maintaining, upgrading, and expanding physical assets effectively throughout their life cycle. It focuses on business and engineering practices for resource allocation and utilization, with the objective of better decision making based upon quality information and well defined objectives" (AASHTO 2020).

III.4.1.3 Managing Infrastructure Assets under a Changing Climate

In addition to the challenges derived from deterioration and congestion, extreme weather events present significant challenges for maintaining existing roads and planning future projects. Extreme weather events, indeed, are rated as the risk with the highest criticality in the management of Texas's state transportation system (TxDOT 2022), with this criticality score combining a quantification of the likelihood of occurrence and impact. Transportation is not only impacted by these extreme events but may also be a contributing factor to these phenomena. Transportation is the nation's biggest emitter of greenhouse gas (GHG) emissions, contributing to 28% of total GHG emissions in 2022 (U.S. EPA 2024). These GHG emissions contribute to climate change, and climate change, in turn, has the potential to negatively affect the transportation system (IPCC 2023). The transportation sector faces increasing risk from climate-related extreme weather.

In the last five years, Texas has experienced 68 extreme weather events classified as billion-dollar disasters, that is, with losses exceeding $1 billion each (NCEI 2025). These events include hurricanes and tropical storms, as well as numerous localized droughts, tornadoes, wildfires, and flood events. Texas has 367 miles of coastline along the Gulf of Mexico and 11 major rivers with extensive tributary systems, making flood risk one of the primary challenges faced by Texas's transportation system (TxDOT 2022). By the year 2100, projections from the National Oceanic and Atmospheric Administration (NOAA) State Climate Summary and Texas U.S. Geological Survey National Climate Change Viewer data indicate that Texas will experience a warmer and drier climate, face more intense extreme weather events, encounter further sea level rise, and endure higher storm surges (TxDOT 2025c) that are expected to impact Texas's transportation system.

III.4.2 HOW CLIMATE CHANGE MAY IMPACT ROADWAY INFRASTRUCTURE

Climate change may affect transportation at local, regional, and national scales by impacting system safety and performance, system cost, and the economy, and reduced access to social services and necessities (U.S. EPA 2022). Some specific examples of how different climate hazards impact the vulnerability of roadways are summarized in Table III.4.1 (Liban et al. 2023). For more details on the impacts of climate stressors on transportation and pavements, the Federal Highway Administration's state of practice report (Muench et al. 2023) serves as a relevant resource.

TABLE III.4.1
Vulnerability of Roadway Infrastructure to Climate Hazards

Climate Hazard	Vulnerability of Roadway Infrastructure
Extreme temperatures	• Cracking, buckling, and rutting from heat and permafrost thaw (Markolf et al. 2019; Transportation Research Board and National Research Council 2008). • Unsafe working conditions. • Increased maintenance frequency.
Storms	• Damage from flooding, erosion, saturated soil, and sea level rise. • Increased risk of landslides. • Reduced life cycle from repeated runoff events (i.e., flow of water over the surface or through the subsurface). • Travel lanes and bus routes blocked or rerouted due to flooding.
Drought	• Reduced pavement integrity due to subsidence, collapsible soils, and increased groundwater pumping. • Reduced slope stability due to decreased roadside seeding uptake. • Emergency bridge maintenance. • Reduced visibility and increased closures from dust storms.
Fire	• Road closures and reduced visibility. • Obstructions and debris flows (i.e., fast-moving landslides that generally occur during periods of intense rainfall or rapid snowmelt). • Increased chance of runoff. • Reduced slope stability due to burn scar.

Source: Liban et al. (2023).

The Texas Department of Transportation (TxDOT) is in the process of completing a study, as part of the Texas Statewide Resiliency Plan (TxDOT 2025a), to assess the vulnerability of the state's transportation system to various hazards and extreme weather events. Recent research, including studies by TxDOT (2025c) and Mostafavi et al. (2023) suggest that extreme weather events, increased flooding, and inundation potential may result in temporary road closures. Road closures may also be needed as these events induce permanent deformations in the pavement (i.e., a phenomenon known as rutting), buckling, and cracking. These studies also suggest that increased risk of power outages may disrupt traffic signals and intelligent transportation systems aimed at enhancing the flow and safety of the transportation network (e.g., cameras to enforce traffic laws, signs that dynamically adjust speed limits depending on conditions, etc.). Finally, warmer and drier conditions may lead to longer wildfire seasons and increased wildfire potential that may result in temporary road closures due to fire, smoke, or limited visibility conditions.

Given the impacts that climate has on infrastructure condition and system performance, accounting for climate stressors in the long-term management of infrastructure becomes crucial. This analysis should not only consider how a changing climate may accelerate infrastructure deterioration, but also how communities may be impacted. It is important to understand how extreme events may impact traffic, the ability of travelers to complete trips, and the increased time and reduced speeds that may further exacerbate the already pressing problem of congestion in the U.S. transportation system in general and Texas in particular.

III.4.3 MOBILE LOCATION DATA TO QUANTIFY IMPACTS OF EXTREME EVENTS ON TRAFFIC

In the past decades, transportation agencies have relied on physical traffic counters to obtain traffic data, such as traffic volume, speed, vehicle classification, and weight. Agencies collect traffic data using either continuous count programs or short-term count programs (FHWA 2024a). Continuous count programs collect data from permanently installed traffic counters that gather traffic data for at least a year. Continuous count programs are helpful to understand daily, weekly, and monthly changes in traffic volumes. Short-term count programs, on the other hand, collect data in shorter periods, often relying on portable traffic counters that collect data for 48 or 72 hours. Short-term count programs are generally used to determine average annual daily traffic (AADT).

Permanent traffic counters are usually installed on major highways, and their number is limited due to their elevated cost. For example, TxDOT has approximately 350 permanent traffic count stations installed across the state of Texas (TxDOT 2025b). Considering that Texas has the largest transportation network in the United States, with a total length exceeding 703,000 lane-miles (TxDOT 2024a), permanent traffic count stations cover only a small fraction of the total transportation network.

Due to the limited coverage of traffic counters, usually limited to major highways, measuring the holistic impacts that extreme weather events – such as hurricanes, floods, and winter storms – have on the highway network is challenging. These events are known to disrupt the normal functioning of transportation networks by reducing traffic volumes (Hu et al. 2024) and speeds (Pregnolato et al. 2017), and by increasing the number of accidents (Kasmalkar and Suckale 2021). Since traditional traffic data collection methods are primarily limited to major highways, using this data to quantify the impacts of extreme weather events on traffic neglects the effects on minor roads, which play a critical role in supporting communities. Understanding the impacts of extreme weather on minor roads and communities is crucial for identifying areas most severely impacted by extreme events that may need infrastructure improvements.

Crowdsourced data has emerged as a promising alternative for obtaining traffic data, as it addresses many challenges of traditional traffic data collection methods. Crowdsourced traffic data is usually derived from devices that use the global positioning system (GPS). Various firms collect GPS data from various providers, aggregate it, and process it to obtain traffic metrics (Yabe et al. 2022). Crowdsourced data provides high coverage, providing traffic data for most roads and areas in the United States, without the need to install and maintain additional infrastructure on roads (Krile and Habtemichael 2024). In addition, crowdsourced data provides historical data, thus enabling the analysis of changes in traffic over time. As an example of the capabilities of crowdsourced data, Hu et al. (2024) used crowdsourced traffic data to evaluate the impacts of adverse weather on the Dallas–Fort Worth area in Texas. Hu et al. (2024) analyzed a total of 19,179 road segments. Using traditional traffic data would have considerably limited the scope of the study, as only 15 traffic counters are available in the Dallas–Fort Worth area.

Various transportation agencies are currently using crowdsourced data, such as the Departments of Transportation in Indiana, Utah, and South Carolina (FHWA 2024b). TxDOT, in particular, is leveraging crowdsourced speed data to adjust traffic signal timing patterns in the Houston district (TxDOT 2024b).

III.4.4 CASE STUDY: IMPACTS OF FLOODING ON MOBILITY IN TEXAS

To demonstrate the capabilities of crowdsourced data to quantify the impacts of extreme weather events on traffic, this section analyzes historical flood events impacting Texas.

III.4.4.1 Flooding Is a Recurring and Increasing Threat to Transportation Infrastructure in Texas

In the past three years (2022–2024), weather and climate disasters have resulted in over $461 billion in economic losses and at least 1,534 deaths (NCEI 2025) in the United States. Texas is the state with the highest number of disasters per year, and this frequency is increasing over time (NCEI 2025).

Flooding poses a considerable threat to transportation networks and people's mobility. Flood events disrupt the normal functioning of transportation systems. Depending on the severity, flooding can reduce traffic speeds or lead to full road closures (Pregnolato et al. 2017). These disruptions have direct consequences for people who must adapt their daily lives to cope with the disruption (Behr et al. 2022). Flooding can limit access to essential services such as hospitals, police, and fire departments (Alabbad et al. 2021), and many travelers may not be able to commute to work, leading to personal economic losses (Behr et al. 2022; Kontou et al. 2017). These experiences also impact the well-being of people, as it has been reported that lack of control, congestion, and uncertainty have been related to higher levels of stress (Chatterjee et al. 2020).

Coastal communities and those located near rivers are at greater risk of flooding. In future years, various communities in Texas – such as Galveston, Corpus Christi, and Rockport – are expected to experience more frequent and severe flooding due to sea level rise (NOAA 2025). For example, by 2100, NOAA projects a sea level rise of 1.5 meters (5 feet) in the Galveston area. As a result, the northern side of Galveston may become permanently inundated, and other areas of Galveston may have increased vulnerability to flood events triggered by high tides, storms, and tropical cyclones.

Crowdsourced data enables a deeper understanding of how different transportation networks and communities are impacted during extreme weather events, such as flooding. Crowdsourced data enables the evaluation of the location, frequency, and magnitude of the disruption extreme events have on traffic. The next section showcases the application of crowdsourced data to evaluate the impacts of different flood events on traffic in various communities in Texas, an assessment that would not be possible using traditional data collection methods.

III.4.4.2 Data

III.4.4.2.1 Locations and Events

The analysis considered 59 census tracts across Texas (Figure III.4.1). Census tracts – subdivisions of counties with populations ranging from 1,200 to 8,000 people (U.S. Census Bureau 2025) – were selected as the unit of analysis because they are among the smallest geographic subdivisions for which traffic data is available. Analyzing smaller areas allows for a more accurate identification

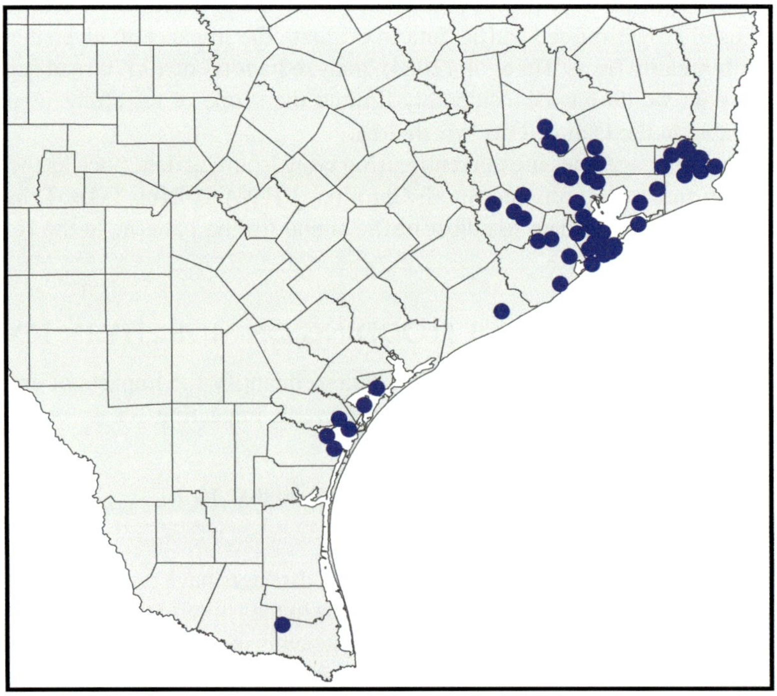

FIGURE III.4.1 Location of the 59 census tracts in Texas selected for analysis.

of communities impacted by flooding. Since the effects of flooding can be highly localized, using larger geographic units – such as counties – might not be appropriate to capture the impacts of flooding (Praharaj et al. 2021).

The analysis considered flood events occurring between 2018 and 2021. Since there is no official database identifying which census tracts have experienced flooding, the analysis relied on two data sources: (1) flood insurance claims; and (2) flood maps. Data from the National Flood Insurance Program (FEMA 2023) was used to identify census tracts with a high number of flood insurance claims. This analysis considered that a greater number of claims corresponded to more severe flooding. Additionally, flood maps from "The Flood Observatory" (Brakenridge, 2023) were used to identify census tracts that flooded during major events such as Tropical Storm Imelda and Hurricane Sally.

III.4.4.2.2 Traffic Data

To measure traffic impacts, the analysis considered the daily number of trips taken within each census tract. This metric was chosen because travel demand metrics are generally more sensitive to disruptive events than other traffic metrics, such as speed, travel time, or trip length (Contreras and Torres-Machi 2025; Donovan and Work 2017; Hu et al. 2024). For example, Donovan and Work (2017) found that taxi trips in New York City during Hurricane Sandy were driving at faster speeds than usual, despite lower taxi demand. Similarly, Contreras and Torres-Machi (2025) observed that while the number of trips declined substantially during flood events, other traffic metrics – such as trip length, travel time, and average speed – did not necessarily decrease. This suggests that, although the number of vehicles on the roads drops significantly, the trips that do occur can maintain their usual duration, distance, and speed.

Based on these results from previous research, this case study used the daily number of trips as the metric to measure the impacts of flooding. The daily number of trips was aggregated at the census tract level, and the data were obtained from StreetLight Data (StreetLight Data 2025), a firm that provides traffic data derived from crowdsourced data.

III.4.4.3 Analysis

Three steps were taken to transform the traffic data into measures of impact: (1) data normalization; (2) identifying times of disruption; and (3) quantifying traffic disruption. For more details on the method, refer to Contreras and Torres-Machi (2025).

1. **Data normalization:** This step involves defining a measure of baseline traffic for each day. This baseline captures normal (i.e., expected) traffic that can be compared to actual traffic metrics to quantify functionality. In practical terms, a functionality value of one represents normal traffic levels, whereas values below one indicate lower-than-usual traffic, and values above one indicate higher-than-usual traffic. For example, a functionality value of 0.5 means that only 50% of the usual number of trips were recorded on a given day. Seasonal and Trend decomposition using Loess (STL) was used to estimate expected daily traffic. STL decomposes the data into three different components – a pattern (which includes weekly and seasonal patterns common in traffic), a trend, and a remainder (Hyndman and Athanasopoulos 2018). The expected number can then be approximated as the sum of the trend and the weekly and seasonal patterns.
2. **Identifying times of disruption:** This step focused on identifying periods of traffic disruption. Recent studies have used outlier detection techniques for this purpose (Ansari Esfeh et al. 2022; Donovan and Work 2017; Yang et al. 2022). In this study, values are classified as abnormal (i.e., outliers) if they fall below two times the median absolute deviation. Similarly to standard deviation, the median absolute deviation measures data dispersion. This metric, however, is less affected by outliers than the standard deviation – a feature that makes this metric more appropriate for this study.

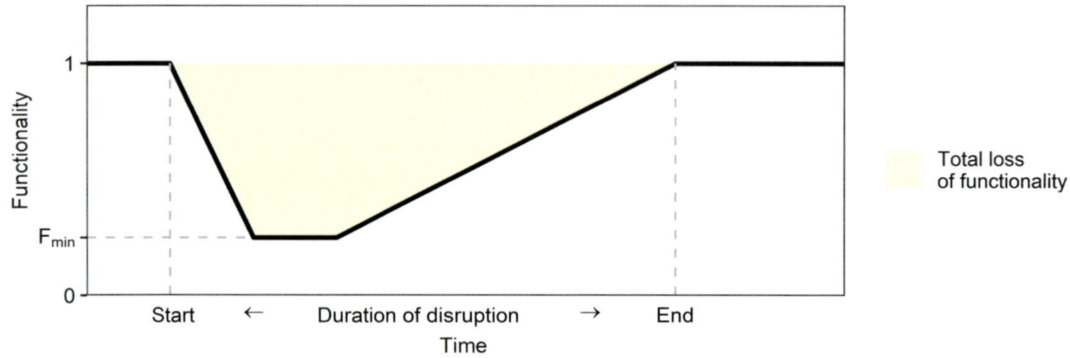

FIGURE III.4.2 Metrics used to quantify traffic disruption.

3. **Quantifying traffic disruption:** To quantify the magnitude of the impacts on traffic, three metrics were used: (i) total loss of functionality; (ii) minimum functionality; and (iii) duration of the disruption. These metrics are depicted in Figure III.4.2. Total loss of functionality is calculated as the area over the curve, minimum functionality is the minimum value of traffic during the flood event (F_{min} in Figure III.4.2), and duration of the disruption is the total number of consecutive days in which abnormal values of traffic are registered. In the context of resilience, particular emphasis is placed on the recovery phase following a disruption event. The selected metrics, also used in recent literature (Aghababaei et al. 2021; Hong et al. 2021; Hu et al. 2024), capture the recovery process by accounting for both the magnitude and duration of the disruption.

III.4.4.4 Results

III.4.4.4.1 Impacts of Flooding on Traffic Are Recurrent in Time and Impact Multiple Areas Across Texas

The analysis shows that flood-induced traffic disruption is a recurring issue for various coastal communities across Texas. A total of 98 events of traffic disruption caused by flooding were found to impact the census tracts under analysis between 2018 and 2021. These flood events were caused by hurricanes – Delta (2020), Hannah (2020), Sally (2020), and Nicholas (2021), Tropical Storm Imelda (2019) – and high-precipitation events occurring throughout the period of analysis (Figure III.4.3).

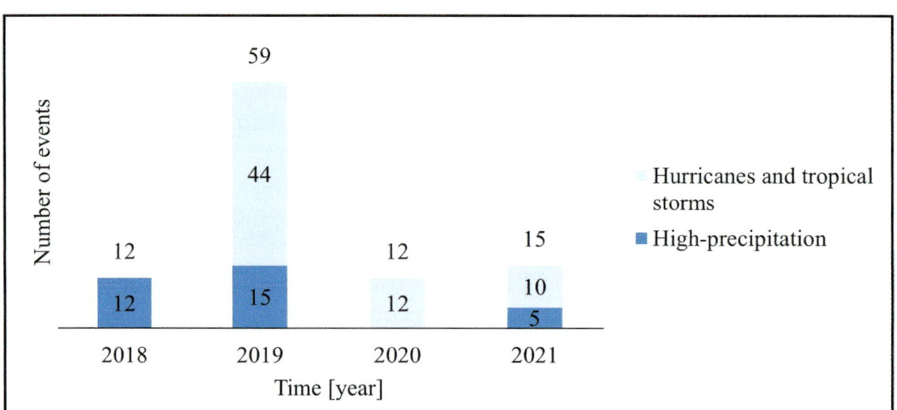

FIGURE III.4.3 Traffic disruption events from flooding from 2018 to 2021 in Texas.

Figure III.4.4 summarizes the geographical location of these disruption events and their evolution over time. In Figure III.4.4, gray dots are used to represent the census tracts under analysis, whereas red dots represent those whose traffic was impacted by flooding. In 2018, there were no hurricanes that impacted traffic (Figure III.4.3). However, several flood events caused by heavy rains resulted in traffic disruptions in 12 census tracts. Coastal communities in Galveston and Rockport were particularly impacted in 2018 (Figure III.4.4a). In 2019, Tropical Storm Imelda and heavy rainfall events led to 59 flood events that disrupted normal traffic (Figure III.4.3). Most of the census tracts suffering traffic disruptions were located in the Houston and Galveston area (Figure III.4.4b). In comparison to 2018, census tracts located farther away from the coast were impacted in 2019. In 2020, hurricanes Delta, Hannah, and Sally were responsible for traffic disruptions in 12 census tracts mainly located in the areas of Galveston and Corpus Christi (Figure III.4.4c). Finally, in 2021, high-precipitation events and Hurricane Nicholas led to traffic disruption events in 15 census tracts located near Galveston, Rockport, and Corpus Christi (Figure III.4.4d).

These results suggest that traffic disruption from flooding is a recurrent threat in Texas, whether caused by hurricanes, tropical storms, or other heavy rainfall events, with coastal areas near Houston, Galveston, Rockport, and Corpus Christi suffering recurring instances of traffic disruption.

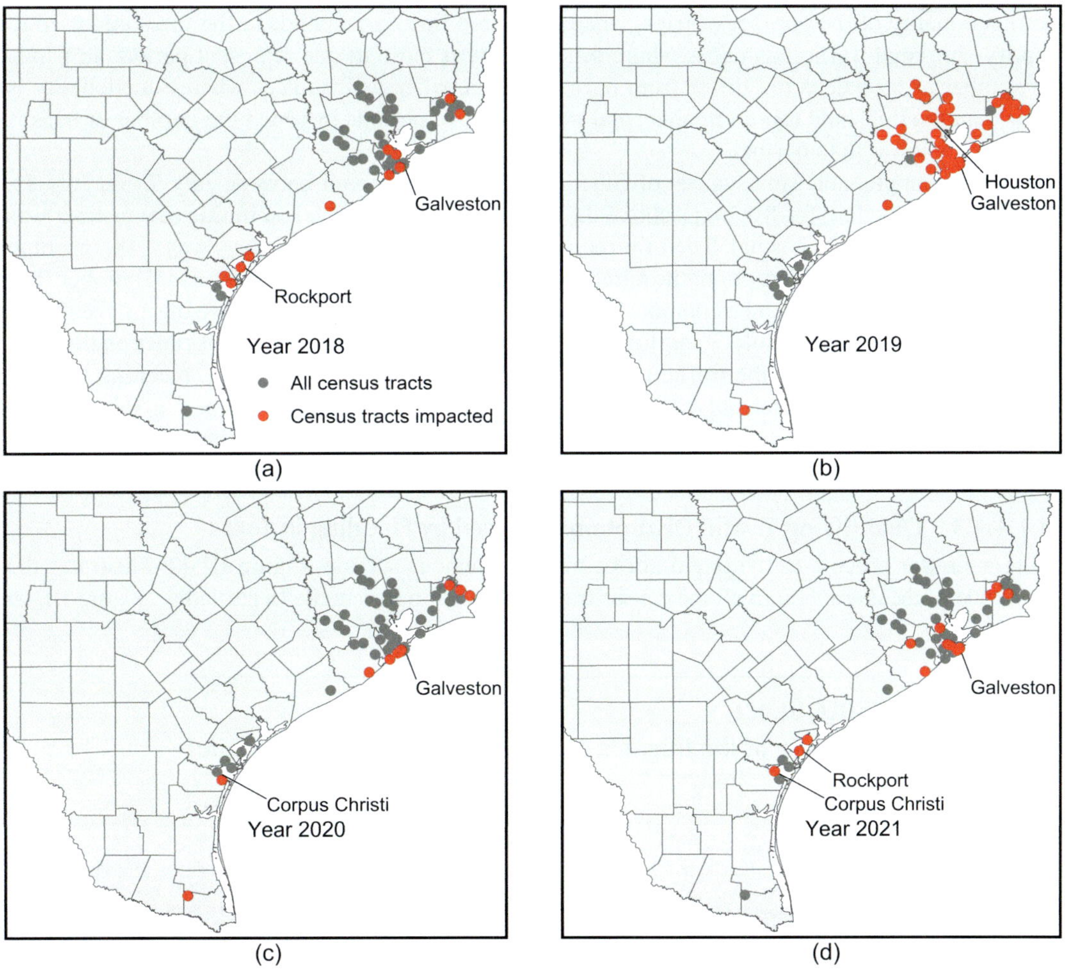

FIGURE III.4.4 Census tracts that experienced traffic disruption from flood events in (a) 2018, (b) 2019, (c) 2020, and (d) 2021.

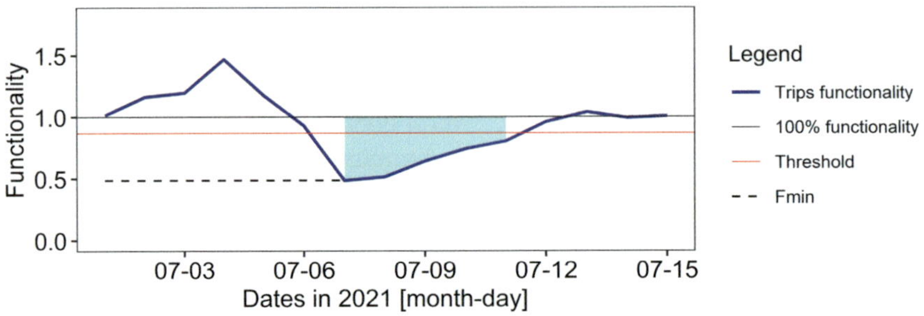

FIGURE III.4.5 Functionality of a census tract in Aransas County for the month of July 2021.

III.4.4.4.2 Quantifying Traffic Impact in a Specific Location

In addition to measuring the location and frequency of traffic disruption events caused by flooding, crowdsourced location data also enables quantifying the magnitude of the impact. As an illustrative example, the functionality of the number of trips of a census tract in Rockport (depicted in Figure III.4.4d as a red dot labeled "Rockport") is shown in Figure III.4.5. The blue line represents the functionality of the number of trips, whereas the red line corresponds to the threshold used to identify abnormal traffic. Instances where functionality is close to one represent periods in which the number of trips does not differ from normal values, whereas instances where functionality is lower than the threshold represent abnormal traffic, that is, periods in which the number of trips is significantly lower than normal.

In this example, abnormal values of functionality were observed between July 7 and July 11, 2021. This means that, for five consecutive days, the number of trips taken in this census tract was significantly lower than normal. The minimum functionality during those days was 0.48, meaning that, on the worst day of disruption, the number of trips was reduced to 48% of its usual number. The total loss of functionality can be computed as the area over the functionality curve during the period of disruption (i.e., between July 7 and July 11, 2021). In this case, the total loss of functionality was 1.81. This value represents the traffic that did not occur due to the flood event, meaning that the equivalent of 1.81 days of normal traffic was lost. Using these values, summarized in Table III.4.2, the impacts of flood events on different communities can be compared to determine which communities are more heavily affected by flooding.

III.4.4.4.3 Quantifying Traffic Disruptions Induced by Flooding in Texas

Using a similar process to the one described in the previous section, Figure III.4.6 illustrates the magnitude of the disruption in all the census tracts analyzed in this study. In particular, Figure III.4.6

TABLE III.4.2
Summary of the Impacts of a Flood Event on a Census Tract

Variables	Values
Location	Rockport, Aransas County, Texas
Start date	July 7, 2021
End date	July 11, 2021
Total loss of functionality	1.81
Minimum functionality	0.48
Duration of the disruption	5 days

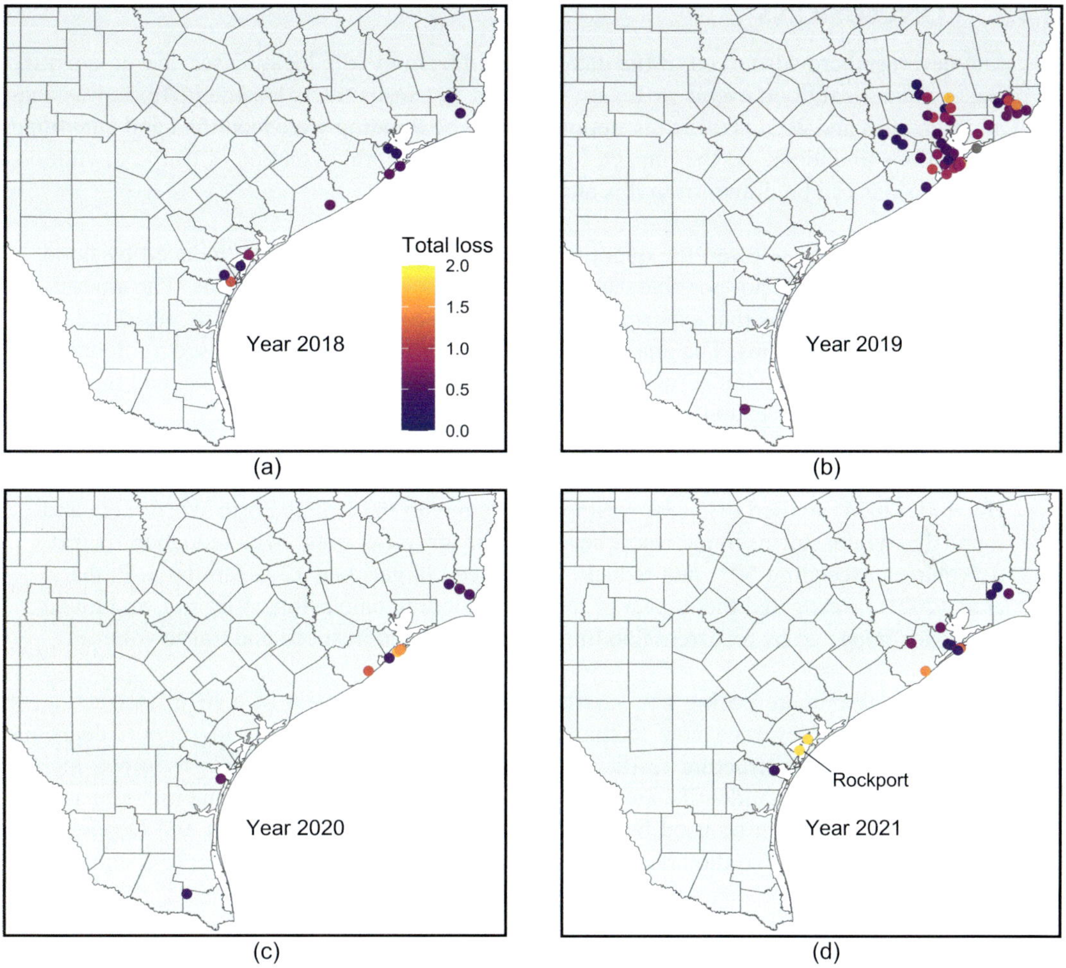

FIGURE III.4.6 Total loss of functionality for (a) 2018, (b) 2019, (c) 2020, and (d) 2021.

depicts the total loss of functionality in impacted census tracts across time. On average, the total loss of functionality was 0.68, meaning that flooding led to a 68% reduction in the number of trips compared to a typical day.

However, some events had a significantly greater impact. For example, in 2021, two areas near Rockport experienced disruptions of almost 2.0 (these instances are plotted as yellow dots in Figure III.4.6d). In these cases, the number of lost trips was equivalent to almost two full days of typical traffic. Higher impacts on traffic ultimately lead to people disrupting their normal lives for longer periods of time, with potentially higher economic losses and well-being impacts. A higher value of total loss of functionality suggests that these communities experienced a longer and more severe reduction in the number of trips taken. Traffic disruption reflects travelers' decisions or ability to travel – suggesting that, during those days, people refrained or were not able to make their usual trips due to extreme weather conditions. Such disruptions can have significant impacts on daily life, limiting access to work, healthcare, and other essential services.

In summary, the magnitude of impacts varies by year and location. The use of crowdsourced data makes it possible to identify communities that are more severely and frequently affected.

III.4.5 CONCLUSIONS

This study leveraged crowdsourced traffic data obtained from mobile location services to quantify the impact of historical flood events in Texas. Traffic in 59 census tracts located in Texas was analyzed between 2018 and 2021. Traffic disruptions caused by flooding were identified and quantified in these locations to understand the impact, location, and temporal evolution of traffic disruption. Several conclusions can be drawn from this analysis:

- Crowdsourced data allowed the quantification of the traffic disruption induced by flood events, an assessment not possible using traditional data collection methods. The analysis identified communities that experienced abnormally low levels of traffic, as measured by the daily number of trips. The analysis revealed that traffic disruption caused by flooding is a recurrent phenomenon across Texas, especially for coastal communities, and that the magnitude of impacts changes over time and location.
- The results obtained in this analysis are aligned with previous work that has used crowdsourced data to evaluate the impacts of extreme weather events on transportation networks. Hu et al. (2024) studied different weather conditions in the Dallas Forth-Worth area and found that winter storms impact more severely than any other event, with reductions in traffic volumes exceeding 50% and reductions of speed larger than 12%. Similarly, Praharaj et al. (2021) investigated the impact of recurrent flooding on Norfolk, Virginia, and found that roads impacted by recurrent flooding have a reduction in speed and traffic volumes.

With a changing climate posing new challenges in the management of civil infrastructure, it becomes imperative to leverage new technologies and advanced methods to support decision-making and enhance infrastructure resilience. This study showcases how crowdsourced mobile location data can be used to identify areas impacted by extreme events and quantify impacts on traffic. This information can be used by system operators (e.g., local agencies and Departments of Transportation) to optimize the allocation of funding aimed at enhancing the resilience of the transportation network.

NOTE

1. Lane-miles are the product of the centerline length (in miles) multiplied by the number of lanes. For example, the one-mile centerline length of a two-lane road equals two lane-miles.

REFERENCES

AASHTO. 2020. *AASHTO Transportation Asset Management Guide.*

Aghababaei, M. T. (Siavash), S. B. Costello, and P. Ranjitkar. 2021. "Measures to evaluate post-disaster trip resilience on road networks." J. Transp. Geogr., 95: 103154. https://doi.org/10.1016/j.jtrangeo.2021.103154.

Alabbad, Y., J. Mount, A. M. Campbell, and I. Demir. 2021. "Assessment of transportation system disruption and accessibility to critical amenities during flooding: Iowa case study." *Sci. Total Environ.*, 793: 148476. https://doi.org/10.1016/j.scitotenv.2021.148476.

Ansari Esfeh, M., L. Kattan, W. H. K. Lam, M. Salari, and R. Ansari Esfe. 2022. "Road network vulnerability analysis considering the probability and consequence of disruptive events: A spatiotemporal incident impact approach." *Transp. Res. Part C Emerg. Technol.*, 136: 103549. https://doi.org/10.1016/j.trc.2021.103549.

ASCE. 2021. "ASCE 2021 Infrastructure Report Card." Accessed June 1, 2021. http://www.infrastructurereportcard.org.

ASCE. 2025a. "ASCE 2025 Report Card for America's Infrastructure." Accessed March 26, 2025. http://www.infrastructurereportcard.org.

ASCE. 2025b. "ASCE 2025 Report Card for Texas's Infrastructure." Accessed March 26, 2025. https://infrastructurereportcard.org/state-item/texas/.

Behr, J. G., W. Yusuf, G. McLeod, S. Stafford, D. Loftis, A. Anuar, and R. Diaz. 2022. "Measuring the economic impact of recurrent flooding on workforce productivity and property." Report Prepared for "COCA/SARP – Assessing and Communicating Economic Impacts and Risks Associated with Water Resource Management Challenges Along the Coast" NOAA Award Number NA19OAR4310309. https://digitalcommons.odu.edu/odurc-presentations/31/

Brakenridge, G. R. 2023. "Global Active Archive of Large Flood Events. Dartmouth Flood Observatory, University of Colorado, USA." Accessed September 1, 2023. http://floodobservatory.colorado.edu/.

Chatterjee, K., S. Chng, B. Clark, A. Davis, J. D. Vos, D. Ettema, S. Handy, and A. Martin, and L. Reardon. 2020. "Commuting and wellbeing: A critical overview of the literature with implications for policy and future research." *Transp. Rev.*, 40 (1): 5–34. Routledge. https://doi.org/10.1080/01441647.2019.1649317.

Contreras, F., and C. Torres-Machi. 2025. "Data-driven methodology to quantify traffic resilience of communities from crowdsourced location data." *Int. J. Disaster Risk Reduct.*, 105219. https://doi.org/10.1016/j.ijdrr.2025.105219.

Donovan, B., and D. B. Work. 2017. "Empirically quantifying city-scale transportation system resilience to extreme events." *Transp. Res. Part C Emerg. Technol.*, 79: 333–346. https://doi.org/10.1016/j.trc.2017.03.002.

FEMA. 2023. "OpenFEMA Dataset: FIMA NFIP Redacted Claims." Accessed September 24, 2023. https://www.fema.gov/openfema-data-page/fima-nfip-redacted-claims-v2.

FHWA. 2024a. "Traffic Monitoring Guide."

FHWA. 2024b. "Crowdsourcing for Advancing Transportation Operations | Case Studies and Fact Sheet." Accessed March 20, 2025. https://ops.fhwa.dot.gov/crowdsourcing/case_studies.cfm.

Hong, B., B. J. Bonczak, A. Gupta, and C. E. Kontokosta. 2021. "Measuring inequality in community resilience to natural disasters using large-scale mobility data." Nat. Commun., 12 (1): 1870. Nature Publishing Group. https://doi.org/10.1038/s41467-021-22160-w.

Hu, S., K. Wang, L. Li, Y. Zhao, Z. He, and Y. Zhang. 2024. "Multi-crowdsourced data fusion for modeling link-level traffic resilience to adverse weather events." *Int. J. Disaster Risk Reduct.*, 112: 104754. https://doi.org/10.1016/j.ijdrr.2024.104754.

Hyndman, R. J., and G. Athanasopoulos. 2018. *Forecasting: Principles and Practice.* OTexts: Melbourne, Australia.

IPCC. 2023. "Intergovernmental Panel on Climate Change - Sixth Assessment Report (AR6)."

Kasmalkar, I., and J. Suckale. 2021. "Traffic accidents and delays present contrasting pictures of traffic resilience to coastal flooding in the San Francisco Bay Area, USA." *Urban Clim.*, 37: 100851. https://doi.org/10.1016/j.uclim.2021.100851.

Kontou, E., P. Murray-Tuite, and K. Wernstedt. 2017. "Duration of commute travel changes in the aftermath of Hurricane Sandy using accelerated failure time modeling." *Transp. Res. Part Policy Pract.*, 100: 170–181. https://doi.org/10.1016/j.tra.2017.04.015.

Krile, R., and F. Habtemichael. 2024. *Guidelines for Obtaining AADT Estimates from Non-Traditional Sources.* United States. Department of Transportation. Federal Highway Administration. Office of Highway Policy Information, Report FHWA-HPL-23-018.Washington, D.C.

Liban, C. B., R. Kafalenos, L. Alessa, S. Anenberg, M. Chester, J. DeFlorio, F. J. Dóñez, A. Flannery, M. R. Sanio, B. A. Scott, and A. M. K. Stoner. 2023. "Ch. 13. Transportation." Fifth National Climate Assessment. Washington, DC, USA.

Markolf, S. A., C. Hoehne, A. Fraser, M. V. Chester, and B. S. Underwood. 2019. "Transportation resilience to climate change and extreme weather events – Beyond risk and robustness." *Transp. Policy*, 74: 174–186. https://doi.org/10.1016/j.tranpol.2018.11.003.

Mostafavi, A., C.-C. Lee, C. Fan, A. Rajput, C.-W. Hsu, H. Farahmand, A. Esmalian, F. Yuan, F. Patrascu, B. Li, J. Ma, C. Liu, and C.-F. Liu. 2023. Establish TxDOT Transportation Planning Scorecard and Best Practices: Technical Report. Austin, Texas.

Muench, S., T. Van Dam, P. Ram, and K. Smith. 2023. "Pavement Resilience: State of the Practice." FHWA: Washington D.C.

NCEI. 2025. "U.S. Billion-Dollar Weather and Climate Disasters."

NOAA. 2025. "Sea Level Rise Viewer." https://coast.noaa.gov/slr/#.

Praharaj, S., T. D. Chen, F. Zahura, M. Behl, and J. Goodall. 2021. "Estimating impacts of recurring flooding on roadway networks: A Norfolk, Virginia case study." *Nat. Hazards*, 107: 1–25. https://doi.org/10.1007/s11069-020-04427-5.

Pregnolato, M., A. Ford, S. M. Wilkinson, and R. J. Dawson. 2017. "The impact of flooding on road transport: A depth-disruption function." *Transp. Res. Part Transp. Environ.*, 55: 67–81. https://doi.org/10.1016/j.trd.2017.06.020.

StreetLight Data. 2025. "StreetLight Data." https://www.streetlightdata.com/.

Transportation Research Board and National Research Council. 2008. "Potential Impacts of Climate Change on U.S. Transportation: Special Report 290." Transportation Research Board: Washington, D.C.

TRIP. 2023. "America's Rolling Warehouses: Opportunities and Challenges on the nation's Freight Delivery System." Washington D.C.

TxDOT. 2022. "Texas DOT Transportation Asset Management Plan."

TxDOT. 2024a. "2023 Roadway Inventory Report."

TxDOT. 2024b. "Adaptive Signal Operations Using Crowdsourced Speed Data."

TxDOT. 2024c. "Roadway Inventory Annual Report 2023."

TxDOT. 2025a. "Statewide Resiliency Plan." Accessed March 26, 2025. http://txdot.gov/en/home/projects/planning/statewide-resiliency-plan.html.

TxDOT. 2025b. "Traffic Count Maps." Accessed March 5, 2025. http://txdot.gov/en/home/data-maps/traffic-count-maps.html.

TxDOT. 2025c. "Statewide On-Road Greenhouse Gas and Climate Change." TxDOT Environmental Affairs Division.

U.S. Department of Transportation. 2024. "Transportation Statistics Annual Report 2024." Washington, DC.

US Census Bureau. 2025. "Glossary." *Census.gov*. Accessed February 14, 2025. https://www.census.gov/programs-surveys/geography/about/glossary.html.

US EPA. 2022. "Climate Change Impacts on Transportation." Overviews and Factsheets. Accessed March 26, 2025. https://www.epa.gov/climateimpacts/climate-change-impacts-transportation.

US EPA. 2024. "Fast Facts on Transportation Greenhouse Gas Emissions." Overviews and Factsheets. Accessed March 21, 2025. https://www.epa.gov/greenvehicles/fast-facts-transportation-greenhouse-gas-emissions.

Yabe, T., N. K. W. Jones, P. S. C. Rao, M. C. Gonzalez, and S. V. Ukkusuri. 2022. "Mobile phone location data for disasters: A review from natural hazards and epidemics." *Comput. Environ. Urban Syst.*, 94: 101777. https://doi.org/10.1016/j.compenvurbsys.2022.101777.

Yang, Z., F. Liu, Z. Gao, H. Sun, J. Zhao, D. Janssens, and G. Wets. 2022. "Estimating the influence of disruption on highway networks using GPS data." *Expert Syst. Appl.*, 187: 115994. https://doi.org/10.1016/j.eswa.2021.115994.

III.5 The Economic Impacts of Climate Change in Texas: A Regional Lesson with National Impact

Lessons for the U.S. Economy

Alla Semenova

III.5.1 INTRODUCTION

Focusing on the state of Texas, this chapter discusses the adverse economic impacts of climate change,[1] drawing lessons for the broader U.S. economy. While Texas is the nation's most weather-vulnerable state (Kinniburgh et al., 2015) with the greatest economic impacts from climate change, increasingly more frequent and severe weather hazards have been changing the economic climate throughout the entire U.S. From Texas to North Dakota, a growing number of communities, businesses, and governments across the U.S. have been feeling the heat of the melting economic conditions in their states. By analyzing the economic impacts of climate change in Texas, this chapter offers a glimpse of the economic costs and consequences of climate change in the nation. While the degrees of impacts vary depending on the state, distinct climate-change driven economic trends can be identified across the U.S. This chapter examines these trends by, first, analyzing the economic impacts of climate change in Texas.

Although Texas is accustomed to heat and humidity, the accelerating pace of climate change has been inflicting "significant and widespread economic harm" on the Lone Star State (Kinniburgh et al., 2015, p. 2). Rising temperatures, changing precipitation patterns, worsening droughts and wildfires, intensifying storms and hurricanes, increased flooding, along with other weather extremes, have been putting Texas at "significant economic risk" (Kinniburgh et al., 2015). While it is not possible to attribute every weather extreme to anthropogenic climate change, the growing frequency, severity, duration, and synchronicity of extreme weather events and conditions have been linked to the Earth's rapidly changing climate (Cho, 2021; Intergovernmental Panel on Climate Change, 2022; Nielsen–Gammon et al., 2024; United States Environmental Protection Agency, 2016).

The negative effects of climate change and extreme weather on the Texan economy have been increasingly discussed by the Federal Reserve Bank of Dallas. The Dallas Fed has emphasized that "weather events and climate-related trends are impacting economic conditions and financial stability" in the state of Texas, the Federal Reserve's Eleventh District,[2] as well as across the nation (Kaplan, 2019, p. 4). The Dallas Fed has cited rising temperatures, changing precipitation patterns, droughts, water shortages, flooding, sea level rise, and increased intensity of hurricanes as important risk factors for the Texan economy (Kaplan, 2019). According to the Dallas Fed analysis, climate change-exacerbated extreme weather "can have a substantial human and economic cost" to the state and the Fed's Eleventh District (Kaplan, 2019, p. 1). The Dallas Fed has further cautioned that "the severity and damage caused by extreme weather events are likely to intensify in the years ahead," with disproportionate impacts on the Texan economy (Kaplan, 2019, p. 1). In the upcoming

DOI: 10.1201/9781003406600-18

decade, the negative effects of climate change on Texas's gross state product[3] (GSP) will likely be twice as pronounced as in the rest of the U.S. (Jayashankar et al., 2023).

While economists have traditionally "looked through" weather events, viewing them as temporary supply shocks (Makhlouf, 2023), the Dallas Fed has stressed the lasting impacts of weather extremes on the state's economy:

> [...], while many of the impacts of these severe weather events have historically been transitory, we are increasingly cognizant of the longer-term risks that are leading several cities and companies to make substantial capital investments in improving infrastructure in order to mitigate the damage of future severe weather.
>
> **(Kaplan, 2019, p. 4)**

With rising greenhouse gas (GHG) emissions and the accelerating pace of climate change, Texas will continue experiencing extreme weather-driven reductions in agricultural crop yields, production shortages and supply-chain interruptions, disruptions to energy supply, surging energy demand, rising energy prices and production costs, transportation disruptions and higher shipping costs, water shortages, rising water rates, elevated consumer prices, declining labor productivity, growing GSP losses, rising insurance and financing costs, increases in the rates of mortality, rising health care spending, growing strains on public budgets, as well as hundreds of billions of dollars in infrastructure and property damages and losses. Climate change will keep threatening the state's infrastructure in the manufacturing, transportation, energy, agriculture, water supply and other key sectors. The state's food supply system remains increasingly vulnerable to climate hazards as well (Barber and Stone, 2022). By undermining the state's food security, climate change will exert disproportionate impacts on Texas's most economically vulnerable communities. Rising food insecurity is one of the ways in which climate change has been exposing and exacerbating economic inequality in Texas.

In what follows, I will discuss these and other economic impacts of climate change in Texas. I will also include a brief overview of the Texan economy, in order to provide a general economic context for this discussion. I will conclude the chapter by analyzing the key economic consequences of climate change for the U.S. households, businesses, and state and local governments.

III.5.2 'THE REPUBLIC OF TEXAS': AN ECONOMIC SNAPSHOT

Driven by oil and natural gas, transportation, agriculture and livestock, and the technology sectors (Kinniburgh et al., 2015), Texas is the second largest economy in the U.S. (after California). In recent years, California, Texas, New York, Florida, and Illinois together accounted for more than 40% of the U.S. gross domestic product (GDP). In 2022, Texas added more than $2.4 trillion to the U.S. nominal GDP of over $25.7 trillion. As such, Texas accounted for almost one-tenth of the U.S. economy. In 2023, Texas contributed more than $2.56 trillion to the U.S. nominal GDP of over $27 trillion. As a result, Texas represented almost one-tenth of the U.S. economy in 2023 again (Bureau of Economic Analysis, 2024). If Texas were a separate country, it would rank as the world's eighth largest economy (Texas Comptroller, 2024), surpassing the national economies of Italy, Brazil, Canada, Russia, Mexico, and Australia (Office of the Governor of California, 2024).

Texas is headquarters to 55 Fortune 500 companies, mainly in the energy and transportation sectors, the greatest number of Fortune 500 companies compared to any other U.S. state (Kinniburgh et al., 2015; Office of the Texas Governor, 2023). Responsible for almost a quarter of the nation's energy supply, Texas is the No. 1 energy-producing state in the U.S. Generating much more energy than it consumes, Texas is also the nation's largest net energy supplier (United States Energy Information Administration, 2023b). With 936,476 energy sector jobs in 2022, Texas's energy sector accounted for 7% of the state's total non-farm employment, representing 11.5% of the nation's energy sector jobs (United States Department of Energy, 2023).

With an estimated total population exceeding 30 million residents in 2022, Texas is the second most populous state in the U.S., home to 9% of the U.S. population (Jansen and Sinha, 2024). Between 2012 and 2022, Texas's population grew by 15.1%, more than double the nation's average population growth rate of 6.2% (Texas Comptroller, 2024). From 2020 to 2021, Texas's population grew ten times as fast as the nation's average, at 1.3% compared to 0.13% nationally (Donald and Grubbs, 2022). In 2022, Texas's population increased by 1.6%, adding about 471,000 new residents to the state (Texas Comptroller, 2024). From 2012 to 2022, population growth in Texas's four largest metropolitan areas exceeded 18%[4] with Austin–Round Rock–San Marcos metropolitan area expanding at 32% (Minton and Alvarado, 2022). Much of Texas's population gains come from people leaving California and New York, a trend that has been amplified by the migration of high-tech companies to Texas (Kerr et al., 2024). The migration of people and businesses to the state has been aiding Texas's GSP growth (Kaplan, 2019).

Between 2017 and 2022, Texas's real GSP increased by 15.4%, exceeding the nation's real GDP growth rate of 11.3% (Texas Comptroller, 2024). In 2023, the Texan economy grew by 5.7%, the second highest growth rate in the nation (after North Dakota). The mining sector was the leading contributor to Texas's GSP growth in 2023 (Bureau of Economic Analysis, 2024). In recent years, the mining industry has been responsible for about 9% of the state's nominal GSP (Thompson and Hines, 2023).

In 2022, Texas accounted for more than half of the U.S. production in the mining, quarrying, and oil and gas extraction industries (Texas Comptroller, 2024). Texas is the No. 1 producer of crude oil and natural gas in the U.S. (Federal Reserve Bank of Dallas, 2024b), responsible for nearly 43% of U.S. crude oil and 28% of U.S. natural gas production in 2023 (Foreman, 2024). If Texas were a separate country, it would rank as the world's third largest natural gas producer, after Russia and the U.S. (Busby et al., 2021). Texas would also rank as the world's fourth largest crude oil producer, after the U.S., Russia and Saudi Arabia (Texas Economic Development Corporation, 2022).

Texas represents more than one-third of U.S. crude oil refining capacity. The state's 32 oil refineries can process more than 5.9 million barrels of crude oil per day (United States Energy Information Administration, 2023b). A top crude oil producing and refining state, Texas is responsible for about 30% of U.S. gasoline supply (Thompson and Hines, 2023). While a top producer, Texas is also the leading petroleum consumer, accounting for 21.6% of U.S. petroleum use. Texas's industrial sector is responsible for almost two-thirds of the state's petroleum consumption. Texas is the nation's largest natural gas consumer as well, accounting for about 15% of the nation's natural gas use. Texas's industrial sector is typically responsible for up to 23% of the nation's industrial sector's natural gas consumption (United States Energy Information Administration, 2023b).

Given its vast oil-refining capacity, Texas produces about 75% of U.S. petrochemicals supply (Thompson and Hines, 2023) and houses the world's largest petrochemicals cluster (Texas Economic Development and Tourism, 2023). In 2021, Texas's petrochemicals industry employed 67,100 Texans, with more than 300,000 Texans indirectly supported downstream in supply chains, construction and maintenance, logistics, engineering and other sectors (Thompson, 2021). Texas's petrochemicals and other chemicals manufacturing accounted for nearly 13% of the nation's chemicals supply in 2023 (Federal Reserve Bank of Dallas, 2024a).

With more than 313,200 miles of public roads, 10,400 miles of railways, 380 airports, and 16 seaports, Texas boasts the largest transportation network in the U.S. (Kaplan, 2019; Texas Economic Development Corporation, 2023). In 2022, Texas's freight systems moved 3.4 billion tons of freight valued at more than $3.1 trillion—the largest freight volume and monetary value compared to any other U.S. state (TRIP, 2023).

A natural resource-rich state, sharing a border with Mexico and housing 16 seaports, Texas is the nation's largest exporter (Thompson and Hines, 2023). Texas generates 22% of U.S. exports (Federal Reserve Bank of Dallas, 2024b), with three-fifths of the state's exports being energy-related (Thompson and Hines, 2023). With more than a trillion-dollar international trade volume[5] in 2022 (Texas Comptroller, 2024), Texas's main international trading partners included Mexico, Canada, the Netherlands, China, and South Korea. In 2023, exported goods accounted for 17.3% of

Texas's GSP. In that year, manufactured products represented 65.5% of the state's goods exports. In addition to petroleum and coal products, the state's manufactured goods exports included chemicals, computer and electronic products, transportation equipment, and machinery (Office of the United States Trade Representative, 2023).

Texas is a key manufacturing state, with the manufacturing sector contributing 11% to the state's nominal GSP (Federal Reserve Bank of Dallas, 2024a) and employing 975,000 Texans in 2023 (Texas Workforce Commission, 2024). In 2023, Texas's contribution to the nation's manufacturing GDP was nearly 11%, as the state ranked second behind California in factory production (Federal Reserve Bank of Dallas, 2024a). Many firms in Texas's services sector are at least partially dependent on spending by the manufacturing (and energy) sector companies and their employees (Thompson and Hines, 2023).

With more than 125 million acres in farmland—14% of the nation's total—Texas is home to 12.3% of the nation's farms and ranches, boasting the greatest number of farms and ranches in the U.S. (Minton and Alvarado, 2022). Texas also ranks first nationwide in forage land used for hay production (Kinniburgh et al., 2015). The state's top agricultural commodities by market value include cattle, poultry and eggs, dairy, corn, cotton, greenhouse, fruits, vegetables and tree nuts, wheat,[6] sorghum, and rice (Texas Department of Agriculture, 2022). Based on cash receipts, Texas ranks as the fourth largest agricultural-producing state in the U.S. Cattle are the Beef State's top agricultural commodity, accounting for 40% of the state's total agricultural cash receipts in 2021, more than three times that of milk (Minton and Alvarado, 2022).

While commonly known for its oil and natural gas, and the agricultural industry, Texas is also a rising leader in the U.S. technology sector (Bridge for Innovation, 2021). After California, Texas boasts the second largest semiconductor workforce in the nation. Texas is home to some of the nation's most promising artificial intelligence (AI) companies (Kerr et al., 2024). The state ranks first and second nationwide for data center growth and data infrastructure, respectively (Bridge for Innovation, 2021). The high-tech industry[7] represents nearly 5% of Texas's GSP, accounting for more than 9% of the state's non-farm employment. Job growth in Texas's high-tech sector has significantly outpaced the overall job growth in the state and in the nation (Kerr et al., 2024). Texas's high-tech sector is especially prominent in the Austin area, where high-tech jobs account for 16.3% of Austin's metro area's total employment, compared to 9% nationwide (as of 2022). In 2022, Austin's high-tech sector employment grew by 9.8%, compared to 5.9% nationally (Kerr, 2023).

Despite its large energy and manufacturing sectors, Texas's employment profile resembles that of the rest of the U.S. (Thompson and Hines, 2023). In 2023, the state's largest industries by employment were trade, transportation, and utilities; professional and business services; and education and health services, accounting for 19.9%, 15.2% and 13.7% of the state's total non-farm employment, respectively. Employment in these industries was followed by employment in the government sector, leisure and hospitality, and manufacturing, which represented 14.7%, 10.5%, and 6.9% of the state's total non-farm employment, respectively. The financial industry was responsible for 6.6% of the state's total non-farm employment, followed by the construction sector, information services, and mining and logging. The latter three sectors accounted for 5.8%, 1.7%, and 1.6% of the state's total non-farm employment, respectively. The remaining 3.4% of the state's non-farm employment took place in other services (Texas Workforce Commission, 2024).

Texas is a high-emitting and a high energy-consuming state, given its reliance on energy- and GHG emissions-intensive industries such oil and natural gas production and mining, manufacturing, and agriculture (Kinniburgh et al., 2015). Nationally, Texas ranks No. 1 in terms of pollution and GHG emissions, producing twice as many emissions as California. The industrial sector accounts for 42.4% of Texas's total emissions, followed by the transportation[8] and the electric power generation industries,[9] which produce 29.8% and 24.4% of the state's total emissions, respectively. The state's industrial sector emissions are primarily driven by the oil and natural gas industries (Skinner et al., 2021).

Consuming about one-seventh of the nation's energy supply, Texas is the largest energy-user in the U.S., leading the nation in energy consumption across all sectors. Texas's industrial sector accounts for more than half of the state's total energy use. Nationally, Texas's industrial sector is responsible for almost a quarter of industrial sector's energy consumption. The transportation sector is the second largest energy consumer in Texas, accounting for one-fifth of the state's total energy use. Texas also ranks sixth nationally in per capita energy consumption (United States Energy Information Administration, 2023b).

The state's rising GSP, growing population, expanding manufacturing facilities and data centers, and the more frequent episodes of extreme weather have contributed to Texas's electricity consumption rising at an annual rate exceeding the nation's average (Golding et al., 2024). The residential sector is the state's leading electricity consumer, responsible for almost two-fifths of the state's electricity sales in 2022. Texas's commercial and industrial sectors consumed one-third and three-tenths of the state's electricity in 2022 (United States Energy Information Administration, 2023b).

While the Texan economy has been growing, the state's economic performance remains increasingly vulnerable to climate change. The state's oil and gas, energy generation, manufacturing, agriculture, transportation, and other vital infrastructure is at risk from growing temperature extremes, rising sea levels, heavier flooding, increased severity of storms and hurricanes, worsening droughts, and other weather hazards. Climate change has been threatening to undermine Texas's economic base, putting the state's productivity and employment at risk. While Texas's population has been growing, the state's changing climate will make it increasingly more difficult to live in the state. The state's long-term economic prosperity will depend on how well Texans prepare for increasingly more damaging weather hazards (Texas 2036, 2021).

Climate change has cost Texans hundreds of billions of dollars in economic losses and damages. As of June 10, 2024, the cumulative cost of post-1980 billion-dollar weather and climate disasters in Texas has been estimated at up to $420 billion (NOAA NCEI, 2024). As the weather hazards affecting Texas are becoming increasingly more damaging and disruptive, recovering from them will prove more and more costly (Texas 2036, 2021). Weather disaster response and recovery will put strains on the state and local government budgets, as extreme weather-related spending will compete with the government's other fiscal priorities.[10]

In what follows, I will provide a snapshot of the economic costs and losses from extreme weather in Texas. As the ensuing discussion will demonstrate, Texas is the natural disaster capital of the U.S., leading the nation in cumulative costs and losses from weather extremes.

III.5.3 THE ECONOMIC COSTS OF WEATHER AND CLIMATE DISASTERS IN TEXAS

Since 1980, Texas has endured the greatest number of billion-dollar weather and climate disasters compared to any other U.S. state (Smith, 2024). As the nation's most weather-vulnerable state (Kinniburgh et al., 2015), Texas accounts for more than 15% of all billion-dollar weather and climate disasters that occurred in the U.S. since 1980 (Leining, 2024). Between 1980 and 2023, Texas incurred $402 billion in direct cumulative costs from billion-dollar weather and climate disasters, the greatest cost in the nation. For the Unites States as a whole, the direct cumulative costs of billion-dollar weather and climate disasters for the 1980–2023 period were estimated at $2.660 trillion (1980–2023) (Smith, 2024).[11] As of June 10, 2024, the nation's cumulative costs of billion-dollar weather and climate disasters climbed to $2.743 trillion (NOAA NCEI, 2024).

The number of and the costs of billion-dollar weather and climate disasters in the U.S. are estimated by the U.S. National Oceanic and Atmospheric Administration's (NOAA) National Centers for Environmental Information (NCEI). NOAA NCEI cost estimates are adjusted for inflation, and account for direct damages and losses from weather and climate disasters, both insured and uninsured. According to the methodology adopted by NOAA NCEI, direct cost estimates of weather and climate disasters include:

- Physical damage to residential, commercial and government buildings;
- Material assets (content) within buildings;
- Time element losses such as business interruption or loss of living quarters;
- Damage to vehicles and boats;
- Public assets including roads, bridges, and levees;
- Electrical infrastructure and offshore energy platforms;
- Agricultural assets including crops, livestock, and commercial timber;
- Wildfire suppression costs, among others (NOAA NCEI, 2024).

The cost estimates provided by NOAA NCEI *do not* account for the following:

- Losses due to natural capital or environmental degradation;
- Mental or physical healthcare-related costs;
- The value of a statistical life;
- Supply-chain, contingent business interruption costs (NOAA NCEI, 2024).

As such, NOAA NCEI cost estimates "should be considered conservative with respect to what is truly lost" as a result of weather and climate disasters (NOAA NCEI, 2024). For example, while the NOAA NCEI cost estimates do not quantify ecosystems damage from weather and climate disasters, ecosystems changes have significant economic consequences, such as reduced food production, diminished water and forestry resources, declining real estate values, and reduced opportunities for recreation and tourism (Hsiang et al., 2023). Further, if we account for additional factors, such as lost productivity, foregone wages, human displacement, and migration, as well as the long-term impacts of weather and climate disasters, then "the assessed cost of direct damages tends to be on the order of about 10 percent of the actual cost" (Hayhoe, 2019, p. 16). Lastly, the direct costs of weather and climate disasters "cannot be completely measured due to a lack of consistently available data" (NOAA NCEI, 2024). The NOAA NCEI cost estimates, therefore, significantly understate the extent of losses from weather and climate disasters. NOAA NCEI cost estimates are, nevertheless, useful, as they account for many important impacts of weather and climate extremes, and allow for cross-state comparisons of cost impacts.

While all U.S. states have witnessed at least one type of a billion-dollar weather or climate event, Texas has been one of the five states to endure seven different types of billion-dollar weather and climate disasters, such as hurricanes, storms, floods, winter storms, freezes, droughts, and wildfires (Lowe and Marx, 2020). Between 1980 and 2024,[12] Texas witnessed 179 billion-dollar weather and climate disaster events, including 19 drought events (10.61%), 9 floods (5.02%), 1 freeze event (0.56%), 118 severe storms (65.93%), 14 tropical cyclones (7.82%), 7 wildfires (3.91%), and 11 winter storms (6.15%) (NOAA NCEI, 2024).

Compared to Louisiana, which is the second most weather-vulnerable state in the U.S., Texas endured 77% more billion-dollar weather and climate disasters during the 1980–2023 period. During that timeframe, Louisiana witnessed 97 billion-dollar weather and climate events, while Florida—the nation's third most weather-vulnerable state—endured 84 billion-dollar weather and climate disasters (Leining, 2024).

The frequency of extreme weather and climate disasters to hit Texas has been accelerating with each decade, with a rapid increase over the past 14 years.[13] The number of billion-dollar weather and climate events endured by Texas during the 2020–2023 period (48 events) was more than three times as great as the number of billion-dollar weather and climate events to hit Texas in 1980–1989 (14 events). The three years with the greatest number of billion-dollar weather and climate disasters in Texas were 2023, 2021, and 2020. From 2010 to 2019, Texas sustained an estimated $200–300 billion in direct costs from billion-dollar weather and climate disasters. For the 2019–2023 period, those costs were assessed at $50–100 billion. In 2023, billion-dollar weather and climate disasters cost Texas an estimated $10–20 billion in direct losses. The three costliest years for Texas were 2008, 2017, and 2021 (NOAA NCEI, 2024).

Most of the state's losses from billion-dollar weather and climate disasters come from tropical cyclones (hurricanes), which account for 57.4% of all losses since 1980 (NOAA NCEI, 2024). By 2030 and 2050, Texas's average annual losses from hurricanes and other coastal storms are likely to increase by up to $222 million and $650 million, respectively. By 2050, the state's annual hurricane damage could amount to $3.9 billion (Kinniburgh et al., 2015). These projections should be viewed as conservative as they do not account for inflation, and assume historical levels of hurricane activity.

The state's losses from tropical cyclones (hurricanes) are followed by losses from severe storms and droughts. Since 1980, severe storms and droughts have been responsible for 21.6% and 10.2% of the state's costs from billion-dollar weather and climate disasters, respectively (NOAA NCEI, 2024).

In what follows, I will discuss the economic costs and impacts of hurricane Harvey (2017)—one of the costliest severe weather events ever to affect Texas. I will then analyze the economic impacts of the winter storm Uri (2021)—the costliest superstorm ever witnessed by the state. These two weather disasters will serve as case studies, providing examples of significant, widespread, and lasting economic damage that climate change can bring to the state.[14]

III.5.4 THE ECONOMIC IMPACTS OF HURRICANE HARVEY

Global warming has been increasing the severity of tropical cyclones affecting the Gulf Coast states (Lowe and Marx, 2020; USGCRP, 2018). When hurricane Ike (category 2) struck Texas in September 2008, it killed 74 people and caused about $30 billion in direct damages to the state (Texas General Land Office, 2009). Ike damaged the state's critical infrastructure such as oil platforms, storage tanks, pipelines, and refineries, causing gasoline and other product shortages in southwestern U.S. (Texas Comptroller, 2018).

Nine years later, a catastrophic 60 inches of rain—Hurricane Harvey (category 4)—caused Texas at least $100 billion in direct damages (Lowe and Marx, 2020). Harvey produced the largest historical rainfall from a single weather event in the state's history (Dix et al., 2023). Attribution studies suggest that global warming made Hurricane Harvey at least three times more likely to occur. The record rainfall that Harvey brought to Houston was made 15 to 38 times more intense by global warming (Lowe and Marx, 2020).

After Hurricane Katrina (2005), Harvey was the second costliest hurricane to hit the U.S. (Texas Division of Emergency Management, 2017).[15] Harvey produced widespread and catastrophic flooding, devastating the Corpus Christi–Houston–Galveston area, with thousands of roads, homes, and businesses affected.[16] Harvey's other immediate impacts included deaths from drowning and injury, claiming the lives of at least 68 Texans (Blake and Zelinsky, 2018). In addition to causing direct death and injury, Harvey affected public health by disrupting the state's vital support systems such as drinking water and wastewater treatment. Surface water contamination caused by Harvey led to a spike in skin and gastrointestinal infections (USGCRP, 2018).

Leaving 300,000 customers in Texas without electricity (USGCRP, 2018) and damaging or destroying more than 178,400 Texas homes (Texas Comptroller, 2018), Harvey forced mass evacuations, with the number of evacuees exceeding the capacity of the state's emergency shelters. As a result, otherwise healthy people or those who needed electricity to power their medical devices had to go to hospitals.[17] The local hospitals' ability to serve their patients was strained, as the hospitals struggled with power shortages and loss of indoor air and temperature controls. Staff shortages posed a significant challenge as well. As a result, at least 15 hospitals evacuated their patients in the aftermath of Harvey. The health care system's ability to serve Texans was also limited as hospitals and pharmacies faced shortages of essential medications (including insulin and respiratory inhalers). Medical suppliers were unable to make deliveries due to the massive flooding caused by Harvey (USGCRP, 2018).

With more than 300,000 homes and businesses flooded (Kaplan, 2019), Harvey caused severe damage to the state's property and infrastructure. NOAA reported Texas's total property damage was valued at more than $51 billion. In the five Texas counties[18] that were most severely affected by the storm, small business real estate and content (inventory, machines, equipment, and leasehold improvements) losses were estimated at more than $3 billion and $1.4 billion, respectively (Lowe and Marx, 2020). Harvey's damage to Texas public property was assessed at $669 million. This included damage to government buildings, roads, bridges, water facilities, and electric utilities, among other public assets. The estimates for car and truck damage in the state varied from a quarter-million to a million vehicles. Texas's crop and livestock losses caused by Harvey were estimated at more than $200 million (Texas Comptroller, 2018).

It must be noted that insured losses represented a relatively small fraction of residential property damage caused by Harvey. During Harvey, a large share of property damage was caused by flooding rather than wind, and flooding is generally not covered under homeowner insurance policies (Texas Comptroller, 2018). Further, as explained by Hayhoe (2019), the majority of flooding caused by Harvey occurred outside the designated flood zone. As a result, many affected homes and buildings were not covered under flood insurance. The Federal Reserve Bank of Dallas, likewise, notes that many homeowners were uninsured, and even those who had flood insurance had to wait a long time to receive compensation from the Federal Emergency Management Agency (FEMA). The amount received from FEMA often did not cover the full cost of rebuilding (Torres, 2022).[19]

Most businesses in the Harvey-affected areas had to shut down in the immediate aftermath of the storm due to flooding, loss of electrical power, employees' inability to get to work, and other factors. As a result, manufacturing, energy, chemicals, retail, and other Texas industries "experienced significant and expensive downtime" (Texas Comptroller, 2018, p. 4). When 23 ports and 700 roads in Houston were closed, regional freight transportation and economic activity was significantly curtailed as well (USGCRP, 2018). The closures of oil refineries in Houston and the upper Texas Gulf Coast had ripple effects, impacting jobs and the flow of economic activity throughout the region (Lowe and Marx, 2020).[20]

In the aftermath of Harvey, 11% of the Gulf of Mexico's oil refining and 25% of the Gulf's oil production capacity went offline. This resulted in actual and anticipated gasoline shortages, causing regional and national gasoline price increases (Kaplan, 2019; USGCRP, 2018). With oil refinery utilization rates falling from 90% to 60%, more than half of U.S. basic petrochemicals production was disrupted by Harvey (USGCRP 2018). Petrochemicals shortages elevated the prices of basic petrochemical products that are used to manufacture a range of final and intermediate goods (Kaskey, 2017).

With winter tourism down by 50%, Texas's coastal tourism industry suffered crippling losses due to Harvey (Texas Comptroller, 2018). Across the Texas Coastal Bend, ecotourism supports 8% of the local workforce, generating over a billion dollars in annual economic value (McPherson et al., 2023). Consequently, the disruptions to the state's coastal tourism industry had significant impacts on the regional economy.

The economic disruptions from Harvey cost Texas an estimated $16.8 billion GSP loss[21] in the first year after the storm. The cumulative loss to Texas GSP in the first three years post-Harvey was estimated at $19.8 billion (Texas Comptroller, 2018). It must be noted that the extent of GSP losses from Harvey was somewhat downplayed by the Texas Comptroller, who emphasized the positive impacts of government spending post-Harvey on the state's economy instead. According to the Texas Comptroller, government spending on post-Harvey relief, reconstruction, and restoration would serve as a major stimulus for Texas's economy, which would ultimately offset the GSP losses caused by Harvey. In the words of the Texas Comptroller:

> The extent and cost of Harvey's destruction were significant. But when balanced against the anticipated increase in business activity due to reconstruction and restoration efforts, combined with an influx of funding from federal aid and insurance payments, the effect of the state's economy may be much less severe than expected. Gains from increased construction activity and associated spending undertaken to repair and rebuild should help offset losses in the months and years to come.
>
> **(Texas Comptroller, 2018, pp. 4–5)**

According to the Texas Comptroller, one year after the storm, the *net* loss to Texas GSP was only about $3.8 billion. This was due to the positive impacts of federal, state, and local government spending,[22] which provided a significant boost to Texas's GSP. The Texas Comptroller (2018) further stressed Harvey's *positive net impact* on Texas GSP in the three years after the storm. Over that timeframe, the gains to Texas GSP resulting from relief, reconstruction, and restoration spending would more than offset the GSP losses caused by the storm (Texas Comptroller, 2018).

While the Texas Comptroller (2018) stressed Harvey's positive *net* impacts on Texas GSP, the Dallas Fed adopted a more cautionary stance, underscoring the structural and long-term impacts of Harvey, such as lost homes and businesses, displaced communities, rising economic inequality, higher insurance costs, and cost impacts on government budgets (Kaplan, 2019).[23]

The public health consequences of Harvey have been significant and long-term. Ten months after Harvey, 16% of Texas Gulf Coast residents affected by the storm reported a new or a worsened health condition. Respiratory ailments such as asthma represented nearly a third of the new health conditions reported. Bacteria and mold growth in flood-damaged homes, along with dust from repair activities, contributed to the rise in respiratory health problems post-Harvey. Chemicals exposure due to chemicals release from industrial accidents could explain the spike in respiratory ailments as well (Constible, 2018). The long-term public health consequences of Harvey will put significant strains on public and private budgets in the state.

Over the long-run, Harvey-affected areas could face the challenge of maintaining their economic attractiveness and competitiveness (Karp et al., 2017). Harvey could negatively affect the Corpus Christi–Houston–Galveston area's perceived economic attractiveness for business relocation (Institute for Regional Forecasting, 2018). This will deter economic investment and population migration into the area, reducing the region's economic potential and undermining the region's and the state's economic development (Karp et al., 2017).

III.5.5 THE ECONOMIC IMPACTS OF WINTER STORM URI

The historic winter storm Uri, which hit Texas on February 12, 2021, killed hundreds of people[24] and caused $20.4 billion in direct damages, rendering Uri "the costliest superstorm ever recorded" (So and Hardin, 2021, p. 1). According to NOAA NCEI (2024), the direct losses from Uri in Texas range from $20 billion to $50 billion. Some estimates suggest that the true costs of Uri range from $80 billion to $130 billion in direct and indirect losses in the state (So and Hardin, 2021).[25]

Driven by global warming,[26] Uri "was a relatively brief 10–day shock" of extremely cold temperatures, which "caused cascading failures across many sectors" and produced long-lasting economic impacts for Texas and its residents (Zamuda et al., 2023). Uri disrupted the state's vital support services such as natural gas production, electricity generation, drinking water and wastewater treatment, health care, and transportation, among others (Castellanos et al., 2023). The storm left 13 million Texans without electricity and/or water, and caused "billions of dollars in abnormally high energy bills" (Castellanos et al., 2023, p. 1) and other damages. The financial repercussions from Uri could last for decades given the exorbitant energy bills that some ratepayers were faced with (Busby et al., 2021).

Among other critical issues, Uri "exposed the inability of the state's energy supply chain to withstand extremely cold temperatures" (Golding, 2023). Texas relies on natural gas for almost half of its electricity generation. In 2021, natural gas-powered plants generated 48.6% of the state's total electricity supply (Nuclear Energy Institute, 2022).[27] When winter storm Uri brought ice and freezing temperatures to Texas, more than one-fifth of the state's natural gas processing plants had to shut down due to freeze-offs,[28] cold weather-related equipment failures, and/or electricity service disruptions. As a result, during February 13–17, 2021 the state's natural gas production plummeted by 45%. Natural gas transportation infrastructure was significantly impacted by the freezing temperatures as well (Golding et al., 2021).

Natural gas shortages and natural gas equipment failures lowered Texas's electric grid's capacity to produce electricity. This happened in a state where 55% of households rely on electricity as a heat source during winter. While the grid's generation capacity shrunk, rigid temperatures across the state pushed electricity demand to record highs (Golding, 2023). To prevent the collapse of the grid amid escalating electricity demand and falling generation capacity, the Electricity Reliability Council of Texas (ERCOT) implemented rolling blackouts across Texas (Castellanos et al., 2023). As a result, nearly 4 million Texasscustomers—representing more than 11 million people—lost electricity (Golding et al., 2021). What was initially announced as rolling blackouts turned into multi-day power outages for many customers (Busby et al., 2021), with disproportionate impacts in low-income and minority-dominated areas (Castellanos et al., 2023). Electricity supply was also inadvertently disrupted to many natural gas production and processing facilities (Golding, 2023), which further amplified natural gas production disruptions and electricity shortages.

Many water treatment facilities across Texas lost electric power as well, resulting in their inability to treat drinking water adequately. As a consequence, 43% of the state's community public water systems issued boil water notices. Power outages at the state's wastewater treatment facilities posed additional risks to public health. With waterpipes bursting in freezing temperatures and water system pressure dropping due to millions of dripping faucets across the state, approximately half of Texans lost water access at some point (Castellanos et al., 2023).

Power outages across the state led to widespread and significant residential and commercial property damage (Golding et al., 2021), such as burst waterpipes, structural water damage to buildings, and water damage to appliances (Castellanos et al., 2023). Residential and commercial insured property losses from Uri were estimated at $10–20 billion (Golding et al., 2021). Because not all insurance claims were approved, these losses underestimate the true extent of property damage from Uri. In addition to interior property damage, Uri produced exterior damage to property due to snow, ice, and fallen trees.

Power outages and hazardous road conditions[29] during Uri shut down economic activity across much of Texas (Castellanos et al., 2023), causing significant losses in foregone business income (Golding et al., 2021). Essential businesses such as food processing plants and grocery stores had to close and face significant losses due to food spoilage (Castellanos et al., 2023; see also Texas Comptroller, 2021). For example, milk processing plants across the state had to shut down due to lack of electricity, natural gas, frozen pipes, or icy road conditions, which prevented milk deliveries. As a result, millions of gallons of milk had to be disposed of (Texas Comptroller, 2021), with some of the state's dairy operations losing as much as $8 million a day (Schattenberg, 2021). Interruptions in food processing, production, transportation, and retail during Uri illustrated the high vulnerability of the state's food supply chains to extreme weather events.

Among other key sectors, Uri disrupted the state's oil refining and petrochemicals manufacturing, producing not just short-term interruptions, but lasting operational damage. Oil refining and petrochemicals plants rely on natural gas as both a heat source and a source of raw materials for their products and processes. Oil refining and petrochemicals production infrastructure is vulnerable to freezing temperatures as well. The combined effect of freezing temperatures, natural gas shortages, and electric power outages led to a rapid shutdown of the state's oil refining and petrochemicals facilities. Decreased oil refining capacity[30] further resulted in lower gasoline and diesel production (Thompson, 2021).

Petrochemical facilities closures interrupted the production of basic goods such as chlorine-based disinfectants, plastic bottles, packaging, fertilizers, and pesticides. The production of many petrochemicals-based intermediate goods was interrupted as well. Production facilities outside of Texas, which were not directly affected by Uri, had to curtail operations due to shortages of intermediate petrochemicals inputs. Because petrochemicals-based intermediate goods produced in Texas enter the supply chains around the world, winter storm Uri further exacerbated supply-chain disruptions caused by the Covid-19 pandemic "leading to price pressures and scores of product shortages" (Thompson, 2021, p. 1). For example, the producer price index (PPI) for plastics and resins increased by 9.1% in March 2021 (Thompson, 2021)

One important shortage was that of polycarbonate resin, a material used to manufacture intermediate products such as car bumpers and headlight lenses. Another key shortage was that of polyvinyl chloride, a petrochemical used to make dashboards and other vehicle parts. These and other petrochemicals-based product shortages adversely affected the global auto industry. Toyota and Honda had either to suspend[31] or slow down production at their facilities in Mexico, the U.S., and Canada (Thompson, 2021).

By disrupting Texas's oil refining and petrochemicals production, winter storm Uri slashed the real value of Texas's petrochemicals, plastics, and rubber products exports[32] by more than 20% in February 2021. In that month, Texas sustained the largest one-month drop in exports since the global financial crisis of 2008 (Thompson, 2021).

Alongside petrochemicals, winter storm Uri dealt a massive blow to the state's agricultural industry. The prolonged periods of freezing temperatures brought by Uri destroyed many fruit and vegetable crops in the state's Rio Grande Valley (Texas Comptroller, 2021). Animal agriculture was affected as well, as the Storm caused statewide losses of livestock grazing supplies such as oats, rye grass, and triticale. As a result, Texas ranchers struggled to procure feed for their livestock. Additionally, many farm animals died during the freezing temperatures (Schattenberg, 2021; see also Texas Comptroller, 2021). Because the storm hit during the calving season, livestock losses included newborn cattle. Consequently, Texas ranchers lost more than one generation of livestock, with longer-term effects on the state's herd size. Agricultural property damage caused by Uri was also recorded (Schattenberg, 2021).

The state's agricultural losses from Uri were estimated at over $608 million (Texas Department of Agriculture, 2024), led by the losses to the state's citrus crop. The freezing temperatures brought by the storm destroyed 200 acres of lemon and lime crops in South Texas. Rio Grande Valley citrus producers lost virtually all of their Valencia orange harvest, and more than 60% of their grapefruit crop.[33] In addition to ravaging the state's citrus harvest, Uri damaged many citrus trees in South Texas. Given that many citrus trees had to be replanted, it could take years for the state's citrus industry to fully recover from Uri (Schattenberg, 2021).

The state's citrus production losses of $230 million[34] were followed by $228 million in livestock losses,[35] and at least $150 million in lost vegetable crops. The vegetable crop losses included over $42 million in lost onion sales, over $27 million in lost leafy greens sales, over a $20 million loss in the sale of watermelons, and more than a $15 million loss in the sale of cabbage. Other vegetable crops adversely affected by Uri included beets, celery, and potatoes, among others. According to Texas A&M AgriLife Extension Service, these agricultural loss estimates should be viewed as conservative (Schattenberg, 2021).

Winter storm Uri had negative impacts on the state's sugarcane industry as well. While the 2021–2022 sugarcane crop was not affected as most of it was harvested before the storm, the freezing temperatures did destroy most of the sugarcane plants in South Texas. As a result, sugarcane producers were forced to start over, which reduced the 2022–2023 sugarcane crop. Corn and grain sorghum planted in South Texas and the state's Coastal Bend prior to Uri had to be replanted as well (Schattenberg, 2021).

With climate change, crop losses and agricultural production disruptions have been accelerating across Texas. While the state's agriculture has always been exposed to weather risks, those risks have been exacerbated by climate change, costing the state billions of dollars in lost agricultural output and productivity. In what follows, I will analyze the impacts of climate change on Texas's agriculture, discussing both direct and indirect impacts, such as supply-chain disruptions and rising food prices.

III.5.6 AGRICULTURAL IMPACTS

Texas's agriculture has been increasingly affected by rising temperatures, changing precipitation patterns, increased severity and frequency of droughts, wildfires, storms, and hurricanes, growing water shortages, and other climate change-driven factors. The 19 severe droughts endured by Texas since 1980[36] have cost the state an estimated $20–50 billion in direct losses (NOAA NCEI, 2024). Texas's agricultural losses from the historic 2011 drought were estimated at more than $7 billion, amounting to about 40% of the state's average annual agricultural production[37] (Kaplan, 2019). The 2010–2014 drought cost Texas an estimated $11.1–15.5 billion (in 2021 dollars) in agricultural industry losses. The state's overall economic losses from the 2010–2014 drought were assessed at $73 billion (in 2021 dollars) (Donald and Grubbs, 2022).

In addition to direct effects, such as agricultural production disruptions and crop losses, extreme weather can lead to *indirect*, secondary impacts by interrupting the flow of economic activity throughout the supply chain between the farmers and the consumers. This supply chain includes various business activities such as farm machinery repair, fertilizer production and sales, food processing and production, transportation, warehousing, wholesale distribution, retail sales and eating establishments. Certain financial and real estate services are linked to agricultural production as well. Agriculture, therefore, has wide-ranging, indirect contributions to the economy (Minton and Alvarado, 2022). When severe weather curtails agricultural production, this results in income losses not just for farmers and harvest laborers, but for farm equipment repair mechanics, fertilizer producers, truck drivers, food processors, and other players in the agriculture-related businesses. Lost incomes throughout the supply chain further lead to reduced consumer spending in the local economy (Kaplan, 2019). For example, when planted acres of rice in Texas's drought-stricken Matagorda county decreased from 22,000 acres to 2,100 acres in 2011, the ripple effects of declining rice agriculture were felt throughout the local economy. Local farm equipment repair shops lost about 70% of their business. Some long-established, family-owned agricultural services businesses had to shut down (Baddour, 2014).

Rising food prices represent additional consequences of severe weather-induced agricultural production disruptions and losses. As an example, consider the price effects of the 2010–2014 drought in Texas. The drought reduced the availability of grazing land and increased livestock feed costs across the state. Faced with higher feed costs, many Texas ranchers were forced to reduce their herd size. This resulted in $3.23 billion in livestock losses, and contributed to "historic" calf and beef prices in Texas. Calf and beef prices remained elevated for years, peaking in late October 2014, despite beef cattle herd recovery by then (Russell, 2022).

The 2022 drought endured by Texas led to similar price effects. By reducing the availability of grazing land, and lowering hay and forage stocks, the drought resulted in higher hay and feed prices (Donald and Grubbs, 2022; Russell, 2022). At the same time, the drought reduced the state's 2022 corn crop to a below-average level, which further contributed to elevated feed prices. Faced with rising feed costs and drought-exacerbated water shortages, many Texas ranchers had to reduce their livestock numbers. Shrinking cattle herds further contributed to higher calf and beef prices across the state. As post-drought herd recovery could take several years, beef prices could remain elevated for years to come (Russell, 2022).

The long-lasting cost effects of droughts have been emphasized by the Dallas Fed economists. Robert Kaplan, the former president of the Federal Reserve Bank of Dallas, has cautioned that the "drought effects are often felt over multiple years due to the resulting pasture damage and herd reduction that take years to recover from" (Kaplan, 2019, p. 4). As a result, the costs of animal feed and other agricultural inputs can remain elevated for several years following a severe drought event (Kaplan, 2019, p. 4).

Alongside higher feed costs, rising pesticide expenditures and declining livestock productivity pose significant challenges to Texas's agriculture. These climate change-related impacts have been discussed by the Dallas Fed:

> Hotter temperatures prompted by climate change also pose additional pest problems for farmers, driving up pesticide use and cost, and hamper livestock productivity – increasing time to grow animals to a given weight and decreasing calving rates and milk production.
>
> **(Kaplan, 2019, p. 4)**

The growing costs of irrigation water pose important challenges to Texas's agriculture as well (Kaplan, 2019; see also Zamuda et al., 2023). Irrigation water use is the largest water use category in Texas, responsible for 55% of the state's total water consumption in 2020 (Texas Water Development Board, 2022a). While rising temperatures and the growing severity of droughts will increase the state's demand for irrigation water (Zamuda et al., 2023), the state's available water supply is projected to dwindle (Texas Water Development Board, 2022b). As a result, irrigation costs across Texas will likely go up (Zamuda et al., 2023).

The rising costs of agricultural production inputs such as irrigation water, animal feed, and pesticides, along with reduced herd size and declining livestock productivity, will keep contributing to rising food prices across the state. These price pressures will be exacerbated by agricultural production disruptions and crop losses due to growing weather hazards.[38]

Extreme weather-driven agricultural crop losses have been mounting across Texas. In 2022, extreme weather cost Texas $6.457 billion worth of agricultural crops, the highest crop loss in the nation, followed by $3.365 billion agricultural crop losses in neighboring Kansas. Texas's 2022 agricultural crop losses were mainly driven by widespread and exceptional drought conditions, May hailstorms, and severe weather in April. The agricultural crop loss sustained by Texas in 2022 included nearly $1 billion in wheat crop damage and $1.7 billion in damages to forage and rangeland (Munch, 2023). In 2022, Texas's peanut production decreased by 12%, falling to its lowest level since 2011 (Sawadgo, 2023). In the same year, the state's cotton production decreased by 74% (Reinsch et al., 2023), a record-setting crop loss of $2.9 billion (Munch, 2023), a consequence of drought and water shortages (Reinsch et al., 2023).

In 2023, natural disasters and severe weather cost Texas more than $4.862 billion in agricultural crop losses (Munch, 2024). 2023 was the warmest year in Texas, with record-setting temperatures of 113–114 degrees observed in Del Rio, Rio Grande, and San Angelo on June 20, 2023 (Powell, 2023). The widespread and exceptional drought conditions endured by Texas in 2023 accounted for about 80% of the state's agricultural crop loss, with the remaining crop damage attributed to the tropical storm Harold, flooding, and hailstorms (Munch, 2024).

In 2023, Texas's agricultural crop loss from drought, wildfire, and heat was estimated at $3.996 billion, the greatest crop loss nationwide. In the same year, the state's agricultural crop loss from hurricanes, flooding, and severe precipitation was assessed at almost $569 million, the second largest crop loss in the nation, following Florida. Texas's 2023 agricultural crop loss from major hailstorms[39] was estimated at almost $300 million, the second largest crop loss in the nation, following Kansas. The state's 2023 agricultural crop loss included $408 million in wheat crop damage, $1.5 billion in damages to forage and rangeland, and $2.3 billion in damages to cotton (Munch, 2024).

Recent reductions in Texas's cotton crop have had global production and price repercussions. The U.S. is the world's leading cotton exporter, and the third largest cotton producer (Reinsch et al., 2023). The state of Texas is the largest cotton producer in the U.S. (Kinniburgh et al., 2015), responsible for 40% of the nation's cotton output[40] (United States Department of Agriculture, 2022). With drought conditions in Texas and other cotton-producing states, the U.S. has been losing billions of dollars in annual cotton production. Cotton production losses in the U.S. have added to the global fiber deficit, which has affected major industries across all sectors, including retail and clothing manufacturing (Donald and Grubbs, 2022). The shortages of cotton have contributed to rising cotton prices across the world. Between February 2022 and May 2022 alone, global cotton prices rose by 18.5%, reaching their 11-year peak.[41] As Texas's cotton output is projected to decline with climate change (Kinniburgh et al., 2015), global cotton prices will likely remain elevated.

Alongside cotton, corn and soybeans are important cash crops in Texas. Their production has been adversely affected by weather hazards as well. Between 2020 and 2039, Texas's annual corn and soybean output could decline by 22% and 17%, respectively, as a consequence of changing temperatures and precipitation. By 2050, Texas's corn crop losses could grow to as much as 39% per year (Kinniburgh et al., 2015). These 2015 projections are likely conservative, as they assume the past trends of GHG emissions.

While agricultural production has inherent weather-related risks, these risks have been amplified by climate change. As a result, climate change could make it more difficult and expensive for Texan farmers to obtain financing and insurance. As crop yields become more volatile due to climate change, the lenders could tighten their lending standards and terms, as they become more concerned about the repayment of agricultural loans. At the same time, Texas's agricultural producers will be faced with higher insurance costs, as insurers will have more extreme weather-related indemnities to pay (Kaplan, 2019). The higher costs of borrowing and insurance in Texas's agricultural industry will likely contribute to rising food prices across the state.

By disrupting the state's agriculture and elevating food prices, climate change has been undermining food security in Texas. About one in every eight Texans is food insecure, and the state's growing population will only exacerbate this issue (Minton and Alvarado, 2022). According to the Texas Comptroller, severe weather and natural disasters "have long posed risks" to the state's food supply (Texas Comptroller, 2021). Climate change has been aggravating those risks, reducing both the quantity and the quality of foods produced in Texas.

III.5.7 DECLINING LABOR SUPPLY, PRODUCTIVITY, AND ECONOMIC GROWTH

In addition to adverse agricultural impacts, there are multiple other ways in which climate change undermines a state's economic performance. One of them is the negative effects of rising temperatures on labor supply. When temperatures rise above 90°F, mortality rates increase by 2%, which causes reductions in labor supply (Jayashankar et al., 2023). According to a 2015 study, extreme heat will likely cause Texas as many as 2,580 additional annual deaths by 2020–2039, rendering Texas the nation's leading state in terms of heat-related deaths. By 2040–2059, Texas will likely see as many as 4,549 additional annual heat-related deaths. By 2050, Texas and Florida combined will likely account for as many as 10,000 additional heat-related deaths per year (Kinniburgh et al., 2015). While the elderly are at the greatest risk of heat-related deaths, the state's working-age population is adversely affected as well. Consequently, rising heat-related deaths are posing a risk to the state's labor supply (Jayashankar et al., 2023; Kinniburgh et al., 2015).

Through their negative effects on human health, rising temperatures pose a threat to the state's labor availability as well. According to the Federal Reserve Bank of Dallas, labor hours decline significantly when daily maximum temperatures exceed 85°F (Jayashankar et al., 2023). With rising temperatures, some counties in Texas are projected to experience more than 6% losses in annual labor hours by the end of the century (USGCRP, 2018).

In addition to reductions in labor supply and availability, rising temperatures can lead to heat-related labor productivity losses (Kinniburgh et al., 2015). Higher temperatures make it more difficult to work outside, as well as in non-climate controlled indoor spaces. Worker productivity can also decrease due to the negative effects of rising temperatures on human health[42] (Jayashankar et al., 2023). While the state's mining, construction, transportation, agriculture, utilities, manufacturing, and leisure and hospitality industries are the most vulnerable (Jayashankar et al., 2023; Kinniburgh et al., 2015), all sectors across Texas will experience heat-related labor productivity losses by 2050. Based on a 2015 study, these losses could cost Texas up to $12.5 billion per year, with a 1-in-20 likelihood of costing the state more than $19.6 billion annually (Kinniburgh et al., 2015). In addition to labor productivity losses, the state's industrial sector can experience equipment failures and production cuts due to the episodes of extreme heat (Jayashankar et al., 2023).

Through their negative effects on labor supply, availability, and productivity, rising temperatures have been reducing employment growth across Texas. According to the analysis by the Federal Reserve Bank of Dallas, a 1°F increase in average summer temperature reduces Texas's mining sector's employment growth by 1.4 percentage points. As mining is one of the most heat-exposed sectors in Texas,[43] employment growth in the state's mining industry is at the highest risk from rising summer temperatures. In the state's construction sector, a 1°F rise in average summer temperature reduces employment growth by 0.5 percentage points. In Texas's manufacturing industry, employment growth declines by 0.25 percentage points for each 1°F increase in the average summer temperature. The negative effects of rising temperatures on the manufacturing sector's employment growth are less pronounced given that most of the state's manufacturing takes place indoors. However, since most factories are not air-conditioned, rising temperatures can lead to falling worker productivity and equipment failures due to excess heat (Jayashankar et al., 2023).

Heat-related declines in labor supply, availability, productivity, and employment growth have had adverse impacts on Texas's GSP growth. According to the Dallas Fed analysis of 2000–2022 data, every 1°F increase in average summer temperature reduces Texas's annual nominal GSP growth by 0.4 percentage points. In 2023, Texas summer temperatures exceeded the post-2000 average by 2.5 degrees. As a result, the state's nominal GSP growth was curtailed by 1 percentage point. In absolute terms, this amounted to a $24 billion nominal GSP loss in 2023. The Dallas Fed analysis of data since the mid-1960s further indicates that the effects of rising temperatures on economic growth are non-linear. The higher the average summer temperature, the greater the negative impact of additional temperature increases on nominal GSP growth (Jayashankar et al., 2023).

Rising summer temperatures can further undermine Texas GSP growth by reducing consumer spending across the state. At higher temperatures, Texas consumers are more likely to stay indoors, or to leave the state for vacations in cooler climates. This slows down economic activity across Texas, leading to revenue losses for the state's businesses. The Federal Reserve Bank of Dallas *2023 Texas Business Outlook Surveys* revealed that 25% of respondents suffered sales revenue and production losses due to the summer heatwave of 2023. The state's leisure and hospitality industry was the most affected, followed by transportation, retail, and manufacturing. According to the *Survey* respondents, the summer heatwave dampened customer demand, reduced production activity in temperature-sensitive worksites, and lowered labor productivity, resulting in revenue losses and production declines (Jayashankar et al., 2023).

III.5.8 RISING PRICES

In addition to lowering growth and employment, climate change can elevate the prices of goods. As has been discussed in the "Agricultural Impacts" section, climate change can exert price pressures by increasing the costs of agricultural production inputs and/or reducing agricultural production volumes. Climate change-driven price increases can last for several years following an extreme weather event.

As another example of climate change-driven price increases, consider peanut and peanut butter price movements in the aftermath of the 2011 drought in Texas. Faced with high irrigation costs, many peanut growers in Texas had to abandon planted peanut acres during the drought. This caused an estimated 26% reduction in Texas peanut production in 2011. Peanut production in other peanut-growing states such as Georgia and Oklahoma was adversely impacted by the drought conditions as well (United States Department of Agriculture, 2011).

As a result of drought-induced peanut production losses in Texas, Georgia, and Oklahoma,[44] U.S. farm peanut prices increased, and remained elevated for about two years after the 2011 drought. Higher farm peanut prices caused an increase in the retail prices of peanut butter. At their peak, average U.S. retail peanut butter prices rose by as much as 21.3%. While farm peanut prices returned to their pre-drought levels in two years, retail peanut butter prices remained elevated for six years after the 2011 drought. Average U.S. retail peanut butter prices were 6.2% higher for four years after farm peanut prices returned to their pre-drought levels (Dong and Astill, 2021). Retail peanut butter prices remained elevated due to the "considerable market and pricing power" at the level of peanut shelling and food retailing, which are both highly concentrated industries (Dong and Astill, 2021, p. 263). As this case study demonstrates, the price effects of climate change can be further exacerbated by the price-setting power of firms in highly concentrated industries. Higher food prices, therefore, represent a significant economic consequence of climate change.

Alongside agriculture and food, climate change can exert price pressures in the oil and gas, and the petrochemicals industries. As was previously discussed, hurricane Harvey (2017) disrupted Texas's oil refining and petrochemicals manufacturing, resulting in gasoline and petrochemicals shortages and price increases. In 2021, winter storm Uri produced petrochemicals shortages and rising petrochemicals prices. Additionally, Uri led to exorbitant increases in the prices of natural gas,[45] which were 33 times higher than average during the winter storm (Dyer, 2023).

"Record-setting natural gas prices" during Uri contributed to "soaring wholesale electricity costs" across the state and other regions (American Public Power Association, 2022, p. 1). In February 2021, Texas wholesale electricity prices hit their regulatory cap of $9,000 per megawatt hour (Mwh) (Everhart and Molnar, 2021). Elevated wholesale electricity prices cost Texas's utility customers billions of dollars in abnormally high electricity bills (American Public Power Association, 2022). It may take years for those excessive electric bills to be repaid (Dyer, 2023). By elevating residential energy bills, the spike in natural gas prices during Uri "created an enormous wealth transfer from consumers to gas sellers" (American Public Power Association, 2022, p. 1).

In addition to energy prices, Uri elevated the service fees paid by Texas's utility customers. Texas utility companies and ERCOT incurred significant costs and losses due to the winter storm. In order to recoup those costs and losses, they have been charging higher service fees to their utility customers (Thompson, 2022).

Alongside cold weather, Texas utility customers have been affected by rising summer temperatures. For example, consider the effects of the 2023 summer heatwave in Texas. As Texas's households and businesses increased their use of air conditioning, Texas's wholesale electricity prices rose above $1,000 per Mwh 182 times. While rising wholesale electricity prices did not immediately affect the state's utility customers with fixed-rate contracts, wholesale price volatility and rising load forecasts do increase forward energy prices. Over time, higher forward energy prices result in retail electricity price increases. For instance, while the state's residential rate per kilowatt-hour (Kwh) ranged from 14 to 17 cents[46] during the 2015–2019 period, the retail rate increased to more than 30 cents per Kwh in 2022, remaining volatile ever since (Golding et al., 2024). As summer temperatures rise and a growing number of Texas's households and businesses increase their use of air conditioning, electricity demand and prices across the state will keep rising (Kinniburgh et al., 2015). Population growth and rising urban density will further increase electricity demand and prices across the state.

With rising electricity demand, Texas's "grid reliability increasingly hinges upon an aging collection" of natural gas and coal plants. These facilities are proving less and less reliable during both extremely hot and extremely cold temperatures (Golding, 2023). For example, during the summer 2023 heatwave, ERCOT warned that rolling blackouts might have to be implemented in order to prevent the collapse of the grid (Golding et al., 2024). During the cold and wind event in December 2022, Texas natural gas production fell by almost 20%. While no power outages occurred in December 2022, the sharp decline in natural gas production is a cause of concern, given how less severe the weather was compared to February 2021 (Golding, 2023).

While winter storm Uri led to new regulations, weatherization standards, and operational changes within the state's energy supply system,[47] "critical gaps" in Texas's energy supply remain. While Texas's electric grid is in better shape compared to 2021, rising electricity demand and accelerating extreme weather continue to pose challenges for the state's energy generation. Texas's energy supply remains vulnerable to power outages during peak demand periods (Golding et al., 2024). As a result, energy prices across the state will remain volatile.

As rising temperatures decrease the efficiency of energy generation, transmission and delivery systems, energy production costs across Texas will rise. Additionally, higher temperatures increase water demand for energy production, as power plants need to be cooled down more often. Increased water usage and rising water rates will further increase energy production costs across Texas. At the same time, the state's available water supply is projected to decrease. Consequently, the state's energy production capacity will likely be lowered, necessitating the construction of addition power generation facilities. The business costs of new construction and transmission will further increase energy prices across Texas, as businesses will have to recuperate their costs (Kinniburgh et al., 2015).

Energy prices across Texas will keep rising despite the sizeable presence of renewable energy infrastructure across the state. While Texas leads the nation in wind-sourced electricity generation (United States Energy Information Administration, 2023), and is the nation's second-largest solar power producer (Texas Comptroller, 2023), the state's renewable energy infrastructure is vulnerable to extreme weather events. The reliability of renewable energy generation and distribution in Texas has been continuously challenged by tropical storms, wildfires, heatwaves, winter weather, flooding, and droughts (McPherson et al., 2023). Further, because wind and solar power are intermittent sources of energy generation, they cannot always be relied upon. For example, due to low wind speeds, Texas could not rely on wind-sourced electric power generation during Uri. Frozen equipment and ice on wind turbine blades posed additional obstacles to wind-sourced electricity generation during Uri (Busby et al., 2021).

Declining wind speeds present a further risk for Texas's renewable energy, as global temperatures and CO_2 levels continue to rise. By 2100, average annual global wind speeds could drop by as much as 10%, posing a significant challenge for wind energy generation. A 10% decline in wind speeds could curtail wind energy production by as much as 30%. As a result, decreased wind speeds could significantly exacerbate energy markets volatility (Robbins, 2022; see also Zamuda et al., 2023).

Texas's transmission line capacity further limits the state's reliance on renewable energy. While Texas's combined wind and solar energy generation capacity could double by 2035, the limitations of the state's energy transmission grid would necessitate increased wind and solar energy curtailments (United States Energy Information Administration, 2023a).

Given the constraints of the state's renewable energy system, rising retail and wholesale electricity prices will keep posing challenges for Texas businesses and households. By reducing consumer spending, higher utility bills will slow down the Texan economy. By making production more expensive, higher energy costs will elevate the prices of consumer goods (Byers, 2022; Congressional Budget Office, 2006; Letourneau, 2022). Even relatively small increases in energy prices can have significant effects on production costs in many Texas's industries (Kinniburgh et al., 2015).

Alongside energy costs, climate change will elevate shipping rates across the state. By disrupting existing transportation routes and/or reducing their efficiency (McPherson et al., 2023; Walewski et al., 2017), climate change makes the shipment of goods more expensive (Walewski et al., 2017). Think of heavy rain, flooding, hurricanes, storms, wildfires, extreme heat, and other weather hazards, which interrupt the shipment of goods, increase transit times, and necessitate cargo load reductions. By increasing transportation costs, climate change will further elevate consumer prices across the state.

In addition to food, energy, petrochemicals, and transportation, climate change can elevate production costs and consumer prices in a range of other sectors. By destroying property and reducing the availability of habitable land, climate change can increase property prices as well. While climate hazards reduce property values in hazard-prone areas, property values appreciate in regions with less exposure to climate risks (Hsiang et al., 2023). By worsening the existing housing shortages, and increasing human migration, the climate crisis will, on average, elevate property values across Texas and the U.S.

Rising prices of food, energy, transportation, and housing will pose significant challenges for Texan consumers, with disproportionate impacts on the state's low-income and minority households. Water rates across Texas will likely go up as well, as the state's water shortages are projected to worsen. The impacts of climate change on the state's water supply and prices will be discussed in the next section.

III.5.9 RISING WATER SHORTAGES AND RATES

With rapid population growth in Texas's urban centers and the worsening severity of droughts, water demand across Texas has been rising (USGCRP, 2018). At the same time, the state's existing water supply is projected to decrease.[48] As a result, Texas's water supply will prove inadequate in meeting the state's water needs in times of droughts. According to the Texas Water Development Board (TWDB), between 2020 and 2070, a severe drought will increase the state's water shortages by 120%.[49] The greatest water shortages will be endured by the state's municipal users, who will account for 46% of the state's water deficit in 2070. In 2020, municipal water shortages accounted for just 7% of the state's water deficit. By 2070, municipal water shortages are projected to become the leading water shortage category in the state, surpassing agricultural irrigation water shortages. Agricultural irrigation water shortages are projected to remain the state's leading water shortage category through 2060 (TWDB, 2022b).

According to the TWDB, the state's challenges of "providing sufficient water supplies at reasonable costs" will continue to amplify (TWDB, 2022b, p. 4). The costs of developing new water supply projects will continue to rise, while the new water supply projects will take years to materialize[50]. The state's failure to meet its growing water needs will adversely affect Texas's households, businesses, industry, and the state's economic development, as well as public health and safety (TWDB, 2022b, p. 4). If Texas does not have a good water supply to sustain the effects of droughts, industry and businesses will become more reluctant to relocate to the state (Donald and Grubbs, 2022). Further, droughts will make the state's water more expensive, which will increase production costs (Cho, 2019).

If Texas fails to increase its available water supply, a severe drought could cost the state an estimated $153 billion in 2070. This loss represents forgone business and labor income due to reduced economic output in agricultural, industrial, and commercial sectors. To put this income loss in perspective, it accounts for 8.5% of Texas's 2018 GSP, valued at around $1.8 trillion. In terms of employment losses, water shortages during a severe drought would cost Texas 1.4 million jobs in 2070. Texas could lose 785,000 jobs and $128 billion worth of income already by 2030, if the state fails to increase its available water supply and a severe drought occurs. By 2050, 1.019 million jobs and $132 billion worth of income could be forgone due to the state's inadequate water supply during a severe drought (TWDB, 2022b).

The above and other loss estimates are presented in Table III.5.1. According to the TWDB, the estimates are conservative, with the actual losses likely significantly larger. Further, the estimates do

TABLE III.5.1
Projected Statewide Economic Impacts from Not Meeting Texas's Water Needs Under a "Drought of Record" (2011) Conditions

Impact Measure	2030	2040	2050	2060	2070
Income Losses (billions)	$128	$128	$132	$140	$153
Job Losses	785,000	883,000	1,019,000	1,179,000	1,371,000

Source: Texas Water Development Board (2022b).

not account for future inflation, as they are measured in 2018 dollars (TWDB, 2022b). Additionally, the economic loss estimates do not account for climate change, as they consider the 2011 drought as the historic drought of record. While the TWDB mentions the challenge of "adequately preparing in the face of continued uncertainty of future droughts that may be worse than the drought of record" (TWDB, 2022b, p. 4), the board's estimates (and water supply planning) do not account for droughts that are worse than the 2011 drought (McPherson et al., 2023).

Water shortages and the rising costs of water will pose significant risks for Texas's oil and natural gas industries, given that fuels production requires large amounts of water (Kinniburgh et al., 2015). As previously mentioned in the "Rising Prices" section, water shortages will adversely affect the state's power generation as well, as power plants require more water for cooling at higher temperatures. Water shortages and rising water rates will strain the state's energy sector, increasing energy production costs, and leading to higher energy prices across the state (Kinniburgh et al., 2015).

In addition to water shortages, climate-related changes in water quality may present a growing challenge for Texas. For example, sea level rise and saltwater intrusion into coastal aquifers can increase groundwater salinity beyond the potable levels (Jasechko et al., 2020; USGCRP, 2018). The high costs of water desalination (McPherson et al., 2023) could potentially increase water rates in the state. Additionally, saltwater intrusion from sea level rise can adversely affect the state's water supply infrastructure (USGCRP, 2018). The added costs of infrastructure repair and maintenance could further contribute to rising water rates in the state. Alongside water supply infrastructure, sea level rise poses significant risks for the state's diverse coastal assets, as will be discussed in the next section.

III.5.10 EROSION OF COASTAL PROPERTY AND INFRASTRUCTURE

Texas' coastal assets have been under the ongoing threat of coastal flooding due to sea level rise (Skinner et al., 2021). Sea level rise poses a risk to Texas's coastal property, roads, hospitals, schools, power plants, drinking water treatment facilities, hazardous waste disposal sites, and other essential infrastructure (Skinner et al., 2021; Strauss et al., 2014; USGCRP, 2018). The costs to repair flood damage to Texas Gulf Coast facilities "are expected to grow substantially" (Kaplan, 2019, p. 2).

Along the Texas Gulf Coast, relative sea level rise is twice as large as the global average (USGCRP, 2018). Between 2005 and 2017, sea level rise eroded $76.4 million in relative property value along the Texas coast (First Street Foundation, 2019). More than 1,000 square miles of land along the Texas coast is located within 5 feet of the high tide line. The property that sits on this land was valued at $9.6 billion in 2014. In 2014, that property included nearly 37,000 homes which housed more than 45,000 people. Further, 2,400 square miles of land along the Texas coast lies less than 10 feet above the high tide line. The property that sits on this land was valued at $33 billion in 2014. In 2014, that property included more than 175,000 homes which housed 320,000 people (Strauss et al., 2014). According to the *Fourth National Climate Assessment* report, up to $20.9 billion in Texas's coastal property is projected to be flooded at high tide by 2030. By 2050, more than $30 billion worth of Texas's coastal property is projected to be below the high-water mark. These estimates should be viewed as conservative as they assume the 2015 trends in GHG emissions (USGCRP, 2018).

As climate change brings heavier rainfall to the state, Texas Gulf Coast cities and towns are increasingly at risk of high-impact floods (Zamuda et al., 2023). According to the Texas State Climatologist, in 2036, Texas will experience 100% more urban flooding compared to 1950–1999, and 50% more urban flooding compared to 2000–2018 (Nielsen-Gammon et al., 2024). The vast impervious surfaces (e.g., parking lots) of coastal metropolitan areas, such as the Greater Houston area, make widespread flooding more likely to occur due to increased runoff. By 2100, under a 3.3 feet of sea level rise along the Texas Gulf Coast, a Category 2 hurricane could cause three to ten times more building damage, and be $10.4 billion more costly (in 2022 dollars) compared to a storm today (Zamuda et al., 2023). Texas's property and infrastructure losses from coastal flooding and storms are projected to exceed the national average (Kinniburgh et al., 2015). By damaging or destroying residential property, frequent flooding in Texas will displace millions of people (Jay et al., 2023).

Sea level rise and coastal flooding pose a threat to the state's oil and natural gas industry, affecting both fuels production and transportation. Texas coastline in the vicinity of Galveston and Texas City is a critical oil refining and transportation hub. The disruptions to coastal oil-refining and transportation facilities can cause ripple effects throughout the region, including fuel shortages and price increases (USGCRP, 2018).

Alongside fuel transportation facilities, sea level rise threatens the state's coastal roadways and highways (McPherson et al., 2023; Walewski et al., 2017). The state's entire transportation infrastructure is increasingly vulnerable to growing climate hazards, as will be discussed in the next section.

III.5.11 DETERIORATING TRANSPORTATION INFRASTRUCTURE

Extreme weather events and conditions can have significant impacts on a state's transportation infrastructure. Rising temperatures, changing precipitation patterns, increased severity and frequency of floods, worsening droughts, and other weather hazards have been accelerating the decay of the nation's transportation assets, shortening the lifespans of the nation's transportation facilities, or rendering the facilities inoperable. Besides the risks to public safety, extreme weather-driven deterioration in the nation's transportation infrastructure has caused emergency repairs and replacements, higher maintenance and labor costs, along with supply-chain disruptions and economic activity losses (Dix et al., 2023). For example, a one-day highway shutdown can cost metropolitan area users up to $76 million in lost time, wages, and productivity. Freight operators lose about $3.4 billion (about 32 million hours) when stuck in weather-related traffic delays in metropolitan areas (Walewski et al., 2017).

Infrastructure design typically does not account for climate change, as infrastructure is built to withstand historical climate extremes (Dix et al., 2023; USGCRP, 2018). Currently, much of Texas's transportation infrastructure "is aging and in significant need of repair," with damage from climate change-driven weather extremes (McPherson et al., 2023, p. 31). For example, extreme heat, and the alternating periods of extreme heat and rainfall, have buckled the state's roads (McPherson et al., 2023; Reader, 2023; Walewski et al., 2017). Heavy rainfall has eroded the foundations of the state's roads and bridges (McPherson et al., 2023; Walewski et al., 2017). Fixing the pavement damage caused by the 2011 heatwave and drought cost Texas Department of Transportation (TxDOT) $26 million. The 2016 report by the U.S. Department of Transportation linked road surface degradation in Texas urban centers to $5.7 billion in annual vehicle operating expenses (USGCRP, 2018).

Many of the state's transportation facilities are constructed on highly expansive (plastic) clay soils, which are highly sensitive to changes in moisture content due to droughts or heavy rain. By causing the soils to contract or expand, changes in soil moisture can cause significant damage to pavements, roadbeds, and buried utilities. These and other weather-related damages pose a threat to the integrity of the state's highways and roadways (Cambridge Systematics, Inc., 2015; Dix et al., 2023).

Extreme weather endangers Texas's railway systems as well. For example, heavy rainfall and flooding can wash out the state's rail lines, while extreme heat can contribute to their thermal misalignment. Thermal rail misalignment can increase the risk of train derailment, cause operational disruptions, and lower train operating speeds (Cambridge Systematics, Inc., 2015).

Along the Gulf of Mexico, Texas has 367 miles of coastal roads, which remain increasingly vulnerable to hurricanes, tropical storm winds, and surging Gulf waters (Dix et al., 2023). Storm surges and strong wave action can displace the state's coastal rail and highway bridge decks (Walewski et al., 2017). When hurricane Ike struck the Texas Gulf Coast in 2018, it cost the state an estimated $131.8 million in transportation system losses, including bridge, highway, railway track, locomotive, and administrative building damage. Additionally, an estimated $2.4 billion was needed for "erosion, waterway dredging and other infrastructure repairs to navigable waterways, ports and coastlines" (Federal Emergency Management Agency, 2008, p. 2). The port of Galveston "sustained extensive damage to its cruise terminal, including sediment deposits and saltwater damage" (Federal Emergency Management Agency, 2008, p. 2).

The thin pavement structure in Texas's coastal metropolitan areas, such as Greater Houston, is especially vulnerable to damage from flooding (Dix et al., 2023). In addition to flooding, water-related stressors for Texas's transportation infrastructure include saltwater intrusion, moisture-damage from humidity, and riverbank erosion due to heavy rainfall (Walewski et al., 2017).

As extreme weather events and conditions are growing in frequency and severity, their adverse impacts on Texas's transportation infrastructure are becoming more damaging and costly. The rising costs of infrastructure maintenance, repair and replacement are posing a growing challenge for state and local government budgets. Those budgets are limited and do not contain adequate funds for infrastructure spending (USGCRP, 2018). In addition to cost impacts, extreme weather (especially heat) poses a health and safety hazard to maintenance and construction workers in the state's transportation sector (Cambridge Systematics, Inc., 2015).

A reliable, efficient, and safe transportation infrastructure is critical to Texas's economic development and prosperity. While the state's freight transportation system has been one of the leading contributors to the strength of the Texan economy (Walewski et al., 2017), this may change, as growing weather hazards threaten to undermine the structural integrity of the state's transportation facilities. Through their adverse impacts on the state's transportation sector, weather extremes threaten to destabilize all other Texass industries.

Further, extreme weather can disrupt and limit access to the state's public transit services. Alongside transportation route disruptions and emergency infrastructure repairs, weather hazards can create operational challenges for the state's fleet vehicles. For example, extreme heat can stress bus air-conditioning systems. Buses without functioning air-conditioning units are typically removed from service. Further, weather hazards such as extreme heat can affect the passengers' ability to walk to bus stops and train stations, and/or wait for transit service at heat-exposed outdoor locations. This can limit public access to the state's transit services (see Cambridge Systematics Inc., 2015), which are already geographically scarce.

Extreme weather-driven disruptions and reduced access to public transit will have significant and disproportionate economic impacts on the state's low-income and minority communities. These communities routinely rely on public transportation in order to commute to work, health care facilities, social services offices, or even grocery stores. By undermining the availability of and access to public transportation, climate change will shut the state's most vulnerable communities out of the jobs and essential goods and services that they need. This will further contribute to rising economic inequality in Texas.

III.5.12 RISING ECONOMIC INEQUALITY

Texas is one of the leading U.S. states in income and wealth inequality (Skinner et al., 2021). The state's 66 billionaires own more wealth than 70% of Texans combined. Texas's richest 1% and 5% of households own 49% and 74% of the state's total wealth, respectively (Cervantes, 2023). The median household income for the richest 5 % of Texans' income-earners is 30 times as high as the median income for the state's lowest-income households (Every Texan, 2023).

The extent of income and wealth inequality in Texas has been exposed and magnified by the effects of climate change (Skinner et al., 2021). This is due to the uneven impacts that climate change has been exerting on the state's most economically vulnerable communities. For example, think of extreme heat, which disproportionately affects the state's low-income populations, especially those in urban areas, "where asphalt is plentiful and trees are rare" (Zamuda et al., 2023, p. 10). Because low-income households do not always have or use air conditioning, extreme heat exposes them to greater health risks (Zamuda et al., 2023, p. 10). Consequently, low-income communities are more likely to face higher health care bills due to heat-related and/or heat-exacerbated illnesses.

As another case in point, consider the inequitable impacts of the hurricane Harvey, which struck the state in August 2017. By wiping out savings, aggravating housing insecurity, and increasing the need for health care and related services, Harvey inflicted disproportionate economic burdens on

Texas's low-income (Kaplan, 2019) and minority communities. Low-income and minority communities often live in poor-quality housing units in flood-prone areas, lacking access to adequate flood infrastructure, which contributes to their increased flood vulnerability and limits their post-flood recovery (McPherson et al., 2023; United States Department of the Treasury, 2023). One year after Harvey, 27% of the state's Hispanic residents, and 20% of the state's Black residents reported that their homes remained damaged. The reported statistic was 11% for Texas's white residents (Environmental Defense Fund, 2020).

Further, low-income and minority communities tend to live in close proximity to petrochemical plants or chemical storage facilities, which puts them at a greater risk of chemicals exposure due to extreme weather-driven industrial accidents. It has been reported that chemical releases after hurricane Harvey tended to be near predominantly Hispanic neighborhoods (McPherson et al., 2023).

As another example of climate change-exacerbated inequities, consider the state's response to the winter storm Uri (February 2021). During ERCOT-ordered blackouts, low-income and minority households were at a greater risk of residential power loss (Castellanos et al., 2023; Golding et al., 2024). Texas's communities with higher shares of racial minorities were more than four times more likely to experience power blackouts, compared to predominantly white areas (Golding et al., 2024). The longest electricity blackouts occurred in minority-dominated areas as well (Castellanos et al., 2023). In Houston, non-White communities disproportionately experienced power outages, regardless of their income level (McPherson et al., 2023). Prolonged blackouts drove many households to unsafe heating methods, leading to carbon monoxide poisoning and deaths (Golding et al., 2024).

Property damage repairs in the aftermath of Uri placed a disproportionate financial burden on Texas's low-income and minority households as well (Castellanos et al., 2023). While faced with limited to no savings, the state's low-income and minority households were also less likely to receive insurance payments in the aftermath of Uri (Hedquist et al., 2023). This is concerning, given that low-income and minority communities "experienced the brunt of storm-related damages and would arguably benefit the most from financial protection" (Hedquist et al., 2023, p. 2). As post-Uri experience has revealed, following an extreme weather event, "Texas insurers are disproportionately denying claims from more socially vulnerable communities" (Hedquist et al., 2023). This further exacerbates economic inequality across Texas, as low-income and minority households find themselves saddled with falling property values and rising debt due to property damage and repairs.

Likewise, the rise in energy bills in the aftermath of Uri had disproportionate financial impacts on the state's low-income and minority households. Faced with significant energy burdens[51] even before Uri, the state's low-income and minority households further saw their disposable incomes erode when confronted with elevated energy bills due to Uri. This impacted the householders' ability to provide for basic necessities such as food and medication (Castellanos et al., 2023).

Low-income and minority householders' energy burdens will likely increase over time as climate change-driven weather hazards continue to intensify (Castellanos et al., 2023). Consequently, more and more of vulnerable Texans will be forced to reduce essential daily spending. Already in 2023, 57% of Texas households earning less than $50,000 per year reported forgoing meals and medication in order to pay their energy bills (Golding et al., 2024). By reducing low-income households' spending on food and medication, higher energy bills will cause adverse public health impacts as well.

By disproportionately affecting low-income and minority communities, climate change will continue to amplify the state's growing economic inequities (Castellanos et al., 2023). Sequential weather events could further compound the adverse impacts of climate change on the state's disadvantaged communities, if recovery has not occurred before the next weather event. For example, the vulnerable communities near Houston have not had a chance to recover fully from hurricane Harvey (2017) when they were struck by the winter storm Uri in 2021 (Zamuda et al., 2023). A lack of full recovery from Harvey further magnified Uri's impacts on Houston area's vulnerable communities.

By increasing economic inequality across Texas, climate change will further undermine the state's economy by reducing consumer spending. The marginal propensity to consume (MPC)[52] is greater at the lower wealth quintiles, while it is lower at the higher wealth quintiles (Fisher et al.,

2019). Conversely, the marginal propensity to save (MPS)[53] is less at the lower wealth quantiles, and greater at the higher wealth quintiles. By transferring wealth to savers and eroding low-income households' disposable incomes, climate change will reduce aggregate consumption across Texas. This will limit the state's economic growth, as consumption spending is a significant driver of economic activity. By exacerbating economic inequality across the U.S., climate change will limit the nation's GDP growth. Rising economic inequality, and other adverse economic impacts of climate change across the U.S. will be discussed in the next section.

III.5.13 CLIMATE CHANGE AND THE U.S. ECONOMY

Alongside Texas, the economic conditions across the U.S. have been increasingly affected by climate change. According to a 2023 report by the U.S. Department of the Treasury, 53% of US counties face elevated exposure to at least one climate hazard. While climate hazards affect "households across income and wealth spectrums, financial burdens are not distributed evenly," as vulnerable communities across the U.S. face a "disproportionate financial strain" from climate change (United States Department of the Treasury, 2023, p.1).

The nationwide economic impacts of climate change outlined in the Treasury's 2023 report align with the above-discussed economic impacts of climate change in Texas. While the magnitude of economic impacts has been greater for Texas, given the size of the state's economy and the state's heightened weather vulnerability, most other U.S. states have been facing similar economic impacts from climate change. What is happening in Texas is indicative of what has been happening across other U.S. economies, although to varying degrees, as the extent of climate change's impacts across the U.S. regions has been uneven (Hsiang et al., 2023).[54]

In particular, the U.S. Department of the Treasury highlights the following economic impacts of climate change on communities across the U.S.:

1. Growing income losses due to:
   - reduced working hours, job loss, or furlough,
   - illness or injury,
   - falling worker productivity;
2. Growing financial strains due to property damage or destruction;
3. Rising relocation costs;
4. Rising costs and reduced availability of property insurance;
5. Rising prices of consumer goods;
6. Rising consumer spending on energy;
7. Rising transportation costs;
8. Reduced availability of public transportation;
9. Rising health care spending;
10. Rising costs and reduced availability of credit;
11. Growing economic inequality (United States Department of the Treasury, 2023).

Additionally, the *Fifth National Climate Assessment* report outlines the following economic impacts of climate change on U.S. households, businesses, and governments:

1. Rising costs of health insurance;
2. Declining property values in climate hazard-affected areas;
3. Rising property values in areas with lower exposure to climate hazards;
4. Increasing commercial and residential mortgage rates;
5. Growing business income losses due to:
   - production and supply-chain disruptions,
   - falling worker productivity,

    - declining consumer spending,
    - accelerating climate hazards abroad;
6. Growing GDP losses;
7. Growing strains on state and local government budgets due to falling tax revenues and increased expenditures;
8. Rising municipal borrowing costs (Hsiang et al., 2023).

When weather hazards damage businesses and key infrastructure, they create unsafe working and operational conditions, necessitating business and infrastructure closures. As a result, businesses and workers lose income, with outsized impacts in weather-exposed industries, such as agriculture, construction, manufacturing, and tourism. These sectors tend to employ lower- to middle-income workers who belong to minority groups (United States Department of the Treasury, 2023). In the near future, climate change is projected to cause significant income losses for U.S. outdoor workers. For example, according to a recent study, reduced working hours due to extreme heat could cost tens of millions of US outdoor workers a collective $55.4 billion loss in annual labor income by 2050 (about $1,700 per worker) (Licker et al., 2022).

By increasing the incidence of work injuries, weather hazards can further slash working hours and incomes of U.S. workers. For example, extreme heat (≥85° day) has been shown to increase work injuries by 5–15% per hot day. By permanently affecting worker productivity, extreme weather-related work injuries can have long-term negative impacts on worker compensation (Hsiang et al., 2023). Overall, prolonged exposure to weather hazards such as extreme heat could adversely affect worker physical or cognitive abilities (regardless of specific injuries), causing productivity and income losses (United States Department of the Treasury, 2023). By damaging the quality of sleep, reducing happiness, harming emotional health, altering outdoor recreational opportunities, and increasing domestic violence, weather hazards such as extreme heat can reduce worker productivity and incomes as well (Hsiang et al., 2023). While the growing severity of weather hazards will have different income implications depending on occupation and geographical location, on average, climate change will reduce American households' income gains compared to what they would earn in the absence of climate change (Hsiang et al., 2023).

In addition to losing income, climate-affected workers could forego health insurance and other employment benefits, if extreme weather conditions necessitate prolonged periods away from work. What is more, U.S. workers in weather-exposed industries could face persistent, long-term unemployment as recurring weather hazards threaten to reduce job availability in certain weather-exposed sectors (United States Department of the Treasury, 2023).

Climate change-related income and employment losses will pose significant financial challenges for U.S. households. According to the U.S. Consumer Financial Protection Bureau, more and more U.S. households are increasingly unprepared for even short-term interruptions in employment. In 2023, 39.7% of U.S. households would be unable to cover regular expenses for more than a month if they lost their main source of income. This statistic is 58% and 54% for Black and Hispanic households, respectively (Consumer Financial Protection Bureau, 2023). A financial shock, such as a climate-related income loss, could put millions of U.S. households in severe financial distress.

In addition to income and employment losses, weather hazards can cause financial strain by damaging or destroying residential property. Alongside property repair and replacement costs, U.S. households may face the financial burden of falling property values due to weather damage. When households lack the resources for property repairs, delayed or incomplete property work may lead to additional, long-term property damage. This can further erode property values, and lead to property abandonment in some cases. When weather hazards make property uninhabitable or unsafe for occupation, affected households can experience significant financial strains due to relocation expenses (United States Department of the Treasury, 2023).

According to a recent study, weather hazards affected more than 14,566 million single- and multi-family homes in the U.S. in 2021, causing an estimated $56.92 billion in residential property damage. The study revealed that about 1 in every 10 U.S. homes was affected by a natural catastrophe in

TABLE III.5.2
Estimated Property Impacts from Major Weather Hazard Events in 2021

Weather Hazard Event	Number of Homes Impacted	Property Damage (billions)
Hurricane	1,233,860	33.00
Winter Storm	12,764,941	15.00
Severe Weather	563,627	7.46
Wildfire	4,101	1.46
Total	**14,566,529**	**56.92**

Source: CoreLogic (2022).

2021. The greatest home damage was caused by hurricanes, estimated at $33 billion. Winter storms created $15 billion in residential property damage, while wildfires cost residential property owners $1.46 billion. Other weather hazards such as tornadoes and hailstorms caused $7.4 billion in residential property damage in 2021 (CoreLogic, 2022). These estimates are presented in Table III.5.2.

Property damage or destruction from weather hazards inflicts disproportionate financial impacts on low-income households, as they have limited savings, lack insurance protection, and often live in poor-quality housing units, which are more prone to damages from weather hazards. Further, low-income households are more likely to live in weather hazard-exposed areas, making them even more vulnerable to property damage (United States Department of the Treasury, 2023).

The negative effects of climate change on property insurance markets will pose additional challenges for U.S. household finances. While property insurance would traditionally protect homeowners against financial losses, insurance markets themselves are being stressed by climate change. Growing weather hazards are already triggering insurer defaults in the U.S.[55]. As climate risks grow, insurers find it increasingly more difficult to offer policies at affordable rates, which has led to "a growing disaster insurance gap" (Hsiang et al., 2023, p. 22). Faced with elevated risks of weather hazards and a growing number of claims, insurers are expected to further increase premiums, reduce coverage, or drop coverage for households in certain areas (United States Department of the Treasury, 2023). Private insurers are already abandoning wildfire- and hurricane-prone locations across the U.S., and are expected to leave more high-hazard areas (Hsiang et al., 2023).

Higher premiums, lowered coverage, and reduced availability of insurance will further undermine low-income households' ability to protect themselves against property damage or destruction due to weather hazards (United States Department of the Treasury, 2023; see also Hsiang et al., 2023). As mentioned earlier, low-income households are more likely to live in hazard-prone areas. As a result, "[i]t remains the case that those most in need of the financial protection of insurance are least able to afford it" (Hsiang et al., 2023).

In addition to higher insurance and property damage repair costs, weather hazards can cause financial strain by reducing property values in hazard-prone areas. A home's exposure to weather risks reduces its value due to potential property damage or destruction. For example, residential properties located in the present-day 100-year floodplain cost about 5% less compared to similar properties outside the floodplain. Homes that are projected to be flooded by 1 foot of sea level rise cost nearly 15% less. Residential properties located near one recent wildfire cost 9.3% less, while those located near two recent wildfires lose 27.7% of their value.[56] Other weather hazards, such as hurricanes, droughts, extreme temperatures, and the overall ecosystem health, can adversely affect property values as well (Hsiang et al., 2023). With falling property values, U.S. homeowners can find themselves with "underwater mortgages," where they owe more on the mortgage than the property is worth. Alongside households, declining property values can adversely affect mortgage markets, and financial institutions' balance sheets, potentially leading to financial instability and crises (Hsiang et al., 2023).

Growing weather hazards can also lead to residential (and commercial) mortgage rate increases, as lenders face higher risks of mortgage delinquencies. For example, for every 10% increase in hurricane damage, commercial mortgage delinquency rates rise by 28%. Some extreme weather events have already triggered widespread mortgage delinquencies in the U.S. (Hsiang et al., 2023). Weather hazards can trigger mortgage delinquencies due to property-owner income loss and/or property repair costs.

In addition to income losses, property damage, falling property values, and rising mortgage rates, climate change can strain household finances by increasing the prices of consumer goods. The U.S. Department of the Treasury cites reductions in agricultural crop yields, interruptions to production and distribution, product shortages, transportation/logistical challenges, and higher shipping/storage costs due to droughts, floods, extreme temperatures, and other weather hazards as price-drivers across the nation. Given the U.S. reliance on imports, American households have been exposed to price increases due to both foreign and domestic weather hazards (United States Department of the Treasury, 2023).

It bears noting that the price impacts of climate change can be exacerbated by the pricing power of the U.S. corporate sector. As extreme weather events and conditions are becoming more frequent, prolonged, and synchronous, prices may rise to levels which are not justified by the production-side impacts of extreme weather. As the U.S. public grows increasingly more accustomed to climate-driven price increases, the corporations may raise the prices of goods by more than what the rise in the costs of production would warrant (Semenova, 2023).

Rising consumer prices have been a growing burden for U.S. households, with disproportionate impacts on U.S. low-income communities. The Unites States Department of the Treasury has identified food as "a spending category of a particular concern," noting the growing impacts of climate change-driven food price increases on food insecurity across the nation (United States Department of the Treasury, 2023, p. 8). Food insecurity disproportionately impacts low-income communities, households led by single mothers, families with children, and households in Southern U.S. states (United States Department of the Treasury, 2023).

In addition to higher grocery bills, U.S. households will face the rising costs of utilities. According to the U.S. Department of the Treasury, energy prices across the U.S. will rise due to climate hazard-related disruptions to fuel production and distribution, interruptions to energy generation, and damage to energy infrastructure. At the same time, extreme weather conditions will necessitate greater household energy use. The increase in energy consumption and prices will put a strain on U.S. household finances. Low-income households are especially vulnerable to energy expenditures increases, as they already face a disproportionately high energy burden (United States Department of the Treasury, 2023).

Extreme weather-related energy security risks faced by the nation were highlighted by the winter storm Uri, which hit Texas in February 2021. Given that Texas is the world's third largest natural gas producer, the state's failure to meet its natural gas needs during the storm was astonishing (Busby et al., 2021). According to Michael Webber, an energy resources expert at the University of Texas at Austin, what happened in Texas "should be a warning to other states: [i]f an energy-abundant, rich state that likes to invest in infrastructure can suffer a significant energy crisis, then it can happen anywhere" (Webber, 2021, p. 1). The 2021 energy disaster in Texas was not an abnormal, one-off event, but "only the beginning" of cascading, extreme weather-driven energy crises that will be unfolding in the U.S. and globally (Ibid.). The recurring energy crises in the United States and around the world will keep energy prices at permanently elevated levels.

In addition to rising energy costs, U.S. households will face higher transportation expenditures, as growing weather hazards are expected to cause increasingly more oil production shortages and gasoline price increases. At the same time, weather hazards will disrupt service or limit access to public transportation across the state. As a result, public transit-dependent households may be forced to adapt to costlier transportation alternatives, such as personal gas-powered vehicles. Rising transportation costs and reduced access to public transit could limit many households' ability to access employment and essential services (United States Department of the Treasury, 2023).

As weather extremes deepen and increase the incidence of many health conditions (Price, 2020), U.S. households' health care expenditures will rise (United States Department of the Treasury, 2023). For example, extreme heat can aggravate respiratory and cardiovascular illnesses, and increase the risk of vector-borne diseases. Further, as the quality of drinking water and the nutritional content of food deteriorates with climate change, water quality- and food quality-related illnesses will exacerbate as well (Price, 2020; see also; Hsiang et al., 2023; McPherson et al., 2023). Rising health care expenditures will strain private and public budgets across the nation (Hsiang et al., 2023; Price, 2020; United States Department of the Treasury, 2023). The growing adverse health impacts of climate change are expected to raise health insurance premiums as well (Hsiang et al., 2023).

Rising health care spending will have disproportionate impacts on the nation's most vulnerable communities. Low-income households, who often do not have access to health insurance, will be struggling to absorb the additional health care costs. Older adults and individuals with pre-existing medical conditions will be disproportionately impacted by climate change-driven increases in health care expenditures as well (United States Department of the Treasury, 2023).

In addition to direct health impacts, such as illnesses caused or exacerbated by extreme weather, climate change can affect public health by reducing the availability of regular health care services. For example, weather hazards such as flooding and hurricanes can cause health care facilities closures, limiting individuals' access to routine medical services (Hsiang et al., 2023; United States Department of the Treasury, 2023). Lack of routine medical care can lead to adverse health consequences, such as the worsening of existing health conditions. This can put additional financial strains on affected individuals, including higher long-term health care costs, and/or increased time away from work. U.S. households in rural and lower-income areas are especially vulnerable to medical facilities closures, as they already struggle with limited geographic access to health care providers (United States Department of the Treasury, 2023).

Faced with limited insurance protection and inadequate savings, many U.S. households may turn to credit in order to manage climate change-related expenditures, such as rising health care costs and mounting home repair bills (United States Department of the Treasury, 2023). According to the U.S. Federal Reserve survey data, 37% of American adults would not be able to cover a $400 emergency using cash or its equivalent[57] in 2022 and 2023 (Board of Governors of the Federal Reserve System, 2023). According to the U.S. Consumer Financial Protection Bureau, almost 30% of U.S. consumers had $500 or less in their checking or savings account in 2023 (Consumer Financial Protection Bureau, 2023). A climate hazard-related financial shock may leave millions of U.S. households with no choice but reliance on credit.

Recent increases in interest rates have made it more expensive to manage financial exigencies with credit cards (Consumer Financial Protection Bureau, 2023). As a growing number of climate-impacted U.S. households will turn to borrowing, credit may become even more costly, and less accessible. Many U.S. consumers already lack access to commonly used consumer credit products (Consumer Financial Protection Bureau, 2023; United States Department of the Treasury, 2023). According to the U.S. Consumer Financial Protection Bureau, 41.5% of U.S. consumers applied for credit in 2023.[58] More than one-third of those who applied were turned down or did not get as much credit as they requested, with disproportionate impacts on Black and Hispanic consumers (Consumer Financial Protection Bureau, 2023). Faced with limited access to commonly used consumer credit products, many climate-affected households may turn to predatory alternatives such as payday and pawn shop loans (United States Department of the Treasury, 2023).

The negative effects of climate change on U.S. household finances will have significant implications for consumer spending across the nation. U.S. consumer spending will fall due to climate change-driven financial impacts such as income loss, higher energy (utilities) spending, rising health care costs, elevated food prices, higher insurance premiums, rising interest rates, mounting property repair bills, rising transportation expenditures, growing economic inequality, declining property values in climate-hazard affected areas, and rising property values in areas with less exposure to climate hazards Hsiang et al., 2023). Consumer spending accounts for 70%of U.S. GDP, representing the main driver of the nation's economy. Climate-change driven reductions in consumer spending will slow down all sectors of the U.S. economy, even those not directly impacted by climate hazards. Faced with falling consumer demand and declining sales revenues, U.S. businesses will be forced to reduce production and employment. This will further stall consumer spending and cool down the U.S. economy.

In addition to falling consumer demand, the negative effects of climate change on U.S. businesses include reductions in worker productivity, output losses,[59] supply-chain disruptions, business activity interruptions, commercial property damage, and rising transportation, energy, and other production costs. Climate change-driven disruptions to global trade and supply chains, along with falling demand for U.S. exports, pose additional challenges for U.S. businesses (Hsiang et al., 2023).

Alongside consumers and businesses, climate change has been impacting state and local governments across the nation. With growing climate hazards, government spending on healthcare, income support, infrastructure repair and replacement,[60] disaster assistance, and public safety[61] has been rising. At the same time, climate change is projected to lower tax revenues across many states, given the negative effects of climate hazards on income, employment, and economic growth. In certain areas, climate change-driven reductions in property values could further reduce local governments' tax receipts. The combination of declining tax revenues and rising expenditures will contribute to rising deficits, increased reliance on borrowing, and higher borrowing costs incurred by state and local governments. The fiscal risks associated with climate change may threaten public programs that vulnerable communities rely upon (Hsiang et al., 2023).

The negative effects of climate change on U.S. households, businesses, financial institutions, and state and local governments could compound over time, generating "large cumulative losses" (Hsiang et al., 2023, p. 17) and "cascading effects" (Hsiang et al., 2023, p. 18), which could cause a "wider failure in the [U.S.] financial system" (Hsiang et al., 2023, p. 18). According to the *Fifth National Climate Assessment* report, "[t]here is growing concern that climate change could pose a systemic risk to financial stability" of the U.S. (Hsiang et al., 2023, p. 18). This concern has been previously voiced by the Federal Reserve Bank of Dallas (Kaplan, 2019).

At the same time, climate change could threaten the nation's overall societal integrity, further undermining the nation's economic stability. The *Fifth National Climate Assessment* report cites increased political turnover, impeded functioning of institutions and governmental organizations, displaced populations, rising violence, and growing population unrest as climate-change driven factors, which could potentially destabilize the U.S. social system (Hsiang et al., 2023).

III.5.14 CONCLUSIONS: CLIMATE CHANGE AND OUR ECONOMIC FUTURE

Climate change has been inflicting significant and widespread economic costs in Texas and across the U.S. The economic impacts of climate change in Texas include agricultural crop losses, production and supply-chain disruptions, rising production costs, higher consumer prices, growing food insecurity, declining labor supply and availability, falling worker productivity, GSP losses, disruptions to energy supply, accelerating energy demand, rising energy insecurity, higher energy prices, growing water shortages, rising water rates, transportation disruptions, higher shipping and storing costs, property damage and destruction, infrastructure damage and losses, rising insurance and financing costs, widening economic inequality, and falling consumer demand, among others. Climate change has cost Texans hundreds of billions of dollars in economic damages and losses. The direct economic costs of the hurricane Harvey (2017) and the winter storm Uri (2021) alone were estimated at more than $100 billion and $20 billion, respectively. As the state with the greatest number of billion-dollar weather and climate disasters, Texas leads the nation in direct cumulative costs and losses from climate change.

If left unmitigated, climate change threatens to undermine the Texan economy. The state's agricultural, manufacturing, transportation, energy, and other key infrastructure remains increasingly at risk from rising temperatures, worsening droughts and wildfires, changing precipitation patterns, rising sea levels, heavier flooding, more frequent storms, stronger hurricanes, and other growing climate hazards. Texas's labor supply, availability, and productivity remain vulnerable to accelerating climate hazards as well. While the state's economy and population have been growing, climate change could reverse the state's productivity, employment, and population gains. Climate change could undo the economic progress and development that the state has achieved.

Alongside Texas, climate change has been inflicting "substantial and increasing" (Hsiang et al., 2023, p. 5) economic costs and losses on communities across the U.S.. According to the *Fifth National Climate Assessment* report (Hsiang et al., 2023) and the United States Department of the Treasury (2023), the economic impacts of climate change across the nation include reductions in agricultural crop yields,[62] rising consumer prices, growing food insecurity, employment and income losses, declining worker productivity, disruptions to production, transportation and supply chains, GDP losses, higher transportation and shipping costs, road and infrastructure degradation, property damage and destruction, disruptions to fuels production and energy generation, rising costs of energy and utilities, higher health care costs, growing health care expenditures, rising costs and reduced availability of credit, rising costs and reduced availability of property insurance, rising mortgage rates, falling property values in weather hazard-prone areas, rising property values in areas with less exposure to weather hazards, growing relocation expenditures, worsening economic inequality, falling consumer spending, and growing strains on public budgets, among others.[63]

As the effects of climate change are becoming increasingly non-linear,[64] climate change will exert increasingly larger negative impacts on the U.S. economy, with disproportionate burdens on the nation's most vulnerable communities. As the adverse economic impacts of climate change compound over time, climate change could pose a systemic risk to the nation's economic and financial stability (Hsiang et al., 2023).

Climate is "a national asset that enables and contributes value to diverse economic activities," from agriculture to real estate and finance (Hsiang et al., 2023, p. 5). In addition to supporting the economy, climate conditions help maintain overall societal stability (Hsiang et al., 2023, p. 5). If left unmitigated, climate change could cause a breakdown of the U.S. society and economy, threatening people's lives and livelihoods.

APPENDIX: SAMPLE FUTURE ECONOMIC IMPACT ESTIMATES OF PROJECTED CLIMATE HAZARDS

Impact Type	Climate Hazard	Impact Estimate
Agricultural yields (maize, soybeans, winter wheat, spring wheat, cotton, and sorghum)	Temperature, moisture changes	12%–29% decrease (2050–2100)* 20%–48% decrease (2050–2100)***
Agricultural yields (maize, soybeans, and cotton)	Temperature, precipitation changes	30%–46% decrease (2070–2099)* 63%–82% decrease (2070–2099)***
Aggregate multisector impact	Temperature increases	–0.1%–1.7% GDP loss* 1.5%–5.6% GDP loss***
Income	Hurricanes	29% GDP loss**
Property tax revenue	Sea level rise	–1.4% (3-foot rise)
Coastal damages	Sea level rise	+$550 billion (optimal adaptation)*** +$2.6 trillion (no adaptation)***
Electricity outages	Temperature, Precipitation changes	+$2.3–$6.8 trillion consumer costs***
Flooding costs	Flooding	+61% annual losses (2050)*
Railroad disruption	Temperature increases	+$30–$55 billion costs from network delays* +$43–$73 billion costs from network delays***
Road degradation	Temperature, precipitation changes	+$116 billion costs***
Urban drainage degradation	Temperature, Precipitation changes	+$29 billion costs***

Source: Hsiang et al. (2023).
Key: * indicates an intermediate scenario; ** indicates a high scenario; *** indicates a very high scenario.

NOTES

1. This chapter does not address the economic impacts of climate change mitigation, including proposed technological responses to climate change.
2. Gross State Product (GSP) is a measurement of a state's annual economic output. It is the sum of value added from all industries in the state. GSP is the state counterpart to the national Gross Domestic Product (GDP).
3. Gross State Product (GSP) is a measurement of a state's annual economic output. It is the sum of value added from all industries in the state. GSP is the state counterpart to the national Gross Domestic Product (GDP).
4. Dallas–Fort Worth–Arlington metropolitan area had a 19.6 percent population growth rate during the 2012–2022 period. During the same time frame, San Antonio–New Braunfels, and Houston–Pasadena–he Woodlands metropolitan areas both registered a population growth rate of 18.7 percent (Texas Comptroller, 2024).
5. In 2022, more than $547 billion was exported and more than $503 billion was imported through the ports of entry located in Texas (Texas Comptroller, 2024).
6. Texas also ranks fifth nationwide in acreage used for wheat production (Kinniburgh et al., 2015).
7. A high-tech sector manufactures or provides services that involve computers, electronics, software, data, and scientific research and development. High-tech sectors have a high concentration of workers in STEM fields: science, technology, engineering, mathematics (Kerr et al., 2024).
8. Texas' transportation emissions have been growing rapidly. The state's major cities such as Dallas and Houston have the largest commute climate impacts in the nation (Skinner et al., 2021).
9. Texas' electric power generation is heavily reliant on fossil fuels (Skinner et al., 2021). The amount of natural gas used for electricity generation in Texas is greater than in any other US state (United States Energy Information Administration, 2023). In 2022, natural gas- and coal-fired plants accounted for about 50 percent and 16 percent of the state's total electricity production, respectively. In the same year, wind power accounted for about 20 percent of the state's total electricity generation. The state's two operating nuclear power plants supplied 8 percent of the state's electricity in 2022. The remaining electricity production came from solar energy. The share of wind in Texas' electricity generation increased from 16.9 percent in 2018 to 20 percent in 2022. In that year, the share of coal fell to 16 percent (from 23.4 percent in 2018) (United States Energy Information Administration, 2023; Skinner et al., 2021).
10. For presidentially declared disasters, the federal government generally covers 75 percent of the eligible Public Assistance (PA) expenditures. PA funding covers debris removal; repair, replacement and restoration of disaster-damaged public facilities; and lives, public health, and public safety hazards mitigation. State and local governments are responsible for the remaining 25 percent of the PA expenditures (United States Government Accountability Office, 2015). Consequently, extreme weather disasters can result in significant costs for state and local government agencies (Walewski et al., 2017). As with other state expenditures, disaster funding is typically part of a state's annual operating budget. As a result, disaster costs compete with other state spending priorities, unless a separately-funded disaster response fund is maintained by a state (United States Government Accountability Office, 2015). When a presidential disaster is not declared, it can be problematic for state and local government agencies to allocate sufficient funds to extreme weather disaster response (Walewski et al., 2017).
11. This estimate is $2.7073 billion according to NOAA NCEI 2024.
12. As of June 10, 2024.
13. From 1980 to 1989, Texas experienced 14 severe weather and climate events. From 1990 to 1999, Texas experienced 21 severe weather and climate events. From 2000 to 2009, Texas witnessed 25 severe weather and climate disasters. From 2010 to 2019, Texas endured 62 severe weather and climate events. Between 2014 and 2023, Texas endured 90 severe weather and climate disasters (NOAA NCEI, 2024b).
14. Although the number of extremely cold days experienced in Texas is expected to decrease with global warming, the state will still experience extreme cold events despite the warming climate. These events will result in considerable costs to the Texan economy (Zamuda et al., 2023).
15. The total direct cost of Harvey has been estimated at $125 billion (Texas Comptroller, 2018). In 2024 dollars, the total direct cost of Harvey has been assessed at $158.8 billion (NOAA, 2024a). The total direct cost of Katrina has been estimated at $161.3 billion (2017 dollars) (Texas Comptroller, 2018). In 2024 dollars, the total direct cost of Katrina has been assessed at $198.8 billion (NOAA, 2024a).
16. Harvey affected eight Texas' Councils of Government (COG) regions including Houston–Galveston, South East Texas, Golden Crescent, Coastal Bend, Brazos Valley, Deep East Texas, and Capital and Alamo area (Texas Comptroller, 2018).

17. Many Texans saw their medical treatments interrupted when they were displaced from their homes (USGCRP, 2018).
18. Those counties include Harris, Jefferson, Fort Bend, Orange, and Galveston.
19. FEMA's National Flood Insurance Program (NFIP) paid out $8.8 billion for flood claims (Lowe and Marx, 2020).
20. Houston and the upper Texas Gulf Coast provide more than 25 percent of US oil and natural gas production and refining capacity (Lowe and Marx, 2020).
21. The $16.8 billion GSP loss includes the $16 billion loss in the Houston–Galveston area, the 13–county region which was hardest hit by the storm (Texas Comptroller, 2018).
22. As an immediate response to Harvey, FEMA provided Texas with over $2 billion in Public Assistance (PA) and over $1.656 billion in Individual Assistance (IA) funds. While PA funds help local governments with debris removal, damaged public facilities repairs, and hazards mitigation, IA funds are aimed at natural disaster victims, such as individuals and families. IA funds help with expenses that are not covered by insurance, such as home repairs and medical bills. During the first year after Harvey, FEMA provided around $14 billion to affected Texas communities and their residents (Lowe and Marx, 2020). According to the Texas Comptroller, federal, state, and local governments, along with private insurers, had spent or committed $31 billion for Harvey-related disaster relief and rebuilding as of November 30, 2017 (Texas Comptroller, 2018).
23. In the aftermath of Harvey, Gulf Coast cities, the state of Texas, and various federal government agencies have been working together on multibillion-dollar investments in reservoir upgrades and flood mitigation infrastructure to better protect the Gulf Coast region in the future (Kaplan, 2019).
24. According to the Texas Department of Agriculture (2024), more than 700 people died as a result of Uri.
25. According to McPherson et al. (2023), the economic losses from Uri in Texas range from $85.6 billion to $139.1 billion (in 2022 dollars).
26. See Cohen et al. 2021; The Meadows Center for Water and the Environment; and Zamuda et al. 2023.
27. While Texas leads the nation in wind-powered electric generation (United States Energy Information Administration, 2023b), the state could not rely on wind power during the winter storm Uri. This was due to low wind speeds, frozen equipment, and ice on wind turbine blades (Busby et al., 2021). The state's multiple coal power plants experienced equipment failures due to freezing temperatures as well (Golding, 2023).
28. Freeze-offs "occur when water and other liquids in the raw natural gas stream freeze at the wellhead or in natural gas gathering lines near production activities" (United States Energy Information Administration, 2021, p. 1). This creates blockages and disrupts the flow of natural gas from the well or through the natural gas transportation facilities (United States Energy Information Administration, 2024). According to the US Energy Information Administration, the decline in natural gas production during the winter storm Uri "was mostly a result of freeze-offs" (United States Energy Information Administration, 2021, p. 1).
29. During Uri, the roads in Texas remained unplowed and unsalted (Busby et al., 2021).
30. As a result of Uri, oil refining in the Gulf Coast Region (Texas, Louisiana and New Mexico) fell by 50 percent in February 2021. The Region accounts for more than half of US operable oil refining capacity. The drop in oil refining during Uri was comparable to the production impacts of hurricanes Ike (2008) and Harvey (2017) (Thompson, 2021).
31. Honda suspended its North American production operations for a week in March 2021 (Thompson, 2021).
32. These products accounted for almost 17 percent of Texas' exports in the three months before Uri (Thompson, 2021).
33. Due to grapefruit crop losses from Uri, and lower grapefruit production in Florida and California, US 2021 grapefruit production hit a 50-year low, down 39 percent from the 2019 – 2020 season. Low grapefruit production put an upward pressure on US grapefruit prices (Kramer et al., 2021).
34. This amount includes the estimated 2022 citrus crop losses resulting from Uri (Schattenberg, 2021).
35. These livestock loss estimates include damage to livestock industry infrastructure and housing facilities; increased heating costs to keep the animals warm; and poultry loss. Beef cattle losses include estimated value of death losses; additional feed used; lost winter small grain grazing; lost weights and feed efficiency in feedlots; and losses due to delayed marketing. Sheep and goat losses include estimated death losses. Dairy losses include cattle death loss; lost milk production; and the value of milk dumped due to milk transportation and processing disruptions (Schattenberg, 2021).

36. Nine out of those 19 droughts took place during the 2010–2024 period (NOAA NCEI, 2024; as of June 10, 2024).
37. Texas Water Development Board references $7.62 billion as the agricultural loss from the 2011 drought, "making 2011 the costliest drought year in history" (Texas Water Development Board, 2022b, p. 35).
38. Even a forecast of bad weather can immediately drive up grocery bills (Horn-Muller, 2024).
39. According to Powell (2023), the May 2023 hailstorm affected 62,000 Texans and resulted in $1.6 billion in total losses.
40. According to Donald and Grubbs (2022), cotton production contributes $2.4 billion to Texas' GSP, supports 40,000 jobs, and generates $1.55 billion in annual labor income.
41. Even though global cotton prices have been falling since May 2022, the price level remains elevated. As of August 2023, global cotton prices were 51 percent higher compared to their April 2020 level (International Monetary Fund, 2023).
42. In addition to labor productivity losses, the adverse health effects of rising temperatures could lead to reductions in labor supply and availability.
43. Most of the oil and natural gas extraction work is performed outdoors (Jayashankar et al., 2023)
44. The decline in peanut production in Texas, Georgia and Oklahoma was estimated at 18.7 percent (United States Department of Agriculture, 2011).
45. Wholesale natural gas spot prices skyrocketed during the winter storm Uri because natural gas prices were completely deregulated (American Public Power Association, 2022).
46. The prices are adjusted for inflation.
47. For example, post-Uri operational protocols at ERCOT require that more thermal power generation is held in reserve (Golding, 2023).
48. Between 2020 and 2070, the state's existing water supply is projected to decrease by 18 percent (Texas Water Development Board, 2022b).
49. This estimate does not account for the growing severity of droughts, and considers the 2011 drought as the historic drought of record (McPherson et al., 2023).
50. The "challenges relating to water availability in the state" have been discussed by the Federal Reserve Bank of Dallas as well (Kaplan, 2019, p. 4)
51. An 'energy burden' is the percentage of annual household income spent on energy bills. For Texas' urban centers, the energy burden is "consistently the highest for low-income and minority groups" (Castellanos et al., 2023, p. 3).
52. The MPC is the fraction of the increase in income that is spent on consumer goods and services.
53. The MPS is the fraction of the increase in income that is saved.
54. The economic impacts of climate change vary by location due to different weather hazards, regional climate change patterns, historical climate and population density (Hsiang et al., 2023).
55. As a result, much of natural disaster insurance coverage is now offered through fully or quasi-public programs (Hsiang et al., 2023).
56. These examples come from specific studies, some of which are not nationally representative (Hsiang et al., 2023).
57. Cash equivalents refer to savings, or credit card balances fully paid off at the next billing statement.
58. At the same time, almost a quarter of US consumers chose not to apply for credit in 2023 for fear of being turned down by the creditors (Consumer Financial Protection Bureau, 2023).
59. In addition to goods-producing industries such as agriculture and manufacturing, these losses apply to service-industries such as finance, real estate, and insurance, among others (Hsiang et al., 2023)
60. The nation's essential infrastructure, such as water, energy, transportation, and communication systems, will be increasingly compromised "by the compounding effects of climate change impacts" (Hsiang et al., 2023, p 14).
61. Income losses and other climate change impacts could lead to rising crime rates across the states (Hsiang et al., 2023).
62. While cold regions across the United States may benefit from low levels of warming, nationwide, climate change impacts will cause a net loss in agricultural production (Hsiang et al., 2023). See Appendix below.
63. The quantitative projections for some of these impacts, such as agricultural production losses, etc., are presented in the Appendix at the end of this chapter.
64. For example, a 2°F warming is projected to cause more than twice the economic harm produced by a 1°F warming (Hsiang et al., 2023).

REFERENCES

American Public Power Association, 2022. Winter storm Uri, extreme winter events, and natural gas reform. Issue brief, January 2022. https://www.publicpower.org/system/files/documents/January%202022%20-%20Winter%20Storm%20Uri.pdf (accessed June 25, 2024).

Baddour D, 2014. During drought, once-mighty Texas rice belt fades away. National Public Radio, August 12, 2014. https://stateimpact.npr.org/texas/2014/08/12/during-drought-once-mighty-texas-rice-belt-fades-away/ (accessed June 25, 2024).

Barber J, Stone A, 2022. Texas food access study. Texas Department of Agriculture, December 31, 2022. https://static.texastribune.org/media/files/aa8d78c575b5e5667f659395cc2e4841/Texas%20Food%20Access%20Study.pdf (accessed June 25, 2024).

Blake E, Zelinsky D, 2018. National hurricane center tropical cyclone report: Hurricane Harvey. https://www.nhc.noaa.gov/data/tcr/AL092017_Harvey.pdf (accessed June 25, 2024).

Board of Governors of the Federal Reserve System, 2023. Report on economic well-being of U.S. households. https://www.federalreserve.gov/consumerscommunities/sheddataviz/unexpectedexpenses.html (accessed June 25, 2024).

Bridge for Innovation, 2021. Technology's impact in Texas. https://bridgeforinnovation.org/impact-map/technologys-impact-in-texas/ (accessed June 25, 2024).

Bureau of Economic Analysis, 2024. Gross domestic product by state and personal income by state, 4th quarter 2023 and preliminary 2023. News Release, March 29, 2024. https://www.bea.gov/data/gdp/gdp-state (accessed June 25, 2024).

Busby J, Baker K, Bazilian M, Gilbert A, Grubert E, Rai V, Rhodes J, Shidore S, Smith C, Webber M, 2021. Cascading risks: Understanding the 2021 winter blackout in Texas. Energy Research & Social Science, 77, doi:10.1016/j.erss.2021.102106.

Byers D, 2022. Energy and inflation: 5 things you need to know. United Stated Chamber of Commerce, June 27, 2022. https://www.uschamber.com/energy/energy-and-inflation-5-things-you-need-to-know-now (accessed June 25, 2024).

Cambridge Systematics, Inc., 2015. Central Texas extreme weather and climate change vulnerability assessment of regional transportation infrastructure. Final Report. Prepared for Capital Area Metropolitan Planning Organization and City of Austin Office of Sustainability. https://www.austintexas.gov/sites/default/files/files/Sustainability/Climate/CAMPO_Extreme_Weather_Vulnerability_Assessment_FINAL.pdf (accessed June 25, 2024).

Castellanos S, Potts, J, Tiedmann H, Alverson S, Glazer Y, Robison A, Russo S, Harmon, D, Ken-Opurum B, Weisz M, Acuna F, Stephens K, Faust K, Webber M, 2023. A synthesis and review of exacerbated inequities from the February 2021 winter storm (Uri) in Texas and the risks moving forward. Prog Energy, 5:012003, doi:10.1088/2516-1083/aca9b4.

Cervantes S, 2023. Proposition 3 will maintain Texas extreme wealth inequality. Every Texan. https://everytexan.org/2023/10/20/proposition-3-will-maintain-texas-extreme-wealth-inequality/ (accessed June 25, 2024).

Cho R, 2019. How climate change impacts the economy. State of the Planet: News from the Columbia Climate School. https://news.climate.columbia.edu/2019/06/20/climate-change-economy-impacts/ (accessed June 25, 2024).

Cho R, 2021. Attribution science: Linking climate change to extreme weather. News From the Columbia Climate School. https://news.climate.columbia.edu/2021/10/04/attribution-science-linking-climate-change-to-extreme-weather/ (accessed June 25, 2024).

Cohen J, Agel L, Barlow M, Garfinkel C, White I, 2021. Linking Arctic variability and change with extreme winter weather in the United States. Science, 373:1116–1121, doi:10.1126/science.abi9167.

Congressional Budget Office, 2006. The economic effects of recent increases in energy prices. https://www.cbo.gov/system/files/2018-10/07-21-energy-dist.pdf (accessed June 25, 2024).

Constible J, 2018. The emerging public health consequences of hurricane Harvey. The National Resources Defense Council. Expert Blog, August 29, 2018. https://www.nrdc.org/bio/juanita-constible/emerging-public-health-consequences-hurricane-harvey#:~:text=Millions%20of%20Pounds%20of%20Air,respiratory%20problems%20such%20as%20asthma (accessed June 25, 2024).

Consumer Financial Protection Bureau, 2023. Making ends meet in 2023. Research Publication No. 2023–8. https://files.consumerfinance.gov/f/documents/cfpb_making-ends-meet-in-2023_report_2023-12.pdf (accessed June 25, 2024).

CoreLogic, 2022. CoreLogic climate change catastrophe report estimates 1 in 10 U.S. residential properties impacted by natural disasters in 2021. Press Release, February 17, 2022. https://www.corelogic.com/press-releases/corelogic-climate-change-catastrophe-report-estimates-1-in-10-u-s-residential-properties-impacted-by-natural-disasters-in-2021/ (accessed June 25, 2024).

Dix B, Wagner H, Vargo A, Proctor G, Varma S, Filosa G, 2023. Addressing resilience to climate change and extreme weather in transportation asset management. Report No. FHWA–HIF–23–010. Prepared for United States Department of Transportation, Federal Highway Administration. https://www.fhwa.dot.gov/asset/pubs/hif23010.pdf (accessed June 25, 2024).

Donald J, Grubbs S, 2022. Drought in Texas: How rain scarcity affects Texans and the economy. Texas Comptroller Fiscal Notes, December 2022. https://comptroller.texas.gov/economy/fiscal-notes/archive/2022/dec/drought.php (accessed June 25, 2024).

Dong X, Astill G, 2021. The short- and long-term costs of a severe drought on retail peanut butter prices and consumers. Journal of Agricultural and Applied Economics, 53(2):259–279, doi:10.1017/aae.2021.6.

Dyer RA, 2023. Texas gas utility customers face 16 years of charges from winter storm Uri consumption. Atmos Cities Steering Committee Blog, November 6, 2023. https://atmoscitiessteeringcommittee.org/texas-gas-utility-customers-face-16-years-of-charges-from-winter-storm-uri/ (accessed June 25, 2024).

Environmental Defense Fund, 2020. Texas and the rising cost of extreme weather. https://www.edf.org/sites/default/files/u76/CostExtremeWeatherOnePager-Texas.pdf (accessed June 25, 2024).

Everhart K, Molnar G, 2021. Severe power cuts in Texas highlight energy security risks related to extreme weather events. International Energy Agency. https://www.iea.org/commentaries/severe-power-cuts-in-texas-highlight-energy-security-risks-related-to-extreme-weather-events (accessed June 25, 2024).

Every Texan, 2023. Income and poverty data brief. https://everytexan.org/wp-content/uploads/2023/12/Income-Poverty-Data-Brief-2023.pdf (accessed June 25, 2024).

Federal Emergency Management Agency, 2008. Hurricane Ike impact report: December 2008. https://www.fema.gov/pdf/hazard/hurricane/2008/ike/impact_report.pdf (accessed June 25, 2024).

Federal Reserve Bank of Dallas, 2024b. Your Texas economy. https://www.dallasfed.org/-/media/Documents/research/econdata/texaseconomy.pdf (accessed June 25, 2024

Federal Reserve Bank of Dallas, 2024a. Texas manufacturing outlook survey. https://www.dallasfed.org/research/surveys/tmos/about (accessed June 25, 2024).

First Street Foundation, 2019. Sea level rise washes away $76.4 million in Texas home values. https://assets.firststreet.org/uploads/2019/04/Texas-Press-Release_042219_NO-embargo.pdf (accessed June 25, 2024).

Fisher J, Smeeding T, Thompson J, 2019. Estimating the marginal propensity to consume using the distributions of income, consumption, and wealth. Federal Reserve Bank of Boston Research Department Working Paper No. 19-4. https://www.bostonfed.org/publications/research-department-working-paper/2019/estimating-the-marginal-propensity-to-consume-using-the-distributions-income-consumption-wealth.aspx#:~:text=Key%20Findings,it%20is%20for%20wealthy%20households (accessed June 27, 2024).

Foreman D, 2024. TXOGA energy economics perspective: Q1 2024. Texas Oil and Gas Association, February 29, 2024. https://www.txoga.org/txoga-energy-economics-perspective-q1-2024/#:~:text=Texas'%20Increased%20Criticality&text=In%202023%2C%20Texas%20accounted%20for,despite%20relatively%20modest%20drilling%20activity (accessed June 25, 2024).

Golding G, 2023. Texas electrical grid remains vulnerable to extreme weather events. Federal Reserve Bank of Dallas, Dallas Fed Economics, January 24, 2023. https://www.dallasfed.org/research/economics/2023/0117 (accessed June 25, 2024).

Golding G, Kumar A, Mertens K, 2021. Cost of Texas' 2021 deep freeze justifies weatherization. Federal Reserve Bank of Dallas, Dallas Fed Economics, April 15, 2024. https://www.dallasfed.org/research/economics/2021/0415#:~:text=Economically%20justified%20measures&text=Using%20our%20VOLL%20estimate%20of,less%20than%20%24430%20million%20annually (accessed June 25, 2024).

Golding G, Perlmeter ER, Kalkunte P, 2024. Addressing Texas grid reliability: Time to go nuclear? Federal Reserve Bank of Dallas, Southwest Economy, May 13, 2024. https://www.dallasfed.org/research/swe/2024/swe2407#:~:text=Thirty%20years%20after%20Texas'%20last,relentless%20but%20not%20without%20precedent (accessed June 25, 2024).

Hayhoe K, 2019. Texas offers perfect setting to study impacts, costs of climate change. Federal Reserve Bank of Dallas Conversation with Katharine Hayhoe, Southwest Economy, Third Quarter 2019. https://www.dallasfed.org/~/media/documents/research/swe/2019/swe1903e.pdf (accessed June 25, 2024).

Hedquist A, Salas RN, Soto M, Orav EJ, Figueroa JF, 2023. Disparities in property insurance relief among socially vulnerable Texas communities after winter storm Uri. Health Affairs Scholar, 1(5):qxad062, doi:10.1093/haschl/qxad062.

Horn-Muller A, 2024. How forecasts of bad weather can drive up your grocery bill. Grist, June 10, 2024. https://grist.org/food-and-agriculture/how-forecasts-of-bad-weather-can-drive-up-your-grocery-bill/ (accessed June 25, 2024).

Hsiang S, Greenhill S, Martinich J, Grasso M, Schuster R, Barrage L, Diaz D, Hong H, Kousky C, Phan T, Sarofim M, Schlenker W, Simon B, Sneeringer S, Avery C, Scheetz, 2023. Ch. 19. Economics. In Fifth National Climate Assessment, Crimmins AR, Avery CW, Easterling DR, Kunkel KE, Stewart BC, Maycock TK, eds. U.S. Global Change Research Program, Washington, DC, USA, doi:10.7930/NCA5.2023.CH19.

Institute for Regional Forecasting, 2018. Harvey in perspective: The Houston economy and hurricanes past and present. University of Houston, February 6, 2018. https://www.bauer.uh.edu/centers/irf/houston-updates-feb18.php (accessed June 25, 2024).

Intergovernmental Panel on Climate Change, 2022. Climate change 2022: Impacts, adaptation and vulnerability. In Contribution of Working Group II to the Sixth Assessment Report of the Intergovernmental Panel on Climate Change, Pörtner H-O, Roberts DC, Tignor M, Poloczanska ES, Mintenbeck K, Alegría A, Craig M, Langsdorf S, Löschke S, Möller V, Okem A, Rama B, eds. Cambridge University Press, Cambridge, doi:10.1017/9781009325844.

International Monetary Fund, 2023. Global price of cotton. Retrieved from FRED, Federal Reserve Bank of St. Louis. https://fred.stlouisfed.org/series/PCOTTINDUSDM (accessed June 25, 2024).

Jansen D, Ghosh Sinha, S, 2024. Top US states by real GDP and per capita real GDP. Texas A&M University Private Enterprise Research Center. https://perc.tamu.edu/blog/2024/02/gdp-vs-per-capita-gdp.html (accessed June 25, 2024).

Jasechko S, Perrone D, Seybold H, Fan Y, Kircher J, 2020. Groundwater level observations in 250,000 coastal US wells reveal scope of potential seawater intrusion. Nature Communications, 11, doi:10.1038/s41467-020-17038-2.

Jay AK, Crimmins AR, Avery CW, Dahl TA, Dodder RS, Hamlington BD, Lustig A, Marvel K, Méndez-Lazaro PA, Osler MS, Terando A, Weeks ES, Zycherman A, 2023. Ch. 1. Overview: Understanding risks, impacts, and responses. In Fifth National Climate Assessment, Crimmins AR, Avery CW, Easterling DR, Kunkel KE, Stewart BC, Maycock, TK, eds. U.S. Global Change Research Program, Washington, DC, USA, doi:10.7930/NCA5.2023.CH1.

Jayashankar A, Kalkunte P, Kumar A, Orrenius P, 2023. Hotter summer days heat up Texans but chill the state economy. Federal Reserve Bank of Dallas, Southwest Economy, October 18, 2023. https://www.dallasfed.org/research/swe/2023/swe2309 (accessed June 25, 2024).

Kaplan RS, 2019. A brief discussion regarding the impact of climate change on economic conditions in the Eleventh District. Federal Reserve Bank of Dallas, June 27, 2019. https://www.dallasfed.org/research/economics/2019/0627b (accessed June 25, 2024).

Karp N, Nava M, Nash-Stacey B, Blazheski F, 2017. Economic impact of hurricanes Harvey and Irma. Banco Bilbao Vizcaya Argentaria's (BBVA) Research. US Economic Watch, September, 18, 2017. https://www.bbvaresearch.com/wp-content/uploads/2017/09/170918_EconomicImpact_Harvey_Irma-1.pdf (accessed June 25, 2024).

Kaskey J, 2017. Harvey disrupts more than one third of US chemical production. Bloomberg Markets, August 28, 2017. https://www.bloomberg.com/news/articles/2017-08-28/harvey-disrupts-more-than-one-third-of-u-s-chemical-production (accessed June 25, 2024).

Kinniburgh F, Greer Simonton M, Allouch C, 2015. Come heat and high water: Climate risk in the southeastern US and Texas. Risky Business Project Report. https://www.cakex.org/documents/come-heat-and-high-water-climate-risk-southeastern-us-and-texas (accessed June 25, 2024).

Kerr B, 2023. High tech industry report. Austin Chamber of Commerce. https://www.austinchamber.com/blog/07-05-2023-high-tech-industry (accessed June 25, 2024).

Kerr E, Leigh R, Torres L, 2024. Texas high tech shakes off post-pandemic slump, readies new growth path. Federal Reserve Bank of Dallas, Southwest Economy, May 3, 2024. https://www.dallasfed.org/research/swe/2024/swe2406 (accessed June 25, 2024).

Kramer J, Simnitt S, Calvin L, 2021. Fruit and tree nuts outlook: September 2021, FTS-373. U.S. Department of Agriculture (USDA), Economic Research Service, September 29, 2021. https://www.ers.usda.gov/webdocs/outlooks/102267/fts-373.pdf?v=8594 (accessed June 25, 2024).

Leining L, 2024. Texas has the most billion-dollar disasters nationwide. Texas 2036. https://texas2036.org/posts/texas-has-the-most-billion-dollar-disasters-nationwide/#:~:text=Texas%20leads%20the%20nation%20in,weather%20and%20climate%20disaster%20events (accessed June 25, 2024).

Letourneau M, 2022. Combating inflation: What's behind high energy prices and what to do about it. United States Chamber of Commerce, May 5, 2022. https://www.uschamber.com/energy/combating-inflation-whats-behind-high-energy-prices-and-what-to-do-about-it (accessed June 25, 2024).

Licker R, Dahl K, Abatzoglou JT, 2022. Quantifying the impact of future extreme heat on the outdoor work sector in the United States. Elementa: Science of the Anthropocene, 10 (1):00048, doi:10.1525/elementa.2021.00048.

Lowe M, Marx R, 2020. Climate change-fueled weather disasters: costs to state and local economies. Datu Research. Prepared for Environmental Defense Fund. https://www.edf.org/sites/default/files/content/report-ClimateChange-FueledWeatherDisasters.pdf (accessed June 25, 2024).

Makhlouf B, 2023. What's up with the weather (and what might it mean for inflation)? Central Bank of Ireland Blog, September 6, 2023. www.centralbank.ie/news/article/blog-whats-up-with-the-weather-and-what-might-it-mean-for-inflation (accessed June 25, 2024).

McPherson RA, Fay PA, Alvarez SG, Bertrand D, Broadbent TL, Bruno T, Fares A, McCullough B, Moore GW, Moorhead B, Patiño L, Petersen A, Smith NG, Steiner JL, Taylor A, Warziniack T, 2023. Ch. 26. Southern Great plains. In Fifth National Climate Assessment, Crimmins AR, Avery CW, Easterling DR, Kunkel KE, Stewart BC, Maycock TK, eds. U.S. Global Change Research Program, Washington, DC, USA, doi:10.7930/NCA5.2023.CH26.

Minton L, Alvarado A, 2022. Agriculture industry grows Texas: Farms, ranches cultivate the state's economy. Texas Comptroller Fiscal Notes, October 2022. https://comptroller.texas.gov/economy/fiscal-notes/archive/2022/oct/agriculture.php#:~:text=According%20to%20the%20U.S.%20Department,includes%20them%20in%20the%20count (accessed June 25, 2024).

Munch D, 2023. New estimates reveal major 2022 weather disasters caused over $21 billion in crop losses. The American Farm Bureau Federation, Market Intel, March 3, 2023. https://www.fb.org/market-intel/new-estimates-reveal-major-2022-weather-disasters-caused-over-21-billion-in-crop-losses#:~:text=The%20National%20Oceanic%20and%20Atmospheric,billion%20and%202005's%20%24244%20billion (accessed June 25, 2024).

Munch D, 2024. Major disasters and severe weather caused over $21 billion in crop losses in 2023. The American Farm Bureau Federation, Market Intel, February 29, 2024. https://www.fb.org/market-intel/major-disasters-and-severe-weather-caused-over-21-billion-in-crop-losses-in-2023#:~:text=The%20assessment%20puts%20total%20crop,programs%20as%20of%20February%202024 (accessed June 25, 2024).

National Oceanic and Atmospheric Association National Centers for Environmental Information, 2024. U.S. Billion-Dollar Weather and Climate Disasters (2024). https://www.ncei.noaa.gov/access/billions/doi:10.25921/stkw-7w73 (accessed June 30, 2024).

Nielsen–Gammon J, Holman S, Buley A, Jorgensen S, Escobedo J, Ott C, Dedrick J, Van Fleet A, 2024. Assessment of historic and future trends of extreme weather in Texas, 1900–2036. Office of the Texas State Climatologist Report. Texas A&M University. https://texas2036.org/wp-content/uploads/2023/06/2024_ClimateReport.pdf (accessed June 24, 2024).

Nuclear Energy Institute, 2022. State electricity generation fuel shares. https://www.nei.org/resources/statistics/state-electricity-generation-fuel-shares (accessed June 24, 2024).

Office of the Governor of California, 2024. California remains the world's 5th largest economy. Governor Gavin Newsom Newsroom, April 16, 2024. https://www.gov.ca.gov/2024/04/16/california-remains-the-worlds-5th-largest-economy/ (accessed June 24, 2024).

Office of the Texas Governor Gregg Abbott, 2023. Texas again leads nation with most Fortune 500 companies. News Release, June 5, 2023. https://gov.texas.gov/news/post/texas-again-leads-nation-with-most-fortune-500-headquarters (accessed June 24, 2024).

Office of the United States Trade Representative, 2023. Texas. https://ustr.gov/map/state-benefits/tx#:~:text=The%20state's%20largest%20market%20was,%2C%20South%20(%2421.1%20billion) (accessed June 24, 2024).

Powell E, 2023. Weather disasters in 2023 cost Texas billions in damage. Texas Farm Bureau, Texas Agriculture Daily. https://texasfarmbureau.org/weather-disasters-in-2023-cost-texas-billions-in-damage/#:~:text=Texas%20Agriculture%20Daily&text=The%20report%20shows%20a%20total,drought%20caused%20the%20most%20damage (accessed June 24, 2024).

Price S, 2020. The temperature's rising: Preparing for the health effects of climate change. Texas Medical Association. https://www.texmed.org/Template.aspx?id=53324 (accessed June 24, 2024).

Reader G, 2023. Explainer: Are the cracks in your road from extreme heat? KXAN Austin, June 23, 2023. https://www.kxan.com/news/local/austin/explainer-are-the-cracks-in-your-road-from-extreme-heat/ (accessed June 24, 2024).

Reinsch A, Thibault D, Kerstens E, 2023. Climate change and US agricultural exports: How future weather patterns will impact US competitiveness. Center for Strategic and International Studies. https://www.csis.org/analysis/climate-change-and-us-agricultural-exports (accessed June 24, 2024).

Robbins J, 2022. Global 'stilling': is climate change slowing down the wind? Yale Environment 360, Yale School of the Environment, September 13, 2022. https://e360.yale.edu/features/global-stilling-is-climate-change-slowing-the-worlds-wind#:~:text=As%20carbon%20dioxide%20levels%20rise%20and%20the,which%20drives%20much%20of%20the%20world's%20climate (accessed June 24, 2024).

Russell A, 2022. Shrinking cattle herd could equal high calf, beef prices. Texas Crop and Weather Report, Agrilife Today, Texas A&M University, July 26, 2022. https://agrilifetoday.tamu.edu/2022/07/26/shrinking-cattle-herd-could-equal-high-calf-beef-prices/ (accessed June 24, 2024).

Sawadgo W, 2023. 2023 peanut market outlook. Southern Ag Today 3 (8.1), https://southernagtoday.org/2023/02/20/2023-peanut-market-outlook/ (accessed June 24, 2025).

Schattenberg P, 2021. Agricultural losses from winter storm exceed $600 million. Texas A&M AgriLife Communications. https://today.tamu.edu/2021/03/02/agricultural-losses-from-winter-storm-exceed-600-million/ (accessed June 24, 2024).

Semenova A, 2023. Rising temperature and rising prices: The inflationary impacts of climate change and the need for degrowth-based solutions to the ecological crisis. Globalizations, 21(1): 103–120, doi:10.1080/14747731.2023.2222482.

Skinner LR, Cha JM, Moskowitz H, Phillips M, 2021. Combatting climate change, reversing inequality: A climate jobs program for Texas. ILR Worker Institute, Cornell University. https://ecommons.cornell.edu/items/3f37053d-17a3-47f6-aa47-8629d5f70029 (accessed June 24, 2024).

Smith B, 2024. 2023: A historic year of U.S. billion-dollar weather and climate disasters. NOAA Climate.gov Beyond the Data Blog, January 8, 2024. https://www.climate.gov/news-features/blogs/beyond-data/2023-historic-year-us-billion-dollar-weather-and-climate-disasters#:~:text=There%20were%2028%20weather%20and,of%20at%20least%20%2492.9%20billion (accessed June 24, 2024).

So K, Hardin S, 2021. Extreme weather cost U.S. taxpayers $99 billion last Year, and it is getting worse. Center For American Progress Report. https://www.americanprogress.org/article/extreme-weather-cost-u-s-taxpayers-99-billion-last-year-getting-worse/ (accessed September 1, 2021).

Strauss B, Tebaldi C, Kulp S, Cutter C, Emrich C, Rizza D, Yawitz D, 2014. Surging sea: A vulnerability assessment with projections for sea level rise and coastal flood risk. Climate Central Research Report. https://sealevel.climatecentral.org/uploads/ssrf/DC-Report.pdf (accessed June 24, 2024).

Texas 2036, 2021. Texas' future depends on extreme weather preparedness, new studies show. https://texas2036.org/posts/texas-future-depends-on-extreme-weather-preparedness-new-studies-show/ (accessed June 24, 2024).

Texas Comptroller, 2018. A storm to remember: Hurricane Harvey and the Texas economy. Texas Comptroller Fiscal Notes, February 2018. https://comptroller.texas.gov/economy/fiscal-notes/archive/2018/special-edition/ (accessed June 24, 2024).

Texas Comptroller, 2021. Supply chains: Food supply chain overview. https://comptroller.texas.gov/economy/economic-data/supply-chain/2021/food.php#:~:text=All%20supply%20chains%20are%20vulnerable,increased%20this%20essential%20sector's%20vulnerability (accessed June 24, 2024).

Texas Comptroller, 2023. Texas comptroller energy tour: Solar energy overview. https://comptroller.texas.gov/economy/economic-data/energy/2023/solar.php (accessed June 24, 2024).

Texas Comptroller, 2024. Statewide report: 2024 regional report. https://comptroller.texas.gov/economy/economic-data/regions/2024/statewide.php (accessed June 24, 2024).

Texas Department of Agriculture, 2022. Top Texas commodities. https://texasagriculture.gov/About/Texas-Ag-Stats (accessed June 24, 2024).

Texas Department of Agriculture, 2024. Opinion: A call to action: Weathering another deep freeze. Commissioner Sid Miller Newsroom. https://texasagriculture.gov/News-Events/Article/9837/OPINION-A-CALL-TO-ACTION-WEATHERING-ANOTHER-DEEP-FREEZE (accessed June 24, 2024).

Texas Division of Emergency Management, 2017. Hurricane Harvey DR – 4332. https://www.tdem.texas.gov/disasters/hurricane-harvey#:~:text=Harvey%20made%20landfall%20three%20times,the%20U.S.%20mainland%20since%201900 (accessed June 24, 2024).

Texas Economic Development Corporation, 2022. Texas energy sector goes beyond oil and gas. CEO Blog, April 4, 2022. https://businessintexas.com/ceo-blog/texass-energy-sector-goes-beyond-oil-and-gas/ (accessed June 24, 2024).

Texas Economic Development Corporation, 2023. Texas' economic strength and growth. https://businessintexas.com/why-texas/economic-strength/ (accessed June 24, 2024).

Texas Economic Development and Tourism, 2023. Texas manufacturing industry. https://gov.texas.gov/uploads/files/business/Manufacturing_Snapshot.pdf (accessed June 24, 2024).

Texas General Land Office, 2009. Hurricanes Ike & Dolly. https://recovery.texas.gov/action-plans/hurricane-ike-and-dolly/index.html#:~:text=Hurricanes%20Ike%20%26%20Dolly,devastating%20hurricanes%20in%20U.S.%20history (accessed June 24, 2024).

Texas Water Development Board, 2022b. 2022 state water plan: Water for Texas. https://www.twdb.texas.gov/waterplanning/swp/2022/index.asp (accessed June 24, 2024).

Texas Water Development Board, 2022a. Texas water use estimates summary for 2020. https://www.twdb.texas.gov/waterplanning/waterusesurvey/estimates/doc/2020TexasWaterUseEstimatesSummary.pdf?d=29438.89999999851 (accessed June 24, 2024).

Texas Workforce Commission, 2024. Texas labor market reached historic highs in 2023. Press Release, January 19, 2024. https://www.twc.texas.gov/news/texas-labor-market-reached-historic-highs-2023 (accessed June 25, 2024).

Thompson J, 2021. Texas winter deep freeze broke refining, petrochemical supply chains. Federal Reserve Bank of Dallas, Southwest Economy, Second Quarter 2021. https://www.dallasfed.org/research/swe/2021/swe2102/swe2102c (accessed June 25, 2024).

Thompson J, 2022. Blame costlier natural gas, rising fees for increasing Texas power bills. Federal Reserve Bank of Dallas, Southwest Economy, Fourth Quarter 2022. https://www.dallasfed.org/research/swe/2022/swe2204/swe2204e (accessed June 25, 2024).

Thompson J, Hines K, 2023. State output remains distinctly Texan, while job mix increasingly resembles the U.S. Federal Reserve Bank of Dallas, Southwest Economy, November 17, 2023. https://www.dallasfed.org/research/swe/2023/swe2313 (accessed June 25, 2024).

Torres L, 2022. Looking back: Harris County five years after hurricane Harvey hit. Federal Reserve Bank of Dallas, Southwest Economy, Second Quarter 2022. https://www.dallasfed.org/research/swe/2022/swe2202/swe2202a (accessed June 25, 2024).

TRIP, 2023. Value of freight shipped in Texas highest in US, with large truck travel and fatal crashes involving large trucks among the highest in nation. News Release, December 5, 2023. https://tripnet.org/reports/freight-texas-news-release-12-05-2023/#:~:text=In%202022%20Texas'%20freight%20system,and%2052%20percent%20by%20weight (accessed June 25, 2024).

United States Department of Agriculture, 2011. Crop production. https://peanutsusa.com/phocadownload/CropProduction/Peanut%20Crop%20Production%20-%20August%202011.pdf (accessed June 25, 2024).

United States Department of Agriculture, 2022. Cotton sector at a Glance. USDA Economic Research Service. https://www.ers.usda.gov/topics/crops/cotton-and-wool/cotton-sector-at-a-glance/#:~:text=Among%20the%20U.S.%20States%2C%20Texas,cotton%20production%20in%20recent%20years (accessed June 25, 2024).

United States Department of Energy, 2023. Texas: US energy and employment report–2023. https://www.energy.gov/sites/default/files/2023-06/USEER23-TX-v2.pdf (accessed June 25, 2024).

United States Department of the Treasury, 2023. The impact of climate change on American household finances. https://home.treasury.gov/system/files/136/Climate_Change_Household_Finances.pdf (accessed June 25, 2024).

United States Energy Information Administration, 2021. Texas natural gas production fell by almost half during recent cold snap. Energy Today, February 25, 2021. https://www.eia.gov/todayinenergy/detail.php?id=46896 (accessed June 25, 2024).

United States Energy Information Administration, 2023b. Texas state energy profile. https://www.eia.gov/state/print.php?sid=TX (accessed June 25, 2024).

United States Energy Information Administration, 2023a. As Texas wind and solar capacity increase, energy curtailments are also likely to rise. Today in Energy, July 13, 2023. https://www.eia.gov/todayinenergy/detail.php?id=57100 (accessed June 25, 2024).

United States Energy Information Administration, 2024. Winter storms have disrupted U.S. natural gas production. Today in Energy, Marh 13, 2024. https://www.eia.gov/todayinenergy/detail.php?id=61563#:~:text=Over%20the%20last%20four%20winters,from%20S%26P%20Global%20Commodity%20Insights (accessed June 25, 2024).

United States Environmental Protection Agency, 2016. What climate change means for Texas. EPA-430-F-16-045, August 16, 2016. https://www.epa.gov/sites/default/files/2016-09/documents/climate-change-tx.pdf (accessed June 25, 2024).

U.S. Global Change Research Program, 2018. Impacts, Risks, and Adaptation in the United States: Fourth National Climate Assessment, Volume II, Reidmiller DR, Avery CW, Easterling DR, Kunkel KE, Lewis KLM, Maycock TK, Stewart BC, eds. Washington, DC, USA, doi:10.7930/NCA4.2018.

United States Government Accountability Office, 2015. Budgeting for disasters: approaches to budgeting for disasters in selected states. A Report to Congressional Requesters, March 2015. https://www.gao.gov/assets/gao-15-424.pdf (accessed June 25, 2024).

Walewski J, Prozzi J, Polunsky S, 2017. How vulnerable is Texas' freight infrastructure to extreme weather events? Final Report No. PRC 16-62 F. Transportation Policy Research Center, Texas A&M Transportation Institute. https://static.tti.tamu.edu/tti.tamu.edu/documents/PRC-16-62-F.pdf (accessed June 25, 2024).

Webber M, 2021. What happened in Texas is only the beginning. Opinion, The Hill, March 1, 2021. https://thehill.com/opinion/energy-environment/540941-what-happened-in-texas-is-only-the-beginning/ (accessed June 25, 2024).

Zamuda CD, Bilello DE, Carmack J, Davis X, Efroymson R, Goff K, Karimjee A, Loughlin D, Upchurch S, Voisin N, 2023. Ch. 5. Energy supply, delivery, and demand. In Fifth National Climate Assessment, Crimmins AR, Avery CW, Easterling DR, Kunkel KE, Stewart BC, Maycock TK, eds. U.S. Global Change Research Program, Washington, DC, USA. doi:10.7930/NCA5.2023.CH5

Section IV

Adaptation and Mitigation

Considerations for Environmental Management

IV.1 Future Water Use in Texas

Stephen Ferencz

IV.1.1 INTRODUCTION

Future water use is subject to numerous uncertainties about how the coupled natural and human systems will evolve (Reed et al., 2022). Here, the natural system refers to the climate and environment, while the human system refers to all human activities and the human-built and managed environment. Sources of uncertainty in the natural system include future climate conditions such as precipitation, temperature, and the frequency, duration, and intensity of drought. Uncertainties in the human system include future population, technological change, and evolution of the regional, national, and global economies. For example, the future population of Texas (Figure IV.1.1) will affect municipal water use, which is currently the second largest water use sector behind agriculture, with no other sector (energy, manufacturing, mining, or livestock) close in magnitude. Therefore, future population will meaningfully affect where and how much water is used and could influence its availability for other sectors.

State water plans (SWPs) for Texas have consistently assumed high population futures where current population growth rates are maintained, while the Texas Demographic Center (TDC) projections report a range of future populations under different migration rates, with the low versus high growth scenarios differing by 7.3 million Texans by 2060. The third set of population projections shown in Figure IV.1.1 are downscaled shared socioeconomic pathways (SSPs) projections for Texas (Zoraghein et al., 2020). The five SSPs shown represent plausible global and country-level trends in the evolution of social and natural systems over the 21st century and demonstrate how dramatically different population could be given different sets of assumptions about the evolution of the coupled human-natural system (O'Neill et al., 2014). The wide range of future populations in Figure IV.1.1 represent very different water use futures in Texas, not only for municipal use but also for other sectors, such as cooling water use for the electricity needs of the population.

In addition to population, there are uncertainties in other factors such as future energy needs and technologies (Liu et al., 2015), cropping decisions (influenced by dietary preferences, population, and available land and water resources) (Calvin et al., 2019; Davis et al., 2016), and climate (Kundzewicz et al., 2018) whose combined effects present a wide range of demands for water resources and implications for water availability. While population is an intuitive example of how an uncertain factor could influence future water use, changes in how water systems are managed (i.e., institutional factors) could be equally important. For instance, cities may prioritize water reuse in response to future water security issues (Lee and Jepson, 2020), implement water conservation policies (Graham et al., 2021), or pursue large infrastructure projects (Siddik et al., 2023) to move water from where it is plentiful to where it is scarce (in Texas, this could be interbasin transfers from eastern to western Texas).

Uncertainties about the many factors influencing water use make it difficult to accurately predict future water use, especially over longer time horizons. Major water utilities have been wildly inaccurate at predicting future demand, in some cases grossly overestimating demand (Heberger et al., 2016; Richter et al., 2020). This could partly be explained by an incentive for utilities to be biased towards conservatively overestimating supply; there is a much lower downside than underestimating supply and water supply projects can have long timelines, so it is difficult to be "reactive" once a water crisis emerges. Other explanations could be uncertainties in population and sectoral growth, emerging technologies, and unforeseen pressures driving adaptation that aren't accounted for at the time the projection was made (such as adapting to more severe droughts). Given the numerous

DOI: 10.1201/9781003406600-20

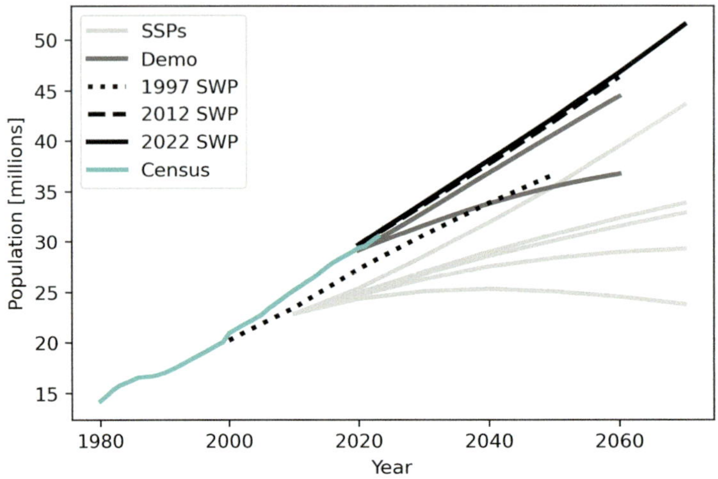

FIGURE IV.1.1 Historical (cyan) and several population projections for Texas show the wide range of uncertainty in future population. Census = U.S. Census data from United States Census Bureau (2025); SWP = State water plan population projections from TWDB (1997), TWDB (2012), and TWDB (2022), demo = Texas Demographic Center population projections under 1 and 0.5 multipliers for 2010–2020 migration rates from Texas Demographic Center (2022), SSPs = Shared socioeconomic pathway projections from Zoraghein et al. (2020).

uncertainties surrounding future water use, the objective of this chapter is not to assign hard numbers or likelihoods to future water use. Instead, the goal is to build an understanding of how water use, including specific water use sectors (i.e., municipal, agriculture) and sources of supply (i.e., groundwater, reuse), may evolve in Texas given current understanding of trends in sectoral water use, near to long-term efforts outlined by state water planning agencies, and possible future climate conditions.

IV.1.2 RECENT AND HISTORICAL TRENDS IN WATER USE

In order to understand how water use could evolve, it is instructive, and arguably essential (McManamay et al., 2021), to examine its recent and historical trends. To that end, this section presents both state-level data on estimates of historical water use as well as county-level data to illustrate the geography of water use in Texas. In addition to reporting water use data, a key aspect of this section is to describe some of the underlying drivers of water use that influence where and how water is used across Texas.

Water use data presented here is sourced from the Texas Water Development Board (TWDB). Available data from TWDB (2025a) includes annual, state-level estimates from 1979–2022 and annual, county-level estimates from 2000–2022 for six water use sectors: municipal, manufacturing, irrigation, steam-electric power, mining, and livestock. The sectoral water use data includes estimated surface water and groundwater for each sector.

The following are brief descriptions of what water use activities each sector captures in the TWDB data:

- Municipal: Water use (including irrigation) at residential (single family and multifamily), commercial (e.g., hotels, restaurants, office buildings), and institutional locations (e.g., schools, hospitals, nursing homes).
- Manufacturing: Large, self-supplied industrial facilities (e.g., chemical, refinery, food processing).

- Irrigation: Water applied by an irrigation system to sustain plant growth for agricultural purposes. TWDB also includes self-supplied water for golf courses in irrigation, although small compared to crops.
- Steam-electric power: Volume of consumed water during the cooling process (not returned to surface water) for power plants that generate electricity with steam-driven turbine generators. These include fossil fuel, nuclear, and biomass plants.
- Mining: Water used for the extraction of solids (e.g., coal, gravel, etc.), liquids (e.g., oil), or gases (e.g., natural gas). Examples include secondary oil recovery, unconventional oil and gas recovery (hydraulic fracturing), milling of mined material, or quarrying.
- Livestock: Water used for livestock feedlots, dairy operations, and other on-farm needs. Includes dairy cows, beef cattle, sheep and lambs, goats, hogs, horses, and poultry.

The county-level data shows that there is large spatial variability in water use across Texas, with specific sectors showing distinct hot spots of high use (Figure IV.1.2). For example, municipal use,

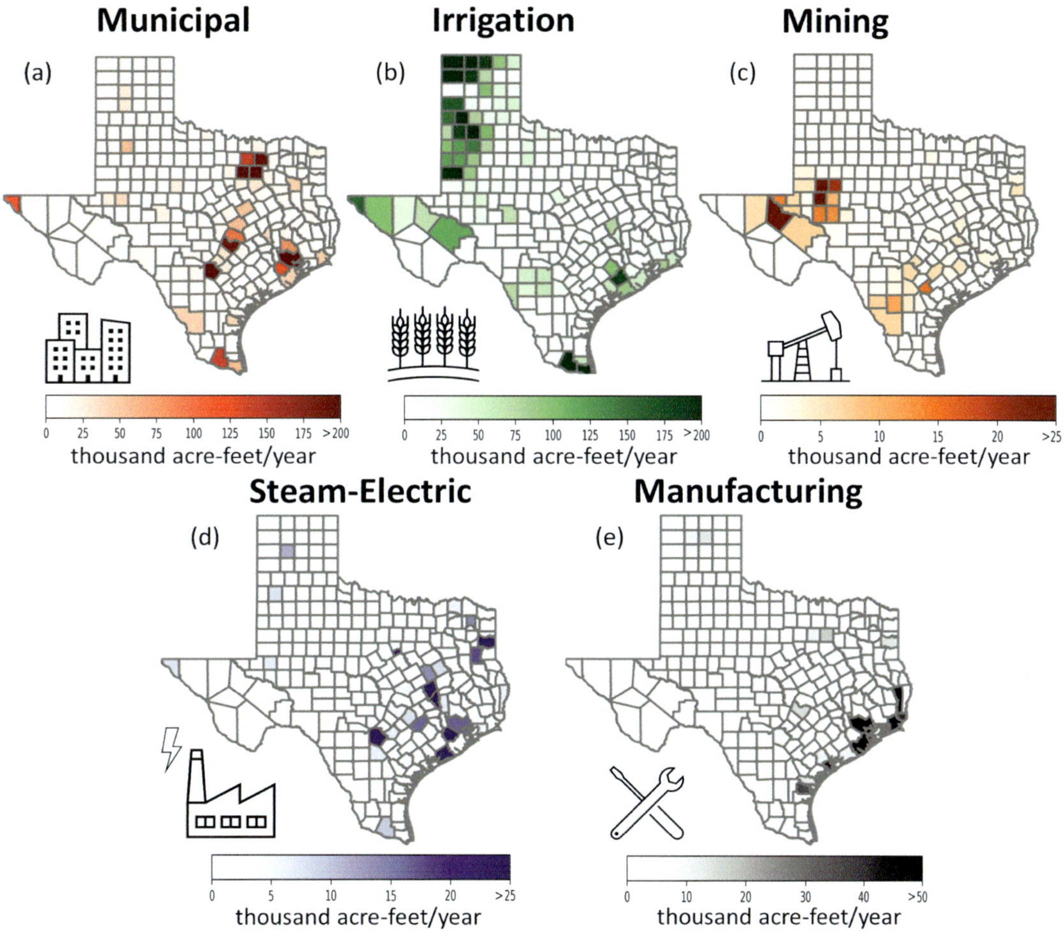

FIGURE IV.1.2 Recent average annual water use in acre-feet (1 acre-foot = 1,233.5 m^3) by county for municipal (a), irrigation (b), mining (c), steam-electric power (d), and manufacturing sectors (e). Livestock was omitted due to its relatively small county-level magnitude compared to other sectors. Values represent averages from 2017 to 2021.

Data from TWDB (2025a).

which roughly mirrors population, is concentrated in counties associated with major cities and their metro areas (Figure IV.1.2a), while manufacturing is largest along the gulf coast (Figure IV.1.2e). It should be noted that the magnitudes of the color scales in Figure IV.1.2 are not consistent across sectors. Additionally, the upper limits for some of the color bars were adjusted such that low to moderate values are discernable.

A handful of counties account for a large fraction of municipal use across Texas's 254 counties. Specifically, these counties are associated with Houston, Dallas/Fort Worth, Austin, El Paso, and San Antonio. In fact, only 10 counties account for nearly 60% of all municipal use, and the highest county (Harris) has more municipal water use (660,000 acre-feet; 1 acre-foot = 1,233.5 m^3) than the lowest 200 counties combined. Irrigation use is more widely distributed than municipal use, and with a few exceptions, high use occurs in counties with low populations and low municipal use. The most noticeable cluster of high irrigation use is in north Texas and is associated with groundwater-sourced irrigation from the Southern High Plains (or Ogallala) Aquifer.

The largest region of mining-related use in west Texas is associated with oil and gas development or production in the Permian Basin (Figure IV.1.2c). Oil and gas development and productions (and associated water use) accelerated in the mid to late 2000s with the emergence of hydrofracturing or "fracking" technologies (Nicot and Scanlon, 2012). While smaller in magnitude, oil and gas operations in the Eagle Ford Shale are also the primary contributor to mining water use in the southern part of Texas (Daher et al., 2019). Fracking can use non-potable water and therefore does not necessarily present a conflict with water availability for other sectoral uses; however, pollution from fracking could potentially impact potable water resources in proximity to fracking operations (Jackson et al., 2014).

Steam-electric water use is predominantly located in the eastern half of the state (Figure IV.1.2d) where surface water resources are much more plentiful, are less drought-prone (i.e., higher supply reliability), and where the majority of the population and manufacturing is located. While there are a small number of counties in the western-half of the state with steam-electric water use, they are much smaller in magnitude. Scanlon et al. (2013a) found that power plants in the western part of the state have switched to lower water demand cooling technologies as an adaptive response to water scarcity concerns. In terms of magnitude, steam-electric water use (often called "thermoelectric" water use) appears small compared to municipal and irrigation. This is because the TWDB quantifies steam-electric water use as the estimated water *consumption* associated with electricity generation – this distinction is not applied to the other sectors where use is equal to withdrawals. Water *withdrawals* for steam-electric power are much larger. The most recent United States Geological Survey (USGS) national water use assessment estimated that steam-electric withdrawals were approximately double municipal withdrawals in Texas (Medalie et al., 2025). However, Scanlon et al. (2013b) points out that while many studies focus on the large water withdrawals associated with steam-electric generation, on average 98% of withdrawals are returned to the source (reservoir, cooling lake, or stream) and are available for subsequent use.

Manufacturing is the most geographically concentrated of all sectors, with a small number of counties in southeastern Texas comprising the vast majority of use (Figure IV.1.2e). The 10 largest manufacturing counties account for almost 80% of all manufacturing use, and the largest county (Harris – also the largest municipal county) is greater than the cumulative amount in 248 of the 254 counties in Texas. Key manufacturing sectors driving water use include chemicals, petroleum refining, paper, and food (Hoffman, 2016). Together these four industries comprised 85% of total manufacturing water use in 2013 with chemical and petroleum refining accounting for over 70% (Hoffman, 2016).

Equally important to where sectoral water use occurs is what water resource is being used. Reuse currently accounts for a small percentage of water use across Texas (TWDB, 2025a) so Figure IV.1.3 focuses on surface water and groundwater. With a few exceptions, areas of high surface water and groundwater use do not overlap (Figure IV.1.3). Areas of large surface water use are concentrated

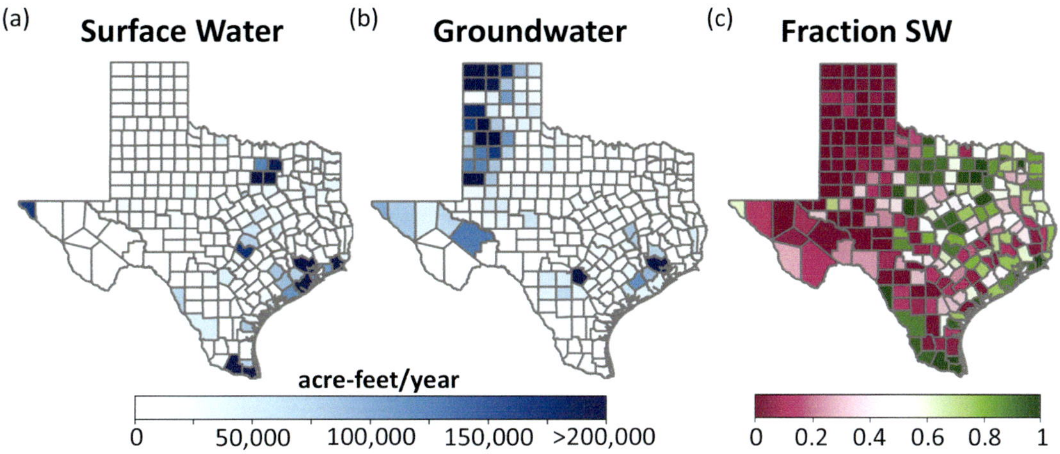

FIGURE IV.1.3 County-level surface water (a), groundwater (b), and percent surface water use (c) averaged over the 2016–2020 period. SW = surface water.

Data from TWBD (2025a).

in the humid, wetter part of the state and also some counties along the Rio Grande River, which defines the southwestern border of Texas (Figure IV.1.3a). In contrast, the majority of high groundwater use counties are in the arid northwestern part of the state and nearly all of this use is associated with agriculture (Figure IV.1.3b). Groundwater and surface water sectoral use data show that manufacturing and steam-electric use the greatest proportion of surface water (both around 80%), municipal uses around 60% surface water, and agriculture (irrigation) only uses 20% surface water (Figure IV.1.4c). Mining is the only sector with a notable trend in water use type, switching from 60% surface water in 2000 to less than 20% surface water in 2020 (Figure IV.1.4c). This trend is due to groundwater being used for oil and gas development in the arid western half of the state with scare surface water resources (Figure IV.1.2c).

Groundwater use in Texas grew enormously during the 1950s and was driven almost entirely by irrigated agriculture. Over the 10-year period between 1945 and 1955, annual use increased fivefold from less than 2 million acre-feet to over 10 million acre-feet (Mace, 2021). This proliferation of groundwater use was enabled by technological progress in well drilling and water pump technology (Green, 1973). Based on available estimates from the USGS and TWDB, groundwater use peaked between 1960 and 1980 with annual extraction during that 20-year period typically larger than 10 million acre-feet and a maximum of over 12 million acre-feet per year (Mace, 2021). Since 2000, groundwater use only exceeded 10-million acre-feet once and averaged 8.7 million acre-feet from 2000 to 2020 (Figure IV.1.4a). While groundwater use has declined in the 21st century, it still remains larger than surface water use, fluctuating between 53% and 63% of combined surface water and groundwater use between 2000 and 2020. Looking at the highest and lowest water use years between 2000 and 2020 shows almost 50% more water was used in the highest year (2011) than in the lowest year (2015). Annual water use can also vary widely from year to year. The best example of this is comparing 2010 versus 2011; 2010 was a normal to wet year, while 2011 was the most severe single drought year in the historical record and affected the entire state. Short-term differences in water use are largely attributable to differences in irrigation water use (both agricultural and municipal) which can be influenced by multiple factors such as weather (irrigation demand and crop feasibility), drought management measures (i.e., water use restrictions), and crop prices.

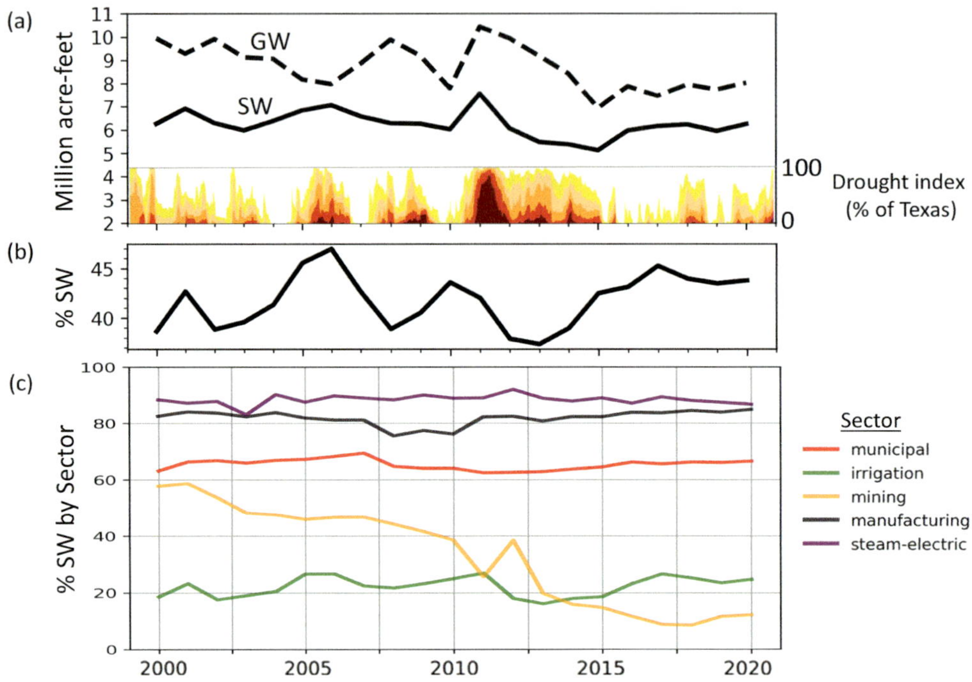

FIGURE IV.1.4 Annual estimated state-level groundwater (GW) and surface water (SW) use and percent of land area experiencing drought conditions (redder = more severe drought conditions) (a), percent surface water use compared to total (groundwater plus surface water) (b), and percent surface water use by sector (c). Note, drought monitor data is sub-monthly resolution while the water use is annual.

Drought index from United States Drought Monitor (2025) and water use from TWDB (2025a).

IV.1.3 FUTURE WATER DEMAND

With a high-level understanding of the geography, magnitude, and trends in historical water use across the major water sectors, this section pivots to examining state water planning projections of future sectoral demand. This section also offers a discussion of factors influencing future sectoral use.

The current water planning process in Texas was mandated by law in 1997 and was created in response to severe drought conditions in 1995 and 1996 (Wurbs, 2015). State water planning dates to as early as 1961, but a top-down approach in earlier plans caused them to be misaligned with actions taken the local and regional levels (Bruun, 2017). Under the current process, state water plans are updated every 5 years, and the most recent plan was published in 2022 (TWDB, 2022). The planning process is coordinated through 16 regional planning groups across Texas. Each planning group develops their own region-specific water plan that informs infrastructure investments needed to meet anticipated water needs over a 50-year planning horizon. The regional process takes a bottom-up approach with each regional water planning group engaging with water user groups within their region. Some examples of water user groups are cities, agricultural districts, or energy utilities. The five-year planning cycle is designed to help water planning in Texas respond to changes in water availability, technology, and sectoral demands (Bruun, 2017; Richter et al., 2024). For example, severe drought during the period from 2008 to 2015 resulted in large changes in estimated water needs and shifts in supply strategies between the 2007, 2012, and 2017 water plans for planning regions within the Colorado River Basin of Texas (Ferencz et al., 2024a).

Demand estimates are a core component of the water planning process in Texas. Future water supply needs are based on evaluating projected demands under water supply scenarios defined by region-specific "drought of record" conditions; drought of record conditions represent the lowest historical water availability conditions experienced for a given planning region. For some regions, the drought of record is the 1950–1957 drought, while for others the recent 2008–2015 drought established a new "worst case" scenario for planning. A potential blind spot or shortcoming of the drought of record approach, noted by Nielsen-Gammon et al. (2020) and Richter et al. (2024), is it overlooks the possibility of more severe drought occurring in the future and doesn't factor in potential changes in water supply due to climate change. However, the drought of record approach does at least evaluate water needs under the most extreme historical drought conditions rather than average conditions.

Demand projections across 25 years of state water plans (1997 to 2022) show substantial differences in anticipated future demand (Figure IV.1.5). Compared to population growth assumed in the state water plans (Figure IV.1.1), projected total demand (Figure IV.1.5a) increases by a comparatively smaller magnitude. For example, in the 2022 SWP, population is projected to increase by 73% by 2070 (compared to 2020) but total demand is only projected to increase by ~9%. This is largely due to projected declines in irrigation demand offsetting some of the assumed growth in municipal (Figure IV.1.5b) and other sectoral demands (Figure IV.1.5d).

Despite substantial differences in magnitude of projected sectoral demand, two consistent trends across all water plans are the anticipated increase in municipal demand (Figure IV.1.5b) and the decrease in irrigation demand (Figure IV.1.5c). Sustained municipal demand growth reflects the assumption of continued long-term growth in population in all plans (Figure IV.1.5). Projected municipal demand in the 1997 plan is considerably less than in the more recent plans due to lower assumed population growth (Figure IV.1.1). Although the 2012 and 2022 plans assume almost identical population growth (Figure IV.1.1), the 2022 plan has lower municipal demand due to revisions in per capita demand informed by increases in municipal water use efficiency. The downward trend

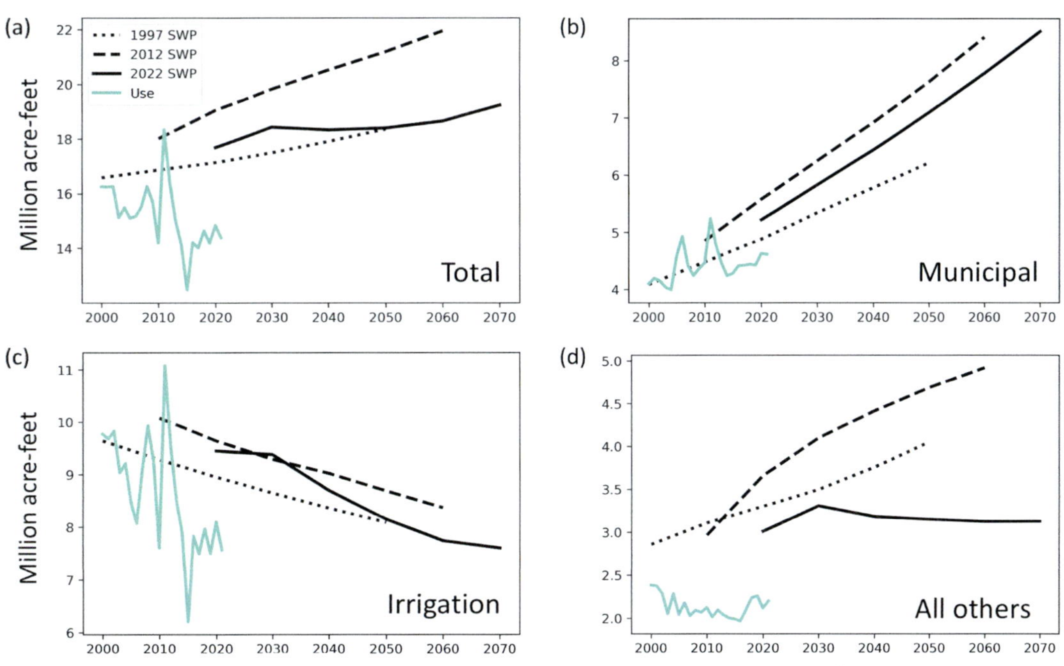

FIGURE IV.1.5 Historical and projected annual total (a), municipal (b), irrigation (c), and water use for all other sectors combined (d). Use = historical use from TWDB (2025a), SWP = state water plan projections from TWDB (1997), TWDB (2012), and TWDB (2022).

in irrigation demand is largely due to assumed reductions in groundwater availability for growing crops in the Southern High Plains Aquifer (region of high irrigation water use in Figure IV.1.2b) from aquifer depletion (TWDB, 2022).

While future municipal and irrigation demands are highly uncertain, both state water planning projections and historical estimated use are consistent in showing a trend of water use shifting from growing crops to meeting the water needs of growing urban areas. However, it should be noted that much of this shift is not a zero-sum transfer of water from agriculture to urban use. As previously mentioned, declining irrigation use is largely expected to occur in the highly agrarian, sparsely populated Southern High Plains Aquifer region in northern Texas, while the majority of municipal growth is expected to occur in the existing population centers and surrounding counties that source their water supplies from different aquifers or whose surface water sources are in entirely different basins. If the trends in the projections hold, the municipal sector could become the largest water use sector in the state – this occurs in 2070 in the 2022 SWP projections (Figure IV.1.5).

Projections for the combined demands of the steam-electric, manufacturing, mining, and livestock sectors vary widely between water plans (Figure IV.1.5d). The most recent water plan projects only a small increase in demand from these sectors over the next 50 years (comparing 2020 to 2070). Reductions in manufacturing and steam-electric demand account for most of the decline in the "all other" sectoral demand in the 2022 plan compared to the earlier plans.

Future manufacturing demands are highly uncertain due to the unpredictable nature of the domestic and global economy 20 or 50 years from now. The 2022 plan reflects the assumption that current rates of manufacturing water demand will continue into the future, while earlier plans were assuming sizeable growth in the manufacturing sector. In reality, water planners have little actionable information about how technological advancements and shifts in domestic and global product demands will influence manufacturing water demands in the coming decades.

The reduction in projected steam-electric demand is due to two factors related to changes in electricity generation in Texas: (1) the conversion of coal-fired to natural-gas fired power which started in the early to mid-2000s (Sanders, 2015; Scanlon et al., 2013a) and (2) the enormous growth of wind and solar energy (Chang and Starcher, 2019). The transition to natural gas from coal reduces water demand due to higher water use efficiency for natural gas plants (Macknick et al., 2012) and wind and solar have no water requirements. Collectively, these two factors suggest that Texas could meet the energy demands of a growing population and economy without large increases in water demand. However, growth in new sectors, like computing data centers could require additional stream-electric supply and increase direct and indirect water demands (Siddik et al., 2021).

The estimated historical use data shows that water use can be very different from assumed demands (Figure IV.1.5). Demand projections can be thought of as more of a moving average of expected water use over longer time scales, while actual year-to-year use can have large annual fluctuations. Importantly, demand projections typically rely on assumptions such as future irrigated acres or population and associated water use factors. Irrigated acreage can fluctuate year-to-year based on market conditions and climate, while population changes more slowly, and water use factors can vary based on short term water management decisions and long-term efficiency improvements. Much of the annual fluctuations in the historical data are due to drought. For example, the large increase in irrigation and municipal use in 2011 is due to the record drought increases water use to compensate for exceptionally dry and hot conditions. While short-term drought can increase water use, prolonged drought can reduce water use due to water use restrictions and adaptation to lower water availability, which is seen in the lower water use from 2012 to 2015 as the drought persisted.

Not shown in Figure IV.1.5 are the sub-annual demand patterns for each sector. While monthly use data are not available, there is large seasonal variation in water use in the municipal and irrigation sectors, primarily driven by increased vegetation water demand for crops and landscapes. Texas's climate is generally drier during the summer, with average monthly precipitation having seasonal highs during the spring and fall (Lee et al., 2017). Efforts to improve irrigation efficiency could reduce the sub-annual variability of water use (Ferencz et al., 2024b). The natural water supply from precipitation and surface water are often lower during the hot summer months when irrigation water demand is high, and reservoirs and groundwater are crucial to providing supply during periods when natural precipitation and streamflow cannot meet sectoral demand.

IV.1.4 CLIMATE IMPACTS ON WATER RESOURCES AND WATER USE

Drier, wetter, or more unpredictable (i.e., switching between dry and wet extremes) future climate conditions could alter water use sources (e.g., surface versus groundwater) and sectoral demands (e.g., irrigated agriculture). Natural internal climate variability (no global warming influence) can produce extended periods of either significantly wetter or significantly drier conditions than experienced in the roughly 100-year recent historical record often used for water planning. Due to its major implications for Western United States (U.S.) water resources, streamflow in the Colorado River Basin (the one encompassing Colorado, Utah, Wyoming, New Mexico, Arizona, Nevada, and California, not the river basin with the same name in Texas) is often cited as an example of how short-term historical records can obscure the full range of possible climate conditions, which can have significant implications for water availability extremes assumed in water planning and management scenarios (Gold et al., 2024; Gupta et al., 2023). In the case of the Colorado River Basin of the Western U.S., the extended historical streamflow record reconstructed using tree ring data shows that much longer and more severe droughts have occurred than observed in the modern historical record (Ault et al., 2016; Woodhouse et al., 2006). These findings suggest that, even without climate change, there is the possibility of more frequent, longer, and more severe drought than the historical streamflow records often used for water planning.

Although there is significant uncertainty in how global warming-driven climate change will affect water resources in Texas, looking at a large number of different climate models (often called an “ensemble”) indicates there is reasonable likelihood climate change could result in hotter temperatures, alter annual average precipitation compared to historical levels, and cause soil moisture to decline (Figure IV.1.6). The Fifth National Climate Assessment (NCA5) is an expert-led synthesis of how climate change could affect different regions of the United States. The NCA5 chapter for the Southern Great Plains Region, which covers Kansas, Oklahoma, and Texas (McPherson et al., 2023), provides a comprehensive but not overly technical summary of possible effects of climate change in Texas.

The question of how much climate change could affect water resources is highly dependent on location, warming scenario, and the time horizon in question (Figure IV.1.6). Warming scenarios are often referred to by representative concentration pathway (RCP) designations, which describe the additional radiative forcing (“warming”) by the year 2100 in units of watts per square meter ($W{\cdot}m^{-2}$) compared to a pre-industrial reference. Climate models can offer useful insight into potential long term (i.e., decadal) changes in factors driving water demand and availability but are not suitable for predicting the timing or severity of specific events. Projected changes in average conditions or extremes can be used to infer possible impacts to sectoral demand (e.g., irrigation) or water availability (e.g., surface water runoff, groundwater recharge).

Projections of average annual precipitation from NCA5 show that there could be both increases and decreases in Texas compared to the 1991–2020 period (Figure IV.1.6a–d). However, the majority of land area shows either no change or reduced precipitation. Due to the projected increase in average temperature for Texas under even the moderate warming scenario (Figure IV.1.6e), the impacts to water availability and demand could be larger than the precipitation changes alone suggest (Venkataraman et al., 2016). This is because warmer temperatures increase evapotranspiration rates, which dry out soils and increase plant water demand, and can lead to more intense drought than would be expected from only precipitation deficits (Payton et al., 2023). Tabari et al. (2021) found increased drought risk across much of Texas for RCPs ranging from 2.6 to 8.5 for the 2070–2100 period. Location-specific impacts to streamflow depend on factors such as seasonal

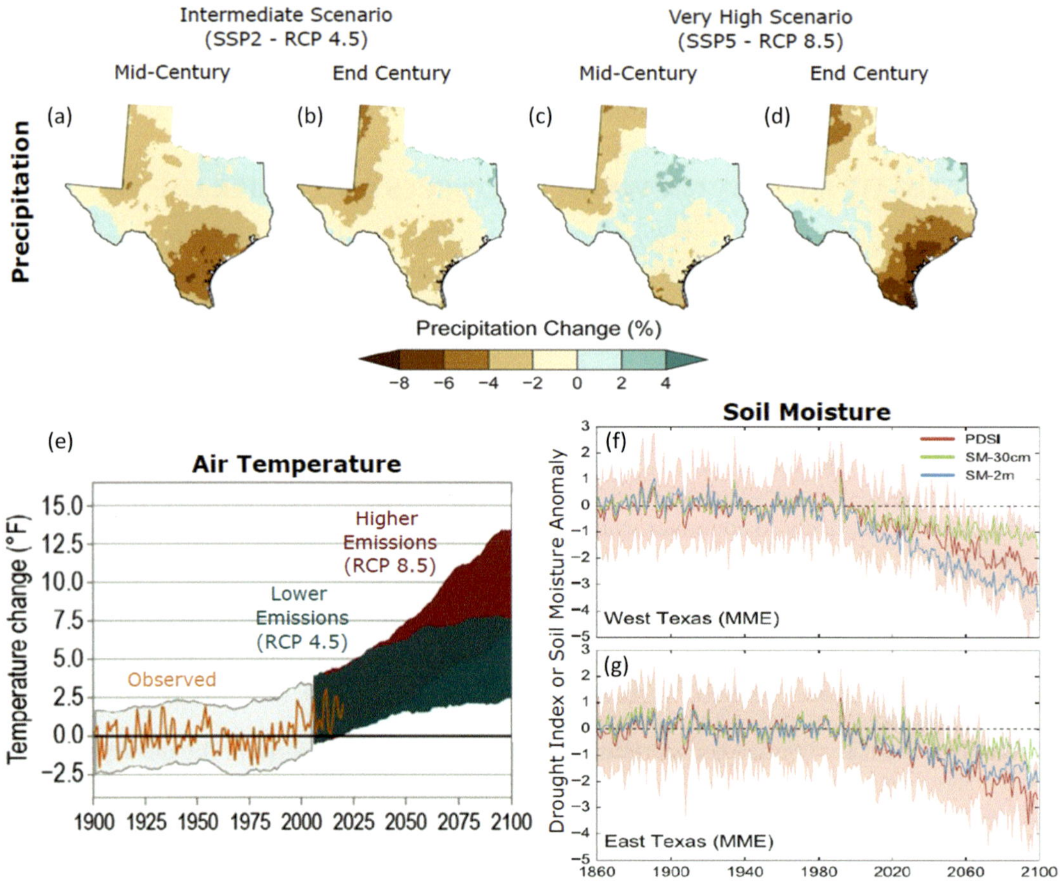

FIGURE IV.1.6 Change in average annual precipitation for two different climate scenarios at the middle (2050) and end of the 21st century (2100) (a–d), change in annual average air temperature across Texas (aerially averaged) (e), and modeled Palmer Drought Severity Index (PDSI) and soil moisture anomaly (SM) for east and west Texas (f–g). Both PDSI and SM anomaly are standardized indexes that enable comparison between different regions and time scales. Values above (below) 0 indicate wetter (drier) conditions. PDSI is an index of drought severity.

Subplots a–e adapted from McPherson et al., 2023; subplots f–g from Nielsen-Gammon et al. (2020).

changes in precipitation and evapotranspiration, the seasonality of the regional climate, and physical properties of the region (e.g., landcover, soil type, groundwater contribution to streamflow) (Van Loon, 2015). Additional complexity in how climate change could influence water demand and availability is that the magnitude and direction (increase versus decrease) of changes in precipitation and evapotranspiration will likely differ between seasons, as shown by Seager et al. (2013), who analyzed seasonal changes in Texas for different future time periods. Nielsen-Gammon et al. (2020) found that *average* soil moisture by 2060 could be as dry as conditions during the recent 2008–2015 severe drought in Texas, and that this long-term drying trend could affect both western and eastern Texas (Figure IV.1.6f–g). If such persistent dry soil conditions were to develop it would likely place profound stress on natural landcover, crops, and water resources.

Warmer temperatures are also expected to increase reservoir evaporation (Banner et al., 2010), which could reduce surface water resource availability during extended drought periods. For example, during the 2008–2015 drought, evaporation from the major reservoirs that provide Austin's water supply exceeded Austin's annual water use (Ferencz et al., 2024a). Under average-to-wet conditions increased evaporative losses may not pose a threat to water supply, but during drought it can accelerate storage declines, which can trigger drought management measures that reduce reservoir releases to downstream water users (Breyer et al., 2018). Reduced reservoir storage can also reduce hydropower production. While Texas has a large number or reservoirs, it generates very little energy from hydropower so reduced reservoir storage from future drought or climate conditions would not place large pressure on hydropower electricity production, which is very different from much of the Western U.S., where hydropower is used more heavily for electricity generation (Sanders, 2015).

In summary, future climate conditions could shift water demands and water availability in both space (spatial-variability across Texas) and time (changes in seasonal climate). Magnitudes of possible near to medium term changes do not differ as much between warming scenarios than over longer time horizons – that is, conditions between different warming scenarios in the next 30 years display less variation than those in the far-future (2070–2100). If there were an increase in aridity or emergence of more frequent drought conditions, it would reduce surface water availability (streamflow and reservoir storage), increase irrigation demands, and could increase reliance on groundwater or non-conventional sources, such as reuse and desalination.

IV.1.5 FUTURE WATER USE ADAPTATION

Estimates of future water supply sources, especially those as far as 50 years into the future, are subject to considerable uncertainty. However, the state water planning data serves as the best "on the ground" information about how water user groups are planning to address future anticipated needs. This section blends an analysis of the most recent state water planning data with grey literature and academic studies on water use adaptation.

The combined cost of all recommended projects in the 2022 state water plan is approximately $80 billion (2018 dollars) (TWDB, 2022). A total of ~2,400 projects are proposed that would provide an additional 7.7 million acre-feet per year of supply by 2070. The 2022 plan has a front-loaded implementation timeline, with 4 million of the 7.7 million acre-feet (52%) proposed by 2030. However, progress towards this objective will not be known until the 2027 state water plan is released; starting with the 2022 SWP, the TWDB requires each planning group to conduct a region-wide survey to track the implementation progress of recommended water management strategies from previous plans (TWDB, 2022).

The four largest sources of additional supply in the 2022 plan, which account for 93% of new supplies by volume, are new surface water (37%), conservation (29%), reuse (15%), and groundwater (12%) (TWDB, 2022). In addition to cost, factors considered by water planning groups when selecting additional supply sources include reliability during drought of record conditions, cost, water quality impacts, and impacts to natural and agricultural resources (TWDB, 2022). An example of a cost-reliability tradeoff is a water user considering reuse versus expanding existing surface water supply. Reuse is more costly (TWDB, 2022) but provides highly reliable supply and also reduces water demand from existing sources since some fraction of demand will be met by recycled water.

Roughly one-third of proposed new surface water is from the construction of new reservoirs. A total of 23 new major reservoirs (> 5,000 acre-feet) are proposed that are projected to increase supply by 866,000 acre-feet per year; however, this volume is small compared to the 31.5 million acre-feet of existing storage capacity (TWDB, 2025b). As of the 2022 SWP, only two of the major reservoirs were under construction, five are in the design and/or land acquisition process, and 14 do not yet have approved water rights. Overall, the planning data suggests that Texas is not planning large increases in reservoir storage over the coming decades. An active area of research is improving reservoir operations strategies to maximize water availability, reliability, and safety under flood and drought conditions (Brekke et al., 2009; Giuliani ct al., 2021). These efforts can help increase the effectiveness of existing storage in meeting demand needs rather than adding new capacity.

The other two-thirds of new surface water supplies span an array of approaches, that include: new small reservoirs (< 5,000 acre-feet), subordination, new infrastructure (e.g., new pipelines to convey surface water from its source to location of demand), reassigning surplus water (these often involve transactions where already developed supplies or water rights are sold to a user with a supply need), or eliminating infrastructure bottlenecks (e.g., increasing water treatment plant capacity for an existing surface connection that is capable of supporting larger water deliveries).

Conservation (often referred to as "demand management"), which is the second largest recommended supply source, has a wide variety of approaches to reduce water use. In addition to being an essential strategy for enhancing water security (Cosgrove and Loucks, 2015), conservation is often one of the least expensive options for addressing water supply needs (Cooley et al., 2019; TWDB, 2022). Conservation increases the sustainability of existing water resources since it reduces demand. Water user groups are required to consider conservation to address identified water needs. The overwhelming majority of proposed conservation efforts, both in number and volume of water savings, are targeted at the irrigation or municipal sectors (TWDB, 2022). Municipal conservation includes programs to reducing outdoor use through landscaping rebate incentives and landscaping irrigation restrictions, water pricing structures that incentivize water savings (e.g., tiered pricing (Cummings and Nercissiantz, 1992)), repairing leaking pipes identified in water loss audits (over 4,000 utilities submit loss audits every five years), improved metering, and efficiency standards for indoor fixtures and appliances (TWDB, 2022). Incredibly, almost 40% of proposed conservation savings come from improved fixtures and appliances alone. Half of the planning groups have municipal use goals of 140 gallons per capita per day (TWDB, 2022). Irrigation conservation strategies include increasing irrigation efficiency (better sprinkler technology, reducing crop water demand through use of lower water demand crops, and better optimizing irrigation timing and volume to reduce excess irrigation (Evett et al., 2020; TWDB, 2022). While urban development is not quantified as a specific conservation strategy, work by Brelsford and Abbott (2017) and Ferencz et al. (2024b) demonstrated that urban development, specifically denser development, can substantially reduce urban water demands through reduction of landscape water demand.

A recent statewide groundwater sustainability assessment by Mace (2021) identified which aquifers are currently being used at unsustainable rates and which have capacity for increased development (Figure IV.1.7a). Their analysis included major and minor aquifers, but here only the major aquifers and associated nearby minor aquifers, which represent most of total statewide groundwater use (>75%), are presented (Figure IV.1.7b). Production (Q) in Figure IV.1.7a is the amount of estimated pumping, modeled available groundwater (MAG) is the annual volume available based on groundwater conservation group desired future condition targets, and maximum sustainable production (QM) is the maximum annual groundwater that could be pumped to prevent decreases in groundwater storage. MAG/QM ratios above 1 indicate aquifers whose desired future conditions do not reflect sustainable groundwater use. Of the major aquifers, only the Edwards and Trinity have values of 1 or lower. The Q/QM ratios show how recent 2019 estimated pumping compares to maximum sustainable production. Only two major aquifers, the Ogallala (Southern High Plains) and Edwards-Trinity, have Q/QM ratios above one indicating unsustainable current groundwater use, while the other four major aquifers could sustainably support increased groundwater production relative to 2019 levels. A concern for groundwater sustainability is the potential for increased use of groundwater if water user groups are concerned with the reliability of surface water (Taylor et al., 2013).

Managed aquifer recharge (MAR) or aquifer storage and recovery (ASR) projects can enhance groundwater sustainability. MAR and ASR both enhance groundwater recharge, typically through surface infiltration basins or injection wells (Dillon, 2005). The difference between the two is the goal of ASR is to recover stored groundwater while MAR only focuses on enhancing recharge. MAR and ASR projects typically use excess surface water (high streamflows that would otherwise not be fully used) or recycled water for recharge. There have been recent efforts to increase the adoption of ASR in Texas. Starting in 2019, regional water planning groups are required to assess ASR as a strategy (Kramer et al., 2019) and the TWDB published a statewide ASR suitability study a)

Aquifer	Max Sustainable Production (QM)	Modeled Available GW (MAG)	MAG/QM	Production 2019 (Q)	Q/QM
Carrizo-Wilcox/Queen City/Sparta/Yegua Jackson	662,542	1,476,571	2.2	441,724	0.7
Edwards (Balcones Fault Zone)	625,986	596,947	1	391,644	0.6
Edwards -Trinity (Plateau)/Pecos Valley Alluvium	236,731	898,597	3.8	261,540	1.1
Gulf Coast Aquifer	791,392	1,797,776	2.3	739,496	0.9
Ogallala	694,526	3,263,351	4.7	4,488,456	6.5
Trinity	386,246	356,844	0.9	190,870	0.5
TOTAL	3,397,423	8,390,086	2.5	6,513,730	1.9

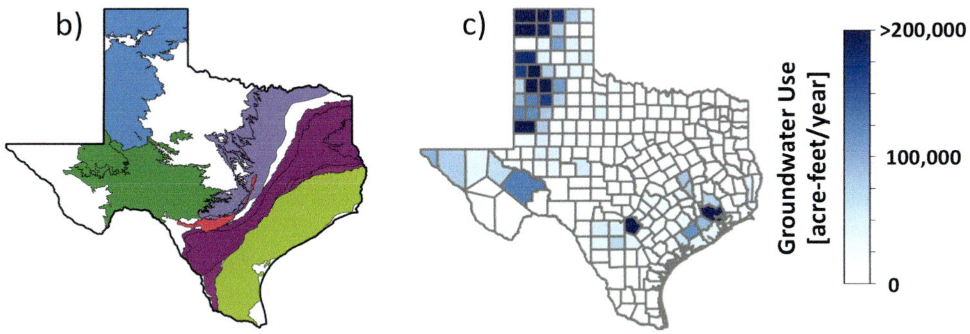

FIGURE IV.1.7 Estimated annual sustainable groundwater production (QM), modeled available groundwater (MAG), and historical 2019 production (Q) for six of the major regional aquifer groups in units of acre-feet from Mace (2021) (a). Aquifer group colors in (a) correspond to aquifer footprint maps in (b). Recent county level groundwater use is shown in (c).

in 2021 (TWDB, 2025c). MAR has a long history and is used in multiple western states contending with groundwater sustainability challenges (Scanlon et al., 2016).

An adaptive response to concerns about reduced reliability and/or sustainability of conventional surface water and groundwater supplies is to supplement a portion of supply with strategies such as ASR/MAR, water reuse, and desalination. There are already examples of municipalities in Texas diversifying their water supplies (Daher et al., 2019; Ferencz et al., 2024a). In the 2022 SWP, seawater desalination (or desal) and ASR both have fairly small proposed volumes (each 190,000 acre-feet/year), and groundwater desalination has slightly less at 157,000 acre-feet/year. While currently small in volume, these strategies could be viewed more favorably and increase in magnitude if future water supplies become less reliable. Reuse has a much larger proposed volume than desal (seawater and groundwater) and ASR. A total of 1.1 million acre-feet by 2070 are proposed across the three reuse categories: direct potable (5.6%), direct non-potable (27.6%), and indirect (67%). Reuse can have lower environmental impacts and is often less costly than desal (Hadjikakou et al., 2019) and can help water service regions work towards "local water independence" where reliance on external imports into a region are reduced due to high levels of reuse (Englehardt et al., 2016). Non-potable and indirect reuse are typically less costly since they do not feed directly into the public drinking water supply and can face less public opposition.

IV.1.6 CONCLUSIONS

Over both the near-term and long-term, Texas will likely continue to rely heavily on traditional surface water and groundwater resources to meet sectoral demands, and the agricultural and municipal sectors will continue to comprise the vast majority of water use. Population growth, conservation, and groundwater management will be key factors influencing future water use from the municipal and agricultural sectors. The over 1.6 million acre-feet of proposed future supply from reuse, desal, and ASR in the 2022 SWP indicates that water user groups are already anticipating a need to augment existing supplies with more sustainable strategies. An emphasis on conservation measures is also a major component of water planning. Conservation has two benefits compared to the other supply strategies: (1) it does not require the development of new resources and (2) it is often the lowest cost option on a per-volume basis. Texas's efforts towards improving long-term groundwater conservation and management are crucial for helping conserve this resource, which in many parts of the state provides almost all of the local water supply. However, in some locations current groundwater conservation plans still exceed natural groundwater recharge. ASR or MAR can help enhance groundwater sustainability, particularly where groundwater is being overutilized, but scarce surface water resources in some regions limit the viability of these approaches for offsetting present levels of groundwater use – the Southern High Plains Aquifer is the best example of this.

Climate change could shift water demands and water availability in both space and time. Increased evaporative demands from warming air temperatures could result in long-term drying of the land surface—potentially increasing irrigation demands, increasing evaporative losses from surface water (streams and reservoirs), and reducing natural groundwater recharge. However, the timing and magnitude of changes in precipitation and temperature are uncertain and the magnitude of potential changes is also dependent on the emission trajectory, which is also uncertain. Vermeulen et al. (2013) advise that for lower magnitudes of climate change (nearer term or lower warming scenarios), the climate signal may be indistinguishable from climate variability, and that efforts towards improving modeling precision and managing known risks could have larger returns than trying to adapt to uncertainties about far future or high warming conditions. Higher magnitudes of climate change, should they come to pass, may require transformative levels of adaptation that exceed current management approaches and could necessitate large-scale innovation and adaptation (Vermeulen et al., 2013).

In closing, a perspective not expressed in previous sections of this chapter but relevant to the future of water use adaptation in Texas is that water supply needs and water resource challenges are highly location-specific (Daher et al., 2019). Wilbanks (2015) advises against trying to apply a rigid set of approaches or solutions due to the location-specific nature of water needs and availability. It is important to recognize that the specific sectors, available water resources, potential climate impacts (both short and longer term), and risk tolerance of the sectors and user groups within a region heavily influence what water supply strategies are most sensible. Three areas of innovation that could improve understanding of future water use and water needs in Texas are: (1) obtaining better data on water use to improve the accuracy of water demand projections (Konar and Marston, 2020); (2) using more sophisticated modeling approaches to help user groups and water planning regions evaluate tradeoffs between investment costs and water supply risks (Gold et al., 2022; Hamilton et al., 2024); and (3) considering a wider-range of water availability scenarios (incorporating natural variability or alternative future climates). Finally, the iterative, five-year planning cycle of the state water planning process will help Texas adapt its water use strategies to changing climate conditions, shifting sectoral water demands, technologies, and advancements in water use data and modeling.

REFERENCES

Ault TR, Mankin JS, Cook BI, Smerdon JE, 2016. Relative impacts of mitigation, temperature, and precipitation on 21st-century megadrought risk in the American southwest. Science Advances, 2(10), e1600873. https://doi.org/10.1126/sciadv.1600873

Banner JL, Jackson CS, Yang ZL, Hayhoe K, Woodhouse C, Gulden L, Jacobs K, North G, Leung R, Washington W, Jiang X, Casteel R, 2010. Climate change impacts on Texas water: A white paper assessment of the past, present and future and recommendations for action. Texas Water Journal. 1(1):1–19. https://doi.org/10.21423/twj.v1i1.1043

Brekke LD, Maurer EP, Anderson JD, Dettinger MD, Townsley ES, Harrison A, Pruitt T, 2009. Assessing reservoir operations risk under climate change. Water Resources Research. 45:W04411. https://doi.org/10.1029/2008WR006941

Brelsford C, Abbott JK, 2017. Growing into water conservation? Decomposing the drivers of reduced water consumption in Las Vegas, NV. Ecological Economics, 133, 99–110. https://doi.org/10.1016/j.ecolecon.2016.10.012

Breyer B, Zipper SC, Qiu JX, 2018. Sociohydrological impacts of water conservation under anthropogenic drought in Austin, TX (USA). Water Resources Research. 54:3062–3080. https://doi.org/10.1002/2017wr021155

Bruun B, 2017. The regional water planning process: A Texas success story. Texas Water Journal. 8(1):1–12. https://doi.org/10.21423/twj.v8i1.7053

Calvin K, Patel P, Clarke L, Asrar G, Bond-Lamberty B, Cui RY, Di Vittorio A, Dorheim K, Edmonds J, Hartin C, Hejazi M, 2019. GCAM v5. 1: Representing the linkages between energy, water, land, climate, and economic systems. Geoscientific Model Development. 12(2):677–698. https://doi.org/10.5194/gmd-12-677-2019

Chang B, Starcher K, 2019. Evaluation of wind and solar energy investments in Texas. Renewable Energy. 132:1348–1359. https://doi.org/10.1016/j.renene.2018.09.037

Cooley H, Phurisamban R, Gleick P, 2019. The cost of alternative urban water supply and efficiency options in California. Environmental Research Communications. 1(4):042001. https://doi.org/10.1088/2515-7620/ab22ca

Cosgrove WJ, Loucks DP, 2015. Water management: Current and future challenges and research directions. Water Resources Research. 51(6):4823–4839. https://doi.org/10.1002/2014WR016869

Cummings RG, Nercissiantz V, 1992. The use of water pricing as a means for enhancing water use efficiency in irrigation: Case studies in Mexico and the United States. In Economics of Water Resources. Natural Resources Journal. 32(4):731–755. https://www.jstor.org/stable/24884504

Daher B, Lee SH, Kaushik V, Blake J, Askariyeh MH, Shafiezadeh H, Zamaripa S, Mohtar RH, 2019. Towards bridging the water gap in Texas: A water-energy-food nexus approach. Science of the Total Environment. 647:449–463. https://doi.org/10.1016/j.scitotenv.2018.07.398

Davis KF, Gephart JA, Emery KA, Leach AM, Galloway JN, D'Odorico P, 2016. Meeting future food demand with current agricultural resources. Global Environmental Change. 39:125–132. https://doi.org/10.1016/j.gloenvcha.2016.05.004

Dillon P, 2005. Future management of aquifer recharge. Hydrogeology Journal. 13(1):313–316. https://doi.org/10.1007/s10040-004-0413-6

Englehardt JD, Wu T, Bloetscher F, Deng Y, Du Pisani P, Eilert S, Elmir S, Guo T, Jacangelo J, LeChevallier M, Leverenz H, 2016. Net-zero water management: Achieving energy-positive municipal water supply. Environmental Science: Water Research & Technology. 2(2):250–260. https://doi.org/10.1039/C5EW00204D

Evett SR, Colaizzi PD, Lamm FR, O'Shaughnessy SA, Heeren DM, Trout TJ, Kranz WL, Lin X, 2020. Past, present, and future of irrigation on the US Great Plains. Transactions of the ASABE. 63(3):703–729. https://doi.org/10.13031/trans.13620

Ferencz SB, Sun N, Turner SW, Smith BA, Rice JS, 2024a. Multisectoral analysis of drought impacts and management responses to the 2008–2015 record drought in the Colorado Basin, Texas. Natural Hazards and Earth System Sciences. 24(5):1871–1896. https://doi.org/10.5194/nhess-24-1871-2024

Ferencz SB, Yoon J, Capone J, McManamay RA, 2024b. Urban morphology and urban water demand: A case study in the land constrained Los Angeles region using urban growth modeling. Environmental Research Letters. 20(1):014048. https://doi.org/10.1088/1748-9326/ada04c

Giuliani M, Lamontagne JR, Reed PM, Castelletti A, 2021. A state-of-the-art review of optimal reservoir control for managing conflicting demands in a changing world. Water Resources Research. 57:e2021WR029927. https://doi.org/10.1029/2021WR029927

Gold DF, Gupta RS, Reed PM, 2024. Exploring the spatially compounding multi-sectoral drought vulnerabilities in Colorado's West Slope River Basins. Earth's Future. 12:e2024EF004841. https://doi.org/10.1029/2024EF004841

Gold DF, Reed PM, Gorelick DE, Characklis GW, 2022. Power and pathways: Exploring robustness, cooperative stability, and power relationships in regional infrastructure investment and water supply management portfolio pathways. Earth's Future, 10(2): e2021EF002472. https://doi.org/10.1029/2021EF002472

Graham NT, Iyer G, Hejazi MI, Kim SH, Patel P, Binsted M, 2021. Agricultural impacts of sustainable water use in the United States. Scientific Reports. 11(1):17917. https://doi.org/10.1038/s41598-021-96243-5

Green DE, 1973. Land of the Underground Rain: Irrigation on the Texas High Plains, 1910–1970. Austin, TX: University of Texas Press.

Gupta RS, Steinschneider S, Reed PM, 2023. Understanding the contributions of paleo-informed natural variability and climate changes to hydroclimate extremes in the San Joaquin Valley of California. Earth's Future. 11:e2023EF003909. https://doi.org/10.1029/2023EF003909

Hadjikakou M, Stanford BD, Wiedmann T, Rowley HV, Kobayashi Y, Ishii S, Gaitan JPA, Johns G, Lundie S, Khan SJ, 2019. A flexible framework for assessing the sustainability of alternative water supply options. Science of the Total Environment. 671:1257–1268. https://doi.org/10.1016/j.scitotenv.2019.03.288

Hamilton A, Reed P, Gupta R, Zeff H, Characklis G, 2024. Resilient water infrastructure partnerships in institutionally complex systems face challenging supply and financial risk tradeoffs. Nature Communications. 15(1):7354. https://doi.org/10.1038/s41467-024-51660-8

Heberger M, Donnelly K, Cooley H, 2016. A Community Guide for Evaluating Future Urban Water Demand. Pacific Institute. http://pacinst.org/wp-content/uploads/2016/08/A-Community-Guide-for-Evaluating-Future-Urban-Water-Demand-1.pdf

Hoffman HW, 2016. Manufacturing and Electric Power Water Use in Texas. Available at: https://savetexaswater.org/resources/doc/Hoffman_Manufacturing_2016.pdf. (accessed March 19, 2025).

Jackson RB, Vengosh A, Carey JW, Davies RJ, Darrah TH, O'Sullivan F, Pétron G, 2014. The environmental costs and benefits of fracking. Annual Review of Environment and Resources. 39(1):327–362. https://doi.org/10.1146/annurev-environ-031113-144051

Konar M, Marston L, 2020. The water footprint of the United States. Water, 12(11), 3286. https://doi.org/10.3390/w12113286

Kramer K, Mullins C, Robbins D, Steinbach SA, Martinsson L, Hightower T, Fowler PL, 2019. Commentary: 86th Texas State Legislature: Summaries of water-related legislative action. Texas Water Journal. 10(1):75–100. https://doi.org/10.21423/twj.v10i1.7100

Kundzewicz ZW, Krysanova V, Benestad RE, Hov Ø, Piniewski M, Otto IM, 2018. Uncertainty in climate change impacts on water resources. Environmental Science & Policy. 79:1–8. https://doi.org/10.1016/j.envsci.2017.10.008

Lee K, Gao H, Huang M, Sheffield J, Shi X, 2017. Development and application of improved long-term datasets of surface hydrology for Texas. Advances in Meteorology. 2017(1):8485130. https://doi.org/10.1155/2017/8485130

Lee K, Jepson W, 2020. Drivers and barriers to urban water reuse: A systematic review. Water Security, 11, 100073. https://doi.org/10.1016/j.wasec.2020.100073

Liu L, Hejazi M, Patel P, Kyle P, Davies E, Zhou Y, Clarke L, Edmonds J, 2015. Water demands for electricity generation in the US: Modeling different scenarios for the water–energy nexus. Technological Forecasting and Social Change. 94:318–334. https://doi.org/10.1016/j.techfore.2014.11.004

Mace R, 2021. Five Gallons of Water in a Ten Gallon Hat: Groundwater Sustainability in Texas. The Meadows Center for Water and the Environment, Texas State University. https://gato-docs.its.txstate.edu/jcr:0abd33a2-0fcf-4af7-8b6b-e888b54508af

Macknick J, Newmark R, Heath G, Hallett KC, 2012. Operational water consumption and withdrawal factors for electricity generating technologies: A review of existing literature. Environmental Research Letters. 7(4):045802. https://doi.org/10.1088/1748-9326/7/4/045802

McManamay RA, Binita, KC, Allen-Dumas MA, Kao SC, Brelsford CM, Ruddell BL, Sanyal J, Stewart RN, Bhaduri BL, 2021. Reanalysis of water withdrawal for irrigation, electric power, and public supply sectors in the conterminous United States,1950–2016. Water Resources Research. 57:e2020WR027751. https://doi.org/10.1029/2020WR027751

McPherson RA, Fay PA, Alvarez SG, Bertrand D, Broadbent TL, Bruno T, Fares A, McCullough B, Moore GW, Moorhead B, Patiño L, Petersen A, Smith NG, Steiner JL, Taylor A, Warziniack T, 2023. Ch. 26. Southern Great Plains. In: Fifth National Climate Assessment. Crimmins AR, Avery CW, Easterling DR, Kunkel KE, Stewart BC, Maycock TK, Eds. U.S. Global Change Research Program, Washington, DC, USA. https://doi.org/10.7930/NCA5.2023.CH26

Medalie L, Galanter AE, Martinez AJ, Archer AA, Luukkonen CL, Harris MA, Haynes JV, 2025. Water use across the conterminous United States, water years 2010–20, chap. D of U.S. Geological Survey Integrated Water Availability Assessment—2010–20: U.S. Geological Survey Professional Paper 1894–D, 56 p., https://doi.org/10.3133/pp1894D

Nicot JP, Scanlon BR, 2012. Water use for shale-gas production in Texas, US. Environmental Science & Technology. 46(6):3580–3586. https://doi.org/10.1021/es204602t

Nielsen-Gammon JW, Banner JL, Cook BI, Tremaine DM, Wong CI, Mace RE, Gao HL, Yang ZL, Gonzalez MF, Hoffpauir R, Gooch T, Kloesel K, 2020. Unprecedented drought challenges for Texas water resources in a changing climate: What do researchers and stakeholders need to know? Earths Future. 8(8):e2020EF001552. https://doi.org/10.1029/2020EF001552

O'Neill BC, Kriegler E, Riahi K, Ebi KL, Hallegatte S, Carter TR, Mathur R, Van Vuuren DP, 2014. A new scenario framework for climate change research: The concept of shared socioeconomic pathways. Climatic Change. 122:387–400. https://doi.org/10.1007/s10584-013-0905-2

Payton EA, Pinson AO, Asefa T, Condon LE, Dupigny-Giroux LL, Harding BL, Kiang J, Lee DH, McAfee SA, Pflug JM, Rangwala I, Tanana HJ, Wright DB, 2023. Ch. 4. Water. In: Fifth National Climate Assessment. Crimmins AR, Avery CW, Easterling DR, Kunkel KE, Stewart BC, Maycock TK, Eds. U.S. Global Change Research Program, Washington, DC, USA. https://doi.org/10.7930/NCA5.2023.CH4

Reed PM, Hadjimichael A, Moss RH, Brelsford C, Burleyson CD, Cohen S, Dyreson A, Gold DF, Gupta RS, Keller K, Konar M, 2022. Multisector dynamics: Advancing the science of complex adaptive human-Earth systems. Earth's Future. 10(3):e2021EF002621. https://doi.org/10.1029/2021EF002621

Richter BD, Benoit K, Dugan J, Getacho G, LaRoe N, Moro B, Rynne T, Tahamtani M, Townsend A, 2020. Decoupling urban water use and growth in response to water scarcity. Water. 12(10):2868. https://doi.org/10.3390/w12102868

Richter SM, Petersen-Perlman JD, Lieberknecht K, 2024. Muddling from the middle out: Multi-scalar governance for place-based and adaptive sustainability in Texas. Journal of Infrastructure, Policy and Development. 8(8):4656. https://doi.org/10.24294/jipd.v8i8.4656

Sanders KT, 2015. Critical review: Uncharted waters? The future of the electricity-water nexus. Environmental Science & Technology. 49(1):51–66. https://doi.org/10.1021/es504293b

Scanlon BR, Duncan I, Reedy RC, 2013a. Drought and the water–energy nexus in Texas. Environmental Research Letters. 8(4):045033. https://doi.org/10.1088/1748-9326/8/4/045033

Scanlon BR, Reedy RC, Duncan I, Mullican WF, Young M, 2013b. Controls on water use for thermoelectric generation: Case study Texas, US. Environmental Science & Technology. 47(19):11326–11334. https://doi.org/10.1021/es4029183

Scanlon BR, Reedy RC, Faunt CC, Pool D, Uhlman K, 2016. Enhancing drought resilience with conjunctive use and managed aquifer recharge in California and Arizona (vol 11, 035013, 2016). Environmental Research Letters. 11:1–15. https://doi.org/10.1088/1748-9326/11/3/035013

Seager R, Ting M, Li C, Naik N, Cook B, Nakamura J, Liu H, 2013. Projections of declining surface-water availability for the southwestern United States. Nature Climate Change. 3(5):482–486. https://doi.org/10.1038/NCLIMATE1787

Siddik MAB, Dickson KE, Rising J, Ruddell BL, Marston LT, 2023. Interbasin water transfers in the United States and Canada. Scientific Data, 10(1), 27. https://doi.org/10.1038/s41597-023-01935-4

Siddik MAB, Shehabi A, Marston LT, 2021. The environmental footprint of data centers in the United States. Environmental Research Letters. 16(6):064017. https://doi.org/10.1088/1748-9326/abfba1

Tabari H, Hosseinzadehtalaei P, Thiery W, Willems P, 2021. Amplified drought and flood risk under future socioeconomic and climatic change. Earth's Future. 9:e2021EF002295. https://doi.org/10.1029/2021EF002295

Taylor RG, Scanlon B, Döll P, Rodell M, Van Beek R, Wada Y, Longuevergne L, Leblanc M, Famiglietti JS, Edmunds M, Konikow L, 2013. Ground water and climate change. Nature Climate Change. 3(4):322–329. https://doi.org/10.1038/Nclimate1744

Texas Demographic Center (TWDB), (2022). Projections of the Total Population of Texas and Counties in Texas, 2020–2060 (0.5 and 1.0 Migration Scenarios). https://demographics.texas.gov/Resources/TPEPP/Projections/2022. (Accessed March 19, 2025).

Texas Water Development Board (TWDB), 1997. 1997 State Water Plan Chapter 3.1 Statewide Future Conditions and Prospective Water Needs 38 pp., 1997. https://www.twdb.texas.gov/waterplanning/swp/1997/index.asp. (Accessed March 19, 2025).

Texas Water Development Board (TWDB), 2012. 2012 State Water Plan. 314 pp., 2012. https://www.twdb.texas.gov/waterplanning/swp/2012/index.asp. (Accessed March 19, 2025).

Texas Water Development Board (TWDB), 2022. 2022 State Water Plan. 202 pp., 2022. https://www.twdb.texas.gov/waterplanning/swp/2022/index.asp. (Accessed March 19, 2025)

Texas Water Development Board (TWDB), 2025a. Historical Water Use Estimates, https://www.twdb.texas.gov/waterplanning/waterusesurvey/estimates/index.asp. (Accessed March 19, 2025).

Texas Water Development Board (TWDB), 2025c. Statewide Survey of Aquifer Suitability for Aquifer Storage and Recovery Projects or Aquifer Recharge Projects. https://www.twdb.texas.gov/innovativewater/asr/index.asp. (Accessed March 19, 2025).

Texas Water Development Board: Texas Reservoirs (TWDB), 2025b. Monitored Water Supply Reservoirs, https://waterdatafortexas.org/reservoirs/statewide. (Accessed March 19, 2025).

United States Census Bureau, 2025. Explore Census Data, https://www.census.gov/data.html. (Accessed March 19, 2025).

United States Drought Monitor, 2025. Drought Severity and Coverage Index, https://droughtmonitor.unl.edu/DmData/DataDownload/DSCI.aspx. (Accessed March 19, 2025).

Van Loon AF, 2015. Hydrological drought explained. Wiley Interdisciplinary Reviews: Water. 2(4):359–392. https://doi.org/10.1002/wat2.1085

Venkataraman K, Tummuri S, Medina A, Perry J, 2016. 21st century drought outlook for major climate divisions of Texas based on CMIP5 multimodel ensemble: Implications for water resource management. Journal of Hydrology. 534:300–316. https://doi.org/10.1016/j.jhydrol.2016.01.001

Vermeulen SJ, Challinor AJ, Thornton PK, Campbell BM, Eriyagama N, Vervoort JM, Kinyangi J, Jarvis A, Läderach P, Ramirez-Villegas J, Nicklin KJ, 2013. Addressing uncertainty in adaptation planning for agriculture. Proceedings of the National Academy of Sciences. 110(21):8357–8362. https://doi.org//10.1073/pnas.1219441110

Wilbanks TJ, 2015. Putting "Place" in a multiscale context: Perspectives from the sustainability sciences. Environmental Science & Policy. 53:70–79. https://doi.org/10.1016/j.envsci.2015.04.009

Woodhouse CA, Gray ST, Meko DM, 2006. Updated streamflow reconstructions for the upper Colorado River Basin. Water Resources Research, 42(5):W05415. https://doi.org/10.1029/2005wr004455

Wurbs RA, 2015. Sustainable statewide water resources management in Texas. Journal of Water Resources Planning and Management. 141(12):A4014002. https://doi.org/10.1061/(ASCE)WR.1943-5452.0000499

Zoraghein H, Nawrotzki R, Jiang L, Dahlke S, 2020. IMMM-SFA/statepop: v0.1.0: initial release (v0.1.0). Zenodo. https://doi.org/10.5281/zenodo.3956703

IV.2 Strategies for Land Management, Conservation, and Nature-Based Solution Support of Climate Action

Forrest Cobb, Alison Lund, Alejandra Rodriguez, and Roel Lopez

IV.2.1 INTRODUCTION

Communities across Texas face mounting challenges from climate change, making the ways in which we manage our land more crucial than ever. Land management, conservation, and nature-based solutions (NBS) will be critical strategies for protecting and sustaining essential water resources, supporting community resilience, and mitigating future climate change. This chapter provides an overview of climate action strategies and explores aspects of the nexus between land and the climate challenges.

The chapter outlines broad ecological and watershed management principles that form the foundation of landscape-scale efforts aimed at building community resilience. We examine the policies and approaches adopted by federal, state, and local governments to encourage beneficial actions, along with specific programs that promote climate action through financial and technical assistance,

FIGURE IV.2.1 Paint Creek overview looking toward Yellow Bluff on Krause Ranch in Real County, Texas, a property permanently conserved through a conservation easement.

DOI: 10.1201/9781003406600-21

incentives, education, regulatory approaches, and partnership initiatives. Lastly, we briefly explore private-led market initiatives that support natural climate solutions, such as carbon stock preservation, increased carbon sequestration, and reduction of greenhouse gas (GHG) emissions.

IV.2.2 CLIMATE CHANGE AS A THREAT MULTIPLIER

Climate change has been described as a "threat multiplier." First coined by Sherri Goodman in 2007, the threat multiplier language is used to illustrate how climate change intensifies or makes existing threats more dangerous, including exacerbation of extreme weather events such as drought, flooding, sea level rise, and wildfire, and increase the spread of life-threatening diseases (Goodman, 2024). Extreme weather events are not new but climate change, the threat multiplier, can and in many cases will intensify these existing challenges. It can alter when, where, and how long these events occur, or create new, overlapping, and co-occurring challenges.

In their 2024 assessment of extreme weather in Texas, the Office of the Texas State Climatologist projected that, by 2036, Texans should be prepared for the number of 100-degree days to quadruple. They estimate that precipitation will decline overall, especially in West Texas, while extreme rain events would increase. Combined with continued periodic drought and increased loss of water from surface water reservoirs due to increased evaporation rates, Texans are predicted to face decreased water availability. Texans may face these challenges in the midst of more frequent and severe wildfires, flooding, coastal subsidence, and storm surges (Nielsen-Gammon et al., 2024). These threats are not new, but the rate of change, uncertainty, and potential new extremes from these threats could overwhelm the state's capacity to withstand them.

The Texas State Water Plan (2022) estimates that the state will face a potential water shortage of nearly 6.9 million acre-feet in another "drought-of-record"—the longest, most severe period of dry conditions, which occurred in the 1950s and lasted seven years. Texas's water needs have grown considerably since that time, especially in the last two decades. With current water infrastructure and water availability, as many as 13,300,000 Texans would have less than half the municipal water supplies they require, agricultural irrigation will face large deficits, and steam power generation will not be able to keep power production capacity (Texas Water Development Board (TWDB), 2022). This threat is further multiplied under future projected climate conditions with increasing temperatures and variable precipitation patterns, which could increase drought severity beyond those of the "drought-of-record" (Nielsen-Gammon et al., 2024).

Increasingly volatile weather events and cycles can cascade and magnify other natural resource issues. Large variable shifts within a year between very dry conditions (< 20th historical percentile rainfall) to very wet conditions (> 80th historical percentile rainfall) are likely to increase (Chen et al., 2022). This hydrological whiplash can magnify water quality, water availability, and fire hazards. For instance, mass vegetation die-off induced by drought can create hazardous fuel loads for wildland fire. If large rainfall events follow a large wildfire, the resultant soil erosion, sedimentation, and nutrient runoff from bare ground can impact water quality as well as reduce surface water storage capacity and groundwater recharge (Paul et al., 2022; TWDB, 2022).

While communities along Texas's coastlines have always been vulnerable to natural hazards, climate change may increase those threats as well. In 2017, Hurricane Harvey caused a cumulative $125,000,000,000 in flooding damage across the Houston Area (Blake and Zelinsky, 2018). Hurricane Harvey was the second costliest hurricane in United States (U.S.) history exceeded only by Hurricane Katrina, with a total of up to 119 cm of rainfall over a four-day period, exceeding the previous Harris County record (from June 1899) by 33.8 cm (Lindner and Fitzgerald, 2018). All 4,700,000 people in Harris County were impacted directly or indirectly during the flooding. All 22 major freeways were flooded, 300,000 vehicles were affected, 60,049 residents had to be rescued by government resources, 34,000 provided with temporary shelters, and 68 flood-related deaths occurred (Lindner and Fitzgerald, 2018). Warming oceans due to climate change are already driving increases in hurricane intensity and this is expected to lead to even, more damaging storms in the future (Kang and Elsner, 2015; Knutson et al., 2021). Sea level rise is accelerating and is very likely to continue through the 21st century, exacerbating these sort of events (Miller and Shirzaei, 2021). In the 2023 Texas Coastal Resiliency Master Plan, the Texas General Land Office identified eight primary vulnerabilities: degraded or lost habitat, erosion of bay and gulf shorelines, storm surges, inland flooding, tidal flooding, degraded water quality, and reduced water quantity (Texas General Land Office, 2023).

Given that climate change is likely to multiply the threats we have faced in the past, our response to this challenge will have to be innovative and similarly require force multipliers that leverage collaboration across as many sectors of economic, public, and community-led initiatives as possible. To address this challenge, we must rethink land management, land stewardship, and conservation's role in the community. It becomes all the more important to galvanize action in natural resources so that conservation of our natural resources and natural spaces can be done within the larger context of helping Texans to withstand and recover from the impacts we will be facing for the foreseeable future. We must treat the lands and waters of our state as vital support tools which must be invested in and sustained so that Texas communities can continue to thrive over this century and into the next.

IV.2.2.1 Climate Action

Actions to address climate change are varied, multifaceted, and spread across many sectors and types of actions. Understanding the overarching goals of climate action is important to understanding what many of these different initiatives are meant to achieve. Climate change action can generally be described as falling within three interconnected approaches to addressing the climate challenge: climate adaptation, climate mitigation, or actions that support climate resilience. The exact definition for these approaches can vary slightly among governmental and non-governmental organizations but generally align with those found in the U.S. Fifth National Climate Assessment (U.S. Global Climate Change Research Program, 2023):

- Mitigation: Measures to reduce the amount and rate of future climate change by reducing emissions of heat-trapping gases (primarily carbon dioxide) or GHGs from the atmosphere;
- Adaptation: The process of adjusting to an actual or expected environmental change and its effects in a way that seeks to moderate harm or exploit beneficial opportunities; and
- Resilience: The ability to prepare for threats and hazards, adapt to changing conditions, and withstand and recover rapidly from adverse conditions and disruptions.

These three approaches or categories of climate action are distinct but mutually supportive and often overlap. Additionally, while mitigation and adaptation generally reflect types of climate action, resilience is often described as a quality to be achieved and is often in a much broader context than climate change alone.

IV.2.2.1.1 Climate Mitigation

Land management and conservation actions that support climate mitigation are those that reduce emissions of GHG or those that remove GHG from the atmosphere (i.e., carbon sequestration). The intent of climate mitigation is to slow the rate of change in future climate conditions by reducing the greenhouse effect. Climate mitigation is sometimes incorrectly used synonymously with "hazard mitigation," which the Federal Emergency Management Agency (FEMA) defines as planning to reduce loss of life and property by minimizing the impact of disasters (FEMA, n.d.). This use is understandable since mitigation plans have been developed for decades for disaster planning from earthquakes to wildland fire planning. The difference is the frame of reference. Hazard mitigation plans are attempting to reduce impacts from an individual and (in the context of the planners) unpreventable hazard such as an earthquake. Actions meant to address the acute and immediate repercussions of climate change, such as increased flooding, drought, and sea level rise would be defined as climate adaptation or resilience actions, addressing the unavoidable impacts of the already changing climate. Examples of climate mitigation actions discussed more in Section II.B would include actions to protect natural carbon stores or to increase soil carbon sequestration rates. In both cases the intended result is fewer GHG in the atmosphere.

IV.2.2.1.2 Climate Adaptation

Climate adaptation actions are those that help reduce vulnerability to the current or expected impacts of climate change, like increased wildfire, drought, flooding, or hurricane events. It can include diverse actions from developing and deploying monitoring and observation systems, establishing funding sources for addressing hazards, creating natural coastal embankments, flood plain management, decision support tools, and other similar actions that help to adjust to current and anticipated future hazards. While climate mitigation actions are intended to prevent future climate change, climate adaptation is adjusting to and reducing the impacts of unavoidable climate-driven hazards.

IV.2.2.1.3 Resilience

Resilience is the capability to avoid, prepare for, minimize the effect of, or adapt to hazards, including extreme weather events. Since resilience is a capability, it can be achieved through many different types of adaptation and mitigation actions. Resilience is also contextual. Community resilience is a quality of the community to include the state and condition of its surrounding lands, warning systems the community has access to, the construction quality of local homes, or the presence of community action groups that organize responses to disasters when they occur. An action that could support community resilience could include agreements and coordination between local and state governments to respond to, and co-fund recovery efforts from major flooding events. The actions that would support resilience will be different when describing a county, a local agricultural sector, a city, or a university.

This is also true if the system being discussed is an ecosystem. The Intergovernmental Science-Policy Platform on Biodiversity and Ecosystem Services (IPBES) encourages the use of ecological restoration to contribute "to longer-term ecosystem resilience," and defines ecosystem resilience as the "capacity of a system to absorb disturbance and reorganize while undergoing change so as to still retain essentially the same function, structure, identity, and feedback" (IPBES, 2019; Walker et al., 2004).

Therefore, when discussing the use of land management, conservation, and NBS to support climate resilience it is essential to define and understand who or what system is meant to achieve resilience from those actions. Conservation actions meant to support community resilience are those which can help local communities to withstand and recover from extreme weather events such as drought or flood events. Conservation actions meant to support ecosystem or species resilience would include those that help wildlife populations to withstand and recover from these events. Ideally, both are achieved but the context and the target of resilience actions is fundamental to understanding the strategies employed in climate action.

IV.2.3 RELATIONSHIP BETWEEN LAND USE, WATER AND CARBON

IV.2.3.1 Land Use and Water

Rivers, reservoirs, aquifers, and other water bodies are intricately linked to the lands and watersheds that surround them. The relationship between land use (the human activities that shape the landscape) and water resources, both in terms of quantity and quality, is critical for effective water management. As the climate changes, the ways in which land is used and managed will have profound implications on water availability and quality, ecosystem health, and the effectiveness of flood mitigation by natural means. Here, we describe how land use influences water resources in three primary ways: water capture, flooding and water storage, and water filtration and purification.

Natural areas like forests, wetlands, and agricultural lands allow rainfall to infiltrate into groundwater systems while reducing surface runoff. Studies have shown that soil infiltration and landscape water retention are key factors in reducing peak storm flows and preventing flooding; the increase in impervious surfaces, such as concrete, pavement, and rooftops, disrupts these essential processes (Konrad, 2003). These surfaces prevent water infiltration, which decreases groundwater recharge, increases runoff, and can overload stormwater systems, increasing flood risk.

Soil also acts as a water filter, removing impurities before water reaches surface water bodies, groundwater, and coastal estuaries. Impervious surfaces prevent water from undergoing this natural filtration process while introducing additional contaminants such as rubber, metal, and engine oil from road runoff. As water moves through soil, it filters through layers of rocks, sand, and clay. Plants also play a vital role in the filtration process, especially in ecosystems like wetlands and estuaries, which slow water down, providing time for plants, soil, and microorganisms to break down pollutants. As the proportion of impervious surfaces increases within a watershed (the land area that channels rainfall or snowmelt into Texas streams, rivers, and lakes), these natural benefits, often referred to as "ecosystem services," are significantly diminished. Ecosystem services are benefits people obtain from ecosystems and are generally described as falling into four categories (Millennium Ecosystem Assessment, 2005):

- Provisioning services, which are tangible resources obtained from natural environments like timber, food, and water;
- Regulating services, which are those that moderate natural phenomena, including climate regulation, flood control, and water purification;
- Cultural services, which provide non-material benefits such as recreation, aesthetic enjoyment, or spiritual fulfillment; and
- Supporting services, which underpin other ecosystem services by maintaining essential processes such as soil formation, photosynthesis, and nutrient recycling.

The loss of undeveloped lands and natural spaces results in loss of these ecosystem services and can result in exacerbation of pollution, increased flooding, and depletion of groundwater resources, further stressing both the environment and communities.

Traditional land management actions and stewardship practices aimed at addressing water issues often draw on Aldo Leopold's five tools of land management, which emphasize maintaining ecological balance and enhancing ecological resilience (Leopold, 1933). Some basic concepts include capturing and/or slowing water flow within a watershed and redirecting and reducing negative impacts such as soil erosion. For example, use of no-graze zones allows for the recovery of overgrazed lands, promoting the regrowth of native plant species and improving soil health, which stabilizes the soil and enhances water retention. Maintaining ground cover is essential for preventing soil erosion, enhancing water infiltration, and promoting biodiversity; strategies such as prescribed burning to maintain healthy woody species cover, and re-seeding with native grasses help foster healthier more resilient ecosystems that support both land and water conservation. Planting windbreaks (rows of trees or shrubs) can protect fragile soils from wind erosion and improve moisture retention in agricultural areas (Smith et al., 2021). Finally, erosion control structures like terraces and check dams can play a vital role in slowing down water runoff, stabilizing soil, and improving water capture, which is crucial for sustaining plant life in arid regions like the High Plains of Texas (Taylor, 1977). These techniques are common land management and conservation tactics that align with Leopold's principles and can be utilized strategically in supporting the resilience of Texas's communities.

IV.2.3.1.1 Nature-based Solutions

Management and conservation of lands are powerful tools that communities can use or leverage to meet the climate challenge. Approaches that focus on conservation, land stewardship, and land management to tackle societal and community challenges are also sometimes referred to as NBS. The International Union for Conservation of Nature (IUCN) defines NBS as actions to address societal challenges through the protection, sustainable management, and restoration of ecosystems, benefiting both biodiversity and human well-being (IUCN, n.d.). While definitions vary (several federal agencies have their own definitions), they tend to all involve providing intentional benefit to people through use, conservation, or management of natural systems.

Activities which are considered NBS range from traditional land management or natural resources practices such as application of forestry and range management to reduce erosion, or reduce impact of forest fires, to the development of hybrid built and natural systems. For example, beaver dam analogues are human-built structures meant to simulate natural beaver dam functions and enhance stream-riparian function (Munir and Westbrook, 2021). These low-tech, permeable structures, often constructed from local materials like branches, mud, and rock, help slow water flow, reduce streambed erosion, improve water quality, and create better habitats. While not sophisticated, this is a human-built and engineered approach that "works with" or "mimics" natural processes to restore or enhance ecological function while mitigating flood or water quality challenges.

One group that has spearheaded hybrid NBS approaches is the U.S. Army Corps of Engineers (USACE), as seen in their Engineering with Nature (EWN) program. The stated intent of EWN is to find beneficial integration of engineered and natural systems, to combine the fields of engineering and ecology for both infrastructure development and environmental management (EWN, n.d.). Some of the earliest NBS work by USACE involved the beneficial re-use of dredged sediment. In 2016, the USACE used dredged sediment from nearby navigation projects to raise the elevation of Pierce Marsh—a coastal marsh area in the Houston-Galveston region of Texas—and created engineered berms to protect and restore these eroded marshlands. The intended benefits of this work were to restore wildlife habitat, provide a buffer against erosion and rising sea levels, and improve water quality (Bridges et al., 2021). Beneficial re-use of dredged sediment may not align with traditional approaches to land stewardship as championed by Aldo Leopold, but it broadly falls within the umbrella of NBS.

IV.2.3.2 Land Use and Carbon

Through photosynthesis, plants convert atmospheric CO_2 into complex carbon molecules (e.g. glucose) to be used as energy or to create plant tissues. Most of the dry mass of plants is carbon sourced from the atmosphere, and globally, plants make up 80% of the Earth's biomass (Bar-On et al., 2018). While carbon can enter ecosystems in other ways, photosynthesis is the primary way in which carbon from the atmosphere enters into the biotic environment. Plant biomass is then consumed by wildlife or enters the soil where, through decay, decomposition, and consumption by the soil microbial community, it becomes soil organic matter. Carbon is also released and emitted through respiration, fire, and other natural processes. Carbon sequestration is the rate at which a mass of carbon is captured, converted, assimilated, and stored in an ecosystem in organic and inorganic forms. A net increase in carbon sequestration within an ecosystem would mean that more carbon is converted into living biomass and stored in a system than is emitted (i.e., lost).

While carbon sequestration is the rate of carbon entering the system, carbon stocks are the amount of carbon already in an ecosystem and are generally described as the mass of carbon in different "carbon pools," the three largest being above-ground biomass (above-ground plant life), below-ground biomass (such as roots), and soil organic matter (which can include living and non-living organic matter incorporated in soil).

The biosphere plays a critical role in mitigating climate change by absorbing and storing a significant portion of anthropogenic CO_2. Since 2010, oceans, vegetation, and soil have sequestered about 44% of the anthropogenic CO_2 emitted into the atmosphere, thus helping slow the rise of atmospheric CO_2 concentrations (Canadell et al., 2021). Of this, oceans have removed 23% and terrestrial ecosystems have removed 31% of emissions. While the majority of emissions, approximately 81–91%, are anthropogenic GHG emissions primarily due to fossil fuel combustion, the remaining portion of emissions is from land-use change and land management. Actions that eliminate or degrade natural systems release carbon stored and thus are sources of emissions (Canadell et al., 2021).

Given its effect on the planet's atmosphere, the management and conservation of ecosystems is usually an important component of climate mitigation strategies. Actions which conserve natural systems, employ land stewardship, or land management practices, with the intent of reducing GHG in the atmosphere are often collectively referred to as natural climate solutions. These strategies are typically beneficial in many ways beyond climate mitigation, but natural climate solutions are evaluated in terms of their capacity to reduce GHG emissions, or to sequester carbon from the atmosphere. In their very influential paper, Griscom et al. (2017) argued that natural climate solutions such as reforestation, forest conservation, improved forest management, grazing management, grassland conservation, and peatland restoration, among others, could provide up to 37% of the cost-effective GHG mitigation needed to hold global warming below 2°C. Debate remains on the durability of natural climate solutions for long-term climate mitigation, as well as to how much they should be relied upon within larger climate mitigation strategies. However, most national, state, and local entities, as well as private actors with climate mitigation plans, include natural climate solutions as a component of their overall mitigation strategy.

IV.2.3.2.1 Natural Climate Solutions

Natural climate solutions and NBS solutions are sometimes used interchangeably, but can have distinct meanings. NBS (Section II.A.1) can refer to actions that address a broad suite of societal challenges beyond climate change. Natural climate solutions, however, usually refer specifically to actions that reduce GHG emissions as a way of mitigating climate change. While natural climate solutions vary, they are generally trying to achieve one of three things: reducing GHG emissions, avoiding GHG emissions, or increasing carbon removal. One way to avoid emissions is to protect ecosystems. Forest ecosystems, for example, contain a large amount of carbon stocks in tree biomass. By sustaining forest cover, or by decreasing the intensity of forest harvesting, emissions are "avoided," since carbon biomass in harvested trees would eventually be re-released into the atmosphere through decomposition, combustion, or other processes. Conservation of existing ecosystems, like wetlands, grasslands, and peatlands, similarly attempts to prevent emissions that would have occurred through ecosystem loss.

Emissions reduction generally refers to implementing actions that lower emissions in comparison to past practices. For example, fertilizer and manure application in agriculture is the primary source of anthropogenic NH_3, NO, N_2O emissions (Pan et al., 2022). Thus, the implementation of efficient fertilization practices, such as adjusting the dosage, timing, type, or method of fertilizer application, can reduce fertilizer volume and is an emission reduction activity (Pan et al., 2022).

Increasing carbon removal can include increasing carbon sequestration rate and carbon stocks in an ecosystem. For example, actively restoring abandoned agricultural land with diverse grassland plants can result in greater rates of soil carbon storage compared to fields undergoing natural succession (Yang et al., 2019). Other practices that can reduce carbon losses from soil or enhance carbon sequestration include the use of cover crops or reduced tillage intensity in croplands and improved grazing practices, such as lower stocking rates and rotational grazing in rangelands and pasturelands.

IV.2.3.3 Land Use Trends

Texas spans nearly 696,241 km^2, with around 81% of the land base designated as *working lands*, comprising privately owned farms, ranches, and forests (Smith et al., 2025). These lands and other undeveloped, open space areas (e.g., local, state, and federal parks) provide critical ecosystem services that are often irreplaceable due to their inherent ecological complexities, or the high costs associated with replicating these services through human-made technologies. For instance, when forests are cleared, the stored carbon in trees and soil is released into the atmosphere as carbon dioxide, contributing to increased GHG emissions. Similarly, removing grasslands disrupts the natural soil structure and diminishes soil organic matter, which further decreases carbon sequestration potential. Draining wetlands not only releases stored carbon but also eliminates their ability to manage excess water and reduce flooding. Furthermore, fragmented landscapes can hinder wildlife migration and reduce genetic diversity, further impairing the ability of ecosystems to adapt to changing climates. As diverse, natural habitat types are lost, the ecological balance is disrupted, making it more difficult for landscapes to recover from climate impacts.

While undeniably important to the state's health and economic well-being, only a small portion (< 4%) of all open space in Texas is permanently protected from development through legal agreements such as conservation easements or similar mechanisms (Smith et al., 2023). Recent land use trends show that Texas's working lands are being fragmented into smaller, less profitable ownerships or converted to non-agricultural uses at an alarming rate—with the threat to Texas agricultural lands ranked the highest in the nation (Freedgood et al., 2020). Land use conversion rarely works in the opposite direction, where development is reversed, and ecological functions are restored. Once lost, the ecological services provided by these lands are very difficult, if not impossible, to restore fully.

According to Texas Land Trends, the state has lost nearly 15,000 km^2 of working lands to development between 1997–2022, with a significant loss of over 7,000 km^2 in the last five-year period (Smith et al., 2025). The loss of working lands in Texas is driven by several overarching and often interconnected factors, including economic development, population growth, and intergenerational land transfer. Over the past decade, Texas has seen a significant influx of high-profile companies, such as Tesla, Oracle, and Hewlett Packard Enterprise, relocating or expanding their businesses in the state. Similarly, the energy industry has expanded dramatically with Texas leading the nation in wind and solar energy production. The expansion of job opportunities driven by these industries, coupled with state characteristics like affordable housing, a generally mild climate, and a younger population, have contributed to Texas's rapidly growing population.

From 2000–2024, the state's population grew from 20,850,000 to 30,700,000 (Texas Demographic Center, 2023; U.S. Census Bureau (U.S. CB), 2010). Looking ahead, the population is projected to increase more than 44%, reaching 44,400,000 by 2060 (Texas Demographic Center, 2022). Currently, 84% of residents live in urban areas, and future migration trends predict continued growth in urban-dominated counties such as Harris, Dallas, Tarrant, Bexar, and Travis (Green, 2023; Texas Demographic Center, 2022). This population growth often leads to increased demand for surrounding rural land to be converted for development.

As demand rises, rural land market values typically increase, resulting in greater risk of fragmentation and conversion of working lands. Similar to traditional home real estate, rural land values fluctuate based on location, land use, property size, and other factors. However, land values tend to be higher closer to major metropolitan areas. As these values rise, landowners may face economic pressures, making it harder to expand operations or tempting them to sell their properties for a higher profit.

These economic dynamics, coupled with Texas's impending largest intergenerational land transfer (over 68% of landowners are currently over the age of 55) raise concerns about the future of working lands (Lund et al., 2020). Many family-owned operations will be passed to younger generations who may have less experience, limited financial resources, or insufficient motivation to maintain family operations, leading to accelerated fragmentation and conversion of these lands.

IV.2.4 FEDERAL, STATE AND LOCAL GOVERNMENT SUPPORT OF CLIMATE ACTION AND COMMUNITY RESILIENCE

While not all of today's climate adaptation and mitigation efforts are government-supported or driven, historically, much of this work has been. Government entities can implement management practices on public lands, but they often depend on the actions of private landowners for scalable impact, especially in states like Texas where the vast majority of undeveloped land is privately owned. To encourage the adoption of beneficial practices beyond public lands, government entities employ a variety of tactics. In this section, we provide examples of legislative actions, programs, and strategies implemented by government agencies and public institutions, which generally fall into the categories outlined below. Some of these initiatives were developed specifically to address climate change, while others are broader programs designed to promote beneficial practices and build resilience within communities. The approaches described can be categorized as follows.

- Financial assistance and incentives: Providing monetary support, credits, or rebates to encourage the adoption of specific land use practices, subsidize actions such as establishing conservation easements, or otherwise reduce the costs of implementing beneficial practices.
- Technical assistance and planning: Offering expertise and guidance for designing and implementing land use practices or NBS, often tailored to specific local needs.
- Education and outreach: Raising awareness about risks, directing landowners and land managers to available resources, and addressing knowledge gaps to encourage the adoption of beneficial practices.
- Regulatory approaches: Requiring or prohibiting certain practices or limiting types of development, for example, through zoning regulations.
- Coordination and partnership: Facilitating collaboration among national, state, or regional stakeholders to identify common priorities, align efforts, and pool resources for the effective deployment of land management, conservation initiatives, or NBS.

IV.2.4.1 Federal Actions

Federal climate change action was particularly prominent during the 117th Congress (2021–2023) and under the Biden Administration (2021–2025). We provide one example of federal legislation as well as some agency initiatives focused on climate action during this period. This section also highlights legislation from earlier administrations and Congresses that support land management conservation and NBS actions. With the election of Donald Trump in 2024, administrative priorities have shifted and so the descriptions below may not reflect current legislation or programs. However, these programs and initiatives still stand as examples of how Congress and federal agencies can attempt to drive land management, conservation, and NBS actions.

IV.2.4.1.1 Climate Focused Federal Legislation

The Inflation Reduction Act (IRA), signed into law in August 2022, became a central component of the Biden Administration's climate strategy and represents the most substantial investment in climate-focused action in U.S. history. The IRA underwent significant revisions before ultimately passing through the budget reconciliation process, which allowed it to pass with a simple majority vote in the Senate. The law addresses a wide range of issues, with a strong focus on incentives for clean energy, emissions reduction technologies, and natural climate solutions.

As part of its approach to addressing climate change, the IRA includes substantial investments in forest resilience, habitat protection, conservation, ecosystem restoration, and climate research. This funding was allocated across various government departments, agencies, offices, and programs, to support projects on federal lands and to provide financial and technical assistance to states, local governments, Indian tribes, private landowners, and other eligible entities. Conservation, forestry, and natural resource investments were primarily distributed through the U.S. Department of Agriculture (USDA), U.S. Department of Interior, U.S. Department of Energy, U.S. Department of Commerce, and U.S. Environmental Protection Agency (EPA).

For example, the IRA provides funding to the USDA for grants through the Forest Legacy Program, which supports the acquisition of land and land interests—such as conservation easements, which are voluntary legal agreements in which landowners give up certain development rights to protect a property's conservation values. Funded projects in 2024 include the Big Thicket–Neches River Corridor project in East Texas, which protects nearly 161 km^2 of Sustainable Forestry Initiative-certified timberland along the Neches River, including almost 97 km^2 of forested wetlands (USDA Forest Service, 2024). This conservation easement expands protected land in the Big Thicket region and is intended to contribute to floodplain protection, support the conservation of various species, and retain working forests under private ownership.

IV.2.4.1.2 Other Federal Legislation and Actions

While the IRA is a climate-focused act, the following two acts address different policy areas and are not directly related to climate action. Nonetheless, actions under these two acts can align with broader climate adaptation and mitigation strategies. Some provisions have been incorporated over time to address climate-related challenges within their respective domains. These acts are representative of recurring legislative actions that often fund initiatives supporting community resilience, even if they do not explicitly target climate change. Because these acts are regularly renewed, the policies, funding allocations, and programs established under each can change over time to reflect evolving needs and the shifting priorities of different Congresses.

IV.2.4.1.2.1 U.S. Farm Bill

Between 1933–2018, the U.S. passed 18 farm bills, multiyear laws that play a crucial role in shaping national agricultural policy. While early farm bills primarily focused on supporting commodity programs, they have evolved to address a wide range of topics, including nutrition assistance, conservation, research, bioenergy programs, forestry, and rural development. The Agriculture Improvement Act of 2018, commonly known as the 2018 Farm Bill, reauthorized and expanded existing conservation programs and created new pilot programs and initiatives. The USDA plays a key role in administering these conservation programs through the Natural Resources Conservation Service (USDA-NRCS) and the Farm Service Agency (FSA).

The USDA-NRCS implements programs funded under the 2018 Farm Bill that help landowners, farmers, and ranchers protect and manage natural resources such as soil, water, air, and wildlife. The Environmental Quality Incentives Program (EQIP), NRCS's flagship conservation program, provides financial and technical assistance to agricultural producers who are interested in conserving these resources and helps them develop conservation plans to address site-specific issues. In 2021, President Joe Biden issued Executive Order 14008, *Tackling the Climate Crisis at Home and Abroad*, which directed the USDA, along with other federal agencies, to address climate change in their respective programs. In response, the USDA-NRCS developed and has continuously updated a list of climate-smart agriculture and forestry mitigation activities eligible for funding through programs such as EQIP (USDA-NRCS, 2024a). The eligible practices varied in focus and intent, including actions aimed at supporting adaptation and resilience, such as restoring field borders to reduce soil erosion and runoff. Other practices focused on climate mitigation benefits, such as increasing carbon stocks in plant biomass and soils by converting cropland to grass-based agriculture or reducing emissions from agricultural equipment by replacing combustion engines with electric motors powered by on-farm renewable energy.

The Conservation Reserve Program (CRP), administered by the FSA, offers financial incentives to landowners to remove environmentally sensitive or highly erodible land from agricultural production and establish vegetative cover. While not intended as a climate-focused program, the CRP's primary goals—to improve soil health and water quality, prevent soil erosion, and provide wildlife habitat—also generate significant co-benefits for climate mitigation and adaptation by increasing carbon sequestration, reducing emissions from agricultural practices, and maintaining ecosystem resilience. The CRP also includes subprograms focused on specific natural resource concerns, such as the Grasslands CRP. This working lands program allows landowners to continue grazing or haying practices that maintain the viability of grasslands, which in turn can support water resilience through increased water infiltration and reduced runoff. As of November 2024, Texas had 8,973 km^2 enrolled in the CRP, making it the fourth-highest enrollment by area behind Colorado, South Dakota, and Nebraska (USDA FSA, 2024a).

In 2021, as part of a broader effort under the Biden Administration to utilize USDA conservation programs for climate change mitigation, the FSA introduced the Climate-Smart Practice Incentive to encourage eligible CRP participants to adopt conservation practices that increase carbon sequestration and reduce GHG emissions (USDA FSA, 2021). The incentive, applied as an additional percentage of the per-acre soil rental rate, varied by practice and cover type, with the highest rate of 10% for woody biomass and the lowest at 3% for grass cover habitat (USDA FSA, 2024b). Eligible practices include tree planting, prairie strips, wetland buffers, and pollinator habitat.

IV.2.4.1.2.2 National Flood Insurance Act

The National Flood Insurance Act of 1968 (42 U.S.C. §§ 4001 et seq.) authorized FEMA to establish the National Flood Insurance Program (NFIP), which must be reauthorized periodically to maintain its statutory authority to operate. The primary purpose of the NFIP is to provide federally backed flood insurance to property owners, renters, and businesses in communities that adopt floodplain management standards set by the program. Communities are not mandated to follow NFIP standards but can voluntarily participate to provide their residents with access to affordable flood insurance. While there are minimum criteria established in federal regulations for community participation, communities are encouraged to adopt higher standards, such as developing comprehensive management plans for flood-prone areas, preserving flood-prone areas as open space, and implementing improved building practices. This program benefits property owners by mitigating some of the financial risk associated with flood damage and benefits the federal government by reducing potential disaster assistance expenditures. The NFIP aims to lower overall flood risk and federal spending through the proactive implementation of floodplain management standards (Congressional Research Service, 2024).

FEMA's mission has continued to expand over time through additional legislation focused on disaster relief, climate preparedness, and community resilience. In 1988, the Robert T. Stafford Disaster Relief and Emergency Assistance Act, commonly known as the Stafford Act, authorized FEMA to create the Hazard Mitigation Grant Program, which provides funding to develop hazard mitigation plans and support rebuilding efforts after a presidentially declared disaster. Over time, FEMA's efforts expanded into a number of grant programs that provide funding to state, local, tribal, and territorial governments to support post-disaster recovery and enhance resilience to climate change through pre-disaster mitigation.

As part of efforts to strengthen the NFIP and mitigate future flood risk, the National Flood Insurance Act was amended in 1994 to establish the Flood Mitigation Assistance (FMA) grant program. FEMA uses this program to support flood risk reduction measures for NFIP-insured properties, particularly those that have experienced repetitive flood losses. Among the communities that have benefited from this program is the City of New Braunfels, which used FMA funding to acquire flood-prone structures and convert the lots near the Guadalupe River into open spaces and public parks, enhancing flood protection and providing lasting community benefits (TWDB, 2023).

IV.2.4.2 Federal Landscape Partnership Programs

There are many federal landscape partnership programs that operate in various forms, including federal-to-federal, federal-to-state or federal-to-local led initiatives that serve a broad range of public, private, and local stakeholders. Below, we provide two examples of federal partnerships that, through coordination, funding, and facilitation, help deploy land management, conservation, and NBS to address resilience challenges at the community and regional levels.

IV.2.4.2.1 Regional Conservation Partnership Program

The Regional Conservation Partnership Program (RCPP) is one of several conservation programs that was funded under the 2018 Farm Bill (USDA-NRCS, 2019). This locally led program relies on strong partnerships between the USDA-NRCS and partner entities (e.g., producer associations, state or local governments, Indian tribes, institutions of higher education, and non-governmental organizations) to accomplish shared conservation objectives. As of 2024, the RCPP had committed over $1 trillion nationally to improve water quality, combat drought, enhance soil health, support wildlife habitat, and protect agricultural viability (USDA-NRCS, 2024c). Through the program, NRCS and its partners co-invest resources to provide conservation assistance (technical and/or funding) to participating producers and landowners within specific watersheds or designated Critical Conservation Areas—regions identified by the Secretary of Agriculture as having shared natural resource priorities. The RCPP structure enables partners to work directly with producers and landowners to develop and implement conservation solutions that are tailored to the unique needs of their area, ensuring greater local relevance and impact.

For example, recognizing the urgent need for conservation and restoration of Texas's Coastal Tallgrass Prairie, a coalition of conservation organizations, supported by the RCPP, formed the Coastal Prairie Additive Conservation Partnership. This initiative focuses on safeguarding the remaining prairies, which historically spanned millions of hectares. It is estimated that over 99% of this ecosystem has been lost to land conversion throughout its range. The remaining remnants are highly fragmented, with approximately 1,000 km^2 scattered throughout South Texas (USDA-NRCS, 2024b). Leveraging partnership resources, the initiative applies several voluntary conservation approaches on private lands, including conservation easements, short-term rental agreements, and comprehensive conservation plans that integrate scientifically proven practices and regenerative, climate-smart agricultural principles.

IV.2.4.2.2 Healthy Watersheds Program

The EPA plays a key role in safeguarding human health and natural resources through broad environmental protections. As part of its mission, the EPA works with states and other stakeholders to fulfill the Clean Water Act's (CWA) primary goal—to restore and maintain the integrity of the nation's waters. Through the Healthy Watersheds Program, the EPA partners with state, local, and tribal governments, and other organizations (i.e., on-the-ground watershed protection practitioners) to identify, maintain, and protect healthy watersheds, ensuring their long-term ecological and economic value (EPA, n.d.b). Its primary objectives are to prevent watershed degradation, promote effective management practices, and raise awareness about the importance of healthy watersheds and how to avoid adverse impacts. The program's partnership approach leverages collective resources and expertise of practitioners, alongside EPA's CWA programs, enabling outcomes that exceed what EPA could achieve alone. Strategies being employed include conducting health and vulnerability assessments, building landscape-level partnerships, and integrating watershed protection into broader environmental initiatives.

While the Healthy Watersheds Program offers an overarching framework for watershed protection, the Healthy Watersheds Consortium Grant Program, established by the EPA, accelerates and expands strategic protection and stewardship efforts (EPA, n.d.a). These grants provide financial support for on-the-ground conservation projects and capacity building by leveraging funds from both government and non-government sources. The Texas Hill Country Conservation Network (THCCN), a coalition of numerous organizations and agencies focused on advancing the protection of natural resources in the rapidly developing 18-county area of Central Texas, utilized the EPA's grant program to build the organizational capacity needed to advance watershed protection initiatives. With this additional staff support, THCCN played a key role in advocating local fundings, helping to secure a $75 million green bond for open space protection in Hays County. Moreover, THCCN facilitated the creation of the Natural Infrastructure Plan, a resource for guiding regional conservation efforts and demonstrating the benefits of NBS to address challenges like flooding (THCCN, 2023).

IV.2.4.3 State Actions

Unlike the U.S. federal government and some states, Texas has never adopted a state-level climate action plan. However, Texas has developed a variety of plans and initiatives aimed at coordinating resilience to extreme weather hazards and events, supporting land use and land planning efforts, and assisting rural and urban communities through conservation and land management. The following are examples of Texas-based initiatives and programs that address water supply challenges, flooding, and a former federally funded, Texas-led climate initiative.

IV.2.4.3.1 Texas Farm and Ranch Lands Conservation Program

The Texas Farm and Ranch Lands Conservation Program is a state-funded and led agricultural conservation easement program designed to conserve privately owned working lands with high agricultural and ecological value, particularly those at risk of development (Texas Parks and Wildlife Department, n.d.). The program leverages state funds to meet cost-share or match requirements of federal conservation programs and is often combined with other funding sources (e.g., municipalities, non-profits, and private entities) to maximize land conservation. By enrolling in a conservation easement, participating landowners agree to permanently restrict certain land uses (e.g., development), thereby safeguarding vital ecosystem services such as carbon sequestration, flood mitigation, and aquifer recharge. From 2016–2024, the program supported conservation easements on nearly 190 km^2 across the state (Lund et al., 2024).

One example is Dreamcatcher Ranch, a 0.85 km^2 working ranch located approximately 0.8 km from the San Marcos city limits in Hays County, Texas (Lund et al., 2020). Situated within the Edwards Aquifer Recharge Zone, the ranch faces intense development pressure due to its proximity to the rapidly growing cities of San Marcos, Austin, and San Antonio. It also lies within the Balcones Escarpment, an environmentally sensitive area that influences both the quality and quantity of water in the San Marcos Springs and the Edwards Aquifer—vital water sources for many surrounding communities. Through the Texas Farm and Ranch Lands Conservation Program, the ranch was permanently protected from development, preserving its ecological value and its contribution to regional water security.

IV.2.4.3.2 Flood Infrastructure Fund

The Flood Infrastructure Fund is a state-funded and led financial assistance program, providing no-interest loans and grants to political subdivisions (i.e., cities, counties, districts, and river authorities) to support flood control, flood mitigation, and drainage projects throughout Texas (TWDB, n.d.). The goal of the fund is to help communities plan, design, and implement flood control measures that improve flood resilience. Eligible projects include a wide range of activities from the Texas State Flood Plan, including the construction and upgrades of traditional flood control structures like detention ponds, levees, and stormwater systems, as well as non-structural and nature-based approaches like floodplain restoration and integration of green infrastructure (TWDB, 2024). Additionally, funds can be used to conduct needed planning assessments and studies, develop comprehensive flood mitigation plans, and support community outreach efforts aimed at raising awareness about flood risks and preparedness.

The North Central Texas Council of Governments (NCTCOG) utilized a grant from the Flood Insurance Fund to finance an integrated transportation and stormwater management study, with the purpose of developing proactive strategies for future population growth and climate challenges in the Upper Trinity River Basin of Texas (NCTCOG, n.d.). This initiative is designed specifically to reduce flood risks, optimize lifecycle costs, and support long-term resilience in an area expected to experience significant growth by 2045. By focusing on comprehensive mapping, modeling, and policy recommendations, the project aims to guide smarter land use decisions that preserve natural floodplains and minimize impervious surface expansion. It will deliver extensive research and data analysis, stakeholder engagement through workshops and public outreach, and integrated infrastructure plans that address stormwater, transportation, and environmental needs. The result will be a roadmap for sustainable urban planning that supports land management practices aligned with water conservation, habitat protection, and community resilience, offering tools and guidelines for replicating these strategies in other regions.

IV.2.4.3.3 Texas Climate-Smart Initiative

The Texas Climate-Smart Initiative was a large-scale pilot project funded by the USDA-NRCS and administered through the Texas State Soil and Water Conservation Board. This initiative was designed to help Texas agricultural producers adopt climate-resilient practices that improve sustainability and productivity of their land while assessing benefits and developing models for voluntary, market-based climate solutions (Texas A&M Agrilife, n.d.). Using a combination of financial incentives, technical planning, and implementation support to introduce commodity specific practices, the program aimed to reduce GHG emissions, conserve water, enhance soil health, and promote long-term agricultural resilience to climate impacts. A unique feature of the program was its integration of research on participating properties, with state universities studying soil health, carbon sequestration, and new market opportunities for climate-smart products (Ledbetter, 2024). As of April 13, 2025, the program was suspended and restructured under the USDA's Advancing Markets for Producers Initiative, which seeks to build on the original goals by expanding climate-smart commodity markets and strengthening the economic viability of sustainable agriculture.

IV.2.4.4 Local Actions

Local county or municipal governments can, and often do, coordinate, fund, and incentivize actions that address water supply and extreme weather resilience challenges through land use, conservation, and the application of NBS. These approaches are varied, and while they are locally administered, the ultimate funding source or enabling authorities may be federal, state, or local. Below, we have included several examples of local government strategies, organized into three relevant topic areas: green infrastructure, land protection, and zoning strategies.

IV.2.4.4.1 Green Infrastructure

Green infrastructure are often hybrid, natural, and built solutions, that replicate natural processes, to reduce flooding, erosion, and runoff. Many cities and counties are increasingly adopting green infrastructure practices to supplement stormwater management capacity of traditional grey infrastructure (e.g. stormwater drains and pipes, concrete channels, and culverts, etc.). While all green infrastructure aims to manage water and improve environmental quality, different types serve distinct functions depending on their design and location. Grassed swales, for example, are large vegetated open channels often employed where land uses or activities produce highly contaminated discharges with pollutant levels higher than those typically found in stormwater, such as in industrial areas (EPA, 2021b). Similarly, smaller bioretention areas are often used in urban settings to capture and treat stormwater runoff from impervious surfaces like streets, sidewalks, and compacted lawns (EPA, 2021a). These generally include rain gardens and planter boxes, which are engineered sunken areas designed to pool water, allowing time for absorption and filtering of pollutants. In coastal areas, living shorelines are used to protect against erosion and provide habitat while also managing stormwater. This strategy consists of strategically placing natural materials like living plants, sand, rocks, and oyster shells to stabilize the coast, reducing wave energy and filtering runoff before it reaches the water (Mississippi-Alabama Sea Grant Consortium, n.d.).

There has been a concerted effort by the City of Dallas, for example, to incorporate green spaces and sustainable infrastructure to address multiple objectives, including flood protection, environmental restoration, recreation, transportation, and economic development. Dallas is one of the largest cities in the U.S., with a population of over 1.3 million people and a land area of about 880 km^2 (U.S. CB, 2020, 2023). As the city continues to grow, there is more need for coordinated, multi-beneficial, and multi-use spaces and infrastructure. The Trinity River Corridor Project is a large-scale example of that effort, spanning over 32 km of the Trinity River that cuts through downtown Dallas (City of Dallas, 2005). As part of the project, a series of constructed wetlands, covering more than 0.5 km^2 throughout the corridor, are being implemented to provide new reservoirs for water storage and a secondary flowage path for floodwater along the west side of the river (Trinity River Corridor Project, n.d.a; Trinity River Corridor Project, n.d.b). Beyond flood control, these man-made wetlands offer numerous ecological and societal benefits, including the enhancement of wildlife habitats and providing additional recreation space for the area (City of Dallas, 2005).

Similarly, regional entities have also played a key role in funding and encouraging beneficial land use and green infrastructure practices. The San Antonio River Authority (SARA) has supported local adoption of low-impact development strategies through a rebate program that offsets construction costs for features like bioswales and rain catchment systems, while also offering technical guidance (SARA, n.d.a). SARA and the City of San Antonio have undertaken major stream and river restoration projects, including the "Mission Reach" restoration of the San Antonio River, which combined ecological restoration with recreational infrastructure such as 25 km of paddling and pedestrian trails (SARA, n.d.b). SARA uses natural channel design techniques which restore the historic functions of streams and rivers for self-sustaining stream systems that reduce flooding, sediment erosion and transport (SARA, n.d.c.). Through its Greenway Trails Program, the City of San Antonio has also invested in an interconnected network of trails, open spaces, and creek corridors, supported by a local sales tax. This citywide effort has created approximately 160 km of accessible trails, 50 trailheads, and preserved 6.5 km^2 of open space (SARA, n.d.b).

IV.2.4.4.2 Land Protection

Strategic land conservation is becoming a common tactic used by communities to safeguard vital natural resources and/or mitigate a range of environmental, social, and economic impacts that arise from urban expansion and changing land uses. This approach goes beyond simply preserving open space and focuses on conserving land that serves specific functions, such as expanding green space buffer, protecting water resources, reducing carbon emissions, or enhancing resilience to natural disasters. Land selection is typically guided by scientific data and comprehensive land use planning, prioritizing the protection of the most valuable and vulnerable areas while considering the overall cost-benefit of the action. The land is protected through either a fee simple acquisition, where the organization purchases the land outright, or a conservation easement, where the landowner retains ownership but agrees to limit certain activities, such as development, to preserve the land's natural value. Both methods offer long-term protection, with fee simple providing complete control over the land, while conservation easements allow landowners to maintain some use of the land while restricting harmful activities. Funding for land conservation implemented at the local level comes from various sources, including dedicated tax revenues, bond measures to raise capital, and grants from federal or state agencies. Additionally, partnerships with non-profits, private landowners, and other entities are common to help cover the necessary funds for conservation efforts.

One notable example of this strategic, locally funded land conservation approach can be found in San Antonio, where safeguarding the Edwards Aquifer has become a top priority. This aquifer is one of the most productive aquifers in the U.S., supplying the majority of the City of San Antonio's water (City of San Antonio, n.d.). Its unique features—such as large sinkholes, sinking streams, and extensive cave networks—facilitate rapid recharge, making the aquifer highly productive and capable of supporting the city's large and growing population. Concern for Edwards Aquifer's protection began to grow in the 1980s as urbanization and water quality issues raised alarms about its vulnerability. In response, the Edwards Aquifer Protection Program was established through a local sales tax initiative, designed to prevent development in the aquifer's recharge and contributing zones, through fee simple purchases and conservation easements. As of 2023, the program has protected nearly 179 km^2 and continues to acquire land and work with landowners to protect additional critical areas (City of San Antonio, n.d.).

IV.2.4.4.3 Floodplain Zoning and Restrictions

In Texas, zoning and building regulations are primarily established at the local level, with municipalities and counties responsible for governing land use and construction standards. While local governments oversee most aspects of zoning and building codes, the federal government plays a key role in floodplain management through the NFIP. As discussed earlier in this chapter, the NFIP is administered by FEMA and provides affordable flood insurance for residents and businesses while encouraging communities to adopt and enforce floodplain management practices that reduce flood risks. To participate in the program, communities must meet minimum regulatory standards set by FEMA. One aspect of the program that has significantly helped local governments in flood-prone areas is the creation of Flood Insurance Rate Maps (FIRMs), which provide detailed information about flood hazards, including flood risk zones and the likelihood of flooding in specific areas.

While zoning and building restrictions can prevent new developments in flood-prone areas, many urban communities still require remedial measures to address flood risks. Strategic home buyout programs serve as a flood mitigation tool, removing homes and other structures that are already built in flood-prone areas. These programs focus on properties that have experienced repeated flood damage or are highly vulnerable to future flooding events. Participants receive compensation, often based on the pre-flood market value of their homes, enabling them to relocate to safer areas and reduce their exposure to future floods. After the home is purchased, they are demolished, and the land is repurposed into green space, parks, or flood retention areas, helping the surrounding community manage stormwater more effectively. A notable example is the Project Recovery Buyout Program launched in the aftermath of Hurricane Harvey (2017) by the Harris County Housing and Community Development Department in partnership with other local agencies. This program offered both voluntary and mandatory buyout options as part of reconstruction efforts, with NFIP requirements and FIRMs guiding the mitigation strategies for the program (Harris County, n.d.). Properties acquired through the buyout program are now owned by Harris County Flood Control District and are permanently dedicated for open space, recreational use, or floodplain and wetland management (Harris County, n.d.).

IV.2.5 MARKET APPROACHES

Of the many efforts and approaches for climate action in the private sector, market mechanisms for funding climate mitigation and adaptation work have attracted a lot of interest in the last decades and continue to drive implementation of climate action. The following section briefly reviews ecosystem service market concepts, types of markets or emerging markets, and then discusses voluntary carbon markets (VCM) as the longest-running and most mature example of these market mechanisms.

IV.2.5.1 Payment for Ecosystem Services Markets

For the last three decades, Payment for Ecosystem Services (PES) ventures have been developed to enable governmental, non-governmental, and private entities to pay for enhancement and preservation of ecosystem services. These market-like mechanisms typically fund or incentivize conservation efforts by compensating landowners for either refraining from developing their properties or implementing specific stewardship and land management practices (Kinzig et al., 2011). PES markets focus on benefits that natural or semi-natural lands and waters provide, such as biodiversity, flood mitigation, water quantity, water quality, and climate regulation (e.g., carbon sequestration). PES markets generally fall into two broad categories:

- Compliance markets, which are regulated by national, regional, or state laws and mandates, or which allow for regulatory relief under those frameworks; and
- Voluntary markets, which are entirely driven by private sector interest and not mandated by law.

Two of the most well-established compliance-based PES market systems in the U.S. are wetland mitigation banking, authorized under Section 404 of the Clean Water Act, and conservation banking, created under the Endangered Species Act. In both systems, developers who impact protected ecosystems are required to purchase credits from approved banks that conserve, restore, or enhance similar ecological resources elsewhere. These mechanisms are administered by federal agencies such as the U.S. Army Corps of Engineers and the U.S. Fish and Wildlife Service, and are designed to ensure ecological equivalency and no net loss of resources.

Texas compliance-based water markets, while not as formalized as those in states like California, exist within regulatory or legal frameworks, often coordinated by governmental bodies. Trading of water rights or allocations occurs in Texas, although it tends to be more limited and region-specific. Water users can apply to the Texas Commission on Environmental Quality for a permit to transfer water rights from one use to another or to sell them to other parties. Groundwater, by contrast, is governed by the "rule of capture," which grants landowners the right to the groundwater beneath their land, although local groundwater conservation districts can regulate and limit extraction to manage aquifer resources. Groundwater use rights can sometimes be bought, sold, or leased to other parties, particularly in areas facing significant water scarcity, such as the High Plains and Central Texas. In some cases, groundwater transactions, if structured with conservation goals, can function similarly to PES market mechanisms. Additionally, interbasin transfers, where water is moved from one river basin to another, primarily support municipal and agricultural water needs but could, under certain frameworks, enable the monetization and transfer of ecosystem services related to water resources. Efforts such as the Texas Water Bank and associated Texas Water Trust, as well as other regional river trusts, provide framework to facilitate the voluntary transfer or deposit of water rights for environmental purposes. These tools support the protection and restoration of in-stream and environmental flows by allowing water rights holders to allocate water for ecological benefits, such as maintaining habitat, improving water quality, and sustaining aquatic ecosystems, often through temporary or permanent dedication of rights. Together, these systems reflect a growing, albeit still emerging, market-based approach to integrating environmental values into Texas water management.

Among the various PES market systems, carbon markets are the most prominent and established by scale, incentivizing activities such as emission reductions (e.g., upgrading to low-emission equipment), avoided emissions (e.g., forest protection), and carbon sequestration (e.g., reforestation or soil enhancement). In the U.S., these markets are primarily voluntary, with most investment driven by corporate sustainability initiatives including net-zero commitments and Environmental, Social, and corporate Governance (ESG) goals rather than federal mandates. State-level exceptions exist—most notably California's Cap-and-Trade Program, a compliance-based carbon market administered by the California Air Resources Board. Under this program, GHG emissions are capped, and allowances can be traded or purchased by regulated entities. While the distinction between VCMs and compliance systems has historically been clear, these boundaries are beginning to blur as some compliance programs begin to integrate or recognize credits from voluntary mechanisms. As discussed in Section IV.B.3, ongoing developments related to the Paris Agreement may further accelerate this convergence internationally. Nevertheless, in Texas and most other U.S. states, there remains no regulatory driver for climate mitigation, and carbon credit transactions continue to operate entirely within voluntary frameworks.

Similarly, voluntary biodiversity credit markets are emerging in the U.S., outside of established compliance-based programs like conservation banking. These voluntary markets are often motivated by ESG goals and alignment with global initiatives such as the Taskforce on Nature-related Financial Disclosures. Voluntary biodiversity projects typically aim to protect, restore, or enhance habitats and species populations, with the goal of improving ecosystem integrity, resilience, and function. This may involve re-establishing native vegetation, creating wildlife corridors, removing invasive species, or implementing conservation-oriented land management practices. In doing so, many of these projects seek not only to stem biodiversity loss but also to provide co-benefits such as water purification, pollination, and climate adaptation. These markets remain nascent and lack established institutional oversight, legal frameworks, and enforcement mechanisms that characterize compliance-based markets.

IV.2.5.2 Voluntary Carbon Markets

IV.2.5.2.1 Overview

Although it is sometimes referred to as "the" voluntary carbon market, there is not a singular entity or marketplace but rather includes a wide range of independent programs, for profit and non-profit entities, all using loosely inter-related but separate standards, protocols, and approaches. While there are many different organizations operating across different VCMs, they are generally dominated by four large standards/registries: Verra, American Carbon Registry, Gold Standard, and Climate Action Reserve.

Different markets operate in a variety of ways; however, most VCMs generally share key components and organizations that fulfill different functions in the crediting process. Registries play a central role by reviewing and certifying the protocols used to generate credits. These protocols, essentially methodologies, stipulate how to develop credits and set criteria that need to be met and standards for how credit development projects will be evaluated. Project developers, often independent for-profit or non-profit entities, then work with landowners or land managers to generate credits under these established protocols. The developers must meet registry protocol criteria, including activity type, monitoring, permanence, and additionality standards, among others. Once the project meets these criteria, the registry, often through a third-party verifier, verifies the project. Upon successful verification, the registry issues credits to the project developer, typically quantified in units of carbon dioxide equivalents (CO_2e), reflecting either avoided emissions or sequestered carbon. Those credits are assigned to a unique serial number which is tracked by the registry. The project developer can then sell or transfer those credits to the purchaser or end-user which can claim a GHG emission offset. When the offset is claimed the registry will "retire" the credits to avoid double counting of credits.

Within the VCM, carbon credits can be developed from non-nature-based or nature-based projects. Non-nature-based credits are typically generated through activities such as renewable energy development or through carbon capture and storage, which uses mechanical or industrial processes to capture and store CO_2, for example by injecting CO_2 into underground geological formations. In contrast, nature-based carbon credits are derived from agriculture, forestry, and other land use (AFOLU) projects, and they have historically represented a large portion of VCM activity (Ecosystem Marketplace, n.d.). Nature-based projects funded through purchase of carbon credits typically incentivize the protection, stewardship, transformation, or restoration of ecosystems with the intent of increasing carbon sequestration, protecting existing carbon stocks, or reducing emissions associated with specific land uses. Forestry projects have dominated the nature-based segment of the market, in part because forests are among the most carbon-dense ecosystems and carbon stocks in these systems can be more easily estimated using established forestry measures and techniques.

Driven by growing pressure from climate-conscious consumers, investors, state governments, and the general public, many U.S. companies have adopted voluntary climate action commitments.

These are often framed as "net-zero commitments," which are publicly announced, non-binding, internal goals to prevent the net increase in GHG emissions from their company's operations by a target year. Strategies to achieve these goals vary but generally involve a combination of direct operational or supply chain emissions reductions, as well as various offsetting strategies which often lean heavily on the purchase of voluntary carbon credits. For example, in 2019, Jeff Bezos announced that Amazon would reach net-zero emissions by 2040 by transitioning to 100% renewable energy and investing $100,000,000 in nature-based carbon off-set projects, developed through the VCM (Thorbecke, 2019).

IV.2.5.2.2 Current State of the Voluntary Carbon Market

Between 2020–2022, carbon credit development and trading soared, with a sharp increase in corporate net-zero and climate action commitments. During this period, annual carbon credit transactions were estimated at a volume between 204,000,000–514,000,000 metric tons of CO_2e. However, high demand also increased the attention from the press, policymakers, climate advocates, and academics. This period saw growing criticism and debate surrounding carbon crediting practices, including allegations of fraudulent credits, disputes over indigenous land rights, overstated conservation claims by certain projects, and challenges related to monitoring, additionality, and verification standards, and a lack of consensus on the utility of carbon credits in meeting net-zero commitments. The result of this was loss of confidence in carbon crediting approaches, accusations of greenwashing, and reputational damage to some environmental organizations that were previously seen as trusted project developers in the marketplace. Finally, some corporate entities, facing these headwinds as well as changing political pressures, just before and after the 2024 U.S. presidential election, either reduced or extended their net-zero commitments, or shifted away from carbon markets altogether.

This period of criticism and uncertainty led to a sharp contraction in the VCM, which declined 56% from its 2021 peak. The steepest reductions in credit volume occurred within the Forestry and Land Use sector (Ecosystem Marketplace, 2024). Nevertheless, 2021 and 2022 represented a substantial increase in carbon trading, and the total value of the 2023 VCM still reached $723,000,000—making it the fourth highest annual total since 2005 (Ecosystem Marketplace, 2024). Nature-based credits continued to represent the majority of credits developed and purchased (Ecosystem Marketplace, 2024).

According to Ecosystem Marketplace's market analysis, buyers in 2023 preferentially sought out credits that had robust quality criteria and clearly demonstrated "additionality." The number of credits with clear reporting on social or nature-positive co-benefits also grew. Together, these trends suggest that while overall demand for carbon credits remained lower than during the 2021 peak, it has continued into 2025. Remaining purchasers have sought more assurances of higher quality credits and ones that create positive benefits in other ways beyond carbon removal (Ecosystem Marketplace, 2024).

IV.2.5.2.3 Private and Public Sector Credit Integrity Initiatives

There have been several initiatives or attempts within federal agencies to see where public-private partnerships could help provide stability and assurances within the VCM. One effort came through the Consolidated Appropriations Act of 2023, in which Congress authorized the USDA to assess the state of U.S. agriculture and forestry carbon markets, provide recommendations, and to stand up a program to support landowner participation. This directive led to the publication of a USDA report on U.S. carbon markets in 2023, followed by formal recommendations in February 2024 for the creation of a Greenhouse Gas Technical Assistance Provider and Third-Party Verifier Program. The recommendations centered on the need to provide technical assistance to landowners attempting to navigate complex carbon markets and identified a potential role for the USDA to act as a trusted third-party verifier. Such a role could help reduce market confusion and support the development of standardized criteria for high-quality carbon credits (USDA, 2024).

There have also been numerous private-sector-led efforts to create criteria for "high-quality" carbon credits. One of the most prominent entities in this space is the Integrity Council for Voluntary Carbon Markets (ICVCM), an international initiative aimed at setting global benchmarks for credit quality. ICVCM released its ten Core Carbon Principles, broad standards intended to identify high-quality carbon credits that represent "real, verifiable climate impact" (ICVCM, n.d.). These principles stipulate the need for mechanisms to address scandals from 2020–2024 and ongoing criticisms such as need for robust third-party verification, for transparency, for additionality, permanence, and for sustainable development benefits among other key principles. While ICVCM focuses on strengthening the integrity of the credits themselves, other groups are working to define how such credits should be used within corporate climate goals. One leading effort is the Science Based Targets initiative, a private-sector coalition that is developing standards for how carbon credits can be credibly incorporated into corporate net-zero commitments.

International processes under the Paris Agreement may also play a growing role in standardization and assurance within carbon markets. The Paris Agreement, an international treaty adopted in 2015 by 196 parties including the United States, seeks to limit global temperature rise to well below 2°C. The Council of Parties (COP) is the primary international decision-making body overseeing implementation of the agreement under the United Nations Framework Convention on Climate Change. At the COP 29 meeting, the adoption of Article 6 established a framework for a global carbon market governed by the United Nations. This included mechanisms for the international transfer of emissions reductions between countries and outlined integrity initiatives intended to align with efforts by organizations such as the ICVCM (Clean Air Task Force, 2024). These developments provide formal avenues for countries to meet their agreed upon actions under the Paris Agreement through use of these crediting mechanisms which could include the VCMs. The details of implementation and the further definition of standards under Article 6 will determine the true impact of this on carbon markets but may potentially increase the amount of global government sponsorship or investment in nature-based carbon credit approaches.

IV.2.5.2.4 Future

The carbon market has evolved rapidly in recent years, and that pace seems likely to continue. With the election of President Trump in 2024, many federal climate actions have been shifted or halted including some aimed at strengthening the VCM. In response to shifting political and public sentiment, many corporate entities have also reduced or paused their participation in the VCM. Despite this contraction, the VCM continues to garner investment and drive land management, conservation, and the application of NBS. Looking ahead, the extent to which federal action will further influence the VCM is uncertain. More likely, the future of the market will continue to depend upon demand from corporate consumers and leadership, confidence in the credibility and quality of carbon credits, the perceived utility of net-zero commitments, and the degree to which international and U.S. state-led efforts drive market engagement and investment.

IV.2.6 CONCLUSION

In the face of climate change—a true threat multiplier—effective, coordinated land management, conservation, and application of NBS is no longer a peripheral concern for Texas communities. It will be central to our collective ability to adapt to, withstand, and recover from extreme weather events while safeguarding the water resources that sustain us. As this chapter has outlined, integrating land use considerations into broader climate action strategies is essential for strengthening community resilience and mitigating climate change. Yet, the complex, interconnected and often unpredictable nature of climate impacts makes coordination across landscapes, levels of government, and between public and private initiatives a significant challenge. Meeting the climate challenge will require natural resource professionals who are deeply engaged with cross-sector efforts and who can work effectively with private landowners and rural communities. Equally important is the need to avoid over-reliance on any single sector or strategy given changing policy priorities, funding streams, and shifting consumer demands and economic conditions. Our hope is that this chapter serves as a foundation for better understanding the critical connection between land use and the climate challenge, and how diverse and innovative strategies can be, and are being, applied at various levels to shape healthy, resilient, and thriving communities into the future.

REFERENCES

Bar-On YM, Phillips R, Milo R, 2018. The biomass distribution on Earth. Proc Natl Acad Sci USA, 115: 6506–6511. https://doi.org/10.1073/pnas.1711842115

Blake ES, Zelinsky DA, 2018. National Hurricane Center tropical cyclone report: Hurricane Harvey. National Hurricane Center. https://www.nhc.noaa.gov/data/tcr/AL092017_Harvey.pdf (accessed April 28, 2025).

Bridges TS, Bourne EM, Suedel BC, Moynihan EB, King JK, 2021. Engineering with Nature: An Atlas, Volume 2. U.S. Army Engineer Research and Development Center. https://doi.org/10.21079/11681/40124

Canadell JG, Monteiro PMS, Costa MH, Cotrim da Cunha L, Cox PM, Eliseev AV, Henson S, Ishii M, Jaccard S, Koven C, Lohila A, Patra PK, Piao S, Rogelj J, Syampungani S, Zaehle S, Zickfeld K, 2021. Global carbon and other biogeochemical cycles and feedbacks. In Climate Change 2021: The Physical Science Basis. Contribution of Working Group I to the Sixth Assessment Report of the Intergovernmental Panel on Climate Change, Masson-Delmotte V, Zhai P, Pirani A, Connors SL, Péan C, Berger S, Caud N, Chen Y, Goldfarb L, Gomis MI, Huang M, Leitzell K, Lonnoy E, Matthews JBR, Maycock TK, Waterfield T, Yelekçi O, Yu R, Zhou B, eds. Cambridge University Press, Cambridge and New York. https://doi.org/10.1017/9781009157896.007

Chen D, Norris J, Thackeray C, Hall A, 2022. Increasing precipitation whiplash in climate change hot spots. Environ Res Lett, 17: 124011. https://doi.org/10.1088/1748-9326/aca3b9

City of Dallas, 2005. Trinity River Corridor comprehensive land use plan. https://dallascityhall.com/departments/pnv/Strategic%20Planning%20Division%20Documents/plans/Trinity%20River%20Corridor%20Comprehensive%20Land%20Use%20Plan%202005.pdf (accessed April 28, 2025).

City of San Antonio, n.d. Edwards Aquifer. www.sanantonio.gov/EdwardsAquifer (accessed April 28, 2025).

Clean Air Task Force, 2024. Article 6 can make or break carbon markets at COP29. Here's all you need to know. https://www.catf.us/2024/10/article-6-make-break-carbon-markets-cop29-heres-all-you-need-know/ (accessed May 1, 2025).

Congressional Research Service, 2024. A brief introduction to the National Flood Insurance Program. https://crsreports.congress.gov/product/pdf/IF/IF10988/25 (accessed April 28, 2025).

Ecosystem Marketplace, n.d. Global Carbon Markets Hub. https://hub.ecosystemmarketplace.com/landing (accessed April 28, 2025).

Ecosystem Marketplace, 2024. State of the voluntary carbon market. https://www.carbon-direct.com/research-and-reports/state-of-the-voluntary-carbon-market-2024 (accessed April 28, 2025).

EPA (Environmental Protection Agency), n.d.a. Healthy Watersheds Consortium Grants. https://www.epa.gov/hwp/healthy-watersheds-consortium-grants (accessed April 28, 2025).

EPA, n.d.b. Healthy Watersheds Protection. https://www.epa.gov/hwp (accessed April 28, 2025).

EPA, 2021a. NDPES: Stormwater best management practices—Bioretention (Rain gardens). https://www.epa.gov/system/files/documents/2021-11/bmp-bioretention-rain-gardens.pdf (accessed April 28, 2025).

EPA, 2021b. NDPES: Stormwater best management practices—Grassed swales. https://www.epa.gov/system/files/documents/2021-11/bmp-grassed-swales.pdf (accessed April 28, 2025).

EWN (Engineering With Nature), n.d. About EWN. U.S. Army Corps of Engineers. https://ewn.erdc.dren.mil/about/ (accessed April 28, 2025).

FEMA (Federal Emergency Management Agency), n.d. Hazard mitigation planning. https://www.fema.gov/emergency-managers/risk-management/hazard-mitigation-planning (accessed April 28, 2025).

Freedgood JM, Hunter M, Dempsey J, Sorensen A, 2020. Farms under threat: The state of the states. American Farmland Trust. https://farmlandinfo.org/wp-content/uploads/sites/2/2020/09/AFT_FUT_StateoftheStates_rev.pdf (accessed April 28, 2025).

Goodman S, 2024. Threat Multiplier: Climate, Military Leadership, and the Fight for Global Security. Island Press, Covelo.

Green D, 2023. Defining rural Texas. Exhibit 1: Urban-rural shares of population and housing in U.S. and Texas. Texas Comptroller of Public Accounts. https://comptroller.texas.gov/economy/fiscal-notes/archive/2023/aug/rural.php#:~:text=While%2084%20percent%20of%20the,Coast%20regions%20(Exhibit%202) (accessed April 28, 2025).

Griscom BW, Adams J, Ellis PW, Houghton RA, Lomax G, Miteva DA, Schlesinger WH, Shoch D, Siikamäki JV, Smith P, Woodbury P, Zganjar C, Blackman A, Campari J, Conant RT, Delgado C, Elias P, Gopalakrishna T, Hamsik MR, … Fargione J, 2017. Natural climate solutions. Proc Natl Acad Sci USA, 114:11645–11650. https://doi.org/10.1073/pnas.1710465114

Harris County, n.d. Buyout program. https://www.harrisrecovery.org/Programs/Buyout (accessed April 28, 2025).

ICVCM (Integrity Council for the Voluntary Carbon Market), n.d. The Core Carbon Principles. https://icvcm.org/core-carbon-principles/ (accessed April 28, 2025).

IPBES (International Science-Policy Platform on Biodiversity Services), 2019. Global assessment report of the intergovernmental science-policy platform on biodiversity and ecosystem services. IPBES secretariat. https://www.ipbes.net/global-assessment (accessed April 28, 2025).

IUCN (International Union for Conservation of Nature and Natural Resources), n.d. Nature-based solutions. https://iucn.org/our-work/nature-based-solutions (accessed April 28, 2025).

Kang N-Y, Elsner JB, 2015. Trade-off between intensity and frequency of global tropical cyclones. Nat Clim Chang, 5:661–664. https://doi.org/10.1038/nclimate2646

Kinzig AP, Perrings C, Chapin FS III, Polasky S, Smith VK, Tilman D, Turner BL, 2011. Paying for ecosystem services-promise and peril. Science, 334: 603–604. https://doi.org/10.1126/science.1210297

Knutson TR, Chung MV, Vecchi G, Sun J, Hsieh T-L, Smith AJP, 2021. ScienceBrief review: Climate change is probably increasing the intensity of tropical cyclones. In: Critical Issues in Climate Change Science, Le Quéré C, Liss P, Forster P, eds. https://doi.org/10.5281/zenodo.4570334

Konrad CP, 2003. Effects of urban development on floods. U.S. Geological Survey-Water Resources. https://pubs.usgs.gov/fs/fs07603/ (accessed April 28, 2025).

Ledbetter K, 2024. AgriLife Research kicks off climate-smart project. Agrilife Today. https://agrilifetoday.tamu.edu/2023/08/04/agrilife-research-kicks-off-climate-smart-project/ (accessed April 28, 2025).

Leopold A, 1933. Game Management. Charles Scribner's Sons, New York.

Lindner J, Fitzgerald S, 2018. Memorandum: Hurricane Harvey—storm and flood information, Harris County Flood Control District. https://www.hcfcd.org/Portals/62/Harvey/immediate-flood-report-final-hurricane-harvey-2017.pdf (accessed April 28, 2025).

Lund AA, Powers GW, Lopez RR, Smith LA, Olson LM, Gregory LF, 2020. Texas Farm and Ranch Lands Conservation Program: 2020 evaluation report. Research report Number 2020–1. Texas A&M Natural Resources Institute. https://nri.tamu.edu/media/3025/tfrlcp2020evaluationreport.pdf (accessed May 1, 2025).

Lund AA, Smith LA, Lopez RR, 2024. Texas Farm and Ranch Lands Conservation Program: 2024 evaluation report. Research report Number 2024-1. Texas A&M Natural Resources Institute. https://nri.tamu.edu/media/nofntula/tfrlcp-2024-evaluation-report.pdf (accessed April 28, 2025).

Millennium Ecosystem Assessment, 2005. Ecosystems and human well-being: Opportunities and challenges for business and industry. World Resources Institute. https://www.millenniumassessment.org/documents/document.353.aspx.pdf (accessed April 28, 2025).

Miller MM, Shirzaei M, 2021. Assessment of future flood hazards for southeastern Texas: Synthesizing subsidence, sea-level rise, and storm surge scenarios. Geophys Res Lett, 48. https://doi.org/10.1029/2021GL092544

Mississippi-Alabama Sea Grant Consortium, n.d. What are living shorelines? https://masgc.org/living-shorelines/what-are-living-shorelines (accessed April 28, 2025).

Munir TM, Westbrook CJ, 2021. Beaver dam analogue configurations influence stream and riparian water table dynamics of a degraded spring-fed creek in the Canadian Rockies. River Res Appl, 37:330–342. https://doi.org/10.1002/rra.3753

NCTCOG (North Central Texas Council of Government), n.d. TSI fact sheet. https://www.nctcog.org/getmedia/06cb610c-fb44-4615-a3d1-48dfd8f5aa78/TSI-4-Page-Fact-Sheet-Digital.pdf (accessed April 28, 2025).

Nielsen-Gammon J, Holman S, Buley A, Jorgensen S, Escobedo J, Ott C, Dedrick J, Van Fleet A, 2024. Assessment of historic and future trends of extreme weather in Texas, 1900–2036: 2024 update. Texas A&M University Office of the Texas State Climatologist. https://texas2036.org/wp-content/uploads/2023/06/2024_ClimateReport.pdf (accessed April 28, 2025).

Pan S-Y, He K-H, Lin K-T, Fan C, Chang C-T, 2022. Addressing nitrogenous gases from croplands toward low-emission agriculture. NPJ Clim Atmos Sci, 5:43. https://doi.org/10.1038/s41612-022-00265-3

Paul MJ, LeDuc SD, Lassiter MG, Moorhead LC, Noyes PD, Leibowitz SG, 2022. Wildfire induces changes in receiving waters: A review with considerations for water quality management. Water Resour Res, 58:e2021WR030699. https://doi.org/10.1029/2021wr030699.

SARA (San Antonio River Authority), n.d.a. Low Impact Development (LID). https://www.sariverauthority.org/services/sustainability/low-impact-development/ (accessed April 28, 2025).

SARA, n.d.b. San Antonio River Walk Mission Reach. https://www.sariverauthority.org/parks-trails/san-antonio-river-walk-mission-reach/ (accessed April 28, 2025).

SARA, n.d.c. Stream restoration. https://www.sariverauthority.org/services/sustainability/stream-restoration/ (accessed April 28, 2025).

Smith LA, Lopez RR, Lund AA, Anderson RE, 2025. Status update and trends of Texas working lands. Texas A&M Natural Resources Institute. https://nri.tamu.edu/media/y04fu4b3/status-update-and-trends-2025-full-report.pdf (accessed April 28, 2025).

Smith LA, Lund AA, Lopez RR, 2023. Conserving Texas: Quantifying ecological return on investment. Texas A&M Natural Resources Institute. https://nri.tamu.edu/media/3672/conserving-texas_quantifying-ecological-return-on-investment.pdf (accessed April 28, 2025).

Smith MM, Bentrup G, Kellerman T, MacFarland K, Straight R, Ameyaw L, 2021. Windbreaks in the United States: A systematic review of producer-reported benefits, challenges, management activities and drivers of adoption. Agric Syst, 187. https://doi.org/10.1016/j.agsy.2020.103032

Taylor CR, 1977. An analysis of some erosion control policies for the High and Rolling Plains of Texas. J ASFMRA. 41:49–52.

Texas A&M Agrilife, n.d. Texas Climate-Smart Initiative. https://climatesmart.tamu.edu/ (accessed April 29, 2025).

Texas Demographic Center, 2022. Projections of the total population of Texas and counties in Texas, 2020-2060 (1.0 Migration Scenario). https://www.demographics.texas.gov/Projections/2022/ (accessed January 30, 2025).

Texas Demographic Center, 2023. Estimates of the total populations of counties and places in Texas for July 1, 2023 and January 1, 2024. https://www.demographics.texas.gov/Estimates/2023/ (accessed January 30, 2025).

Texas General Land Office, 2023. Texas Coastal Resiliency Master Plan. https://www.glo.texas.gov/sites/default/files/resources/glo/coast/coastal-management/coastal-resiliency/resources/files/2023-tcrmp-book.pdf (accessed April 28, 2025).

Texas Parks and Wildlife Department, n.d. Texas Farm and Ranch Lands Conservation Program. https://tpwd.texas.gov/landwater/land/private/farm-and-ranch/ (accessed April 28, 2025).

THCCN (Texas Hill Country Conservation Network), 2023. Land, Water, Sky, and Natural Infrastructure Plan. https://ourtxhillcountry.org/wp-content/uploads/Natural-Infrastructure-Plan-Digital.pdf (accessed April 29, 2025).

Thorbecke C, 2019. Amazon vows to meet the Paris climate agreement requirements 10 years early. *ABC News*. https://abcnews.go.com/Business/amazonvows-meet-paris-climate-agreement-requirements-10/story?id=65720544 (accessed April 29, 2025).

Trinity River Corridor Project, n.d.a. Chain of wetlands. https://trinityrivercorridor.com/flood-control/chain-wet-lands (accessed April 29, 2025).

Trinity River Corridor Project, n.d.b. Habitat creation and environment. https://trinityrivercorridor.com/environment/habitat-creation-environment (accessed April 29, 2025).

TWDB (Texas Water Development Board), n.d. Flood Infrastructure Fund (FIF). https://www.twdb.texas.gov/financial/programs/FIF/index.asp (accessed April 28, 2025).

TWDB (Texas Water Development Board), 2022. State Water Plan. https://www.twdb.texas.gov/waterplanning/swp/2022/docs/SWP22-Water-For-Texas.pdf (accessed April 29, 2025)

TWDB (Texas Water Development Board), 2023. Summary of Texas Flood Mitigation (FMA) grant program. https://twdb.maps.arcgis.com/apps/dashboards/7d734b7041d54d83a7f7654bcc10eb63 (accessed April 29, 2025).

TWDB (Texas Water Development Board), 2024. 2024 State Flood Plan. https://www.twdb.texas.gov/flood/planning/sfp/doc/2024_State_Flood_Plan_Volume_I.pdf (accessed April 28, 2025).

U.S. CB (United States Census Bureau), 2010. Table 3. Intercensal estimates of the resident population by sex, race, and Hispanic origin for Texas: April 1, 2000 to July 1, 2010. https://www.census.gov/data/tables/time-series/demo/popest/intercensal-2000-2010-state.html (accessed April 29, 2025).

U.S. CB (United States Census Bureau), 2020. QuickFacts: Dallas city, Texas. https://www.census.gov/quickfacts/fact/table/dallascitytexas/LND110220 (accessed April 29, 2025).

U.S. CB (United States Census Bureau), 2023. QuickFacts: Dallas city, Texas. https://www.census.gov/quickfacts/fact/table/dallascitytexas/PST045223#PST045223 (accessed April 29, 2025).

U.S. Global Climate Change Research Program, 2023. Fifth National Climate Assessment. US Global Climate Change Research Program. https://doi.org/10.7930/NCA5.2023.

USDA Forest Service, 2024. Forest Legacy 2024 funded projects. https://www.fs.usda.gov/managing-land/private-land/forest-legacy/program/funded-projects (accessed April 29, 2025).
USDA FSA, 2021. USDA expands and renews Conservation Reserve Program in effort to boost enrollment and address climate change. https://www.fsa.usda.gov/news-events/news/04-21-2021/usda-expands-renews-conservation-reserve-program-effort-boost (accessed April 28, 2025).
USDA FSA (Farm Service Agency), 2024a. Conservation Reserve Program monthly summary – November 2024. https://www.fsa.usda.gov/documents/crpmonthlynovember2024withpagenumbers-0 (accessed April 29, 2025).
USDA FSA, 2024b. FSA Handbook: Agricultural Resource Conservation Program, 2-CRP. https://www.fsa.usda.gov/Internet/FSA_File/2-crp_r06_a08.pdf (accessed April 29, 2025).
USDA (United States Department of Agriculture), 2024. Justification Report: USDA intent to establish the Greenhouse Gas Technical Assistance Provider and Third-Party Verifier Program. https://www.usda.gov/sites/default/files/documents/GCSA-JustificationReport.pdf (accessed April 28, 2025).
USDA-NRCS, 2019. Regional Conservation Partnership Program. https://www.nrcs.usda.gov/sites/default/files/2022-06/NRCS-RCPP_FarmBill2018_Factsheet.pdf (accessed April 28, 2025).
USDA-NRCS (Natural Resources Conservation Service), 2024a. Climate-Smart Agriculture and Forestry (CSAF) mitigation activities list for FY2025. https://www.nrcs.usda.gov/sites/default/files/2023-10/NRCS-CSAF-Mitigation-Activities-List.pdf (accessed April 29, 2025).
USDA-NRCS, 2024b. Regional Conservation Partnership Program projects in Texas. https://www.nrcs.usda.gov/sites/default/files/2024-11/RCPP%20Fact%20Sheet_11212024.pdf (accessed April 29, 2025).
USDA-NRCS, 2024c. USDA invests $1.5 billion in 92 partnership projects to advance conservation and climate-smart agriculture as part of the Biden-Harris investing in America agenda. https://www.nrcs.usda.gov/news/usda-invests-15-billion-in-92-partnership-projects-to-advance-conservation-and-climate-smart (accessed April 28, 2025).
Walker B, Holling CS, Carpenter SR, Kinzig A, 2004. Resilience, adaptability and transformability in social-ecological systems. Ecol Soc, 9. https://doi.org/10.5751/ES-00650-090205.
Yang Y, Tilman D, Furey G, Lehman C, 2019. Soil carbon sequestration accelerated by restoration of grassland biodiversity. Nat Commun, 10:718. https://doi.org/10.1038/s41467-019-08636-w.

IV.3 The Great Texas Energy Experiment

The Uncertain Road to Electricity Decarbonization in the Lone Star State

Ryan A. McManamay

IV.3.1 INTRODUCTION

Energy, like all things in Texas, is done the Texan way. The story of renewable energy growth in the Lone Star State, specifically the wind and solar industries, is partly due to pioneering novel approaches to energy legislation and market design. In many ways, Texas was (and still is) an experiment on the frontier of policies, balancing legislation with market-driven approaches aimed at promoting competition among conventional and renewable energy technologies. It is important to keep in mind that Texas's current oil and gas production exceeds the entire remainder of the United States and all of Canada's production. Yet, the paradox is that Texas now has among the United States' largest renewable energy portfolios, dominated by wind energy and a rapidly growing solar industry. The primary motivation for adoption of renewables in the state is an even more interesting enigma – this was not driven, at least predominantly, by an environmentalist movement to reduce CO_2 emissions, but rather by taking economic advantage of an untapped resource fostered by increasing energy market competition (Guiney, 2018). Nonetheless, the Texas experience provides valuable lessons for other states and countries traveling down the road of decarbonization. In many ways, Texas offers an experiment, first an opportunity to examine changes to the past energy system to tease apart the relative roles of several drivers of renewable energy growth – the legislative actions, energy market design, and overcoming the technoeconomic hurdles for high penetration of utility-scale renewables in the electricity system. Second, the Texas renewable energy industry, in and of itself, is also an ongoing experiment, where the future role of renewables in the energy system is still being tested. As with any experiment, the results are uncertain.

IV.3.2 AN ENERGY PRIMER

Before our deep dive into wind and solar dynamics in Texas, it is important to first differentiate "energy" and "electricity" – as modern society uses these terms interchangeably. Electricity is one form of energy, but energy is a much broader concept. Energy is technically defined as the ability to do work - it cannot be created or destroyed, but it can be converted from one form to another. Indeed, our society heavily relies upon converting energy from a primary, raw source to a final usable form. An important consideration is that no energy conversion is 100% efficient – that is, in any conversion, at least some energy is "lost" (not destroyed) as heat. Primary energy sources, either renewable or non-renewable, can be used immediately to produce heat, whereas secondary sources, such as electricity, require conversion from primary energy sources.

DOI: 10.1201/9781003406600-22

This chapter is acutely focused on electricity and the diversity of factors influencing renewable energy expansion in Texas. The primary motivation for renewable energy expansion among the environmentalist community, again not necessarily reflective of Texas energy objectives, is related to mitigating against elevated concentrations of greenhouse gas (GHG) emissions in the troposphere, which through radiative forcing, can induce global warming and influence climate change. Additionally, other byproducts of fossil fuel combustion, such as particulates and aerosols, also contribute to health concerns, particularly respiratory and heart disease. Energy is stored in fossil fuels (petroleum, natural gas, and coal) in the form of hydrocarbons that when combusted, react with oxygen to form CO_2, water (H_2O), and heat as products. Heat from this reaction can be harnessed to do a great deal of work for society, such as direct heat in homes and industries, conversion to mechanical energy in combustion engines to motorize vehicles, or conversion to electricity via turbines through a process known as electromagnetic induction. Although methane and other gases have far higher Global Warming Potential, i.e., amount of heat absorbance, CO_2 is far more abundant and longer lasting in the atmosphere and global cycles. Climate sensitivity to increasing CO_2 emissions is estimated using global climate models (GCM), which although heavily tested and parameterized, can produce uncertain projections (although it should be noted that numerous independent GCM modeling communities produce convergent projections of long-term outcomes of increasing CO_2 levels in the troposphere).

Regardless of the certainty or uncertainty of GCM projections, our society accurately accounts for emissions and their sources because these are directly tied to economic activities. Fossil fuel combustion comprises 74% of total anthropogenic greenhouse gas emissions in the United States, primary as CO_2 for electricity, heat, and transportation (EPA, 2025). Confusion can arise, however, in differentiating amongst the relative magnitude of GHG emissions depending on how economic sectors are divided. Different agencies partition economic sectors differently and use various accounting methods, which require careful examination as to what role "energy" plays. For instance, the Environmental Protection Agency (EPA) reported that electric power generation comprised 25% of all U.S. GHG emissions in 2022, whereas, for other energy sectors, transportation comprised 28%, industries 23%, residential and commercial 13%, and agriculture 10% (US EPA, 2025a). In comparison, the Energy Information Administration (EIA) reported that 31% of GHG emissions in 2023 for the U.S. were from the electric power sector, 37% from transportation, 20% from industrial, and 12% from residential and commercial (agriculture is captured in other sectors, EIA, 2024d). In other words, 25% to 31% of CO_2 emissions are from the electric power sector, whereas 69% to 75% of CO_2 emissions are from other non-electric energy sectors. The most obvious strategy to mitigate against GHG emissions in the electric power sector is by transitioning to power generation from carbon-neutral sources, i.e., renewables, nuclear, or carbon-capture fossil strategies. A more immediate and cost-effective strategy is to reduce total electricity demand as much as possible through efficiency measures (e.g., improved insulation, high-efficiency appliances). However, both the above strategies only tackle 31% of the problem. Mitigation strategies for other sectors, i.e., 69% of the problem, can also include increasing efficiency but primarily focus on increasing carbon-neutral electrification of those sectors (e.g., electric vehicles, heat pumps). Electrification of these sectors, however, dramatically increases the burden on electricity production and transmission, requiring substantial new infrastructures and strategies for load management. This level of decarbonization, termed "deep decarbonization," would require significant investment and deployment in renewable energy and comes with its own challenges. For example, 40% of Texan households use non-electricity (natural gas or propane) sources for primary heat (Ssembatya et al., 2024). As an alternative, heat pumps can be used to cool and heat homes. While the efficiency of standard heat pump models is sufficiently high in hot conditions, efficiency drops considerably in colder temperatures, especially below 30°F. If all non-electrified households adopted heat pumps, severe peak electricity demand during cold extremes, as opposed to hot extremes, could create grid reliability issues (Sembatya et al., 2024).

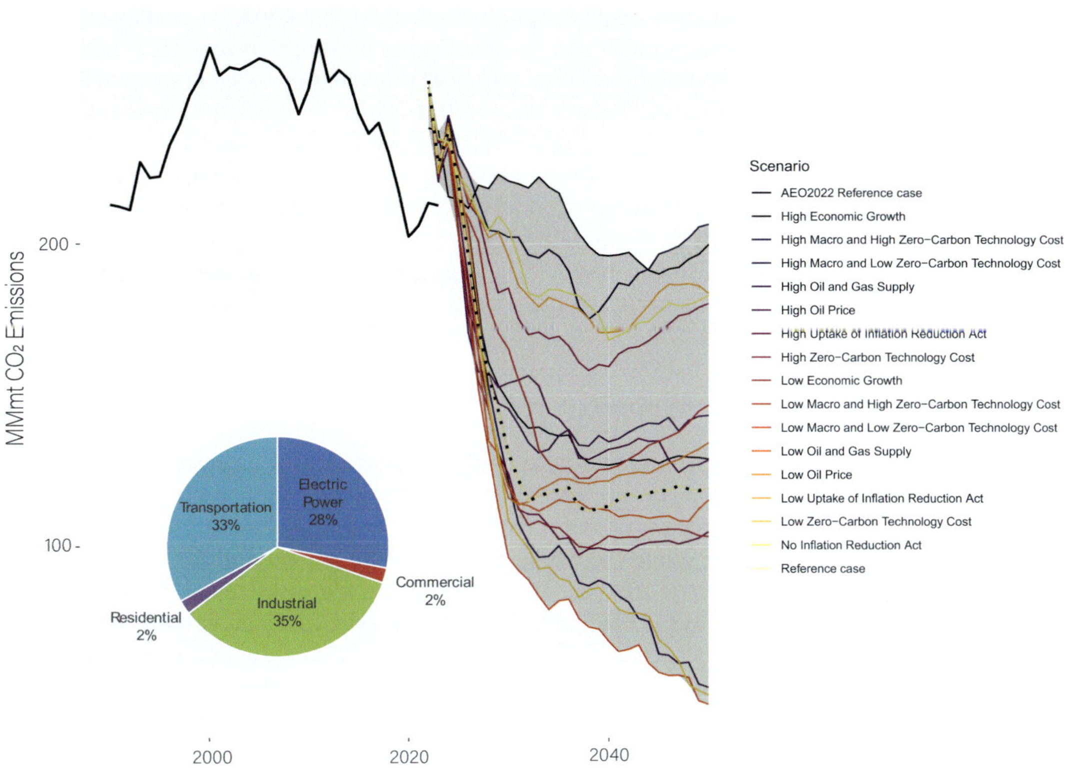

FIGURE IV.3.1 Past and future projections of CO_2 emissions (million metric tons) in Texas related to the electric power sector. The pie chart depicts the relative percentages of emissions from various sectors in Texas. Future scenarios were provided by the 2023 Annual Energy Outlook provided by the Energy Information Administration. Scenarios are described in the AEO 2023 (EIA, 2023a).

As you can imagine, future CO_2 emissions from the electric power sector in Texas are uncertain and due to many factors, including the adoption of mitigation strategies mentioned above, primarily the level of transition to carbon-neutral sources (Figure IV.3.1). This transition is influenced by legislation, economic incentives, financing, natural gas pricing, public support, or opposition, and overcoming technological and economic hurdles. We will explore each of these items in this chapter.

IV.3.3 THE MODERN ELECTRICITY GRID AND MANAGEMENT

Another important topic to review is the modern electricity grid, a complex infrastructural network operated to instantaneously balance electricity production with dynamic and uncertain consumption, all while maintaining strict adherence to operational parameters, such as voltage, frequency, and capacity. The network is comprised of power sources, transmission lines, substations, distribution lines, transformers, and end consumers. Electricity services, including management and provision of electricity, transmission, and power production, are regulated by large utility companies or independent system operators (ISOs). In the southeastern and western United States, large utility companies are vertically integrated monopolies solely responsible for power generation, transmission, and distribution – this generally describes a regulated wholesale electricity market. In contrast, in a deregulated wholesale market, ISOs and regional transmission organizations (RTOs) are independent from entities that provide power production and distribution, but manage the generation, delivery of electricity, and promote competition among power producers. ISOs and RTOs work with state Public Utility Commissions (PUCs) to plan transmission to avoid monopolization and manipulation by power producers. In these cases, utility companies are only responsible for electricity delivery to customers by purchasing electricity from producers. However, retail electricity markets may still be regulated by utilities, who set electricity prices among customers and autonomously purchase electricity from any producers pending legislation mandates requiring purchase from renewable sources. In a regulated market, utilities serving a given geographical area commonly have "captive" customer bases, meaning that customers do not have a choice in electricity provider, the price of electricity, or the source of electricity generation – this is simply determined by where you live.

IV.3.4 THE TEXAS ENERGY EXPERIMENT

Understanding how Texas came to an ideal case study in renewable energy growth requires a brief examination of Texas renewable energy policy history. Motivated to increase wholesale and retail efficiency and reduce electricity rates for Texans, the Texas legislature passed two bills, SB 373 in 1995 (74th Leg) to deregulate the wholesale generation market and SB 7 in 1999 (76th Leg) to deregulate the retail electric market, which eventually led to the establishment of the Electric Reliability Council of Texas (ERCOT). ERCOT manages loads and power supplies in the Texas electric grid interconnection, as well as governing the wholesale and retail electricity markets. Deregulation of the retail electricity market also established the "price-to-beat" concept, where residential and commercial customers can elect contracts with electricity providers of their choice on a month-by-month basis. Currently, there are 145 retail electricity providers in Texas (PUCT, 2025). ERCOT's wholesale and retail electricity market presents among the most aggressive forms of energy deregulation in the U.S., comparable to the open-energy market in Germany (Mormann et al., 2016). In addition to market deregulation, SB 7 also established one of the earliest Renewable Portfolio Standard (RPS) in the nation, designating a "goal for renewable energy" aimed to install 2,000 MW of renewable generating capacity by 2009 (Texas Administration Code §25.173) – the intention then was that most of this energy would likely be comprised by wind generation. However, at the time, the wind industry was young, and the capital costs of new installations were far higher than other conventional technologies. To incentivize integration of wind and support power purchase agreements (PPAs), the bill also established a Renewable Energy Credit (REC) program, requiring load-serving entities, or utilities, to either own renewable generating capacity or source some of their portfolio from renewable sources. In 2005, Texas legislature passed SB 20 (79th Leg) to expand the RPS to require 5,800 MW of installed renewable generating capacity by 2015 and set an aspirational goal of 10,000 MW by 2025. It also established a goal specifying that 500MW should be generated from renewable sources other than wind. While the state was rich in wind energy resources, most of these areas are located far from the most populated areas (and thus the demand centers) and existing transmission infrastructure was insufficient to bridge this gap. Hence, SB 20 established the Competitive Renewable Energy Zone (CREZ) initiative, a directive for the Public Utility Commission of Texas to design new transmission infrastructure expansion plans, including identifying zones to connect areas with high wind energy capacity with population centers. In order to carry out the CREZ plans, the SB20 tasked ERCOT and stakeholders to develop new transmission infrastructure.

As indicated above, many factors influence the growth of renewables, including legislation, energy markets, financing and economics, grid operation and reliability, and of course, the availability of renewable resources. While 19 other states have deregulated electricity markets and 31 have established an RPS or a clean energy standard (CES) (Barbose, 2023), Texas presents an exceptional opportunity to explore the major drivers of renewable growth in the state. More specifically, Texas provides a convenient and controlled experiment to examine the relative role of these drivers. First, Texas's electricity grid is an independent system disconnected from the Eastern and Western interconnections of the remainder of the United States. Hence, the entire grid interconnection for Texas, as well as the ISO controlling the electrical power system, ERCOT, is completely contained within the state, and therefore, is subject to state regulations and somewhat shielded from outside markets. Secondly, Texas has the among the most aggressively deregulated wholesale and retail markets among the U.S. states, with 90% of consumers having a choice of their retail utility (ERCOT, 2025); thus, the energy market aspects of Texas energy experiment have a clear "treatment effect." Furthermore, the social and technoeconomic landscape of Texas also complements the full spectra of challenges and opportunities for renewable growth found in other states and countries. For instance, Texas has the highest electricity generation and consumption among the U.S. states and the 11th largest energy market in the world (EIA, 2025b). As mentioned earlier, Texas also represents an odd paradox of energy endpoints, both serving as a globally significant producer of oil and natural gas, yet a leader in wind and solar generation. And significant potential for renewable growth remains. Texas is only second to Alaska in potential wind resource capacity (1,300 GW) when considering near-future turbine technologies (EIA 2025b) and has the highest utility-scale photovoltaic solar potential (20,500 GW) in the U.S. (Lopez et al., 2012). Hence, renewable growth in Texas is not limited by resource potential, thereby confining our exploration to understanding the other potential variables: legislation, technoeconomics, and society. This chapter explores the energy "transition" in Texas, the past and potential future growth of renewables, and the potential constraints, challenges, and opportunities associated with continued renewable expansion on the road to decarbonization.

IV.3.5 GROWTH OF THE WIND AND SOLAR INDUSTRY IN TEXAS

As of 2024, Texas boasted 40,865 MW of installed wind capacity, producing almost 125 million MWh annually (Figure IV.3.2). This generation is roughly equivalent to all wind generation in Iowa, Oklahoma, Kansas, and California combined. Since 2005, wind generation in Texas has increased 6 million MWh per year, on average, totaling a 30-fold increase. Growth in the solar photovoltaic (PV) industry in Texas has lagged that of wind production by almost 15 years (Figure IV.3.2). Solar, however, has rapidly increased, with current (2024) installed capacities of 24,675 MW and almost 40 million MWH of annual generation, only 8 million MWh less than California, which leads the nation. Most projections suggest that Texas will outpace California in the next 1–2 years (EIA, 2023a). The most noticeable growth in wind started after 2005, whereas solar growth did not escalate until 2016.

In August 2015, the Obama administration announced the Clean Power Plan, implemented by the EPA under the Clean Air Act, Section 111(b), 42 U.S.C. § 111(b). The amendments included New Source Performance Standards (NSPS) for fossil-fuel power plants, with the goal of reducing CO_2 emissions from the power sector by a third by 2030 (based on 2005 levels). The NSPS criteria for emissions (lbs CO_2 per MWh) from combustion turbines proved too restrictive for many of the nation's coal-fired facilities, leading widespread decommissioning of power plants, specifically a 30% reduction in coal generating capacity in Texas. Because natural gas-fired steam and combined cycle units emit roughly half of the CO_2 than coal power plants, natural gas electric generating units rapidly expanded, filling the void left by the coal industry. Simultaneously, however, total electricity demand has continued to increase, providing opportunities for new developments. The share of wind energy generation increased from 6% in 2010 to 22% in 2024, whereas solar generation was minimal in 2010 and is now 6% of generation (Figure IV.3.2).

Wind farms in Texas have generally been developed in the panhandle, north central, and coastal portions of the state distally located from the most populated areas (Figure IV.3.3). As many as 221 wind farms are in operation, whereas 18 are under various phases of construction, planning, or regulatory approval for development in the next few years (3837 MW). Facilities under construction

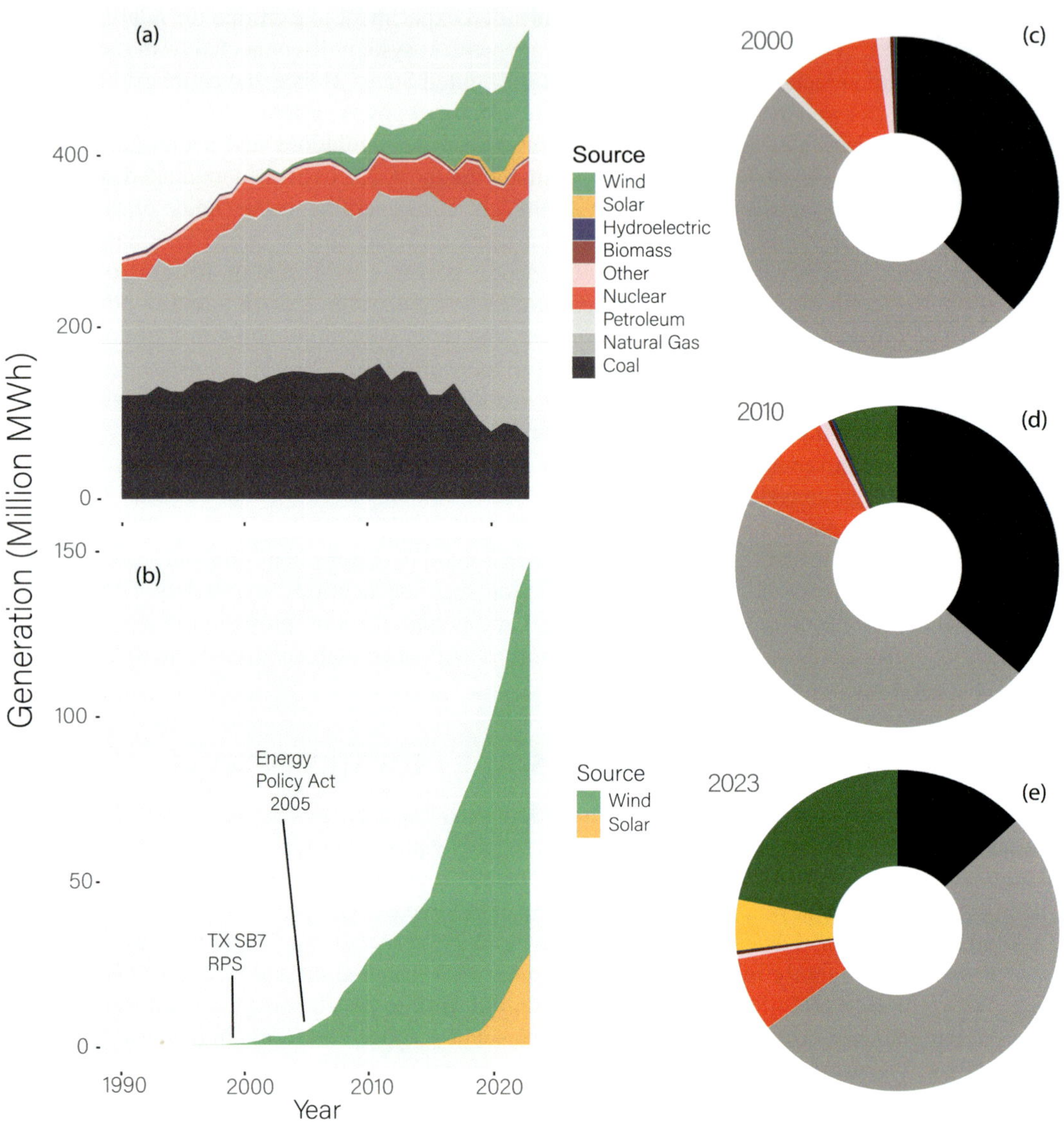

FIGURE IV.3.2 Electricity generation (million megawatt hours) by source in Texas. (a) Electricity generation over time according to all sources and (b) only wind and solar. Shares of electricity generation by source are provided for multiple select years (c-e). The timing of major legislation is shown for reference in panel b.

Data from EIA (2024c).

or planned represent a 9.4% increase in total generating capacity, which if fully developed, would be roughly equivalent to the average growth rate since 2010 (9.7%). However, new developments in 2023 were the lowest since 2013, potentially suggesting a senescence of the massive growth phase of the wind industry. In contrast to wind, solar farms are spread throughout the state and in proximity to dense populations, lessening the transmission distances (Figure IV.3.3). Currently, an estimated 185 solar farms are in operation with an additional 119 facilities under various forms of construction or planning, potentially constituting a doubling of solar generating capacity within the next 2–3 years.

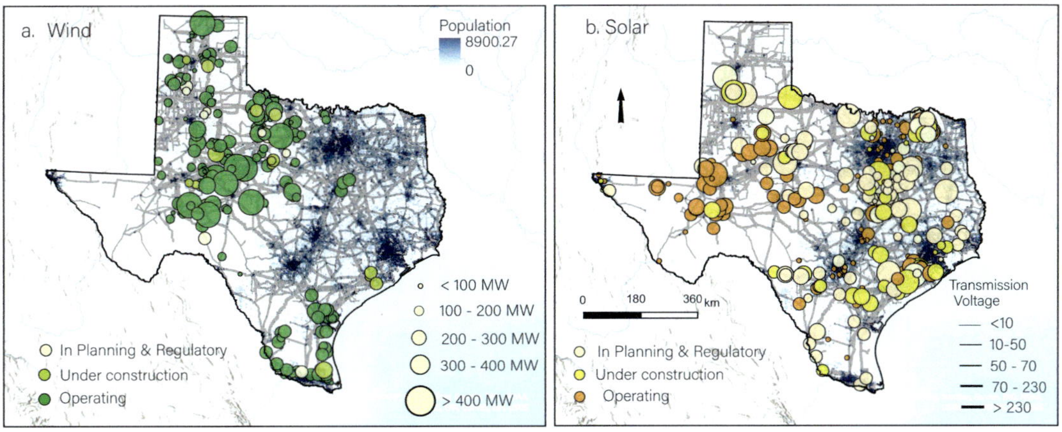

FIGURE IV.3.3 Maps of (a) wind and (b) solar facilities in Texas that are operating, under construction, or in planning or undergoing regulatory processes.

Data from EIA (2024b).

IV.3.6 TECHNOLOGICAL AND ECONOMIC CHALLENGES TO ADOPTION OF WIND AND SOLAR

Electric grids are managed for reliable and continuous provision of electricity to customers. Generally, the ISO forecasts future electricity loads at both seasonal and daily levels, and schedules baseload and peak power facilities in advance to meet demands, and deviations in load and generation must be managed in real-time, using both automated techniques but also manual controls. Deviations in load and generation can be addressed through power producers and load management strategies that provide adaptive responses, including ancillary services. Ancillary services, i.e., the ability of a generator or consumer to increase or decrease the supply of energy in minutes to seconds, can be provided by dispatchable sources, such as fast-ramping generation (e.g., natural gas combined cycle) or through energy storage (e.g., batteries and pumped storage).

In contrast to dispatchable technologies that can quickly respond to load imbalances, wind and solar technologies are Variable Renewable Energy (VRE) sources, which generate electricity only intermittently based on ideal environmental conditions; this creates challenges for integrating these sources into grid operations to ensure reliability (Deetjen et al., 2017). As an example, total electricity load and integrated solar and wind generation are visualized for several days in the summer and winter of 2024 (Figure IV.3.4). During the summer, load profiles are predictable with daily peaks coinciding with late afternoon maximum temperatures and elevated air-conditioning demands (Figure IV.3.4a). In Texas, prevailing winds, however, generally increase in the evening and peak during overnight hours, creating a mismatch with peak load versus peak generation. Solar generation more closely aligns with peak load, although even peak solar generation also starts to wane before loads peak each day. Load is far more unpredictable during winter, which creates even more challenges for integrating VREs (Figure IV.3.4b). Generation by both technologies can provide a more balanced source of baseload power, although dispatchable sources are required to fill deficiencies. There are solutions to the challenges presented by VREs that will be discussed later, but first, let us look at how these challenges are manifested in a deregulated market.

Increased in generation from VREs, especially at high penetration levels, can create challenges for managing grid reliability, specific creating supply-demand imbalances and grid congestion due to transmission constraints (Newbery, 2023). In other situations, market conditions may lead to negative electricity prices, making power production uneconomical. In these situations, the grid

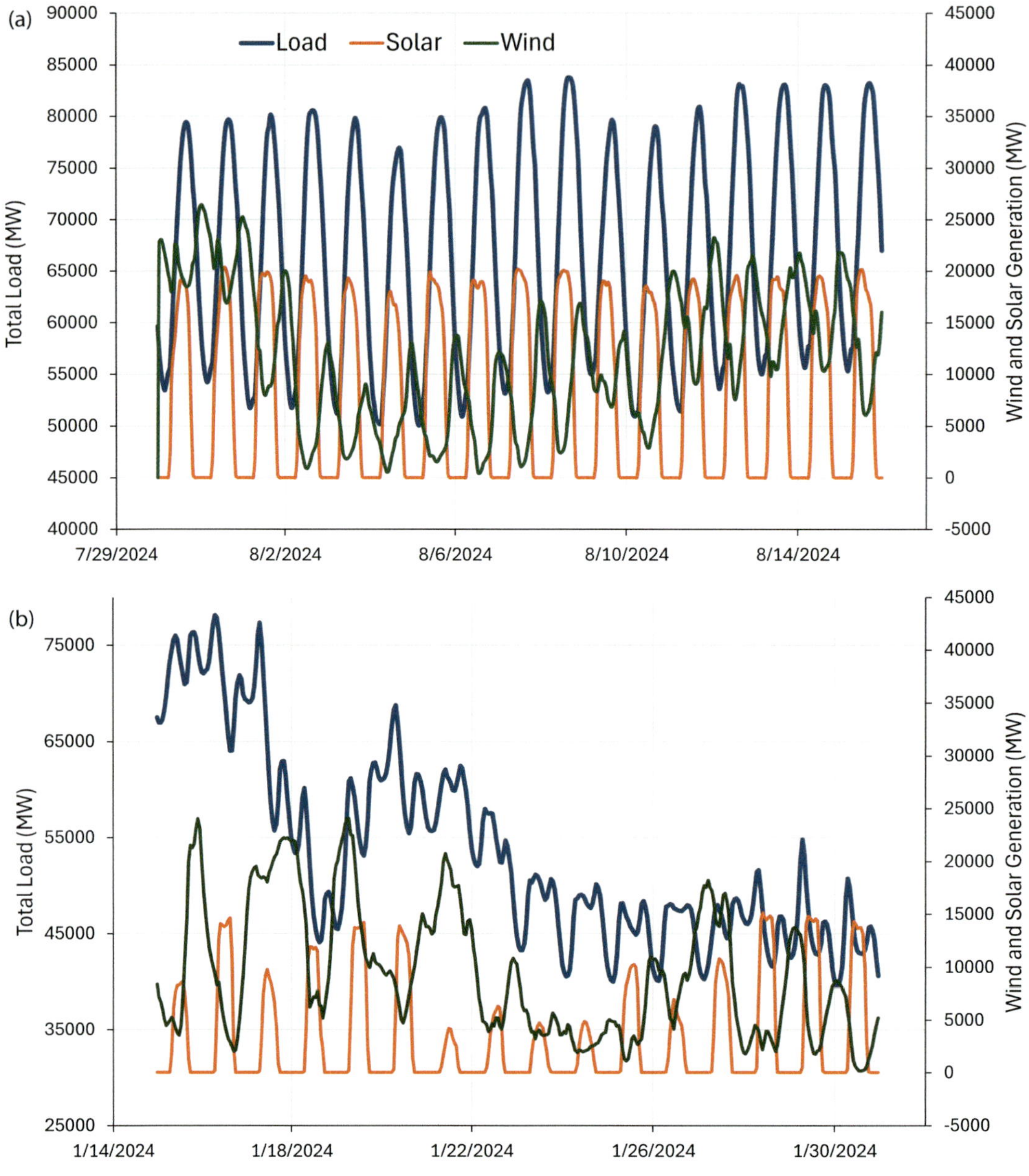

FIGURE IV.3.4 Total ERCOT electricity load relative to solar and wind generation at 15-minute intervals. (a) Wind and solar generation, as well as load, are more predictable during late summer periods, whereas (b) load and generation are far more unpredictable during winter.

Data from EIA (2025d).

manager may intentionally reduce power production, termed curtailment, from renewable sources even though generation is available (Newbery, 2023). In the ERCOT market, this can lead to hundreds of millions of dollars in revenue losses, as well as operational costs from wind and solar producers (Tables IV.3.1 and IV.3.2). Under low penetration of renewables, curtailment may be minimal; however, as VRE penetration has increased, so has curtailment (Tables IV.3.1 and IV.3.2). Although VREs can create challenges to reliability, grid managers can accommodate high amounts of VREs through multiple strategies (Woo et al., 2023), including transmission expansion (Joskow,

TABLE IV.3.1
Annual Curtailment and Revenue Losses for Wind Generation in Texas

Year	% VRE penetration	% Curtailment (total generation)	Curtailed GWh	Market Price ($/MWh)	Revenue Losses ($ Million)
2007	2.2	1.2	109.39		
2008	4.0	8.4	1487.88	$42.77	$63.63
2009	5.0	17.1	3942.07	$23.37	$92.13
2010	6.4	7.7	2087.77	$30.18	$63.00
2011	7.0	8.5	1625.58	$29.05	$47.22
2012	7.5	3.8	1158.64	$21.37	$24.76
2013	8.3	1.2	928.33	$32.29	$29.97
2014	9.2	0.4	928.34	$37.79	$35.08
2015	10.1	1.0	463.77	$23.86	$11.06
2016	12.8	1.6	927.55	$19.63	$18.20
2017	15.3	2.6	1855.88	$19.48	$36.15
2018	16.5	2.5	2319.66	$21.67	$50.28
2019	18.2	2.7	2315.70	$22.94	$53.13
2020	21.3	4.7	3708.60	$12.67	$46.98
2021	23.7	5.1	5333.39	$35.06	$186.98
2022	26.1	4.8	5562.10	$30.37	$168.91
2035	52.7	13.1	18087.14	$22.89	$414.06

VRE refers to variable renewable energy. Curtailment and VRE penetration data obtained from EIA (2023b) and US DOE (2017). Market price data obtained from LBNL (2025).

2021) and demand-response management with ancillary services. During the rapid stages of wind development in 2009, curtailment of wind energy in the ERCOT market peaked at 17%, but curtailment dropped substantially in subsequent years due to the CREZ initiative, i.e., transmission expansion in areas of high wind development, as well as more-efficient wholesale market designs (Sioshansi and Hurlbut, 2010; Wiser et al., 2015, 2016) (Tables IV.3.1 and IV.3.2). However, VRE

TABLE IV.3.2
Annual Curtailment and Revenue Losses for Solar Generation in Texas

Year	% VRE penetration	% Curtailment (total generation)	Curtailed GWh	Market Price ($/MWh)	Revenue Losses ($ Million)
2017	15.3	0.0	0.00		
2018	16.5	1.8	940.76	$47.99	$45.15
2019	18.2	5.3	705.88	$73.77	$52.07
2020	21.3	7.7	706.30	$29.69	$20.97
2021	23.7	8.4	1410.11	$113.00	$159.34
2022	26.1	8.5	1645.40	$74.31	$122.26
2035	52.7	19.4	28234.47	$67.60	$1,908.51

VRE refers to variable renewable energy. Curtailment and VRE penetration data obtained from EIA (2023b) and US DOE (2017). Market price data obtained from LBNL (2025).

penetration is expected to increase to over 50% of generation by 2035 based on projections by the EIA (EIA, 2023b). The EIA also projects increasing curtailment up to 13% for wind and almost 20% for solar by 2035 (EIA, 2023b), which is valued at $400 million and almost $2 billion in revenue losses, respectively (Tables IV.3.1 and IV.3.2). Continued transmission grid expansion, beyond that achieved by CREZ, can not only reach areas of higher renewable energy potential, but also, naturally dampens grid volatility by diversifying energy sources and demand points, thereby reducing the influence of localized variability. Combining wind and solar facilities with batteries also aids in storing surplus energy during times of over production for later use.

Like most new technologies, early adoption and investments are generally slow at first because capital costs and risks to revenue are higher until the industry and technologies mature. Without financial assistance and subsidies, development can be hindered. Construction costs of projects, also termed overnight installation costs (excluding interest and long-term financing) are generally provided on a per-unit generation capacity basis ($/KW) to support comparison among technologies. Construction costs have declined over time for wind and solar where the national averages for 2022 were $1451/KW and $1588/KW, respectively, for new installations; however, these values tend to be lower in Texas, $1395/KW and $1310 for wind and solar, respectively (EIA 2024a,b). Even so, these values are roughly double the 2022 installation cost of combined-cycle, natural gas plants (CC-NG) ($722/KW) (EIA, 2024a,b). Advanced combined cycle facilities (ACC-NG) combine both gas and steam turbines to generate electricity, leading to significant increases in efficiency (Smith et al., 2001). ACC-NG plants have been in use for over 20 years, are modular, and can be constructed relatively rapidly to serve as dispatch technologies for grids (Smith et al., 2001). In many ways, the ACC-NG technology represents the most competitive conventional technology and the price-to-beat for renewables.

The association between renewables and natural gas, however, is complicated. As stated earlier, natural gas power plants emit roughly half the CO_2 emissions of coal-fired generation facilities; therefore, natural gas is rightfully considered a "cleaner alternative" to conventional coal technology. As such, the natural gas has served as a "bridge" fuel for renewable penetration, as it provides a flexible source of energy that can be rapidly dispatched to fill in gaps left by VREs (Pless et al., 2015). However, the natural gas industry goes beyond serving as a mere bridge fuel for renewables but fills a much larger niche of the Texas and U.S. electricity market, supporting over 51% of electricity generation share in Texas and 43% in the United States (EIA, 2025a). For renewables to occupy a larger portion of the energy mix currently served by natural gas, energy storage solutions are required. Current installation costs for batteries have also plummeted over time but are still costlier ($1,205/KW) than natural gas.

Construction and other capital costs, however, may not capture all the factors associated with return on investment, such as the costs of fuels, operations and maintenance, and longevity of a project. Alternatively, the levelized cost of electricity (LCOE) was presented as an overall measure of the pricing competitiveness of different generation technologies, representing a per unit of generation cost of building, operating, and maintaining a plant over its lifetime (Shen et al., 2020; Tran and Smith, 2018). LCOE can be calculated using eq. IV.3.1.

$$LCOE = \frac{sum\ of\ total\ costs\ over\ lifetime}{sum\ of\ electrical\ energy\ produced\ over\ lifetime} = \frac{\sum_{t=1}^{n} \frac{I_t + M_t + F_t}{(1+r)^t}}{\sum_{t=1}^{n} \frac{E_t}{(1+r)^t}} \quad \text{(IV.3.1)}$$

Where I_t is the initial investment expenditures, M_t is the maintenance and operations expenditures, F_t is the fuel expenditures, E_t is the electricity produced, r is a discount rate, for the t years out of a total expected lifespan of n years.

In comparison to initial capital costs, LCOE provides investors with more long-term perspective to inform decisions among multiple competing technologies. LCOE values ($/MWh) for wind and solar have declined considerably over time as the industry has matured (Figure IV.3.5).

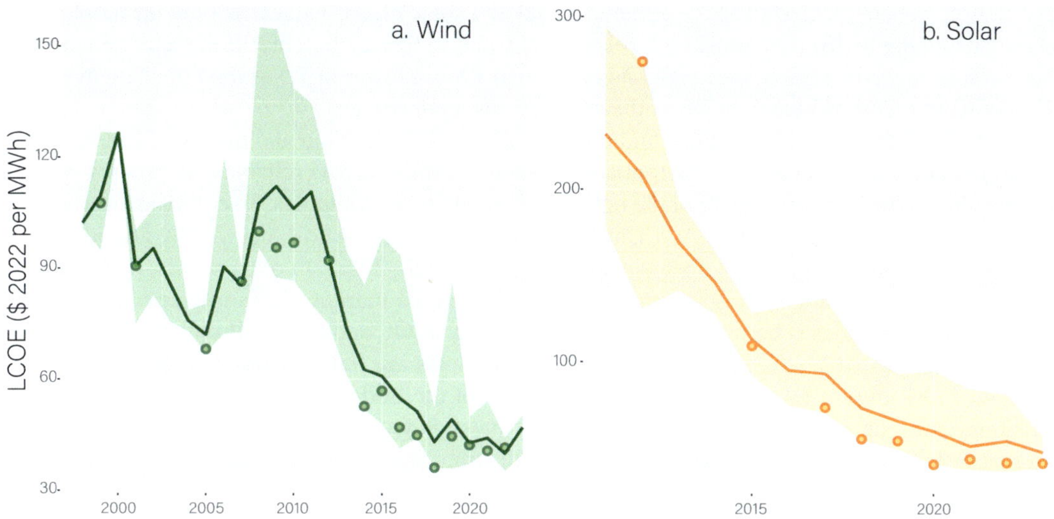

FIGURE IV.3.5 Levelized Cost of Electricity (LCOE) values decreasing over time for (a) wind and (b) solar technologies. Shaded areas represent entire range across different U.S. regions, whereas lines represent the average. Points represent values specific to Texas.

Data from LBNL 2025.

Although wind and solar installation costs are higher than CC-NG plants, the LCOE values are far more competitive. As of 2023, the U.S. average LCOE estimates for PV solar and onshore wind were $41.22/MWh and $50.87/MWh, respectively, without including any tax credits (EIA, 2023a). If storage options are included, LCOE values increase. For example, solar PV projects with batteries are currently estimated at $54/MWh. In contrast, LCOE (2023) values for CC-NG plants are $42.72/MWh (Figure IV.3.6). This does not consider financial subsidies, which are covered in the

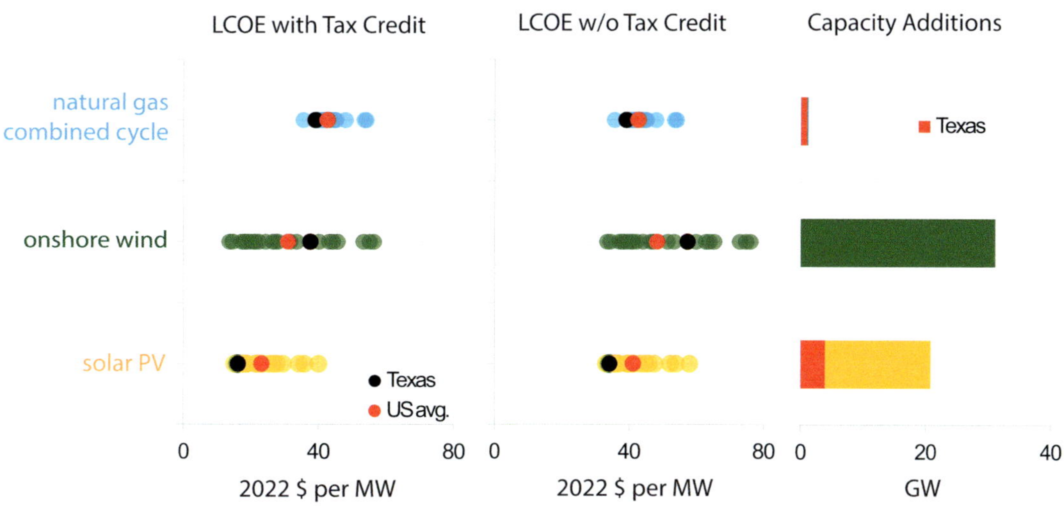

FIGURE IV.3.6 Levelized Cost of Electricity (LCOE) values for three technologies with and without recent Inflation Reduction Act (2022) Tax Credits and capacity additions. Points represent all regions of the U.S., the U.S. average, and specific to Texas region.

Data from EIA (2023a).

next section. When tax credits are considered, wind and solar LCOE values become far more competitive, dropping to \$31.07 and \$23.22, respectively, and to \$36.27 for solar if storage is included. These values range by region, where CC-NG and solar PV are lower in Texas than the U.S. average, whereas wind LCOE is higher in Texas than the U.S. average (Figure IV.3.6).

There are several noted limitations and uncertainties of LCOE (Tran and Smith, 2018). A major limitation is that LCOE does not differentiate between intermittent and reliable generation sources. This provides a limited perspective of the potential revenue of technologies. To compensate, the levelized avoided cost of electricity (LACE), as the name suggests, is the cost incurred by providing the same electricity had the plant not been constructed (Loth et al., 2022). In other words, LACE is the cost of electricity and ancillary services if purchased from the market. Economic competitiveness of different technologies is better explained by comparing LCOE and LACE. Projects are economically attractive if LACE exceeds LCOE. Very few examples are found where LACE dramatically exceeds LCOE (Figure IV.3.7); however, the majority of the buildouts of technologies have been situations in which LACE is approximately equal or slightly exceeding LCOE (Figure IV.3.7). Examples of solar and CC-NG buildouts in Texas illustrate this dynamic (Figure IV.3.7). Here, we see the virtue of the LCOE–LACE comparison. Even though CC-NG plant LCOE values are higher than wind and solar, LACE estimates are also higher, suggesting higher electricity value from dispatchable energy. An economically attractive opportunity occurs where LACE approximates the LCOE, illustrated by the examples of solar PV and CC-GN buildouts in Texas. In contrast, CC-NG appears economically unattractive for other regions across the United States (Figure IV.3.7).

Although solar and wind industries have rapidly grown, capacity additions for conventional technologies, particularly CC-NG, are still ongoing. Even though LCOE and LACE provide long-term perspectives, investors who desire to engage in PPAs with power producers may be unwilling to be bound in long-term agreements. Low-capital fossil-fuel plants, such as CCNG facilities, tend to be favored in markets supporting short-term PPAs (Beck and Martinot, 2004; Painuly, 2001).

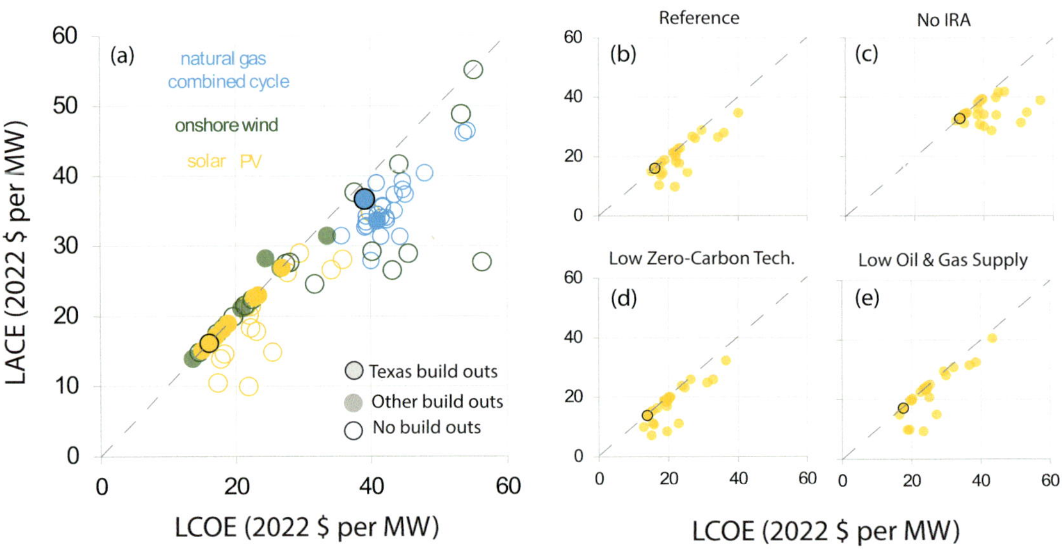

FIGURE IV.3.7 Comparison of Levelized Cost of Electricity (LCOE) values versus Levelized Avoided Cost of Electricity (LACE) for three technologies: (a) Regions with and without buildouts are noted, including buildouts in Texas; (b–e) LCOE versus LACE values for solar PV across regions among different scenarios within the Energy Information Administration's 2023 Annual Energy Outlook. Texas is indicated by points with a solid black outline.

Data from EIA 2023a.

With lower upfront capital costs and low gas prices, ACC-NG IPPs have attracted investors and short-term PPAs, and hence, have expanded rapidly in Texas in the last 10 years. Beyond providing base load power, natural gas power plants participate in the ancillary market (peak demand) and take advantage of higher electricity rates. LACE values may not capture all the additional grid benefits from CC-NG technologies. Going forward, it is difficult to predict how natural gas and renewables will continue to coordinate or compete in Texas's electricity market, pending various influences of subsidies and market dynamics. For these reasons, it is imperative to review the influence of legislation, financial incentives, and deregulation of the electricity market on renewable energy growth in Texas.

IV.3.7 LEGISLATIVE MANDATES AND FINANCIAL INCENTIVES

The wind and solar industries' impressive growth in Texas has far exceeded the goals established by the RPS, with current renewable generating capacities at almost 7x the standard set for 2025. Such a dramatic exceedance of the standards questions the role of the RPS as the sole impetus behind renewable growth. More likely, the RPS, in concert with multiple other legislative and financial incentives, both at state and federal levels, as well as technological and operational measures, are responsible for fostering growth. Discerning which among these factors had the most influence is extremely difficult. However, we can explore the timing of different legislative measures and provision of incentives, or lack thereof, relative to industry growth.

Measurable wind development can be observed following the original establishment of the RPS in 1999; however, a perfect storm of legislation and initiatives in 2005 were responsible for the aggressive growth in the wind industry from 2006 to 2009 (Figure IV.3.8). Specifically, expansion of the RPS and establishment of the CREZ under SB20, supplemented further by federal tax incentives, such as those provided by the Energy Policy Act of 2005, were followed by an upsurge in wind generating capacity (see Figure IV.3.2). As a part of SB 20, the RPS mandated that utilities incorporate roughly 30–35% of wind-generated electricity into their sales through Renewable Energy Credits. At the most fundamental level, this means retail utilities are obligated to create either power purchase agreements (PPAs) or accept day-ahead bids (Bhaskar and Singh, 2014; Sioshansi and Hurlburt, 2010; Van Kooten, 2016) from wind farm operations, regardless of the cost per unit energy. Simultaneously, the new CREZ initiative supported expansion of new high voltage transmission to areas of high wind production in western and northern areas of Texas, enabling wind development to continue. Prior to this time, these areas were limited by substation and transmission capacity.

Since 1992, federal legislation has provided financial incentives to renewables through two measures, specifically the Production Tax Credit (PTC) and the Investment Tax Credit (ITC). As the name suggests, the PTC provides a tax credit for every kwh of renewable energy produced. The ITC allows taxpayers to reduce their tax liability by a percentage of the cost of renewable energy investments. Generally, the PTC, although generally available to most renewable technologies, had significant influence on the wind industry, particularly creating volatility as the policies lapsed and then were extended multiple times (Frazier et al., 2019; Mai et al., 2016) (Figure IV.3.8). In contrast, the ITC has been far more influential on the solar industry (Aslani and Wong, 2014; Frazier et al., 2019; Stokes and Breetz, 2018) (Figure IV.3.8). However, it should be noted that the US-wide growth and subsidence patterns in the wind industry are not necessarily mirrored in Texas. This makes it difficult to understand how much federal policies have influenced wind in Texas relative to other factors.

The Energy Policy Act (1992)[1] originally provided a PTC of 1.5 cents/kwh for qualifying electricity production and was implemented for 6.5 years, after which Congress let it expire in 1999. It was extended twice, but not before expiring each time (3 total lapses). Finally, the Energy Policy Act of 2005[2] extended the PTC for 2 years, which was further extended by the Emergency Economic Stabilization Act of 2008[3] and the American Recovery and Reinvestment Act of 2009[4]. Collectively,

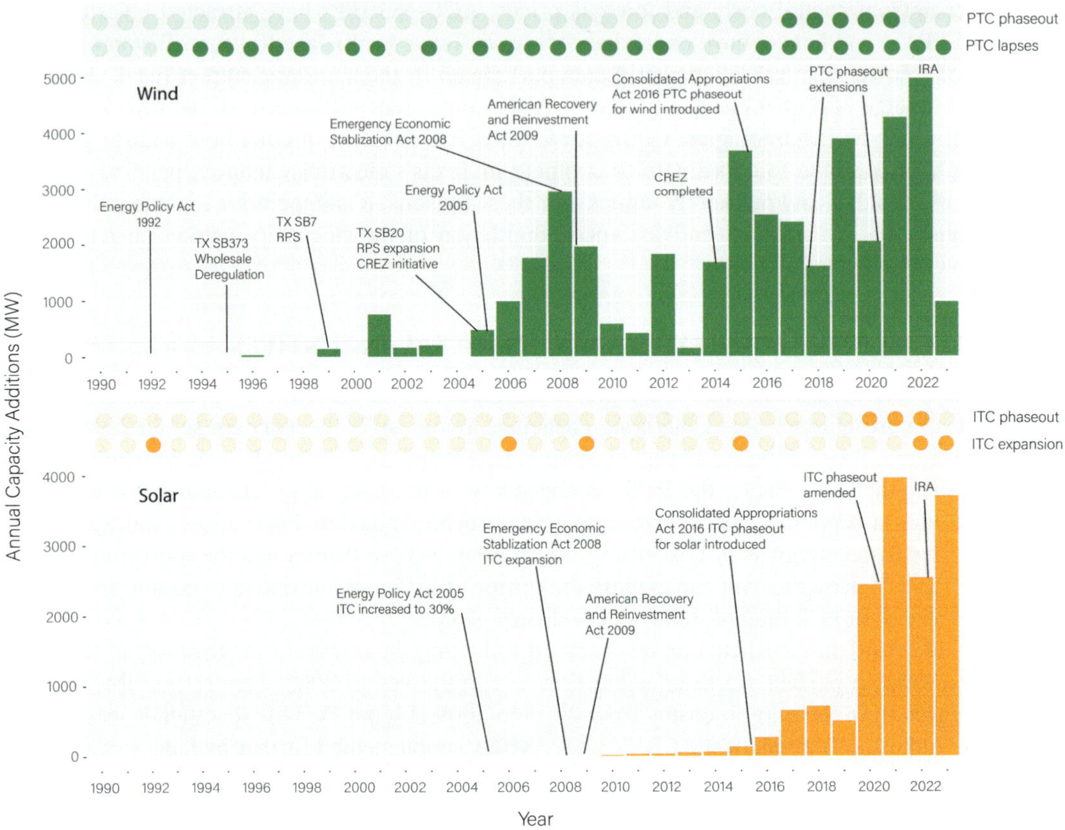

FIGURE IV.3.8 Annual capacity additions of wind and solar in Texas in relation to major federal and state legislation. Solid green dots indicate years of availability in the production tax credit (PTC) for wind production whereas lapses in PTC coverage are indicated by unfilled dots. Green dots also indicate either phaseout introduction or modification for the PTC. Solid Orange dots indicate investment tax credit (ITC) phaseout or years of expansion in relation to solar growth.

Data on capacity additions from EIA Form-860 (EIA, 2024b).

these legislation measures provided the wind industry with a long period of policy stability, fostering large expansion until 2013, when the PTC was allowed to expire again (Figure IV.3.8). Since its inception in 1992, the PTC has expired at least 5 times, each time requiring extensions. The uncertainty of policies and financing during this time led to market volatility, characterized by boom-or-bust cycles of growth (Frazier et al., 2019; Mai et al., 2016). The Consolidated Appropriations Act of 2016[5] provided the wind industry with another long forecast of policy certainty. The Act also increased the PTC to 2.3 cents per kwh but also introduced a phaseout, where projects commencing construction in 2016 would receive the full value, projects started in 2017 would receive 80% of PTC value, 60% for 2017, and 40% for 2019. However, Congress extended these phase outs twice, providing an extended lifeline to continued wind production.

The ITC has a far longer history initiated by the Energy Tax Act of 1978[6], where both solar and wind received 10% refundable tax credit. The tax credit undulated between 10 and 15% from 1980 to 1992, until the 10% credit was made permanent by the Energy Policy Act of 1992. The solar industry received a significant incentive with the Energy Policy Act of 2005 where the ITC was raised to 30% and residential systems were made eligible. Although tax credits provided new financial incentives for investors, developments in Texas remained extremely slow during the recession

(2007–2009) (Figure IV.3.8). The American Recovery and Reinvestment Act of 2009 overcame the financing obstacles by enabling the Treasury Department to issue cash grants in lieu of tax credits, which voided the need for tax equity financing. Growth was minimal after this time until the Consolidated Appropriations Act of 2016, which provided a multi-year extension of the ITC, but also introduced a phaseout, mirrored by the PTC (30% through 2019, then 26% in 2020, 22% in 2021, and 10% in 2022 onward). Noticeable growth in the solar industry was observed in Texas after 2016, whereas substantial expansion occurred in 2020 through current (Figure IV.3.8). Part of this growth was due to the amendment of the phaseout in late 2020.

The Inflation Reduction Act (IRA) of 2022[7] changed the renewable energy policy landscape in several major ways (IRS, 2025). First, the IRA shifted the ITC and PTC away from technology-specific credits to technology neutral credits with zero or negative greenhouse gas emissions. Additionally, the IRA also provided flexibility by affording projects a choice between the ITC or PTC and transferability, and hence monetization, of tax credits. The IRA also provided long-term stability to the ITC and the PTC by making the expiration of the credits tied to achieving reductions in greenhouse gas emissions (75% reduction from 2022) rather than arbitrary dates. Finally, the IRA provided opportunities for bonus credits and labor provisions. Companies could receive bonus credits if using domestically produced equipment, providing energy in low-income areas, and ensuring workforce training and higher wages. Final rules for the IRA were issued by the Department of Treasury and the Internal Revenue Service in January 2025. A summary of the PTC and ITC under the IRA 2022 is provided in Table IV.3.3.

Wind capacity currently makes up 22% of the total energy generation portfolio in Texas, which is a significant achievement given the Department of Energy's Wind Vision of obtaining 20% of the nation's electrical demand from wind by 2030 and 35% by 2050 (US DOE, 2017). Additionally, the growth of wind and solar has exceeded the RPS's goals for generating capacity, which questions the continued role of the RPS. In spring 2015, state senators voted to repeal the RPS, claiming the legislation had done its job (TX SB 931, 84th Leg). The assumption was that wind has a strong foothold in Texas and should be able to compete in Texas's open electricity market without continued subsidies. The repeal failed to get legislative approval. Had it gone through, the Texas wind industry would not have been left empty-handed, given the extensions in PTC. However, the RPS does mandate RECs, which ensure PPAs with retailers and larger profitability of wind production in the

TABLE IV.3.3
Production Tax Credit and Investment Tax Credit Structure and Provisions under the Inflation Reduction Act (2022)

Project Size	Base Incentive	Energy Community Bonus	Domestic Content Bonus	Total Possible
	Production Tax Credit (¢ per kwh)			
< 1 MW	2.75¢	0.275¢	0.275¢	3.30¢
> 1 MW	0.5¢	0.055¢	0.055¢	0.66¢
> 1 MW that meets prevailing wage and apprenticeship requirements	2.75¢	0.275¢	0.275¢	3.30¢
	Investment Tax Credit			
< 1 MW	30%	+10%	+10%	50%
> 1 MW	6%	+2%	+2%	10%
> 1 MW that meets prevailing wage and apprenticeship requirements	30%	+10%	+10%	50%

Taken from US DOE (2025) and US EPA (2025b).

deregulated market beyond the PTC. The continued penetration of wind and solar has and will be partially dependent upon the economic viability apart from legal mandates and subsidies in Texas's deregulated market, which I discuss in the next section.

IV.3.8 DEREGULATION OF ENERGY MARKETS: ENERGY COMPETITION

Deregulation of electricity markets in Texas, like deregulation elsewhere, is hypothesized to provide openings for smaller-scale energy producers, including renewables, through market competition. Debate continues, however, over the success of energy market deregulation in achieving its primary objectives, principally in ensuring reliability, efficiency, and providing cheaper electricity rates for customers (Borenstein and Bushnell, 2000). Most agree that efficiency has been gained (Borenstein and Bushnell, 2000; MacKay and Mercadal, 2024); however, Woo et al. (2006) suggest that deregulation can lead to many challenges, including high set-up costs, complicated market designs, price volatility, inefficient or de-incentivized investments, stranded costs, and unequal distribution of benefits. Furthermore, the role of market competition in promoting renewable energy growth versus within regulated markets is just as unclear (Beck and Martinot, 2004; Eun Kim et al., 2016; Painuly, 2001). To understand the potential influence of deregulated electricity markets on renewable energy competition, an overview of ERCOT's market structure and renewable technologies is beneficial.

Legislative changes in Texas in 1995 and 1999 can be thought of as establishing two phases of deregulation in Texas's electricity market. The first phase – the deregulation of the wholesale energy market – separated investor-owned utilities into distinct entities: retail utilities, energy operators, and transmission services. The second phase of deregulation, i.e., restructuring of the vertical retail energy market, removed the captive customer bases from retail utilities, which allowed customers to select their own retail utility.

Deregulation of the wholesale and retail electricity market required establishing an independent entity, i.e., ERCOT, to manage the grid and retail market. ERCOT is a nonprofit organization that operates the wholesale electricity market for 90% of Texas and schedules power on an electric grid including 46,500 miles of transmission and almost 600 generation units (Figure IV.3.9). ERCOT

FIGURE IV.3.9 Electric Reliability Council of Texas (ERCOT) control room.

Courtesy of ERCOT. https://www.ercot.com/news/mediakit/corp_images

cannot own any generation assets but procures and deploys electricity and ancillary services. In a deregulated market, power producers compete to provide electricity to willing buyers at a low rate. Therefore, ERCOT operates the grid by procuring energy through auction or direct contracts for reliability (Sioshansi and Hurlbut, 2010).

Since the early stages of deregulation and with increasing transmission system evolution, ERCOT has changed its grid management strategy to ensure reliability and efficiency, all while accommodating VREs. Originally, ERCOT operated as a zonal market in 2001, where zones were defined based on commercially significant transmission pathways and areas of congestion (Sioshansi and Hurlburt, 2010). This price structure did not appropriately represent location-variant electricity and market values. Even so, under this structure, ERCOT provided provisions and flexible allowances for VREs, specifically wind producers, allowing extreme deviations (up to 50%) in real-time versus scheduled generation, whereas deviations remained highly restrictive (1.5%) for other resource entities (Sioshansi and Hurlburt, 2010). However, the liberal allowance for deviation created a perverse incentive for wind producers to overschedule generation and receive compensation for the difference under situations of curtailment (Sioshansi and Hulrburt, 2010). Although several corrective measures were implemented to disincentivize this behavior, the zonal market was eventually transitioned into a locational marginal pricing (LMP) model for nodes, or specific locations in the power network where the market settlement price for a generation resource varies to reflect the value of power at a given location. From here on, the remainder of the discussion of deregulated markets reflects the current LMP model.

ERCOT's nodal market is described by Sioshansi and Hurlburt (2010) and consists of four main elements: 1) a day-ahead market (DAM), which is purely a financially binding market for buying and selling electricity, 2) a day-ahead reliability unit commitment (DRUC) to ensure commitments of sufficient generating capacity, 3) an hourly RUC for resolving short-term forecasts in load and generation, and 4) a real-time market, where generation or load-management is dispatched in real-time. Qualified Scheduling Entities (QSEs) submit bids and offers on behalf of resource entities, i.e., power producers, and load serving entities, such as retail electric providers. QSEs schedules resources, procures ancillary services to serve anticipated loads, settles financial agreements and prices, and handles errors. The QSE submits scheduled generators and outputs for each node to ERCOT. ERCOT's DAM is a voluntary, financially binding energy market that matches willing buyers and sellers, subject to constraints and optimization of ancillary services and prevention of grid congestion. Power producers, including renewable sources, can participate in the DAM by submitting bids to sell electricity to willing buyers. Outside of the DAM, producers can engage in long- or short-term PPAs where buyers can purchase electricity from a producer at a fixed price, rather than an outcome of the DAM. Ancillary services are handled in a DAM with generally higher price structures due to the critical nature of the service provided to balance the grid; hence, participation in the ancillary market can be a lucrative strategy for power producers. Under the current market, only registered power producers, capable of dispatchable power production (not VREs), can participate in the ancillary market following tests ensuring they can meet the required operating characteristics. Apart from the DAM, ERCOT conducts the DRUC to ensure sufficient generation capacity and reserves are committed to serve the forecasted base load and ancillary needs; the hourly RUC process is similar but conducted on an hourly basis.

IV.3.8.1 Wholesale Energy Market Deregulation and Renewable Competition

In terms of wind and solar penetration in the energy market, wholesale energy market regulation could ideally benefit independent-power-producers (IPPs), such as renewable plants, by removing monopolies among retailers and operations and allowing wind and solar technologies to compete with other sources of energy (Beck and Martinot, 2004; Painuly, 2001). For instance, unbundling operators from utilities removes biases against renewable technologies by eliminating the price structures and economic advantages of monopolized large power plants, wider client bases, and controls over transmission (Borenstein and Bushnell, 2000; 2015). Restructuring tends to favor IPPs that are smaller, distributed, and constitute lower investment risks than large power plants (Beck and Martinot, 2004; Painuly, 2001). Historically, utilities were also less likely to invest in renewables because of steep costs of developing transmission lines to regions with high wind capacity (Milstein and Tishler, 2011). With the CREZ initiative, however, the cost burden of transmission is now offset to retail utilities and, subsequently, customers (Milstein and Tishler, 2011; Zarnikau, 2005; more on this below). Thus, IPPs compete purely on the grounds of capital cost of facilities developed in closest proximity as possible to grid installments. However, as seen in Figure IV.3.8, wind development was not necessarily responsive to wholesale deregulation in 1995. Capacity additions were minimal post-1995 and were not noticeably higher until after the RPS (1997), increasingly noticeably after 2005. Even though the electricity market was open for competition, the early stages of wind industry growth may not have been economically attractive without subsidies, mandated RECs, and without lower transmission burdens (CREZ). While deregulation should accommodate competition, this assumes that any IPP can equally compete in an open market under status quo financial and operational conditions. Indeed, there are obstacles presented by an open energy market that can also stymie renewable growth. To fully consider the challenges, we need to envision a situation in which Texas's deregulated market occurred in the absence of other state and federal legislative mandates, such as the RPS and tax credits.

Left purely to an open wholesale electricity market, the penetration of wind and solar would be a matter of return on capital investment and competitive pricing ($ per MWh). As discussed, the LCOE and LACE values suggest that, when considering federal subsidies, wind and solar technologies can compete with natural gas. However, LCOE and LACE values consider the capital and operational costs across the life of the project, which may not match the timelines needed to attract investors. In the early stages of deregulation, IPPs with higher initial capital costs, such as large renewable installations, have difficulty competing in markets driven solely by pricing as they tend to require procuring longer-term PPAs (Beck and Martinot, 2004; Painuly, 2001). Under uncertain financial conditions, short-term PPAs may be preferred, leading to investments in low-capital installations that provide higher returns in short order, such as ACC-NG (Beck and Martinot, 2004) despite having higher marginal costs over the long-term (i.e., purchasing fuel). However, natural gas PPAs are volatile and subject to fluctuations in gas prices. Therefore, one advantage of renewable PPAs is that they are constant for the entire term of the agreement. In a regulated market, lower LCOE and LACE could influence investment toward renewables. In a deregulated market, however, the question is whether the potential of increasing gas prices is sufficient to reduce the utility's perceived financial risk to engage in long-term PPAs with renewables or continue to play their odds in a hedge-betting strategy.

Wind and solar producers have responded by providing competitive pricing. Nationwide PPA prices for the wind and solar industries peaked in 2009 and have since continued to fall (Figure IV.3.10). PPA prices in ERCOT are among ethe most competitive wholesale price in conterminous United States (LBNL, 2025; Wiser and Bolinger, 2019). As of 2022, the US-average PPA price for wind was 2.65 cents per kwh and 3.22 cents per kwh for solar (Figure IV.3.10). Wind PPA price in ERCOT in 2017 was 2.37 cents/kwh, whereas solar as of 2018 was 3.22 cents per kwh. Recall that under the Inflation Reduction Act, the current PTC is up to 3.30 cents per kWh for qualifying facilities, whereas the base ITC provides 30% in credit. Therefore, wind and solar producers roughly double their revenue (or half their liability) on account of tax credits.

Federal subsidies, however, can also lead to perverse behaviors in deregulated markets and cascading implications on all participants. As mentioned above, retail suppliers in deregulated markets may be reluctant to engage in long-term PPAs, as many rely on day-ahead pricing to buy energy (Sioshansi and Hurlbutt, 2010; Van Kooten, 2016). In this case, wind and solar producers can still participate in ERCOT's DAM to seek willing buyers. During periods of excess supply, wind and solar operators place negative day-ahead bids to ensure their energy is purchased and then rely on PTCs to recover their production value (Deng et al., 2014; Huntowski et al., 2012; Zarnikau, 2011). In other words, wind operators pay to generate energy and obtain less than half the income that

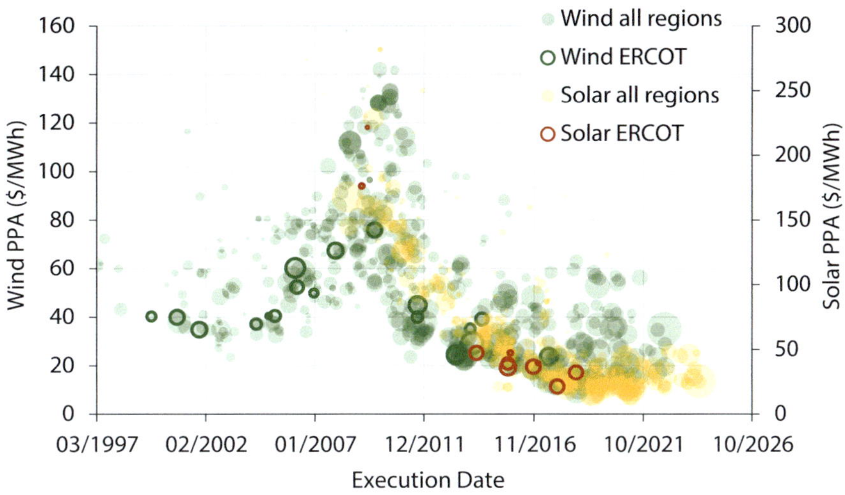

FIGURE IV.3.10 Wind and solar power purchasing agreements (PPA) for entire US and in Texas according to the execution date. Size of points represents relative project size in megawatts generating capacity.

Data from LBNL (2025).

wind operators with long-term PPAs receive. The problem that arose with this strategy is that it distorted price signals in the electricity market, leading to price volatility, reduced incentives for dispatchable generation units, and undermining grid reliability (Huntowski et al., 2012; Woo et al., 2023; Zarnikau, 2011). In any competitive economic market, price signals are important components of balancing supply and demand; when demand is high, prices increase, and when demand is low, prices decrease. In a balanced market, excess supplies are discouraged from market entry because of price reduction. In this case, the federal PTC creates false signals, distorted by incentives provided to some market participants and not others. Over time, this price reduction created directional behaviors and grid reliability challenges, particularly reducing investment incentives in dispatchable generation projects (Woo et al., 2023). ERCOT has since made market adjustments to prevent perverse behaviors and incentivize dispatchable technologies (ERCOT, 2024).

Even in the absence of negative bidding, high penetration of VREs, like wind and solar, can cause challenges. Wind and solar have low marginal costs of operation because they do not rely on purchasing fuel sources, i.e., the lower operating and maintenance costs reflected in LCOE values. Competitive electricity markets will prioritize the lowest electricity production costs for generation, ranking generators based on marginal costs, i.e., a merit order system. Therefore, in deregulated markets, wind and solar are ranked higher for provision of low-cost electricity, leading to merit order effects, i.e., lower ranking and eventual displacement of conventional (and dispatchable) power producers (Quint and Dahlke, 2019). Merit order effects lead to a reduction in market clearing prices, lowering wholesale electricity prices and reducing incentives for conventional dispatchable producers (Pena et al., 2022), eventually leading to grid reliability challenges. To compensate, ERCOT has increased the value of "price adders" that add value to wholesale prices for real-time reserves to continue incentivizing dispatchable technologies (Woo et al., 2023). The reduction in market prices from high penetration of wind and solar, while initially increasing competition with other technologies, can eventually lead to "cannibalization," i.e., undermining the remuneration of their own market over the long term (Pena et al., 2022; Prol et al., 2020). Hence, the stability of the market, including the long-term viability of renewables, will require adaptive adjustments.

Because VREs have no ability to react to short-term market demands and signals, it seems inappropriate for renewable power production to participate in day-ahead deregulated markets (Klessmann et al., 2008). In part, this is due to an inability to make accurate hourly wind forecasts, which if improved, would reduce risks in balancing costs due to unforeseen intermittency and provide better grounds for day-ahead bidding (Holttinen, 2005). Another option used commonly in Europe is a feed-in tariff scheme, which would set aside wind production from other market participants and guarantee a fixed price and sold volume (Hiroux and Saguan, 2010). Feed-in tariffs could also include rules where wind power producers could bear the responsibility of costs incurred through imbalances, which are computed as deviations from the producer's projected day-ahead load-profile (Hiroux and Saguan, 2010).

ERCOT's wholesale market prices have generally been lower than the average of other ISOs across the U.S., with the exception of 2021, due to winter storm Uri (Table IV.3.4). Wholesale market value ($/MWh) of wind and solar generation can be estimated based on generation, nodal location, and capacity in the market (LBNL, 2025). Wholesale market values, apportioned to wind generation, although displaying fluctuations, have decreased 19.5%, on average, over the past 20 years (Figure IV.3.11). In contrast, solar market prices have, on average, remained constant

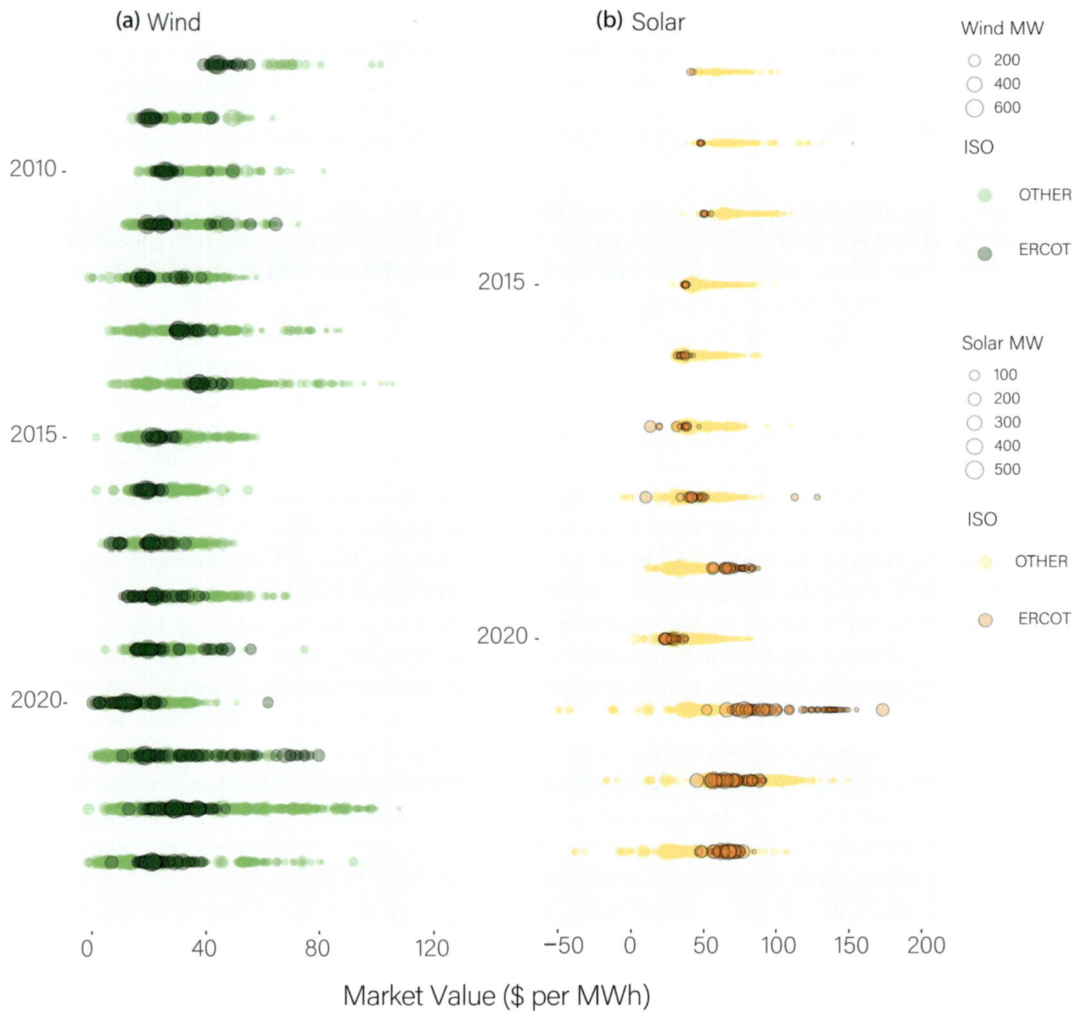

FIGURE IV.3.11 Wholesale market values are estimated for (a) wind and (b) solar generation for all ISO regions, including ERCOT. Generating capacity is provided by size of bubble.

Data from LBNL (2025).

TABLE IV.3.4
Average Annual Wholesale Market Prices ($/MWh) for all ISOs in the US

Year	CAISO	ISO-NE	MISO	NYISO	PJM	SPP	ERCOT
2012	$38.72	$47.23	$33.68	$48.64	$42.68		
2013	$51.09	$71.28	$38.40	$63.23	$46.38		
2014	$58.44	$79.10	$45.82	$69.55	$59.94		$42.40
2015	$38.77	$51.32	$31.32	$45.26	$41.59	$28.26	$30.86
2016	$34.31	$36.03	$30.50	$34.33	$34.20	$26.65	$27.75
2017	$35.83	$41.01	$32.96	$36.33	$35.93	$27.60	$31.79
2018	$40.89	$51.74	$35.59	$45.44	$42.60	$30.69	$38.71
2019	$40.60	$36.17	$28.73	$31.32	$30.81	$25.73	$45.28
2020	$31.29	$27.28	$23.21	$24.54	$24.18	$20.75	$26.96
2021	$45.61	$50.23	$39.99	$44.08	$42.80	$47.42	$168.19
2022	$76.74	$89.39	$61.20	$82.09	$75.97	$46.27	$66.55
2023	$54.05	$36.24	$28.63	$31.65	$29.73	$26.47	$50.15
2024	$35.12	$39.09	$26.47	$36.60	$30.48	$22.97	$28.43

Data from Millstein et al. (2025).

despite fluctuations (Figure IV.3.11). This suggests that higher solar prices may help to stabilize wholesale market prices for other technologies and mitigate against loss of incentives for dispatchable technologies and subsequent reliability issues. ERCOT has and will continue to adapt its policies to ensure market stability and incentives.

IV.3.8.2 Retail Energy Market Deregulation

Providing customers with the option of selecting their electric retail utility can potentially influence renewable energy penetration in two ways. First, most directly, retail deregulation allows customers to make "green" choices or select retailers sourcing a certain amount of their electricity from renewables, providing a market-driven demand for renewable energy (Bae et al., 2014). Secondly, through economic efficiencies and lower marginal costs from renewable generation, deregulation could lower retail electricity prices, potentially cultivating more support for renewables (Bae et al., 2014). Currently, 120 different retail utilities are available for over 90% of Texas's residential customers, i.e., those residing within ERCOT's market area. Of these, 35 providers offer 100% renewable energy sourced electricity plans, 63 provide at least 24% renewable energy sourced plans, and the remainder provide at least 6% renewable energy sourced power (Power to Choose https://www.powertochoose.org/). Although the number of customers adopting different plans is not currently available, pricing among plans can be evaluated. In reviewing all electricity providers, there is no apparent relationship between the percentage of electricity sourced from renewable energy sources and the residential price of electricity. As of April 2025, the average residential price for providers offering 100% renewable-energy-sourced plans is 15.01 cents per kWh, whereas 14.88 cents per kWh is the average price for providers offering 33% or less renewable energy.

Debate continues over the effects of deregulation on retail electricity prices (Borenstein and Bushnell, 2015; Hartley et al., 2019; Zarnikau and Whitworth, 2006). Without even considering retail deregulation, wholesale market deregulation could influence retail prices, under the assumption that savings from lower marginal costs of generation and increased economic efficiencies are passed along to the customer. Additionally, it could be implied that vertically integrated utilities have traditionally overcharged captive consumer bases for electricity. However, rates in regulated markets are generally lower than deregulated markets (American Public Power Association, 2022), although this could be reflective of a regional pattern since the price differences among regulated and deregulated states have remained constant since the 1990s, i.e., before widespread adoption of electricity markets (Hartley et al., 2019).

Trends in retail electricity prices in Texas have further obscured the debate. Immediately following retail deregulation, retail electricity prices increased 44% (Figure IV.3.12). These increases were far higher in competitive markets in Texas than non-competitive ones (Co-operatives and municipalities) (Zarnikau and Whitworth, 2006), suggesting increased prices were reflective of whole market volatility from increased marginal generation costs. To be fair, the increase in prices at that time was largely driven by exogenous drivers, particularly increased natural gas prices (Borenstein and Bushnell, 2015; Hartley et al., 2019). Indeed, deregulated electricity markets are more susceptible to rapid changes in marginal costs of generation from shifts in fuel prices. However, residential electricity prices in Texas have remained high since 2009 despite the large declines in gas prices, whereas commercial and industrial have since declined and stabilized, with the exception of 2021 during winter storm Uri (Figure IV.3.12). Assuming residential electricity prices were simply a linear response to natural gas prices, observed residential prices 2009 to 2025 are far higher than predicted. Increasingly, residential rates, although still reflective of natural gas price signals, have disproportionately increased relative to other electricity sectors. For example, most recently in 2024, natural gas prices hit a record low, yet residential electricity rates have continued to climb. It is extremely important to note that, over the same period, U.S. retail prices, on average, have also increased, albeit those averages reflect 29 other states with unregulated electricity markets,

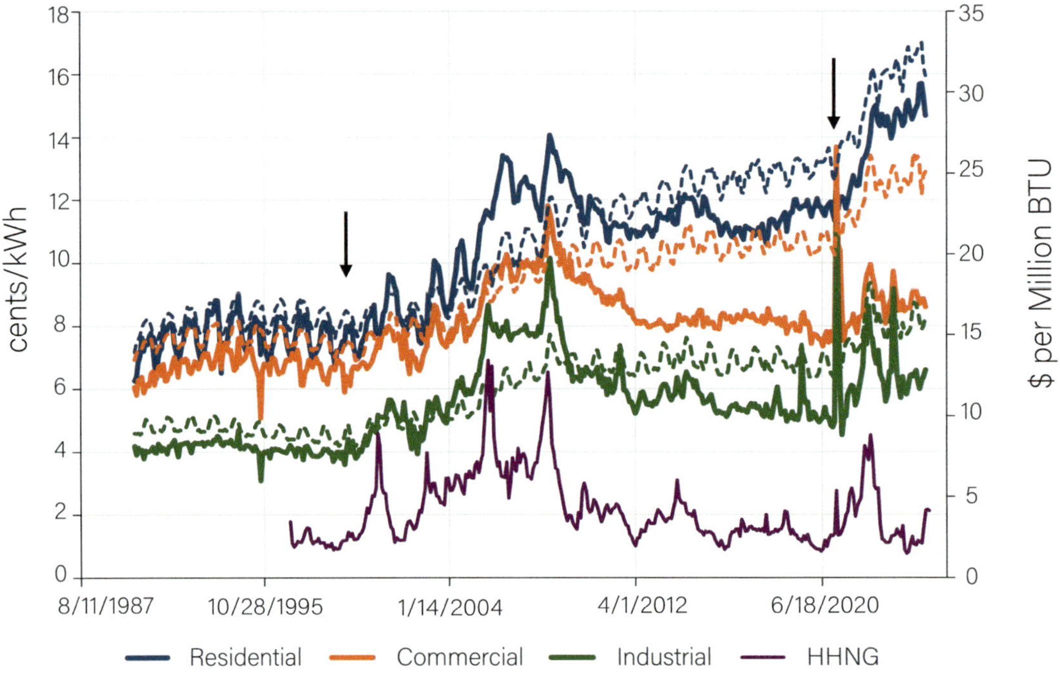

FIGURE IV.3.12 Monthly retail electricity prices in Texas for residential, commercial, and industrial sectors. Monthly Henry Hub Natural Gas spot prices ($per million British Thermal Units) are provided for reference. Solid lines represent prices in Texas whereas dashed lines show the U.S. average. The first arrow depicts the restructuring of retail market in 1999, whereas the second arrow depicts the occurrence of Winter Storm Uri in 2021.

Data from EIA (2025a).

further suggesting that deregulation has not decreased retail prices. However, Texas retail prices have become noticeably lower than the U.S. average post-2009 and have since continued to dissociate. What is more peculiar about Texas, however, is the degree of difference between residential and the other sectors, suggesting an unequal distribution of benefits of deregulation among sectors (Borenstein and Bushnell, 2015; Woo et al., 2006).

There are many potential reasons for increased retail prices following deregulation. Milstein and Tishler (2011) suggested that increases in renewable penetration lead to increased volatility in electricity prices and average electricity rates. Apart from renewable growth, deregulation of wholesale electricity markets inherently creates price volatility due to price swings in response to uncertainty in supply and demand (Milstein and Tishler, 2011). As ERCOT has matured, prices have continued to evolve due to market adjustments. For instance, noticeable increases in prices in 2022 and 2023 are predominantly due to market structure changes in ERCOT in response to the grid failure from winter storm Uri in 2021 (see Chapter IV.4, Energopolitics, D. Boyer).

Borenstein and Bushnell (2015) provide a 20-year retrospective evaluation of the political motivations and economic implications of electricity restructuring in the U.S. Despite the promise of competition-driven lower electricity rates by political advocates, rates have increased abroad, increasingly in deregulated markets. The authors suggest that electricity rates have not been driven by restructuring, but rather exogenous factors, such as generation technologies and natural gas fluctuations. Although there is evidence that competition has increased the economic efficiencies and better grid coordination, most efficiency improvements (that would lead to lower cost to customers) are more so related to technological advancement, such as new grid forecasting techniques and scheduling. In contrast, others suggest that market changes under deregulation are directly responsible for increases prices (MacKay and Mercadal, 2022, 2024). In regulated markets, utilities own generator assets and directly bear the cost of generation, distributing that to customers (Mackay and Mercadal 2024). Once markets were deregulated, retailers were purchasing power at the higher market price, even though generational costs were unchanged, undoubtedly passing higher costs on to customers. Similarly, higher consumer prices following deregulation are also documented in an empirical analysis by MacKay and Mercadal (2022), where wholesale market prices increased. In deregulated markets, developing RTOs, such as ERCOT, require significant start-up costs, as these organizations assume responsibilities for adjusting supply of electricity to meet changing demand, a role historically born by utilities (Milstein and Tishler, 2011; Woo et al., 2006). Although lower wholesale prices can be passed to savings for retail customers (Brown et al., 2020), high VRE increases the need (and cost) for ancillary services and transmission (Woo et al., 2023). Hence, the more probable reason for increased prices is that the burden of RTO development and state-wide transmission expansion (e.g., CREZ) have been passed on to residential customers (Milstein and Tishler, 2011; Zarnikau and Whitworth, 2006).

Hartley et al. (2019) pose a counter argument, suggesting that the deregulation in Texas actually led to declines in rates, arguing that most studies have only looked at short-term temporal changes within Texas and not comparing competitive markets with non-competitive providers (Co-operatives and municipalities). In an evaluation from 2005 to 2016, the authors found that retail prices were more reflective of wholesale prices and marginal costs of generation and have declined post-deregulation. As of 2016, the authors note that the commercial price within competitive markets was 1 cent/kwh lower than non-competitive markets and residential prices within competitive markets have declined to the point of being similar to non-competitive markets.

Even considering the counter-analysis provided by Hartley et al. (2019), the clear benefits of deregulation on retail electricity rates, at least for most Texas customers, have been marginal at best. Other factors, such as exogenous drivers and transmission expansion, have likely diminished any measurable benefit of deregulation, at least for residential customers. Research shows that Texans are willing to pay increased rates for renewable energies (Farhar, 1999), specifically $16.00 (2025 dollars) annually for every 1% increase in the fuel mix from renewable sources (Murakami et al., 2015). The average Texas household currently uses approximately 14,000 kwh annually (EIA, 2025b) at a price of $0.147 per kwh, which translates to $2,055 annually for electricity compared to $1120 in 2002 (0.08 per kwh). Wind and solar energy currently make up 28% of the overall generation in the state compared to 0.5% in 2002, suggesting Texans are willing to pay an additional $440 annually for the current level of renewables. However, this does not account for natural increases in electricity rates over that time period, which, on average, have increased 64% to 83% (an additional $735 annually) in Louisiana and Mississippi, both neighboring regulated states with less than 3% renewable energy each (EIA, 2025c). Texans' electricity rates have increased 85%, paying an additional $965 annually in 2024 than in 2002, roughly $226 more than customers experienced in MS and LA. This estimate is over $200 less than the willingness-to-pay value cited above, which suggests Texans are getting utility scale wind and solar energy for a good deal. All in all, however, it is difficult to conclude as to whether public perceptions in Texas will influence market investment in wind energy, especially with rates increasing.

IV.3.9 FUTURE PROJECTIONS OF WIND AND SOLAR FROM A TECHNOECONOMIC PERSPECTIVE

To briefly review the contents of this chapter thus far, the future development of wind and solar depends on numerous factors including but not limited to the following: i) legislative support and financing, ii) the ability of the electricity market to accommodate high VRE penetration while ensuring reliability, iii) reduction in costs to attract investment, iv) continued transmission expansion, and v) overcoming intermittency through technological obstacles, such as storage. These factors are important considerations in models that project the future viability and expansion of technologies. The EIA produces an Annual Energy Outlook (AEO) that provides projections of long-term energy trends in the United States. The AEO is produced using the National Energy Modeling System that considers technological, economic, and demographic assumptions, as well as considering operational system constraints, laws, and regulations to produce projections. Here, the results of the 2023 AEO for Texas's electrical grid are presented, which provide projections through 20250 including 17 different cases (or scenarios), described in EIA (2023a). The scenarios vary in economic growth, oil and gas prices and supply, and varied technology costs (EIA, 2023a); however, as you will find, scenarios are based on current technoeconomic assumptions, which can shift dramatically altering the anticipated trajectory of different technologies (e.g., see examples of battery installations and natural gas pricing).

In agreement with the number of planned solar and wind facilities, AEO projections suggest continued rapid expansion of solar, whereas wind increases are also expected, but far slower (Figure IV.3.13a, b). Under the AEO, installed solar capacity in Texas could increase by over 5 times the current amount by 2050. This high level of penetration is largely due to the role of increasing grid-scale battery storage solutions (Figure IV.3.13c). Battery installations have grown exponentially since 2010, with current capacities approximated at 15 GW (EIA, 2024b). In 2023, 3.6 GW of battery storage were installed, equivalent to the solar capacity additions (3.7 GW) to Texas's electricity grid that same year. As of 2024, battery installations either installed, under development, or planned totaled 8.8 GW of additional capacity, more than a doubling the 2023 values. This pattern of growth has been unanticipated as is evident by the actual and planned installations (2023–2028) outpacing AEO projections (Figure IV.3.13c, red dashed line). Over the next three years, planned storage installations are expected to increase total capacities to 25 GW, already equivalent to the total capacity projected by the AEO for 2050. The rate and magnitude of continued growth in wind projects is uncertain, as investments in solar have been substantial.

Also, in congruence with current patterns, decommissioning of coal-fired power plants is expected to continue under the AEO 2023 projections (Figure IV.3.13d). Deployment of CC-NG facilities, however, is expected to increase conservatively by 60% by 2050 (Figure IV.3.13e). The role of natural gas in Texas and the US electricity sector remains uncertain. AEO projections of natural gas fuel prices to the electric sector further demonstrate the uncertainty in economic

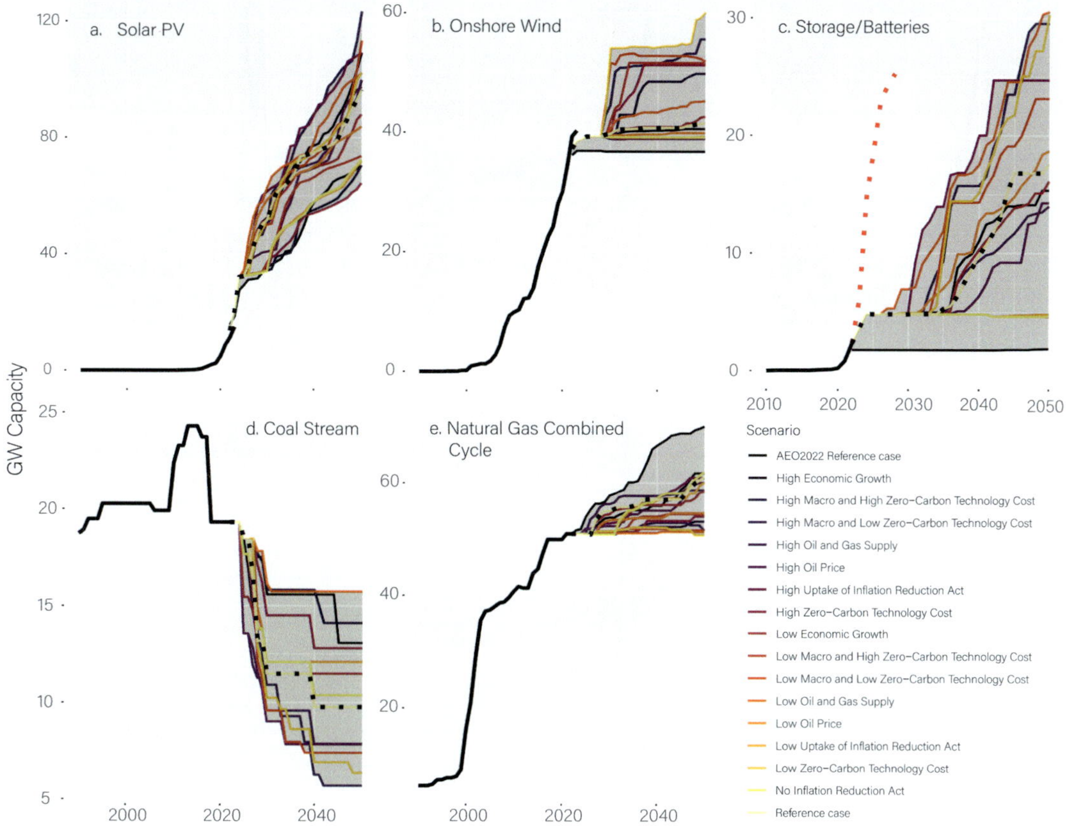

FIGURE IV.3.13 Installed capacity in gigawatts (GW) for five technologies in Texas through 2050 according to projections for various scenarios provided by Energy Information Administration's 2023 Annual Energy Outlook (EIA 2023). Black lines represent observed values in Texas. Red dashed line in panel c represents observed and planned battery installations for 2022 to 2028.

Data from EIA (2023a, 2024b).

assumptions and policies that support technological projection models. The AEO 2023 projections were developed in 2022, when gas prices had peaked at $6 per MMBTU (Million British Thermal Units). These projections are agnostic to the dramatic decline in prices to the 2024 price of $2.19 per MMBTU (red dotted line, Figure IV.3.134c). In contrast, AEO 2025 projections depict far lower natural gas prices, with 2050 values for the reference scenario close to $2 per MMBTU.

Installation costs, projected under the AEO 2023, are expected to continue to decrease and incentivize investments in wind, solar PV, and storage options (Figure IV.3.14, b, d). Costs in Texas and the south region are generally lower than the remainder of the US. The wind industry has led solar by at least 15 years, meaning costs have been lower for some time; however, installation costs for wind are still expected to decline below $1,200/kW. Installation costs for solar PV (with and without storage) have dropped 60% over the last 15 years, with costs expected to be below $1,000/kW by 2050. LCOE and LACE cost assumptions under various scenarios for the AEO 2023 are provided in Figure IV.3.7, depicting costs for projects entering service in 2028. Again, here the lower costs in Texas can be noted. Under the low zero-carbon technology-cost scenario and low-oil-and-gas scenario, solar LCOE and LACE values are competitive, especially in Texas, where LACE values approximate LCOE, indicating ideal economic conditions. The cost of batteries has been erratic but are also expected to decline to $500/kw by 2050 (Figure IV.3.14d). Likewise, CC-NG installation

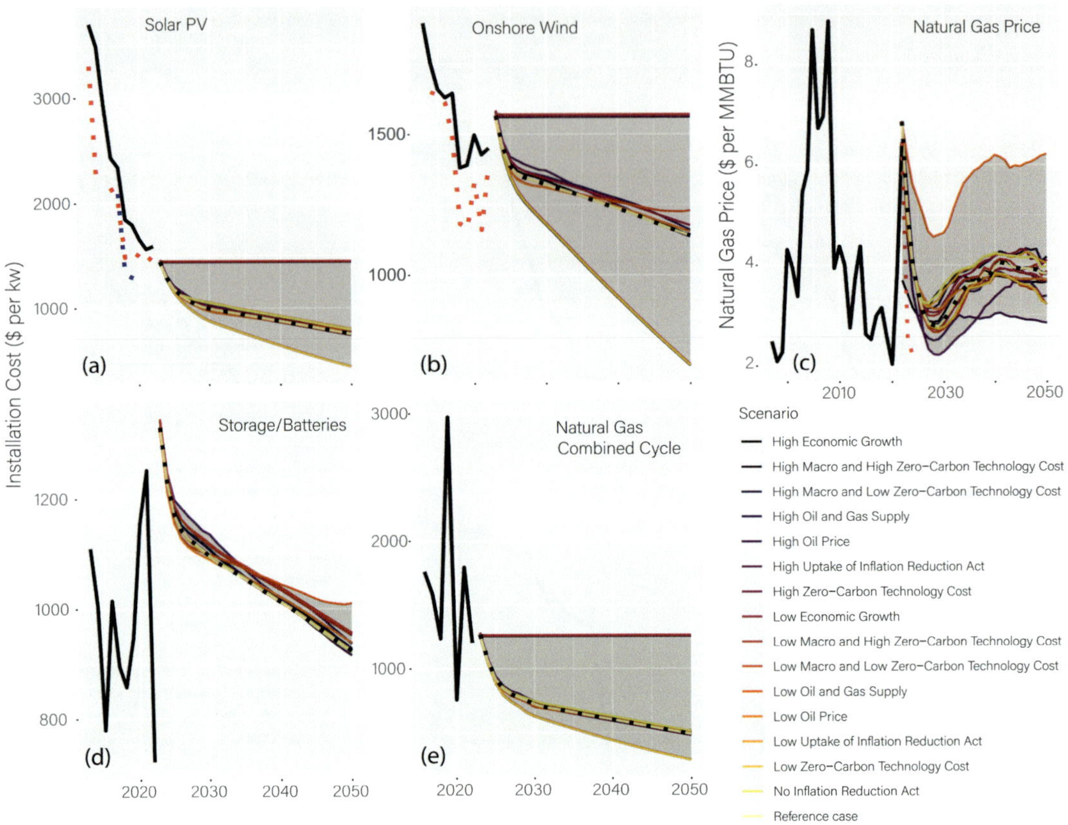

FIGURE IV.3.14 Installation costs ($per kilowatts) projections for four technologies and natural gas prices ($ per million British Thermal Units) through 2050 in the US. Scenarios provided by Energy Information Administration's 2023 Annual Energy Outlook (EIA, 2023a). Black lines represent observed values of U.S. average installation costs. Red dashed line in panels a and b represents observed values for wind and solar developments in the Southern U.S., whereas panel c shows observed U.S. averages for natural gas. Blue dashed line in panel a represents observed values in Texas.

Data from EIA (2023a, 2024b).

costs have fluctuated dramatically over the past decade; however, costs plummeted to $722/kw in 2022 (Figure IV.3.14e). These costs are obviously not reflected in the AEO 2023 projections, which assumed cost increases were going to remain constant with 2021 trajectories. Here again, illustrates the uncertainty of the role of natural gas and CC-NG in the future.

IV.3.10 PUBLIC OPPOSITION, ENVIRONMENTAL CHALLENGES, AND POLITICAL UNCERTAINTY

The Texas political and ideological landscape cannot be overlooked when examining the past and present support and opposition to renewable energy. As wind and solar installations have expanded rapidly over the last 25 years, opposition has arisen at multiple scales, from localized resistance to single developments to state-level hostility manifested as anti-renewable legislation (Taylor, 2023). Opposition to renewables occurs for many reasons – at electoral levels, this may occur based on voter reactions to federal climate policies (Stokes, 2016; Taylor, 2023), whereas at local community levels, most opposition stems from perceived cost-benefit imbalances (Andrews et al., 2025) or a lack of open participatory processes (Susskind et al., 2022). Localized public opposition to renewables can be a significant barrier to continued growth (Susskind et al., 2022). Although 72% of Texans support funding renewable technologies (Howe et al., 2015; Marlon et al., 2022), there remains a social gap in individualistic public behaviors, particularly the "not in my backyard" syndrome (Bell et al., 2005). As one example, the 200-MW Byers and Bluegrove project, an 11,000-acre wind farm, was eventually abandoned by Innergex Renewable Energy, Inc. due to public opposition and financial viability (Susskind et al., 2022). Local opposition can also be a bi-partisan issue, where stakeholders in liberal-leaning jurisdictions oppose impacts on wildlife, aesthetics, or property value (Taylor, 2023). Hence, understanding and legitimizing the reasons for opposition, rather than wholesale dismissal of concerns, is critical for not only gaining public support (Trandafir et al., 2023), but also ensuring long-term sustainability of well-planned projects.

As stated previously, much of the motivation for renewable energy developments in Texas and elsewhere has been for economic, not environmental, reasons (Guiney, 2018). Renewable projects are expensive, yet they attract investment because they generate revenue for developers, operators, and providers who sell electricity to end customers. The broader benefits to the public, which need to be weighed against the costs, include positive externalities and economic returns, including tax income to counties, financial support for local school districts, and benefits to landowners. Rhodes (2023) estimates that the current and planned fleet of wind, solar, and battery storage developments in Texas are expected to generate $20.2 billion in tax revenues to local communities over the life of the projects. Projects can also generate significant revenue for landowners that lease their property to wind and solar developers. It is estimated that Texas landowners will receive $15.1 billion over the lifetime of existing projects, and over $29.5 billion when considering both current and planned projects. Although Rhodes (2023) suggests that over 75% of Texas counties are expected to receive tax revenues from renewable and storage developments, significant opposition can arise from inequitable distribution of benefits, whether perceived or real (Kerr et al., 2017; Mueller and Brooks, 2020).

In a review of 53 utility-scale wind, solar, and geothermal projects in the U.S. that were delayed or blocked by social opposition, only 13% were eventually completed (Susskind et al., 2022). The authors reported that predominant reasons for opposition included concerns over impacts to property value (62%), environmental impacts (60%), and concerns over health or lack of public participation process (28%) (Susskind et al., 2022). Most utility-scale developments have occurred in rural areas, where siting concerns are elevated as development will convert natural or agricultural lands. Proximity to residential properties to large scale solar facilities can elicit significant concern, suggesting the need for appropriate buffer sizes (Carlisle et al., 2016). Although many landowners and community members have directly received benefits, others may only perceive the costs, such as loss in property value or bearing the burden of transmission expansion to support developments. In a recent survey of Texan attitudes toward wind developments, only 39% to 44% of respondents felt that they had a say in local planning processes, and only 48% to 54% felt the project developer acted transparently (Andrews et al., 2025).

As with any land development activity, ensuring appropriate consideration of local and regional environmental impacts are important. Currently, there are no state mandates for permitting and siting wind or solar developments in Texas, including environmental impact statements (EISs). Wind companies are only required to apply for a Federal Aviation Association permit in Texas; any other requirements were at the discretion of the landowner in lease agreements, or third parties involved in power purchase agreements. Solar companies in Texas are required to apply for a Stormwater Prevention Plan permit from Texas Commission for Environmental Quality (TCEQ) and a Pollution Prevention Plan permit from the U.S. Army Corps of Engineers. Other than that, no other permits or EIS are required. Not surprisingly, Texas has become the wild frontier open to rapid and untethered expansion of renewable projects. This seemingly contradictory environmentalism, subsidized by federal liberal policies, likely created localized and statewide opposition to renewable energy installations.

Here, it is critical that political proponents of climate mitigation strategies understand the measurable environmental tradeoffs that come with global or national-scale initiatives for widespread renewable energy development. Although large-scale renewable energy deployment is one carbon-neutral pathway to achieve decarbonization of the energy sector to mitigate climate change, renewable technologies require significant land areas for generation facilities (van de Ven et al., 2021), but also mining metals and materials, manufacturing, transportation of materials, and decommissioning, collectively termed a "life cycle" (UNECE, 2022). Decarbonization from renewable energy development, if achieved, could increase the U.S. energy footprint from 81 million acres to 250 million acres by 2050 (Merrill, 2021; Van Zalk and Behrens, 2018). In particular, the land area required for solar PV life cycles are over 10X that of natural gas and nuclear technologies in the US (Sturtevant et al., 2022). The potential risks to biodiversity loss from global-scale renewable deployment can be significant, but highly variant depending on the location and nature of development (McManamay et al., 2021). Although bird and bat mortalities of wind facilities can be high in some regions (Cabrera-Cruz et al., 2020), other regions report far smaller estimates that are minimal compared to other sources of mortality (Loss et al., 2015). Impaired wildlife migrations (Levin et al., 2023) and vegetation changes (Xu et al., 2024) are realized concerns associated with utility-scale solar development. To compensate, wind and solar have adopted land-sharing strategies, such as co-occurrence of wind and agricultural land (Morris and Blekkenhorst, 2017), agrivoltaics (Stid et al., 2025), and pollinator habitat enhancement (Walston et al., 2018).

Political opposition to wind and solar development in Texas has been ongoing for some time. In the previous 88th legislative session (2023), eight house bills (HB) and five senate bills (SB) were proposed, each addressing renewable energy. The opposition to renewables, however, hit a climax during the latest 89th session (2025) where 60 proposals on increasing regulations, restrictions, and requirements on wind and solar were circulated in the house and senate. The most significant of these, SB 819, is a reincarnation of SB 634, which failed in the 88th legislation session. SB 819 was passed by the Texas Senate on April 15, 2025, in a 22:9 vote, but died one week later in the House State Affairs Committee. Had SB 819 became law, it would have imposed several new requirements on new and existing facilities greater than 10 MW including obtaining environmental permitting requirements, undergoing EIS review and subsequent environmental monitoring, environmental impact fees, removal of tax abatements, fostering public participation in meetings, and maintaining setbacks of at least 3,000 feet from adjacent property lines. SB 819 was expected to result in significant costs and potential attrition of planned facilities. SB 388 also passed the senate but later died in the House; it stipulated that 50% of additional generating capacity installed after January 1, 2026, be sourced from dispatchable generation by requiring power generating companies (including renewables), municipalities, or cooperatives, to either own dispatchable generation units or purchase dispatchable generation credits. SB 714, if passed, would have required ERCOT to adopt rules and procedures to eliminate or compensate for distortion in electricity pricing due to federal Texas credits. SB 3278, also failed, called for studies on the environmental, economic, and infrastructural effects of solar and wind facilities, including direct effects of land use on habitats and ecosystems, environmental effects and challenges of decommissioning, and the indirect consequences of renewable penetration on infrastructure expansion to ensure grid stability and reliability. Among the bills that successfully passed into law, HB 3229 followed up on previous SB 1290 (88th session) requiring oversight on renewable energy decommissioning by requesting wind and solar project owners fully cover the costs of removal and land restoration, including recycling and disposal costs. HB 3229 passed in May 2025 and became effective September 1, 2025.

Although many of the items in the SBs above address legitimate concerns for environmental protection and sustainable development, some of the political impetus may again be unrelated to environmental reasons, but more so backlash from federal policies (Stokes, 2016), particularly the IRA of 2022 (Taylor, 2023). Much of the support or opposition for renewable policies vary based on political ideologies rather than the technology per se (Hess et al., 2016). For instance, policies in the form of mandates tend to gain less support in conservative states, compared to tax reductions or regulatory reductions (Hess et al., 2016). The IRA could be interpreted as more of a mandate than tax reduction, because its life expectancy relates to sufficient reductions in GHG emissions rather than the economic vitality of the wind and solar industries (i.e., typical motivation for past tax credits).

IV.3.11 CONCLUSIONS

After almost 30 years in the making, the Texas energy experiment provides many real-life lessons for decarbonization of the electricity sector by way of renewable deployment. Transitioning the electric sector to carbon neutrality is not simply the outcome of direct government build-out programs as in past eras, e.g., Civilian Conservation Corps, Tennessee Valley Authority. Nor is it the outcome of philanthropic support or ambitious global sustainability targets idealized by the Conference of Parties (COP) or United Nation's Convention on Climate Change, although these international agreements do help mobilize domestic legislation to a degree. The electric sector is a highly political, technologically complex, and economically motivated infrastructural and social system that supports the livelihoods of millions of Americans. Any sizeable shift in the operations or fabric of this system requires overcoming entrenched sociopolitical, economic, and technological inertia, including changes in legislation, financial subsidies, market adjustments, and technological advancements. Perhaps most importantly, it requires finding a common set of political ideological values. Texas energy provides a counter argument to the conventional view of the politics behind renewable energy growth. Wind and solar expansion in Texas was not entirely due to pro-environmental policies, although federal legislation for subsidies certainly influenced development. The story of renewable energy growth in Texas challenges the sociopolitical clichés commonly associated with renewable energy adoption. Furthermore, Texan social and political values will continue to shape the energy landscape in the future.

Here, the major points of the chapter are summarized below.

1. Legislative mandates and incentives have forged the integration of wind and solar into the Texas energy grid and made it economical, at least in the short term. Across the U.S. and Europe (Hiroux and Saguan, 2010), tax credits, subsidies, RPSs, CESs, and analogous frameworks have provided economic incentives for renewable development. Within Texas, federal PTCs and ITCs, at times, have made up anywhere from 50% to 100% of the gross revenues for wind and solar. The CREZ initiative expanded transmission to remote areas of high wind generation that have incentivized development and reduced transmission congestion. These initiatives in conjunction with low regulatory oversight, by far, were the predominant drivers of major wind and solar expansion in Texas. Likewise, by extending the PTC and ITC, the IRA will continue to incentivize financial investment in renewables in the U.S. for the foreseeable future.
2. The extent to which wholesale and retail market deregulation, alone, have supported or jeopardized the investment and long-term economic vitality of renewable penetration is unclear. On one hand, the unbundling of utilities from transmission services has resulted in substantial investments in IPPs, including renewables but also small ACC-NGs. On the other hand, the deregulation of the Texas retail market shifted, at least initially, toward a short-term, low-price bidding system, which removes the incentive of investors to engage in long-term PPAs with renewables. Wholesale deregulation, in association with PTCs, at least originally encouraged perverse behaviors by renewable producers. Without state mandates and federal incentives, wind and solar, at times, were not economically competitive in Texas's energy market, a conclusion shared by others (Eun Kim et al., 2016). Furthermore, lower marginal prices of generation for wind and solar created merit order effects, disincentivizing market participation by reliability units, subsequently creating grid instability issues; this required amendments and incentives from ERCOT for dispatchable generators. Ultimately, merit order effects not only influence markets for other competing technologies but also risk cannibalization of wind and solar's own energy markets.

Despite these challenges, wind and solar industries have entered a new phase of large-scale mature deployment and surpassed that of marginal technologies. Such a level of VRE integration implores a new look at the design of current deregulated electricity markets (e.g., Europe and the U.S.), specifically whether they are appropriately equipped to encourage future renewable investments and if adjustments are needed (Hiroux and Saguan, 2010). While there is likely ongoing debate over rationale for different market designs, the moral of the story is that a discussion regarding new electricity market designs accommodating large-scale renewable integration is needed. However, dispatchable technologies also need incentives to participate and ensure reliability of electricity grids.

3. Continued technological advances, such as grid-scale energy storage, transmission expansion, and grid operations are required to assist with high VRE integration while ensuring reliability and stability. Additionally, improved accuracies of forecasting hourly wind and solar conditions are needed to prevent load imbalances and encourage short-term investments from market participants (Holttinen, 2005). Lower installation costs will continue to incentive investment in renewables. However, as long as natural gas is cheap, gas will likely remain a strong competitor, and wind and solar will have to be even cheaper to compete. The important caveat is whether gas will remain cheap.
4. Social and political ideologies of Texas, particularly at odds with centralized federal mandates, have and will create challenges for the future of wind and solar industries in Texas. Additionally, community-level opposition is growing in many areas, as the past era of wind and solar development could be argued as being, at times, inequitable to all community members (Mueller and Brooks, 2020). Growing environmental concerns required for continued renewable energy development should not, however, be dismissed as a mere political tool. The level of development required for deep decarbonization, including the necessary transmission expansion to accommodate high VRE penetration, could constitute considerable landscape modification. Indeed, concerns have escalated recently due to Texas's wild-west frontier policies fostering rapid development including a complete lack of regulatory oversight and any formalized environmental permitting processes. Overcoming some of these challenges will require wind and solar developers to conduct adequate planning and engage the community in that process.

NOTES

1. Energy Policy Act of 1992, Pub. L. No. 102-486, 106 Stat. 2776 (1992)
2. Energy Policy Act of 2005, Pub. L. No. 109-58, 119 Stat. 594 (2005)
3. Emergency Economic Stabilization Act of 2008, Pub. L. No. 110–343 (2008)
4. American Recovery and Reinvestment Act of 2009, Pub. L. No. 111–5 (2009)
5. Consolidated Appropriations Act of 2016, Pub. L. No. 114-113, 129 Stat. 2243 (2015)
6. Energy Tax Act of 1978, Pub. L. No. 95-618, 92 Stat. 3174 (1978)
7. Inflation Reduction Act of 2022, Public Law No. 117-169 (P.L. 117–169), H.R. 5376 (2022, amended 2024).

REFERENCES

American Public Power Association, 2022. Retail electric rates in deregulated and regulated states: 2021 update. https://www.publicpower.org/resource/retail-electric-rates-regulated-and-deregulated-states-2021-update (accessed May 7, 2025).

Andrews TM, Atkinson-Palombo C, Bayulgen O, Gallaher A, Scruggs L, 2025. The winds of change? Attitudes toward wind projects and their electoral implications in Texas. Energy Policy. 202:114608.

Aslani A, Wong KFV, 2014. Analysis of renewable energy development to power generation in the United States. Renewable Energy. 63:153–161.

Bae M, Kim H, Kim E, Chung AY, Hwangnam K, 2014. Toward electricity retail competition: Survey and case study on technical infrastructure for advanced electricity market system. Applied Energy. 133:252–273.

Barbose G, 2023. U.S. state renewables portfolio & clean electricity standards: 2023 status update. Lawrence Berkeley National Laboratory. https://emp.lbl.gov/publications/us-state-renewables-portfolio-clean (accessed May 6, 2025).

Beck F, Martinot E, 2004. Renewable energy policies and barriers. In Encyclopedia of Energy, Cleveland CJ, ed. Elsevier. https://doi.org/10.1016/B0-12-176480-X/00488-5. ISBN 9780121764807

Bell D, Gray T, Haggett C, 2005. The 'social gap' in wind farm siting decisions: Explanations and policy responses. Environmental Politics. 14(4):460–77.

Bhaskar K, Singh SN, 2014. Wind power bidding strategy in a day-ahead electricity market. In 2014 IEEE PES General Meeting Conference & Exposition 2014 Jul 27 (pp. 1–5). IEEE. DOI: 10.1109/PESGM.2014.6939112

Borenstein S, Bushnell J, 2000. Electricity restructuring: Deregulation or regulation. Regulation, The Cato Review of Business and Government. 23(2):46–52.

Borenstein S, Bushnell J, 2015. The US electricity industry after 20 years of restructuring. Annual Review of Economics. 7(1):437–63.

Brown DP, Tsai CH, Woo CK, Zarnikau J, Zhu S, 2020. Residential electricity pricing in Texas's competitive retail market. Energy Economics. 92:104953.

Cabrera-Cruz SA, Cervantes-Pasqualli J, Franquesa-Soler M, Muñoz-Jiménez O, Rodríguez-Aguilar G, Villegas-Patraca R, 2020. Estimates of aerial vertebrate mortality at wind farms in a bird migration corridor and bat diversity hotspot. Global Ecology and Conservation. 22:e00966.

Carlisle JE, Solan D, Kane SL, Joe J, 2016. Utility-scale solar and public attitudes toward siting: A critical examination of proximity. Land Use Policy. 58:491–501.

Deetjen TA, Rhodes JD, Webber ME, 2017. The impacts of wind and solar on grid flexibility requirements in the Electric Reliability Council of Texas. Energy. 15(123):637–54.

Deng L, Hobbs BF, Renson P, 2014. Negative bidding by wind: A unit commitment analysis of cost and emission impacts. International Association for Energy Economics. 37th IAEE International Conference Proceedings 45–47. http://www.iaee.org/en/publications/proceedingssearch.aspx

EIA (Energy Information Administration). 2023a. Annual Energy Outlook AEO2023. https://www.eia.gov/outlooks/archive/aeo23/ (accessed May 6, 2025)

EIA (Energy Information Administration). 2023b. A Case Study of Transmission Limits on Renewables Growth in Texas. July 2023. https://www.eia.gov/todayinenergy/detail.php?id=57100 (accessed May 6, 2025).

EIA (Energy Information Administration). 2024a. Construction cost data for electric generators installed in 2022. https://www.eia.gov/electricity/generatorcosts/ (accessed May 6, 2025)

EIA (Energy Information Administration). 2024b. Form EIA-860 detailed data with previous form data (EIA-860A/860B). https://www.eia.gov/electricity/data/eia860/ (accessed May 6, 2025).

EIA (Energy Information Administration). 2024c. Form EIA-923 detailed data with previous form data (EIA-906/920). https://www.eia.gov/electricity/data/eia923/ (accessed May 6, 2025).

EIA (Energy Information Administration). 2024d. U.S. Energy-Related Carbon Dioxide Emissions, 2023. https://www.eia.gov/environment/emissions/carbon/ (accessed May 6, 2025).

EIA (Energy Information Administration). 2025b. Texas State Profile and Energy Estimates. https://www.eia.gov/state/?sid=TX (accessed May 6, 2025).

EIA (Energy Information Administration). 2025c. U.S. States. https://www.eia.gov/state/ (accessed May 7, 2025).

EIA (Energy Information Administration). 2025d. Wholesale Electricity Market Portal. https://www.eia.gov/electricity/wholesalemarkets/ (accessed May 6, 2025).

Energy Information Administration. 2025a. Electricity Data. https://www.eia.gov/electricity/data.php (accessed May 7, 2025).

ERCOT (Electric Reliability Council of Texas), 2024. Strategic Plan 2024–2028. https://www.ercot.com/files/docs/2023/08/25/ERCOT-Strategic-Plan-2024-2028.pdf (accessed May 7, 2025).

ERCOT (Electric Reliability Council of Texas), 2025. Company Profile. https://www.ercot.com/about/profile (accessed May 6, 2025).

Eun Kim S, Yang J, Urpelainen J, 2016. Does power sector deregulation promote or discourage renewable energy policy? Evidence from the states, 1991–2012. Review of Policy Research. 33(1):22–50.

Farhar BC, 1999. Willingness to pay for electricity from renewable resources: a review of utility market research. National Renewable Energy Laboratory NREL/TP.550.26148. July 1999. http://www.nrel.gov/docs/fy99osti/26148.pdf

Frazier AW, Marcy C, Cole W, 2019. Wind and solar PV deployment after tax credits expire: A view from the standard scenarios and the annual energy outlook. The Electricity Journal. 32(8):106637.

Guiney N, 2018. The only green that matters is the green in your pocket: Advocating for renewable energy in Red States. Ecology Law Quarterly. 45(2):163–86.

Hartley PR, Medlock III KB, Jankovska O, 2019. Electricity reform and retail pricing in Texas. Energy Economics. 80:1–1.

Hess DJ, Mai QD, Brown KP, 2016. Red states, green laws: Ideology and renewable energy legislation in the United States. Energy Research & Social Science. 11:19–28.

Hiroux C, Saguan M, 2010. Large-scale wind power in European electricity markets: Time for revisiting support schemes and market designs? Energy Policy. 38:3135–3145.

Holttinen H, 2005. Optimal electricity market for wind power. Energy Policy. 33:2052–2063.

Howe P, Mildenberger M, Marlon J, Leiserowitz A, 2015. Geographic variation in opinions on climate change at state and local scales in the USA. Nature Climate Change. 5:596–603. https://doi.org/10.1038/nclimate2583

Huntowski F, Patterson A, Schnitzer M. 2012. Negative electricity prices and the production tax credit. The NorthBridge Group. 2012 Sep. http://graphics8.nytimes.com/news/business/exelon.pdf

Internal Revenue Service. 2025. Inflation Reduction Act of 2022. https://www.irs.gov/inflation-reduction-act-of-2022

Joskow PL, 2021. Facilitating transmission expansion to support efficient decarbonization of the electricity sector. Economics of Energy & Environmental Policy. 10(2):57–92.

Kerr S, Johnson K, Weir S, 2017. Understanding community benefit payments from renewable energy development. Energy Policy. 105:202–11.

Klessmann C, Nabe CH, Burges K, 2008. Pros and cons of exposing renewables to electricity market risks—A comparison of the market integration approaches in Germany, Spain, and the UK. Energy Policy. 36(10):3646–3661.

Lawrence Berkley National Laboratory. 2025. Energy Markets & Policy. Tools & Data. https://emp.lbl.gov/tools-data

Levin MO, Kalies EL, Forester E, Jackson EL, Levin AH, Markus C, McKenzie PF, Meek JB, Hernandez RR, 2023. Solar energy-driven land-cover change could alter landscapes critical to animal movement in the continental United States. Environmental Science & Technology. 57(31):11499–509.

Lopez A, Roberts B, Heimiller D, Blair N, Porro G, 2012. U.S. Renewable energy technical potentials: A GIS-based analysis. National Renewable Energy Laboratory Technical Report NREL/TP-6A20-51946. July 2012.

Loss SR, Will T, Marra PP, 2015. Direct mortality of birds from anthropogenic causes. Annual Review of Ecology, Evolution, and Systematics. 46(1):99–120.

Loth E, Qin C, Simpson JG, Dykes K, 2022. Why we must move beyond LCOE for renewable energy design. Advances in Applied Energy. 8:100112.

MacKay A, Mercadal I, 2022. Deregulation, Market Power, and Prices: Evidence from the Electricity Sector, MIT CEEPR Working Paper 2022-008, April 2022. https://ceepr.mit.edu/deregulation-market-power-and-prices-evidence-from-the-electricity-sector/ (accessed May 7, 2025).

MacKay A, Mercadal I, 2024. Do Markets Reduce Prices? Evidence from the US Electricity Sector. Harvard Business School, Working Paper: 21-095.

Mai T, Cole W, Lantz E, Marcy C, Sigrin B, 2016. Impacts of federal tax credit extensions on renewable deployment and power sector emissions. National Renewable Energy Lab. (NREL), NREL/TP-6A20-65571. Golden, CO (United States); 2016 Feb 1.

Marlon JR, Wang X, Bergquist P, Howe P, Leiserowitz A, Maibach E, Mildenberger M, Rosenthal S, 2022. Change in US state-level public opinion about climate change: 2008–2020. Environmental Research Letters. 17(12):124046.

McManamay RA, Vernon CR, Jager HI, 2021. Global biodiversity implications of alternative electrification strategies under the shared socioeconomic pathways. Biological Conservation. 260:109234.

Merrill D, 2021. The U.S. will need a lot of land for a zero-carbon economy. Bloomberg, Apr. 29, 2021, https://www.bloomberg.com/graphics/2021-energy-land-use-economy/. (accessed May 7, 2025).

Millstein D, O'Shaughnessy E, Wiser R, 2025. Renewables and Wholesale Electricity Prices (ReWEP) tool. Lawrence Berkeley National Laboratory. Version 2025.1 https://emp.lbl.gov/renewables-and-wholesale-electricity-prices-rewep (accessed May 7, 2025).

Milstein I, Tishler A, 2011. Intermittently renewable energy, optimal capacity mix and prices in a deregulated electricity market. Energy Policy. 39:3922–3927.

Mormann F, Reicher D, Hanna V, 2016. A tale of three markets: Comparing the renewable energy experiences of California, Texas, and Germany. Stanford Environmental Law Journal. 2016(35):55.

Morris DW, Blekkenhorst N, 2017. Wind energy versus sustainable agriculture: An Ontario perspective. The Journal of Rural and Community Development. 12(1):23–33.

Mueller JT, Brooks MM, 2020. Burdened by renewable energy? A multi-scalar analysis of distributional justice and wind energy in the United States. Energy Research & Social Science. 63:101406.

Murakami K, Ida T, Tanaka M, Friedman L, 2015. Consumers' willingness to pay for renewable and nuclear energy: A comparative analysis between the US and Japan. Energy Economics. 50:178–189.

Newbery DM, 2023. High renewable electricity penetration: Marginal curtailment and market failure under "subsidy-free" entry. Energy Economics. 126:107011.

Painuly JP, 2001. Barriers to renewable energy penetration: A framework for analysis. Renewable Energy. 24:73–89.

Pena JI, Rodríguez R, Mayoral S, 2022. Cannibalization, depredation, and market remuneration of power plants. Energy Policy. 167:113086.

Pless J, Arent JD, Logan J, Cochran J, Zinaman O, Stark C, 2015. Pathways to decarbonization: natural gas and renewable energy. Lessons learned from energy system stakeholders. Joint Institute for Strategic Energy Analysis. Technical Report NREL/TP-6A50-63904. April 2015.

Prol JP, Steininger KW, Zilberman D, 2020. The cannibalization effect of wind and solar in the California wholesale electricity market. Energy Economics. 85:104552.

Public Utility Commission of Texas (PUCT). 2025. Electric Industry. Market Directories: Electric Companies Serving Texas. https://www.puc.texas.gov/industry/electric/directories/ (accessed May 6, 2025).

Quint D, Dahlke S, 2019. The impact of wind generation on wholesale electricity market prices in the midcontinent independent system operator energy market: An empirical investigation. Energy. 169:456–466.

Rhodes JD, 2023. The economic impact of renewable energy and energy storage in rural Texas. IdeaSmiths LLC, Austin, TX. 2023 Jan. https://www.ideasmiths.com/s/Economic-Impact-of-Renewable-Energy_JAN2023.pdf (accessed May 7, 2025).

Shen W, Chen X, Qiu J, Hayward JA, Sayeef S, Osman P, Meng K, Dong ZY, 2020. A comprehensive review of variable renewable energy levelized cost of electricity. Renewable and Sustainable Energy Reviews. 133:110301.

Sioshansi R, Hurlbut D, 2010. Market protocols in ERCOT and their effect on wind generation. Energy Policy. 38(7):3192–3197.

Smith RW, Polukort P, Maslak CE, Gardiner BD, 2001. Advanced technology combined cycles. GE Power Systems GER-3936A. Schenectady, NY. 16 pg.

Ssembatya H, Kern JD, Oikonomou K, Voisin N, Burleyson CD, Akdemir KZ, 2024. Dual impacts of space heating electrification and climate change increase uncertainties in peak load behavior and grid capacity requirements in Texas. Earth's Future. 12:e2024EF004443. https://doi.org/10.1029/2024EF004443

Stid JT, Shukla S, Kendall AD, et al., 2025. Impacts of agrisolar co-location on the food–energy–water nexus and economic security. Nature Sustainability. 8:702–713. https://doi.org/10.1038/s41893-025-01546-4

Stokes LC, 2016. Electoral backlash against climate policy: A natural experiment on retrospective voting and local resistance to public policy. American Journal of Political Science. 60:958–974. https://doi.org/10.1111/ajps.12220.

Stokes LC, Breetz HL, 2018. Politics in the U.S. energy transition: Case studies of solar, wind, biofuels and electric vehicles policy. Energy Policy. 113:76–86.

Sturtevant J, McManamay RA, DeRolph CR, 2022. US national water and energy land dataset for integrated multisector dynamics research. Scientific Data. 9(1):183.

Susskind L, Chun J, Gant A, Hodgkins C, Cohen J, Lohmar S, 2022. Sources of opposition to renewable energy projects in the United States. Energy Policy. 165:112922.

Taylor ME, 2023. Peril and promise: The challenge of siting renewable energy and infrastructure. Texas Journal of Oil, Gas and Energy Law. 18(1):42–60.

Tran TT, Smith AD, 2018. Incorporating performance-based global sensitivity and uncertainty analysis into LCOE calculations for emerging renewable energy technologies. Applied Energy. 216:157–71.

Trandafir S, Thomas P, Bidwell D, Rezendes R, 2023. Community benefit agreements for solar energy: Examining values, preferences and perceived benefits in the United States using a discrete choice experiment. Energy Research & Social Science. 106:103305.

UNECE (United Nations Economic Commission for Europe), 2022. Carbon neutrality in the UNECE Region: Integrated life-cycle assessment of electricity sources. Geneva, 2021.

US DOE (U.S. Department of Energy), 2017. Wind Vision. Office of Energy Efficiency and Renewable Energy. https://energy.gov/eere/wind/wind-vision (accessed May 7, 2025)

US DOE (U.S. Department of Energy), 2025. Inflation Reduction Act Tax Credit Opportunities for Hydropower and Marine Energy. https://www.energy.gov/eere/water/inflation-reduction-act-tax-credit-opportunities-hydropower-and-marine-energy (accessed May 7, 2025).

US EPA (U.S. Environmental Protection Agency). 2025a. Overview of Greenhouse Gases. https://www.epa.gov/ghgemissions/overview-greenhouse-gases (accessed May 6, 2025).

US EPA (U.S. Environmental Protection Agency), 2025b. Summary of Inflation Reduction Act provisions to renewable energy. https://www.epa.gov/green-power-markets/summary-inflation-reduction-act-provisions-related-renewable-energy (accessed May 7, 2025).

van de Ven DJ, Capellan-Peréz I, Arto I, et al, 2021. The potential land requirements and related land use change emissions of solar energy. Scientific Reports. 11:2907. https://doi.org/10.1038/s41598-021-82042-5

Van Kooten GC, 2016. The economics of wind power. Annual Review of Resource Economics. 8(1):181–205.

Van Zalk J, Behrens P, 2018. The spatial extent of renewable and non-renewable power generation: A review and meta-analysis of power densities and their application in the US. Energy Policy. 123:83–91.

Walston LJ, Mishra SK, Hartmann HM, Hlohowskyj I, McCall J, Macknick J, 2018. Examining the potential for agricultural benefits from pollinator habitat at solar facilities in the United States. Environmental Science & Technology. 52(13):7566–7576.

Wiser R, Bolinger M. 2019. 2018 Wind Technologies Market Report. Lawrence Berkeley National Laboratory. https://escholarship.org/uc/item/2tc1819q (accessed May 7, 2025).

Wiser R, Bolinger M, Barbose G, Darghouth N, Hoen B, Mills A, Rand J, Millstein D, Porter K, Widiss R, Oteri F, Tegen S, Tian T, 2016. 2015 Wind Technologies Market Report. https://doi.org/10.2172/1339558

Wiser R, Lantz E, Mai T., Zayas J, DeMeo E, Eugeni E, et al, 2015. Wind vision: A new era for wind power in the United States. The Electricity Journal. 28(9), 120–132.

Woo CK, King M, Tishler A, Chow LC, 2006. Costs of electricity deregulation. Energy. 31(6–7):747–68.

Woo CK, Zarnikau J, Tsai CH, Cao KH, Qi HS, Li R, 2023. Regional revenues of solar and wind generation in Texas. Energy Policy. 178:113586.

Xu Z, Li Y, Qin Y, Bach E, 2024. A global assessment of the effects of solar farms on albedo, vegetation, and land surface temperature using remote sensing. Solar Energy. 268:112198.

Zarnikau J, 2005. A review of efforts to restructure Texas' electricity market. Energy Policy. 33(1):15–25.

Zarnikau J, 2011. Successful renewable energy development in a competitive electricity market: A Texas case study. Energy Policy. 39(7):3906–13.

Zarnikau J, Whitworth D, 2006. Has electric utility restructuring led to lower electricity prices for residential consumers in Texas? Energy Policy. 34(15):2191–200.

IV.4 Texas Energopolitics

Dominic Boyer

IV.4.1 INTRODUCTION

This chapter focuses on Texan energy politics or "energopolitics" from the late 19th century until the present day and its relationship to climate change. Superficially, this might seem to be a simple story. Since 1901, Texas has served in one form or another as the epicenter of the American oil industry. Texas thus has contributed massively to anthropogenic climate change: burning fossil fuels generates 74% of total greenhouse gas emissions nationally and 91% globally. Indeed, this chapter focuses on the emergence and consolidation of a Texan petrostate, by which is meant the assemblage of the oil, gas, and petrochemical industries and their political operations (Boyer, 2023). However, since the 1990s Texas has also been exceptionally effective at renewable energy deployment and generation, which operates for the most part via electrical infrastructure. Although seemingly paradoxical, the Texan enthusiasm to develop any market-friendly energy form makes sense when one investigates its deep roots in the capitalist resource politics and liberal (in the Lockean sense) settler ideology of the state. Looking forward toward the middle of the 21st century it may well be the case that the Texan model of "all of the above" energy development remains a dominant pathway toward climate mitigation. Although climate scientists increasingly warn of the immediate need to "rapidly and equitably phase out fossil fuel extractions" (Ripple et al., 2024), it is possible that the Texan model of exuberant energy development could help the U.S. to reach national decarbonization goals, even despite frequent and vocal disavowal of environmental protection and climate change threats on the part of Texan political leadership.

The concept of energopolitics offers a helpful lens through which to explore such apparently paradoxes in energy politics. Energopolitics offers an analytical framework through which one can explore how the material qualities and political arrangements of energy impact and shape other kinds of political institutions, practices and even ideas (Boyer, 2014, 2019; Rogers, 2015; Folch, 2019; Daggett, 2019). Taking inspiration from historian Michel Foucault's concept of "biopolitics" and its genealogy of the modern governance of bodies and populations, energopolitical analysis seeks to illuminate how energy infrastructures create certain conditions of possibility for political life. Timothy Mitchell's "Carbon Democracy" project (2009, 2011) exemplifies the energopolitical approach. Mitchell investigates the history of coal and oil to surface the dependency of modern democratic ideas and institutions upon fossil energy systems. Mitchell (2009, p. 407) shows, for example, how the consolidation of social democracy in the late 19th century crucially depended on the material character and related infrastructures of coal. Coal extraction required a great deal of dangerous labor that created exceptional political solidarity among workers. Coal's weight and uneven geographic distribution also meant that the vital fuel of steam-powered modernity had to be moved across a rail network that created potential chokepoints where militant workers could disrupt fuel supplies to assert power over mine owners, governments, and society as a whole. Mitchell believes that without this critical energopolitical leverage it would have been impossible for the working classes to win reforms to labor conditions and wages more broadly. Mitchell views the gradual shift from coal to oil as the dominant fossil fuel form over the first half of the 20th century as a part of an effort by the political interests of capital to diminish the political power of militant coal labor. Oil extraction involved different kinds of labor relations, less union affiliation, and fewer chokepoints. And, oil-powered internal combustion engines pioneered in the late 19th century promised a revolution in mobility (not only terrestrial, but also in the air and across water) with both military and civilian applications. Especially after the discovery of massive petroleum resources in

DOI: 10.1201/9781003406600-23

the East Texas and Baku oilfields, oil was reconceptualized as a fuel form with few of energopolitical limits and many new energopolitical advantages for empowering a new era of industrial and military activity. Mitchell even describes the idea of economic growth itself as a "petroknowledge" (Mitchell, 2009, p. 417) of the mid-20th century, anchored by the cornucopian promises of seemingly limitless oil to infinitely expand economic activity. Energopolitical analysis, in this way, looks to the material and infrastructural politics of energy as a way of generating novel insights into the historical evolution of politics and society.

IV.4.2 SECTION I: TEXAN PETROPOLITICS

IV.4.2.1 Early Texan Petropolitics

Legends favor the colonizer and the history of Texan oil is no different. Many sources (e.g., Warner, 2007, p. 19) date the first Texan experience with oil to the remnants of Hernando de Soto's murderous 1540s expedition across *La Florida* washing ashore on the Texas coast south of Beaumont and utilizing asphaltic resin discovered there to repair their boats en route to Mexico. The truth is that the Indigenous peoples who lived in what would become the Texas colony harvested petroleum from tar seeps and sour springs for a variety of medicinal and practical purposes for centuries if not millennia (Froh, 1984, p. 2).

What is more certain is that Texas's first modern oil well was dug in 1866 near Oil Springs in Nacogdoches County, discovering petroleum at a depth of just over 100 feet. As with many other early Texan wells, the drilling equipment and expertise came largely from western Pennsylvania, where America's first oil boom had begun in 1859. Producing a mere 10 barrels a day, the Melrose Petroleum Oil Company, Texas's first commercial venture, soon went out of business (Froh, 1984, p. 23), and Nacogdoches County would not be drilled again for two decades. The market proposition for Texan oil at the time was rather limited. The early American petroleum industry evolved with a focus on providing fuel for artificial illumination, with most petroleum being refined into kerosene as a replacement for dwindling stocks of whale blubber and more expensive derivatives from coal, and as an industrial lubricant. However, by the 1890s a more robust market for fuel oil had developed for multiple purposes including heating, and providing fuel for shipping and locomotives. The advance of internal combustion engine technology in the 1890s led to petroleum increasingly being used for automobiles, both as fuel and via asphalt for road making (Hanieh, 2024, Ch. 2).

The first two decades of Texan petropolitics were challenging for material reasons too and the industry had great trouble sustaining itself: "Between 1870 and 1890 several companies drilled wells, none deeper than roughly a hundred feet, and either bailed or pumped what oil they found. As many as one hundred wells may have been completed, but there were no gushers among them. Worse yet, from the point of view of attracting wider attention, the oil was high in sulfur and low in paraffin, which meant that it would not be desirable for refining into kerosene, the petroleum product then most in demand" (Olien and Olien, 2002, p. 3). With an 82% rural population in 1900 (Weaver, 2010, p. 139), late 19th century Texas remained dominated by agricultural and ranching interests, drilling focused mostly on water, where the appearance of oil was usually considered a "grave disappointment and a waste of money" (Olien and Olien, 2002, p. 4).

Test drilling for water eventually led to the accidental discovery of a larger oil field at Corsicana in Navarro County in 1894. There was enough oil in these wells to interest local entrepreneurs, who recognized that the capitalistic prospects for oil had improved substantially since the 1860s. They hired a veteran wildcatter—as oil prospectors were called—from Pennsylvania to drill additional test wells and before long there were 43 productive wells in the areas yielded 66,000 barrels of oil per year. Although still modest by national standards, Corsicana laid the groundwork for future booms while also revealing Texas's infrastructural inadequacy: "Corsicana producers had oil, but they did not have pipelines, refineries, or marketing systems. Their crude oil could yield a high percentage of illuminating oil, but there was no refinery within affordable distance. To sell the crude for fuel oil required shipment out of Corsicana; there were railroads, but they had an inadequate supply of tank cars" (Olien and Olien, 2002, p. 5). Cotton and lumber remained vastly more important to local economies in Texas than oil.

All this was to change a few years later with the discovery of the largest oilfield in the United States at Spindletop near Beaumont in January 1901. The Spindletop gusher lasted for nine days, blowing as high as two hundred feet into the area. As historian Joseph Pratt describes the social impact of the discovery of Spindletop, "the Gulf Coast oil fields were much larger than those in Pennsylvania, and the rush to pump oil from the ground had even more extreme impacts in Texas than in Pennsylvania. Texas experienced unfettered capitalism squared, as the 'normal' cutthroat competition of the era was heightened by conditions peculiar to the oil industry and to the region. Under the rule of capture, the legal doctrine governing oil production in the United States until the 1930s, any leaseholder above any underground oil reservoir had the legal right to pump oil from the field. Thus, once oil was discovered, the race was on to acquire a lease over the field and pump as much oil as rapidly as possible. Extreme waste from gross overproduction, considerable evaporation from makeshift storage, dangerous fires, and abundant runoff of oil into surrounding waterways resulted" (Warner, 2007, p. 27). It might seem odd that the development of these giant oil fields was guided by relatively inexperienced local capitalists. Indeed, within a few years, two of the oil industry's future Seven Sisters—Gulf Oil and Texaco—formed because of the abundance of the eastern Texas fields. Texas was insulated from the influence of the national oil behemoth, Standard Oil, by a fortuitously timed U.S. Supreme Court judgment in March 1900, which upheld Texan court rulings against Walker-Pierce Oil Company that it was part of a secret trust orchestrated by Standard with the intent "to control and monopolize the petroleum industry in the United States and the several states thereof, and the business of manufacturing, refining, selling, and transporting petroleum and its products." Eleven years later a federal antitrust suit would lead to the breakup of the Standard trust into 34 competitive entities. The anti-monopolist fervor of the era empowered local wildcatters and local capitalists in Texas, as did Texas's distance from the petropolitical power centers and associations of the Northeast (Cleveland, Titusville, New York, Philadelphia).

The rule of capture sought legal analogy for the mysterious (at the time) subterranean fluidity of oil. In 1889, the Pennsylvania Supreme Court found this analogy in the "fugitive and wandering existence" of hunted animals, with ownership based ultimately upon who capture them. Since subsurface geology was largely a matter of conjecture, the oil belonged to whoever could pump it first. As Adam Hanieh writes: "With echoes of John Locke, what mattered most was whose labour brought oil to the surface – capture conferred possession" (2024, Ch. 2). The rule of capture indeed helped lay the groundwork for the chaotic boom/bust dynamics as wells would be overexploited to wring as much oil out of them as quickly as possible, deflating market price. But the material character of oil reservoirs mattered too; pressurization by gas and water was critical to petroleum extraction in the early decades of the industry. Wells always produced best at the beginning and then tapered over time which helped to structure the boom/bust cycle of the business; moreover, "Oilmen were not the only ones who knew that production was often short-lived; bankers quickly learned that no prudent lender extended a loan on the basis of oil production. On the broader industry level, these production realities meant violent cyclical swings between times of boom and bust. When new fields came in, markets flooded with crude, purchasers had more than they could handle, prices fell, producers were pinched for cash, and local refiners turned crude purchased at distress prices into handsome profits. But when production dropped off and prices rose, these local refiners often found themselves short of crude and forced to buy at unprofitably high prices to stay in business" (Olien and Olien, 2002, p. 19).

The rhythm of booms and busts also informed the worldviews of those who came to work in the oil patch. Oil work could be quite lucrative even for the unskilled, certainly much more financially rewarding than farm work or factory work, although much more dangerous too. With Mitchell's argument about carbon democracy in mind, the boom/bust cycle of oil work seemed to disincentivize union affiliation. "Although the rig builders seemed to have embraced unionization, other oilfield occupations, particularly the drilling industry, never gave it more than a passing glance. The transient nature of almost all the contract labor in the oil patch mitigated against unionization of any type. Their independent mind-set combined with the unsettled nature of the work that depended upon a boom-and-bust cycle worked against unionization. In their mind having a union only saddled them with another entity of control. From their point of view if they were dissatisfied with the job, they could just whip the driller and move on to another job that would probably pay more money anyway" (Weaver, 2010, p. 163). The material character of oil exploration, in other words, reinforced settler Lockean principles of the rewards of individual labor as a pathway to both wealth accumulation and divine grace. The ethics of the oil patch surely helped to cement these principles in Texan regional culture, which remains to this day very enthusiastic about resource development of all kinds, but much less enthusiastic about governmental supervision of that development, let alone environmental regulation.

IV.4.2.2 The Making of the Texan Petrostate

The path to governmental engagement with the Texan oil industry was a circuitous and fraught one. Its first focus was another venerable Lockean concern, the problem of waste. California and Oklahoma were the first states to begin to regulate the petroleum industry with a focus on controlling overproduction to reduce price volatility and oilfield waste, and to guarantee more equitable access among producers to the primary transportation networks of pipelines and railroads. Their experiments in the 1910s offered prototypes for regulatory ventures elsewhere (Childs, 2005, p. 152). World War I also created a national conversation about reducing oil waste, as it was recognized that new elements in the modern machinery of warfare like airplanes, automobiles and tanks depended upon petroleum to function. In Texas, there was also mounting pressure among oil producers for state action to help manage market forces. In 1917, on the heels of a severe drought, Texans approved the Conservation Amendment to the Texas Constitution, creating a state responsibility for managing natural resources. Although the amendment's focus is clearly on water management its language is expansive enough to include energy resources as well. Although historian William Childs concludes that Texas lawmakers never imagined a "comprehensive approach to oil-industry regulation" (Childs, 2005, p. 155) they did pass an Oil and Gas Conservation Law in 1919, which gave the Texas Railroad Commission (TRC) the authority to establish rules and regulations for the purposes of conservation and the reduction of waste. Among other things, "the act defined 'waste' as letting 'commercial quantities' of natural gas produced from a gas formation simply flow into the air; letting water drown out a natural gas formation; burning off natural gas; and using gas wastefully, as in flambeaus or in daytime lighting" (Olien and Olien, 2002, p. 92). The TRC was granted this power to regulate waste because of the interrelated nature of energy and transportation industries, because of the congressional lack of enthusiasm for creating a bespoke regulatory agency focused on oil, and because an investigatory committee determined that the TRC commissioners were "underworked" (Childs, 2005, p. 156).

None of three TRC commissioners had any experience in oil, which made its first decade or more of conservation regulation mostly a lost cause. In its early years, the TRC passed laws regulating waste—for example, by mandating a certain distance between wells—that were largely ignored or undermined by oil producers; but by the end of the 1920s TRC agents had gained more legitimacy providing expertise and performing an umpire-like function to resolve disputes. This led to early experiments with regulatory proration—in which the TRC would set and monitor production limits—but these only succeeded where the rule-making followed requests from producers themselves (Childs, 2005, p. 166). The interests of producers in Texas were no longer singular, but rather increasingly multiple and polarized between larger corporate interests (the majors) and the inheritors of the wildcatting tradition (the independents).

The discovery of a massive new oil field sprawling over five counties in eastern Texas in 1930 weaponized this polarization, unleashing a crisis (the so-called "Texas oil wars") from which the

TRC would eventually emerge with cemented regulatory status. At its peak, this one field would produce 60 percent of all Texas oil, 37 percent of the oil in the United States, and 22 percent of the oil in the world. Its discovery was a system shock in more ways than one. Recent U.S. oilfield discoveries and rising international competition was depressing oil prices (Hanieh, 2024, Ch. 3) but land and labor was still cheaper. "With so many men out of work because of the Depression, East Texas soon swarmed with cheap labor, both experienced and completely green, ready to take jobs for minimal wages" (Olien and Olien, 2002, p. 171). Independent producers mushroomed to take advantage of the massive field, producing oil at rock bottom prices. By the middle of 1931 it was clear that a market disaster was looming. The Depression had reduced energy demand and with the market flooded now with cheap East Texas oil, the market price for oil dropped 78% in only a year (Olien and Olien, 2002, p. 180). The TRC tried to intervene to set a production quota for East Texas but the independents decried this move as regulation in favor of the majors and rallied to oppose the quota. By July 10, a barrel of oil was selling for as little as two cents. In August, the governor declared that there were East Texas producers "in a state of insurrection against conservation laws" and sent in the national guard to impose martial law (Olien and Olien, 2002, p. 186). The TRC continued to try to issue proration orders—19 of them in 1932—but they were ignored or challenged in court; and, courts eventually overturned every one of the TRC proration orders (Childs, 2005, p. 212). Although martial law was overturned by the U.S. Supreme Court in December 1932, the situation remained unsettled. Fearing a federal takeover, the Texas legislature passed three bills to help strengthen the TRC's regulatory authority, including making violating a TRC order a felony punishable by jail time. However, the crucial guarantor of TRC authority came with the passage of the Connally Hot Oil Act by the United States Congress in February 1935, which prohibited the transportation of oil across state lines that had been produced in violation of state laws. This reciprocally validated the regulatory powers of state agencies like the TRC to set proration orders for the purposes of stabilizing the market and reducing waste.

A few years later, World War II and its unprecedented demands for military fuel meant that the problem of petroleum excess in the 1930s turned into a surplus that could enable unprecedented war machines and scales of global conflict. Texas contributed greatly to fueling WWII. "Indeed, the eventual Allied victory in the war is sometimes credited – with only a hint of exaggeration – to the two enormous Inch pipelines built between 1942 and 1943, connecting the oil fields of Texas to refineries on the New Jersey seaboard and providing 500,000 barrels of petroleum each day for Allied forces in Western Europe" (Hanieh, 2024, Ch. 5). World War II was reciprocally revolutionary for the petroleum industry by accelerating the need for petrochemicals and petroplastics, creating huge new markets for petroleum derivatives. The 1940s also saw natural gas byproducts of oil extraction becoming a crucial infrastructure for heating and electricity production. "The natural gas industry expanded at a phenomenal rate during the 1940s and was competing vigorously with hydroelectricity and coal. By the end of the decade, natural gas was the sixth largest industry in the United States" (Childs, 2005, p. 250).

IV.4.2.3 Petropolitics in a Time of Climate Change

After WWII, Texas raised state taxes on oil and gas production to near their current rates (4.6% for oil, 7.5% for natural gas). This move cemented the Texan petrostate complex in which a light hand of governance and regulation was rewarded with exceptional tax revenue (including a record $26.3 billion in 2023) to fund Texas public schools and universities, roadways, first responders, and a variety of other critical biopolitical functions. This largesse explains the outsize political influence oil and gas retains in Texas despite the fact that oil and gas has declined as a share of the Texas GDP from its peak of 21% in 1981 to only 8% in 2021 as the Texan economy has grown and diversified.

But, the national, indeed international, impact of Texas oil remains indisputable. When the formation of OPEC and oil shocks in the Middle East transformed global resource politics in the 1970s, Texas oil and the expansion of Gulf offshore drilling played vital roles in mitigating energy crisis

(Parra, 2004). There is even evidence that the TRC served as the original model for OPEC (Parra, 2004, p. 89) with the idea that the organization's primary function was to manage supply so as to guarantee sufficiently high prices to allow larger companies to earn huge profits while also covering the production costs of smaller independent ventures, thus incentivizing the wildcatting network that had historically guaranteed a steady supply of new oil finds.

The gradual depletion of Texan oil production from 2.5 million barrels/day (bpd) to just under 1 million bpd in 2009 mirrored the rise of American "peak oil" anxieties in the first decade of the 21st century. Although the peak oil movement was about much more than simply energy (Schneider-Mayerson, 2015), it served as a lightning rod for apocalyptic imagination of what would become of America after its oil reserves depleted. A few years later, peak oil fears were unceremoniously swept away by the rise of so-called "unconventional" oil production techniques like enhanced oil recovery, horizontal drilling, and hydraulic fracturing, which revitalized depleted fields and opened production of new shale oil and shale gas fields. Again, Texas led the way in the so-called "shale revolution"; Texan oil production surged back to 2.5 bpd in less than five years (2014) and then broke production record after record, all the way to 5.5 bpd in 2023, 42% of the national total, helping to make the U.S. the largest producer of crude oil, ever.

The oil and gas production renaissance of the past fifteen years might be a cause for celebration were it not for the inconvenient fact of climate change being driven mostly by fossil fuel use. 2023 was also a year in which Texas led the nation with 16 weather-related billion-dollar disaster events, the intensification of which by anthropogenic climate change has been confirmed by a variety of scientific studies. According to NOAA data, between 1980 and 2024 Texas experienced 188 confirmed weather/climate disaster events with losses exceeding $1 billion, including 19 drought events, 9 flooding events, 1 freeze event, 126 severe storm events, 15 tropical cyclone events, 7 wildfire events, and 11 winter storm events (see Figure IV.4.1). That is by far the greatest number of weather/climate disasters experienced by any state during this period, constituting nearly half of the 400 events experienced by the United States as a whole during the 1980–2024 period. Worse still, both the state and the country have seen a noticeable acceleration of billion-dollar disaster events over the past five years, more than double the long-term trend.

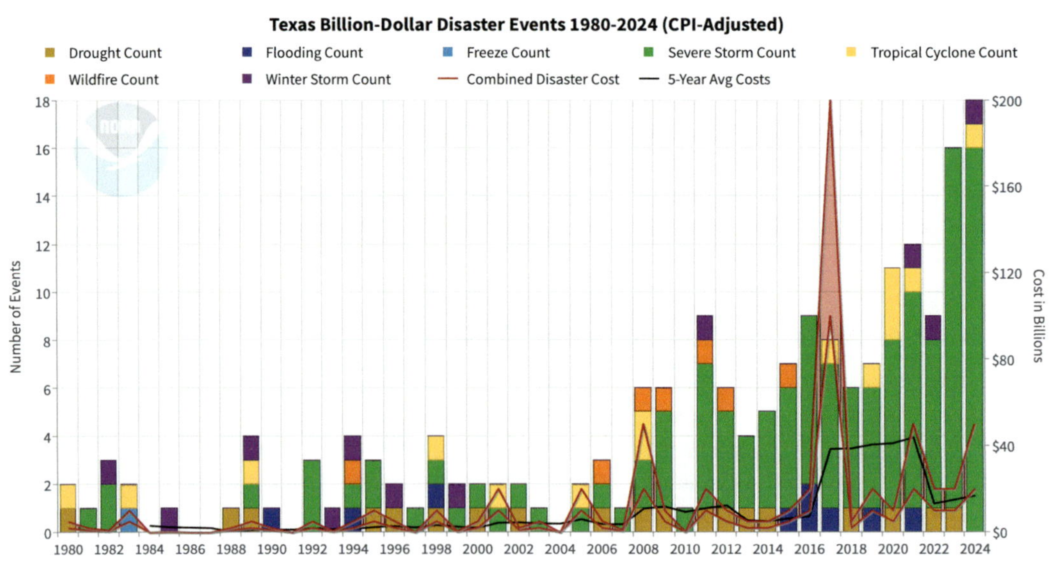

FIGURE IV.4.1 Billion dollar weather/climate disaster events in Texas, 1980–2024.

Source: NOAA.

Texas currently finds itself at a crossroads both in terms of energy and climate. In 2021, Texas produced 1.741 billion barrels of crude oil. The rest of the United States together produced 1.716 billion barrels. But at the same time Texas had the greatest installed capacity of wind power, the same as the next four largest states put together. And, also in 2021, Texas led the nation in terms of residential solar installations. This combination of fossil and renewable energy leadership makes sense in terms of the venerable Lockean ideology that is deeply woven into Texan frontier culture. In its logic, the whole world is nothing more than a bundle of resources to be developed and transformed by humans as a matter of divine right and dispensation. Oil and gas are resources that should be extracted from the earth, no matter the environmental or social cost. But, likewise, the wind and the sun are resources to be harvested lest they lie fallow and go to waste. Texas's considerable renewable energy achievements have rarely been driven by environmental, even conservation, concerns. Instead, they have been dictated by the settler liberal teleology of industrial, technological progress (Boyer, 2023). And, more recently, by the falling price of renewable energy. But, taking a long-term view on the matter, we are likely witnessing the first phase of a transitional energopolitics between an oil-based economy and an economy centered on renewable energy and electricity. Yet, Texas's islanded electrical grid is more precarious than might appear at first glance. For this reason, it makes sense to start an investigation of Texan electropolitics with is arguably the most significant grid event in Texan history, the so-called "Texas freeze" caused by winter storm Uri (February 13–17, 2021).

IV.4.3 SECTION II: TEXAN ELECTROPOLITICS

IV.4.3.1 The Verge of Failure: The Texas Freeze of 2021

Uri would eventually become the most expensive disaster in Texas history causing over $195 billion in damages, 56% more than the $125 billion price tag of the previous record setter, 2017's Hurricane Harvey. But, unlike Harvey, which had elicited much anticipatory doomsaying in the days before the storm's landfall, the cascading implications of Uri's arrival seemed to catch Texas unawares. The advance weather coverage of Uri seems rather tame in retrospect, focused mostly on hazardous icy roads and the need to protect exposed pipes from what were correctly predicted to be the most frigid temperatures Texas had seen since 1989. No one in broadcast media seems to have grasped the enormous challenge the storm would pose to the state's islanded electrical system. How much government knew and when is also a matter of debate. Although hardly the hero of the story, Governor Greg Abbott issued a preemptive disaster declaration for the entire state on February 12, after icy roads in the northern part of the state led to a 133-car pile-up and six deaths near Fort Worth. Yet Abbott's declaration made no specific mention of an anticipated loss of electrical power. ERCOT—the system operator of the Texas grid—held an emergency conference call the same day with the state's electrical utilities and rural electrical cooperatives to advise them that several hours of rolling blackouts were likely coming (Gold 2022).

A deep freeze spread across the state on February 13, flirting with record temperature lows in Dallas (–2 F/–19 C) and Houston (11 F/–12 C), and, although the information was not publicized, electricity generation began to falter as natural gas wellheads froze and critical equipment was disabled by freezing temperatures at coal, nuclear, and wind power generators. No energy source was untouched, although the breakdown in natural gas supply became the greatest destabilizer of the bulk power system (see Figure IV.4.2).

On February 14, at 837p ERCOT announced the new winter peak demand record of 69,150 MW, more than 3.2 GW higher than the previous winter peak in January 2018 and 2 GW worse than the worst-case scenario ERCOT had gamed out for winter demand (Gold, 2022). However, the message seemed strictly informational, appearing to convey that the record demand was being managed. ERCOT's next tweet at 1217a was a level 1 Energy Emergency Alert (issued when reserves dip under 2.3 GW for more than 30 minutes). The situation quickly unraveled from there. At 112a

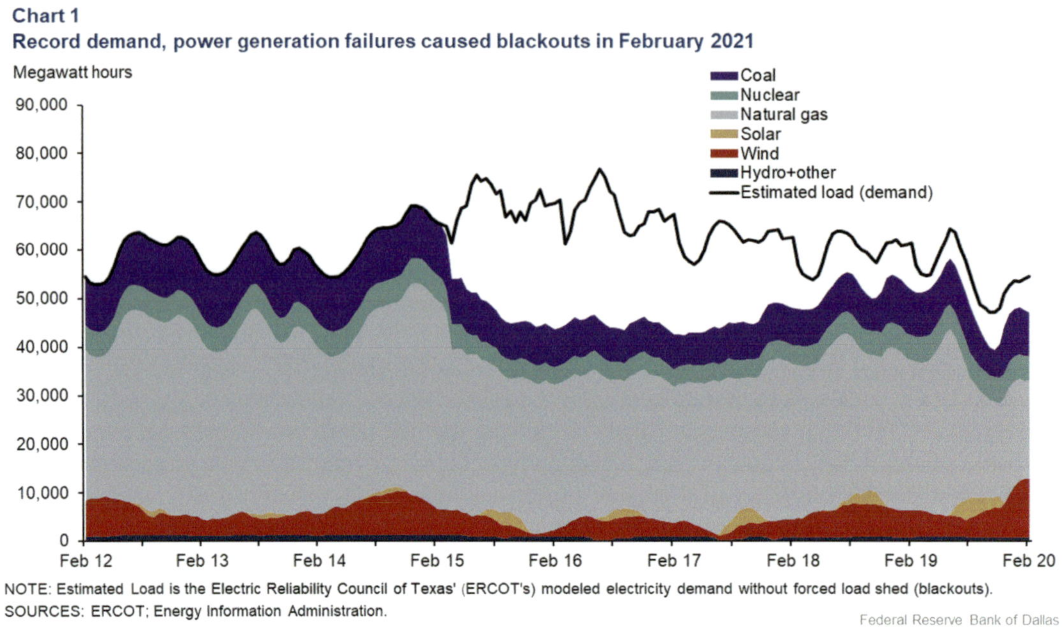

FIGURE IV.4.2 Generation failures during Texas freeze.

Source: ERCOT.

a level 2 EEA was announced, indicating reserves under 1.75 GW and dropping grid frequency. Thirteen minutes later, ERCOT issued a level 3 EEA in which "rolling outages" were announced.

In reality, those rolling outages were emergency load shed that would persist in many Texan neighborhoods for the next four days. What had gone wrong? And why did ERCOT seem to have underestimated the dangers so badly? Less than a half hour after ERCOT entered EEA 3 the grid frequency dropped below 59.4 Hz for four minutes and 23 seconds. At nine minutes below 59.4 Hz, a wide swath of additional generating units would have automatically tripped offline, creating a real threat of a systemic collapse of electrical supply in the context of the coldest weather Texas had seen in over 30 years. The grid frequency actually dropped as low as 59.302 Hz, .002 Hz away from widespread failsafe automatic shutdowns. Facing the prospect of an excruciating weeks or months long "black start" scenario for a customer base of 27 million Texans, the actual 10.5 GW of emergency load shed that took place between 123a and 159a on February 15th, 2021 (impacting 3.6 million homes and businesses) might seem like a victory for the engineers who barely held the state's electrical supply together. Over the next two days, ERCOT averaged only 34 GW of generation, less than half of its peak winter electric load. It was, according to a subsequent Federal Energy Regulatory Commission analysis, "the largest controlled firm load shed event in US history" (FERC, 2021, p. 9). Meanwhile, leading Texas Republicans like Abbott, Lieutenant Governor Dan Patrick, and former Governor Rick Perry took to cable television news shows to blame the crisis (falsely) on the intermittency of renewables and on the specter of the "Green New Deal" (Gold, 2022). The agents of the Texan petrostate saw an opportunity to distract from the greater failure of natural gas at a fragile electropolitical juncture when the interconnection queue for ERCOT is already almost entirely wind, solar, and battery storage projects.

Without denigrating the obvious grace under pressure of ERCOT's engineers in the early hours of February 15, it is hard to interpret the unfolding of the Texas freeze as anything other than an epic failure of Texan electropolitics writ large. In total, 69% of Texans lost power during the storm, with outages averaging 42 hours. At least 246 people (and possibly as many as 700) died because of hypothermia, carbon monoxide poisoning, and problems resulting from the loss of power to critical medical equipment. 75% of Texans had difficulty obtaining food and 63% had difficulty obtaining water. Electricity bills soared, especially for those customers whose bills were tied to real-time power prices. One customer of the rapidly bankrupted firm Griddy received an electric bill for $17,000 for five days' service. Meanwhile, gas suppliers enjoyed about $11 billion in windfall profits despite their poor service record: a 50% regional decline in natural gas production between February 8 and February 17 (FERC 2021, p. 12). Kinder Morgan alone made $1.41 billion, its best quarter ever (Gold, 2022). Within Texas's islanded electricity market, "power companies aren't required to produce enough electricity to get the state through crises ... In fact, they are incentivized to ramp up generation only when dwindling power supplies have driven up prices" (Schwartz et al., 2021).

Jane Bennett wrote of the 2003 North American power disruption that "the electrical grid, by blacking out, lit up quite a lot" (Bennett, 2005, p. 462). Indeed, blackouts tend not to be the neatly bounded events media and politicians portray them to be, but rather ragged assemblages of infrastructural politics often many years in the making. In the next section we look backward to highlight the electropolitical circumstances that made the "fraying wires" (Bakke, 2017) of the Texas freeze and the near collapse of the ERCOT ISO zone possible.

IV.4.3.2 Texan Electropolitics, Past and Present

To understand what happened and did not happen in Texan electropolitics after Uri, it is important to understand the idiosyncratic history of the making of the islanded Texan electric grid itself. As Thomas Hughes (1993) has argued convincingly, energy security looms large in grid emergence and design: grids tend to emerge in the context of wartime industries as a redundancy hedge against power outages from rapidly expanding demand. The Texas Interconnected System (TIS) was founded in 1941 with just such a purpose to guarantee adequate power supply to wartime aluminum smelters along the Gulf Coast (Dyer, 2011, p. 18). But the TIS also materialized in opposition to the 1935 Federal Power Act, which authorized the federal government to regulate the transfer of electricity between states. Recognizing that this would sooner or later become a challenge to Texan energy sovereignty (as well as opening the Texas electricity market to external competitors), most of the Texas utility companies agreed not to send any electricity out of state, severing their interstate lines and creating the core of what would later become the ERCOT zone. "Freedom from federal regulation was a cherished goal—more so because Texas had no state regulation before the 1970s. So eager were the Texas utilities to maintain traditional Texas independence that they memorialized the policy of isolation into a written agreement, binding themselves to intrastate operation" (Cudahy, 1995; Cohn and Jankovska, 2020).

The massive November 1965 blackout in the Northeast catalyzed the 1968 creation of the North American Electric Reliability Corporation (NERC), a voluntary membership of electrical utilities organized to create standards and guidelines for reliability and adequacy in the bulk power system. In keeping with NERC guidelines, TIS created ERCOT as an independent not for profit corporation with advisory rather than regulatory functions. ERCOT's initial staff consisted of only two retired utility employees (Dyer, 2011, 19). A Public Utilities Commission of Texas (PUC) followed in 1975 with regulatory authority over electricity rates and services. TIS transferred its operational functions to ERCOT in 1981. Although ERCOT remained a voluntary membership organization (now composed of 80 investor-owned, municipal, and cooperative utilities), it became the technical operator and rule-maker for a wholesale electricity market that involved of 85% of Texan electricity generation, including 40,000 miles of transmission lines and more than 22 million customers. In 1996, ERCOT became the first independent system operator (ISO) in the United States.

The mid-1990s saw a national push for the "deregulation" of electricity markets that found very warm reception in Texas. In 1995, Texas Senate Bill 373 allowed independent wholesale generators, power marketers and utility affiliates to compete to supply wholesale power. In 1999, Texas

Senate Bill 7 allowed electric retailers to compete for business and ERCOT's functions expanded to include switching customers between retailers on a timely basis, and overseeing transactions in the wholesale spot market for energy. The transition to the deregulated market in the early 2000s was difficult and marred with technical problems and organizational scandals at ERCOT (Dyer, 2011). In keeping with the many failed promises of neoliberal policy, deregulation also pushed power prices higher. Before 1999 Texans had electricity prices consistently below the national average. By 2008, Texans taking power from competitive suppliers paid 29.3 percent above the national average (Dyer, 2011, p. 75). At the same time, rolling blackout became more frequent as ERCOT's market fundamentalism led to the lowest reserve margins of any region covered by the North American Electric Reliability Corporation. As elsewhere under neoliberalism, "deregulation" meant in practice regulation in favor of elite wealthy and corporate interests, usually insulating them in technical opacity and fringing them with noisy populist distractions.

ERCOT's low reserve margins, combined with Texas's enthusiasm for deregulated electricity markets, created important energopolitical conditions of possibility for the 2021 crisis. An investigation led by the Texas Tribune concluded that the PUC, which oversees the state's electric and water utilities, "repeatedly ignored, dismissed or watered down efforts to address weaknesses in the state's sprawling electric grid" (Schwartz et al., 2021), especially the lack of weatherization of energy supplies. According to consumer advocate and environmental activist Tom Smith, "The fault on this one is at the feet of the Legislature and the regulators for their failure to protect the people rather than profits, the utility companies, rather than investing millions of dollars in weatherization that had been recommended in review after review of these kinds of incidents" (Schwartz et al., 2021). Instead of acknowledging these planning and regulatory deficits, the state government sought to activate electropolitical conflicts over the most reliable sources of electricity supply. This speaks to the seemingly paradoxical energopolitics of Texas as an oil and gas petrostate with one of the most impressive renewable energy sectors in the country.

Although the ERCOT zone design emerged in a social environment dominated by petropolitics, it has been amenable to low carbon energy sources so long as they abide the deeper ideological commitments of Texan energopolitics to the decomposition of democratic accountability and the strengthening of corporate power. Confusion has by and large ruled the roost since 2021. Regulatory reform over energy providers is scarcely mentioned and indeed Governor Abbott continues to press for theatrical displays of electropolitical competence over real safety and reliability measures. This is not to underestimate the seriousness and fierceness of the struggle within transitional electropolitics. The plummeting cost of solar PV and wind power have cancerously eaten away at the dominant petropolitical interests of the ERCOT zone, causing them to sacrifice coal in order to secure a natural gas future. But now that future is itself under assault from solar, wind, and battery storage, summoning a crisis for the Texan petrostate, which earned (at the time) a record $24.7 billion in tax revenues from oil and gas sales in 2022.

In July 2024, Governor Abbott called for a $10 billion fund to support new natural gas generation to meet ERCOT's growing demand needs. He also lent his support to a 2023 ERCOT plan to revive so-called "zombie plants," coal-fired facilities that had recently been shuttered for cost and pollution reasons. The plan, fortunately enough, found no support from power providers and not a single zombie was revived. Abbott also strongly supports adding more "dispatchable generation" to the ERCOT zone, known also as "peaker plants," natural gas facilities that fire only when there is high demand on the grid (and high profits to be made). Without mandating the weatherization of such facilities, of course, they would likely suffer similar outages to those that occurred during Uri. Once again, we see the doubling down on petrologics interrupting the potential emergence of a post-fossil ERCOT zone.

The heads of ERCOT and the PUC were fired within days of Uri, but nearly four years on from the crisis of February 2021, the weatherization of energy facilities in the ERCOT zone remains dubious, guaranteeing systemic precarity in the face of climate change and its extreme weather swings. A routine cold snap in December 2022 brought nearly 30 GW of coal and gas-powered electrical supply offline, roughly equivalent to the fossil power losses during Uri (Lewin, 2023) with ERCOT once again drastically underestimating power demand. The anticipation of blackouts has now become a frequent and very vivid feature of Texan life. As one of my interviewees commented: "It used to be the case that we worried about power outages only for a few days each summer. Now, I find myself worried in the winter too. It's like the whole system is breaking down."

Could the near catastrophic failure of the Texas grid happen again today? Texas grid expert Doug Lewin estimated in November 2023 that "If we do see another Winter Storm Uri next year, the likelihood of at least some rolling outages is probably close to 100%" because of state agencies' failure to require the gas industry to weatherize its supply system (Lewin, 2023a). Yet Lewin noted with more optimism that ERCOT's 15-fold increase in battery storage and 10 additional GW of solar capacity since 2021 would likely make the duration and magnitude of Uri-style outages less severe. Lewin sees progress in transitional energopolitics toward a more reliable and resilient system through added renewable and battery resources; "the grid is not fixed; but we're in better shape."

Yet every unhappy grid crisis is unhappy in its own way. The most recent catastrophic episode in Texan electropolitics arrived in the form of Hurricane Beryl, which struck Houston in early July 2024, knocking out power to over 60% of Houston's homes for at least 24 hours with some areas taking nearly two weeks for power to be restored. Houston solar developer Jonathan R. described Beryl in an interview as a "death by a thousand cuts. Because of the direct hit on the city with high winds, every single branch along the infrastructure went down. Hard to reach, hard to repair."

Public wrath was visited this time on Houston's major electrical utility, CenterPoint, and a graffiti meme "CenterPointless" went viral in the days that followed. Abbott's administration castigated CenterPoint leadership publicly while quietly allowing the company to recoup losses from Beryl through securitization, by which ratepayers will be charged to pay the infrastructural costs for climate catastrophes for many years or even decades to come. Lewin commented, "This has become the state's de facto climate policy: rather than fight climate change or prepare for its impacts, the state is simply passing the costs to future generations" (Lewin, 2024).

IV.4.4 CONCLUSION

As we look ahead toward the middle of the 21st century, the energopolitical complexity of Texas shows no signs of abating. Given the general dismissal of climate action by the incumbent political regime, it seems highly unlikely that the state will act to curtail fossil fuel production until the consumer market withers. At the same time, there has been impressive decarbonization progress in terms of the ERCOT grid. Viewed solely through the lens of electricity generation, many states (and countries) could do worse than to follow the Texan lead. But there are significant concerns about Texan readiness to endure the intensifying weather-related disasters being brought about because of climate change. For example, despite experiencing the worst power outage failure event in state history, there has been surprisingly little action taken to prevent its repetition. Ignoring massive public outcry for accountability after Uri, as well as expert calls for greater regulation of power providers to force them to weatherize their equipment, the state government opted only to institute a voluntary weatherization program. Even though natural gas supply emerged as the biggest culprit for the near-collapse of bulk power during the storm, the Texas Governor and legislature have subsequently sought to leverage the crisis to defame renewables and foster even greater dependency upon gas within the ERCOT zone. They may well succeed. Renewables meanwhile continue to expand in Texas at the expense of coal—which could be understood as a transition of sorts—but only because their low cost fits well within the market fundamentalism of Texan politics. Notably, Texas has received the second most funding of any state from Biden government infrastructure investment programs ($33.2 billion), many of which favor post-fossil policies. However, under the single party rule of a political party for which climate denial has become an article of faith, for the foreseeable future Texan electropolitics will very likely continue to embrace natural gas as a so-called "transition fuel," thus quite literally freezing a complete transition away from fossil fuels.

REFERENCES

Bakke, Gretchen. 2017. *The Grid: The Fraying Wires between Americans and Our Energy Future*. Bloomsbury Publishing.

Bennett, Jane. 2005. "The Agency of Assemblages and the North American Blackout." *Public Culture* 17 (3): 445–466.

Boyer, Dominic. 2014. "Energopower: An Introduction." *Anthropological Quarterly* 87 (2): 309–34.

Boyer, Dominic. 2019. *Energopolitics*. Duke University Press.

Boyer, Dominic. 2023. *No More Fossils*. University of Minnesota Press.

Childs, William R. 2005. *The Texas Railroad Commission: Understanding Regulation in America to the Mid-Twentieth Century*. Texas A&M Press.

Cohn, Julie, and Olivera Jankovska. 2020. "Texas CREZ Lines: How Stakeholders Shape Major Energy Infrastructure Projects." Center for Energy Studies, Rice University. https://www.bakerinstitute.org/research/texas-crez-lines-how-stakeholders-shape-major-energy-infrastructure-projects

Cudahy, Richard D. 1995. "The Second Battle of the Alamo: the Midnight Connection." *Natural Resources & Environment* 10 (1): 56–87.

Daggett, Cara. 2019. *The Birth of Energy: Fossil Fuels, Thermodynamics and the Politics of Work*. Duke University Press.

Dyer, R.A. 2011. "The Story of ERCOT." Report published by the Steering Committee of Cities served by Oncor & the Texas Coalition for Affordable Power. http://tcaptx.com/downloads/THE-STORY-OF-ERCOT.pdf

FERC. 2021. "Cold Weather Outages in Texas and the South Central United States." Federal Energy Regulatory Commission. https://ferc.gov/media/february-2021-cold-weather-outages-texas-and-south-central-united-states-ferc-nerc-and

Folch, Christine. 2019. *Hydropolitics: The Itaipu Dam, Sovereignty, and the Engineering of Modern South America*. Princeton University Press.

Froh, Riley. 1984. *Edgar B. Davis: Wildcatter Extraordinaire*. Luling Foundation.

Gold, Russell. 2022. "The Texas Electric Grid Failure Was a Warm-up." *Texas Monthly*. https://www.texasmonthly.com/news-politics/texas-electric-grid-failure-warm-up/

Hanieh, Adam. 2024. *Crude Capitalism: Oil, Corporate Power and the Making of the World Market*. Verso.

Hughes, Thomas. 1993 *Networks of Power: Electrification in Western Society, 1880-1930*. Johns Hopkins University Press.

Lewin, Doug. 2023. "What ERCOT knows and isn't telling us." Substack. January 18. https://www.douglewin.com/p/what-ercot-knows-and-isnt-telling

Lewin, Doug. 2023a. "What does ERCOT's Failed Procurement Mean?" Substack. November 21. https://www.douglewin.com/p/what-does-ercots-failed-procurement

Lewin, Doug 2024. "What Happens Next." Substack. July 30. https://www.douglewin.com/p/what-happens-next

Mitchell, Timothy. 2009. "Carbon Democracy." *Economy and Society* 38 (3): 399–432.

Mitchell, Timothy. 2011. *Carbon Democracy: Political Power in the Age of Oil*. Verso.

Olien, Diana Davids and Roger M. Olien. 2002. *Oil in Texas: The Gusher Age, 1895-1945*. University of Texas Press.

Parra, Francisco. 2004. *Oil Politics: a modern history of petroleum*. I.B. Tauris.

Pratt, Joseph A. 2007. "A Mixed Blessing: Energy, Economic Growth, and Houston's Environment." In *Energy Metropolis*, eds. Martin Melosi and Joseph Pratt. University of Pittsburgh Press, 21–51.

Ripple, William J, Christopher Wolf, Jillian W Gregg, Johan Rockström, Michael E Mann, Naomi Oreskes, Timothy M Lenton, Stefan Rahmstorf, Thomas M Newsome, Chi Xu, Jens-Christian Svenning, Cássio Cardoso Pereira, Beverly E Law, and Thomas W Crowther. 2024. "The 2024 State of the Climate Report: Perilous Times on Planet Earth." *BioScience* 74 (12): biae087. https://doi.org/10.1093/biosci/biae087

Rogers, Douglas. 2015. "Energopolitical Russia: Corporation, State, and the Rise of Social and Cultural Projects." *Anthropological Quarterly* 87 (2): 431–452.

Schneider-Mayerson, Matthew. 2015. *Peak Oil: Apocalyptic Environmentalism and Libertarian Political Culture*. University of Chicago Press.

Schwartz, Jeremy, Kiah Collier, and Vianna Davila. 2021 "'Power companies get exactly what they want': How Texas repeatedly failed to protect its power grid against extreme weather." *Texas Tribune*. February 22. https://www.texastribune.org/2021/02/22/texas-power-grid-extreme-weather/

Warner, Charles A. 2007. *Texas Oil and Gas since 1543*. Copano Bay Press.

Weaver, Bobby D. 2010. *Oilfield Trash: Life and Labor in the Oil Patch*. Texas A&M Press.

IV.5 Adaptation and Mitigation to Climate Change and Resource Use in the Texas Transportation Sector

Aparajita Datta and Ramanan Krishnamoorti

Transportation is essential for maintaining and growing quality of life, and enabling access to socioeconomic opportunity and mobility, healthcare, education, resources, and markets. Its interdependencies are complex, multidimensional, and dynamic (Figure IV.5.1). Despite its decentralized nature, its function and scope are local and global, and a variety of stakeholders, including businesses, individuals, communities, and governments, influence the interdependencies. Changes to any element create impacts that are felt both within and across the economy.

IV.5.1 A TEXAS-SIZED CHALLENGE – THE EMISSIONS IMPACT OF THE TRANSPORTATION SECTOR

In Texas, the transportation sector plays a pivotal role. It serves as a crucial driver and provider of mobility for people and goods, socioeconomic activity, commerce, and trade. Texas relies heavily on its transportation sector, given its vast geographic expanse of nearly 270,000 sq. miles (Texas Parks and Wildlife, 2025), a booming population of over 31 million (U.S. Census Bureau, 2025), major international trade corridors, and the presence of multiple global ports and logistics hubs that contribute to its gross state product of $2.7 trillion. The state has one of the nation's most extensive and multimodal transportation systems, including vast roadways, bridges, railroads, public transit, sidewalks, bikeways, airports, seaports, waterways, and national and international border crossings. In 2024, the sector employed 2.8 million people in direct, indirect, and induced jobs, and contributed about 20% to the state's gross domestic product (GDP) (Texas Workforce Commission, 2025).

The outsized impact of the state's transportation sector also makes it the second-largest source of greenhouse gas (GHG) emissions and the second-largest energy user in the state (U.S. EIA, 2025). As evidenced by Figure IV.5.2, energy-related emissions in Texas almost doubled between 1970 and 2022. On the one hand, while other high-emitting states have lowered or steadied their energy-related emissions since 2005, emissions in Texas have continued to grow. On the other hand, emissions related to the transportation sector have continued to grow among all states, except New York, between 1970 and 2022. The most significant growth can be observed in Texas, with a 2.6 times increase between 1970 and 2022. In 2022, the transportation sector in the state contributed over a third of the total emissions impact in the state at 220 million metric tons of CO_2 or $MMtCO_2$. The significant emissions impact of the transportation sector in Texas can be predominantly attributed to its population, long commute distances, low cost of transportation fuels, in-state trade dependent on road and rail transportation, and consumer preferences for light-duty trucks over passenger cars (Krishnamoorti and Datta, 2023).

What makes the challenge truly Texas-size are the attributes that are exclusive to transportation in the state: its distinctive mobility patterns and travel habits. The state consistently ranks among the highest in total vehicle miles traveled. Figure IV.5.3 presents the annual total vehicle miles (VMT) traveled against population in all the states and D.C., based on 2022 data. A few states, including

DOI: 10.1201/9781003406600-24

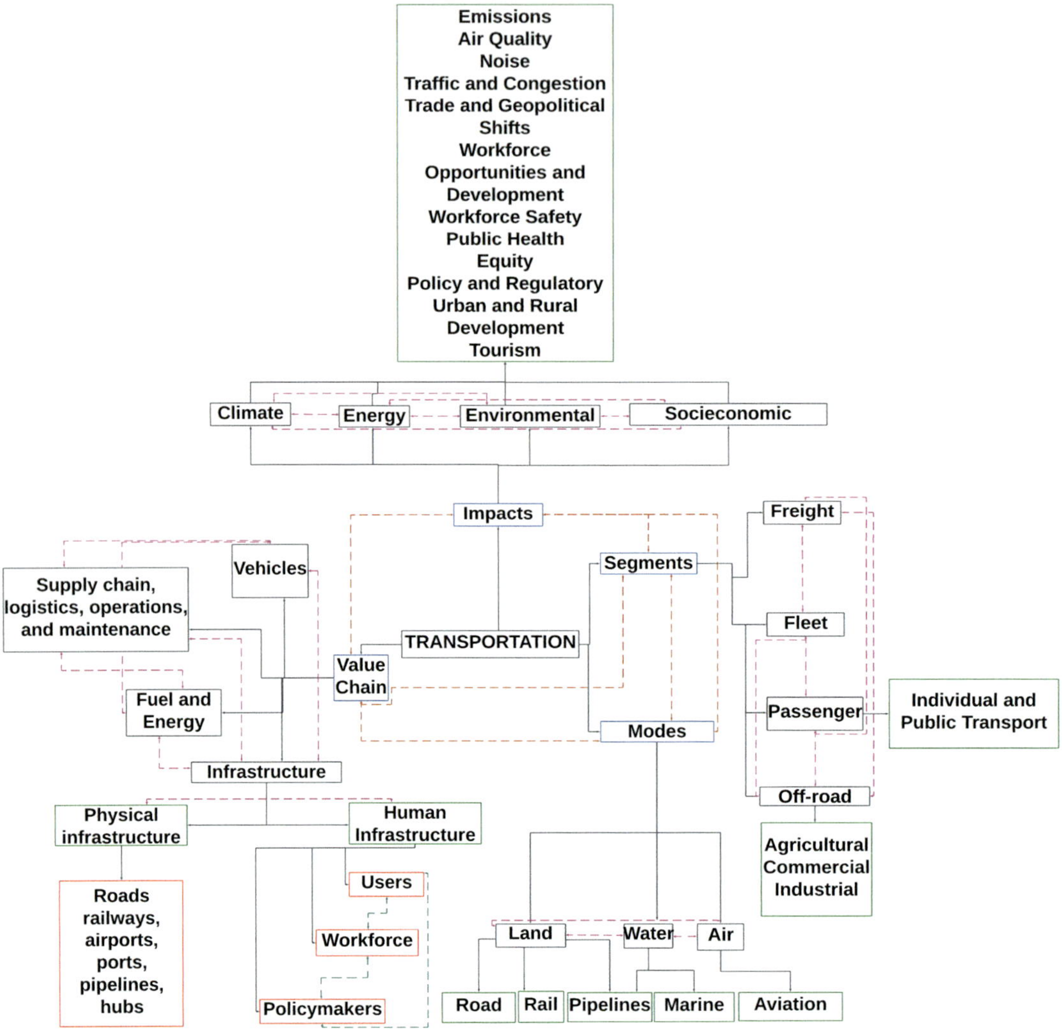

FIGURE IV.5.1 The complex interdependencies within the transportation sector highlight primary, secondary, and tertiary elements of its value chain, segments, modes, and direct and indirect impacts.

Texas, where drivers drove over 288 billion miles, stand out. Among different sectors of transportation and vehicle types, light-duty vehicles (LDVs) contribute the most to the emissions impact, at about 56%, followed by medium- and heavy-duty vehicles at 26%. Air, rail, marine, and all other modes contribute 8%, 2%, 2%, and 6%, respectively. Hence, the predominant focus of the discussion in this chapter will be on the on-road fleet and infrastructure related to light-duty, medium-duty (MDVs), and heavy-duty vehicles (HDVs) (emissions-related details are presented in section 3).

IV.5.1.1 Emerging Concerns and Challenges for the Transportation Sector

While the trend in VMT is often attributed to Texans' love for driving, there are a few underappreciated and connected contributing factors. The preference for driving long distances in the state and its car-centric culture is driven by the sprawl of urban metropolitan areas, the necessity of long-distance travel between urban, suburban, and rural regions, and the lack of accessibility and alternatives to public transit.

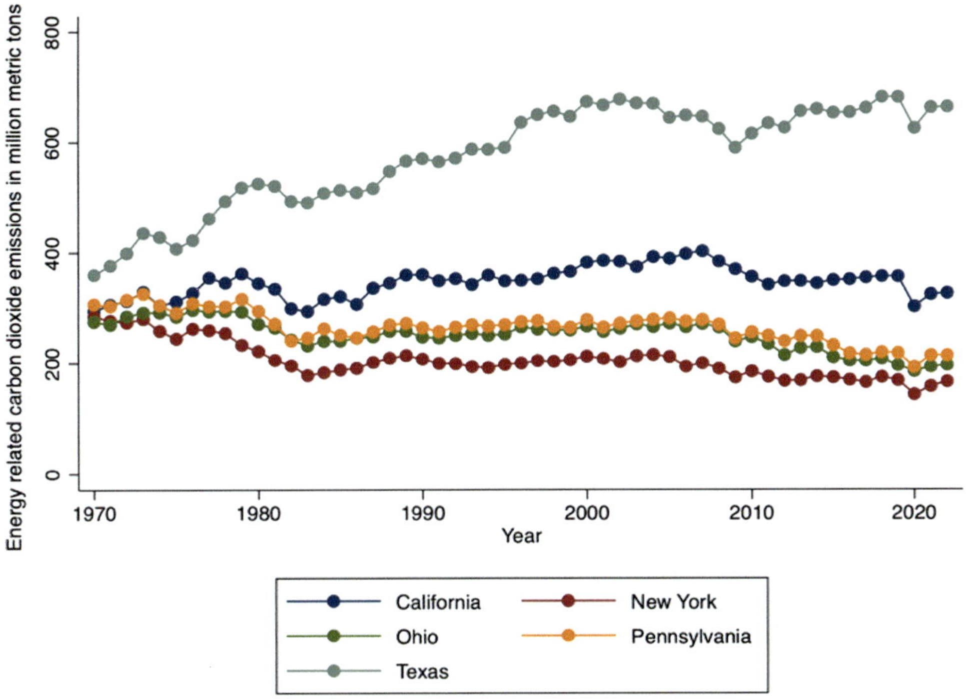

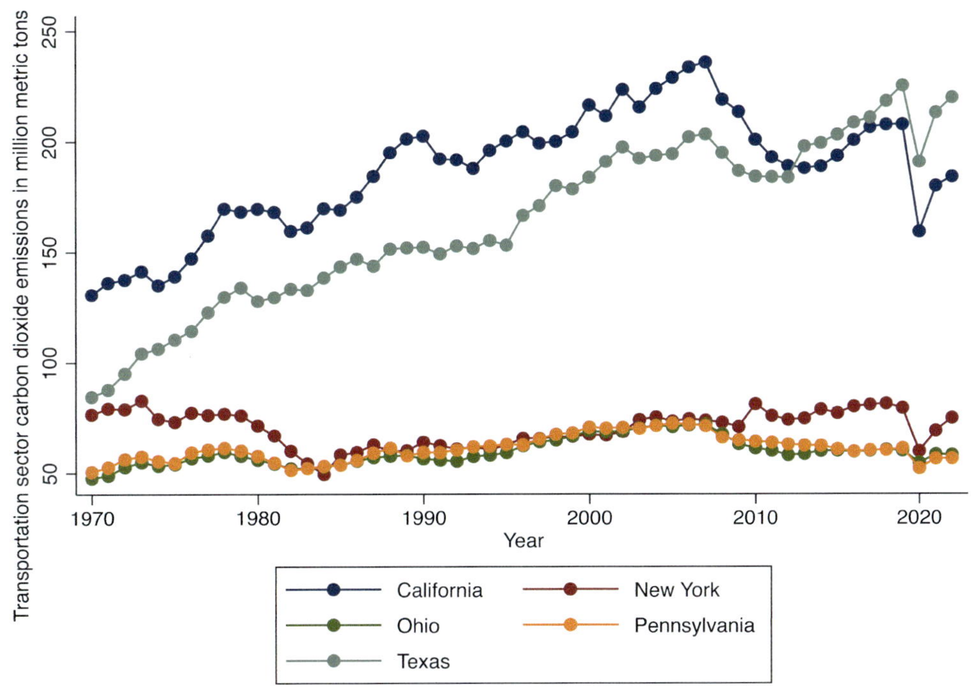

FIGURE IV.5.2 Energy-related emissions (a) and transportation-related emissions (b) in $MMtCO_2$ from 1970 to 2022, sectoral contributions (c) for the top five emitters (total energy-related emissions) in 2022. *(Continued)*

Source: U.S. EIA.

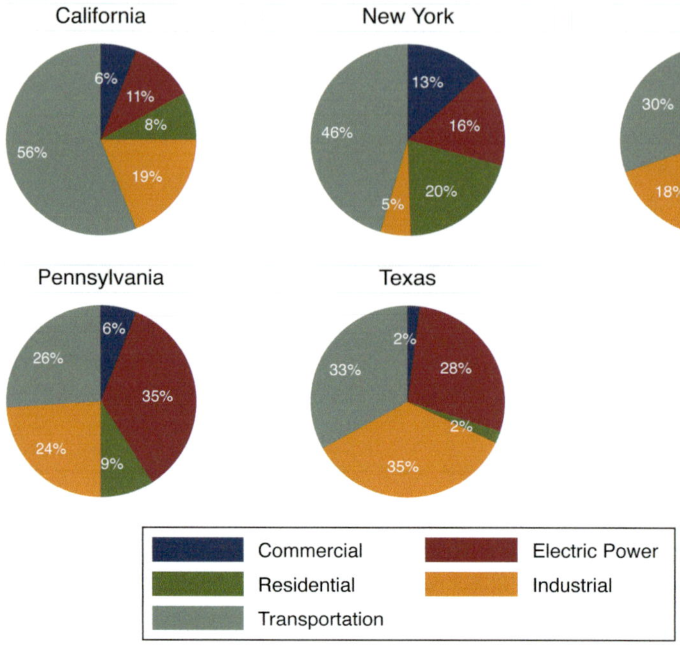

FIGURE IV.5.2 (*Continued*)

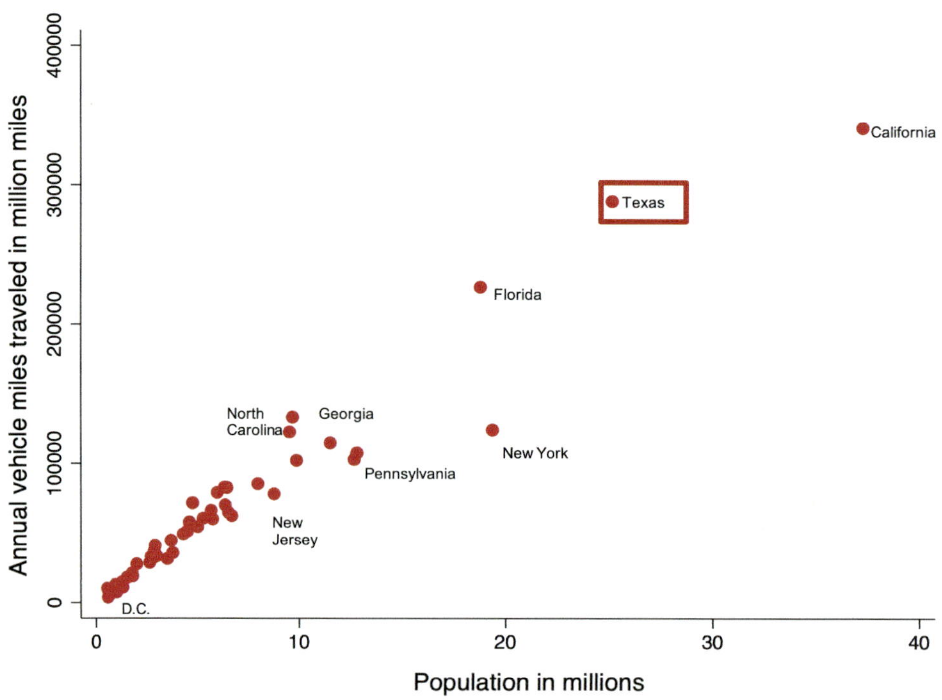

FIGURE IV.5.3 Vehicle miles traveled and population in millions in U.S. states and D.C.

Source: **U.S. Department of Transportation Federal Highway Administration.**

Since 1997, the state's population has almost doubled. The growth has been the highest in the Texas Triangle megaregion, encompassing the Dallas-Fort Worth Metroplex to the north, Houston to the southeast, and San Antonio to the southwest. The region is home to approximately 65% of the state's population (Texas Department of Transportation, 2025a). Major highways connect these urban centers, including Interstate 35 (which runs from Dallas–Fort Worth to San Antonio), Interstate 10 (from San Antonio to Houston), and Interstate 45 (from Houston to Dallas–Fort Worth). These highways pass through urban areas such as Austin, Waco, and College Station. Estimates suggest that the Texas Triangle, formed by the state's four main urban centers, Austin, Dallas-Fort Worth, Houston, and San Antonio, and which is the most rapidly urbanizing part of Texas, has sprawled between 49% and 173% (Kolankiewicz et al., 2017). The number of all registered vehicles in the state has increased by 47% over this period, with only marginal increases in roadways and highway capacity, resulting in congestion. In addition to the time loss, congestion leads to lost productivity, wasted fuel, and increased delivery times, translating into billions in annual economic losses. Resultantly, individuals and businesses bear the brunt of inefficient mobility. Two areas of the Texas Triangle are among the top 15 most congested very large urban areas in the country, *i.e.*, areas that house over 3 million people. It is noteworthy that the Greater Houston area ranks third for excess fuel consumption because of its congestion and urban sprawl. On average, the Texas Triangle and Austin lose about 50–80 hours per year due to traffic congestion (Table IV.5.1).

TABLE IV.5.1
Characteristics of Very Large Congested Urban Areas in the U.S. in 2022

Urban Area	Annual Person-Hours of Delay per Commuter	Daily Vehicle-Miles of Travel (Freeway & Arterial)	Annual Excess Fuel Consumed	Annual Congestion Cost per Commuter	Total Congestion Costs
	Hours	Miles	Gallons	2022 $	2022 million $
Los Angeles-Long Beach-Anaheim CA	122	221,454	114,655	3,214	2,839
San Francisco-Oakland CA	109	209,679	78,938	3,148	2,743
New York-Newark NY-NJ-CT	92	98,903	299,412	2,239	1,937
Washington DC-VA-MD	85	125,087	322,395	1,953	1,702
Atlanta GA	82	51,963	47,031	1,927	1,615
Seattle WA	82	108,772	61,950	1,903	1,470
Miami FL	79	108,281	68,093	1,874	1,404
Boston MA-NH-RI	73	127,214	71,812	1,852	1,492
Chicago IL-IN	72	80,984	94,104	1,675	1,545
Houston TX	69	81,767	126,751	1,664	1,474
Phoenix-Mesa AZ	69	87,454	85,430	1,645	1,564
Dallas-Fort Worth-Arlington TX	68	53,274	70,698	1,528	1,521
Philadelphia PA-NJ-DE-MD	68	82,566	70,787	1,523	1,543
San Diego CA	66	75,079	85,729	1,441	1,180
Detroit MI	61	56,782	57,007	1,277	1,058

Source: Texas A&M Transportation Institute (2022).

While cities struggle with congestion, rural areas in the state face inadequate infrastructure investments, limiting mobility and the available economic opportunities. Additionally, many highways, bridges, and transit systems require modernization. The American Society of Civil Engineers (ASCE) has identified infrastructure deficiencies that need urgent attention, giving the state's infrastructure a grade of "C" in its latest assessment (The American Society of Civil Engineers, 2025).

The state of public transit in Texas is worse off. The state ranks among the lowest in terms of quality of access, based on measures including the share of jobs accessible by public transit and the distribution of the average household's access to transit within a state (U.S. Environmental Protection Agency, 2023). A longstanding policy history of prioritizing highway and road construction over public transit, with state funding favoring these projects, has resulted in a general lack of alternatives, access, and reliable last-mile connectivity through public transit. The COVID-19 pandemic further reduced public transit ridership for buses and rails. In 2023, the ridership recovered to about 75% of 2019 levels. Much of this recovery can be attributed to ridership in Small Urban Transit Districts (92% of 2019 levels) and Rural Transit Districts (82% of 2019 levels) (Texas Department of Transportation, 2023). However, safety incidents, including fatalities, injuries, and/or property damage costing above $25,000, continue to have regressive effects on ridership in Metropolitan and Large Urban Transit Districts, with a reported increase of 1.8 incidents per million miles between 2019 and 2023.

IV.5.2 CLIMATE CHANGE AND THE TRANSPORTATION SECTOR IN TEXAS

The transportation sector's emissions impact is foundational to understanding its complex and cyclical relationship with climate change impacts. As discussed in Section 1, the sector is a significant source of decentralized and hard-to-abate greenhouse gas emissions. While the emissions footprint directly amplifies climate change impacts, the sector's cyclical relationship with climate change is exacerbated by the dependencies created by urban sprawl, congestion, and infrastructural limitations. At the same time, rising temperatures, extreme weather, and sea-level rise damage infrastructure, disrupt supply chains, create hazards for public health and human safety, and increase maintenance costs for vehicles and infrastructure. Hence, adaptation and mitigation efforts for the transportation sector must be unique to the sector's nature and needs. Addressing the emissions from the transportation sector and the interdependencies with climate change impacts will require distinctly different pathways and approaches than those being adopted for the industrial and power generation sectors in Texas. Any changes in transportation modalities and segments will have wide-reaching consequences in the state and beyond.

IV.5.2.1 Understanding the Feedback Loop between Climate Change and Transportation

The circular relationship between climate change impacts and the transportation sector can be best understood through observed ongoing impacts in the state. Transportation-related emissions exacerbate climate change impacts like extreme weather events. Rising temperatures, which are driven by increasing emissions, have led to increased evaporation rates, resulting in reduced soil moisture and worsening droughts. In March 2025, 14% of the state faced "exceptional drought" conditions, which is the most extreme category, with an expansion of drought area, as well as increasing drought intensity (Texas Water Development Board, 2025). The state's water supply is also strained, with surface water sources and underground aquifers replenishing at a lower rate due to prolonged dry periods and the increasing water demand in Texas. A combination of these factors will result in a 9% increase in water demand by 2070, while the supply is expected to decrease by 18% (Texas Water Development Board, 2023). Increased emissions from the transportation sector also contribute to higher ground-level ozone, negatively impacting agricultural yields, which is further stressing water resources in the state (U.S. Environmental Protection Agency, 2001; U.S. Department of

Agriculture, 2013). When high temperatures coincide with drought, subsidence can impact transportation infrastructure, creating a feedback loop. Roads and airport runways are vulnerable, as prolonged heat can soften and degrade asphalt, leading to structural damage, while extreme heat can cause rail lines to buckle, increasing the risk of derailments (U.S. Department of Homeland Security, 2015). Persistent drought conditions can disrupt port and waterway transportation by reducing navigable routes and restricting cargo capacity, leading to higher transportation costs (U.S. Department of Homeland Security, 2015).

Drier conditions, changes in precipitation, and higher temperatures related to increasing emissions and climate change have also led to ideal conditions for wildfires, altering when and where wildfires are most severe. The state has experienced a 50% increase in annual wildfires over the past two decades, with over 238,000 wildfires affecting 12.6 million acres of land (Texas A&M Forest Service, 2025). The Smokehouse Creek Fire, which occurred in February and March of 2024 in the Texas Panhandle and parts of Oklahoma, became the largest wildfire in the state's history, consuming over one million acres and causing record-breaking agricultural losses (Texas House of Representatives, 2025). Between 2005 and 2021, 85% of Texas wildfires ignited within two miles of the affected community, in a zone known as the Wildland-Urban Interface (WUI), which represents the transition area between unoccupied land and human development, where both natural and man-made structures are at risk (Texas Comptroller of Public Accounts, 2025). The Texas WUI spans 11 million acres and contains 3.2 million housing units, ranking second in the U.S. behind California, and has outpaced growth in 42 of the 48 contiguous states between 1990 and 2020. As a result, over 25% of Texas homes are in the WUI, increasing the risk of property damage from wildfires. Wildfire-prone areas are particularly notable in Northeast Houston, outside Dallas-Fort Worth, between Austin and San Antonio, and along the Texas–Mexico border. Wildfires release additional carbon dioxide and air pollutants, creating a feedback loop that exacerbates the impacts of climate change. Often, roadways and highway infrastructure are at risk, with major fires damaging roadways, bridges, and utility lines, leading to costly repairs. Excessive debris flow after a fire increases the risk of functional and structural failures in roadway infrastructure. Severe wildfires can lead to road closures, heightened maintenance demands, and expensive reconstruction efforts. In rural areas, where alternative routes and transportation options are limited or nonexistent, such failures can have even greater consequences for restricting individual mobility and access to essential goods and services (Fraser et al., 2022). The extensive road infrastructure and vehicle congestion in urban areas create urban heat islands due to fuel combustion in vehicles, which releases heat. This also contributes to making urban areas significantly warmer than surrounding rural areas (Zhu et al., 2017; Rizvi et al., 2024).

The urban heat island effect is also exacerbated by intense heat waves in Texas. The jet stream, a key atmospheric current that helps regulate weather patterns and a fast-flowing air current that travels west-to-east around the globe's Northern hemisphere, is shifting due to climate change (Arctic Council, 2024). This has led to prolonged heatwaves in Texas as warm air becomes trapped in parts of the state. In 2023, 334 people in Texas died from heat-related causes, which is more than twice the number recorded in 2011, which held the record until 2021 (Texas Department of Health and Human Services, 2023). Given the vulnerabilities of the state's power grid, power outages and hot weather are a dangerous mix in Texas, with compounding risks for heat-sensitive populations. Between 2000 and 2023, Texas had the most heat season outages of any other state (Climate Central, 2024). These 107 events accounted for 61% of all such events across the southern part of the country. The feedback between heatwaves and the transportation sector results in infrastructure damage like buckling roads and rails, increasing vehicle breakdowns, and potentially requiring reduced speed limits or service disruptions (U.S. Environmental Protection Agency, 2017). Research has also found that heatwaves have led to shifts in travel patterns across the country, with people opting for air-conditioned vehicles and reducing public transit use (Batur et al., 2024).

As emissions have increased, warmer sea surface temperatures have amplified the intensity of hurricanes, and the associated destructive rainfall, storm surges, and winds. Since the Category 4 Great

Galveston Hurricane of 1900, which is the deadliest recorded hurricane in U.S. history, Texas has been affected by another 20 major hurricanes (classified as Category 3 or higher) (Texas Comptroller of Public Accounts, 2018; Weinkle et al., 2018). Hurricane Harvey, a Category 4 storm with unprecedented rainfall of up to 60 inches in parts of Houston, caused historic flooding and infrastructure damage of over $125 billion in 2017 (U.S. National Oceanic and Atmospheric Administration, 2018). Although Hurricane Ike (2008) and Hurricane Beryl (2024) were both Category 2 storms, they caused massive infrastructural damage in the state of nearly $40 billion and $4 billion, respectively (Hagerty, 2024). Hurricanes create a feedback loop with the transportation sector in the state as widespread damage to infrastructure results in road closures, port closures, and disruptions to air and rail travel. For example, the Houston Ship Channel, which is one of the largest ports in the U.S. in terms of waterborne tonnage, experienced operational disruptions, causing economic losses of more than $300 million per day during Hurricane Harvey (Krishnamoorti and Datta, 2022). It also impacted 10% of all trucking in the U.S. due to road closures, led to airport and rail facilities closures, and shut down of warehouses and storage facilities (AFP Global Logistics, 2025).

IV.5.2.2 Primary, Secondary, and Tertiary Impacts of the Feedback Loop between Climate Change and Transportation

The impacts of the feedback loop between climate change and the transportation sector do not occur in isolation but create ripple effects throughout the economy and society. The direct damage to infrastructure, including roads, bridges, railways, and ports, due to extreme weather events is a primary impact on the transportation sector. In contrast, the supply chain disruptions, increased maintenance costs, travel delays, and decreased reliability and safety of transportation systems result in secondary impacts that disrupt commerce and increase costs for producers, suppliers, and consumers in areas beyond those experiencing the extreme weather event. Lastly, tertiary impacts on the transportation sector are experienced in the form of threats to safety and well-being, economic losses, workforce displacement, and environmental damage that have long-term societal effects. These can manifest as increased health risks to truck drivers, airline workers, and dockworkers from day-to-day exposure to extreme heat, productivity loss, job disruptions, and loss of accessibility to public transit to get to work during an extreme weather event. Ultimately, socioeconomically vulnerable populations, including the low-income and communities of color, many of whom may be employed by the transportation sector or related sectors of the economy, also live in areas of heightened risks associated with climate change impacts. This makes emergency preparedness and hazard mitigation interventions for them more pressing, but they are rarely prioritized from a policy perspective.

IV.5.2.3 Other Impacts on Public Health and Safety from Transportation-Related Emissions

Transportation emissions, primarily through vehicle exhaust and freight operations, have significant consequences on public health, the environment, and overall quality of life. In addition to carbon dioxide, vehicle emissions include pollutants like nitrogen oxides (NOx) and volatile organic compounds (VOCs), which contribute to respiratory and cardiovascular diseases, adverse birth outcomes, and increased mortality rates. The exposure is particularly harmful to children, the elderly, and those with preexisting conditions. Nationwide, exposure to ambient particulate matter (PM) has led to approximately 131,000 premature deaths, of which about 21,000 deaths can be attributed specifically to transportation-related emissions (Caiazzo et al., 2013). Specifically in Houston, a study found that transportation-related noise and crashes account for 4% of premature deaths (Sohrabi and Khreis, 2020). Houston households with lower median income, living in the central business district and the vicinity of highways and airports, were found to be at greater risk. In contrast, a statewide study found that reductions in vehicle exhaust emissions over a 20-year period resulted in modest improvements in birth outcomes (Hystad et al., 2025).

IV.5.2.4 The Lack of Direct Mitigation Pathways in the Transportation Sector

The lack of direct mitigation pathways makes abating and adapting to the impacts of the transportation sector on climate change and reducing the intensity of the feedback loop even more complex. The distributed nature of the sector's emissions makes the application of scalable carbon abatement technologies like carbon capture, utilization, and storage (CCUS) to transportation processes impossible. Moreover, the fuel intensity, travel distance range, and the long lifecycles of and upfront investments required for global infrastructure for heavy-duty trucking, freight shipping, and aviation make the sector's emissions particularly hard to abate. Hence, electrification, low-carbon fuels, and efficiency standards are essential to mitigation and adaptation measures for the sector.

In Q3 of 2024, Texas had more than 300,000 registered electric vehicles (EVs) and an expanding network of public fast-charging stations, supported by more than $400 million in federal funding for the Texas National Electric Vehicle Infrastructure (NEVI) Planning Program at the time (Texas Department of Transportation, 2024a). EVs in the state have increased by over 16 times, now representing over 1% of all LDVs (Figure IV.5.4) (U.S. Energy Information Administration, 2024; U.S. Department of Energy, 2025), but EV adoption for M/HDVs remains low. The state currently has 12,315 alternative fueling stations, including those for compressed and liquefied natural gas (CNG and LNG), hydrogen, biodiesel, renewable diesel, ethanol, propane (LPG), and electric charging stations. Texas is also making significant advances in renewable natural gas (RNG) and hydrogen for the medium- and heavy-duty vehicles (M/HDV) segment, electric trains for public transit and connecting urban centers, sustainable aviation fuel for aviation, and hydrogen and ammonia for shipping.

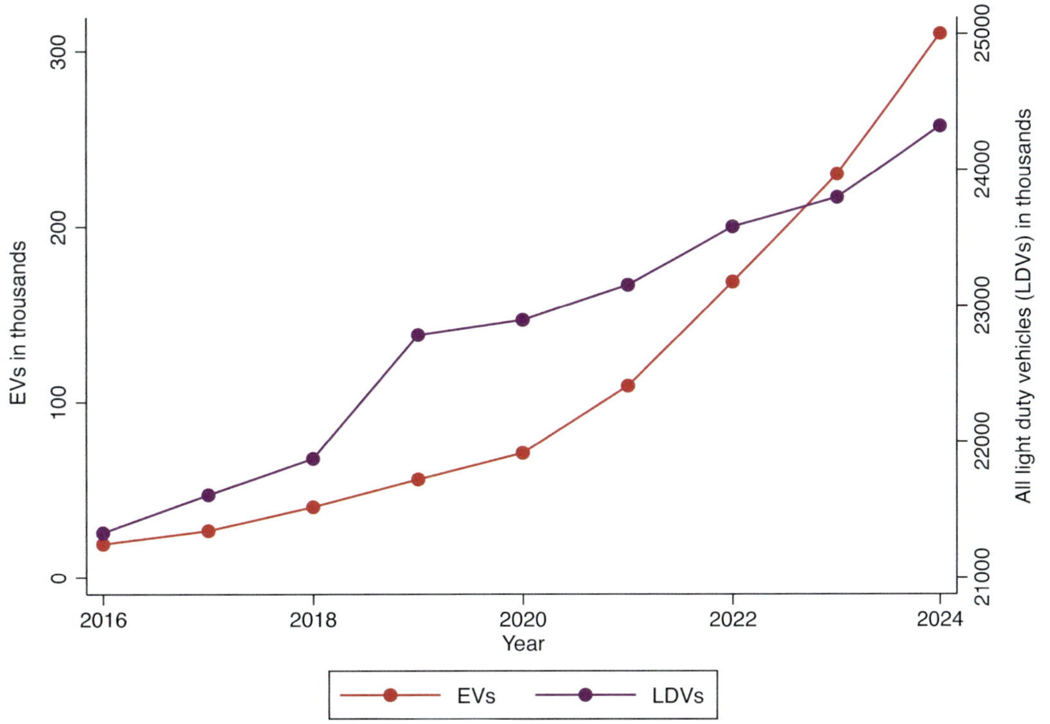

FIGURE IV.5.4 EVs and all light-duty vehicles (LDVs) in Texas in thousands, 2016–2024.

Source: **U.S. Energy Information Administration and U.S. Department of Energy**

IV.5.3 THE STATE OF THE FLEET AND INFRASTRUCTURE, AND NEEDED INVESTMENT FOR RESILIENCE

IV.5.3.1 On-Road Vehicles

The dominant vehicle types in the on-road transportation fleet in the state are gasoline-powered LDVs. Among M/HDVs, about 80% are diesel-based and can be categorized into 17 market segments based on their weight class, from class 2B (heavy-duty pickup and van) to class 8 (regional and long-haul tractor, transit bus, refuse hauler, freight box truck, and dump truck). Regional and long-haul class 8 vehicles represent about 15% of the M/HDV segment but account for about 60% of the emissions impact, while class 2B heavy-duty pickup vans and trucks make up the majority of the remaining emissions impact from M/HDVs (Krishnamoorti and Datta, 2023). In 2024, about 3% of all on-road vehicles were alternately fueled vehicles (AFVs), including EVs, while the remaining were non-AFVs. The growth in AFVs and non-AFVs since 2016 and their emissions impact across highways and roads in the state are presented in Table IV.5.2. Undisclosed and unknown categories are included because of the standard vehicle identification number (VIN) coding process in the automotive sector.

Texas has over 703,600 miles of lane miles (rural roads – 62.3% and urban roads representing 37.7%) and over 314,600 miles of public roads (The American Society of Civil Engineers, 2025). The 45 interstate highways in Texas cover over 3,200 miles. Additionally, the state has more than 55,000 bridges that carry vehicular traffic, which is about 28,000 more bridges than any other state and more than the combined inventories of 17 states (Texas Department of Transportation, 2022). Of these, 33 are international highways that connect to Mexico. The average age of a bridge in the state is 48 years, while that of the highway system is 34 years.

Roads in Texas received a grade of C- and bridges received a grade of B- (the highest for Texas) in ASCE's 2025 assessment (The American Society of Civil Engineers, 2025). The assessment recommends that the state allocate $2 billion annually for bridge expansions and $1.8 billion annually to address the backlog of deficient bridges over the next 10 years to meet future capacity demands and maintain the existing network. It must also incorporate resilience as a criterion in the design of new bridges, invest in a full-time workforce for bridge expansions, and in research and implementation of advanced technologies such as AI, machine learning, and digital twins. Additionally, it must prioritize public safety by analyzing crash data, conducting regular inspections, and implementing new technologies and methods to enhance the safety and efficiency of Texas roads and bridges. The assessment also recommends mitigating the high cost of deteriorated roads by raising tolls, securing additional government funds, and revising the motor fuel tax to adequately fund roads and highways. Additionally, it recommends the state create dedicated cycling infrastructure, invest in infrastructure to optimize traffic flow, and provide more transportation options to reduce single-occupancy vehicles, mitigate the strain on existing infrastructure, enhance public safety and health, and reduce emissions. For public transit, which received a grade of D+, ASCE recommends that Texas establish legislation for sustained investments and secure private equity for transit (bus and rail) transportation, invest in workforce development, leverage lessons from other state entities for public transit, implement the Federal Transit Administration's (FTA) requirements for facility inspections, and quantify the benefits of public transit for the state.

IV.5.3.2 Railways

Texas has a total of 14,771 miles of rail tracks, the most in the U.S. Passenger rail systems in the state are categorized into high-speed rail, intercity rail, commuter and regional rail, light rail, trolleys, and tourism rail (Texas Department of Transportation, 2025b). High-speed rail operates at speeds exceeding 110 mph on dedicated tracks. The long-delayed high-speed rail between Dallas and Houston was taken over by Amtrak in August 2024, and the project received a $64 million federal

TABLE IV.5.2
All Vehicle Types in Texas by Fuel Type and Associated Emissions Based on Truck and LDV Segments for the National Highway System (NHS) and All Other Roads in Texas, 2016–2024

Fuel Type	2016	2017	2018	2019	2020	2021	2022	2023	2024
Hybrid (Gasoline)	199,096	217,084	233,645	225,212	201,629	256,654	346,471	410,330	512,256
Electric	8,397	11,724	18,990	29,540	36,418	60,528	105,807	175,497	248,096
Natural Gas	3,889	3,901	3,063	3,284	2,836	3,366	3,412	3,610	4,271
Propane	1,038	1,276	1,310	1,517	1,451	1,835	2,168	2,171	2,221
Ethanol	-	-	54	238	200	211	194	172	156
Hybrid (Diesel)	-	-	4	6	4	7	10	12	17
Methanol	-	-	4	3	3	2	1	1	1
Hydrogen	-	-	-	-	-	-	1	1	13
Subtotal, AFV	**212,420**	**233,985**	**257,070**	**259,800**	**242,541**	**322,603**	**458,064**	**591,794**	**767,031**
Gasoline	16,622,760	17,237,827	17,510,554	16,845,463	15,024,613	17,244,692	18,589,078	18,825,861	19,065,744
Flexible	2,127,669	2,215,878	2,231,880	2,168,253	1,945,689	2,067,804	2,063,603	1,986,004	1,861,906
Diesel	1,327,585	1,369,414	1,258,596	1,276,164	1,188,617	1,343,897	1,454,229	1,592,323	1,605,598
Convertible	6,549	5,756	4,473	3,830	2,926	2,811	2,562	2,225	2,074
Undisclosed	555	715	336,527	329,708	291,046	275,206	256,783	224,073	71,473
Unknown	3,734,699	3,084,014	3,012,625	4,214,381	5,319,377	1,570,235	415,115	305,236	260,371
Trailers	–	–	–	–	–	2,393,190	2,386,158	2,377,167	2,331,365
Subtotal, non-AFV	**23,819,817**	**23,913,604**	**24,354,655**	**24,837,799**	**23,772,268**	**24,897,835**	**25,167,528**	**25,312,889**	**25,198,531**
Total	**24,032,237**	**24,147,589**	**24,611,725**	**25,097,599**	**24,014,809**	**25,220,438**	**25,625,592**	**25,904,683**	**25,965,562**
NHS: Truck CO_2	31,945,721	32,514,246	33,937,027	35,735,825	36,037,235	31,987,680	36,539,978	37,165,457	38,886,871
NHS: LDV CO_2	67,587,487	69,603,143	69,800,518	70,687,430	71,540,504	60,992,569	69,656,312	71,302,702	73,030,679
NHS: Total CO_2	**99,533,208**	**102,117,390**	**103,737,544**	**106,423,256**	**107,577,739**	**92,980,249**	**106,196,289**	**108,468,159**	**111,917,550**
All Roads: Truck CO_2	53,139,315	54,310,954	56,949,059	60,229,211	60,959,479	55,146,842	61,539,165	62,123,540	64,715,126
All Roads: LDV CO_2	112,426,723	116,263,285	117,130,880	119,136,696	121,015,719	105,151,344	117,312,366	119,185,302	121,536,895
All Roads: Total CO_2	**165,566,037**	**170,574,238**	**174,079,938**	**179,365,907**	**181,975,198**	**160,298,185**	**178,851,531**	**181,308,842**	**186,252,021**

Source: Texas Department of Motor Vehicles, 2024, Texas Department of Transportation, 2024a.

grant for its permitting processes. Additionally, the North Central Texas Council of Governments (NCTCOG) received a federal grant in 2023 to conduct a feasibility study for high-speed rail between Fort Worth, Dallas, and Houston. Intercity rail primarily provides long-distance travel at lower speeds than high-speed rail and is supported by Amtrak along three routes over 1,540 miles of track and 19 stations: the Texas Eagle (San Antonio to Chicago), the Sunset Limited (Los Angeles to New Orleans), and the Heartland Flyer (Fort Worth to Oklahoma City). Commuter and regional rail serve daily passengers traveling between suburban and urban areas, often operating on shared freight corridors. Light rail typically runs within urban areas on dedicated tracks using specialized equipment to serve commuters efficiently. Texas also leads the nation in freight railway tracks, at over 14,000 freight track miles (about 8% of the U.S. capacity) that move 410 million freight tons worth $405 billion to, from, and within Texas annually, and another 132 million freight tons worth $369 billion that pass through the state (Texas Department of Transportation, 2025b). The predominant categories of freight include energy products, agricultural products, and automotive and consumer goods.

Rails and related infrastructure received a grade of C from ASCE in 2025 (The American Society of Civil Engineers, 2025). The assessment recommends that the state invest in rail infrastructure and expand its rail services to reduce congestion and support urban development. To achieve this, the state must establish a more effective, equitable, and reliable funding source to support long-term rail expansion, which is capital-intensive.

IV.5.3.3 Aviation

Texas is a hub for both domestic and international air travel, as well as air freight, with 90 million passengers and three million tons of cargo moved in 2022. The state is home to six of the nation's 50 busiest airports. Among these are Dallas Fort Worth International Airport (DFW): 3rd, Houston George Bush Intercontinental Airport (IAH): 15th, Austin-Bergstrom International Airport (AUS): 29th, and Dallas Love Field (DAL): 32nd. Overall, airfield infrastructure remains in good condition with 9,100 aircraft based across Texas and 5.7 million operations annually, contributing approximately $2.5 billion to the state's economy. The state has 26 commercial service airports and 266 general aviation (GA) airports.

Aviation received a grade of B from ASCE in 2025 (The American Society of Civil Engineers, 2025). The assessment recommends that, as passenger traffic rebounds to or surpasses pre-pandemic levels and as terminals face an increased strain, Texas must prioritize capital investments through increased state funding, technology upgrades, and decarbonization solutions for facilities and the fleet, resilience, and sustainability, and expand intermodal and multimodal airport connections for passenger movements and 'nearshoring' for cargo logistics.

IV.5.3.4 Marine

Texas has 23 seaports and an extensive network of waterways across 1,020 miles of navigation channels. Additionally, the Gulf Intracoastal Waterway (GIWW-T), spanning 379 miles within the state and connecting to Louisiana, Mississippi, Alabama, and Florida, facilitates the movement of energy products, consumer goods, and commercial shipments in the U.S. and globally. Texas handles about 750 million tons of foreign and domestic freight annually, which is the highest in the nation. Additionally, about 80% of the containerized cargo passing through Texas seaports is either produced or consumed by Texans (Texas Department of Transportation, 2024b).

Ports received a grade of C from ASCE in 2025 (The American Society of Civil Engineers, 2025). The assessment recommends that the state increase funding for port infrastructure and ship channel expansion, support innovation and reduce carbon emissions, promote effective public-private partnerships, and establish asset management programs.

IV.5.4 ADAPTATION AND MITIGATION STRATEGIES FOR THE TRANSPORTATION SECTOR IN TEXAS

Broadly, the transportation sector's adaptation and mitigation efforts involve fuel switching, technological advancements to make the fleet more fuel efficient, more stringent fuel and engine regulations, infrastructural upgrades and resilience, and incentivizing changes in public behavior and acceptance for low-carbon solutions. For on-road vehicles, strategies can also include travel demand management (TDM) and traffic system management (TSM). Table IV.5.3 highlights the current adaptation and mitigation strategies in the transportation sector, and the role of federal, state, and

TABLE IV.5.3
Fleet and Infrastructure-Related Policies and Programs in the Transportation Sector as They Relate to Climate Change Impacts

Program	Target and Scope	Agency
Federal Policies and Programs		
Infrastructure Investment and Jobs Act	$30 million in funding for efficiency upgrades and retrofits; $ 577 million for bridge replacement, rehabilitation, preservation, protection, and construction program; $641 million to support projects transportation emissions reduction, $729 million for the Transformative, Efficient, and Cost-saving Transportation (PROTECT) Program; $25.4 billion in Federal-Aid Highway apportioned programs; $69 million in construction of ferry boats and ferry terminal facilities; $3.7 billion for Public Transit; $1.2 billion Aviation Infrastructure Grants; $105 million in Clean School Bus Program for alternative fuels and zero-emission school buses	Multiple federal agencies
Bipartisan Infrastructure Law	$153 million funding for highway safety traffic programs; $3.4 billion for Public Transit, compete for $10 billion for rail improvement and safety grants, and $5.5 billion for grade crossing safety improvements; $408 million over five years to support the expansion of an EV charging network.	Multiple federal agencies
FHWA Resilience Pilot Program	Provides grants for states to integrate climate resilience into transportation planning. Texas is in its third round of funding, with the Houston-Galveston Area Council (H-GAC) participating as one of 11 pilots.	Federal Highway Administration (FHWA)
Inflation Reduction Act	$2.8 billion in Environmental and Climate Justice Block Grants for "community-led" projects that reduce GHGs in disadvantaged communities in Texas; $690 million for Home Energy Rebates programs; $9.9 billion in announced investments from associated federal grants and loans, $17.17 billion in clean energy and transportation investments.	Multiple federal agencies
FTA Low-No Emission Grant Program	Provides grants to state and local governments for the purchase or lease of zero-emission and low-emission transit buses: $36 million for Texas.	U.S Department of Transportation (U.S. DOT)
Clean Air Act & Corporate Average Fuel Economy (CAFE) Standards	A comprehensive federal law that regulates air emissions from stationary and mobile sources; CAFE standards aim to improve the average fuel economy of cars and light trucks sold in the U.S.	U.S. Environmental Protection Agency (U.S. EPA), U.S Department of Transportation (U.S. DoT)

(Continued)

TABLE IV.5.3 (*Continued*)
Fleet and Infrastructure-Related Policies and Programs in the Transportation Sector as They Relate to Climate Change Impacts

Program	Target and Scope	Agency
State Policies and Programs		
Texas Resiliency Plan (TRP)	Enhancing infrastructure durability against extreme weather events like hurricanes, flooding, and heat waves. Key components include (1) flood mitigation measures: expansion of stormwater drainage systems and flood-resistant roadways; (2) heat-resistant infrastructure: development of cool pavements and materials that minimize heat absorption; (3) port and rail resilience: hardening Texas's ports and rail systems against hurricanes and sea-level rise; (4) emergency response upgrades: smart traffic management systems to facilitate evacuation and disaster response.	Texas Department of Transportation (TxDOT)
Texas Emissions Reduction Plan (TERP)	Grants for businesses and individuals to replace high-emission vehicles with cleaner alternatives.	Texas Commission on Environmental Quality (TCEQ)
Seaport and Rail Yard Emissions Reduction Grants and Heavy-Duty Vehicle and Equipment Grants	Grants to eligible entities to replace, repower, or purchase drayage and cargo handling equipment. Eligible projects include heavy-duty on-road vehicles with a gross vehicle weight rating of over 26,000 pounds, off-road yard trucks, and other cargo-handling equipment. Eligible engines must be powered by electricity or meet federal emissions standards and reduce NO_x by at least 25% compared to the engine being replaced. Eligible replacement on- and off-road vehicles must be powered by diesel, natural gas, propane, or electricity.	Texas Commission on Environmental Quality (TCEQ)
Light-Duty Motor Vehicle Purchase or Lease Incentive Program	CNG and propane vehicles, including bi-fuel vehicles, are eligible for a rebate of up to $5,000. EVs and PHEVs are eligible for a rebate of up to $2,500.	Texas Commission on Environmental Quality (TCEQ)
Natural Gas Vehicle Grant	Grants to replace existing medium- and heavy-duty vehicles with new, converted, or repowered natural gas or propane vehicles that operate in one or more of the eligible counties for at least 75% of the activity life. Qualifying vehicles must be on-road vehicles with a gross vehicle weight rating of more than 8,500 pounds, operate on at least 60% natural gas or propane, and be certified to current federal emissions standards.	Texas Commission on Environmental Quality (TCEQ)
Clean Fleet Grants	Grants to replace existing fleet vehicles with alternative fuel vehicles (AFVs) or hybrid electric vehicles (HEVs). An entity that operates a fleet of at least 75 vehicles and commits to placing 20 or more qualifying vehicles in service for use in the Clean Transportation Zone may be eligible. Qualifying AFV or HEV replacements must reduce emissions of nitrogen oxides or other pollutants by at least 25% as compared to baseline levels and must replace vehicles that meet operational and fuel usage requirements.	Texas Commission on Environmental Quality (TCEQ)
Clean School Bus Grants	Any public school district or charter school may receive the grant to pay for the incremental costs to replace school buses or install diesel oxidation catalysts, diesel particulate filters, emission-reducing add-on equipment, and other emissions-reduction technologies in qualified school buses.	Texas Commission on Environmental Quality (TCEQ)

TABLE IV.5.3 (*Continued*)
Fleet and Infrastructure-Related Policies and Programs in the Transportation Sector as They Relate to Climate Change Impacts

Program	Target and Scope	Agency
Diesel Fuel Blend Tax Exemption	The biodiesel or ethanol portion of blended fuel containing taxable diesel is tax-exempt.	Texas Comptroller of Public Accounts
Authorization of the Governmental Alternative Fuel Fleet Grant Program	Grants for the purchase or lease of a new vehicle and the purchase, lease, or installation of alternative fueling equipment. Eligible alternative fuels include natural gas, propane, hydrogen, and electricity. State agencies and political subdivisions are eligible to apply for a grant under the program if the entity operates a fleet of more than 15 vehicles. Mass transit and school transportation providers are also eligible.	Texas Commission on Environmental Quality (TCEQ)
Alternative Fuel Vehicle (AFV) Registration Tracking Program	The Texas Department of Motor Vehicles collects data on the number of AFVs registered in the state and must submit an annual report to the Texas Legislature detailing the results each year.	Texas Department of Motor Vehicles (DMV)
Alternative Fuel Use and Vehicle Acquisition Requirements	State agency fleets with more than 15 vehicles, excluding emergency and law enforcement vehicles, may not purchase or lease a motor vehicle unless the vehicle uses natural gas, propane, ethanol, or fuel blends of at least 85% ethanol (E85), methanol, or fuel blends of at least 85% methanol (M85), biodiesel, or fuel blends of at least 20% biodiesel (B20), or electricity (including plug-in hybrid electric vehicles). Covered state agency fleets must consist of at least 50% of vehicles that can operate on alternative fuels and use these fuels at least 80% of the time the vehicles are driven.	Texas Commission on Environmental Quality (TCEQ)
Funding for Hydrogen Program	TxDOT may seek funding from public and private sources to acquire and operate hydrogen vehicles and establish and operate publicly accessible hydrogen fueling stations. TxDOT must ensure that data on emissions from the vehicles, fueling stations, and related hydrogen production are monitored and compared with data on emissions from control vehicles with internal combustion engines that operate on fuels other than hydrogen	Texas Department of Transportation (TxDOT), Texas Commission on Environmental Quality (TCEQ)
Fuel Dispenser Labeling Requirement	All equipment used to dispense motor fuel containing at least 1% ethanol or methanol must be clearly labeled.	Texas Department of Licensing and Regulation
Neighborhood Electric Vehicle (NEV) Access to Roadways	NEVs are defined as vehicles that can attain a maximum speed of 35 mph and that must comply with the safety standards in Title 49 of the U.S. Code of Federal Regulations, section 571.500. NEVs may only be used on roadways that have a posted speed limit of 45 mph or less, except to cross at an intersection.	Texas Department of Motor Vehicles (DMV)
Texas's National Electric Vehicle Infrastructure (NEVI) Planning	Submit an EV Infrastructure Deployment Plan to the DOT and U.S. Department of Energy (DOE) describing how the state intends to distribute federal NEVI funds	Texas Department of Transportation (TxDOT)

(*Continued*)

TABLE IV.5.3 (*Continued*)
Fleet and Infrastructure-Related Policies and Programs in the Transportation Sector as They Relate to Climate Change Impacts

Program	Target and Scope	Agency
Clean Vehicle and Infrastructure Grants – Emissions Reduction Incentive Grants (ERIG) Program and Rebate Grants Program as part of the Texas Emissions Reduction Plan (TERP)	Grants to improve air quality in the state's nonattainment areas and other affected counties. Eligible projects include those that involve replacement, retrofit, repower, or lease or purchase of new heavy-duty vehicles; alternative fuel dispensing infrastructure; idle reduction and electrification infrastructure; and alternative fuel use. The Rebate Grants Program provides grants to upgrade or replace diesel heavy-duty vehicles and non-road equipment. Qualifying projects must reduce emissions of NO_x or other pollutants by at least 25% as compared to baseline levels and must meet operational and fuel usage requirements.	Texas Commission on Environmental Quality (TCEQ)
Governmental Fleet Grants	Up to 10% of awarded funds may be granted for the purchase, lease, or installation of refueling infrastructure or equipment, or refueling services in conjunction with an eligible vehicle purchase or lease for the purchase or lease of new vehicles powered by natural gas, propane, hydrogen, or electricity.	Texas Commission on Environmental Quality (TCEQ)
Transportation-Related Components of Selected Local Policies and Programs		
Houston Climate Action Plan and METRO Next Plan	Focuses on expanding EV charging stations, enhancing public transit, improving walkability, and implementing flood-resistant road designs. 7.5 billion plan to expand bus and rail services to underserved areas.	City of Houston
Austin's Urban Heat Resilience Program and Project Connect	Implements reflective pavement and tree cover expansion to combat urban heat island effects. $7.1 billion investment in light rail, bus rapid transit (BRT), and transit hubs.	City of Austin
Dallas's Smart Infrastructure Initiative and Comprehensive Environmental Plan	Uses real-time sensors to monitor road conditions and structural integrity; aims to mitigate climate change impacts by reducing greenhouse gas emissions.	City of Dallas
Port of Houston's Shore Power Initiative	Reduces emissions by electrifying docked vessels, aiming for carbon neutrality by 2050, including a renewable electricity contract, LED lighting, hybrid-electric cranes, and EVs.	Port of Houston Authority

local decision-makers and agencies in shaping these strategies. Based on this, Figure IV.5.5 presents a general overview of the overlap and coordination of policies and programs related to the transportation sector and climate change, highlighting relevant federal, state, and selected local agencies.

IV.5.4.1 Overlap with Policies and Programs on Land Use, Urban Planning, Resilience, and Energy Diversification

Land use policy, including urban planning, plays a crucial role in shaping transportation outcomes by influencing the location of destinations, the necessity of travel, the feasibility of alternative transportation modes, and the demand for infrastructure investments. Given the challenges of urban sprawl and congestion, the necessity to travel long distances between urban and rural areas in the state, and the land impacts associated with transportation infrastructure siting, land use policies, and urban planning are especially relevant to mitigation and adaptation in the transportation sector in Texas. Similarly, the transportation sector in Texas cannot be decarbonized without diversifying the state's electricity mix and ensuring power grid resilience.

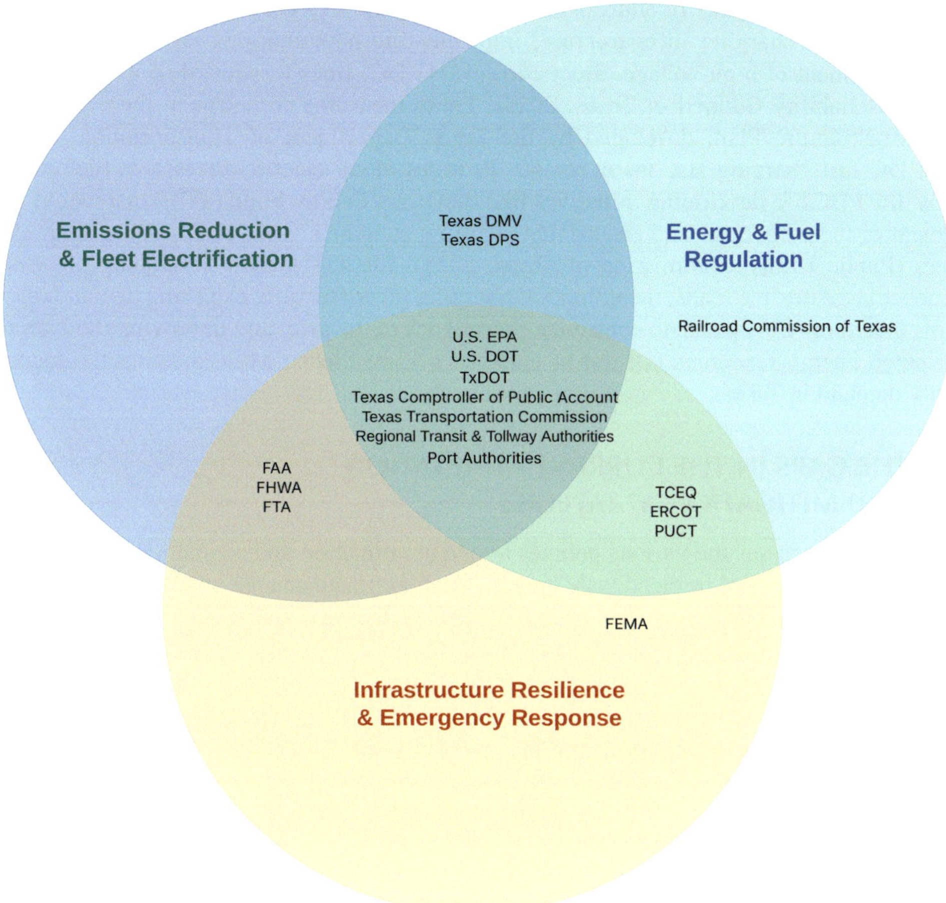

FIGURE IV.5.5 Priority areas and targets for policies and programs related to the transportation sector as they relate to climate change, and the associated federal, state, and local agencies, highlighting areas of policy overlap and coordination.

The policy blueprint, which is expected to serve as the foundation for multimodal transportation planning, incorporating roads, transit, freight, and other mobility options with land use planning and zoning, is the Statewide Long-Range Transportation Plan (SLRTP): Connecting Texas 2050 plan (Texas Department of Transportation, 2024c). It is designed to create a more sustainable and efficient transportation system by integrating various modes of travel and enhancing regional mobility. Developed with input from multiple key stakeholders, including communities, it aims to improve safety, reliability, resilience, and connectivity in Texas. Similarly, the Texas Metropolitan Mobility Plan empowers local communities to implement land-use policies that enhance mobility, pedestrian and cyclist safety, and public transit accessibility through mixed-use, walkable neighborhoods, and reduced reliance on cars (Texas Department of Transportation, 2025c). These policies are also designed to help urban areas adapt to climate-related risks like flooding and extreme heat. Transit-oriented development (TOD) in the state, particularly in Houston, with a TOD ordinance, is encouraging higher-density communities near transit hubs to reduce their dependence on private vehicles and alleviate congestion (North Central Texas Council of Governments, 2025).

As EV adoption continues to rise in the state, both the Electric Reliability Council of Texas (ERCOT) and the Public Utility Commission of Texas (PUCT) are facing challenges in balancing increased electricity demand. To address these concerns, ERCOT is collaborating with stakeholders to enhance EV charging infrastructure, implementing new charging strategies, and exploring the development of high-voltage direct current (HVDC) lines to strengthen grid connectivity (Electricity Reliability Council of Texas, 2023). These measures complement the Texas Electric Vehicle Infrastructure Plan, developed by the Texas Department of Transportation, that aims to install DC fast charging stations every 50–70 miles along electric alternative fuel corridors. Similarly, the PUCT is developing initiatives like the Texas Energy Fund (TEF) that would aid the deployment of these chargers and ensure that charging stations are not classified as retail electric providers (Public Utility Commission of Texas, 2025). ERCOT is also developing processes to interconnect large flexible loads, including EV charging infrastructure, exploring demand response programs that allow EV owners to contribute power back to the grid, and improving the integration of distributed energy resources (DERs) to enhance grid resilience while meeting the increasing electricity demand in Texas.

IV.5.5 THE GAPS IN THE CURRENT ADAPTATION AND MITIGATION STRATEGIES

While Section 4 discusses the various policies aimed at mitigation and adaptation for the transportation sector, Texas remains tremendously unprepared for climate change impacts due to structural, financial, and policy-related gaps. The interplay of the state's sheer size, longstanding policy history of favoring highway expansion over transit investments, economic dependence on transportation, and extreme climate vulnerability creates challenges for adaptation and mitigation. These are compounded by the lack of direct mitigation pathways for the transportation sector discussed in Section 2. Despite recent advances, policies remain fragmented and lack adequate, dedicated funding. Moreover, government agencies are yet to completely develop a unified strategy across segments of the transportation sector, regions in Texas, and levels of the government. Given the size and scale of the transportation industry in Texas, these failings have outsized national and global implications on adaptation and mitigation efforts. These include implications for the Texas workforce and economic impacts across the state, key statistics for which are presented in Figure IV.5.6 and Table IV.5.4. Employment in the trade, transportation, and utilities sectors, which roughly maps on to direct, indirect, and induced jobs related to the transportation sector in the state, grew by 20% over the last decade while weekly wages grew by 25%.

Therefore, without proactive adaptation and mitigation strategies for the transportation sector, Texas risks economic losses, supply chain failures, and worsening inequality and climate impacts locally and globally. The magnitude of the gap in adaptation and mitigation strategies for the transportation sector can be understood through a 2023 analysis focused on guiding the state to net-zero emissions by 2050 (Krishnamoorti and Datta, 2023). Based on population growth estimates from the 2010 Census, four scenarios were modeled in this analysis, including a "business as usual" scenario that relied on no substantial policy or market deviations from the current state to boost the electrification of the Texas fleet of LDVs, MDVs, and HDVs. Alternative scenarios assume EVs make up 100% of all new sales by 2040 or 2050, respectively. Alternate pathways through electrification and fuel switching were analyzed for rail, air, and marine transportation. The analysis highlighted that:

- Given current growth rates, an additional 2.5 million LDVs will be on Texas roads between 2022 and 2050, assuming vehicles have an average lifespan of 15 years and travel about 11,500 miles each year. Electric vehicles (EVs) in the light-duty fleet are projected to rise from 8% of new car sales in 2022 to approximately 63% in 2050 under a business-as-usual scenario. At the same time, internal combustion engine vehicles (ICEVs) will comprise 60% of the total fleet under this scenario, with 32% in the 2050 scenario and 7% in the 2040 scenario. Additionally, there will be about 82,000 medium and heavy-duty vehicles

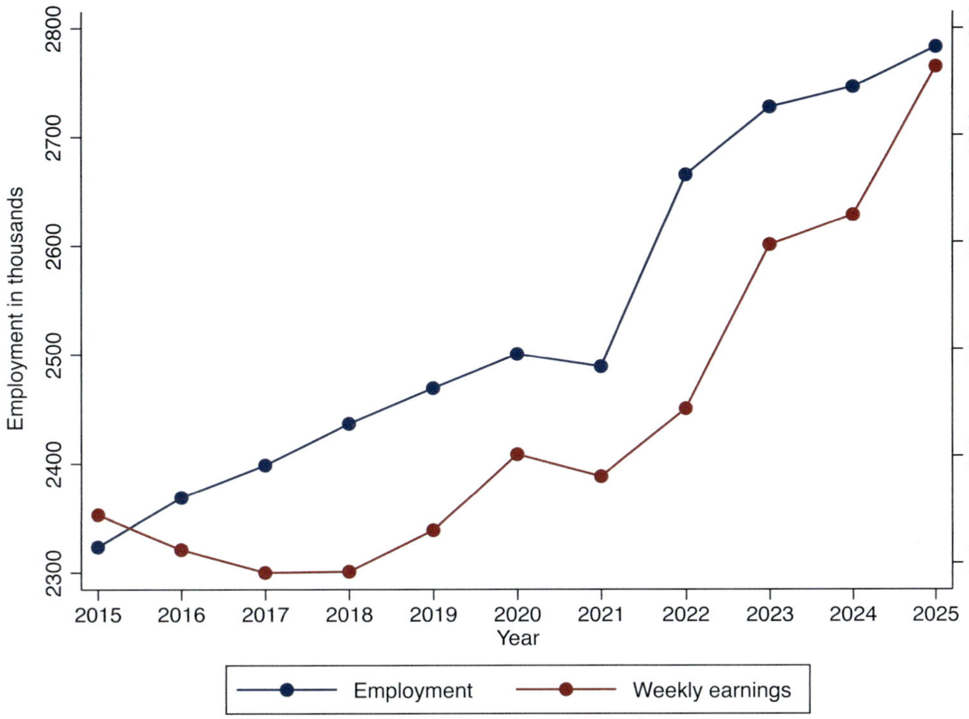

FIGURE IV.5.6 Employment in trade, transportation, and utilities in thousands and average weekly earnings of all employees in dollars per week in Texas, 2015–2025. Values are not seasonally adjusted. *Source*: Federal Reserve Bank of St. Louis (2025).

on Texas roads by 2050, assuming an average lifespan of 12 years and an annual vehicle miles traveled of 25,500.

- Texas will be unable to reach net zero emissions from the transportation sector even with its most aggressive decarbonization goals, due to the current electricity mix, slow EV adoption, and the large number of ICEVs expected to still be on the road in 2050.

TABLE IV.5.4
The Economic Impact of the Transportation Industry in Texas

Segment or Mode	Impact on Direct Jobs	Impact on Labor income (billions)	Impact on Gross State Product (billions)	Impact on Local and State Revenue (billions)
Aviation	778,955	$30.1	$94.3	Data unavailable
Aviation (freight only)	146,000	$18.1	$35.6	$0.2
Railways (freight only)	294,000	$2.5	$83.6	$2.5
Shipping (freight only)	384,000	$42.1	$153.9	$2.3
Warehouse and Distribution	173,000	$8.8	$13.0	$2.4
Freight (overall)	4	$302.0	$572.0	$102.0
Trucking and Public Roads	4.3 million	$289.1	$562.6	$16.2
Pipelines	548,000	$70.4	$180.4	$6.3

Source: Based on a 2018 analysis by the Texas Department of Transportation.

- Electrifying the rail, aviation, and marine sectors or producing carbon-neutral fuels using carbon-neutral electricity could reduce emissions from these modes by as much as 99% by 2050.
- In addition to the cost of replacing the on-road fleet, the electrification of the road fleet will be expensive, requiring about $250 million to $640 million for Level-2 (L2) charging stations and between $500 million and $1.3 billion for DC Fast Charging (DCFC) stations. The cumulative cost for charging infrastructure ranges from $19 billion to $35 billion up to 2050, without the price of the land or other infrastructure required to deliver electricity to the stations. Additionally, the switch from ICEVs to EVs will result in the retirement of many gas stations, requiring expenditures of between $2 billion and $5 billion for environmental remediation.
- Based on current battery technology, the total volume of lithium and cobalt necessary for the number of EVs expected in Texas by 2050 would exceed the current worldwide production of both minerals. There will also be a significant increase in the consumption of materials, including copper, manganese, and graphite.
- Lastly, accounting for the job growth from the electrification of the on-road fleet and adjusting for the job losses associated with the retirement of the conventional fleet, will add more than 130,000 jobs to the Texas economy by 2050 under the business-as-usual scenario. These include direct, indirect, and induced jobs in the transportation, electricity, advertising, retail, data and networking, and maintenance sectors. The policy target of all new sales being EVs by 2050 will add about 40,000 jobs, while the aggressive policy target of all new sales being EVs by 2040 will add about 180,000 jobs by 2050.

In addition to these findings, the Texas Department of Transportation's assessment of funding and investments needed in support of maintaining the state's transportation infrastructure is presented in Table IV.5.5. All values have been adjusted for inflation since the 2014 assessment, but do not include many of the new externalities and costs the state's transportation infrastructure will experience because of climate change. The funding needed per year is calculated based on a 2025 to 2040 timeline. Even when all the announced and planned funding presented in Table IV.5.5 and the Texas Department of Transportation's investment plans are accounted for, the needed funding is more than twice the available and expected funding.

TABLE IV.5.5
Texas Funding Needs by Transportation Mode through 2040 and Per Year, in Billions of 2025 Dollars

Mode	Funding Needs through 2040	Funding Needs/Year
Highways – Expansion	320.6	21.4
Highways – Pavement	138.9	9.3
Highways – Bridges/Culverts	53.6	3.6
Transit Systems (Excluding Passenger Rail)	135.6	9.0
Passenger Rail	28.9	1.9
Aviation	27.3	1.8
Intelligent Transportation Systems**	17.4	1.2
Non-Highway Freight***	7.6	0.5
Bicycle and Pedestrian	2.9	0.2
Total	732.9	48.9

Note: ** Equipment and software used to improve traffic safety and mobility, such as dynamic messaging signs, "smart" signals, video cameras, traffic detection devices, etc. *** Includes freight rail, air cargo, pipelines, ports, and waterways.
Source: Texas Department of Transportation (2018).

IV.5.6 PUBLIC-PRIVATE PARTNERSHIPS: THE CASE STUDY OF EVOLVE HOUSTON

Evolve Houston, a non-profit created in 2019 as a result of a partnership between the City of Houston, Houston METRO, Shell, CenterPoint Energy, and the University of Houston, is committed to accelerating electric vehicle (EV) adoption in the Greater Houston area (Evolve Houston, 2020). Since its inception, Evolve has worked on a pilot program assessing over 15,000 fleet vehicles in Houston, identifying $28 million in economic savings and 17,000 $MMtCO_2$ of annual emissions reductions (Evolve Houston, 2025). The organization has made significant strides in expanding EV infrastructure and promoting equitable mobility, with its focus on equity being a key driver of this progress. Its eMobility Microgrant Program, which focuses on underserved communities by supporting projects related to EV infrastructure, education, workforce development, and micro-transit access, awards $10,000–15,000 per project and is now in its third cycle. The program complements Evolve's Community Connector initiative, with the Houston METRO and the City of Houston, which helps underserved communities with on-demand micro-transit services for last-mile connectivity using EVs in places where public transit access is limited. It has also partnered with RYDE to provide free EV rideshare services in the underserved Third Ward neighborhood in Houston, with plans to expand to other neighborhoods in the future. Evolve's impact in Houston highlights that public-private partnerships can help the state accelerate the adoption of equitable, safer, reliable, and low-carbon mobility alternatives, as well as support infrastructural expansions.

Despite these successes in Houston, the lack of overarching policy support and a coordinated framework for the transportation sector means that Texans are still facing barriers to EV adoption. Our survey of 124 Houston residents who attended the Auto Show in 2023 revealed that only about 15% owned an EV. When the respondents were asked about the most important considerations for purchasing an EV, access to fast charging at public charging, environmental impact, and technology were found to be the most important (very or extremely important), while access to charging at home/multifamily housing, discounts from electric utilities for home charging, and discounts to install home charging were the least important (not at all or slightly important). A tabulation of the results is presented in Table IV.5.6.

TABLE IV.5.6
Share of Respondents and Ranking of Considerations for Buying an EV Based on a Survey of 124 Respondents at the 2023 Houston Auto Show

Consideration for Buying an EV	Not at all important	Slightly important	Moderately important	Very important	Extremely important
Access to fast charging at home/multifamily housing	20.0%	26.7%	13.3%	26.7%	13.3%
Access to fast charging at public charging stations	7.1%	14.3%	14.3%	28.6%	35.7%
Access to fast charging at work	26.7%	6.7%	33.3%	13.3%	20.0%
Battery range	7.7%	0.0%	38.5%	23.1%	30.8%
Discounts from electric utilities for home charging	26.7%	20.0%	20.0%	20.0%	13.3%
Discounts for installing home charging	28.6%	14.3%	14.3%	28.6%	14.3%
Environmental impact	14.3%	14.3%	21.4%	35.7%	14.3%
Performance	0.0%	21.4%	35.7%	21.4%	21.4%
Preferential parking space	21.4%	35.7%	7.1%	35.7%	0.0%
Price	0.0%	14.3%	42.9%	42.9%	0.0%
Technology	7.1%	14.3%	28.6%	21.4%	28.6%
Vehicle Design	0.0%	6.7%	46.7%	26.7%	20.0%

IV.5.7 OUTLOOK AND CONCLUSIONS

Texas faces an unprecedented challenge in managing the emissions from the transportation sector and its interactions and interdependencies with climate change. The entrenched nature of the sector in the state's economy and way of life, the sheer scale of the industry's emissions impact, the lack of direct mitigation pathways, and the distributed nature of these emissions make the challenge particularly complex. The bulk of the emissions from the transportation sector in the state arise from its on-road fleet, with a disproportionate share of the emissions from light-duty vehicles. While consumer preference for light-duty trucks over passenger cars contributes to the sector's enormous emissions impact, the state's geographic sprawl, rapidly urbanizing and growing metropolitan areas, and the need to drive long distances between urban and rural parts of the state due to the absence of other efficient transit options contribute to why Texans drive as much as they do. With some of the highest vehicle miles traveled annually in the country and a policy history of prioritizing road development over public transit, the state's reliance on its on-road fleet and infrastructure has resulted in associated challenges of severe congestion and infrastructure strain. At the same time, the transportation sector is a major contributor to the state's economy, with millions of jobs tied to freight, aviation, trucking, and public roads, and billions of dollars in personal income, contributions to the state's GDP, and revenue. Any disruptions in the transportation sector due to extreme weather or infrastructure failures have widespread effects on supply chains, businesses, communities, and households, with primary, secondary, and tertiary effects extending beyond the borders of Texas.

These concerns and challenges are intensified by the feedback loop between transportation and climate change. Transportation-related emissions compound the effects of climate change, and contribute to extreme weather events, including hurricanes and flooding, worsening droughts and wildfires, and urban heat island effects that the state is experiencing. These climate impacts, in turn, risk transportation fleets and infrastructure by causing quicker degradation and resulting in costly repairs. Ultimately, these impacts severely threaten public health and safety and the quality of life in Texas, and the consequences are expected to mount as the severity and frequency of climate change impacts increase. By not scaling mitigation and adaptation strategies to the state's needs, Texas is also risking economic stagnation, increased socioeconomic inequality, and a loss of status as a global leader in trade, transportation, energy and utilities, healthcare, and innovation.

Several federal, state, and local policies are attempting to address the complex interdependencies of climate change and the transportation sector, and its integration with other policy areas like land-use and urban planning, community resilience, and decarbonization of the Texas power grid. However, in the absence of a coordinated cross-sectoral and cross-agency effort, the lack of a unified strategy across levels of the government, and a significant gap in the available funding and the scale of the investment needed, Texas is falling behind in addressing its transportation emissions, the effects of the feedback loop with climate change, and safeguarding its future against climate risks.

Given the state of current mitigation and adaptation strategies, Texas is unlikely to be able to fully decarbonize its on-road fleet by 2050, which is the timeline and goal outlined in several current federal policies, even if it undertakes the most aggressive policy shifts. Technological advancement and fuel switching in the rail, aviation, and marine sectors can significantly reduce the emissions impact of these modes over the same timeline, but the emissions impact of the on-road segment will continue to dominate. Any mitigation and adaptation effort through policy shifts will result in enormous costs to Texans, impact multiple sectors of the economy, and require trading off several costs and benefits. These include the scale and pace of environmental remediation, associated environmental impacts, and the carbon intensity of emerging technologies, changes in fuel and road taxes, workforce impacts, economic impacts on other sectors of the economy, public health and safety, emergency preparedness and response, community resilience, and socioeconomic equity.

Ultimately, Texas will need to prioritize cohesive, feasible, multipronged mitigation and adaptation strategies that can simultaneously and rapidly deliver the most benefits for many groups.

However, with the impacts of climate change becoming more severe and frequent, the associated costs for any strategy will increase exponentially if the state fails to act now. As evidenced by the case study in this chapter, innovative funding mechanisms and solutions for bringing public and private entities and community stakeholders together can help scale mitigation and adaptation strategies in Texas. However, this will require an urgent, coordinated policy effort across levels of government and among all stakeholders, which Texas has not been able to fully achieve so far. The outlook for Texas in being able to address the intertwined challenges of climate change and the transportation sector is not entirely optimistic. At the same time, technological and business solutions exist, but their success will depend on policy pathways and priorities, and what Texans are willing to pay for the transition.

REFERENCES

AFP Global Logistics. (2025). The Impact of Hurricane Harvey on Transportation Logistics. https://afplus.com/impact-hurricane-harvey-transportation-logistics/

Arctic Council. (2024). Shifting Winds: How a wavier polar jet stream causes extreme weather events. https://arctic-council.org/news/shifting-winds-how-a-wavier-polar-jet-stream-causes-extreme-weather-events/#:~:text=But%20climate%20change%20is%20disrupting,the%20North%20and%20the%20South

Batur, I., Alhassan, V. O., Chester, M. V., Polzin, S. E., Chen, C., Bhat, C. R., & Pendyala, R. M. (2024). Understanding how extreme heat impacts human activity-mobility and time use patterns. Transportation Research Part D: Transport and Environment, 136, 104431.

Caiazzo, F., Ashok, A., Waitz, I. A., Yim, S. H., & Barrett, S. R. (2013). Air pollution and early deaths in the United States. Part I: Quantifying the impact of major sectors in 2005. Atmospheric Environment, 79, 198–208.

Climate Central. (2024). Heat Season Power Outages. https://www.climatecentral.org/climate-matters/heat-season-power-outages

Electricity Reliability Council of Texas. (2023). ERCOT EV Allocation Study: Methodology for Determining EV Load Impact at the Substation Level. https://www.ercot.com/files/docs/2023/08/28/ERCOT-EV-Adoption-Final-Report.pdf

Evolve Houston. (2020). A Year with Evolve Houston. https://evolvehouston.org/news-reports/a-year-with-evolve-houston

Evolve Houston. (2025). Electric Fleet Study. https://evolvehouston.org/electric-fleet-study.

Federal Reserve Bank of St. Louis. (2025). Current Employment Statistics (Establishment Survey): Trade, Transportation, and Utilities. https://fred.stlouisfed.org/series/USTPU

Fraser, A. M., Chester, M. V., & Underwood, B. S. (2022). Wildfire risk, post-fire debris flows, and transportation infrastructure vulnerability. Sustainable and Resilient Infrastructure, 7(3), 188–200.

Hagerty, M. (2024). Beryl Cost Houston Billions, but Was a Fraction of Ike or Harvey's Impact. Houston Public Media. https://www.houstonpublicmedia.org/articles/news/hurricane/2024/08/13/496407/beryl-cost-houston-billions-but-was-a-fraction-of-ike-or-harveys-impact/

Hystad, P., Hill, E. L., Larkin, A., Schrank, D., Harleman, M., Volkin, E., … & Willis, M. D. (2025). Changes in traffic-related air pollution exposures and associations with adverse birth outcomes over 20 years in Texas. International Journal of Epidemiology, 54(1), dyae178.

Kolankiewicz, L., Ruark, E., Beck, R., & Manetas, A. (2017). Population Growth and Sprawl in Texas: How an Exploding Population Consumes Natural Habitat and Agricultural Land in the Lone Star State. https://www.numbersusa.com/wp-content/uploads/2023/12/2017-Texas-Sprawl-Study_FINAL.pdf

Krishnamoorti, R., & Datta, A. (2022). Climate Resilience in Houston: Implications of the IPCC's Sixth Assessment Report for Houston's Climate Action Plan. https://uh.edu/uh-energy-innovation/uh-energy/energy-research/white-papers/white-paper-files/uh-energy-white-paper-series-climate-resistance.pdf

Krishnamoorti, R., & Datta, A. (2023). Net Zero Texas by 2050: The Role of Transportation. https://uh.edu/uh-energy-innovation/uh-energy/energy-research/white-papers/white-paper-files/uhe-transportation.pdf

North Central Texas Council of Governments. (2025). Transit-Oriented Development. https://www.nctcog.org/trans/plan/land-use/tod

Public Utility Commission of Texas. (2025). The Texas Energy Fund. https://www.puc.texas.gov/industry/electric/business/texas-energy-fund/

Rizvi, S. H. A., Agrawal, P., Batra, S., & Singh, V. (2024). Assessing urban heat island intensity and emissions with compressed natural gas (CNG) in non-commercial vehicles. TIDEE: TERI Information Digest on Energy and Environment, 23(1/2), 39–39.

Sohrabi, S., & Khreis, H. (2020). Burden of disease from transportation noise and motor vehicle crashes: Analysis of data from Houston, Texas. Environment International, 136, 105520.

Texas A&M Forest Service. (2025). Wildfires and Disasters. https://tfsweb.tamu.edu/TWPP/

Texas A&M Transportation Institute. (2022). Congestion Data for Your City. https://mobility.tamu.edu/umr/congestion-data/

Texas Comptroller of Public Accounts. (2018). A Review of the Texas Economy. https://comptroller.texas.gov/economy/fiscal-notes/archive/2018/special-edition/history.php

Texas Comptroller of Public Accounts. (2025). Texas communities face elevated wildfire threats. https://comptroller.texas.gov/economy/fiscal-notes/infrastructure/2025/fire-cost/

Texas Department of Health and Human Services (2023). State health officials provide hot weather precautions. https://www.dshs.texas.gov/news-alerts/state-health-officials-provide-hot-weather-precautions

Texas Department of Transportation. (2018). Transportation Infrastructure. https://comptroller.texas.gov/economy/fiscal-notes/archive/2018/may/transportation.php#:~:text=According%20to%20the%20Federal%20Aviation,hangar%20rents%20and%20fuel%20sales.

Texas Department of Transportation. (2022). Bridges FY 2022. https://ftp.txdot.gov/pub/txdot-info/library/pubs/bus/bridge/facts-22.pdf

Texas Department of Transportation. (2023). 2023 Texas Transit Statistics. https://www.txdot.gov/content/dam/docs/division/gov/texas-transit-statistics-report-2023.pdf

Texas Department of Motor Vehicles. (2024). Alternativey Fueled Vehicle Report. https://www.txdmv.gov/sites/default/files/report-files/FY-2024-Alternatively-Fueled-Vehicle-Report.pdf

Texas Department of Transportation. (2024a). Texas electric vehicle planning. https://www.txdot.gov/projects/projects-studies/statewide/texas-electric-vehicle-planning-03-22-22.html?utm_source=local%20profile&utm_campaign=local%20profile%3A%20outbound&utm_medium=referral

Texas Department of Transportation. (2024b). Anchoring Prosperity: The Vital Role of the Texas Maritime System. https://www.txdot.gov/content/dam/docs/maritime/2024-mrd-brochure.pdf

Texas Department of Transportation. (2024c). Statewide Long-Range Transportation Plan. https://www.txdot.gov/projects/planning/ttp.html

Texas Department of Transportation. (2025a). Statewide Report. https://comptroller.texas.gov/economy/economic-data/regions/2024/statewide.php#:~:text=Texas%20had%20an%20estimated%20total,San%20Marcos%20(8.1%20percent).

Texas Department of Transportation. (2025b). 2025–2026 Educational Series. https://www.txdot.gov/content/dam/docs/government/edu/edu-rail-2025.pdf

Texas Department of Transportation. (2025c). Texas Metropolitan Mobility Plan: Breaking the Gridlock. https://www.dot.state.tx.us/btg/tmmp.pdf

Texas House of Representatives. (2025). Investigative Committee on The Panhandle Wildfires. https://www.house.texas.gov/pdfs/committees/reports/interim/88interim/House-Interim-Committee-on-The-Panhandle-Wildfires-Report.pdf

Texas Parks and Wildlife. (2025). https://tpwd.texas.gov/education/hunter-education/online-course/wildlife-conservation/texas-ecoregions#:~:text=Due%20to%20its%20size%20and,to%20Alaska%20in%20land%20area.

Texas Water Development Board. (2023). Texas Water: Present and Future Needs 2023. https://comptroller.texas.gov/about/media-center/news/20230905-texas-water-present-and-future-needs-2023-1693338176814

Texas Water Development Board. (2025). Water Weekly. https://www.twdb.texas.gov/newsmedia/drought/doc/weekly_drought_report.pdf

Texas Workforce Commission. (2025). Texas Labor Market Growth Continues with Over 37,000 Jobs Added in December. https://www.twc.texas.gov/news/texas-labor-market-growth-continues-over-37000-jobs-added-december

The American Society of Civil Engineers. (2025). Texas 2025 Report Card. https://infrastructurereportcard.org/state-item/texas/

U.S. Census Bureau. (2025). Quick Facts: Texas. https://www.census.gov/quickfacts/fact/table/TX/PST045224

U.S. Department of Agriculture. (2013). Climate Change and Agriculture in the United States: Effects and Adaptation. https://www.usda.gov/sites/default/files/documents/CC%20and%20Agriculture%20Report%20(02-04-2013)b.pdf

U.S. Department of Energy. (2025). Vehicle Registration Counts by State. https://afdc.energy.gov/vehicle-registration

U.S. Department of Homeland Security. 2015. Drought Impacts to Critical Infrastructure. https://content.govdelivery.com/attachments/USDHSFACIR/2015/04/30/file_attachments/386553/Drought+Impacts+to+Critical+Infrastructure.pdf

U.S. Energy Information Administration. (2024). State Energy Profile. https://www.eia.gov/state/data.php?sid=TX

U.S. Environmental Protection Agency. (2001). Ground-level Ozone Basics. https://www.epa.gov/ground-level-ozone-pollution/ground-level-ozone-basics#:~:text=air%20emission%20sources.-,How%20does%20ground%2Dlevel%20ozone%20form?,can%20experience%20high%20ozone%20levels.

U.S. Environmental Protection Agency. (2017). Latest Findings on National Air Quality: 2000 Status and Trends. 2017 U.S. billion-dollar weather and climate disasters: a historic year in context. https://www.epa.gov/sites/default/files/2017-11/documents/trends_brochure_2000.pdf

U.S. Environmental Protection Agency. (2023). Smart Location dataset. https://www.epa.gov/smartgrowth/smart-location-mapping

U.S. National Oceanic and Atmospheric Administration. (2018). https://www.climate.gov/disasters-2017#:~:text=Hurricane%20Harvey:,inflation%20adjustment%20to%202017%20dollars

Weinkle, J., Landsea, C., Collins, D., Musulin, R., Crompton, R. P., Klotzbach, P. J., & Pielke, R., Jr. (2018). Normalized hurricane damage in the continental United States 1900–2017. Nature Sustainability, 1(12), 808–813.

Zhu, R., Wong, M. S., Guilbert, É., & Chan, P. W. (2017). Understanding heat patterns produced by vehicular flows in urban areas. Scientific Reports, 7(1), 16309.

IV.6 Climate Adaptation from the Regional Planning Perspective

Notes on Institutional Reality, Anthropogenic Responses, Phronesis, and Incremental Pragmatism

Mukesh Kumar, Mark Keeley, and Ryan A. McManamay

IV.6.1 INTRODUCTION

The phenomenon of climate change, by itself, is unsurprising. The science behind climate change is indisputably obvious in its broad claims, whose concerning aspects have more to do with our (in)ability to respond effectively to cope with the consequences and effects. Responses as practical matters can be classified into the categories of individual single unit responses and the collective institutional responses. Individual responses are the selections of actions from a set of choices available to an individual that can be modeled well within the rational choice frameworks. Collective responses often involve civic morality, institutional ethics, and the context where they are more or less likely to be viable. In this chapter, we concern ourselves with an analysis of our institutional responses rather than the phenomenon of climate change itself or individual responses such as household recycling, composting, and reusable grocery bags.

Analyses of institutions, whether conceptualized via sociologists' *social structures* (Parsons, Skocpol, Giddens), or neoclassical economists' *rational choice theory* (originating with Samuelson, Von Neuman, and Morgenstern), have largely relied on the idea of voluntarily aggregated choices made by individual actors that define the operations of institutions, be they political, economic, or social. These concepts neatly fall into what Grafstein describes as the "conventionalist" paradigm. A conventionalist analysis considers institutions as products of deliberate individual choices. In this reading, institutions are voluntary creations, whose rules and processes are effective only to the extent that they reflect some aggregate of deliberate individual choices of rational actors. Such institutions can be reformed, recast, and revised. Grafstein offers a realist interpretation of institutions, in which individual rationality is constrained by institutional dependencies (Grafstein, 1992).

Hall and Taylor (1996), in their seminal paper on institutional perspective, clarify the significant differences among historical institutionalism, rational choice institutionalism, and sociological

DOI: 10.1201/9781003406600-25

institutionalism. The authors further explain why, despite their common interest in explaining the role of institutions in socio economic and political life, these three approaches developed largely independently and have distinct analytical lenses regarding how institutions affect behavior, what institutions do, and why they persist or change. In the current environment, sociological institutionalism with its focus on identities and preferences offers a more realistic perspective than historical and rational choice institutionalism.

Despite renewed and vigorous interest in institutionalist perspectives on social, political, and economic behavior, there are several claims on how there exists a fundamental institutional failure, especially in the context of climate change and global warming. Oriordan and Jordan (1999) argue that institutions help to define climate change both as a problem and a context, through such socialized devices as the use of scientific knowledge, culturally defined interpretation of scientific findings, and politically tolerable adaptation strategies (O'Riordan and Jordan, 1999). Yet, they are based on fundamentally contradictory assumptions of human behavior, and hence the relative ineffective institutional responses that have made little difference to meaningful responses to climate change and global warming. Even the new institutionalism – further divided among historical institutionalism, rational choice institutionalism, and sociological institutionalism – fails to resolve the central dilemma of climate adaptation strategies. That it is a wicked problem (Head, 2022) and even though the classical five ways of relating to other individuals (Thompson et al., 1990) that shape the structures of the institutions may be a useful framework, it still yields only incomplete responses to the problem of climate change.[1]

A more pragmatic challenge for responses to climate change as a wicked problem is brought forth by Crowley et al. (2020) and further extended by Head (2022) that, even in fortunate situations where robust solutions have been successfully negotiated, the implementation process is often subject to the constraints of the governance and resource challenges. Head (2022) further breaks down institutional responses to wicked problems (including environmental challenges) into seven categories: (1) avoidance, denial, and minimal responsibility; (2) coercive controls; (3) compartmentalized micro-management; (4) technocratic problem-solving; (5) incremental and pragmatic adjustment; (6) stakeholder collaboration; and (7) coping and prevention policies.

In McLennan County, Texas, most local responses could be classified under the categories of incremental pragmatic adjustment and stakeholder collaboration. Similar to Agrawal and Agrawal (2008), we recognize that adaptation to climate change is inherently local and that local institutions play a critical role in shaping how communities respond to climate impacts. A government can only respond to future concerns with laws and policies which guide development and generations based on the information currently available. As the local government is the primary witness and influencer of the Code of Ordinances, strategic development guidelines and policies will help to shape the community through new development and redevelopment that supports an overall future vision for issues of critical concern to public health and wellbeing, which can, of course, include future implications of climate change.

Numerous strategies are available for municipalities and regional planning organizations to plan and mitigate for climate-related risks (United Nations, 2014). The United Nations provides a stepwise assessment for urban planners when initiating new climate planning exercises by facilitating planners and community members to ask four key questions: (1) What is happening? (2) What matters most? (3) What can we do about it?, and (4) Are we doing it? Here, we are starting at questions 3 and 4. For the urban planner, instruments of enacting adaptations or mitigations are largely achieved through existing policies and regulations. In other words, realistically operationalizing actions should understand the constraints and purview of existing environmental regulations, related to the municipality or county government. In the context of public health and wellbeing, this rapidly narrows to considerations of future urbanization, transportation planning, and land zoning and resource demands. These factors are particularly important for their implications on air and water quality, in conjunction with climate change, perhaps mitigating greenhouse gas (GHG) emissions through smart design and planning.

How is the demand of a growing population met with a desire for more housing, available jobs, and expanded transportation networks, all while considering the impacts of those developments on the most important natural resources? How and where land is manipulated, what materials are used in the development process, and enforcement of water and air quality requirements can help to establish priorities. One challenge is that these factors are not just influenced by the immediate jurisdiction. It is of vital importance to pay attention to the source watershed and air quality influences within and outside jurisdictional boundaries when considering development regulations. Air quality is frequently less sensitive to local adaptation (except point-source emission) than water due to its wider geographical movements, but in significant ways, it is easier to improve at larger spatial scales. Additionally, local institutions often have significantly fewer tools to influence air quality directly but are better equipped to manage water quality.

Local institutional responses to climate change are better articulated and implemented as adaptation and resilience strategies rather than long-term solutions to reverse or at least improve the negative trends. The local democratic processes further make incremental decision-making a far more pragmatic way forward than creating options for larger-scale and longer-term solutions. The local adaptation and resilience strategies are also better communicated through the day-to-day challenges of water availability and ambient air quality as lived experience of citizens. Most importantly, local decision-making is further constrained by the legal and political framework at state and national levels that allow for a limited set of options available for local institutions.

In this chapter, an institutional perspective is provided on how climate-related risks are addressed indirectly, somewhat incrementally, through the purview of urban planning related to air, water quality, and land use change. The City of Waco, Texas is used as a case study, providing the local context of regulation, challenges, and pragmatic considerations. A brief overview of air, water, and land use regulations are discussed, followed by approaches for implementing protective measures.

IV.6.2 COMPLEXITY OF PLANNING AND POLICY RESPONSES

Planning for air and water quality has been often characterized as one of the many wicked problems (complex, ill-defined, and resistant to simple solutions) in planning and policy literature. Lazarus (2009) further characterizes climate change with characteristics of: (1) time is running out; (2) no central authority; (3) actors seeking to solve the problem are also causing it; and (4) policies discount the future irrationally. Proposing asymmetric precommitment strategies, it further argues calls for robust, resilient, and forward-looking legal and institutional frameworks for climate change that can withstand political fluctuations and ensure sustained action toward a livable future.

An essentialist view on the “tragedy of the commons” treats air and water quality, and to some extent land, issues largely as conventional commons that are readily accessible to all and thus prone to misuse, underinvestment, and free riding (Ansari et al., 2013). Ansari, Wijen, and Gray offer an institutional perspective on the non-linear process of the emergence of a “commons logic.” Using a 40-year longitudinal analysis, they highlight three conditions necessary for a commons logic to emerge: recognition of interconnected fates of key actors regarding a problem issue, increased perception of the actors’ role in contribution to the problem, and steps taken toward collective action and institutional development. While the exact process that allows diverse actors to achieve a working consensus and avert the predictable tragedy is unpredictable, Ansari et al. identify five key mechanisms that facilitate frame shifts among actors: collective theorizing, issue linkage, active learning, legitimacy seeking, and catalytic amplification. Calling climate change as a “grand challenge,” Gümüsay, Claus, and Amis contend how deeply intertwined they are to the prevailing institutional logics – the shared, socially constructed beliefs, values, and practices that guide action in different societal sectors (e.g., market logic, professional logic, state logic) (Gümüsay et al., 2020).

Property rights and jurisdictional boundaries have generally precluded land from being considered a “common.” Land resources can be clearly partitioned and privatized, the boundaries of which are well defined and mutually exclusive. Perhaps stemming from an ideological standpoint of privatization and individual freedoms, land management has been a localized, rather than centralized, regulatory process, generally controlled through local ordinances under the purview of state laws. This has also led to its own complexities, which are further complicated by drivers of land use change. Land use change is a place-based phenomenon – the outcome of multiple interactions between sociopolitical and cultural factors, natural and spatial forces, but especially economical and market drivers (Hersperger et al., 2018). Land use changes are driven by spatially proximate and immediate processes as well as regional market forces and even continental and intercontinental teleconnections from distal, non-adjacent socioeconomic pressors (McManamay et al., 2022; Seto et al., 2012).

IV.6.3 CURRENT KNOWLEDGE

IV.6.3.1 Water Regulations

Water regulations are reactive. Water quality regulation in the United States was only enacted after the recurrence of several dangerous and severe fires caused by the industrial pollution of the Cuyahoga River in Ohio (USEPA, 2025a). It is hard for us to fathom that we will ever run out of water due to its vastness around us. Are we not the blue planet? What are we to do with all this water? As many already know, the vast majority of water available on the planet is not suitable for consumption in its current condition: either it is salinized (97%), or it is frozen (2%). Of the remaining 1% left as available "fresh" water, much is considered polluted and not suitable for swimming, fishing, or consumption. According to the 2017 National Water Quality Inventory Report to Congress by the United States Environmental Protection Agency (EPA), 46% of U.S. river and stream miles are in poor biological condition and 21% of the nation's lakes are hypereutrophic (USEPA, 2017).

From this lens, now understanding how little water we have, I ask again what are we to do with all this water around us? How do we protect it in an ever-growing environment of a consumer group in the U.S. who doesn't understand how precious water is or that civilizations have grown and thrived because of water (Hosseiny et al., 2021)? Arguably the most powerful historical empire, the Romans, fell by conquest due to the complete disruption of a fresh water aqueduct system by the Goths (Procopius, 1919).

Aside from availability, we also face localized shortages with increased temperatures and the effects of climate change on the remaining available water with extreme swings in weather patterns and drought-flood dynamics. This is exacerbated by urban sprawl and population densities.

So how do we respond, we as a species who have our own existence at stake? Yes, we are worried about the endangered salamander and the biota of polluted waters, but we are talking about human existence.

In our bureaucratic way, we have attempted to improve on the initial Clean Water Act of 1972 multiple times. Regulations for water use and protection trickle down from Federal to State to Local Municipalities with a suite of agencies performing individual portions of these roles. However, this is all laughable: have you ever had a conversation with water to explain jurisdictional boundaries? By its very nature, water is cyclical and does not follow jurisdictional boundaries; instead, it follows watersheds that extend well beyond a traditional governmental territory of control. This concept was understood by John Wesley Powell, former head of the U.S. Geological Survey, in 1879 when he proposed political jurisdictional boundaries be drawn to match watersheds in water limited environments (Letsinger, 2019). It is essential to have a wholistic approach to water protection and conservation.

The current water crisis revolves around source water quality, availability, treatment, affordability, and insecurity. We will focus on water quality and availability as they are the fundamental components that lead to the ability for water to be treated and distributed to tackle the affordability and insecurity concerns and are those most influential by government in an incremental pragmatic adjustment, and stakeholder collaboration approach.

In many cities, source water quality is so poor from pollution that treating it for consumption is becoming very difficult. The EPA is still defining how to deal adequately with microplastics and just released in 2024 the final per- and polyfluoroalkyl substances (PFAS) national primary drinking water regulations. PFAS, or "forever chemical," is of particular concern due to its persistence, bioaccumulation tendencies, and toxicity (Krafft and Riess, 2015) and has been found exceeding these new EPA limits in several large Texas cities including most of the Texas Triangle (Martinez, 2024). Water quality is impacted by point source and non-point source pollution, and sedimentation and erosion which can be influenced by development due to changes in the velocity and volume of stormwater runoff.

Availability of water is not only the presence of water in your community, but the treatment of that water and the management of the volume to meet the demands of the population in your area. Items that affect water availability include weather patterns, drought dynamics, reservoir access, water rights, and difficulty of treatment to make that water potable. Texas is a "first-in-time, first-in-right" state, meaning rights are governed under "prior appropriation." Allocations from senior right holders are satisfied first, followed by junior right holders – this can become critical in times of drought. Both surface and groundwater rights can be sold, making this a complex process to navigate.

How can society respond to this question of managing water quality and availability? What can we do now to ensure not only our access to water, but future access for our children and their children?

IV.6.3.2 Air Quality Regulations

Air pollution[2] is defined by the World Health Organization as contamination of the air by various agents that change the atmosphere's natural characteristics. Human activities contribute to air quality issues through two main sources: stationary sources, like power plants, and mobile sources, like vehicles. Emissions from stationary sources can be easily monitored and regulated, while those from mobile sources are more challenging. The Clean Air Act of 1970 and its subsequent amendments (Rogers, 1990) in the United States regulate air emissions and allows the EPA to set National Air Quality Standards (NAAQS) to protect public health and the environment. States must monitor air quality to ensure compliance with these standards. The transportation sector is linked to many harmful emissions from fossil fuel combustion, including particulate matter (PM), carbon monoxide (CO), sulfur dioxide (SO_2), nitrogen oxides (NO_2, NO_3), volatile organic compounds (VOCs), ground-level ozone (O_3), and carbon dioxide (CO_2). Most of these substances can have serious health effects over time, leading to respiratory issues, heart-lung diseases, and even death (Manisalidis et al., 2020). The NAAQS are summarized in Table IV.6.1.

Particulate matter results from the incomplete burning of fuels. For instance, diesel fuel emits more PM because of its dense fuel-to-air ratio. Other harmful emissions arise from high-temperature reactions during combustion, leading to nitrogen oxides, which can create ground-level ozone. While ozone is natural in the upper atmosphere, it becomes dangerous at ground level. EPA had lowered the level of NAAQS for ozone from 0.08 parts per million (ppm) to 0.075 ppm in 2008. This level was further lowered to 0.070 in 2015. The last update to the NAAQS was made in 2024 that lowered the level for PM 2.5 from 12.0 to 9.0 micrograms per cubic meter ($\mu g{\cdot}m^{-3}$).

Similar to the requirement for most states, the Texas Commission on Environmental Quality (TCEQ) maintains the Texas State Implementation Plan (SIP) that serves as the state's comprehensive plan to clean the air and meet the federal air quality standards. If a county (or counties as part of a metropolitan area) violates these standards, it may enter "nonattainment," requiring measures to reduce emissions, particularly from new sources or existing pollutants. This can affect future transportation projects, potentially causing delays until emission standards are met.

Air quality can vary significantly due to various factors, including human activities and weather conditions. Localized ambient air quality conditions are complex and can be influenced by numerous factors and can change dramatically over time for a given location. Temperature and air currents can influence local air quality considerably; hence, ambient air quality may be influenced more by advection

TABLE IV.6.1
EPA NAAQS, Including Units of Measure. Parts Per Million (ppm) by Volume, Parts Per Billion (ppb) by Volume, and Micrograms Per Cubic Meter of Air (µg·m^{-3})

Pollutant [links to historical tables of NAAQS reviews]		Primary/ Secondary	Averaging Time	Level	Form
Carbon Monoxide (CO)		primary	8 hours	9 ppm	Not to be exceeded more than once per year
			1 hour	35 ppm	
Lead (Pb)		primary and secondary	Rolling 3 month average	0.15 µg·m^{-3}	Not to be exceeded
Nitrogen Dioxide (NO_2)		primary	1 hour	100 ppb	98th percentile of 1-hour daily maximum concentrations, averaged over 3 years
		primary and secondary	1 year	53 ppb	Annual Mean
Ozone (O_3)		primary and secondary	8 hours	0.070 ppm	Annual fourth-highest daily maximum 8-hour concentration, averaged over 3 years
Particle Pollution (PM)	PM2.5	primary	1 year	9.0 µg·m^{-3}	annual mean, averaged over 3 years
		secondary	1 year	15.0 µg·m^{-3}	annual mean, averaged over 3 years
		primary and secondary	24 hours	35 µg·m^{-3}	98th percentile, averaged over 3 years
	PM10	primary and secondary	24 hours	150 µg·m^{-3}	Not to be exceeded more than once per year on average over 3 years
Sulfur Dioxide (SO_2)		primary	1 hour	75 ppb	99th percentile of 1-hour daily maximum concentrations, averaged over 3 years
		secondary	3 hours	0.5 ppm	Not to be exceeded more than once per year

Primary standards provide public health protection, including protecting the health of "sensitive" populations such as asthmatics, children, and the elderly. Secondary standards provide public welfare protection, including protection against decreased visibility and damage to animals, crops, vegetation, and buildings. Taken from https://www.epa.gov/criteria-air-pollutants/naaqs-table

of pollution from distant sources, such as big cities located upstream, rather than local emissions. For instance, air pollution in Waco often originates from Dallas–Fort Worth (DFW) and Houston–Galveston-Brazoria (HGB) metropolitan areas. GHG emissions, notably CO_2 from transportation, contribute to climate change and are also regulated by the EPA, with transportation being the largest source of GHGs in the U.S. Planning for transportation projects must consider these air quality impacts.

IV.6.3.3 Land Use Regulations

Federal regulations for land use are not as extensive as those for air and water; land governance is largely handled at local and regional levels and property rights regime do not allow land ownership to guarantee unlimited freedom to develop due to varying local restrictions. Local economies, market values, and population growth significantly influence land development in the U.S., and relatively higher land prices in urban areas create two opposing trends (Albouy et al., 2018; Sun et al., 2020). *Intensification* (densification/infilling) involves making efficient use of existing space, but can lead to issues like overcrowding, traffic, and public health concerns if not managed with tools like zoning. *Extensification* (sprawl) implies new development occurring in suburban and peri-urban areas. While offering more space, it transforms natural landscapes, impacts agriculture, creates infrastructure inefficiencies, and increases transportation emissions due to longer commutes and the conversion of carbon-sequestering vegetation to impervious surfaces (McManamay et al., 2024).

To manage these development patterns, *land use planning* and *zoning regulations* are employed. On the one hand, land use planning involves systematically assessing and creating sustainable future uses for land, aiming to balance economic, health, and environmental demands. It helps prevent overcrowding, ensures suitable land use, minimizes hazard exposure, and facilitates efficient infrastructure and transportation. Zoning regulations, on the other hand, are regulatory tools that allow municipalities to control or restrict private property use to ensure livable, safe, and sanitary cities (Dahlstrom, 2013). While different from land use planning, zoning can implement the strategic vision set by comprehensive plans (Nadigroup, 2025).

The *U.S. Standard Zoning Enabling Act of 1926* provided the framework for local zoning regulations, giving municipalities power to control density, building types, and other factors, and emphasized that zoning should align with a comprehensive plan (Dahlstrom, 2013). However, highly restrictive zoning and market forces can push development outward. Containment policies, like urban growth boundaries (UGBs), aim to prevent this extensification by limiting development outside a defined area (Nelson and Moore, 1993). While adopted in places like Oregon, their effectiveness has been mixed, often failing significantly to enhance densification or curb sprawl against strong market pressures (Lewis and Parker, 2021; Nelson and Moore, 1993; Schuster Olbrich et al., 2022). This suggests that UGBs and zoning may not be sufficient on their own. Integrating market-based approaches, containment policies, and zoning, along with regional coordination, is suggested as a more effective strategy for managing urban growth (Fertner et al., 2016).

IV.6.4 LOCAL CONTEXT AND MOTIVATION

IV.6.4.1 Water Issues

When discussing water management and quality, we must first define local. As stated previously, water does not follow manmade jurisdictional boundaries. In the past century, Texas surface water supply storage increased almost 600% from 254,000 acre-feet of water in 1917 to 1.67 million acre-feet in 2012 amongst 188 major water supply reservoirs (TSHA, 2018). These major water supplies are catchments in localized watersheds. A watershed is an area of land which is divided by natural ridges within which all surface waters from rains flow to one point. A watershed can be as small as a puddle to as large as the Mississippi river basin covering a large portion of the central United States. Texas has 15 major river basins or watershed which flows to the Gulf of Mexico (TWDB, n.d.a).

These watersheds are divided in scale by the U.S. Geological Survey Hydrologic Unit Codes (HUCs), which delineate watersheds in 22 regions (2-digit) through subregions, basins, subbasins, watersheds, and subwatersheds (12-digit). Most of Texas is in HUC Region 12, an area comprised of 17 river and coastal basins in Texas, including the Sabine, Neches, Trintiy, San Jacinto, Brazos, Colorado, Lavaca, Guadalupe, San Antonio, and Nueces river basins (TCEQ, 2025a; USGS, n.d.a; USGS, n.d.b). The Brazos river basin, which cuts through much of Central Texas, is the second largest river basin in Texas and has a total drainage area of 45,573 square miles (TCEQ, n.d.; TWDB, n.d.b). To put that into perspective, the Brazos river basin is larger in size than 17 U.S. states, the largest being Ohio (USCB, 2021). For this chapter, we will focus on the HUC subbasins within the Brazos river basin impacting the McLennan County Area for water availability and quality, primarily the Bosque River Watershed, HUC12060203 and 12060204, which flow into the largest surface water reservoir in McLennan County, Figure IV.6.1.

Lake Waco is the primary surface water source for McLennan County serving roughly 206,000 central Texans directly or through wholesale and water exchange resale groups from the City of Waco Water Utilities Department and is projected to grow to nearly 300,000 by 2060; see

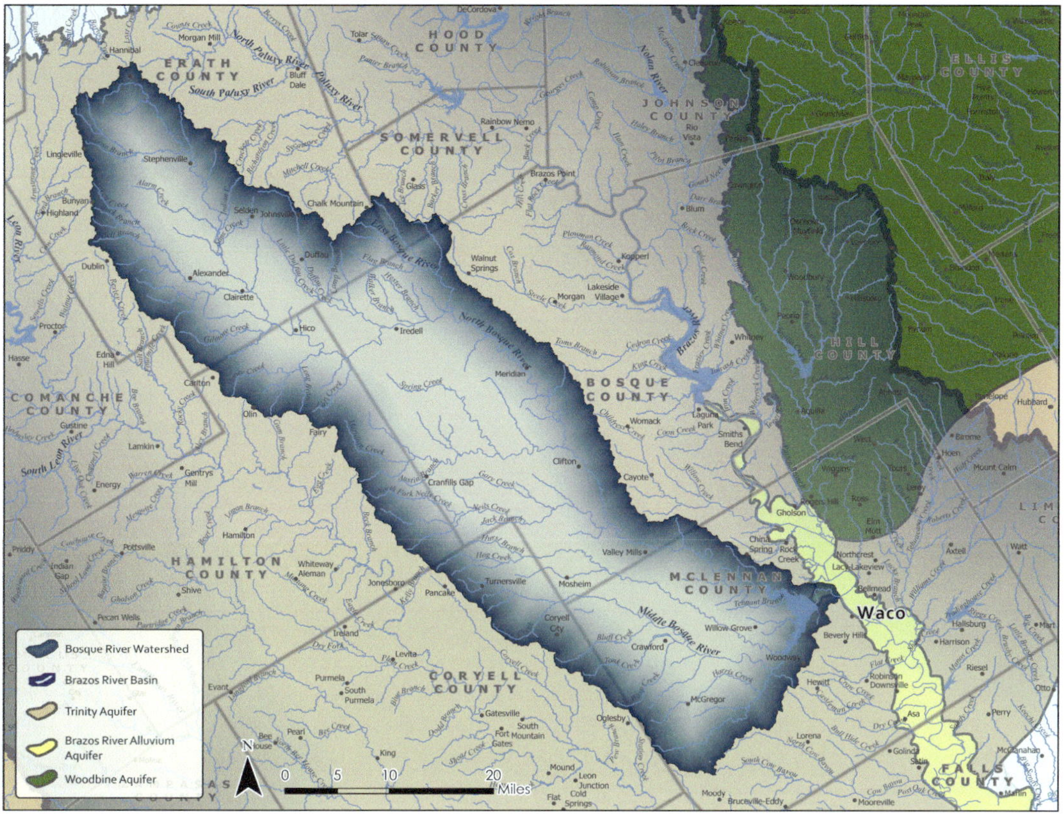

FIGURE IV.6.1 McLennan County regional water supply dynamics and sources (Scanes, 2025).

Figure IV.6.2 (COW, 2024). Initial damming of the Bosque River in 1930 established Lake Waco securing 39000 acre-feet of water and provided flood mitigation for the City of Waco and downstream portions of the Brazos. Through foresight and planning the City of Waco began planning for water supply expansion in 1950 leading to the second and current dam and expansion of compacity to 98000 acre-feet of water completed in 1964. The next expansion planning began in 1979 resulting in a determination to elevate the lake level by 7 feet in 1987. This elevation did not take place until 2003, which expanded water holding capacity to the current 189,773 acre-feet of water at a normal lake elevation of 462 feet above mean sea level (USACE, n.d.).

Over the past 25 years, McLennan County has been considered drought free for only eight years and ten months; meaning that for 65% of the past quarter century the county has experienced drought, four of those years being labeled as extreme or exceptional drought (NOAA, 2025). In that same timeframe, 2000–2025, Lake Waco has been below conservation pool capacity nearly 70%; dipping as low as 55.3% of capacity in October of 2023, dropping 17% (32,089 acre-feet of water) in just over four months (WDT, 2025). Without rain, the storage of the lake would have been exhausted, at this rate of consumption, in less than two more years. Without the elevation of the conservation pool in 2003 the lake level would have been just over 13,000 acre-feet of water, 13.5% of the pre-2003 capacity and less than 7% of the current pool.

Aside from surface water there is a large use of groundwater from the Trinity and Brazos River Alluvium aquafers in McLennan County, Figure IV.6.1. The annual usage of ground water as of 2019 in McLennan County was 13,823 acre-feet of water, 13,401 from the Trinity Aquifer and 422 from the Brazos River Alluvium Aquifer (STGCD, 2021). This ground water use accounted for roughly 20% of the 2020 water demand within the county (STGCD, 2021). The two largest uses of

FIGURE IV.6.2 Waco water service area (COW, 2024).

groundwater are industrial/mining and manufacturing, followed by rural well and municipal conservation (TWDB, 2024).

The Southern Trinity Groundwater Conservation District was created by the 80th Texas legislature in 2007 with the primary purpose to conserve, preserve, protect, recharge, and prevent the waste of groundwater and to control subsidence caused by groundwater withdrawals (STGCD, 2021). The purpose of conserving, preserving, protecting, and preventing waste of a water resource seems obvious, but subsidence, the sinking of the ground because of underground material movement caused by the removal of water or other natural below ground resources, is something many do not consider when dealing with ground water over use (NOAA, 2024). In a recent subsidence study, the cities within the Texas Triangle (Houston, Dallas, and Fort Worth) have all been found to be sinking at a rate greater than 4 mm per year (Ohenhen et al., 2025).

IV.6.4.2 Air Pollution

The last Air Quality Model completed for Waco area was undertaken in 2021 for the Heart of Texas Council of Government (HOTCOG) and TCEQ (McGaughey et al., 2012). This report updated the conceptual model for understanding high ozone concentrations in the Waco, Texas area, incorporating data from 2010 to 2012. The model focused on the annual, seasonal, and daily variability of ozone, linking it to large-scale weather patterns and local meteorological conditions. It also identified potential upwind source regions for background ozone affecting Waco.

For the region the CAMS 1037 monitoring station, located northeast of downtown Waco, is the primary data source, and has been operational since 2007. The report concluded that while Waco area was in compliance with the 75 ppb 8-hour ozone National Ambient Air Quality Standard (NAAQS) at that time, it would exceed a potentially lowered standard of 60–70 ppb, which was proposed in 2010 but later withdrawn.

The key findings of the report were organized around four primary considerations of ozone occurrence and precursors, meteorological conditions and background ozone, source regions and transport, and large-scale weather patterns.

1. For ozone occurrence and precursors, the report highlighted the point that high ozone days (≥ 75 ppb) occurred 17 times between 2007 and 2012, with no clear annual trend; these days were predominantly from June to September in the Waco area, with August having the most occurrences; Nitrogen Oxides (NOx) showed a slight increase during the period from 2007 to 2012, with higher concentrations on high ozone days; and, Sulfur Dioxide (SO_2) plumes contributed to ozone on a few high ozone days, but were not the dominant factor for the 8-hour average maximums.
2. For meteorological conditions and background ozone, the key points were that high ozone days were typically characterized by warm temperatures (average 97°F), light winds (6.1 miles per hour), and moderate humidity (40%); winds often come from the north-northeasterly through south-southeasterly directions; CAMS 1037 is positioned to measure regional background ozone entering eastern McLennan County rather than ozone formed from local Waco emissions; and, Background ozone concentrations are a significant factor, with regional values indicating that even small increases in local emissions or future background levels could lead to NAAQS exceedances.
3. In the area of source regions and transport, it made important observations that inter-state potential source regions for background ozone frequently included areas to the north or northeast of Texas, such as eastern Oklahoma, Louisiana, Arkansas, and Missouri, and parts of the Ohio/Mississippi River Valleys; intra-state upwind regions within Texas most often included areas northeast, east, or southeast of Waco; local upwind regions identified by ground-level wind data included eastern McLennan County and parts of surrounding counties; and, Ozone from the HGB area likely contributed to at least one high ozone day, and the DFW area contributed to at least three.
4. For large-scale weather patterns, the report highlighted that high ozone episodes were primarily driven by large-scale weather patterns favorable for ozone transport, formation, and accumulation; that these often involved a surface ridge of high pressure extending into Texas from the Central Plains or eastern U.S., leading to clear skies, warm temperatures, and light winds; the inflow of continental air from outside Texas, likely carrying elevated

background ozone and its precursors, is a necessary condition; and, many episodes are initiated by a cold front passage through Waco, which can lead to increased solar radiation and drier air.

To summarize, the report concluded that high ozone in Waco is largely a regional phenomenon, influenced by specific meteorological conditions and the long-range transport of ozone and its precursors from broader continental and intra-state source areas, rather than primarily local Waco emissions. It also brought to light that there had been several days of exceeding the ozone level even at the older standard of 0.075 ppb.

Since 2007, ambient air quality monitoring by TCEQ has occurred for the greater Waco area at the Mazanec Station (TCEQ, 2024a). Parameters monitored at the station include CO, O_3, PM2.5, SO_2, along with other meteorological variables, including solar radiation, temperature, and wind direction and speed. NAAQS standards for O_3 are calculated as the fourth highest daily 8-hour maximum concentration occurring every year, averaged as a 3-yr moving average. Hence, instantaneous daily and hourly concentrations may exceed 70 ppb; however, prolonged concentrations over 70 ppb lead to exceedances of the NAAQS standard. Long-term trends suggest O_3 concentrations exceeded the NAAQS standard prior to 2013 and subsequently remaining in attainment until most recently, as seen in Figure IV.6.3. Primary NAAQs standards for PM2.5 are calculated as a three-year

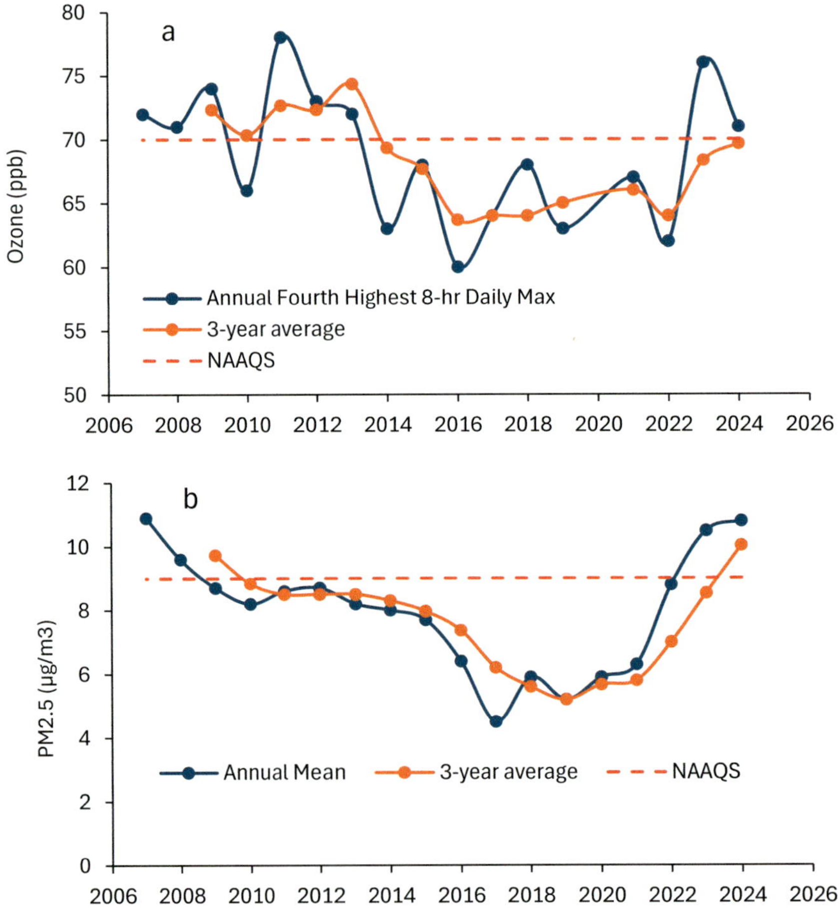

FIGURE IV.6.3 Annual (a) ozone (o_3) concentrations and (b) PM2.5 concentrations at the Waco Mazanec air quality monitoring station relative to NAAQS.

moving average of annual means. PM2.5 values tracked similar patterns in O_3. PM2.5 concentrations exceeded the NAAQS standard in 2009, and subsequently dropped into attainment until 2024. Potential non-attainment status will depend upon concentrations for the remainder of 2024 and 2025. As mentioned previously, this could have implications for future transportation projects.

IV.6.4.3 Land Change

Waco has observed both urban intensification and extensification, the majority following the latter style of growth. From 1970 to 2010, population growth within the city was far lower (31%) than the remainder of McLennan County (59%), suggesting the majority of new growth was occurring outside city limits, leading to dispersed development. In the past 40 years, urban lands have increased 33% in McLennan County, 18% just in the last 20 years (Figure IV.6.4). Much of this growth occurred in the southern portion of the county where rapid suburbanization expanded the communities of Woodway, Hewitt and Lorena. In turn, this new growth has placed stress on existing infrastructures, particularly requiring new wastewater treatment plant development in Lorena

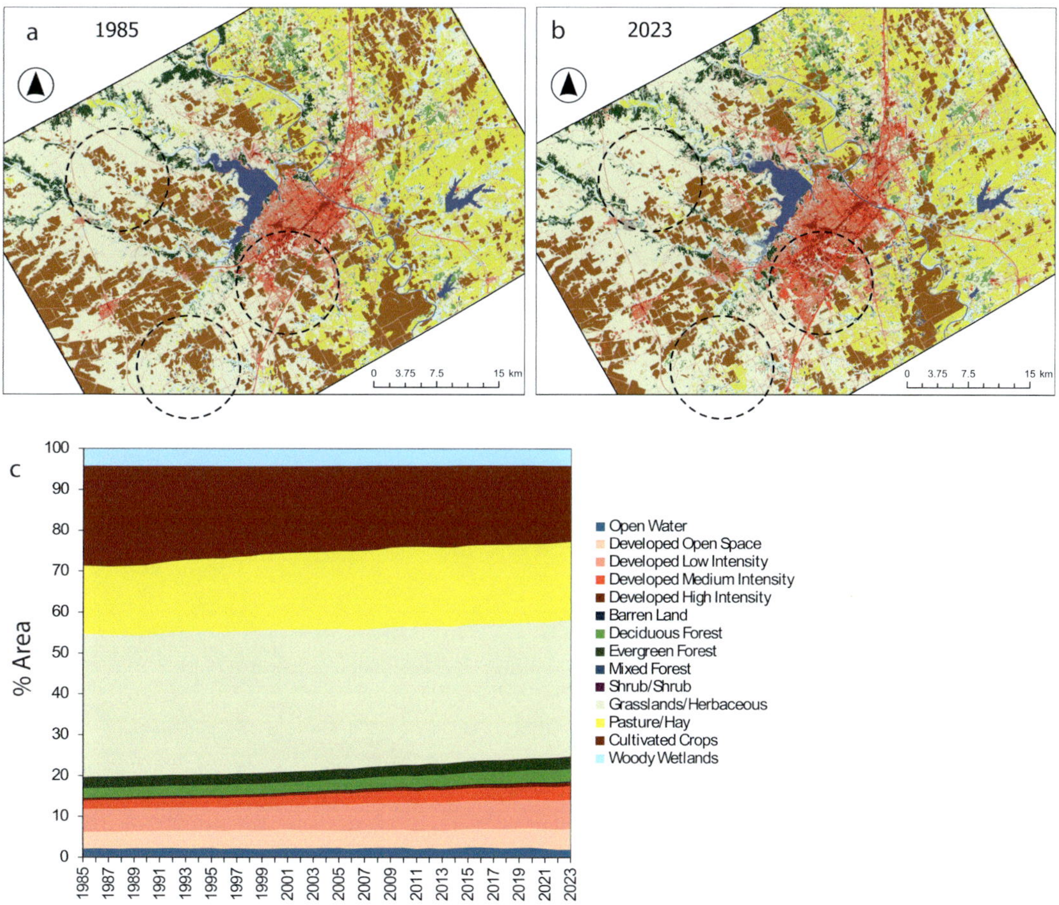

FIGURE IV.6.4 National land cover dataset annualized land cover maps for McLennan County, TX in (a) 1985 and (b) 2023, and (c) 40-year changes in land cover proportions. The dashed circles show areas of changes including increased suburban expansion as well as shifts in agricultural uses between pasture/hay, grassland, and cultivated crops.

Data from Multi-Resolution Land Characteristics Consortium, https://www.mrlc.gov/viewer/.

(Jatko et al., 2024). Expansion of urban lands have primarily compromised cultivated crop land, grasslands, and pasture lands, reducing several important services, including reductions in rainfall interception, heat mitigation, and carbon sequestration capacity. Reduction of cultivated lands and pasture lands proximate to Waco and other cities has led to secondary offsets, or compensation, by shifting agricultural uses in the rural outlying regions of McLennan County. Open space urban development, primarily low-growth green spaces within the urban area have decreased, being replaced with intensified forms of urban development.

Future projections of urban land expansion within Waco's Metropolitan Statistical Area, according to numerous sources, are highly uncertain (Figure IV.6.5). Conservative estimates suggest an additional 500 km² of urban lands by the end of century, a 67% increase from current values. However, high rates of urbanization could lead to an additional 1,500 km² of urban lands, depending on future economic growth, population, and growth policies.

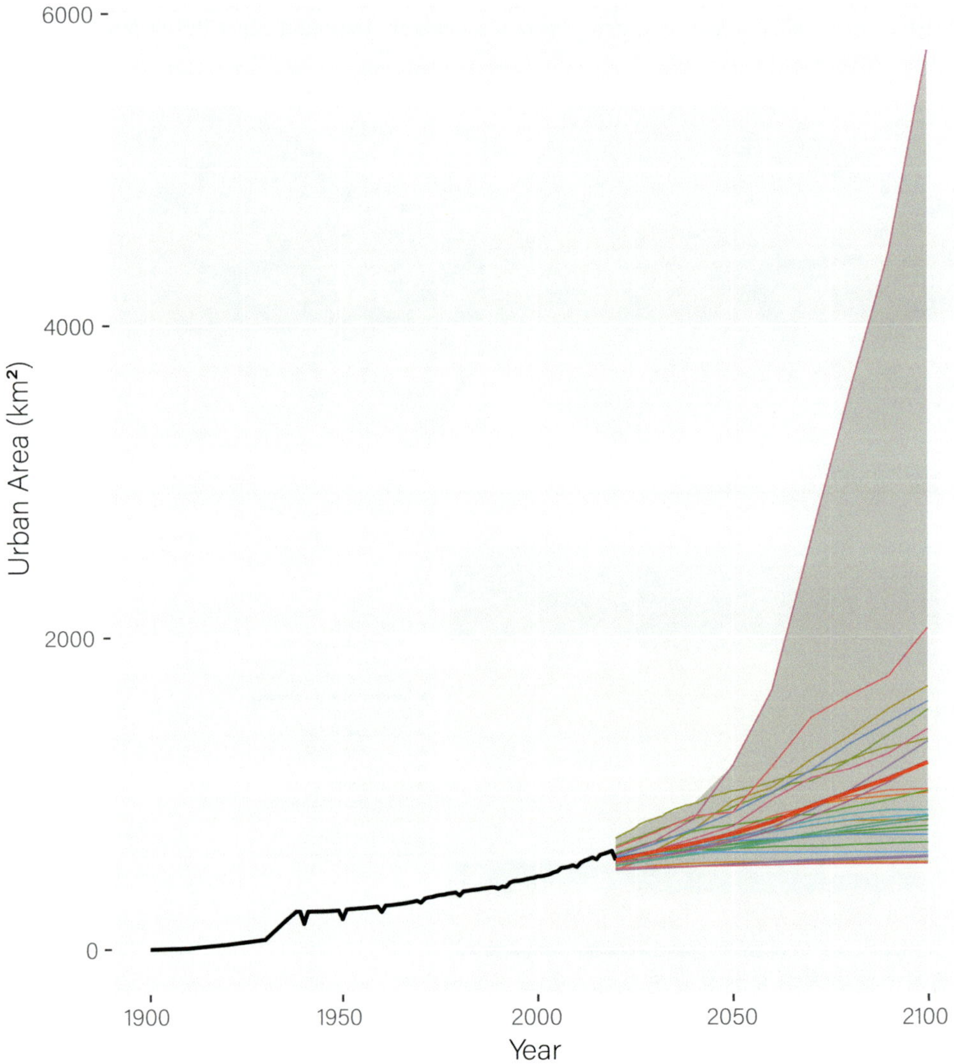

FIGURE IV.6.5 Urban land (km²) changes in Waco metropolitan statistical area and future urban land projections according to numerous models (Chen et al., 2020; Dornbierer et al., 2021; Gao and O'Neill, 2020; Li et al., 2021; USEPA, 2009).

Land planning efforts will be needed to manage uncertain drivers of urban land extensification in the future. However, the direct relationship between comprehensive plans, i.e. land use planning, and eventual implementation is not always clear and varies by municipality. Indeed, this leads to uncertainties in observed land changes versus land planning. To more clearly link land planning to urban development, Hersperger et al. (2018) suggest that more effort should be devoted to relate the following elements: (1) the intentions expressed in the plans; (2) the means of implementing plans through regulation; and (3) the role of external factors (i.e., markets, etc.) influencing implementation. In Texas, areas outside the city boundaries are termed "extraterritorial jurisdiction zones" (ETJs). Local governments can regulate some activities within the ETJ but not zoning, which suggests that rural land change and urbanization could be driven largely by market forces.

IV.6.5 PRACTICAL INSTITUTIONAL RESPONSES TO CLIMATE CHALLENGES

IV.6.5.1 Water

IV.6.5.1.1 Water Availability

Over the past century, suggestions to improve water availability primarily included increasing the holding capacities of Lake Waco with heavy investment in dam infrastructure expansion and improvements. There are additional sources of water that have been and can be acquired along with conservation efforts to increase the resiliency of local water resources. In the state of Texas, surface water access is regulated by right and is owned by the state and held in trust for the citizens (TCEQ, 2025b). TCEQ oversees this program and enforces rights to the water using a first in time, first in right approach, where the earlier date by which the right was established provides order of seniority to access. This means in times of drought the older, or more senior, rights are allowed to draw their allotted quantity prior to a more recent or younger established right. Acquiring new additional surface water rights in Texas can be difficult and requires meeting several guidelines provided in the Texas Water Code (TWC) §11.134; however, you can purchase older existing rights which are often tied to land (TWDB, n.d.d). In an effort to shore up senior water rights and prevent mandated releases from Lake Waco downstream, the City of Waco purchased 628 acres of land with a Brazos River 1921 diversion water right of 701 million gallons per day in 2022 (Saegert, 2022).

Groundwater is regulated differently and follows a geographic approach where the landowner owns the water beneath the property. With groundwater, the Rule of Capture applies; if water can be sourced beneath the property, then the property owner has right to that water, even if it impacts neighbors (Dowell, 2013). There are some protections in place in the TWC which restrict pumping when water is wasted, causes malicious drainage, or leads to negligent subsidence (TWC §36.002). There is also legislative authority provided to local conservation districts, like the Southern Trinity Groundwater Conservation District, for high use permitting to help protect the general resource, however exemptions exist. A permit is not required under a few circumstances including, if the well is on a 10-acre or larger tract, is for domestic or livestock use, draws < 25,000 gallons a day, or if the well is in active use supporting drilling or exploration for oil or gas, (TWC §36.117). For non-exempt, typically larger uses, a permit is required to meet local conservation district requirements; however, under this permit, landowners can pump and sell their water to others, which provides an opportunity to redistribute groundwater to areas where local availability is unable to meet demand (Dowell, 2013). San Antonio is an example of this redistributed ground water strategy, where a 142-mile pipeline transects seven counties to move up to 50,000 acre-feet of water from the Carrizo Aquifer to the city. This project was implemented to alleviate stress on the Edwards Aquifer, while providing water for future economic development and population growth (PDE, n.d.; SAWS, 2025).

TABLE IV.6.2
Annual City of Waco Water Rights in Gallons by Source as of 2024

Water Supply Source	Source Type	Total Gallons
Lake Waco	Surface	25,700,000,000
Brazos River	Surface	2,509,052,700
Trinity Aquifer	Ground	289,168,943

Source: **COW (2024).**

Uniquely, the water in Lake Waco, a United States Army Corps of Engineers (USACE) flood mitigation lake, is owned by the United States government. Additionally, as it flows into the Brazos River, access to the water must be shared with the Brazos River Authority (BRA) to allow for downstream discharge requirements and downstream water rights (TWDB, n.d.c). As the only USACE lake located entirely within a single municipal jurisdiction, City of Waco has priority water rights to Lake Waco (Conry, 2010). With access to the Trinity Aquifer and the Brazos River, Waco has accumulated water rights to 28,498,221,643 gallons annually, as seen in Table IV.6.2, nearly 140,000 gallons per citizen served (COW, 2024).

IV.6.5.1.2 Water Conservation

With water acquired it is important for a local jurisdiction to conserve that water. Conservation is where incremental pragmatic adjustment and stakeholder collaboration are fully employed, where stakeholder support and buy in help to shape the actual use of the available water. In the Texas Water Development Board (TWDB) 2022 State Water Plan, many water management strategies have been identified to help conserve water by reducing overall demand: agricultural conservation, municipal conservation, drought management, and industrial conservation (TWDB, 2021).

This call for conservation is important as the population of Texas is anticipated to increase by 73% in the year 2070 with a water shortage estimate of 6.9 million acre-feet of water if a drought of record occurred in that same year leading to 25% of Texans having less than half their needed water supply, as can be seen in the demand and existing supply depiction overtime in Figure IV.6.4 (TWDB, 2021). While advances in water use efficiency have led to a reduced per person consumption rate, 21% decrease in use from 2000–2015 – which causes the reduction in slope for the demand line in Figure IV.6.6 -, we cannot keep up with the overall demand with the expected increase in population (Walker, 2017).

At the local level this could mean the implementation of water restrictions. In 2023 the City of Waco implemented Stage 2 of their four-stage drought contingency plan from July 2022 to October 2023 and was nearly in Stage 3 until a major flooding event occurred in the Bosque Watershed and replenished the lake in two days (Kyle, 2024). After experiencing the drought of 2022 and 2023, a look at the drought plan spurred new changes that were introduced in May of 2024 with updates to the contingency plan and the introduction of a long-term water conservation plan (Kyle, 2024). The Waco plan calls for public participation in the reduction of water usage and took input from citizens and Council members before taking final form (COW, 2024).

As of January 2025, 42 of the public water systems in Texas were operating under a base-level water restriction as a means of wider conservation efforts (TCEQ, 2025c). These restrictions can include lawn watering rules and timing assignments, pressure washing of hard surfaces elimination, pool and spa construction permit reductions or eliminations, and high use water business operational restrictions (COW, 2024).

Municipal Water Conservation Plan (WCP) guidelines have been laid out by TCEQ in Title 30 of the Texas Administrative Code Chapter 288. The guidelines require any surface water right

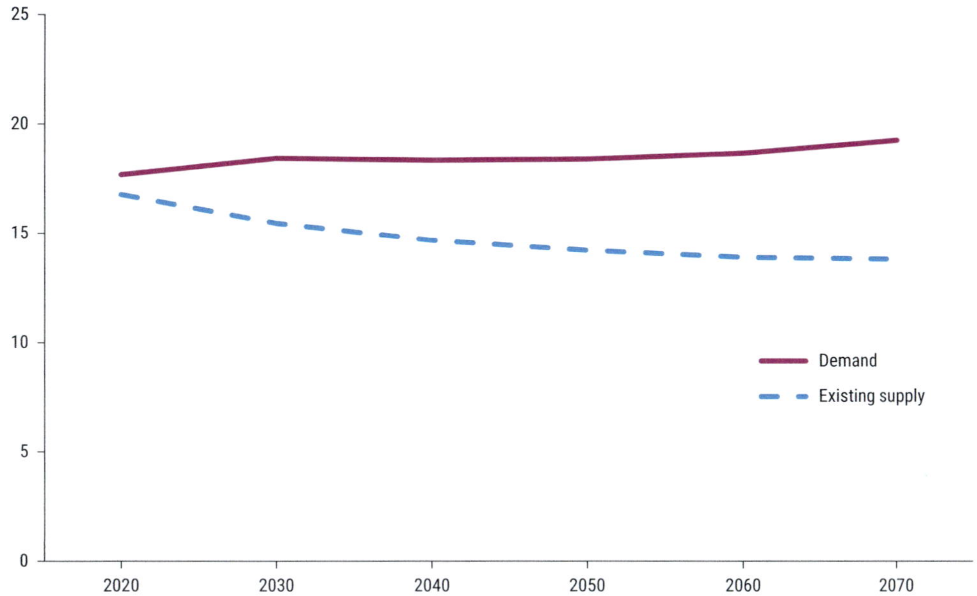

FIGURE IV.6.6 Projected total annual water demand and existing water supply for all sectors in Texas (millions of acre-feet) (TWDB, 2021).

holder with rights to more than 1,000 acre-feet of water for municipal, industrial, and/or other non-irrigation uses to develop a WCP and to submit water conservation implementation reports. These guidelines encourage public involvement in development and require ongoing public outreach and education of the plan.

IV.6.5.1.3 Water Quality

The general protection of this precious resource must be considered a local climate change action. As the resource becomes scarcer, the protection of that resource becomes ever more important to maintain or improve its quality. While we have already discussed many regulating bodies geographically (TWDB, TCEQ, BRA, USACE), we have not discussed the direct regulation of water quality.

Water quality is monitored and enforced locally with federal implications. Texas has the second largest environmental agency on the globe in TCEQ only second to the EPA (Mazur, 2022). Under a memo of understanding, the EPA delegated authority to TCEQ in 1998 to administer the National Pollutant Discharge Elimination System (NPDES) within the boundaries of Texas as the appropriately named Texas Pollutant Discharge Elimination System (TPDES) which has federal authority over discharges of pollutants to Texas surface waters, with additional authorities granted from EPA in 2021 (TCEQ, 2024b).

Under the TPDES program 16 different water quality general permits are available associated with stormwater, oil and gas, and wastewater discharges to surface waters of Texas, as well as many individual permits for specific discharges like industrial and municipal domestic wastewater (TCEQ, 2025d). These permits issue specific guidelines and thresholds for allowable discharges into the surface waters of the state in an attempt to maintain or improve surface water quality as it relates to Texas Surface Water Quality Standards. Within the jurisdictional boundaries of a municipality the government is issued enforcement and inspection authority over stormwater discharges through the Municipal Separate Storm Sewer System (MS4) permit, which allows for direct involvement in management of local waters through state issued requirements to permittees.

The Federal Emergency Management Agency (FEMA) also plays a vital role in water quality management through locally enforced floodplain ordinances. Unfortunately, some of the cheapest lands available is the land that manages and mitigates water flows and quality, floodplains. While less expensive they endanger those who build in them, particularly with the flash flood dynamics of Central Texas. Development in floodplains directly removes nature's own ability to manage water quality and recharge groundwater, while also limiting other environmental benefits like habitat protection and biological productivity (FEMA, 2024).

Urban development falls under these general permits and requirements of TCEQ, FEMA, and any local code or policies with the Construction General Permit, Multi-Sector General Permit, and the National Floodplain Insurance Program. The Brazos River suffers water quality degradation from urban development influence like increased bacteria levels, reduced dissolved oxygen, and turbidity (BRA, 2022; USEPA, 2025b). Urban development also influences groundwater recharge with the increased impervious surfaces promoting rainwater run-off and reducing infiltration and recharge eventually lowering the water table (USGS, 2018).

External jurisdictional influences of rural development do impact the Brazos River; however it is the primary impact on the Bosque River watershed facing water quality degradation from nitrogen and phosphorus, and bacterial contamination primarily due to dairy operations and other human activity (BRA, 2022; Terfera et al., 2024). TCEQ again regulates this through concentrated animal feeding operation (CAFO) permits but has increased requirements by establishing a total maximum daily load (TMDL) explicitly for the Bosque as a designated major sole-source impairment zone (TCEQ, 2025e). The City of Waco directly influences and support stakeholders in this watershed by managing and participating in the Bosque River Coalition, an organization comprised of landowners, businesses, and the City of Waco focusing on water quality, conservation, habitat restoration, and responsible waste management (COW, 2025). The Bosque watershed has seen improvements after the implementation of the TMDL program in 2001 to reduce phosphorous with reductions of soluble reactive phosphorus of up to 50% at some monitoring sites (TCEQ, 2024c).

IV.6.5.2 Air Quality – Transportation System Planning

Given the exceedances of air quality attainment thresholds, the process of integrating considerations of potential emissions and potential reduction strategies, such as mitigation techniques are under way via the 25-year Metropolitan Transportation Plan. In planning for future transportation projects, several guidelines can be considered and used in evaluating alternative projects. National averages of emission profiles (grams per mile) from vehicles vary according to vehicle types (e.g., passenger, light-duty vehicles, heavy-duty vehicles, buses, etc.) (BTS, 2024), and can provide the basis for estimation of potential emission effects from transportation projects. Additionally, projections of future vehicle emission rates, due to technological changes, are available (BTS, 2024) and can be used to provide updated estimates (Figure IV.6.7).

Vehicle emission profiles could provide generalized metrics for evaluating projects on the basis of changes in total mileage per vehicle types. Considerations for projections could use a simplified checklist, relying on considerations that extend beyond mileage estimates, such as congestion, travel quality, and the potential for public transportation or forms of non-vehicle travel (Table IV.6.3). Potential mitigation options that could stand to ultimately reduce net emissions could also be considered (Table IV.6.4). More holistic considerations could include whether transportation projects are situated in residential versus commercial or industrial settings, leading to growth or transitions in these areas. Traffic densities will vary in residential developments of various intensities compared to commercial developments, such as hotels, restaurants, or retail destinations, leading to very different future emission estimates. In contrast, industrial facilities could have their own regulated or permitted emissions contributing to ambient air quality apart from transportation.

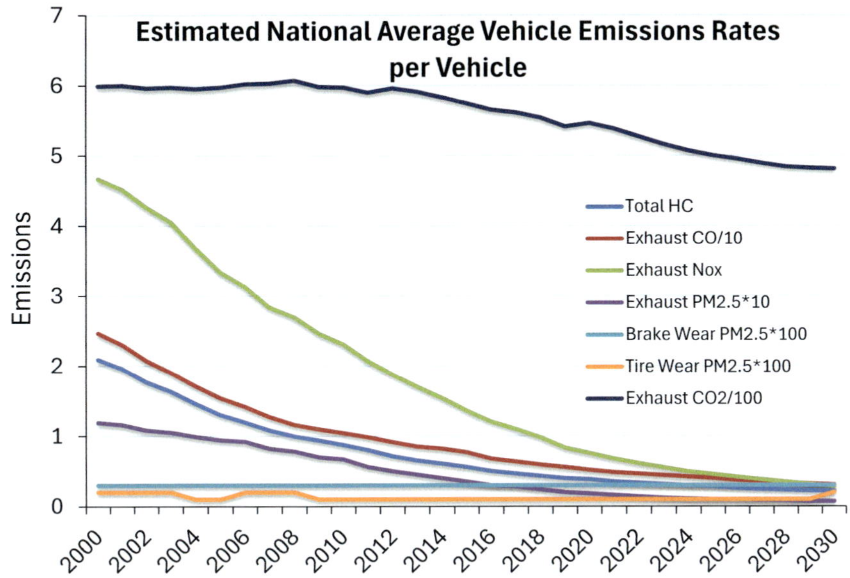

FIGURE IV.6.7 Estimated past and future vehicle emission rates for various pollutants. Taken directly from Bureau of Transportation Statistics (Source: https://www.bts.gov/product/national-transportation-statistics).

TABLE IV.6.3
Example of a Checklist for Transportation Project Considerations Related to Impacts to Emissions and Air Quality

Travel Attribute	Increase/Decrease?	Reason/Justification
Person trips	Is your project likely to increase or decrease each factor? If so, by how much (%)?	Be prepared to justify or reason through your estimates and be transparent with assumptions.
Average trip length		
Vehicle-miles traveled		
Vehicle-hours traveled		
Congestion indicators		
Travel quality, speed		
Overall traffic volume		
Public transit		
Walking and bicycling		

TABLE IV.6.4
Potential Mitigation Options for Offsetting CO_2 Emissions from Transportation Projects

Mitigation Options	Proposed Idea	Sufficient to mitigate impacts?
Tree plantings	e.g., type, age, amount	Estimated CO_2 sequestration
Renewable energy offsets	Size of solar power system (kW)	Offset CO_2 from fossil fuel power plants
Local monitoring station	Devote resources to long-term air quality monitoring station	Provides an ability to directly measure rather than estimate emissions
Public transportation	Reduce personal vehicle usage through bus route	Estimated reduction in emissions
Electric vehicle charging station	Increased likelihood of electric vehicles on route	Estimated reduction in emissions

IV.6.5.3 The City Plan – Waco Comprehensive Plan 2040

Similar to most urban areas across the country, most cities in Texas maintain a comprehensive plan that outlines the land use policies and sets the enabling framework for adopting and implementing zoning regulations. The notable comprehensive plans in the Waco region include the ones adopted by the Cities of Waco, Hewitt, Woodway, McGregor, Robinson, Bellmead, and Lacy-Lakeview. Of all the current comprehensive plans, the one adopted by the City of Waco in 2016 makes the most explicit attempt to address climate resilience and natural resource management. The City Plan, Waco Comprehensive Plan 2040 includes guidance for economic development, management of growth, and addresses issues related to transportation, utilities, housing, and the environment (COW, 2016). This plan aimed to tackle challenges like aging infrastructure, limited funding for new utility expansion, and increasingly spread-out development. These challenges also presented opportunities to direct growth more strategically, especially by leveraging subsidized utility access. While not explicitly stated, the plan's focus on efficiency and regulated growth indirectly contributed to climate resilience.

The plan highlighted the need for major repairs and replacements to Waco's roads, water, sewer, and stormwater systems. It proposed *mixed-use developments* within the city to increase efficiency in providing utilities and services, while also generating revenue for infrastructure improvements. Given insufficient funds for continued suburban sprawl, the plan suggested managing rural development with two primary land uses: large-lot single-family homes or higher-density clustered single-family homes, with the latter reducing construction and maintenance costs, as well as per-capita energy and water footprints. The City Plan (Figure IV.6.8) adopted different strategies for the city limits and its extra-territorial jurisdiction (ETJ), mainly because Texas municipalities cannot enforce zoning in the ETJ. The overall goal was sustainable development through maximizing efficiency.

Within city limits, the plan promoted mixed-use categories (e.g., mixed flex space, combined commercial-residential) to foster a more urban lifestyle, including walkability and bike paths, to lessen transportation burdens. This approach aimed to attract more residents, increasing population density sufficiently to support public transit and affordable housing, and encourage infill development to stabilize neighborhoods and support diverse incomes. About a third of the city's space was designated for existing suburban and urban residential areas, and 21% was preserved as open space, including significant areas like Cameron Park and Lake Waco Wetlands.

In the ETJ, the plan focused on preserving the rural character and minimizing the need for new subsidized infrastructure. It aimed to foster clustered development on smaller lots, preserving the remaining land as open space, parks, recreation, and agriculture. Alternatively, oversized lots (mini-ranch style) were supported to maintain the rural aesthetic. These strategies reduce infrastructure requirements for streets, drainage, and utilities, and ease the burden on public services.

It is important to emphasize that land use plans are not synonymous with zoning regulations, and inconsistencies between the plan and existing zoning regulations can exist (Figure IV.6.8). While zoning change requests can address non-conformities, they can also move land uses in directions that contradict the plan. Therefore, a combination of instruments and approaches is needed to effectively use land use planning for climate resilience. Locally, two of the limiting constraints are due to (1) widely varying considerations of climate change, resilience, and natural resources across municipalities, and (2) lack of land use regulatory authority for the county. Since a significant portion of McLennan County is still rural and continually developing urbanized areas are subject to different municipalities, it is challenging to use land use regulations as a tool to address climate change concerns alone.

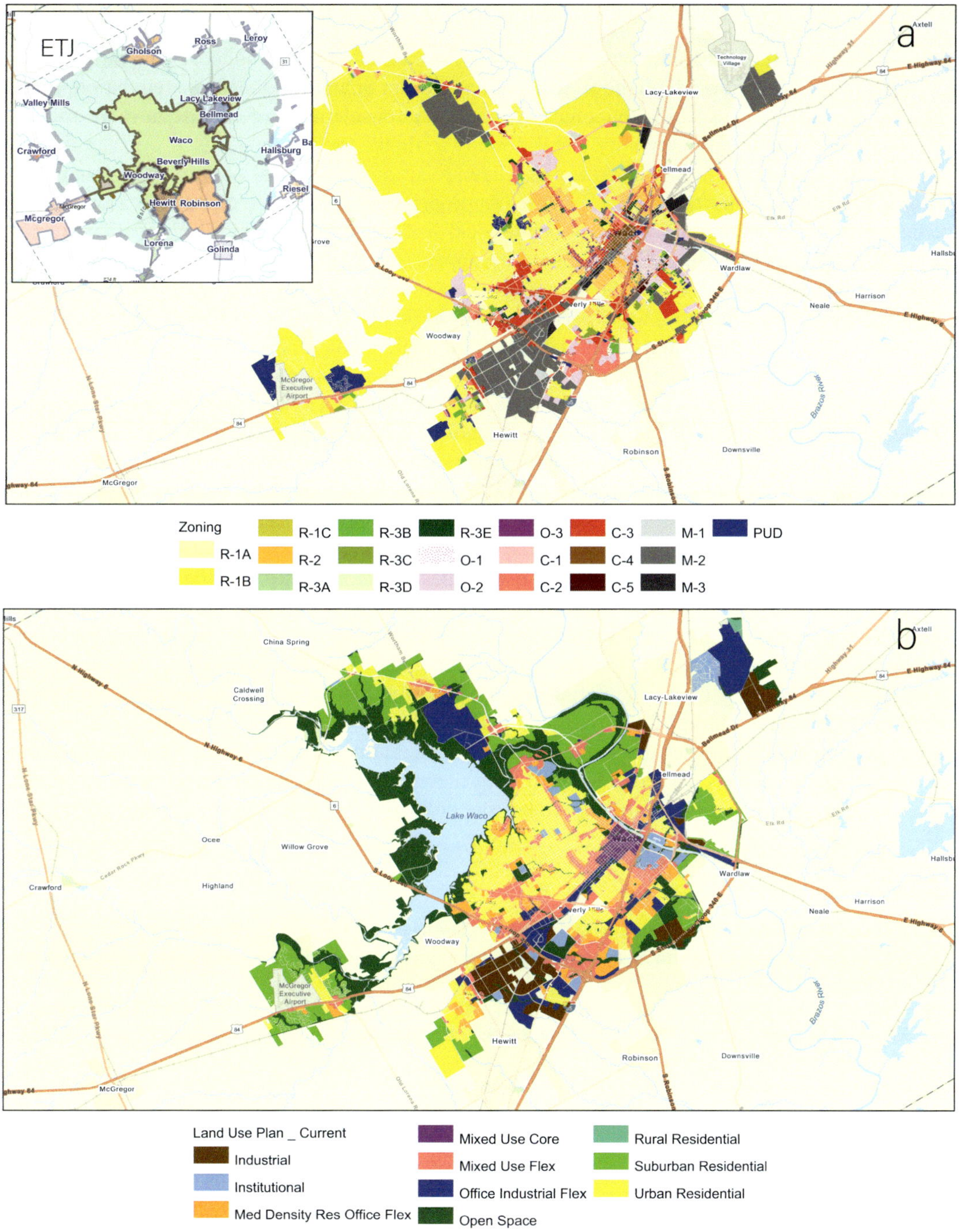

FIGURE IV.6.8 Land uses for the City of Waco, including (a) current land zoning ordinances, and (b) the land use plan included within the City's Waco 2040 comprehensive plan. The inset in panel a depicts the extraterritorial jurisdiction (ETJ) boundary and other incorporated areas outside of the City of Waco limits.

IV.6.6 EXPECTED OUTCOMES AND CONCLUDING REMARKS

Within the central Texas region, there are multiple institutions representing local governments that play a partial yet important role in the future of air and water quality, and land use regulation. In the near term we expect two specific outcomes: framing the consideration of climate in more specific terms of ensuring air and water quality that is directly available for current needs as well as expected future demands, and integrating the concerns related to air and water quality into the regional planning processes for transportation and land use. Appropriately framing the climate consideration is crucial to building the adaptive capacity of the local institutions. More frequently, policy-making and planning process regarding climate adaptation appeal to normative value preferences privileging one generation over another. While the temporal aspect of climate adaptation strategies is integral to understanding the phenomenon, we expect the action to be driven by more immediate and tangible concerns of water availability and health impacts of air quality. Appropriate framing is also essential to improving the planning processes to integrate climate adaptation into decisions on related areas. Quite often climate adaptation considerations are classified as costs and weighed against other value considerations such as growth, development, economic development, and efficiency as tradeoff. Appropriate valuations of costs and benefits associated with climate adaptation strategies will be essential to integrating them into decisions that affect the air and water quality in the region.

In the last decade numerous studies have explored the cognitive barriers to climate adaptation strategies (Acciarini et al., 2021; Dodds, 2021; Golman et al., 2017; Marshall, 2015; Ranney and Clark, 2016; Zhao and Luo, 2021). Most of these studies focus on identifying the biases inherent in individual processing of climate information and how these biases influence their decision-making. Within the frameworks of historical institutionalism, and rational choice institutionalism, an understanding and recognition of these cognitive biases offer valuable insights into institutional responses as scaled up aggregated climate adaptation strategies. Using the framework of sociological institutionalism, such understanding allows for incremental pragmatic adjustment, and stakeholder collaboration as immediate strategies that offer a platform for creating a path forward toward sustainable institutional responses. As we move forward with these short-term strategies, we hope to be able to overcome cognitive biases toward a more holistic collective response toward climate adaptation.

NOTES

1. The five ways refer to fatalists, hierarchists, individualists, egalitarians, and the hermits. Each category defines the problem of nature in its own unique way under different contexts. For details see O'Riordan and Jordan (1999).
2. Much of this section on air quality was earlier prepared for inclusion into the Convey 2050: Shaping Regional Mobility Across McLennan County (the 2050 Metropolitan Transportation Plan of the Waco Metropolitan Planning Organization).

REFERENCES

Acciarini C, Brunetta F, Boccardelli P, 2021. Cognitive biases and decision-making strategies in times of change: A systematic literature review. Management Decision. 59(3):638–652.

Agrawal A, Agrawal A, 2008. The role of local institutions in adaptation to climate change 17, 28274. Washington, DC: World Bank.

Albouy D, Ehrlich G, Shin M, 2018. Metropolitan land values. Review of Economics and Statistics. 100:454–466.

Ansari S, Wijen F, Gray B, 2013. Constructing a climate change logic: An institutional perspective on the "tragedy of the commons". Organization Science. 24(4):1014–1040.

BRA (Brazos River Authority), 2022. Brazos River Basin Summary Report 2022. https://brazos.org/Portals/0/Documents/CRP/Brazos-River-Basin-Summary-Report-2022.pdf

BTS (Bureau of Transportation Statistics), 2024. Estimated U.S. Average Vehicle Emissions Rates per Vehicle by Vehicle Type Using Gasoline, Diesel, and Electric. https://www.bts.gov/content/estimated-national-average-vehicle-emissions-rates-vehicle-vehicle-type-using-gasoline-and (accessed May 29, 2025).

Chen G, Li X, Liu X, Chen Y, Liang X, Leng J, Xu X, Liao W, Qiu Y, Wu Q, Huang K, 2020. Global projections of future urban land expansion under shared socioeconomic pathways. Nature Communications. 11:537. https://doi.org/10.1038/s41467-020-14386-x

Conry TM, 2010. Lake Waco comprehensive study: Background and overview. Lake and Reservoir Management. 26(2):74–79. https://doi.org/10.1080/07438141.2010.494131

COW (City of Waco), 2016. The city plan: Waco comprehensive plan 2024. https://www.waco-texas.com/Departments/Development-Services/Planning-Services/City-Plan (accessed June 3, 2025).

COW, 2024. Utility profile & water conservation plan. https://www.waco-texas.com/files/sharedassets/public/v/5/departments/water-utility/documents/city-of-waco-conservation-drought-contingency-plan.pdf

COW, 2025. Bosque river coalition. https://www.waco-texas.com/Departments/Water-Utility-Services/Bosque-River-Coalition#section-1 (accessed May 29, 2025).

Crowley K, Stewart J, Kay A, Head BW, 2020. Reconsidering Policy: Complexity, Governance and the State. Bristol, UK: Policy Press.

Dahlstrom W, 2013. Chapter 4, zoning regulations in Texas in A guide to urban planning in Texas Communities 2013. American Planning Association Texas Chapter. September 7, 2012. https://txplanningguide-ojs-utexas.tdl.org/txplanningguide (accessed June 9, 2025).

Dodds J, 2021. The psychology of climate anxiety. BJPsych Bulletin. 45(4):222–226.

Dornbierer J, Wika S, Robison C, Rouze G, Sohl T, 2021. Prototyping a methodology for long-term (1680–2100) historical-to-future landscape modeling for the conterminous United States. Land. 10(5):536.

Dowell T, 2013. Texas water; basics of groundwater law. Texas A&M AgriLife Extension. https://agrilife.org/texasaglaw/2013/10/22/texas-water-basics-of-groundwater-law/ (accessed May 15, 2025).

FEMA (Federal Emergency Management Agency), 2024. Benefits of natural floodplains. https://www.fema.gov/floodplain-management/wildlife-conservation/benefits-natural (accessed May 29, 2025).

Fertner C, Jorgensen G, Sick Nielsen TA, Bernhard Nilsson KS, 2016. Urban sprawl and growth management—drivers, impacts and responses in selected European and US cities. Future Cities and Environment. 2:9.

Gao J, O'Neill BC, 2020. Mapping global urban land for the 21st century with data-driven simulations and Shared Socioeconomic Pathways. Nature Communications. 11:2302. https://doi.org/10.1038/s41467-020-15788-7

Golman R, Hagmann D, Loewenstein G, 2017. Information avoidance. Journal of economic literature. 55(1):96–135.

Grafstein R, 1992. Institutional Realism: Social and Political Constraints on Rational Actors (pp. 89–112). New Haven, CT: Yale University Press.

Gümüsay AA, Claus L, Amis J, 2020. Engaging with grand challenges: An institutional logics perspective. Organization Theory. 1(3):2631787720960487.

Hall PA, Taylor RC, 1996. Political science and the three new institutionalisms. Political studies. 44(5):936–957.

Head BW, 2022. Wicked Problems in Public Policy: Understanding and Responding to Complex Challenges (p. 176). Cham, Switzerland: Springer Nature.

Hersperger AM, Oliveira E, Pagliarin S, Palka G, Verburg P, Bolliger J, Grădinaru S, 2018. Urban land-use change: The role of strategic spatial planning. Global Environmental Change. 51:32–42.

Hosseiny SH, Bozorg-Haddad O, Bocchiola D, 2021. 9-water, culture, civilization, and history. In Economical, Political, and Social Issues in Water Resources, Bozorg-Haddad O, ed. Elsevier. https://doi.org/10.1016/C2020-0-03830-2(978-0-323-90567-1)

Jatko JT, McManamay RA, Woodyard MT, Roush LK, Back JA, King RS, Matson CW, 2024. Differentiating between point and non-point source nutrient loadings of wastewater in an agriculturally impacted area using a hybrid statistical model. Science of The Total Environment. 912:169553.

Krafft MP, Riess JG, 2015. Selected physiochemical aspects of poly- and perfluoroalkylated substances relevant to performance, environment and sustainability – Part one. Chemosphere. 129:4–19. https://pubmed.ncbi.nlm.nih.gov/25245564/

Kyle M, 2024. Waco's three-day watering limits to take effect in July under updated drought plan. Waco Tribune-Herald. https://wacotrib.com/news/local/government-politics/article_9f00c4c2-0d85-11ef-9738-476ec5f40dd0.html

Lazarus RJ, 2009. Super wicked problems and climate change: Restraining the present to liberate the future. Cornell Law Review. 94:1153.

Letsinger B, 2019. The United Watershed States of America: A Biography of John Wesley Powell. https://brandonletsinger.com/biography/the-united-watershed-states-of-america-a-biography-of-john-wesley-powell/#:~:text=John%20Wesley%20Powell%20felt%20so,rest%20of%20his%20public%20career. (accessed May 21, 2025).

Lewis R, Parker R, 2021. Exurban growth inside the urban growth boundary? An examination of development in Oregon cities. Growth and Change. 52:885–908. https://doi.org/10.1111/grow.12481

Li X, Zhou Y, Hejazi M, Wise M, Vernon C, Iyer G, Chen W, 2021. Global urban growth between 1870 and 2100 from integrated high resolution mapped data and urban dynamic modeling. Commun Earth Environ. 2:201. https://doi.org/10.1038/s43247-021-00273-w

Manisalidis I, Stavropoulou E, Stavropoulos A, Bezirtzoglou E, 2020. Environmental and health impacts of air pollution: A review. Frontiers in public health. 8:14.

Marshall G, 2015. Don't Even Think About It: Why Our Brains Are Wired to Ignore Climate Change. New York, NY: Bloomsbury Publishing USA.

Martinez A, 2024. Dozens of Texas water systems exceed new federal limits on "forever chemicals". The Texas Tribune. https://www.texastribune.org/2024/04/16/texas-pfas-forever-chemicals-public-water-systems-epa-limit/

Mazur JB, 2022. Texas 2036: Supporting agency capacity for effective regulation. https://texas2036.org/wp-content/uploads/2022/10/TCEQ-LAR_One-Pager_9.22.pdf

McGaughey G, Durrenberger C, McDonald-Buller E, 2012. Conceptual model for ozone for the Waco area. Report Prepared for Heart of Texas Council of Governments & Texas Commission on Environmental Quality.

McManamay RA, Brinkley C, Vernon CR, Raj S, Rice JS, 2022. Urban land teleconnections in the United States: A graphical network approach. Computers, Environment and Urban Systems. 95:101822.

McManamay RA, Vernon CR, Chen M, Thompson I, Khan Z, Narayan KB, 2024. Dynamic urban land extensification is projected to lead to imbalances in the global land-carbon equilibrium. Communications Earth & Environment. 5(1):70.

Nadigroup, 2025. Do you know the difference between land-use planning and zoning? https://www.nadigroup.com/insights/do-you-know-the-difference-between-land-use-planning-and-zoning (accessed June 9, 2025).

Nelson AC, Moore T, 1993. Assessing urban growth management: The case of Portland, Oregon, the USA's largest urban growth boundary. Land Use Policy. 10(4):293–302.

NOAA (National Oceanic and Atmospheric Administration), 2024. National ocean service: What is subsidence? https://oceanservice.noaa.gov/facts/subsidence.html (accessed May 15, 2025).

NOAA, 2025. National integrated drought information system: Drought conditions for McLennan County. https://www.drought.gov/states/texas/county/McLennan (accessed May 15, 2025).

Ohenhen LO, Zhai G, Lucy J, Werth S, Carlson G, Khorrami M, Onyike F, Sadhasivam N, Tiwari A, Ghobadi-Far K, Sherpa SF, Lee J, Zehsaz S, Shirzaei M, 2025. Land subsidence risk to infrastructure in US metropolises. Nature Cities. https://doi.org/10.1038/s44284-025-00240-y

O'Riordan T, Jordan A, 1999. Institutions, climate change and cultural theory: towards a common analytical framework. Global Environmental Change. 9(2):81–93.

PDE (Pape-Dawson Engineers), n.d. Vista ridge regional water supply project. https://www.pape-dawson.com/services/water-resources/water-wastewater/water-supply-infrastructure-design/vista-ridge-regional-supply-project/ (accessed May 15, 2025).

Procopius, 1919. Procopius with an English Translation. In Seven Volumes, Vol 3: History of the Wars, Book V and VI, Translation by H.B. Dewing. Cambridge: Harvard University Press.

Ranney MA, Clark D, 2016. Climate change conceptual change: Scientific information can transform attitudes. Topics in Cognitive Science. 8(1):49–75.

Rogers PG. 1990. The clean air act of 1970. *EPA Journal* 16, 21.

Saegert R, 2022. Waco buying Brazos River water right dating to 1921. Waco Tribune-Herald. https://wacotrib.com/news/local/govt-and-politics/waco-buying-brazos-river-water-right-dating-to-1921/article_6b7ad29e-8ec2-11ec-baad-b3f048852dc5.html

SAWS (San Antonio Water System), 2025. Diversifying our water supply. Securing water for the future. https://www.saws.org/your-water/management-sources/vista-ridge-pipeline2/about-this-project/ (accessed May 15, 2025).

Scanes D, 2025. McLennan County regional water supply dynamics and sources. Scale: 1:600,000. Data sources: ESRI basemap, and data from Texas Geographic Information Office, TIGER Census, TxDOT, Texas Water Development Board, USGS. Created using ArcGIS Pro 3.4.3, Redlands, CA: Esri Inc. (May 29, 2025).

Schuster Olbrich JP, Vich G, Miralles-Guasch C, Fuentes L, 2022. Urban sprawl containment by the urban growth boundary: The case of the Regulatory Plan of the Metropolitan Region of Santiago of Chile. Journal of Land Use Science. 17(1):324–338. https://doi.org/10.1080/1747423X.2022.208631

Seto KC, Reenberg A, Boone CG, Fragkias M, Haase D, Langanke T, Marcotullio P, Munroe DK, Olah B, Simon D, 2012. Urban land teleconnections and sustainability. Proceedings of the National Academy of Sciences. 109(20):7687–7692.

STGCD (Southern Trinity Groundwater Conservation District), 2021. Groundwater management plan. https://southerntrinitygcd.org/wp-content/uploads/2024/06/STGCD_MP_2021_07_15_Final.pdf

Sun L, Chen J, Qinglan L, Huang D, 2020. Dramatic uneven urbanization of large cities throughout the world in recent decades. Nature Communications. 11:5366.

TCEQ, n.d.. Basin 12 Brazos River. https://www.tceq.texas.gov/assets/public/waterquality/swqm/assess/02twqi/basin12.pdf

TCEQ, 2024b. What is the "Texas Pollutant Discharge Elimination System (TPDES)"? https://www.tceq.texas.gov/permitting/wastewater/pretreatment/tpdes_definition.html (accessed May 15, 2025).

TCEQ, 2024c. TCEQ publication AS-204/23 - Status report: Implementing TMDLs to reduce phosphorus. https://www.tceq.texas.gov/downloads/water-quality/tmdl/north-bosque-river-narrative-criteria-06/as-204-23-bosque-river-status-report-final.pdf

TCEQ, 2025a. Hydrography data and map-based Daa viewers. https://www.tceq.texas.gov/waterquality/tmdl/hydromaps.html (accessed May 1, 2025).

TCEQ, 2025b. Am I regulated? Water rights in Texas. https://www.tceq.texas.gov/permitting/water_rights/wr-permitting/wr_amiregulated.html (accessed May 15, 2025).

TCEQ, 2025c. List of Texas PWSs Limiting water use to avoid shortages. https://www.tceq.texas.gov/drinkingwater/trot/droughtw.html (accessed May 15, 2025).

TCEQ, 2025d. Permitting, registrations, and reporting: Wastewater and stormwater. https://www.tceq.texas.gov/permitting/wastewater (accessed May 15, 2025).

TCEQ, 2025e. Concentrated animal feeding operation (CAFO) individual permit: Am I regulated? https://www.tceq.texas.gov/permitting/wastewater/cafo/cafo_individual.html (accessed May 29, 2025).

TCEQ (Texas Commission on Environmental Quality), 2024a. TAMISWeb v5.4.6 - Site details - Basic information. https://www17.tceq.texas.gov/tamis/index.cfm?fuseaction=report.view_site&siteAQS=483091037 (accessed May 29, 2025).

Terfera GW, Ray RL, Singh VP, 2024. Surface water quality under climate change scenarios in the Bosque watershed, Central Texas of United States. Ecohydrology & Hydrobiology. Lodz, Poland: Elsevier. https://doi.org/10.1016/j.ecohyd.2024.07.008

Thompson M, Ellis R, Wildavsky A, 1990. Cultural Theory. Boulder, CO: Westview Press

TSHA (Texas State Historical Association), 2018. Environment: Texas' water resources. In Texas Almanac 2018–2019, Alvarez EC, Harch R, Plocheck R, eds. Dallas: Quad Williamson.

TWDB, n.d.a. River basins. https://www.twdb.texas.gov/surfacewater/rivers/river_basins/index.asp (accessed May 1, 2025).

TWDB, n.d.b. Brazos River basin. https://www.twdb.texas.gov/surfacewater/rivers/river_basins/brazos/index.asp (accessed May 1, 2025).

TWDB, n.d.c. A Texan's guide to water and water rights marketing. https://www.twdb.texas.gov/publications/reports/infosheets/doc/WaterRightsMarketingBrochure.pdf

TWDB, n.d.d. Lake Waco (Brazos River Basin). https://www.twdb.texas.gov/surfacewater/rivers/reservoirs/waco/index.asp (accessed May 15, 2025).

TWDB (Texas Water Development Board), 2021. 2022 State Water Plan. https://www.twdb.texas.gov/waterplanning/swp/2022/index.asp

TWDB, 2024. McLennan County water supply planning information & resources. https://www.twdb.texas.gov/waterplanning/rwp/outreach/doc/mclennan.pdf

UN (United Nations), 2014. Planning for Climate Change: A strategic, values-based approach for urban planners. United Nations Human Settlements Programme (un-habitat), 160 pp.

USACE (United States Army Corps of Engineers), n.d.. History of Lake Waco. https://www.swf-wc.usace.army.mil/waco/Information/History.shtml (accessed May 1, 2025).

USCB (United States Census Bureau), 2021. State area measurements and internal point coordinates. https://www.census.gov/geographies/reference-files/2010/geo/state-area.html (accessed May 1, 2025).

USEPA (United States Environmental Protection Agency), 2009. Land-use scenarios: National-scale housing-density scenarios consistent with climate change storylines (Final Report). EPA/600/R-08/076F. https://ordspub.epa.gov/ords/eims/eimscomm.getfile?p_download_id=489947

USEPA, 2017. National water quality inventory: Report to congress. https://www.epa.gov/sites/default/files/2017-12/documents/305brtc_finalowow_08302017.pdf
USEPA, 2025a. Watershed academy web: Introduction to the clean water act. https://cfpub.epa.gov/watertrain/pdf/modules/IntrotoCWA.pdf
USEPA, 2025b. Urbanization – Water and sediment quality. https://www.epa.gov/caddis/urbanization-water-and-sediment-quality (accessed May 29, 2025).
USGS, n.d.a. Hydrologic Unit Code (HUCs) explained. https://nas.er.usgs.gov/hucs.aspx (accessed May 1, 2025).
USGS, n.d.b. Hydrologic Units of the United States. https://water.usgs.gov/themes/hydrologic-units/ (accessed May 1, 2025).
USGS (United States Geological Survey), 2018. Urbanization and water quality. https://www.usgs.gov/special-topics/water-science-school/science/urbanization-and-water-quality#overview (accessed May 29, 2025).
Walker J, 2017. Texas has an epic story to tell about water conservation. Texas Living Waters. https://texaslivingwaters.org/texas-good-story-water-conservation/
WDT (Water Data for Texas), 2025. Reservoirs: Waco Lake. https://waterdatafortexas.org/reservoirs/individual/waco (accessed May 15, 2025).
Zhao J, Luo Y, 2021. A framework to address cognitive biases of climate change. Neuron. 109(22):3548–3551.

IV.7 Texas Climate Policy Considerations

Julie A. King

IV.7.1 INTRODUCTION

Each semester for over a decade, I have surveyed my *Environmental Policy* students on the first day of class, asking the question: "What is the most pressing environmental problem we face today?" Overwhelmingly, each and every semester, students identify climate change as the most pressing issue. Pollution, closely interrelated with climate impacts, is always identified second. Additionally, student responses related to the climate change issue have evolved over time to include issues related to inaction, politicization, and lack of understanding, revealing insights into our growing social problems, perhaps incognizance, rather than a pure environmental phenomenon. In fact, many students express significant concern and consternation over climate change politicization.

In *Environmental Policy*, we use a working definition of public policy as "a course of government action or inaction in response to social problems" (Kraft et al., 2025). Some social problems are difficult or impossible to solve because of their interconnected nature. They are particularly complex, value-laden, and involve multiple stakeholders with competing interests. These are considered "wicked problems" not because they are inherently bad, but due to their complex socioecological nature, insoluble by traditional means or methods. Addressing wicked problems require sustained contributions from industry, every level of government, academic institutions, foundations, and philanthropists, really all of us. Specific to climate change, the challenge is finding affordable, equitable ways to bring every aspect of the global economy to net-zero carbon no later than 2050. At the same time, society must adapt to the effects of climate change that are already happening.

In the context of U.S. climate policy, we have experienced both action and inaction at the local, state, and federal level. This dichotomy has been particularly true in Texas. Some aspects and implications of the climate change problem are unique to Texas. This chapter seeks to address recent action and inaction on climate change in Texas and the subsequent policy implications. Additionally, I will explore policy as a tool for addressing climate change, specifically policies that incorporate climate change governance, climate litigation, and other strategies to address complex, wicked problems.

IV.7.2 THE NATURE OF THE PROBLEM – SCIENCE AND SOCIETY

As we consider the overwhelming scientific evidence of climate change and the scope and scale of solutions, it is critical to reflect on the nature of the problem. If the problem were driven by natural forces, policy would focus on human and biological adaptation. The study of climate change was rooted in addressing and answering the question of why the earth goes in and out of ice ages. Those studies found that atmospheric levels of carbon dioxide have been rising steadily since the beginning of the Industrial Revolution when fossil fuels first came into widespread use. In 1800, the carbon dioxide level of the atmosphere was about 290 parts per million (ppm). By 1958, when Charles Keeling first began collecting samples of the air atop Mount Mauna Loa in Hawaii and in Antarctica, the level had risen to 315 ppm. In 2007, Keeling's five-year average showed a level of 382 ppm (Weart, 2003).

The most significant cause of rising carbon dioxide levels is the burning of fossil fuels, which puts about six billion metric tons of carbon dioxide into the atmosphere each year. Natural processes

DOI: 10.1201/9781003406600-26

that tend to regulate atmospheric carbon dioxide levels – plant photosynthesis, absorption by soil and water – remove only about half that amount. These natural processes are generally busy recycling "natural" sources of carbon dioxide, such as the air we exhale. Research shows that higher levels of atmospheric carbon dioxide are associated with warmer periods. One explanation for this association is that rising carbon dioxide levels raise temperature slightly, which allows the atmosphere to hold more water vapor. Water vapor is a more powerful greenhouse agent than carbon dioxide, so its presence exacerbates the temperature problem.

In the 1980s, climate scientists reached a consensus that the atmosphere had warmed significantly in the past century and that industrial processes are largely responsible. Using climate models, scientists projected future carbon dioxide levels and temperatures, evaluating the accuracy of their premise regarding human-generated carbon dioxide. They expected those models to tell them whether they were on the right track around the year 2000. Their expectations were met.

Now the consensus is that greenhouse gases (GHGs) are the most significant cause of the current warming. In August 2021, the Intergovernmental Panel on Climate Change (IPCC), a global group of preeminent climate scientists, released its Working Group I report, *Climate Change 2021: The Physical Science Basis.* The most notable results confirmed with greater certainty that human influence has caused climate change and detailed the accelerating rate of warming and its effects. The report states unequivocally that human activities have warmed the atmosphere, ocean, and land. IPCC's best estimate is that humans are responsible for approximately 1.07°C of the observed 1.09°C warming between 1850–1900 and 2011–2020 (IPCC, 2021). Now, much of the uncertainty and discussion in the scientific community centers on variations within that premise – for example, how fast will the earth warm, what will change as it warms, will new feedbacks bring the temperature back down or make it rise more rapidly.

Once it was established that human activity was the significant contributor to a warming climate, the problem becomes, "How do we alter the human drivers of climate change?" Optimistically, solutions exist. Climate change can be mitigated by significantly reducing carbon dioxide and other GHGs produced by industrialization and the burning of fossil fuels for energy. Another option includes adding more natural sources of carbon absorption. Identifying anthropogenic factors as the driving force behind climate change led to policies addressing how to mitigate those drivers. Initial policy formulation focused on allocating the cost of carbon dioxide to the largest contributors. Incentive-based policy concepts, such as carbon pricing and carbon taxation, are based on the "polluter pays principle" where the societal cost is allocated to the responsible industrial sources, making it more costly for businesses to pollute. This can take the form of a carbon tax and carbon pricing or an emissions trading system (ETS), such as cap and trade. The goal is to encourage businesses to reduce their carbon footprint by internalizing the costs of emissions, leading to lower emissions and potentially faster progress towards climate goals. Some states are pursuing programs incorporating carbon pricing, specifically with cap and trade. However, Texas is not one of them.

In addition, during the second Trump administration, the federal government is signaling the potential for undoing the finding of endangerment for carbon dioxide issued by EPA in 2009 following the Supreme Court's landmark case addressing climate change, *Massachusetts v. EPA* (*Massachusetts v. EPA*, 549 U.S. 497 (2007)). On December 7, 2009, the EPA Administrator signed two distinct findings, an "endangerment finding" and a "cause or contribute finding" regarding GHGs under section 202(a) of the Clean Air Act (Final Rule, 2009). Under the endangerment finding, the administrator found that the current and projected concentrations of the six key well-mixed GHGs – carbon dioxide (CO_2), methane (CH4), nitrous oxide (N_2O), hydrofluorocarbons (HFCs), perfluorocarbons (PFCs), and sulfur hexafluoride (SF6) – in the atmosphere threaten the public health and welfare of current and future generations. In addition, under the "cause or contribute finding", the administrator found that the combined emissions of these well-mixed GHGs from new motor vehicles and new motor vehicle engines contribute to the GHG pollution that threatens public health and welfare. This was the first action taken by EPA to begin to address GHG reductions from new motor vehicles. It led to joint rule-making by EPA and DOT in 2012 to increase fuel-efficiency for new passenger cars, light-duty trucks, and medium-duty passenger vehicles and decrease carbon dioxide emissions. During the first Trump Administration, EPA successfully weakened Obama-era standards, reducing the required annual increase in fuel efficiency for vehicles from 5% to 1.5%. However, President Trump did not seek to undermine the finding of endangerment during his first administration. President Trump's Executive Order 14154, entitled Unleashing American Energy, signaled that the new administration has taken it under stronger consideration (Executive Order, 2025).

On August 1, 2025, U.S. Environmental Protection Agency (EPA) published a proposed rule "proposing to repeal all greenhouse gas (GHG) emission standards for light-duty, medium-duty, and heavy-duty vehicles and engines to effectuate the best reading of Clean Air Act (CAA) section 202(a)" (90 Fed. Reg. 36288 (2025)). The summary of the proposed rule states EPA's position on its lack of authority under the Clean Air Act in stating "we propose that CAA section 202(a) does not authorize the EPA to prescribe emission standards to address global climate change concerns and, on that basis, propose to rescind the Administrator's prior findings in 2009 that GHG emissions from new motor vehicles and engines contribute to air pollution which may endanger public health or welfare. We further propose, in the alternative, to rescind the Administrator's prior findings in 2009 because the EPA unreasonably analyzed the scientific record and because developments cast significant doubt on the reliability of the findings" (90 FR 36388, 2025). The comment period for the Proposed Rule closed on September 22, 2025. After consideration of the comments, EPA will promulgate a Final Rule. It is not anticipated that there would be significant change in the Final Rule from the Proposed Rule. However, in the Rulemaking process, EPA must review and consider all comments. For this Proposed Rule, comments are voluminous. After the Final Rule is promulgated, EPA is likely to face legal challenges to rescinding the endangerment finding. Scientific data on the impacts of GHGs on climate change has only increased since the original 2009 finding.

In 2023, planet-warming pollution in Earth's atmosphere hit the highest levels in human history, and global temperatures were on track to hit yet another record high. Public expectation of another record high year has become the norm. Concentrations of carbon dioxide, the most important driver of global warming, are now growing faster than at any time since our species evolved, according to the World Meteorological Organization's (WMO) annual Greenhouse Gas Bulletin (WMO, 2024). The increase can be traced back to stubbornly high rates of fossil fuel consumption, the report said, as well as ecosystems that are becoming net emitters, potentially less capable of absorbing excess carbon.

Levels of the potent GHGs methane and nitrous oxide also hit all-time highs in 2023 (WMO, 2023). The total heat-trapping potential of the atmosphere is now 51.5% higher than in 1990, when United Nations Intergovernmental Panel on Climate Change (IPCC) scientists first warned the world was on track for catastrophic climate change. The concentration of carbon dioxide in Earth's atmosphere last year exceeded 420 parts per million – a level not seen since the Pliocene Epoch more than three million years ago. At that time, global temperatures were 2 to 3 degrees Celsius (3.6 to 5.4 degrees Fahrenheit) warmer, sea levels were 30 to 60 feet higher, and *Homo sapiens* did not yet exist. Most of the recent growth comes from people burning coal, oil, and gas, but the WMO researchers also found evidence that human-driven warming has caused natural systems to release more GHGs and jeopardize the Earth's ability to absorb emissions (WMO, 2023).

The hike in carbon dioxide concentrations in 2024 also coincided with the largest-observed spike in carbon monoxide – a related gas that is produced when trees burn. During the 2023–2024 fire season, global carbon emissions from forest fires were 16% above the average of the 20 previous years, as Australia endured a historic drought, and Canada saw a record 37 million acres of forest go up in flames (Jones et al., 2024).

Surging levels of methane may also be traced to degraded ecosystems, data suggests. Chemical analysis of the gas, which traps 28 times more heat than carbon dioxide over a 100-year timeframe, suggests that it is increasingly coming from microbial activity, rather than fossil fuel burning (EPA, 2025). Although some of that increase can be attributed to bacteria living in landfills and the guts of cows, researchers worry it is also being produced by warming tropical wetlands and thawing Arctic permafrost.

IV.7.3 TEXAS VULNERABILITY

In Texas, the scope of the problem and projections of the impact of climate change is particularly significant. Climate change models predict that Texas will face increasing temperatures, more frequent extreme weather events, and shifting precipitation patterns (see Chapter I.2, "The Changing Climate of Texas"). These changes have direct consequences for public health (Chapter III.2, infrastructure (Chapters I.5 and III.4), agriculture (Chapter III.1), and economic stability (Chapter III.5). In an evaluation of Texas climate impacts, Banner et al. concluded that "under essentially all climate model projections, Texas is susceptible to significant climate change in the future. Most projections for the twenty-first century show that with increasing atmospheric greenhouse gas concentrations, there will be an increase in temperatures across Texas and a shift to a more arid average climate" (Banner et al., 2010). Economic analysis of the cost of inaction is critical to the financial commitment that is critical to addressing the problem. "This investment should support development of 1) research programs that provide policy-relevant science; 2) education programs to engage future researchers and policymakers; and 3) connections between policymakers, scientists, water resource managers, and other stakeholders" (Banner et al., 2010).

IV.7.4 PUBLIC SUPPORT FOR CLIMATE POLICY

Public understanding of the scope of the problem is key to action. Clearly, public awareness and understanding, co-dependent upon climate education and political rhetoric, is a critical component of voicing public concern about a problem to bring that issue to a state or federal governmental body. Public understanding of the problem of climate change is related to climate education and political rhetoric. A crucial factor driving action on a problem is public support. Yale Center for Climate Communication provides extensive data on public perception of the problem of climate change. The annual survey addresses beliefs, perceptions, and policy support, all key driving forces for policy mobilization. The Yale Center for Climate Communication provides extensive data on Texans' perception of the problem of climate change. According to Yale's 2024 data, 72% of Texans believe global warming is happening, and 58% believe it is mostly caused by human activity. Two-thirds of Texas voters support government action to address climate change. As many as 70% of Texans support regulating carbon dioxide as a pollutant. About half of Texans believe there is a link between climate change and severe weather events, 70% of Texans believe global warming will harm future generations (Yale Climate Opinion Maps, 2024).

The Pew Research Center also studies the public's view of the problem of climate change. A Pew Research Center survey of 9,593 U.S. adults conducted in October 2024, takes stock of Americans' views on climate change, including its impact on their own lives, their support for policies to address the issue and the emotional reactions they have to climate news. Among Republicans and those who lean to the Republican Party, 56% think climate policies usually hurt the economy, whereas 52% of Democrats and those who lean Democratic say they usually help.

Public opinion and political views on the issue factor significantly into climate policy initiatives. Overall, 69% of Americans say large business and corporations are doing too little to help reduce the effects of global climate change. 60% also say state elected officials are doing too little on climate (Pew Research Center, 2024). Overall, **a** majority of registered voters (55%) said federal agencies should increase their efforts to protect people from health problems related to global warming**,** outnumbering those who said federal agencies should decrease these efforts (15%) by more than 3 to 1. An additional 20% of registered voters said agencies should do about the same amount as they are doing now (Fine et al., 2025).

Congressional activity has focused on incentives such as tax credits and rebates and voluntary programs like Energy Star and LEED building certification to maintain consumer and corporate autonomy based on public input. A significant majority of Americans favor tax credits for those who improve their home's energy efficiency (83%) and tax credits for businesses that develop carbon capture technologies (79%). Notably, majorities of both Republicans and Democrats support these policy actions (Kennedy and Tyson, 2024).

Americans also recognize and disdain the political divide on this topic. Eight-in-ten Americans say climate news makes them feel frustrated about the level of political disagreement on the issue. A large share (73%) also say climate news has made them feel sad about what is happening to the Earth. At the same time, 51% of U.S. adults say they have felt suspicious of the groups pushing for action on climate change (a view expressed by 75% of Republicans) (Kennedy and Tyson, 2024). 64% say climate change currently affects their local community either a great deal or some (Kennedy and Tyson, 2024). Among Republicans, 41% say climate change is currently impacting their own community. For Democrats, a large majority (86%) holds this view (Kennedy and Tyson, 2024). These wide partisan differences in perceptions exist across geographic regions. Relatively few expect to make major sacrifices in their lifetime due to climate change. Three-quarters of U.S. adults expect they will have to make at least minor sacrifices in their everyday life during their lifetime because of global climate change. However, the share who expect to make major sacrifices remains modest, at 23% (Kennedy and Tyson, 2024).

IV.7.5 MITIGATION POLICIES

Climate policy has focused on two important approaches to address the problem: Mitigation and Adaptation. Mitigation focuses on first acknowledging the connection between carbon dioxide and other GHG emissions as the primary driver behind the warming climate system. Then, mitigation involves systemic changes that target sectors and sources of carbon dioxide and other GHGs, particularly working to reduce those initial emissions and capture or sequester additional emissions. Amongst the wealth of bad news on climate change, mitigation is generally perceived in a positive light by the public, at least from the standpoint of research on climate change. If climate change were an outcome of purely exogenous phenomenon, such as shifts in solar output, with no connection to anthropogenic drivers, then our collective focus would be on adaptation rather than addressing the cause. Mitigation, however, holds promise because we now understand human activities are at the center of the problem. Mitigation provides an opportunity to alter the course of climate change. The challenge, however, is not necessarily technological. It is a social problem. Changing behaviors in society takes focused discipline and public support.

Since 1970, environmental policy has been significantly administrative and regulatory. It is also a demonstration of federalism, where both the federal and state governments concurrently share jurisdiction, with federal supremacy to arbiter any conflicts when states are in conflict with the federal government. The Environmental Protection Agency (EPA), in partnership with state agencies, such as Texas Commission on Environmental Quality (TCEQ), play a significant role in formulating and implementing climate action and addressing air pollution and emissions that significantly contribute to climate change.

States also have significant policy authority in the state legislature. Some states have comprehensive climate action plans. However, during the Abbott administration, there has been significant inaction on the issue of climate change. A recent proposed House Bill on climate action did not pass in the Texas Legislature. In 2021, Texas State Representative James Talarico introduced Texas House Bill 87(R) HB 2206 entitled “The Texas Climate Action Act” (Texas House Bill, 2021). The Preamble states that: “The aim of the Texas Climate Action Act is to combat climate change by establishing greenhouse gas emissions reductions targets, energy efficiency standards, and a resiliency plan with a framework for local involvement” (Texas House Bill, 2021).

Another Texas bill seen as anti-climate action, Senate Bill 1860, targeted a proposed “climate charter” in El Paso (Tex. S.B. 1860). Proposition K, which El Paso residents voted down in 2025, would have amended the city’s charter to create aggressive renewable energy goals and make controlling carbon emissions a cornerstone of major city decisions. Senate Bill 1860 was signed by Governor Abbott and became effective on September 1, 2023. It requires cities to get permission from the Texas Legislature before approving changes to their charters that purport to address climate change. Texas climate policy had become more local, but the Texas Legislature aims to keep Texas cities from acting without their approval. Austin, Houston, Dallas, and San Antonio have initiated Climate Action Plans aiming for carbon neutrality by 2050. These plans encompass strategies for renewable energy adoption, energy efficiency, sustainable transportation, and green space preservation, contributing to broader biodiversity and environmental health goals.

Texas is home to eight of the top 50 facilities with the most GHG emissions in the country, according to a new analysis from Frontier Group (Ridlington et al., 2023). A new interactive map from Environment Texas Research and Policy Center shows every facility in Texas which reports GHG pollution to the EPA (2022 data – the most recent). Texas emits more GHGs than any other state, emitting more than twice as much as runner up, California. Texas emitted 873 million metric tons of GHGs in 2022, California emitted 393 million metric tons (Ridlington et al., 2023). Most of Texas's GHG emissions come from the electric power, industrial, and transportation sectors.

Climate inaction in Texas is largely due to the critical importance of the energy sector to the state's economy, but also the nature of regulatory controls, particularly that of oil and gas, in the state. Climate mitigation policies target the energy sector the most, therefore, having the most consequences (US EPA, 2022). There is much at stake in Texas because the energy industry has long been an important contributor to the state's economy. According to the 2025 U.S. Energy and Employment Report, Texas had 990,052 energy workers statewide in 2024, representing 11.7% of all U.S. energy jobs. Of these energy jobs, 310,185 in Fuels; 71,818 are in Electric Power Generation (EPG); 213,076 in Transmission, Distribution, and Storage (TDS); 182,506 in Energy Efficiency; and 212,467 in Motor Vehicles and Component Parts (MV & CP). Energy in Texas represents 7.1% of total state employment (DOE, 2025).

In addition to its significant economic contribution, Texas' energy industry is unparalleled. Texas is the undisputed national leader in energy production and energy consumption, producing twice as much energy as Florida, only second to Texas. Texas leads the nation in energy consumption across all sectors. Texas produces the most oil and gas in the U.S., averaging 42% of crude oil and 27% of marketed natural gas According to the U.S. Energy Information Administration (EIA), Texas is the top oil and gas state in the U.S., producing 42% of crude oil and 27% of marketed natural gas (US Energy Information Administration, 2025). Texas has 32 petroleum refineries (more than any other state), which can process more than 5.9 million barrels of crude oil per day (32% of U.S. refining capacity). Texas also leads the nation in energy consumption across all sectors and is the largest energy-consuming state in the nation. Texas produces more electricity than any other state, generating more than 12% of the nation's total and produces about 26% of all U.S. wind-powered electricity generation, the most of any state (US Energy Information Administration, 2025).

States have government agencies and commissions that are responsible for regulating the oil and gas industries in their states, which can often have a wider impact. States have authority over pipeline transportation that occurs within one state. Texas regulatory oversight structure is unique among the states. The Railroad Commission of Texas (RRC) is the state agency with primary regulatory jurisdiction over the oil and natural gas industry, pipeline transporters, natural gas and hazardous liquid pipeline industry, natural gas utilities, the LP-gas industry, critical natural gas infrastructure, and coal and uranium surface mining operations. RRC exists under the provisions of the Texas Constitution and exercises its statutory responsibilities under state and federal laws for regulation and enforcement of the state's energy industries. RRC was established in 1891 under a constitutional and legislative mandate to prevent discrimination in railroad charges and establish reasonable tariffs. RRC has distinct regulatory oversight of fossil fuel energy in the state. It is the oldest regulatory agency in Texas and one of the oldest of its kind in the nation. RRC no longer has any jurisdiction or authority over railroads in Texas, a duty which was transferred to other agencies, with the last of the rail functions transferred to the Texas Department of Transportation in 2005.

This sole control vested in one agency, while generating significant economic benefit to Texas' natural gas industry, could be challenged to provide significant environmental oversight of fossil fuel operations in Texas. RRC also has regulatory and enforcement responsibilities under federal environmental law including the Surface Coal Mining Control and Reclamation Act, Safe Drinking Water Act, Pipeline Safety Acts, Resource Conservation Recovery Act, and Clean Water Act. Therefore, RRC's actions and oversight are crucial for addressing climate change mitigation in Texas. Recently, RRC became involved in challenging federal climate action.

During the Biden administration, EPA issued a 2023 methane rule which required oil and gas producers to stop methane leaks from their equipment and end the practice of routine flaring of excess methane (EPA, 2023). EPA describes methane as a climate "super pollutant" that is many times more potent than carbon dioxide and is responsible for approximately one third of the warming from GHGs occurring today (EPA, 2024). This rule applied to both newly built equipment as well as existing facilities and would significantly reduce methane emissions from the oil and gas industry with big climate benefits (US EPA, 2024). EPA touted that the "final rule leverages the latest cost-effective, innovative technologies and proven solutions to prevent an estimated 58 million tons of methane emissions from 2024 to 2038, the equivalent of 1.5 billion metric tons of carbon dioxide – nearly as much as all the carbon dioxide emitted by the power sector in 2021" (EPA, 2024).

After the new methane rule was promulgated, it was up to each state to develop a State Implementation Plan to enforce the rule within their borders where permitting and monitoring happen. However, on January 30, 2024, the three-member Texas Railroad Commission voted to formally ask Texas Attorney General Ken Paxton to sue the EPA to stop implementation of the rule, a move that was confounding given support of the rule by oil and gas producers as well as environmental advocates.

Industry representatives supported the new methane rule, pointing out that industries need a predictable and consistent regulatory environment in which to operate, which the new methane rule would provide. BP America, a major natural gas producer in the United States, with a significant presence in both traditional onshore shale and renewable natural gas (RNG) issued a statement in favor of the new methane rule. Orlando Alvarez, Chair and President of BP America said "BP welcomes the finalization of a federal methane rule for new, modified and - for the first time - existing sources. A well-designed rule will help drive material methane emission reductions this decade and beyond. We appreciate the collaborative way EPA, NGOs, and industry worked together on this rulemaking. We congratulate the Administration on this important milestone and look forward to working together on the next phases of implementation" (EPA, 2024). The American Petroleum Institute also stated its support for the methane rule, with its chief executive saying that "this is a new position for API, but we think given where the industry is at this time and the continued importance of reducing methane, it was critical we update this position as the administration changes" (API, 2023).

The methane rule required emissions reporting and also included a more controversial 'methane tax' provision, the waste emissions charge (WEC), which imposed a fee on large sources of methane emissions, designed to incentivize reduced emissions from facilities. The fee started at $900 per metric ton of methane in 2024, increasing to $1,200 in 2025 and was set to increase to $1,500 in 2026 and beyond. As mentioned, the State of Texas opposed the new rule through the TRC. In January 2025, Texas Attorney General Ken Paxton, joined by a 22-state coalition, filed a lawsuit EPA to block the implementation of the methane emissions charge. In March of 2025, the methane rule was disapproved by a joint resolution of Congress pursuant to the Congressional Review Act. President Trump signed the Joint Resolution of Disapproval relating to the Waste Emissions Charge for Petroleum and Natural Gas Systems on March 14, 2025. EPA's website indicates it is "currently evaluating options and obligations for implementing Clean Air Act section 136(c-g) relating to methane and will provide additional information to the regulated community at an appropriate time" (EPA, 2025).

During the Biden Administration, there was also more significant climate action at the federal level that flowed to the state level due to the federal government's approach to incentivize action and coordination among the states on the issue of climate change by funding state initiatives to measure and reduce GHG emissions. Texas participated in this climate action. TCEQ received an EPA grant for statewide planning to reduce climate pollution as part of the Climate Pollution Reduction Grants (CPRG) program. Under this grant, TCEQ is developing two plans – a priority action plan and a comprehensive action plan. These plans will include an inventory of GHG emissions for Texas and ways to reduce air pollution and greenhouse gases together (TCEQ, 2025).

The CPRG program is an EPA program created as part of the Inflation Reduction Act. The CPRG program consists of two phases. As part of Phase I, Texas has received grant funding to develop a plan to reduce greenhouse gases within the state. Six Texas metropolitan areas have also received Phase I grants. Texas will be able to apply for competitive implementation grants in Phase II. Texas' plan focuses on incentivized, voluntary measures with co-pollutant reductions from the three largest GHG emitting sectors in Texas: industry, transportation, and electric power (TCEQ, 2025).

Texas Commission on Environmental Quality's "Climate Pollution Priority Action Plan for the State of Texas" addresses the highest emitting sectors of the total Greenhouse Gas (GHG) Emissions by economic sector. In 2021, three sectors accounted for 87% of the total emissions. "Fossil fuel combustion is the largest individual source of GHG emissions in Texas and accounts for about 67% of GHG emissions no matter the economic sector" (TCEQ, 2024). Fossil fuel combustion accounts for 99% of GHG emissions in the electricity sector, 97% in the transportation sector, and 46% in the industrial sector (TCEQ, 2024). Within the electricity sector 98% of GHG from fossil fuel combustion is from coal and natural gas. For the transportation sector, 91% of the fossil fuel production is from petroleum combustion. Measures for this sector focus on zero emission light-, medium-, and heavy-duty vehicles, infrastructure, and decarbonization strategies for ports, airports, and railways (TCEQ, 2024). Estimates show that implementation of these measures could reduce GHG emissions in Texas by 174 million metric tons (MMT) from 2025 through 2030 and 592 MMT from 2025 through 2050. Co-pollutants could also be reduced by 0.6 MMT from 2025 through 2030 and 3.0 MMT from 2025 through 2050.

TCEQ's Action Plan also addresses incentivized voluntary "priority measures." Most of these will focus on the three economic sectors that produce the most GHG emissions in the state: industry, transportation, and electric power.

The Climate Pollution Reduction Plan consists of three essential phases:

1. Priority Climate Action Plan (Due March 1, 2024)
    - Designed for communities committed to making a difference.
    - Focuses on immediate actions with flexibility.
    - Targets specific sectors for maximum impact.
    - Emphasizes a foundational approach that will build on past environmental planning efforts.

    Texas met this first deadline in March 2024. The Texas Priority Climate Action Plan (PAP), developed by the Texas Commission on Environmental Quality (TCEQ) as part of the U.S. Environmental Protection Agency's (EPA) Climate Pollution Reduction Grants (CPRG), focuses on incentivized, voluntary measures to reduce GHG emissions and co-pollutants, particularly in the industry, transportation, and electric power sectors. These three sectors account for nearly 90% of Texas's GHG emissions. The plan aims to reduce GHG emissions by 174 million metric tons from 2025 to 2030 and 592 million metric tons from 2025 to 2050 (TCEQ, 2024).
2. Comprehensive Climate Action Plan (Due in Summer 2025)
    - Expands on the groundwork of the Priority Climate Action Plan.
    - Aiming for substantial, long-term impact.
    - Encompasses a wide array of measures to reduce GHG emissions across all sectors.
    - Requires comprehensive analyses for robust implementation.

3. Status Report (Fall 2027)
   - Initiates a continual improvement process.
   - Requires a thorough review and analyses of elements from the earlier plans.
   - Documents progress and next steps for key metrics.
   - Provides confirmation that pollution reduction strategies are on target.

The second and third essential phases of the CPRG may not be met. Texas' Comprehensive Climate Action Plan was due by July 2025. At the time of publication, Texas has not published its Comprehensive Climate Action Plan. With the transition to the new Trump administration underway at EPA, it is anticipated that the last two phases of CPRG may not be concluded. A recent news report indicates the Trump administration's shift away from climate action: "On March 12, 2025, the Environmental Protection Agency terminated grant agreements worth $20 billion issued by the Biden administration under a so-called green bank to finance clean energy and climate-friendly projects. (Associated Press, 2025) The action comes weeks after the EPA froze the grants, which EPA Administrator Lee Zeldin has characterized as a 'gold bar' scheme marred by conflicts of interest and potential fraud" (Associated Press, 2025). However, as of June 10, 2025, the CPRG plan remains intact on EPA's website. At the time of publication, details on the CPRG plan and Texas' implementation are present on TCEQ's website (TCEQ, 2025).

IV.7.6 ADAPTATION POLICIES

Adaptation recognizes the effects of climate change and addresses society's attempts to adapt to the changes by planning and making changes, such as agricultural changes in planting zones and harvesting cycles. Adaptation is the component of policy that is not dependent on acceptance of anthropogenic drivers to climate change, but as a necessary response to the impacts of climate change.

Texas has not focused significantly on adaptation measures to address climate change, such as tree cover and green infrastructure. In fact, Texas has not developed a statewide adaptation plan, or a statewide climate action plan that would necessarily include adaptation. However, in June 2019 the State of Texas established the Texas Infrastructure Resiliency Fund (TIRF), a new special fund in the state treasury for the purpose of financing flood mitigation and protection projects and related planning efforts. TIRF is administered by the Texas Water Development Board (TWDB) and includes four separate accounts: a federal matching account, a floodplain management account, a flood implementation account, and a Hurricane Harvey account.

In 2019, the State of Texas established the Texas Flood Infrastructure Fund (TFIF), which provides financial support to communities for drainage, flood mitigation, and flood control projects. Administered by the Texas Water Development Board (TWDB), the TFIF fund provides financial assistance to political subdivisions (cities, counties, or state-established districts or authorities) in the form of grants and zero-interest loans. The funding can be used to support planning, design, construction, and rehabilitation of flood projects, whether structural or non-structural (including nature-based) (Houston-Galveston Area Council, 2024).

The Texas Climate-Smart Initiative (TCSI) is part of the Partnerships for Climate-Smart Commodities, a $3.1 billion investment across the U.S. administered by the U.S. Department of Agriculture to support the production and marketing of climate-smart commodities. The Texas State Soil and Water Conservation Board (TSSWCB) has partnered with Texas A&M AgriLife to work directly with producers to implement new climate-smart practices. TSSWCB will provide producers with planning and implementation support for climate-smart practices along with financial incentives for successfully executing the practices on their land. A cross-functional team of researchers at Texas A&M, Prairie View A&M, Tarleton State University, and University of Texas Rio Grande Valley will be working with a select group of producers to do extensive research on soil health, carbon sequestration, and new market opportunities for climate-smart products. Over a five-year period, TCSI will put $38.8 million in the pockets of producers while helping them improve the long-term sustainability, productivity, and profitability of their operations.

IV.7.7 TEXAS CLIMATE LITIGATION

In May 2011, an environmental group, Texas Environmental Law Center, filed a rule-making petition requesting that the TCEQ adopt rules to reduce statewide GHG emissions from fossil fuels pursuant to the public trust doctrine. This was part of a larger movement to utilize the common law theory of resources held in the public trust, the public trust doctrine, and extend this theory to atmospheric trust and the rights of younger populations and future generations. Texas Environmental Law Center's petition was part of a nationwide campaign by Our Children's Trust and iMatter, public interest groups that seek to combat climate change on behalf of future generations. On June 27, 2011, The Texas Commission on Environmental Quality denied the petition, stating that Texas was currently in litigation with EPA concerning the regulation of GHGs, and that the use of the public trust doctrine in the state had been limited to waters and did not extend to GHGs.

Atmospheric trust litigation, a theory pioneered by Mary Christina Wood at the University of Oregon, calls upon the judicial branches of government worldwide to force carbon reduction on the basis of their fiduciary responsibility to protect the public trust (Wood & Woodward, 2016). The theory itself branches from the public trust doctrine, a doctrine that holds natural resources are held in a "trust" for the citizens of the world. Within this public trust, the atmosphere is an asset, and as such, no sovereign nation is permitted to degrade it. Any carbon pollution endorsed on behalf of these governments thus breaks their fiduciary duty to the trust and can result in litigation. While the public trust doctrine is a sound legal theory for its near global applicability, its stability in the United States federal government is a bit less clear. In 2012, the Supreme Court reiterated "the public trust doctrine remains a matter of state law" (*PPL Montana, LLC v. Montana* (2012)).

Texas state courts have addressed atmospheric public trust. In July 2011, Texas Environmental Law Center filed a lawsuit in Texas state court on behalf of a young adult and minor children challenging the denial of their rulemaking petition seeking state action to reduce GHG emissions (*Bonser-Lain v. TCEQ*, 2012). In July 2012, the judge hearing the case issued a letter order indicating that the court would hold that the Commission's conclusion that the public trust doctrine is limited to waters was legally invalid, and that the doctrine includes all natural resources of the state (*Bonser-Lain v. TCEQ,* 2012). The judge also held that the Commission's conclusion that it was prohibited from regulating air quality pursuant to section 109 of the Clean Air Act was also legally erroneous because the Clean Air Act was a floor, not a ceiling (Triana, 2012). The Court said, however, that it would find that because the "legal landscape is uncertain" due to ongoing state and federal cases concerning federal and state authority to regulate GHGs, the court would find that the Commission's refusal to exercise its regulatory authority was a reasonable exercise of discretion (Triana, 2012).

On appeal, the Texas Court of Appeals ruled on June 14, 2014, that the district court erred in concluding that it had subject matter jurisdiction. The appellate court concluded that neither the Texas Administrative Procedure Act nor the Texas Water Code waived sovereign immunity for judicial review of denials of rule-making petitions (*TCEQ v. Bonser-Lain*, 2014).

IV.7.8 CONCLUSION

The future of climate policy action in Texas, and throughout the United States, focuses on climate resiliency. Climate resiliency is the ability of ecosystems, societies, and businesses to anticipate, prepare for, and respond to the impacts of climate change. It is about building the capacity to withstand, recover from, and even thrive in the face of climate-related risks. It requires a multi-faceted approach, involving government, businesses, communities, and individuals working together to build a more sustainable and adaptable future. Much of this work will require significant capital investment and innovative policy initiatives.

As Texas' population is expected to continue to grow rapidly, climate resilience will be key. By 2050, the population is projected to reach 40 million, with most people clustered in urban centers including Dallas–Fort Worth, Houston, San Antonio, and Austin (Texas Demographic Center, 2022). Population growth will contribute significantly to climate change and stress natural resources.

Effective action on climate resiliency is necessary at the local level to address specific local impacts of climate change. Local governments and interest groups can provide site-specific strategies which are critical for effective resilience to climate impacts. State funding for these efforts can come from the aforementioned TFIF fund which provides financial assistance to political subdivisions within Texas.

Across Texas, there are unique opportunities for resiliency development in green infrastructure with the capacity to address the significant issues of stormwater runoff and carbon absorption. Currently, there is significant local action on climate resiliency across Texas. For example, Austin's initiatives focus on community preparedness, resilient infrastructure, and sustainable solutions. Coastal communities are also preparing along the Texas Gulf Coast for extreme storm events and flooding. The Texas Living Waters Project is a coastal resiliency model for other coastal states. It is a collaboration of conservation groups working together to ensure that Texas has the water it needs for thriving communities and healthy fish and wildlife. The Texas Living Waters Project is working to equip local leaders and decision-makers in the Texas Mid-Coast with locally relevant information about the area's future related to climate change impacts, enabling them to make greater investments in natural assets to shore up coastal environments for wildlife, local economies, and protection of their communities (Pathak and Fuller, 2021). Municipal and regional planning incorporating climate resiliency is imperative for the climate challenges facing Texas now and in the future.

REFERENCES

American Petroleum Institute (API, 2023) Press Release, API Statement on EPA Methane Rule, https://www.api.org/news-policy-and-issues/news/2023/12/02/api-statement-on-epa-methane-rule

Associated Press, March 12, 2025. EPA terminates Biden-era climate grants worth $20 billion. https://apnews.com/article/epa-green-bank-zeldin-climate-trump-5168d11b7f63aeaf72001b50221c3c19.

Banner JL, Jackson CS, Yang ZL et al., 2010. Climate change impacts on Texas water: A white paper assessment of the past, present and future and recommendations for action. Texas Water Journal. 1(1):1–19.

Bonser-Lain v. Texas Commission on Environmental Quality, 2012. (201st Dist. Ct., Travis County, Texas.

Department of Energy, 2025 United States Energy & Employment Report, (DOE 2025), https://www.energy.gov/media/348936.

EPA, 2023, EPA Press Release, Biden-Harris Administration Finalizes Standards to Slash Methane Pollution, Combat Climate Change, Protect Health, and Bolster American Innovation, https://www.epa.gov/newsreleases/biden-harris-administration-finalizes-standards-slash-methane-pollution-combat-climate.

EPA, 2024, EPA Press Release, Biden-Harris Administration Announces Final Rule to Cut Methane Emissions, Strengthen and Update Greenhouse Gas Emissions Reporting for the Oil and Gas Sector, https://www.epa.gov/newsreleases/biden-harris-administration-announces-final-rule-cut-methane-emissions-strengthen-and

EPA, 2025. Inventory of U.S. greenhouse gas emissions and sinks. https://www.epa.gov/ghgemissions/overview-greenhouse-gases#methane.

Exec. Order No. 14154, 90 Fed. Reg. 8353 (Jan. 20, 2025).

Final Rule, 2009. 40 CFR Chapter I endangerment and cause or contribute findings for greenhouse gases under section 202(a) of the clean air act.

Fine J, Yue Y, Ettinger J, et al., 2025. A majority of registered voters want federal agencies to increase their efforts to protect people from the health harms of global warming. Yale University and George Mason University. New Haven, CT: Yale Program on Climate Change Communication.

Houston-Galveston Area Council, 2024. Climate Pollution Reduction Grant (CPRG) Program. https://www.h-gac.com/climate-pollution-reduction-grant-program.

IPCC, 2021. Climate change 2021: The physical science basis, the working group I contribution to the sixth assessment report. Cambridge University Press, Cambridge, UK.

Jones MW, Kelley DI, Burton CA et al., 2024. State of wildfires 2023–2024. Earth Syst. Sci. Data. 16: 3601–3685. https://doi.org/10.5194/essd-16-3601-2024

Kennedy B, Tyson BA, 2024. How Americans view climate change and policies to address the issue: Mixed views of how climate policies impact the U.S. economy, Pew Research Center. https://www.pewresearch.org/science/2024/12/09/how-americans-view-climate-change-and-policies-to-address-the-issue/html

Kraft ME, Rabe BG, Vig NJ, 2025. Environmental Policy: New Directions for the Twenty-First Century, 12th ed. CQ Press.

Massachusetts v. EPA, 2007. 549 U.S. 497.

Pathak A, Fuller A, 2021. Vulnerability and Adaptation to Climate Change: An Assessment for the Texas Mid-Coast. Austin, TX: National Wildlife Federation.

Pew Research Center. (2024, December 9). How Americans View Climate Change and Policies to Address the Issue. https://www.pewresearch.org/science/2024/12/09/how-americans-view-climate-change-and-policies-to-address-the-issue/

PPL Montana, LLC v. Montana, 2012. 565 U.S. 576.

Reconsideration of 2009. Endangerment Finding and Greenhouse Gas Vehicle Standards, 90 FR 39345 (August 15, 2025).

Ridlington E, Frank L, Scarr A, 2023. Who are the top climate polluters in the country? https://frontiergroup.org/resources/who-are-the-top-climate-polluters-in-the-country/

S.B. 1860, 88th Leg., Reg. Sess. (Tex. 2023). A bill to be entitled an act relating to the adoption of a climate policy in a municipal charter.

TCEQ, March 1, 2024. Climate Pollution Reduction Grants Priority Action Plan for the State of Texas. https://www.tceq.texas.gov/agency/climate-pollution-reduction-grants

Texas Commission on Environmental Quality (TCEQ), 2025. Climate pollution reduction grants program. https://www.tceq.texas.gov/agency/climate-pollution-reduction-grants

Texas Commission on Environmental Quality v. Bonser Lain, 2014. 438 S.W.3d 887.

Texas Demographic Center, 2022. Projections of the Total Population of Texas and Counties in Texas, 2020-2060 (0.5 Migration Scenario). Retrieved February 21, 2024, from https://demographics.texas.gov/Resources/TPEPP/Projections/2022/2022statetotsex_mig100.csv

Texas House Bill, 2021. 87(R) HB 2206 entitled "The Texas Climate Action Act". (Talarico, H.B. No. 2206, introduced).

Triana GD, 2012. Judge 200th District Court, Letter to Counsel, *Bonser-Lain* v. Texas Commission on Environmental Quality, (201st Dist. Ct., Travis County, Texas).

US Energy Information Administration, 2025. "Texas State Profile and Energy Estimates." https://www.eia.gov/state/?sid=TX

US Environmental Protection Agency (US EPA), 2022. Inventory of U.S. Greenhouse Gas Emissions and Sinks.

US EPA, 2024. 40 CFR Part 60, Subpart OOOOb Standards of Performance for Crude Oil and Natural Gas Facilities for Which Construction, Modification or Reconstruction Commenced After December 6, 2022.

Weart SR, 2003. The Discovery of Global Warming. Cambridge, MA: Harvard University Press.

WMO Greenhouse Gas Bulletin, 2024.

Wood MC, Woodward CW IV, 2016. Atmospheric trust litigation, and the constitutional right to a healthy climate system: Judicial recognition at last, 6 wash. Journal of Environmental Law and Policy. 633.

Yale Program on Climate Change Communication, 2024. Yale Climate Opinion Maps. https://climatecommunication.yale.edu/visualizations-data/ycom-us/

Section V

The Future of Texas in a Changing Climate

V.1 The Future of a Changing Climate in Texas

Mona Wells

In one way or another, most chapters in this book reference evidence indicating that Texas is already experiencing irreversible climate shifts. As noted in the chapter on the changing climate of Texas, the single indisputable direct effect of climate change on Texas is that the state's average temperature has risen and will continue to rise, with urban areas experiencing even greater increases due to heat island effect. As noted in the preface, 2023 was the hottest summer on record for all of the major cities in Texas. With the increasing heat on average (globally), ice reservoirs are melting and sea level is rising such that the Texas Gulf Coast is retreating, with concomitant increased tidal flooding and storm surges. Another consequence of increased heat is that there is more energy in the ocean. Accordingly, this brings increasing severe weather. Hurricane Harvey (occurring in 2017) was the most significant rainfall event in U.S. history, and Winter Storm Uri (occurring in 2021) was devastating to the Texas power grid. Extremely worrying is the prospect that increasing heat and disruption of weather patterns has had and is projected to continue to have on record-breaking droughts in Texas, as in 2011, being the driest year in state history. These physical changes play their part in affecting land cover, which, combined with changes to water availability, are altering the natural landscapes and biodiversity of Texas in a manner not only destructive to natural and cultural heritage, but also deleterious to human well-being. The marine ecosystem is also affected in myriad deleterious ways, due to the sensitivity of this system to temperature, coupled with the additional substantial stressor of ocean acidification caused by ever increasing levels of atmospheric carbon dioxide.

Associated with the direct and indirect effects of climate change are impacts of which this book, as a single volume, is perforce limited to a representative subset. As illustrated by the chapters herein, however, impacts of climate change will increasingly affect every aspect of daily life: increasing extreme events such as wildfires, drought, and dust, flooding, disruptions in transport, disruptions in the power grid, food insecurity, power insecurity, decreasing air quality, decreasing ecosystems and human health, extinctions, crop failures, economic damage, and increasing difficulties in dealing with all impacts due to lack of sufficient policy and political agility. Exacerbating the climate change impacts from any single stressor is the ubiquitous circumstance that all impacts are inextricably intertwined. As an example, with less water available due to drought, evapotranspiration decreases, making heatwaves even more intense. Less moisture in the soil means less evaporation, which normally cools the air, hence reduced transpiration exacerbates high temperatures. Attendant vegetation loss reduces shade and cooling effects, making heatwaves more intense, and dry ground is hot(ter) ground. Vegetation loss also entails that wind and waves are able to erode coastlines faster, while the reduced river flows result in less sediment transport to coastal areas, also rendering coastal areas more vulnerable to wave erosion. Depressed freshwater levels allow saltwater to intrude inland, weakening coastal soils, and augmenting the process whereby, in conjunction with other effects from climate change, coastlines sink and erode faster.

This example of interconnected and intensifying impacts resulting from drought illustrates why the result of climate change intensification has been likened to a chain reaction beyond control, whereby each impact exacerbates others in a chaotic and never-ending series of feedback loops, moreover, with climate change representing only one part of the system. As discussed in the chapter on planetary boundaries, it is a misdirection to consider climate change as "the" existential threat or global catastrophic risk in need to solving, at the expense of other Earth systems, and, in fact, while

DOI: 10.1201/9781003406600-28

results presented in that chapter show that climate change is likely to be the greatest environmental challenge in Texas, globally, it takes a back seat to others, most notably, the effect that humans have had in launching the sixth great extinction event in just under 450 million years of the Earth's history.

Earth's system, incorporating an interplay between the nine subsystems comprising the Earth's biophysical Planetary Boundaries, functions productively to support life on the planet because of being in balance. This balance is maintained in a manner that can be likened to a chemical buffer, whereby if the natural state of balance or buffer is perturbed, equilibrium rapidly reestablishes. Such systems, however, are not infinitely resilient, and over time, may be disrupted past the point of return. Examples from everyday life that might be relatable include the strategy and dexterity game Jenga, where 54 wooden blocks are stacked three to a layer in alternating directions to form a tower, and players take turns removing a block from the tower without causing it to collapse. In a typical game, more than half the blocks can be removed without destroying the tower, but, inevitably, collapse, when it comes, is instantaneous. Another analogy would be the health effects of high blood pressure. It is not medically unusual that a person might have high blood pressure for years with no major issues, only to die suddenly of a stroke or heart attack. These phenomena, the game of Jenga, the sudden medical effects from years of high blood pressure, Earth's biophysical system balance, are all examples of action-reaction systems that can be referred to as mathematically stiff. While there are regions of system change/input whereby a discrete input (pulling a Jenga block, eating a single high salt meal, killing off 200,000 species or so) results in only modest change, there are other system states whereby a single input results in irrevocable systems collapse, and this is the danger in our studied and obstinate neglect of systems thinking with respect to the Earth's environment and our effect on it. Climate change, and other changes to Earth's subsystems as represented by planetary boundaries, represent a continuous erosion of system stability, leading to a gradual fracturing of predictability whereby once-stable systems become erratic, making planning difficult to impossible. As well, unlike most crises, there is a deep time perspective – effects do not span years, or even decades.

In considering the future of Texas in a changing climate, and recognizing that climate change is only a part of the overall Earth biophysical system, let us take a step back, and revisit one of the earliest, certainly the most ambitious, efforts at modeling the economic and social outcomes of how humans interact with our environment. The Limits to Growth (LtG) study, funded by the Club of Rome and published in 1972 by a group of prominent MIT systems dynamics researchers, was a groundbreaking study that used systems modeling to explore the consequences of continued economic and population growth on a finite planet (Meadows et al., 1972). The study constructed and utilized the World3 computer model, shown in Figure V.1.1, to simulate various future scenarios, looking at the effects of demographic trends (population), industrial output/resource extraction, food production, use of nonrenewable resources, and effects of pollution on various metrics related to human social and economic well-being. The study calibrated the model based on historical data, and then used the calibrated model to forecast results for multiple future scenarios under various implementations of human stewardship, only one of which, the "stabilized world," did not lead to economic and societal collapse. Collapse, as the study framed it, is not synonymous with human extinction and could instead be thought of as an undesirable/overly sudden rate of diminution in the human standard of living as a result of overexploitation of the global biophysical system (cumulative resource depletion, environmental degradation, and food shortages). The key to the stabilized world scenario's lack of collapse was to reduce population growth and resource use in time to protect the Earth system's ability to remain in balance.

The results of the LtG study, notwithstanding the very high profile of the investigators who led the study, were highly controversial at the time of publication – stridence and munificence might be modestly said to have triumphed over the factual basis of most objections. The brouhaha died down relatively rapidly, however, with the results and implications of the LtG study consigned to obscurity until the early 2000s, when various researchers started to revisit the model. In particular, much attention has been devoted to revisiting two scenarios from the original study, the business as usual (BAU) and Technological World scenarios. The BAU scenario was based on humanity's continuing to pursue economic growth and unsustainable resource production, as in the 1970s, and

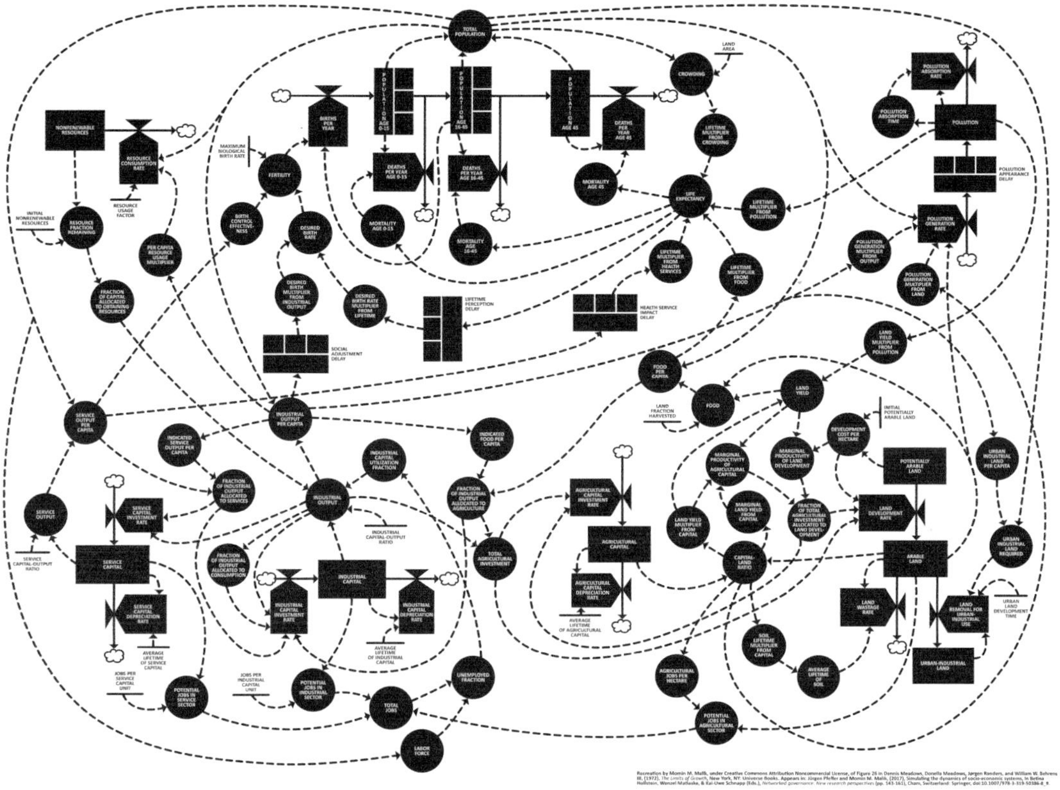

FIGURE V.1.1 Recreation of the schematic representation of the World3 model used in the LtG study. Physical quantities that are measured directly are represented by rectangles, rates that influence the quantities measured directly are indicated by a valve symbol, and auxiliary variables that influence rates are represented as circles. Time delays in feedback loops are shown as rectangles sectioned with white lines. Real flows of people, goods, money, and other quantities are shown by arrows, and causal relationships are shown by broken arrows. Clouds represent sources or sinks that do not affect overall model behavior. (Courtesy of Momin Malik under Creative Commons Attribution Noncommercial License (https://www.creativecommons.org/licenses/by-nc/4.0/deed.en), reproduced from Pfeffer and Malik (2017) as adapted from the LtG (Meadows et al., 1972).)

Technological World offered a nod to Jevon's paradox[1] in meretriciously assuming implementation of technological innovations to capture resource use efficiencies, i.e., offsetting resource depletion, when a large body of literature indicates this is very frequently not the case. As limited as the model was given computing resources at the time (metrics based on aggregate data, with data limitations as they existed at the time), over 20 years since the first revisitation of the original study in 2004, it is startling how well the BAU and Technological World scenarios have predicted (by now) over 50 years of human effects on the planet, as discussed to date by a growing list of highly prominent scientists (Bardi, 2011; Hall and Day, 2009; Herrington, 2021; Meadows et al., 1972; Meadows et al., 2004; Pereira et al., 2012; Randers, 2012; Rockström et al., 2009; Turner, 2008; Turner, 2012).

Central questions raised by the original LtG study were "when will collapse happen?"; i.e., how accurate were model predictions with respect to timing of collapse, and "what will collapse look

[1] Jevons Paradox is based on studies of the 19th century economics by economist William Stanley Jevons, who found a generalizable trend that the increasing the efficiency with which a resource is used leads, rather than to a reduction of resource use from the captured efficiency, instead to a paradoxical overall increase in consumption of that resource. In Jevons's various studies, efficiency makes a resource cheaper per unit of output, and lower cost boosts demand, such that total overall consumption rises. The paradox entails that technological fixes alone not only do not guarantee sustainability – without broader changes in behavior, incentives, or policy, technology is more likely to scale up resource use and therefore impacts faster.

like to those encountering its development?" The surmised best-guess answers to these questions were that signs of collapse might begin to be apparent as early as 2020 to observers aware of the underlying systems theory. Because the LtG model was based on aggregated data, the predictions for change in various metrics were given as smoothed curves, however, the original investigators noted that a more likely outcome would be that some manifestations would occur as a series of intensifying shocks. Earmarks of collapse included, but were not limited to, economic instability, job insecurity, inflation, financial crises, political and social unrest, challenges maintaining and ultimate deterioration of roads, power grids, and water systems, large-scale migration of populations under one or another form of stress, biodiversity loss, increasing water scarcity and competition/conflicts over water, energy insecurity, increasing pollution and impacts on public health, and increasing supply chain difficulties (affecting access to and pricing of essential goods).

In addition to the number of studies revisiting LtG, an increasing number of publications have also predicted societal collapse based on data and the increasing number of indicators serving as harbingers of collapse (Bendell, 2018; Diamond, 2005; Homer-Dixon, 2006; Korowicz, 2012; Arnscheidt et al., 2025; Motesharrei et al., 2014; Ophuls, 2012; Read and Alexander, 2019; Servigne and Stevens, 2020; SRC, 2025; Steffen et al., 2007; Tainter 1988; Tainter 2000; Turchin, 2016). These are not publications by modern-day Nostradamus(es), or colorfully turbaned psychics prognosticating doom, but rather, publications that represent scholarly, evidence-based research. Some of the researchers studying collapse conclude that it is too late to avoid collapse, which would be consistent with LtG findings, however, all of these researchers conclude that there is an urgent need for adaptation – not piece-wise adaptation, but radical adaptation.

Returning now to the discussion of the future of Texas in a Changing Climate, what is Texas actually doing now (?), as that will drive the future. First and foremost, Texas has become famous for wind and solar power generation, with the largest installed wind capacity in the United States (U.S. EIA, 2023), and Texas ranks second in the U.S. for solar power generation, trailing only California. Between 2020 and 2022, Texas was scheduled to add over 10 GW of utility-scale solar capacity, one-third of all new utility-scale solar capacity in the U.S. (U.S. EIA, 2021). This is certainly a positive development with respect to energy generation; however, while wind and solar energy are "renewable," wind turbines, solar panels, and associated infrastructure, are not. Very large amounts of energy and materials are required to manufacture wind turbines and solar panels, which also have a limited lifespan of 20 to 25 years. One way to assess the sustainability of an energy supply is to look at the energy return on investment, or EROI. EROI is the ratio of energy obtained in relation to that energy needed to realize the energy obtained (Hall et al., 2014). For petrochemicals, for example, for every volume of fuel taken at the pump, the raw resource must be found, extracted, refined, transported, etc., and this is energy that must be expended to exploit the resource, but is not available for use. As a general rule of thumb, current estimates are that wind and solar have marginal EROIs; put another way, they do not generate enough energy for us to have a surfeit to use, while also relying only on wind and solar energy to build and instate more wind turbines and solar panels (after mining materials, fabricating materials, building wind and solar farms, and decommissioning and landfilling or recycling of materials). Wind and solar power, in different realizations, have some of the lowest EROIs of any energy source and the highest material consumption in tonnes per TWh of electricity produced (see, for instance, Figure V.1.2, Håkansson, 2021; U.S. DOE, 2015). EROI is, like any metric, imperfect, and while some criticize its use, it is conceptually grounded not only in biophysical reality, but in thermodynamics. Finding the correct EROI of a given energy source, is however, a valid concern; a widely accepted (as having integrity) comparison of EROIs is shown in Figure V.1.3 (Weißbach et al., 2013). To assume a higher service life or higher production efficiency will artificially inflate wind and solar EROIs, as will not including factors relating to the inability to use all wind and solar energy (storage, maintenance of grid stability). Wind farm EROIs may be inflated by more than a factor of three (Weißbach et al., 2013). Given that climate change is not the most pressing environmental challenge of our day, that the largest remaining tall grass prairie in Texas has now been paved with solar panels (Fears, 2022), and

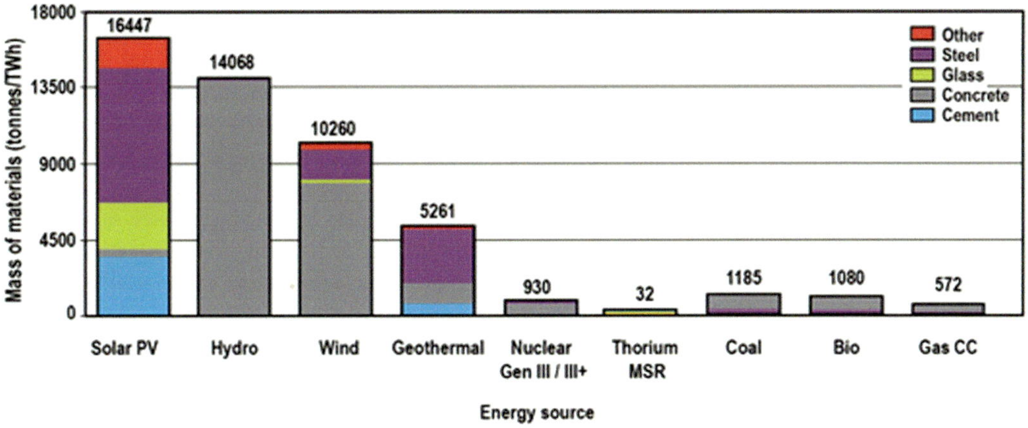

FIGURE V.1.2 Material consumption in tonnes per TWh of electricity produced for different energy. (Based on Håkansson, 2021 and the U.S. DOE, 2015, reproduced courtesy of Peter Rudling (https://energyeducation.se/wind-and-solar-energy-are-neither-renewable-nor-sustainable/).)

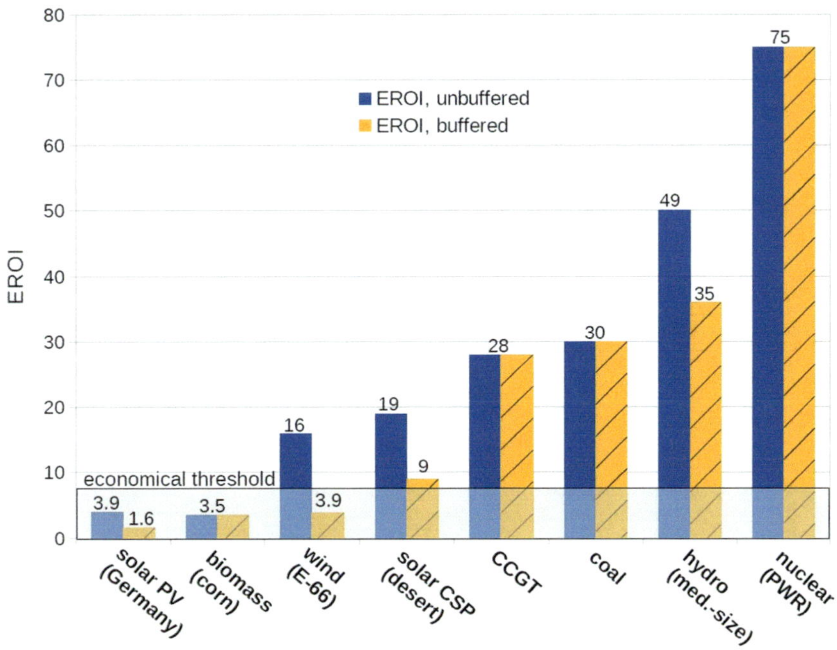

FIGURE V.1.3 EROIs of different energy types shown with the with economic "threshold" EROI, i.e., the EROI at which the energy source is estimated to become independently economically viable (without subsidies). For coal, transportation not included (resulting in an artificially high EROI) and for solar CSP the grid connection is not included (artificially increasing EROI). Buffered versus unbuffered refers to whether an energy source includes storage or backup systems to compensate for its intermittency – especially relevant for renewables like wind and solar, that is, unbuffered assumes energy storage is not included. (Reproduced from Weißbach et al., 2013, with permission.)

that Sweetwater Texas has now become, in the words of *Texas Monthly*, an outsized wind turbine dump (Gold, 2023), the truth as to what environmental benefit might be captured from wind and solar remains, therefore, elusive.

This wind and solar adaptation response of Texas to the need to reduce fossil fuel use is, like Texas, bigger than most other places, and at the same time, it serves as a metaphor. The LtG study, and all of the plenitude of studies cited herein that have followed, stressed the need for reduced consumption. We cannot change the laws of thermodynamics, and, we don't need to. Some studies have shown that reduced consumption is possible. In particular, two recent studies have examined the relationship between energy consumption and human well-being globally (Jackson et al., 2022; Millward-Hopkins et al., 2020). Across diverse metrics of human well-being, including health, happiness, access to food, and others, the research showed that many rich countries use grossly disproportionate amounts of energy to little or no increase in well-being metrics (Figure V.1.4). Reducing energy consumption in rich countries would provide scope for poorer countries to continue to improve their lot, all in a manner that would drastically reduce the global ecological footprint to a sustainable resource use level.

So, why don't we do it? Not just for energy, but for other resources as well? Other than having lived in Texas for a snapshot of time in my youth, then again from 2022 to 2024, I have lived overseas for much of my life, and I find the circumstance of resource consumption, in most places, but certainly in the U.S., and in Texas in particular, baffling and surprisingly illogical. Where I live, most of the cars on the street are light passenger cars, fuel efficient at that, not trucks approaching the size of the Gulf of America. Many houses still rely on rain water collection for household use, hence water use is much less, and much more sustainable. Cycle ways are ubiquitous in any city. If one values one's life, no problem, cycling to work is quite possible – and safe (many people are fit in consequence – I am routinely smoked by people a decade or more older than I). Rather than having 24/7 climate control to within a hermetically sealed 1–2 degree range of temperature, people "rug up" in the winter and moderate heat use, with no climate control at all in summer (in Texas the analogous response would be to moderate air conditioning in summer). More illogical and baffling

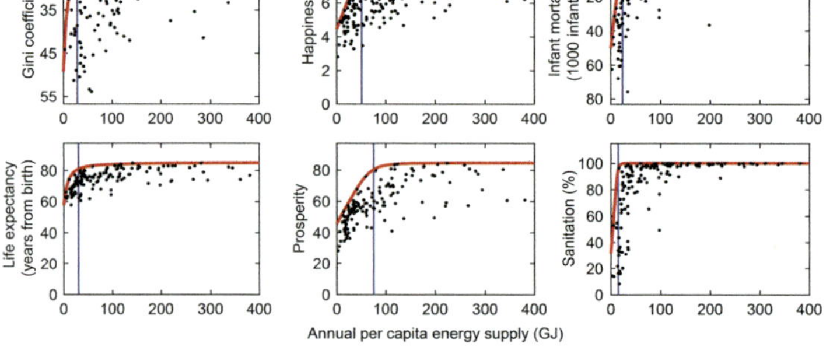

FIGURE V.1.4 Nine metrics used as indicators of human well-being plotted against annual national per capita total energy supply (gigajoules) by country. Black points are data, red lines are fitting functions, and blue lines mark the point of diminishing returns, beyond which much more energy is expended to little gain in human well-being. NB: The Gini coefficient is a measure of income inequality, with 0 indicating that everyone in a population has the same income or wealth, and 1 the opposite. Happiness is based on world polling. (Reproduced from Jackson et al., 2022, with permission (https://creativecommons.org/licenses/by/4.0/).)

still to me is the political nature of the discourse. At least climate change deniers are true to their beliefs in rejecting the idea of climate change as a threat, and accordingly taking no action. In contrast, my Texas friends and associates who are "pro-environment" I find most intractable. An office mate could not abide that I suggested setting the thermostat at 80°F in the summer (a temperature, even coming from my cool country, to which I was readily able to adapt at home during my recent stint in Texas). Another co-worker came to work upset one day, in an insouciant fit of cognitive dissonance, complaining to me that parents taking their children to the school which her children also attended would not "save fuel" by turning off their engines during the traffic jams that ensue in these modern times when children are no longer allowed/expected to walk, cycle, or bus to school. But the matter of the children is what I find most perplexing of all – it is said to change a habit, i.e., resource overconsumption, it is first necessary to identify the habit, and next to have *a motivation to change*. I do not have children, but most people do, including most Texans. These children are the future of Texas, yes, under climate change. Would there be any stronger motivator than this?

REFERENCES

Arnscheidt CW, Beard SJ, Hobson T, Ingram P, Kemp L, Mani L, Marcoci A, Mbeva K, Ó hÉigeartaigh SS, Sandberg A, Sundaram LS, Wunderling N, 2025. Systemic contributions to global catastrophic risk. Global Sustainability. 2025:8:e19.

Bardi U, 2011. *The Limits to Growth Revisited*. Springer: Berlin.

Bendell J, 2018. Deep adaptation: A map for navigating climate tragedy. University of Cumbria Institute for Leadership and Sustainability, Occasional Paper 2.

Diamond J, 2005. *Collapse: How Societies Choose to Fail or Succeed*. Viking: New York.

Fears D, 2022. *Texas Prairie Clash Shows Climate Solutions can Threaten Conservation Goals*. The Washington Post, 8 March 2022. https://www.washingtonpost.com/climate-solutions/2022/03/08/texas-prairie-solar-panels-climate-change/.

Gold R, 2023. Thousands of old wind turbine blades pile up in West Texas. Texas Monthly, 24 August 2023. https://www.texasmonthly.com/news-politics/sweetwater-wind-turbine-blades-dump/.

Håkansson A, 2021. Energy for the greens. *Kvartal*, 11 March 2021. https://kvartal.se/artiklar/energi-for-grona/.

Hall CAS, Day JW, 2009. Revisiting the limits to growth after peak oil. American Scientist. 97:230–237.

Hall CAS, Lambert JG, Balogh SB, 2014. EROI of different fuels and the implications for society. Energy Policy. 64:141–152.

Herrington G, 2021. Update to limits to growth: Comparing the World3 model with empirical data. Journal of Industrial Ecology. 25:614–626.

Homer-Dixon T, 2006. *The Upside of Down: Catastrophe, Creativity, and the Renewal of Civilization*. Island Press: Washington, DC.

Jackson RB, Ahlström A, Hugelius G, Wang C, Porporato A, Ramaswami A, Roy J, Yin J, 2022. Human well-being and per capita energy use. Ecosphere. 13:e3978.

Korowicz D, 2012. *Trade-Off: Financial System Supply-Chain Cross-Contagion—A Study in Global Systemic Collapse*. Feasta: Dublin.

Meadows DH, Meadows DL, Randers J, Behrens WW III. 1972. *The Limits to Growth: A Report for the Club of Rome's Project on the Predicament of Mankind*. Universe Books: New York.

Meadows DH, Randers J, Meadows DL. 2004. *Limits to Growth: The 30-Year Update*. Chelsea Green Publishing: White River Junction, VT.

Millward-Hopkins J, Steinberger JK, Rao ND, Oswald Y, 2020. Providing decent living with minimum energy: A global scenario. Global Environmental Change. 65:102168.

Motešarrei S, Rivas J, Kalnay E, 2014. Human and nature dynamics (HANDY): Modeling inequality and use of resources in the collapse or sustainability of societies. Ecological Economics. 101:90–102.

Ophuls W, 2012. *Immoderate Greatness: Why Civilizations Fail*. CreateSpace: North Charleston, South Carolina.

Pereira HM, Navarro LM, Martins IS, 2012. Global biodiversity change: The bad, the good, and the unknown. Annual Review of Environment and Resources. 37:25–50.

Pfeffer J, Malik MM, 2017. Simulating the dynamics of socio-economic systems. In *Networked Governance: New Research Perspectives*, Hollstein B, Matiaske W, Schnapp K-U, eds. Springer: Cham, Switzerland.

Randers J, 2012. *2052: A Global Forecast for the Next Forty Years*. Chelsea Green Publishing: White River Junction, VT.

Read R, Alexander S, 2019. *This Civilization Is Finished: Conversations on the End of Empire and What Lies Beyond*. Simplicity Institute: Melbourne.

Rockström J, Steffen W, Noone K, Persson Å, Chapin FS III, Lambin EF, Lenton TM, Scheffer M, Folke C, Schellnhuber HJ, Nykvist B, de Wit CA, Hughes T, van der Leeuw S, Rodhe H, Sörlin S, Snyder PK, Costanza R, Svedin U, Falkenmark M, Karlberg L, Corell RW, Fabry VJ, Hansen J, Walker B, Liverman D, Richardson K, Crutzen P, Foley JA, 2009. A safe operating space for humanity. Nature. 461:472–475.

Servigne P, Stevens R, 2020. *How Everything Can Collapse: A Manual for Our Times*. Polity Press: Cambridge.

Steffen W, Crutzen PJ, McNeill JR, 2007. The Anthropocene: Are humans now overwhelming the great forces of nature. Ambio. 36:614–621.

Stockholm Resilience Centre. (n.d.). Planetary boundaries. Stockholm Resilience Centre. https://www.stockholmresilience.org/research/planetary-boundaries.html. Science. 461:472–475.

Tainter JA, 2000. Problem solving: Complexity, history, sustainability. Population and Environment 22:3–41.

Tainter JA, 1988. *The Collapse of Complex Societies*. Cambridge University Press: Cambridge.

Turchin P, 2016. *Ages of Discord: A Structural-Demographic Analysis of American History*. Beresta Books: Chaplin, CT.

Turner GM, 2008. A comparison of the limits to growth with thirty years of reality. Global Environmental Change. 18:397–411.

Turner GM, 2012. Is global collapse imminent? An updated comparison of limits to growth with historical data. GAiA - Ecological Perspectives for Science and Society. 21:116–124.

United States Department of Energy (US DOE), 2015. Quadrennial Technology Review – An assessment of energy technologies and research opportunities – Chapter 10: Concepts in integrated analysis. In *Quadrennial Technology Review 2015*. US Department of Energy: Washington, DC.

United States Energy Information Administration (US EIA), 2023. Texas renewable electricity profile. https://www.eia.gov/renewable/state/Texas/.

US EIA, 2021. Texas likely to add record utility-scale solar capacity in the next two years. https://www.eia.gov/todayinenergy/detail.php?id=47636.

Weißbach D, Ruprecht G, Huke A, Czerski K, Gottlieb S, Hussein A, 2013. Energy intensities, EROIs, and energy payback times of electricity generating power plants. Energy. 52:210–221.

Index

Note: Page numbers in **bold** and *italics* refer to tables and figures, respectively. Page numbers followed by n refer to notes.

C

D

E

F

G

Q

R

U

V

W

X

Y

Z

For Product Safety Concerns and Information please contact our
EU representative GPSR@taylorandfrancis.com Taylor & Francis
Verlag GmbH, Kaufingerstraße 24, 80331 München, Germany